MW01268057

Logic Circuit Design

Saunders College Publishing Series in Electrical Engineering, formerly The Holt, Rinehart and Winston Series in Electrical Engineering

M.E. Van Valkenburg, Senior Consulting Editor
Adel S. Sedra, Series Editor/Electrical Engineering
Michael R. Lightner, Series Editor/Computer Engineering

ALLEN/HOLBERG
CMOS Analog Circuit Design

BELANGER/ADLER/RUMIN
Introduction to Circuits with Electronics: An Integrated Approach

BOBROW
Elementary Linear Circuit Analysis, 2/e

BOBROW
Fundamentals of Electrical Engineering

BUCKMAN
Guided-Wave Photonics

CHEN
Analog and Digital Control System Design

CHEN
Linear System Theory and Design

CHEN
System and Signal Analysis

COMER
Digital Logic and State Machine Design, 2/e

COMER
Microprocessor-Based System Design

COOPER/McGILLEM
Probabilistic Methods of Signal and System Analysis, 2/e

GHAUSI
Electronic Devices and Circuits: Discrete and Integrated

GUNGOR
Power Systems

GURU/HIZROGLU
Electric Machinery and Transformers

HOSTETTER/SAVANT/STEFANI
Design of Feedback Control Systems, 3/e

HOSTETTER
Digital Control System Design

HOUTS
Signal Analysis in Linear Systems

ISHII
Microwave Engineering

JONES
Introduction to Optical Fiber Communication Systems

KENNEDY
Operational Amplifier Circuits: Theory and Application

KUO
Digital Control Systems, 2/e

LASTMAN/SINHA
Microcomputer-Based Numerical Methods for Science and Engineering

LATHI
Modern Digital and Analog Communication Systems, 2/e

LEVENTHAL
Microcomputer Experimentation with the IBM PC

LEVENTHAL
Microcomputer Experimentation with the Intel SDK-86

LEVENTHAL
Microcomputer Experimentation with the Motorola MC6800 ECB

McGILLEM/COOPER
Continuous and Discrete Signal and System Analysis, 3/e

MIRON
Design of Feedback Control Systems

NAVON
Semiconductor Microdevices and Materials

PAPOULIS
Circuits and Systems: A Modern Approach

RAMSHAW/VAN HEESWIJK
Energy Conversion

SADIKU
Elements of Electromagnetics

SCHWARZ
Electromagnetics for Engineers

SCHWARZ/OLDHAM
Electrical Engineering: An Introduction, 2/e

SEDRA/SMITH
Microelectronic Circuits, 3/e

SHAW
Logic Circuit Design

SINHA
Control Systems

VAN VALKENBURG
Analog Filter Design

VRANESIC/ZAKY
Microcomputer Structures

WARNER/GRUNG
Semiconductor Device Electronics

WOLOVICH
Robotics: Basic Analysis and Design

YARIV
Optical Electronics, 4/e

Logic Circuit Design

Alan W. Shaw
Utah State University

SAUNDERS COLLEGE PUBLISHING
Harcourt Brace Jovanovich College Publishers
Fort Worth Philadelphia San Diego New York Orlando Austin
San Antonio Toronto Montreal London Sydney Tokyo

Text Typeface: Times Roman
Compositor: Monotype Composition
Senior Acquisitions Editor: Emily Barrosse
Assistant Editor: Laura Shur
Managing Editor: Carol Field
Project Editor: Anne Gibby
Copy Editor: Mary Patton
Manager of Art and Design: Carol Bleistine
Art Director: Doris Bruey
Cover Designer: Lawrence R. Didona
Text Artwork: Vantage Art, Inc.
Director of EDP: Tim Frelick
Production Manager: Joanne Cassetti
Marketing: Monica Wilson

Cover Credit: © 1991 Bruce Forster/All Stock, Inc.

Printed in the United States of America

Library of Congress Catalog Card Number: 92-056726

LOGIC CIRCUIT DESIGN

ISBN: 0-03-050793-6

3456 016 987654321

Book printed on **recycled** paper containing a minimum of 10% post-consumer waste and 50% pre-consumer waste.

Cover printed on **recycled** paper containing a minimum of 10% post-consumer waste and 40% pre-consumer waste.

To Joan

Preface

This text is intended to teach logic circuit design to sophomore-level engineering and computer science students. The subject matter requires no prerequisites other than algebra. Enough material is contained in the text for a two-quarter sequence (and even a two-semester sequence) of courses. Normally I use this material to teach a one-quarter course in combinational logic circuit design and a one-quarter course in sequential logic circuit design. These courses are offered to students in electrical engineering, mechanical engineering, and computer science. Electrical engineering students usually follow these courses with a one-quarter course specifically dealing with the design of digital systems using programmable logic devices. In a two-semester sequence this additional material on programmable devices could be included in the second semester.

APPROACH

The text introduces logic circuit design to the student who knows nothing about the subject and carries that student through to the design of a small stored-program computer. This small computer is designed and simulated in Chapter 12, ''Digital Systems Design.''

Since the text is intended for courses in which students first encounter the concept of design, one purpose is to present a design methodology using logic circuit design. Computer aids to design, in particular computer simulation, are introduced early and stressed throughout the text. Most circuits in the text are drawn using *OrCad*[1] schematic capture software and the remainder of the circuits are drawn using

[1]*OrCad* is a registered trademark of OrCad Systems, Inc.

Viewlogic[2] schematic capture software; all simulations use *Viewlogic* simulation software. PLD designs use PALASM[3] software. Descriptions of standard digital devices are given as they're needed throughout the text, but it's not the purpose of this text to introduce the student to a large repertory of digital devices but rather to teach design methodology.

The text is well supplied with worked examples illustrating methods introduced in each section. These examples illustrate the use of Karnaugh maps, circuit diagrams, state transition diagrams, and computer simulations as applicable.

STYLE

The text is written in an easy, conversational tone. Concepts are first introduced by example; they are then stated concisely; finally, more examples are presented and worked to their logical conclusion. Key terms are listed at the end of each chapter, as are a variety of exercises employing the skills learned.

NOTES ABOUT NOTATION

The text uses a notation called *polarized mnemonics*, which differs from the notation used in many (even most) logic design and analysis texts. I've found that a beginner who knows nothing about logic circuits has no difficulty learning this notation. Moreover, it can easily be specialized to the conventional notation. For the student who already knows something of logic circuits, I've found that the notation may at first appear difficult, but it offers advantages in design simplification and clarity of documentation which makes it well worth the effort to learn.

The text uses a notation for state transition diagrams which differs somewhat from conventional notation. The notation is less cryptic than the standard notation and offers the advantage of clarity at the expense of more writing. The relation of this notation to standard notation is clearly indicated.

ORGANIZATION

Organization of the text departs from the standard in which analysis is presented first with design following. In this text, design is presented first and analysis follows. This order allows a systematic design process—which leads to easily documented, easily analyzed circuits—to be presented before circuits of more haphazard design are analyzed.

---

[2]*Viewlogic* is a registered trademark of Viewlogic Systems, Inc., of Marlboro, Massachusetts.

[3]PALASM is a copyrighted program of Advanced Micro Devices Incorporated, © 1991.

EXERCISES

An ample selection of exercises can be found at the end of each chapter. There is enough of a selection in most chapters to allow a different set of exercises to be assigned for several quarters (or semesters) running. Available as ancillary material to the text will be a solutions manual containing solutions to these exercises.

END-OF-TEXT MATERIAL

The appendix describes a type of state transition diagram called a low control diagram. This type of state transition diagram is not used in the text, but it enjoys enough popularity to warrant its inclusion for completeness.

ACKNOWLEDGMENTS

In the preparation and writing of this text, I've received a great deal of help through comments and critical readings. I would first like to thank the good nature, professionalism, and extreme patience of the staff at Saunders College Publishing/Harcourt Brace Jovanovich, especially Emily Barrosse, Senior Acquisitions Editor, Laura Shur, Assistant Editor, Marc Sherman, Senior Project Editor, and Janet Nuciforo and Anne Gibby, Project Editors. I would also like to acknowledge with gratitude the excellent reviews and suggestions I received from reviewers Bruce Black, Rose Hulman Institute of Technology; Charles Doty, University of North Carolina, Charlotte; and Jay L. Smith, Weber State University.

This text was originally conceived four years ago with the expectation of writing it with my colleague and friend, Ronald L. Thurgood, Utah State University. Overwhelmed by other commitments, Ron was forced to drop out of the project before actual writing began, but I've benefited greatly by his notes, suggestions, and most of all, his constant encouragement. I would also like to acknowledge the help and encouragement of Paul Israelsen and Paul Wheeler for their critical comments and suggestions, and Todd K. Moon and Doug Woodhead who taught classes from the text and found for me many of the inevitable errors. I would like to acknowledge my colleague, Bill Fletcher, whose text, *An Engineering Approach to Digital Design*, has been for many years the classic text in digital design using polarized mnemonics and has served for this text as a logical takeoff point.

Special thanks are due my students, who have suffered with fortitude and good will the various drafts of this text.

Finally, I'd like to express my love and gratitude to my wife, Joan Shaw, a technical editor with nevertheless little love for circuits and transistors, who read and edited and suffered with me over the preparation of these pages. I couldn't have finished *Logic Circuit Design* without her.

Alan Shaw
December 18, 1992

Contents

1 Introduction

Electronic circuits can be divided into analog circuits and digital circuits. Analog circuits are those in which the current and voltages can take on a continuous range of values. These circuits are used in radio, television, telephone, and sound systems where the quantities involved, such as sound and light, must take on values from soft to loud or dark to light. They can also be used in a type of computer, called an analog computer, where numbers are represented by a continuous range of voltages.

In digital circuits, which are used in digital computers and related equipment, voltages and currents do not take on continuous values, but instead take on a discrete set of values. In almost all applications, this set comprises just two values. Digital circuits *count*, and in most cases they count only two values, high and low; or they *control* by using the same two values.

The purpose of this book is to describe these two-valued digital circuits. In their most basic form, two-valued circuits are called logic circuits. The development of logic circuits has been phenomenally swift during the last couple of decades.

1.1 HISTORY OF COMPUTATION

In the lab for which I'm currently consulting, we're designing a prototype digital image compression system for commercial television. Digital images have much higher quality than conventional analog images, but their commercial transmission without compression is not economically feasible. Image compression will allow more of these digital TV channels to be transmitted within a satellite's restricted bandwidth. Because satellite transmission cost is based on bandwidth, compression is essential. Image compression technology is expected to replace cable entirely, raining compressed signals from satellites above us to decompression devices built into home and business television receivers on the ground. Several labs across the country are working on compression-decompression technology. The pressure to produce well and produce early is intense.

Electrical engineers are primarily designers, and designing in a highly dynamic field carries with it a certain amount of this type of pressure, but also excitement. Electrical engineers often work with technology that has been developed only in the last year or two. In turn, the technology these engineers develop for one device may be used by later designers to develop or improve other devices. So technology marches on, swelling in size as one advance builds upon another. In fact, the acceleration of this march has been so impressive during this past century that we're likely to forget the slow and stumbling first efforts of human beings to perform such elementary operations as counting their fingers. Counting, and a search for aids to counting and computations, is where technology began.

1.1.1 Primitive Counting. Judging by the practice of present-day primitive societies in Australia, Brazil, and New Guinea, early humanity's need for technology was limited. Most of today's primitive tribes have no names for numbers beyond two or three; some go to twenty, using fingers and toes or fingers and ears, elbows, or eyes. For a long time, historians propose, numbers came to no more than this—a range of one to twenty: one to twenty bison, one to twenty wolves. Anything beyond twenty was simply uncountable, mind-boggling, a vast horde. Slowly, counting systems expanded as people accumulated possessions, stockpiled reserves, or gathered herds. Fingers and toes were no longer adequate, so they kept track of numbers with piled stones, notches on a stick, or marks on a piece of leather. Here were operations with numbers on the most primitive level; they led eventually from arithmetic to logic.

By the time agrarian societies formed around 11,000 years ago, people needed to keep track of days and seasons and to compute the amount of food and seed grain to store. As civilizations advanced, governments needed to count their citizens and levy taxes; architects needed a system of measurements for temples, palaces, and tombs; seamen needed a system of navigation as they sailed from port to port, trading between newly formed nations. Egyptians, Babylonians, Greeks, and Romans all developed number systems to fill these needs, and all were cumbersome. Then someone discovered that if movable beads or disks were attached to a board, computation was much faster than when it was done by piling stones or marking immovable figures on leather or papyrus; thus the abacus evolved.

1.1.2 The Abacus. The Babylonians are generally credited with the invention of the abacus, and the Chinese are credited with the refined instrument we have today: a counting frame with beads that slide back and forth on wires or in slots. Used for computation by experts, the abacus is an amazingly fast mechanical aid for performing addition and subtraction, even today.

1.1.3 The Indispensable Zero. The Hindus developed a nine-digit numbering system some time before 876. Much later they added their most remarkable invention, the zero, which represented the empty column in the abacus. This accomplishment led to a method of notation, still used today, in which any number is represented in the base 10 system by an ordered sequence of digits, lending a remarkable ease to all types of computations. The Hindus began an important era of mathematics by adding that zero to the decimal system; their leadership lasted until around 1200 and led to major advances in algebra.

The Arabs, in their considerable commercial dealings with the Indians, adopted both the decimal system and the zero. In fact, while on an extended visit to India, the Arab mathematician al-Khawarizmi wrote a book, *Al-jabr wa'l muqabalah*, from whose title our word *algebra* evolved.

The invention of the printing press in the fifteenth century made it possible to teach arithmetic to the masses, particularly merchants. Modifications of the Arabic system in the sixteenth and seventeenth centuries included the use of the addition sign (+) and of decimal fractions.

1.1.4 Counting Machines. In the meantime, advances in the physical sciences created urgent needs for computational aids, particularly for division and multiplication, which the abacus, the earliest counting machine, could not deal with easily. In the early seventeenth century, logarithm tables were developed by the Englishman Henry Briggs, and the slide rule was invented by his countryman William Oughtred, a clergyman. But the most fascinating will o' the wisp for mathematicians remained—some kind of mechanical counting device that would take the drudgery out of arithmetic.

Among the inventors of such machines was Blaise Pascal, a Frenchman who produced a computer in the seventeenth century, primarily to assist his father, a tax collector. Several of Pascal's counters, essentially adding machines, survive in museums in England and France. Later in the century, the German Gottfried Wilhelm von Leibniz produced a mechanical multiplier (incidentally advocating the binary numbering system, which would one day become a cornerstone of computer technology). The Leibniz machines eventually went into production and over a sixty-year span numbered approximately 1500; they were used primarily by merchants.

In the meantime, more and more complex mathematical tasks confronted scientists, and the search for adequate counting machines continued unabated into the nineteenth century, when modern technology had its birth and machines became kings.

1.1.5 Charles Babbage. An eccentric English mathematician, engineer, and inventor, Charles Babbage, conceived the idea that not only mathematical computations but logic and even analysis could be mechanized. Babbage's *Difference Engine* began with six wheels and was finally scuttled in 1833 for an all too familiar reason, lack of funds, after a decade of bitter battles with government agencies controlling funds for research.

The Difference Engine was succeeded by Babbage's *Analytical Engine*, a forerunner of today's computer. It was to take up approximately 50 square feet and be equipped with concentric toothed wheels; some of these wheels measured 2 feet in diameter. It was to have a *Store* (what we would now call memory) and a *Mill* (what we would now call the central processing unit, or CPU). There would also be wheels and gears that transferred numbers back and forth between the mill and the store and, most interesting of all, *punch cards* to direct the nature of the operations to be performed. The machine was to be powered by steam.

Babbage continued to perfect his Analytical Engine, meanwhile keeping up an interest in a bewildering variety of subjects—steam locomotive engineering, the ''occulating'' lighthouse, politics, secret codes, crime, an ophthalmoscope, and various crusades, most notably against street musicians. He was still working on the plans and funding for the Analytical Engine when he died at 71.

The trouble with Babbage, biographers say, was that except for his autobiography and a book on operations research, he never finished anything he started, instead jumping from one idea to another, quarreling with heads of government and scientific establishments, and dropping friends and honors alike because of imagined slights. He ended his life lonely and unknown except for his eccentricities. Remarkably, most of the scientists who, a century later, took up where Babbage had left off were not even aware he—or his plans for a mechanical computer—had ever existed.

1.1.6 Boolean Algebra. In 1854 an Englishman named George Boole, a self-taught mathematician and Babbage's contemporary, developed Boolean algebra.[1] Boole proposed that anything, including postulates in logic, could be described in equations using a binary, or two-valued, representation. In Boole's system, any given thing either *is* (1) or *is not* (0)—a profound advance in mathematical thought. In Boolean algebra, + means *or*; therefore,

$$\textit{is} + \textit{is} = \textit{is}$$

$$\textit{is not} + \textit{is not} = \textit{is not}$$

$$\textit{is} + \textit{is not} = \textit{is}$$

or, in present-day Boolean algebraic terms,

$$1 + 1 = 1$$

$$0 + 0 = 0$$

$$1 + 0 = 1$$

Also, a dot (·) means *and*, so

$$\textit{is} \cdot \textit{is} = \textit{is}$$

$$\textit{is not} \cdot \textit{is not} = \textit{is not}$$

$$\textit{is} \cdot \textit{is not} = \textit{is not}$$

[1]Later in the book we'll discuss Boolean algebra as an aid in giving machines information they can understand.

or, in present-day Boolean algebraic terms,

$$1 \cdot 1 = 1$$

$$0 \cdot 0 = 0$$

$$1 \cdot 0 = 0$$

1.1.7 Claude Shannon. In the 1930s Claude Shannon, an engineer for Bell Laboratories, developed a communications theory, based on Boolean algebra, to help improve the reliability of long-distance telephone and telegraph lines, and for this work he is justly famous. His lesser known work in digital switching systems, an outcome of his communications theory, is one of the bases of modern logic systems and computer technology. In 1938, while at MIT, Shannon published a paper in which he proposed a machine that could be built to handle Boolean logic and calculations using telephone on-off switching techniques. The paper was a significant contribution to the development of digital switching systems. Here was a binary system *making decisions—on* or *off*. From a binary machine that simply counted numbers mechanically, technology had entered the realm of *binary logic* as a means of *control*.

Around this same time, Bell Laboratories scientist George R. Stibitz built a computer that used relays to control either the presence or absence of current, which represented binary numbers—an idea similar to Shannon's. Stibitz's machine converted decimal numbers to binary numbers, made calculations using binary numbers, and then reconverted them to decimal answers. His machine was called the *Complex Number Computer* because it was designed to make calculations of complex numbers. It turned out to be a dead end because of the mechanical nature of its technology, but was significant for its use of a binary, or two-state, logic.

1.1.8 Herman Hollerith. Back in the 1880s, another part of early computer technology found its way into computation: the punch card. During this period America developed a very real problem: the Census Bureau needed to count people. In a nation where the population was increasing by approximately a third every ten years, where the tabulating of vital statistics and occupations threatened to take longer than the time elapsing between censuses, some kind of mechanical counting device was critically needed. The man who met this need was a young mining engineer named Herman Hollerith who was working on a survey of power usage for the U.S. Census Office. Hollerith wound up reinventing Babbage's punch card and adding a sorter, which organized the cards according to the information represented by the holes, such as birthday and occupation.

Eventually Hollerith changed the hand-punched holes to machine-punched holes; this advance marked not only the start of modern data processing but the large-scale employment of women as office workers. Although Hollerith's work has only peripheral importance to logic circuits, his move into the commercial field of business computation was important to logic circuit development later in the century, for his tabulating machine company, through incorporation and reorganization, evolved into the International Business Machine Company, now known as IBM.

In 1944, Howard Aiken, working at IBM after graduation from Harvard, modified the standard punch card technology into IBM's *Mark I*, the first automatic computer. Both Stibitz's Complex Number Computer and the IBM Mark I were basically electromechanical computers, and were driven into obsolescence; by then scientists were already working on computers and logic systems based on electronic devices.

1.1.9 Vacuum Tubes. By the 1940s computer technology was rapidly reaching what seemed like a utopia: the production of an electronic, general purpose computer that used a binary system for both computation and control. John Mauchly and J. Presper Eckert are generally given credit for the development of the first of these machines, finished in the late 1940s: the ENIAC, a room-size behemoth housed at the University of Pennsylvania. The ENIAC used the vacuum tube, a device perfected in 1907 by Lee De Forest, an American inventor. Storage of information in vacuum tubes greatly increased speed of operation, and the ENIAC used 18,000 vacuum tubes.

In the late 1940s and early 1950s, Mauchly and Eckert worked on the world's first *commercial* general purpose electronic computer, the UNIVAC. With memory storage in mercury delay lines, this computer used a fraction of the vacuum tubes needed in the ENIAC and made its owner, Remington-Rand, the leading manufacturer of commercial computers. Vacuum tubes, however, produced excessive heat and were proving too large and unreliable for use in large computers. As one writer put it,

> *Troubleshooting was a nightmare. If one tube blew, engineers usually could not find the offender quickly and sometimes had to resort to pulling the computer apart and crawling inside. Moreover, the tubes suffered from what the engineers called ''morning sickness,'' the inability of cold, shut-down tubes to sense the first electric impulse they received when the machine was turned on in the morning.*[2]

1.1.10 The Transistor. The stage was set for a compact storage device, and the *transistor* came on the scene in 1947, again at Bell Laboratories. Only eight years later Bell Laboratories was producing *integrated circuits*, in which transistors and diodes imbedded in silicon chips performed both arithmetic and logic functions. By the 1960s the behemoths of the 1940s had shrunk to a manageable size, and the way was opened for the microprocessors and microcomputers of the 1970s. Solid state technology had now been developed to the point where a personal computer sitting on a desktop was capable of many times more computing and decision-making capability than the room-size ENIAC of the 1940s.

1.1.11 Modern Machines. The computer industry has grown at an incredible rate since the development of solid state technology during the late 1940s and 1950s.

---

[2]Joel Shurkin, *Engines of the Mind* (New York, London: W. W. Norton, 1984), 301.

During the 1960s computers were produced largely for scientific applications. But in 1965 Digital Equipment Corporation introduced the PDP-8. Its relatively low price, $50,000, helped to move the computer out of sophisticated research laboratories and into manufacturing plants across the country, marking the beginning of the minicomputer industry.

In 1969 Datapoint Corporation contracted with Intel and Texas Instruments to design an elementary computer on a single logic chip. Intel succeeded, but the design executed instructions more slowly than Datapoint had specified, and the company decided against buying it. Intel was left with a single-chip, computer-like logic device on its hands, which it eventually decided to market, in 1972, in the Intel 8008, a desk-size computer. This innovation marked the beginning of the microcomputer industry.

1.1.12 Integrated Circuits. The integrated circuits with which we work today would have appeared unbelievable to the scientists developing computation and logic functions only a decade ago. These circuits now provide engineers with huge numbers of logic functions crammed onto a single chip. In fact, the VSLIs (Very Large Scale Integrated Circuits) being manufactured right now can hold up to 1.5 million transistors on a chip approximately a half-inch square.

The design of logic functions using integrated circuits is what this book is about. By the end of the course of study set out here, students should be able to design and construct a moderately complicated digital system. Chapter 2 reviews some of the ideas involved in numbers systems and develops the idea of a two-valued, or binary, number system. The remainder of the book is divided into two parts. The first deals with combinational logic circuits, circuits that allow the electronic evaluation of logic expressions. In Chapter 3, the mathematics to describe logic expressions, called Boolean algebra, is described. In Chapter 4, first the design and then the analysis of circuits that can evaluate logic expressions are explained. Design is introduced using more conventional discrete logic, but the easy extension to design using Field Programmable Gate Arrays is also made. Chapter 5 discusses important examples of combinational logic circuits. Chapter 6 introduces combinational logic design using fixed architecture programmable devices such as PROMs, PALs, and PLAs.

The second part of the book deals with sequential logic circuits, circuits that involve a sequence of logic evaluations usually controlled by some type of clock. Chapter 7 discusses the basic ideas of sequential design, develops the basic model of a sequential machine, and describes the distinction between synchronous and asynchronous sequential machines. The use of computers as sequential machines is considered briefly. Chapter 8 introduces the design of sequential machines using logic circuits. It does so by considering the design of two-state state machines, or flip-flops. Although these initial designs have limited usefulness, more practical flip-flops, to be used in later chapters, are also described.

Chapter 9 expands the design concepts of Chapter 8 to sequential machines with more than two states. Designs use conventional discrete logic and Field Programmable Gate Arrays. Practical problems such as asynchronous inputs, logic simplification, and output glitches are discussed. Chapter 10 introduces sequential designs with register PROMs, register PALs, and register PLAs, unconventional sequential design, analysis, and asynchronous design. Chapter 11 offers examples of sequential machines and shows how to make very large sequential machines using counters and PROMs. Chapter 12 describes the design of digital systems (often and unfortunately called controllers), which use a sequential machine as the basic control element.

2 Decimal, Binary, Octal, and Hexadecimal Numbers

2.1 COUNTING AND NUMBERS

As already mentioned, early humans most likely first counted objects by using their fingers, so when they developed a number system it was only natural to base the system on 10, the number of fingers on a human hand. By using the fingers, it was possible to count up to ten objects. To count beyond 10 required some innovation, although the idea of 10 + 10 and 10 + 10 + 10 most likely came early.

With the development of a written language it became convenient to use symbols to represent numbers. Here we're concerned with numbers used for counting (one, two, three), or *cardinal numbers*, as distinct from numbers used for ordering (first, second, third), or *ordinal numbers*. One not very convenient system of numbers was the system of Roman numerals. It had symbols to represent one (I), five (V), ten (X), and so on, and a number was represented by grouping these symbols in order of descending values, starting on the left. For example, the number 26 was written as XXVI, that is, 10 + 10 + 5 + 1. If a symbol representing a lower value was written before one of higher value, it was subtracted from the one of higher value. For example, the number 34 was written as XXXIV, that is, 10 + 10 + 10 + 5 − 1. This type of number system made computations such as addition and subtraction particularly difficult. Computations only became easy with **positional numbers**.

2.2 DECIMAL NUMBERS

In a positional number system, any positive integer is represented by a juxtaposition of symbols called **digits**. The number N is thus written in the form

$$N = d_n d_{n-1} \cdots d_1 d_0$$

where the d_i are the digits. In a system based on 10, the digits have ten values ranging from 0 to 9. Notice that in a positional number system it's necessary to have a digit that represents nothing (0). One can count from 0 to 9 using the number N with only one digit. To use the number N to count higher than 9, one must use the positional features of the number. In a positional numbering system, each digit has a value based on its position as well as an intrinsic value. This can best be seen by writing the number N as

$$N = d_n \times 10^n + d_{n-1} \times 10^{n-1} + \cdots + d_1 \times 10^1 + d_0 \times 10^0$$

The digit on the extreme right is called the least significant digit because it has the lowest positional value of 1 (for integers). The next digit to the left of this least significant digit has the positional value of 10; the next, 100; and so on. The digit on the extreme left is called the most significant digit because it has the highest positional value.

Using the idea of positional numbers, we see that the number N = 375 means

$$N = 3 \times 100 + 7 \times 10 + 5 \times 1$$

The digit 5 is a units digit, the digit 7 is a tens (10s) digit, and the digit 3 is a hundreds (100s) digit—all because of their respective positions.

The use of positional numbers can be extended to fractions by adding a decimal point and letting digits to the right of the decimal represent 10ths, 100ths, and so on, depending on position. Thus,

$$N = d_n \cdots d_0 \,.\, d_{-1}d_{-2} \ldots \qquad \textbf{(2.2-1)}$$

means

$$N = d_n \times 10^n + \cdots + d_0 \times 10^0 + d_{-1} \times 10^{-1} + d_{-2} \times 10^{-2} \ldots \qquad \textbf{(2.2-2)}$$

2.3 GENERAL POSITIONAL NUMBERS

The positional system we normally use is based on 10, but a number system can have any **base**. The base of a number system is also called the **radix**. Some common bases used in digital systems are 2 (binary), 8 (octal), and 16 (hexadecimal).

If we generalize the idea of a positional number to any base b, then

$$N = d_n \cdots d_0.d_{-1}d_{-2} \ldots \qquad \textbf{(2.3-1)}$$

means

$$N = d_n \times b^n + \cdots + d_0 \times b^0 + d_{-1} \times b^{-1} + d_{-2} \times b^{-2} \ldots \qquad \textbf{(2.3-2)}$$

In this number representation the d_i are the digits, b is the base or radix, and the "."

is the **radix point**. Sometimes we need to explicitly specify the base to be used with a number. This is done using the notation

$$N_b$$

where the subscript denotes the base. Thus,

$$N = 356_{10}$$

means 356 is a number with base 10.

2.4 BINARY NUMBERS

Most electronic storage devices are most easily placed in two distinct different states, charged and discharged. Thus they can be used to store two different values. It's natural that systems containing these devices use a number system based on only two digits. In this **binary** (two-valued) **number** system, the digits are zero (0) and one (1) and the radix is two (2). An example of an integer binary number is

$$N = 10110101_2 = 10110101$$

where the subscript denoting the base is usually left off if the meaning is clear without it. The digits of a binary number are often called **bits**, so this is an 8-bit number. It can be evaluated in terms of base 10 numbers (decimal numbers) using the formula for the general positional number, Eq. (2.3-2):

$$\begin{aligned} N &= 1 \times 2^7 + 0 \times 2^6 + 1 \times 2^5 + 1 \times 2^4 + 0 \times 2^3 \\ &\quad + 1 \times 2^2 + 0 \times 2^1 + 1 \times 2^0 \\ &= 128 + 32 + 16 + 4 + 1 = 181 \end{aligned}$$

A more convenient way to do this computation is to write the positional value above each of the digits and then add the positional values corresponding to nonzero digits (1). If we want to evaluate the preceding number we write the positional value above each digit

128	64	32	16	8	4	2	1
1	0	1	1	0	1	0	1

and then sum

$$\begin{array}{r} 128 \\ 32 \\ 16 \\ 4 \\ \underline{1} \\ 181 \end{array}$$

The preceding binary number is an 8-bit (digit) number. The largest 8-bit binary number is the number with all bits 1. Using the above technique, the decimal value of this number is found to be

128	64	32	16	8	4	2	1
1	1	1	1	1	1	1	1

$$\begin{array}{r} 128 \\ 64 \\ 32 \\ 16 \\ 8 \\ 4 \\ 2 \\ \underline{1} \\ 255 \end{array}$$

An 8-bit binary number can be used to count from 0 to

$$2^8 - 1 = 255$$

or

$$2^8 = 256$$

different values, including zero. In general, an n-bit binary number can be used to count from 0 to

$$2^n - 1$$

or

$$2^n$$

different values.

Example 2.4-1

Evaluate 11110111 as a decimal number.

128	64	32	16	8	4	2	1
1	1	1	1	0	1	1	1

$$\begin{array}{r} 128 \\ 64 \\ 32 \\ 16 \\ 4 \\ 2 \\ \underline{1} \\ 247 \end{array}$$

Example 2.4-2

Evaluate 100001 as a decimal number.

32	16	8	4	2	1
1	0	0	0	0	1

$$\begin{array}{r} 32 \\ \underline{1} \\ 33 \end{array}$$

Binary numbers can be used to represent fractions by means of a radix point (called a binary point for binary numbers). The evaluation of binary fractions as decimal numbers is a simple extension of the method just used. The positions beyond the binary point have positional values $2^{-1} = \frac{1}{2} = 0.5$, $2^{-2} = \frac{1}{4} = 0.25$, $2^{-3} = \frac{1}{8} = 0.125$, and so on.

Example 2.4-3

Evaluate 1100.101 as a decimal number.

8	4	2	1	.5	.25	.125
1	1	0	0 .	1	0	1

$$\begin{array}{r} 8 \\ 4 \\ .5 \\ \underline{.125} \\ 12.625 \end{array}$$

Example 2.4-4

Evaluate 10.0011 as a decimal number.

2	1	.5	.25	.125	.0625
1	0 .	0	0	1	1

$$\begin{array}{r} 2 \\ .125 \\ \underline{.0625} \\ 2.1875 \end{array}$$

Now suppose we have a decimal number and wish to change it to a binary number. It's possible to do this by using the general formula for positional numbers, Eq. (2.3-2), but all the computations would have to be done in binary arithmetic. The following computational technique is much easier, and all the computations use decimal arithmetic. Suppose we wish to find the binary equivalent of 213. We repeatedly divide 213 by 2 and save all the remainders.

```
2)213
2)106  1
2) 53  0
2) 26  1
2) 13  0            Read
2)  6  1  LSB        Up
             piPS1246/
2)  3  0  ART/
2)  1  1  chapt2/
             arrow.up
    0  1  MSB
```

Here the repeated divisions have been arranged in a convenient array with the quotient of each division below the number being divided and the remainder at the right. The binary number equivalent to 213 is the number formed from the remainders, with the most significant (the left-most bit) at the bottom of the array:

$$213_{10} = 11010101_2$$

As a check, we can evaluate the binary number as a decimal number:

128	64	32	16	8	4	2	1
1	1	0	1	0	1	0	1

$$\begin{array}{r} 128 \\ 64 \\ 16 \\ 4 \\ \underline{1} \\ 213 \end{array}$$

The procedure works because we're removing 1 bit at a time starting with the least significant bit. First we remove the least significant bit of 213 (the 1s bit) by dividing by 2:

$$213 = 11010101_2$$
$$\frac{213}{2} = 106 + 1(\text{rem}) = 1101010 + 1(\text{rem})$$

Division by 2 for the binary number causes a shift one position to the right. Now we remove the least significant bit of 106 (the 2s bit of 213) by dividing by 2:

$$106 = 1101010_2$$
$$\frac{106}{2} = 53 + 0(\text{rem}) = 110101 + 0(\text{rem})$$

We continue this process until the quotient is zero. In each case the remainder is the value of the bit that was removed.

Example 2.4-5

Find the binary number equivalent to 63 decimal.

```
2)63
2)31  1
2)15  1
2) 7  1
2) 3  1
2) 1  1
   0  1
```

$$63_{10} = 111111_2$$

Checking,

32	16	8	4	2	1
1	1	1	1	1	1

```
32
16
 8
 4
 2
 1
--
63
```

Example 2.4-6

Find the binary number equivalent to 163 decimal.

```
2)163
2) 81   1
2) 40   1
2) 20   0
2) 10   0
2)  5   0
2)  2   1
2)  1   0
    0   1
```

$$163_{10} = 10100011_2$$

Checking,

128	64	32	16	8	4	2	1
1	0	1	0	0	0	1	1

```
128
 32
  2
  1
---
163
```

If we want to convert a decimal number with a fractional part (decimal point) to a binary number, we do it in two parts. The integer part of the number (before the decimal) is computed as just described, but the fractional part (after the decimal) needs a different (but similar) computation. Suppose we wish to convert 0.625 decimal to binary. The computation uses repeated multiplication by 2 with a truncation of the integer part of the product at each multiplication.

```
 .625
   ×2
-----
1.250
   ×2
-----
0.500
   ×2
-----
1.000
```

The binary number is formed from the integer values starting at the top:

$$.625_{10} = 0.101_2$$

We've again put the computation into a convenient array. In each of the multiplications, only the fractional part is multiplied by 2; the integer part is ignored. We can check this binary number by finding its equivalent decimal number:

.5	.25	.125
1	0	1

$$\begin{array}{r} .5 \\ \underline{.125} \\ .625 \end{array}$$

The array used in the computation can be shortened by leaving out the $\times 2$s and writing,

$$\begin{array}{r} .625 \\ 1.250 \\ 0.500 \\ 1.000 \end{array}$$

where each row is twice the fractional part of the row above. In the conversion of a decimal fraction to a binary number, the continued multiplication may never end. In this case, we carry as many bits as we need for the accuracy we want.

The procedure works because we're removing 1 bit at a time starting with the most significant bit. Each time we multiply by 2, the most significant bit in the fraction is changed to an integer, and this is reflected as a 1 or a 0 in the integer position. For example,

$$.625_{10} = .101_2$$

Multiplying by 2,

$$1.25_{10} = 1.01_2$$

where multiplying a binary number by 2 shifts all its bits one position to the left. Now the integer parts of the two numbers must be equal, so the integer part of the decimal number is the first bit of the binary fraction. Continuing this for the fractional part of 1.25_{10},

$$.25_{10} = .01_2$$

Multiplying by 2,

$$0.50_{10} = 0.1_2$$

and the second bit is 0. Continuing again for the fractional part of 0.50_{10},

$$.50_{10} = .1_2$$

Multiplying by 2,

$$1.00_{10} = 1.0_2$$

and the third bit is 1.

Example 2.4-7

Find the binary number equivalent to 37.28 decimal.

First consider the integer part of the number:

$$
\begin{array}{rl}
2\overline{)37} & \\
2\overline{)18} & 1 \\
2\overline{)\ 9} & 0 \\
2\overline{)\ 4} & 1 \\
2\overline{)\ 2} & 0 \\
2\overline{)\ 1} & 0 \\
0 & 1
\end{array}
$$

The integer part of the number is 100101.

Now consider the fractional part of the number:

$$
\begin{array}{r}
.28 \\
0.56 \\
1.12 \\
0.24 \\
0.48 \\
0.96 \\
1.92 \\
1.84
\end{array}
$$

The fractional part of the number is .0100011 to an accuracy of 7 binary bits ($2^{-7} = 0.0078125$). A more accurate binary fraction could be found by continuing the process of multiplication. The complete binary number, with both integer and fractional parts, is

$$37.28_{10} = 100101.0100011_2$$

Checking the integer part,

32	16	8	4	2	1
1	0	0	1	0	1

$$
\begin{array}{r}
32 \\
4 \\
\underline{1} \\
37
\end{array}
$$

Checking the fractional part,

.5	.25	.125	.0625	.03125	.015625	.0078125
0	1	0	0	0	1	1

$$\begin{array}{r} .25 \\ .015625 \\ \underline{.0078125} \\ .2734375 \end{array}$$

Notice the inaccuracy due to roundoff.

2.5 OCTAL AND HEXADECIMAL NUMBERS

Two common bases for numbers used in digital systems are 8 (octal numbers) and 16 (hexadecimal numbers). The **octal number** system requires eight digits. The digits used are 0 through 7, borrowed from decimal numbers. The **hexadecimal number** system requires 16 digits. The first ten of these are 0 through 9, borrowed from decimal numbers. The remaining digits are the letters A through F, with the values

$$A = 10$$
$$B = 11$$
$$C = 12$$
$$D = 13$$
$$E = 14$$
$$F = 15$$

The value of an octal or hexadecimal number can be found using the formula Eq. (2.3-2).

Example 2.5-1

Evaluate the octal number 174 as a decimal number.

$$\begin{aligned} 174_8 &= (1 \times 8^2) + (7 \times 8^1) + (4 \times 8^0) \\ &= (1 \times 64) + (7 \times 8) + (4 \times 1) \\ &= 124_{10} \end{aligned}$$

Example 2.5-2

Evaluate the *hex* (abbreviation for *hexadecimal*) number B78 as a decimal number.

$$\begin{aligned} B78_{16} &= [11 \times (16)^2] + [7 \times (16)^1] + [8 \times (16)^0] \\ &= (11 \times 256) + (7 \times 16) + (8 \times 1) \\ &= 2936_{10} \end{aligned}$$

Octal and hexadecimal numbers are used in digital systems because of their close relationship to binary numbers. In a binary number, each set of 3 bits represents an octal digit and each set of 4 bits represents a hex digit. The binary-to-octal conversions are

Binary	Octal
000	0
001	1
010	2
011	3
100	4
101	5
110	6
111	7

The binary-to-hexadecimal conversions are

Binary	Hex
0000	0
0001	1
0010	2
0011	3
0100	4
0101	5
0110	6
0111	7
1000	8
1001	9
1010	A
1011	B
1100	C
1101	D
1110	E
1111	F

To convert a binary number to octal, form it into groups of 3 bits, starting at the binary point and working to both the left and the right. Then translate the 3-bit groups according to the preceding binary to octal table. To convert a hexadecimal, form into groups of 4 bits, again starting at the binary point, and translate using the binary to hex table.

Example 2.5-3

Convert the binary number 101001101.110 to an octal number and to a hexadecimal number.

Forming into groups of 3 bits and using the table,

$$101\ 001\ 101.110_2 = 515.6_8$$

Forming into groups of 4 bits and using the table,

$$1\ 0100\ 1101.1100_2 = 14D.C_{16}$$

To convert from octal or hexadecimal to binary, just reverse the process.

Example 2.5-4

Convert 7251_8 and $B396_{16}$ to binary.

$$7351_8 = 111\ 011\ 101\ 001_2$$

$$B396_{16} = 1011\ 0011\ 1001\ 0110_2$$

To convert from decimal to octal or hexadecimal, it's easiest to first convert to binary using the procedure we've found, and then convert to either octal or hexadecimal. It's also possible to use repeated division by 8 or 16 in a process similar to that used to convert from decimal to binary.

Example 2.5-5

Convert 123 decimal to octal and hexadecimal.

First convert to binary using repeated division by 2:

$$\begin{array}{rl}
2\underline{)123} & \\
2\underline{)\ 61} & 1 \\
2\underline{)\ 30} & 1 \\
2\underline{)\ 15} & 0 \\
2\underline{)\ \ 7} & 1 \\
2\underline{)\ \ 3} & 1 \\
2\underline{)\ \ 1} & 1 \\
0 & 1
\end{array}$$

$$
\begin{aligned}
123_{10} &= 1111011 \\
&= 1\ 111\ 011 = 173_8 \\
&= 111\ 1011 = 7B_{16}
\end{aligned}
$$

This can also be done using repeated division by 8 and repeated division by 16.

```
8)123
8) 15   3
8)  1   7
    0   1
```

$$123_{10} = 173_8$$

```
16)123
16)  7   11
     0    7
```

$$123_{10} = 7B_{16}$$

where $11_{10} = B$.

The use of repeated division by 8 is not too difficult, but repeated division by 16 can be computationally challenging, as you can see.

2.6 ADDITION

Now let's look at computations using positional numbers. We said that computations were made much simpler by positional numbers. All of us have used this simplification without thinking. Look at an example of addition of decimal numbers:

```
     1
  2367
 +1326
 -----
  3693
```

To do this addition, we line up the 1s digits, the 10s digits, the 100s digits, and the 1000s digits; then we add columns, because we're adding like-valued positional digits. If the result of a column addition is greater than a single-digit number (as is the double-digit 13 in the right column), we carry the higher order digit (1) to the next higher column and add it to the like-valued positional digits. The carry is shown above the upper number in the example.

Addition of binary numbers is much the same as addition of decimal numbers, except perhaps easier. First we make a table defining all possible results of 1-bit additions:

$$\begin{array}{r} 0 \\ +0 \\ \hline 0 \end{array} \qquad \begin{array}{r} 0 \\ +1 \\ \hline 1 \end{array} \qquad \begin{array}{r} 1 \\ +0 \\ \hline 1 \end{array} \qquad \begin{array}{r} 1 \\ +1 \\ \hline 10 \end{array}$$

Note that when we add 1 and 1, we get a binary 10, or a 0 with 1 carried to the next higher position. The carry here is just like the carry in decimal addition. Now, by using this table, we can do a binary addition. For example,

$$\begin{array}{r} 11\ \ 11 \\ 10111001 \\ +00110011 \\ \hline 11101100 \end{array} \qquad \begin{array}{r} \\ 185 \\ +51 \\ \hline 236 \end{array}$$

In this addition, the carry bits are shown above the upper number. Note that when we add the sixth column from the right, we have a carry of 1 to add to 1 and 1, which gives us an 11, or 1 and 1 carry. The equivalent decimal numbers are shown at the right as a check on the addition.

Example 2.6-1

$$\begin{array}{r} 1\ 1111 \\ 10100011 \\ +00101101 \\ \hline 11010000 \end{array} \qquad \begin{array}{r} \\ 163 \\ +45 \\ \hline 208 \end{array}$$

Example 2.6-2

$$\begin{array}{r} 1 \\ 10101010 \\ +10000100 \\ \hline 100101110 \end{array} \qquad \begin{array}{r} \\ 170 \\ +132 \\ \hline 302 \end{array}$$

Notice that in the last example, the result of adding two 8-bit binary numbers is a 9-bit binary number. If we were working in a system that could handle only 8-bit numbers (which is the case with some small computers), then the 8-bit answer obtained in Example 2.6-2 (100101110), without the carry (00101110), would be incorrect. The carry-out of the high order bit indicates a carry error.

2.7 SUBTRACTION

In subtraction of positional numbers, we use the same ideas as in addition. Look at an example of decimal number subtraction:

```
   3
  472
 -391
 ----
   81
```

Like-valued positional digits are placed in columns and subtracted. If, as in the second column, the digit in the upper number is smaller than that in the lower number, a higher order digit is borrowed from the next column, and the higher order digit in that column is reduced by 1. In the preceding example, the digit 4 is reduced to 3, the digit 7 is replaced by 17, and the subtraction is completed.

Like addition, subtraction of binary numbers is much the same as subtraction of decimal numbers. First we'll make a table defining all possible results of 1-bit subtractions:

```
  0     1)0     1     1
 -0     -1     -0    -1
 --     --     --    --
  0      1      1     0
```

Note that in order to subtract binary 1 from 0, we need to borrow 1 from the next higher position. The borrow here is just like the borrow in decimal subtraction. Now, using this table, we can do a binary subtraction. For example,

```
    01
  10111001       185
 -00110011       -51
 ---------       ---
  10000110       134
```

In this subtraction the bits written above the upper number are values resulting from a borrow. Note that when we subtract in the second column from the right, we need to borrow from the next higher position. This position, however, contains a zero, so we must borrow in turn from the next higher position. The equivalent decimal numbers are shown at the right as a check on the subtraction.

Example 2.7-1

```
  01011
  10100011       163
 -00101101       -45
 ---------       ---
  01110110       118
```

Example 2.7-2

```
    0
 10101010        170
-10000100       -132
 --------       ----
 00100110         38
```

Example 2.7-3

```
-1)  01
  01110010              114
 -10000110             -134
  --------             ----
  11101100  (236)       -20
```

In the last example, the number in parentheses is the decimal equivalent of the result of the binary subtraction. It's obviously incorrect. This is because we're subtracting a larger number from a smaller number. The result should be a negative number, but we have not extended our binary number system to handle negative numbers yet. To complete the subtraction as shown, we had to borrow in the high order bit position. A borrow in this position indicates an error.

In this last example, we're left with a borrow from the bit position with positional value 256. If we consider the -1 caused by the borrow in the 256-bit position, we'll get the correct solution for our subtraction:

$$-256 + 236 = -20$$

Later we'll use a similar idea to represent negative numbers.

2.8 MULTIPLICATION

We've all used the simplifications of positional numbers to do multiplication without thinking about it, just as we do in addition and subtraction. Look at an example of decimal multiplication:

```
   147
  ×265
  ----
   735
  882
 294
 -----
 38955
```

Multiplication is a two-step process. We find the product of the top number and each of the digits of the bottom number, then add these partial products. In the example, we first multiply the top number by 5 (with the appropriate carries, as in addition) and write down the partial product. Second, we multiply the top number by 6 and write this second partial product under the first, but displaced one column to the left because each digit represents ten times as much as the corresponding digit in the first partial product. Third, we multiply the top number by 2 and write this third partial product below the second partial product, again displaced one digit to the left because each digit represents ten times as much as the corresponding digit in the second partial product. Then we add the partial products.

As with addition and subtraction, multiplication of binary numbers is much the same as multiplication of decimal numbers. First we'll make a table defining all possible results of 1-bit multiplications. Note that the multiplication of binary bits never generates a carry.

```
  0     0     1     1
×0    ×1    ×0    ×1
 --    --    --    --
  0     0     0     1
```

Now, using this table, we can do a binary multiplication. For example,

```
   1011      11
  ×1001       9
  -----      --
   1011      99
  0000
 0000
1011
-------
1100011
```

Notice that the product of these two 4-bit binary numbers is a 7-bit binary number. In some cases, such as the following example, the product can be as large as an 8-bit number.

```
     1110     14
    ×1100     12
    -----     --
     0000     28
    0000     14
   1110      ---
  1110       168
--------
10101000
```

In general, the product of two n-bit numbers is a number with, at most, twice as many bits. The product of an m-bit number and an n-bit number is, at most, an (n + m)-bit number.

Example 2.8-1

```
        10110011          179
       ×10100110          166
        00000000         1074
       10110011         1074
      10110011          179
     00000000          29714
    00000000
   10110011
  00000000
 10110011
 111010000010010
```

Example 2.8-2

```
        11000110          198
       ×11010001          209
        11000110         1782
     11000110           3960
   11000110            41382
 11000110
 1010000110100110
```

In this last example, we've simplified the computation by leaving out all the lines of zeros, which add nothing to the result.

2.9 DIVISION

Again, we've all used the simplification of positional numbers to do division without thinking about it. Look at an example of decimal division:

```
       2149
 123)264329
     246
      183
      123
       602
       492
       1109
       1107
          2
```

In division we determine the digits of the quotient, starting with the high order digit and using a repeated divide, multiply, and subtract. We divide 123 into the first three digits of 264329 and write the quotient 2 above the 4. This establishes the 2 in the 1000s position in the quotient. Now we multiply 123 by 2 and write the result so that it lines up with the 1000s digit position. Finally we subtract 246000 from 264329 to remove 2000 × 123. For simplification, we do this without writing out all the trailing zeros. The entire process is now repeated with the numbers remaining, to find the next lower order digit. This procedure is repeated until all the digits of 264329 are exhausted. The final difference (2) is the remainder.

As with addition, subtraction, and multiplication, division of binary numbers is much the same as division of decimal numbers. Look at the example:

```
        10001            17
 1101)11101001        13)232
      1101               13
      --------           ---
         11001           103
          1101            91
         -----           ---
          1100            12
```

First we divide 1101 into the first four bits of 11101001. Here the division is easier than for decimal numbers because the quotient is either 1, if the number formed from the first four bits is larger than 1101, or 0 if it's smaller. Since the partial quotient is 1, we copy 1101 below the first four digits of 11101001 and subtract. For binary division the multiplication is easy because it's a multiplication by either 1 (copy the number) or 0. When the quotient is 0 we do not show the subtraction. Notice that the 3 bits following the high order 1 are all zero, so no subtraction is shown. The last difference (1100) is the remainder.

Notice also that we divided an 8-bit number by a 4-bit number and the result was a 5-bit number. In general, if we divide an m-bit number by an n-bit number, the result is a number with m − n or m − n + 1 bits. This point can be important when we consider computers, which store numbers of only a certain length.

Example 2.9-1

```
            1010101            85
 11101)100110100101        29)2469
       11101                  232
       ------------           ---
        100110                149
         11101                145
        -------               ---
          100101                4
           11101
          -------
            100001
             11101
            ------
               100
```

Example 2.9-2

```
                 10011101                  157
        10111001)111000110010111       185)29079
                 10111001                  185
                  101010001                1057
                   10111001                 925
                   100110000               1329
                    10111001               1295
                     11101111                34
                     10111001
                      11011011
                      10111001
                        100010
```

2.10 TWOs COMPLEMENT BINARY NUMBERS

Because binary numbers are used for machine calculations, negative binary numbers are not handled in the same way as negative decimal numbers. The use of **2s complement numbers** allows additions involving both positive and negative numbers to be handled as simple additions. Their use also simplifies subtraction.

Twos complement numbers are simply numbers in which the high order bit position is assigned a negative positional value. For an n-bit, 2s complement binary number N,

$$N = b_{n-1}b_{n-2} \ldots b_0 \quad (2.10\text{-}1)$$

means

$$N = -b_{n-1} \times 2^{n-1} + b_{n-2} \times 2^{n-2} + \ldots + b_0 + 2^0 \quad (2.10\text{-}2)$$

Thus, in an 8-bit, 2s complement number, the high order bit has a positional value of −128 rather than 128. To see the effect of this assignment, we'll find the decimal values for several different 8-bit, 2s complement binary numbers. We'll do this just as with positive numbers, by writing the positional value above the bit. Evaluating N = 01110101,

−128	64	32	16	8	4	2	1
0	1	1	1	0	1	0	1

```
  64
  32
  16
   4
   1
 ---
 117
```

Evaluating N = 10110100,

−128	64	32	16	8	4	2	1
1	0	1	1	0	1	0	0

$$\begin{array}{r} -128 \\ 32 \\ 16 \\ 4 \\ \hline -76 \end{array}$$

As we can see from these examples, the presence of a 1 in the high order bit means the number is negative; however, although this 1 is sometimes called a sign bit, it's not a sign bit. Notice that if a number is in 2s complement form, we must specify its length because we need to know which is the bit with negative positional value. It is instructive to list some representative 8-bit positive numbers and their decimal values, as follows.

Binary	Decimal
01111111	127
01111110	126
...	...
00000011	3
00000010	2
00000001	1
00000000	0
11111111	−1
11111110	−2
11111101	−3
...	...
10000010	−126
10000001	−127
10000000	−128

The positive numbers look much like normal binary numbers except that the high order bit is always 0. The negative numbers are quite different, but they're related to the positive numbers of the same values. We'll soon see how to form them from the positive numbers. One unusual feature of these numbers is the asymmetry between the positive and negative values. For 8-bit numbers the largest number is 127, and the smallest number is −128.

A 2s complement number need not be 8 bits long; it could be 16 bits long, for example. In this case the high order bit would have a positional value of

$$-2^{15} = -32768$$

and the largest positive number that could be represented would be 32767, whereas the smallest number would be −32768. In general, the largest number that can be represented in an n-bit, 2s complement number system is

$$2^{n-1} - 1$$

and the smallest number is

$$-2^{n-1}$$

Example 2.10-1

Evaluate the 12-bit, 2s complement binary number 011011001011 as a decimal number.

−2048	1024	512	256	128	64	32	16	8	4	2	1
0	1	1	0	1	1	0	0	1	0	1	1

$$\begin{array}{r} 1024 \\ 512 \\ 128 \\ 64 \\ 8 \\ 2 \\ \underline{1} \\ 1739 \end{array}$$

Example 2.10-2

Evaluate the 12-bit, 2s complement binary number 111011001011 as a decimal number.

−2048	1024	512	256	128	64	32	16	8	4	2	1
1	1	1	0	1	1	0	0	1	0	1	1

$$\begin{array}{r} -2048 \\ 1024 \\ 512 \\ 128 \\ 64 \\ 8 \\ 2 \\ \underline{1} \\ -309 \end{array}$$

2.11 ADDITION OF TWOs COMPLEMENT NUMBERS

Twos complement numbers can simply be added directly, ignoring the carry-out of the high order bit—independent of whether the numbers are both positive, both negative, or one negative and one positive. The only problem that can occur is an arithmetic overflow or underflow. The sum may be too large or too small to be represented as a 2s complement number of the length used in the calculation.

To see how this addition works, look at some examples.

$$
\begin{array}{rcr}
01101101 & & 109 \\
+00010010 & & 18 \\
\hline
01111111 & & 127
\end{array}
$$

$$
\begin{array}{rcr}
01011011 & & 91 \\
+01000100 & & 68 \\
\hline
10011111 & (-97) & 159
\end{array}
$$

The number in parentheses is the decimal equivalent of the binary sum. It's obviously incorrect.

$$
\begin{array}{rcr}
10010110 & & -106 \\
+01100001 & & 97 \\
\hline
11110111 & & -9
\end{array}
$$

$$
\begin{array}{rcr}
11111000 & & -8 \\
+11111101 & & -3 \\
\hline
11110101 & & -11
\end{array}
$$

In this example, we ignore the carry-out of the high order bit.

$$
\begin{array}{rcr}
10000011 & & -125 \\
10011111 & & -97 \\
\hline
00100010 & (34) & -222
\end{array}
$$

Here we ignore the carry also, but the binary sum is incorrect. The decimal equivalent of the binary sum is shown in parentheses.

Notice that in all of the above examples involving 8-bit, 2s complement binary numbers, simple addition gives the correct answer unless the answer is too large or too small to be represented as an 8-bit, 2s complement binary number (arithmetic overflow or underflow). This kind of error can occur only if we add two positive numbers or two negative numbers. For mixed numbers, simple addition always gives a correct sum. When the addition is not mixed, there is a simple test to determine whether the sum is correct. If we add two positive numbers and the sum is positive, the sum is correct. If we add two negative numbers and the sum is negative, the sum is correct.

Also notice that the arithmetic overflow is not determined by the carry out of the high order bit. It can be determined by looking at both the carry out of the high order bit and the carry out of the bit next to the high order bit. To see how this is done, first consider the sum of two positive 8-bit numbers:

$$\begin{array}{r} 0XXXXXXX \\ \underline{0XXXXXXX} \end{array}$$

where the Xs represent arbitrary bits (0s or 1s). If there is no carry from the seventh to the eighth order bit, the sum is positive and therefore correct. If there is a carry from the seventh to the eighth order bit, the sum is negative and therefore incorrect. In either case, there is no carry out of the eighth order bit. Now consider the sum of a positive 8-bit number and a negative 8-bit number:

$$\begin{array}{r} 1XXXXXX \\ \underline{0XXXXXX} \end{array}$$

The sum here is always correct, but notice that if there is a carry from the seventh order bit to the eighth order bit, there is also a carry out of the eighth order bit. Finally, consider the sum of two negative numbers:

$$\begin{array}{r} 1XXXXXX \\ \underline{1XXXXXX} \end{array}$$

If there is no carry from the seventh order bit to the eighth order bit, then the sum is positive and incorrect. If there is a carry from the seventh order bit to the eighth order bit, then the sum is negative and correct. In both cases, there is a carry out of the eighth order bit. We can summarize these results quite succinctly: *Considering both the carry from the seventh to the eighth order bit and the carry-out of the eighth order bit, zero or two carries means a correct sum, and one carry means an incorrect sum.* This is the way computers determine arithmetic overflow. In hand calculations, however, we'll continue to use the rules in the previous paragraph.

Example 2.11-1

Add the following 12-bit, 2s complement binary number and state whether the binary sum is correct.

$$\begin{array}{rr} 010010011001 & 1177 \\ \underline{+111110100000} & \underline{-96} \\ 010000111001 & 1081 \end{array}$$

The sum in a mixed addition is always correct.

Example 2.11-2

Add the following 12-bit, 2s complement binary number and state whether the binary sum is correct.

$$\begin{array}{rlr} 100010011001 & & -1895 \\ +110110100000 & & -608 \\ \hline 011000111001 & (1593) & -2503 \end{array}$$

The sum is incorrect. The sum of two negative numbers results in a positive number.

2.12 SUBTRACTION OF TWOs COMPLEMENT NUMBERS

In hand calculations, 2s complement numbers can be subtracted directly, with a high order borrow if needed—independent of whether the numbers are both positive, both negative, or one negative and one positive. The only problem that can occur is an arithmetic overflow or underflow. The difference may be too large or too small to be represented as a 2s complement number of the length used in the calculation.

To see how this subtraction works, let's look at some examples.

$$\begin{array}{rr} 01101101 & 109 \\ -00010010 & -18 \\ \hline 01011011 & 91 \end{array}$$

$$\begin{array}{rlr} 01011011 & & 91 \\ -10000100 & & -(-124) \\ \hline 11010111 & (-41) & 215 \end{array}$$

In the preceding example, we need a borrow in the high order bit. The number in parentheses is the decimal equivalent of the binary difference. It's obviously incorrect.

$$\begin{array}{rlr} 10010110 & & -106 \\ -01100001 & & -97 \\ \hline 00110101 & (53) & -203 \end{array}$$

In the preceding example, the number in parentheses is the decimal equivalent of the binary difference. It's obviously incorrect.

$$\begin{array}{rr} 11111000 & -8 \\ -11111101 & -(-3) \\ \hline 11111011 & -5 \end{array}$$

In the preceding example, we need a borrow in the high order bit.

10000011	−125
− 10011111	−(−97)
11100100	−28

In the preceding example, we need a borrow in the high order bit.

Notice that in all of the foregoing examples involving 8-bit, 2s complement binary numbers, simple subtraction gives the correct answer unless the answer is too large or too small to be represented as an 8-bit, 2s complement binary number (arithmetic overflow or underflow). This kind of error can occur only if we subtract mixed numbers because this is the same as adding two positive or two negative numbers. If subtraction is not mixed, the difference is always correct. When the two terms have different signs, we can determine whether or not the difference is correct by changing the sign of the bottom term and applying the foregoing rule for addition. This means, more simply, that the difference must have the same sign as the top term in order to be correct.

Using this idea on the second subtraction example, we find that the subtraction is equivalent to the addition of two positive numbers, which should give a positive sum. In the next example, the subtraction is equivalent to the addition of two negative numbers, which should give a negative sum.

Example 2.12-1

Subtract the following 12-bit, 2s complement binary number and state whether the binary difference is correct.

010010011001	1177
− 111110100000	−(−96)
010011111001	1273

This subtraction is equivalent to the addition of two positive numbers, so the result is a positive number and is therefore correct. To put it another way, the result is the same sign as the top term and is therefore correct.

Example 2.12-2

Subtract the following 12-bit, 2s complement binary number and state whether the binary difference is correct.

100010011001	−1895
− 110110100000	−(−608)
101011100001	−1187

This subtraction is the difference of two like numbers, and the difference is always correct.

2.13 COMPLEMENTS AND SUBTRACTION

Although we find it convenient to make hand calculations using the ideas of subtraction presented in the last section, computers do not subtract. Computers negate (change the sign of) the second term in a subtraction and add. Negating a 2s complement binary number is much like complementing the number. So let's look first at the complement and then at the negation of a 2s complement binary number.

The simple complement of a binary number (sometimes called the 1s complement) is formed by changing all the 0 bits to 1 and all the 1 bits to 0. Some examples of binary numbers and their complements follow, with the complements under the binary numbers.

10101100	10000110	11111111	10000001
01010011	01111001	00000000	01111110

Notice that in each case, each number is the complement of the other. The complement of the complement of a binary number is the binary number itself.

Now let's find how to negate a 2s complement binary number. Suppose we have an 8-bit, 2s complement binary number:

$$N_2 = b_7b_6b_5b_4b_3b_2b_1b_0$$

To negate this number, we can subtract it from zero:

$$\begin{array}{r} 0\ 0\ 0\ 0\ 0\ 0\ 0\ 0 \\ -\,b_7\ b_6\ b_5\ b_4\ b_3\ b_2\ b_1\ b_0 \\ \hline \end{array}$$

Rather than do a straight subtraction, we'll manipulate the zero top term to simplify the procedure. Subtracting 00000001 from the 8-bit zero,

$$\begin{array}{r} 00000000 \\ -\,00000001 \\ \hline 11111111 \end{array}$$

So

$$00000000 = 11111111 + 1$$

Now, going back to the negation, we can write it as

$$\begin{array}{r} 1\ 1\ 1\ 1\ 1\ 1\ 1\ 1\ \ +1 \\ -\,b_7\ b_6\ b_5\ b_4\ b_3\ b_2\ b_1\ b_0 \\ \hline \overline{b}_7\ \overline{b}_6\ \overline{b}_5\ \overline{b}_4\ \overline{b}_3\ \overline{b}_2\ \overline{b}_1\ \overline{b}_0\ \ +1 \end{array}$$

where $\overline{b}_i$ is the simple complement of the bit b_i, because $1 - 1 = 0$ and $1 - 0 = 1$. To negate, we complement and add one. The preceding demonstration is for an 8-bit, 2s complement binary number. The same procedure works for a number of any length. This process of number negation is called taking the *2s complement* of the number. Be sure to understand the distinction between the 2s complement *of* a number and a *2s complement number.* The first is the *negative of a number.* The second is a *way of representing negative (or positive) binary numbers.*

Example 2.13-1

Negate the 8-bit, 2s complement binary number 10110100.

Complementing each bit and adding 1, we get

10110100	−76
01001011	
+1	
01001100	76

Example 2.13-2

Negate the 8-bit, 2s complement binary number 00100000.

Complementing each bit and adding 1, we get

00100000	32
11011111	
+1	
11100000	−32

Example 2.13-3

Negate the 8-bit, 2s complement binary number 10110101.

Complementing each bit and adding 1, we get

10110101	−75
01001010	
+1	
01001011	75

These examples show a pattern that allows the addition of 1 to be performed simply. Starting at the right end of the number, we change all the 1s to 0s until we find the first 0, which we change to 1. With practice we can complement and add 1 in one step.

Example 2.13-4

Negate the 12-bit, 2s complement number 010010110100.

Complementing each bit and adding 1, we get

010010110100	1204
101101001011	
+1	
101101001100	−1204

Notice that we change the last two 1s to 0s and the next 0 to 1.

Example 2.13-5

Negate the 12-bit, 2s complement number 110011100100. Complementing each bit and adding 1, we get

110011100100	−796
001100011011	
+1	
001100011100	796

Notice that we change the last two 1s to 0s and the next 0 to 1.

Negation gives us an easy way to find the value of a 2s complement negative number without using the negative positional idea. We can simply negate the number and find the value of the equivalent positive number. For example,

$$10001001 = -01110111 = -119_{10}$$

Suppose we want to change from negative decimal numbers to 2s complement binary numbers. We do this in two steps, using the 2s complement. First, ignore the sign of the decimal number and change it to binary, using the repeated division by 2 already described (multiplication by 2 for fractions). Second, negate the number (take the 2s complement). For example, to find the 8-bit, 2s complement binary representation of −100, first find the binary representation, ignoring the sign.

2)100	
2) 50	0
2) 25	0
2) 12	1
2) 6	0
2) 3	0
2) 1	1
2) 0	1

$$100_{10} = 1100100_2$$

Second, since we want an 8-bit, 2s complement binary representation, we need to pad with zeros on the left until we have 8 bits and then take a 2s complement.

$$\begin{array}{r} 01100100 \\ 10011011 \\ +1 \\ \hline 10011100 \end{array}$$

$$-100_{10} = 10011100_2$$

Example 2.13-6

(a) Find the 8-bit, 2s complement binary equivalent of −91.

(b) Find the 12-bit equivalent.

(c) Find the 16-bit equivalent.

(a) First, find the binary equivalent of 91.

$$\begin{array}{rl} 2\underline{)91} & \\ 2\underline{)45} & 1 \\ 2\underline{)22} & 1 \\ 2\underline{)11} & 0 \\ 2\underline{)\ 5} & 1 \\ 2\underline{)\ 2} & 1 \\ 2\underline{)\ 1} & 0 \\ 0 & 1 \end{array}$$

$$91_{10} = 1011011_2$$

To find the 8-bit, 2s complement representation of −91, pad with zeros out to 8 bits and take the 2s complement:

$$\begin{array}{r} 01011011 \\ 10100100 \\ +1 \\ \hline 10100101 \end{array}$$

$$-91_{10} = 10100101_2$$

(b) To find the 12-bit representation, pad to 12 bits and take the 2s complement.

$$\begin{array}{r} 000001011011 \\ 111110100100 \\ +1 \\ \hline 111110100101 \end{array}$$

$$-91_{10} = 111110100101_2$$

(c) To find the 16-bit representation, pad to 16 bits and take the 2s complement.

$$\begin{array}{r} 0000000001011011 \\ 1111111110100100 \\ +1 \\ \hline 1111111110100101 \end{array}$$

$$-91_{10} = 1111111110100101_2$$

This example shows us how to extend the length of a negative 2s complement binary number: pad with 1s on the left.

Now that we know how to negate a number, we can perform subtraction like machines. To perform the 8-bit, 2s complement binary subtraction,

$$\begin{array}{r} 10101100 \\ -01111011 \\ \hline \end{array}$$

we perform the equivalent addition

$$\begin{array}{r} 10101100 \\ +10000101 \\ \hline 00110001 \end{array}$$

But this is an incorrect answer (arithmetic underflow) because the sum of two negative numbers gave a positive sum.

Example 2.13-7

Do the 8-bit, 2s complement subtraction.

$$\begin{array}{r} 11110100 \\ -10001111 \\ \hline \end{array} \qquad \begin{array}{r} 11110100 \\ +01110001 \\ \hline 01100101 \end{array}$$

This answer is correct because the sum of mixed numbers is always correct.

Example 2.13-8

Do the 12-bit, 2s complement subtraction.

$$\begin{array}{r} 101111110100 \\ -\,101100001111 \\ \hline \end{array} \qquad \begin{array}{r} 101111110100 \\ +\,010011110001 \\ \hline 000011100101 \end{array}$$

The answer is correct because the sum of mixed numbers is always correct.

We will not discuss techniques for multiplying and dividing 2s complement binary numbers. If we need to multiply or divide, we can always change all the numbers to positive numbers and then assign a sign to the answer based on the signs of the two factors involved. This procedure is the same as with decimal numbers. If we take the product of two numbers with like signs, the answer is positive. If the signs are different, the answer is negative. If we take the quotient of two numbers with like signs, the answer is positive. If the signs are different, the answer is negative.

2.14 CODING

In this section we're going to put a name to some of the procedures we've already learned. We've designed a system of numbers that can be used in machines; this is called **coding.** We've designed a particular code for positive numbers only, and another code for both positive and negative numbers (n-bit, 2s complement numbers). The codes we've designed come naturally (a 1 is always a 1), but there are many other ways we can code numbers. One rather arbitrary code for representing numbers is the **unit distance code**. It's designed so that only one bit changes with each number change. A unit distance code for the numbers 0 through 15 could be

Number	Code	Number	Code
0	0000	8	1100
1	0001	9	1101
2	0011	10	1111
3	0010	11	1110
4	0110	12	1010
5	0111	13	1011
6	0101	14	1001
7	0100	15	1000

This kind of code is useful in some digital operations—in particular, state machines, which will be discussed later.

Moreover, codes are not used only for representing numbers. A standard 7-bit code is used for representing alphanumeric characters and symbols; it is called the **ASCII** (American Standard Code for Information Interchange) **code.** The ASCII code for common alphanumeric characters and symbols is given in Table 2.14-1. Included is the equivalent hexadecimal for each of the binary codes. The hexadecimal form is much more compact.

Table 2.14-1 ASCII Code

Character	Code	Hex	Character	Code	Hex	Character	Code	Hex
Space	0100000	20	@	1000000	40	‘	1100000	60
!	0100001	21	A	1000001	41	a	1100001	61
″	0100010	22	B	1000010	42	b	1100010	62
#	0100011	23	C	1000011	43	c	1100011	63
$	0100100	24	D	1000100	44	d	1100100	64
%	0100101	25	E	1000101	45	e	1100101	65
&	0100110	26	F	1000110	46	f	1100110	66
’	0100111	27	G	1000111	47	g	1100111	67
(	0101000	28	H	1001000	48	h	1101000	68
)	0101001	29	I	1001001	49	i	1101001	69
*	0101010	2A	J	1001010	4A	j	1101010	6A
+	0101011	2B	K	1001011	4B	k	1101011	6B
,	0101100	2C	L	1001100	4C	l	1101100	6C
–	0101101	2D	M	1001101	4D	m	1101101	6D
.	0101110	2E	N	1001110	4E	n	1101110	6E
/	0101111	2F	O	1001111	4F	o	1101111	6F
0	0110000	30	P	1010000	50	p	1110000	70
1	0110001	31	Q	1010001	51	q	1110001	71
2	0110010	32	R	1010010	52	r	1110010	72
3	0110011	33	S	1010011	53	s	1110011	73
4	0110100	34	T	1010100	54	t	1110100	74
5	0110101	35	U	1010101	55	u	1110101	75
6	0110110	36	V	1010110	56	v	1110110	76
7	0110111	37	W	1010111	57	w	1110111	77
8	0111000	38	X	1011000	58	x	1111000	78
9	0111001	39	Y	1011001	59	y	1111001	79
:	0111010	3A	Z	1011010	5A	z	1111010	7A
;	0111011	3B	[	1011011	5B	{	1111011	7B
<	0111100	3C	\	1011100	5C	¦	1111100	7C
=	0111101	3D	]	1011101	5D	}	1111101	7D
>	0111110	3E	^	1011110	5E	~	1111110	7E
?	0111111	3F	_	1011111	5F	Delete	1111111	7F

Notice that there are codes for the digits 0 through 9 and that these are different from the binary number codes. Very often the ASCII code is stored in machines that have provisions for storing codes in 8-bit groups (bytes); when this is the case, each ASCII code is padded with a 0 on the left. The code for 'Alan' would thus be

01000001 01101100 01100001 01101110

ASCII is designed with a simple difference between capital and lowercase letters. The code corresponding to the capital has a 0 in the sixth bit from the right; the code corresponding to the lowercase letter has a 1. This makes it easy for machines to handle the shift key on a keyboard.

Example 2.14-1

Translate the following ASCII code into characters and symbols.

(a) 01001000 01100101 01101100 01110000
H e l p

(b) 00110011 00110111 00111000 00101110 00110101
3 7 8 . 5

Example 2.14-2

Write the following characters in an ASCII code.

(a) Code: 01000011 01101111 01100100 01100101

(b) 261.3: 00110010 00110110 00110001 00101110 00110011

In these examples we have inserted spaces between the code groups to clarify what we're doing. In actual fact, an ASCII code would not have spaces.

2.15 BCD NUMBERS

Let's look at one last way of coding numbers for machines. In this system, we separately code each decimal digit as a binary number. This gives the code its name, **Binary Coded Decimal** (BCD). To code the number 37, we code 3 as 0011 and 7 as 0111, so 37 is coded as 00110111. Each digit is coded in 4 bits because 4 bits are needed for the digits 8 and 9. To convert a BCD number to a decimal number, we separate it into groups of 4 bits, and then convert each group to a decimal digit. In the following table the conversion from decimal digits to BCD is given for every digit from 0 to 9.

Decimal Digit	BCD Code
0	0000
1	0001
2	0010
3	0011
4	0100
5	0101
6	0110
7	0111
8	1000
9	1001

Notice that this code is not as efficient as the straight binary code, because we use 4 bits to code ten different digits. But 4 bits can code as many as 16 different digits, so some combinations of 4 bits are not valid codes for decimal digits.

To convert a BCD number to a decimal, separate it into 4-bit groups starting from the right, and use the preceding table to find the decimal digits for each group.

Example 2.15-1

Convert the following BCD numbers to decimal.

(a) 100101010010 = 1001 0101 0010 = 952_{10}

(b) 1000011000010100 = 1000 0110 0001 0100 = 8614_{10}

(c) 110001010111 = 1100 0101 0111 This is not a BCD number; the first group is not a decimal digit because it's the number 12.

To convert from decimal numbers, we use the reverse process, representing each decimal digit with a 4-bit binary code.

Example 2.15-2

Convert the following decimal numbers to BCD.

(a) 324 = 0011 0010 0100

(b) 1065 = 0001 0000 0110 0101

(c) 32791 = 0011 0010 0111 1001 0001

In this example, we have again left a space between the code groups to clarify what we're doing. An actual BCD number would not have spaces.

Now let's look at the addition and subtraction of BCD numbers. If we simply add BCD numbers as we did binary numbers, we do not always get the correct answer. Consider the example

0011 1000	38
+0100 0001	+41
0111 1001	79

In the decimal addition shown on the right, we can see that simple binary addition gives a correct sum. Now consider the example

1000 0111	87
+1001 0100	+94
1 0001 1011	181

Here, binary addition does not give a correct BCD sum. In fact, the code group 1011 for the least significant digit is not even a decimal digit, and the code group 0001 for the next significant digit is simply not correct. In both cases the problem occurs because the sum of the 4-bit numbers is greater than 9. It's easy to correct the answer in order to arrive at a proper BCD number. If we add 0110 (6) to each of the two incorrect digits, then

```
  1 0001 1011
  +0110 0110
  1 1000 0001
```

and the sum is correct. To see why adding 0110 corrects the answer, let's suppose we have 1010 (10_{10}) for the result of a BCD addition. In decimal addition, this should give us a 0 digit and a carry of 1. But adding 0110 (6_{10}) to 1010 (10_{10}) will do just this

```
   0110
  +1010
  10000
```

which generates a carry into the next digit and a 0 digit. We can further see that adding 0110 to 1011 (11_{10}) generates a carry and a 1 digit and that, in general, adding

6 to any digit that is 10 or larger or that generates a carry corrects the digit and generates a carry when needed.

The procedure just outlined allows us to state a rule for BCD addition:

1. Do a normal binary addition.
2. Correct by adding 0110 to those digits (4-bit code groupings) in which the digit is too large *or* a carry was generated into the next digit. When we make the correction, we start on the right and make normal carries from one 4-bit code (decimal digit) to the next.

BCD subtraction can be done using simple binary subtraction and then correcting by subtracting 0110 from the digits that caused a borrow from the next higher order digit. The correction in this case is for the fact that when we borrow, we actually borrow 16 rather than 10. As we can see, the correction rule for subtraction is simpler than that for addition.

Example 2.15-3

Do the following BCD additions.

(a)

```
   1000 0110 0111        867
  +0011 0010 1000       +328
  ---------------       ----
   1011 1000 1111       1195
  +0110      0110
  ---------------
 1 0001 1001 0101
```

(b)

```
   0011 0101 1000        358
  +0010 0110 0010       +262
  ---------------       ----
   0101 1011 1010        620
  +     0110 0110
  ---------------
   0110 0010 0000
```

(c)

```
      1
   0110 1000 0110        686
  +0011 1000 0001       +381
  ---------------       ----
   1010 0000 0111       1067
  +0110 0110
  ---------------
 1 0000 0110 0111
```

Example 2.15-4

Subtract the following BCD numbers.

(a)

BCD	Decimal
011 001	
1000 0110 0111	867
−0011 0010 1000	−328
0101 0011 1111	539
−0110	
0101 0011 1001	

(b)

BCD	Decimal
0 10 01	
0011 0101 1000	358
−0010 0110 0010	−262
0000 1111 0110	96
−0110	
0000 1001 0110	

(c)

BCD	Decimal
0 0	
0101 1000 0110	586
−0011 1000 0001	−381
0010 0000 0101	205

We haven't yet addressed the problem of subtraction of BCD numbers when the lower number is larger than the upper number, resulting in a negative difference. In fact, we haven't addressed the problem of the coding of negative BCD numbers. A common way to code negative BCD numbers is to use a 10s complement, which is similar to a 2s complement number but defined for decimal numbers. We will not develop 10s complement numbers in this book.

KEY TERMS

Positional Numbers Numbers composed of symbols called digits in which the digits have a value based on position as well as an intrinsic value.

Digits Symbols used in positional numbers.

Base or Radix The difference in positional value between two digits which are next to each other in a positional number system.

Radix Point The point that separates digits with a positional value greater than 1 from those with a positional value less than 1 (a general decimal point).

Binary Numbers Positional number with base 2.
Bits Digits of a binary system. These may have value 1 or 0.
Octal Number Positional number system with base 8.
Hexadecimal Number Positional number system with base 16.
Twos Complement Numbers A binary number system modified to handle negative as well as positive numbers.
Twos Complement A process for negating a 2s complement number.
Coding A processing of assigning some sequence of bits (0s and 1s) to an entity.
ASCII Code A standard coding for alphanumeric characters (alphabetic characters, symbols, and decimal digits).
Binary Coded Decimal (BCD) A coding for numbers in which each decimal digit is coded as a 4-bit binary number.

EXERCISES

1. Convert the following binary numbers to decimal numbers.

(a) 101101 (b) 11101010
(c) 1011.1101 (d) 101111.01
(e) 1000000000 (f) 0.00000001
(g) 11010110 (h) 111101100
(i) 10101010 (j) 1101.1101
(k) 111.00111 (l) 0.0011011
(m) 101011.001 (n) 1001.1011
(o) 110001101 (p) 1100.0011

2. Convert the following decimal numbers to binary numbers.

(a) 201 (b) 4013 (c) 50.25
(d) 0.32 (e) 21.61 (f) 153
(g) 132 (h) 89 (i) 827
(j) 112 (k) 1237 (l) 236
(m) 62 (n) 194 (o) 1000
(p) 71.035 (q) 61.23 (r) 0.3721
(s) 0.654 (t) 120.32 (u) 49.6

3. Convert the following binary numbers to octal, hexadecimal, and decimal equivalents.

(a) 101101011111 (b) 101101101
(c) 111111111111 (d) 101010110

(e) 101101011001 (f) 110110010
(g) 110110111101 (h) 101100011
(i) 111101011000 (j) 110100111
(k) 101110100110 (l) 101001101
(m) 111000101011 (n) 110101110
(o) 100000000001 (p) 100100100

4. Convert the following octal numbers to decimal.

(a) 721 (b) 632 (c) 301
(d) 777 (e) 125 (f) 651
(g) 261 (h) 332 (i) 361
(j) 173 (k) 101 (l) 732
(m) 221 (n) 350 (o) 435

5. Convert the following hexadecimal numbers to decimal.

(a) AA (b) 632 (c) 3D1
(d) AB7 (e) 9D6 (f) FF0
(g) 7FC (h) 23D (i) 832
(j) 259 (k) 7F3 (l) 5EF
(m) 22E (n) 261 (o) 348

6. Convert the following decimal numbers to both octal and hexadecimal.

(a) 33 (b) 267 (c) 125
(d) 72 (e) 165 (f) 89
(g) 277 (h) 300 (i) 100
(j) 133 (k) 175 (l) 210
(m) 151 (n) 152 (o) 182

7. Add the following positive integer binary numbers. Check each answer by converting to decimal numbers. State whether the answer is correct if it's limited to 8 bits.

(a) 10110111
    01001101
(b) 10111101
    10101100
(c) 11010011
    01110110

(d) 10011011
    11011001
(e) 11111101
    01101101
(f) 11010011
    11100011

(g) 11011000
    10110001
(h) 01111111
    10000110
(i) 11010111
    10010110

8. Subtract the following binary numbers as indicated. Check by converting to decimal. State whether the answer is correct if it's limited to 8 bits.

(a) $\begin{array}{r} 10111011 \\ -01110001 \\ \hline \end{array}$ (b) $\begin{array}{r} 01011011 \\ -10000011 \\ \hline \end{array}$ (c) $\begin{array}{r} 11010011 \\ -01110110 \\ \hline \end{array}$

(d) $\begin{array}{r} 10011011 \\ -11011001 \\ \hline \end{array}$ (e) $\begin{array}{r} 11111101 \\ -01101101 \\ \hline \end{array}$ (f) $\begin{array}{r} 11010011 \\ -11100011 \\ \hline \end{array}$

(g) $\begin{array}{r} 11011000 \\ -10110001 \\ \hline \end{array}$ (h) $\begin{array}{r} 01111111 \\ -10000110 \\ \hline \end{array}$ (i) $\begin{array}{r} 11101110 \\ -01011010 \\ \hline \end{array}$

9. Multiply the binary numbers in exercise 7. Check by converting to decimal.

10. Divide the following binary numbers. Check by converting to decimal.

(a) $10101111\,\overline{)1001101011101111}$

(b) $10001110\,\overline{)1001011011110111}$

(c) $10110110\,\overline{)1101100000101001}$

(d) $11010011\,\overline{)1001001001001110}$

(e) $10010110\,\overline{)1101110101001100}$

(f) $11010011\,\overline{)1110001000011000}$

11. Convert the following 8-bit, 2s complement numbers to decimal.

(a) 01101101 (b) 11101111 (c) 10000011

(d) 10011001 (e) 00111011 (f) 11111110

(g) 10110010 (h) 01011101 (i) 01101100

(j) 11101101 (k) 10100011 (l) 01011011

(m) 10111011 (n) 01110110 (o) 10111010

12. Convert the following 12-bit, 2s complement numbers to decimal.

(a) 010111110001 (b) 100010111001

(c) 100110010010 (d) 010100011110

(e) 100111011101 (f) 101001110110

(g) 010111101110 (h) 011110001101

13. Convert the following decimal numbers to 8-bit, 2s complement numbers.

(a) 105 (b) −28 (c) 76

(d) −25 (e) −101 (f) 127

(g) −62 (h) 87 (i) 56

(j) −77 (k) 110 (l) −1

(m) 6 (n) −23 (o) 63

14. Convert the following decimal numbers to 12-bit, 2s complement numbers.

(a) 1050	(b) −281	(c) 763
(d) −25	(e) −1013	(f) 127
(g) −627	(h) 876	(i) 560
(j) −77	(k) 110	(l) −1
(m) 6	(n) −235	(o) 631

15. Negate the following 8-bit, 2s complement numbers.

(a) 01111111	(b) 10000010	(c) 00001000
(d) 11110000	(e) 01011001	(f) 11111110
(g) 10110010	(h) 01011101	(i) 01101100
(j) 11101101	(k) 10100011	(l) 01011011
(m) 10111011	(n) 01110110	(o) 10111010

16. Negate the following 12-bit, 2s complement numbers.

(a) 010111110001	(b) 100010111001
(c) 100110010010	(d) 010100011110
(e) 100111011101	(f) 101001110110
(g) 010111101110	(h) 011110001101

17. Add the following 8-bit, 2s complement numbers and indicate whether or not the sum is a correct 8-bit, 2s complement answer.

(a) 10010011 01101101	(b) 10011011 10011011	(c) 01011011 01011010
(d) 10101101 11101101	(e) 11111101 01101101	(f) 11010011 11100011
(g) 11011000 10110001	(h) 01111111 10000110	(i) 11010111 10010110

18. Subtract the following 8-bit, 2s complement binary numbers as indicated. State whether the difference is a correct 8-bit, 2s complement answer.

(a) 10010011 −01101101	(b) 10011011 −10011011	(c) 01011011 −01011010
(d) 10011011 −11011001	(e) 11111101 −01101101	(f) 11010011 −11100011
(g) 11011000 −10110001	(h) 01111111 −10000110	(i) 11101110 01011010

19. Translate the following ASCII code into characters.

(a) 01010100 01101000 01101001 01110011 00100000 01101001 01110011 00100000 01110100 01101000 01100101 01000000 01110000 01101001 01110100 01110011

(b) 01001000 01011001 01101100 01101100 01101111 00101100 00100000 01110011 01110000 01100101 01100101 01101011 00100000 01000001 01010011 01000011 01001001 01001001

20. Convert the following BCD codes to decimal numbers.

(a) 0101 0111 1000
(b) 1001 0011 0100
(c) 1000 1001 0110
(d) 0101 0011 0110
(e) 1001 0001 0111
(f) 1000 0101 0110
(g) 0101 0110 0011
(h) 1001 0110 0100

21. Convert the following decimal numbers to BCD.

(a) 100
(b) 372
(c) 25
(d) 89
(e) 721
(f) 138
(g) 459
(h) 898
(i) 504
(j) 1063
(k) 2785
(l) 1057

22. Add the following BCD numbers.

(a)
```
0111 0101 1001 0000
0010 1001 0011 0101
-------------------
```
(b)
```
0011 0110 0001 0111
0111 0100 0110 1000
-------------------
```
(c)
```
1001 0101 1000 0111
0011 0101 0110 0010
-------------------
```
(d)
```
0110 1000 0011 0101
0100 1000 0101 1001
-------------------
```
(e)
```
1000 0100 0110 0101
0111 0011 1000 0111
-------------------
```
(f)
```
0111 0001 0010 0110
1001 0100 1001 0011
-------------------
```
(g)
```
1000 0101 0100 1000
0011 0110 0010 1000
-------------------
```
(h)
```
0011 1001 0010 0101
0001 1000 0111 0101
-------------------
```

23. Subtract the following BCD numbers as indicated.

(a)
```
  0111 0101 1001 0000
- 0010 1001 0011 0101
---------------------
```
(b)
```
  0011 0110 0001 0111
- 0001 0100 0110 1000
---------------------
```
(c)
```
  1001 0101 1000 0111
- 0011 0101 0110 0010
---------------------
```
(d)
```
  0110 1000 0011 0101
- 0100 1000 0101 1001
---------------------
```
(e)
```
  1000 0100 0110 0101
- 0111 0011 1000 0111
---------------------
```
(f)
```
  0111 0001 0010 0110
- 0001 0100 1001 0011
---------------------
```
(g)
```
  1000 0101 0100 1000
- 0011 0110 0010 1000
---------------------
```
(h)
```
  0011 1001 0010 0101
- 0001 1000 0111 0101
---------------------
```

3 Boolean Functions

3.1 EXAMPLE LOGIC CIRCUIT PROBLEM

Let's look at an example of a simple problem we might want to solve with a logic circuit. Suppose we plan to launch a missile from a launch facility. The decision to launch is partly under the control of a launch safety officer (called S), who usually signifies a "launch-ready" condition by throwing a switch. Suppose we do not entirely trust the launch safety officer, so we assign two other individuals to help him; call these individuals A and B. A and B also have switches to indicate a launch-ready condition, but we give them less authority than we do the launch safety officer. If either A or B signifies a launch-ready condition and the launch safety officer also signifies a launch-ready condition, then we have a launch-ready condition.

To design a logic circuit that will perform the function described in the problem, we first express the function as an equation. Such an equation, describing a launch-ready condition, might be

$$L = (A \text{ or } B) \text{ and } S$$

Notice that we need some way to denote switch-on conditions for A, B, and S, and we need some operators to show the relationship between the various switch-on conditions and the launch-ready condition.

Once we have an equation describing a launch-ready condition, we need to design an electronic circuit that will perform the operations of the equation. In this chapter we'll develop the mathematics needed to describe logic circuits. In the next chapter we'll show how to turn our descriptions into electronic circuits. The circuits we design from the logic expressions developed in this chapter will be simple or complex, depending on whether the logic expressions are simple or complex.

There are a number of sound engineering reasons for producing simple designs, including cost, ease of construction, speed, and even aesthetics. We will therefore devote a major portion of this chapter to developing methods to simplify logic expressions and becoming proficient in using those methods so we can produce simple logic circuits.

3.2 BOOLEAN ALGEBRA

In **Boolean algebra**, all variables and relations are two-valued. Because this algebra was first developed to deal with logical propositions that could be *true* or *false*, those are the two values a variable or relation can have in Boolean algebra. They are normally written as 1 and 0, with one (1) representing *true* and zero (0) representing *false*. If A is a **Boolean variable**, then

$$A = 1$$

means A is true, and

$$A = 0$$

means A is false. In our application of Boolean algebra to logic circuits, we'll find that *asserted* and *not asserted* are better names for the two values a logic variable can have.

Boolean algebra has three operators:

1. The unitary operator *not*, symbolized by an overbar. Thus, for the Boolean variable A, $\overline{A}$ means *not A*. The *not* operator is also called the complement, so $\overline{A}$ is the complement of A. The *not* operator is defined by the relations

$$\overline{1} = 0 \qquad (3.2\text{-}1)$$

$$\overline{0} = 1 \qquad (3.2\text{-}2)$$

2. The binary operator *and*, symbolized by a dot. The *and* of the Boolean variables A and B is written $A \cdot B$. Usually, if the meaning is clear, the dot is omitted and $A \cdot B$ is written as AB. We'll follow this latter notation. The *and* operator is defined by the relations

$$0 \cdot 0 = 0 \qquad (3.2\text{-}3)$$

$$0 \cdot 1 = 0 \qquad (3.2\text{-}4)$$

$$1 \cdot 0 = 0 \qquad (3.2\text{-}5)$$

$$1 \cdot 1 = 1 \qquad (3.2\text{-}6)$$

3. The binary operator *or*, symbolized by a plus sign. The *or* of the Boolean variables A and B is written $A + B$. The *or* operator is defined by the relations

$$1 + 1 = 1 \qquad (3.2\text{-}7)$$

$$1 + 0 = 1 \qquad (3.2\text{-}8)$$

$$0 + 1 = 1 \tag{3.2-9}$$

$$0 + 0 = 0 \tag{3.2-10}$$

Notice that in the definitions of the operators, the result of the operator operating on every possible combination of values can be explicitly given. This is characteristic of Boolean algebra.

The relations that define the *and* operation are duals of the relations that define the *or* operation. This means that the one relation can be formed from the other relation by changing ones (1) to zeros (0), zeros (0) to ones (1), pluses (+) to dots (·), and dots (·) to pluses (+).

The foregoing definitions allow us to write the following identities for the variable A.

$$1 + A = 1 \qquad 0 \cdot A = 0 \tag{3.2-11}$$

$$0 + A = A \qquad 1 \cdot A = A \tag{3.2-12}$$

$$A + A = A \qquad A \cdot A = A \tag{3.2-13}$$

$$A + \overline{A} = 1 \qquad A \cdot \overline{A} = 0 \tag{3.2-14}$$

$$\overline{\overline{A}} = A \tag{3.2-15}$$

To prove these identities, we can try every possible value of the variable A and show that the equality is true for every value. For example, if we substitute 1 for A in the first identity,

$$1 + 1 = 1$$

and if we substitute 0 for A,

$$1 + 0 = 1$$

so the identity is true.

The **commutative** and **distributive** laws hold for this algebra.

Commutative law:

$$A + B = B + A \tag{3.2-16}$$

$$AB = BA \tag{3.2-17}$$

Distributive law:

$$A + (BC) = (A + B)(A + C) \tag{3.2-18}$$

$$A(B + C) = (AB) + (AC) \tag{3.2-19}$$

Notice that both *or* (+) and *and* (·) are distributive in Boolean algebra, but only multiplication (·) is distributive in ordinary algebra. We'll sometimes write expressions like these without some of the parentheses, as

$$A + BC = (A + B)(A + C)$$
$$A(B + C) = AB + AC$$

When there are no parentheses, the order of operations is *and* before *or* (just as in ordinary algebra, where we multiply before adding).

The following three theorems will be useful: the absorption theorem, logic adjacency theorem, and DeMorgan's theorem. They can be proved by using the commutative and distributive laws and the preceding identities; or, if all else fails, they can be proved by substituting all possible values for the variables and showing that the relation is always true (perfect induction).

Absorption theorem:

$$A + AB = A \tag{3.2-20}$$

$$A(A + B) = A \tag{3.2-21}$$

Logic adjacency theorem:

$$AB + A\overline{B} = A \tag{3.2-22}$$

$$(A + B)(A + \overline{B}) = A \tag{3.2-23}$$

DeMorgan's theorem:

$$\overline{A + B} = \overline{A}\,\overline{B} \tag{3.2-24}$$

$$\overline{AB} = \overline{A} + \overline{B} \tag{3.2-25}$$

Other relations:

$$\overline{AB + AC} = (\overline{A} + \overline{B})(\overline{A} + \overline{C}) \tag{3.2-26}$$

$$AB + \overline{A}C + BC = AB + \overline{A}C \tag{3.2-27}$$

$$(A + B)(\overline{A} + C)(B + C) = (A + B)(\overline{A} + C) \tag{3.2-28}$$

$$AB + BC + CA = (A + B)(B + C)(C + A) \tag{3.2-29}$$

DeMorgan's theorem can be generalized for any expression. The complement of any expression is equal to the expression with each of the variables complemented, + replaced by ·, and · replaced by + throughout. For example,

$$\overline{A + BC} = \overline{A + (BC)} = \overline{A}(\overline{B} + \overline{C})$$

$$\overline{(A + B)(C + D)} = (\overline{A}\,\overline{B}) + (\overline{C}\,\overline{D})$$

Example 3.2-1

Prove the absorption theorem.

$$A + AB = A$$

$$A(A + B) = A$$

Proof: Using identity (3.2-12), the distributive law (3.2-19), and identities (3.2-11) and (3.2-12) in succession,

$$A + AB = A \cdot 1 + AB = A(1 + B) = A \cdot 1 = A$$

Using identity (3.2-12), the distributive law (3.2-18), and identities (3.2-11) and (3.2-12),

$$A(A + B) = (A + 0)(A + B) = A + (0 \cdot B) = A + 0 = A$$

Notice that the second proof is the exact dual of the first.

Example 3.2-2

Prove Theorem (3.2-29).

$$AB + BC + CA = (A + B)(B + C)(C + A)$$

Using the distributive law (3.2-19) and the absorption theorem (3.2-20) to manipulate the left side of the expression,

$$\begin{aligned}
(A + B)(B + C)(C + A) &= (AB + AC + BB + BC)(C + A) \\
&= (AB + BB + AC + BC)(C + A) \\
&= (AB + B + AC + BC)(C + A) \\
&= (B + AC + BC)(C + A) \\
&= (B + AC)(C + A) \\
&= BC + BA + ACC + ACA \\
&= BC + BA + AC + AC \\
&= BC + BA + AC \\
&= AB + BC + CA
\end{aligned}$$

Example 3.2-3

Prove DeMorgan's theorem.

$$\overline{A + B} = \overline{A}\,\overline{B}$$

$$\overline{AB} = \overline{A} + \overline{B}$$

These expressions will have to be proved by substitution of all possible values of the variables (perfect induction). For the first expression,

$$\overline{A + B} = \overline{0 + 0} = \overline{0} = 1 = 1 \cdot 1 = \overline{0} \cdot \overline{0} = \overline{A}\,\overline{B}$$

$$\overline{A + B} = \overline{0 + 1} = \overline{1} = 0 = 1 \cdot 0 = \overline{0} \cdot \overline{1} = \overline{A}\,\overline{B}$$

$$\overline{A + B} = \overline{1 + 0} = \overline{1} = 0 = 0 \cdot 1 = \overline{1} \cdot \overline{0} = \overline{A}\,\overline{B}$$

$$\overline{A + B} = \overline{1 + 1} = \overline{1} = 0 = 0 \cdot 0 = \overline{1} \cdot \overline{1} = \overline{A}\,\overline{B}$$

This expression is true for all possible values of A and B. For the second expression,

$$\overline{AB} = \overline{0 \cdot 0} = \overline{0} = 1 = 1 + 1 = \overline{0} + \overline{0} = \overline{A} + \overline{B}$$

$$\overline{AB} = \overline{0 \cdot 1} = \overline{0} = 1 = 1 + 0 = \overline{0} + \overline{0} = \overline{A} + \overline{B}$$

$$\overline{AB} = \overline{1 \cdot 0} = \overline{0} = 1 = 0 + 1 = \overline{1} + \overline{0} = \overline{A} + \overline{B}$$

$$\overline{AB} = \overline{1 \cdot 1} = \overline{1} = 0 = 0 + 0 = \overline{1} + \overline{1} = \overline{A} + \overline{B}$$

Example 3.2-4

Simplify the expression $AB + A\overline{B}\,\overline{C} + \overline{A}BC$.

If we expand the first term ($AB = ABC + AB\overline{C}$),

$$AB + A\overline{B}\,\overline{C} + \overline{A}BC = ABC + AB\overline{C} + A\overline{B}\,\overline{C} + \overline{A}BC$$

Now, grouping together the first and fourth terms and using logic adjacency in A,

$$ABC + \overline{A}BC = BC$$

and grouping together the second and third terms and using logic adjacency in B,

$$AB\overline{C} + A\overline{B}\,\overline{C} = A\overline{C}$$

so

$$AB + A\overline{B}\,\overline{C} + \overline{A}BC = BC + A\overline{C}$$

Example 3.2-5

Simplify the expression $\overline{B}\,\overline{C}\,\overline{D} + B\overline{C}\,\overline{D} + A\overline{C}D$.

If we just combine the first two terms using logic adjacency (shown graphically), we will not get the simplest expression.

$$\overline{B}\,\overline{C}\,\overline{D} + B\overline{C}\,\overline{D} + A\overline{C}D = \overline{C}\,\overline{D} + A\overline{C}D$$

To simplify further, we need to now expand the $\overline{C}\,\overline{D}$ term in A and $\overline{A}$, then simplify:

$$\overline{C}\,\overline{D} + A\overline{C}D = A\overline{C}\,\overline{D} + \overline{A}\,\overline{C}\,\overline{D} + A\overline{C}D = A\overline{C} + \overline{C}\,\overline{D}$$

Example 3.2-6

Simplify the expression $A + B + C + \overline{ABC}$.

To simplify here, use DeMorgan's theorem:

$$A + B + C + \overline{A} + \overline{B} + \overline{C} = 1$$

3.3 TRUTH TABLES

Boolean functions are functions of Boolean variables that can have the values true (1) and false (0), depending on the values of the variables. Because we can enumerate all combinations of the values of the variables in a function, we can define a function by using a table of all possible combinations of variables and the resulting function values. Consider the example from Section 3.1. The *truth table* for this function is

A	B	S	L
0	0	0	0
0	0	1	0
0	1	0	0
0	1	1	1
1	0	0	0
1	0	1	1
1	1	0	0
1	1	1	1

In this table we list all possible combinations of the variables A, B, and S. An easy way to do that is to list the binary numbers from 0 to 7. For each combination of variables we determine, from the description of the problem, whether or not the launch condition (L) is true and enter a 1 or a 0 accordingly. The launch condition is true only if A and S are true, B and S are true, or A, B, and S are all true.

For another example of truth tables, consider the addition of two 2-bit binary numbers,

$$\begin{array}{c} C\,A \\ \underline{D\,B} \\ F_2F_1 \end{array}$$

where A, B, C, and D represent the bits of the numbers to be added, and F_1 and F_2 represent the bits of the sum. Here we need two truth tables to define the functions F_1 and F_2. The first is a function of the two binary variables A and B. The second is a function of the four binary variables A, B, C, and D because of the carry from the low to high order bits.

A	B	F_1
0	0	0
0	1	1
1	0	1
1	1	0

A	B	C	D	F_2
0	0	0	0	0
0	0	0	1	1
0	0	1	0	1
0	0	1	1	0
0	1	0	0	0
0	1	0	1	1
0	1	1	0	1
0	1	1	1	0
1	0	0	0	0
1	0	0	1	1
1	0	1	0	1
1	0	1	1	0
1	1	0	0	1
1	1	0	1	0
1	1	1	0	0
1	1	1	1	1

In these two examples we've seen how to define a function using a truth table. We produced the tables directly from the specifications of the problem. Sometimes we want to form a truth table to describe a Boolean expression or function. Before we do that, however, let's learn to formulate a Boolean expression to represent a function described in a truth table.

3.4 MINTERMS AND MAXTERMS

Any Boolean function can be written in terms of simple expressions called **minterms** and **maxterms**. This is done most easily from the truth table for the function. Consider a simple truth table for a two-variable function, F.

A	B	F
0	0	1
0	1	0
1	0	0
1	1	1

The table contains four entries (rows). The minterm for each entry is formed by *anding* A and B or their complements, $\overline{A}$ and $\overline{B}$, in such a way that the minterm expression is true (1). For instance, the minterm for the first entry in the table,

$$m_0 = \overline{A}\,\overline{B}$$

is true (1) when $A = 0$ and $B = 0$. The minterms are given the symbol m_i, with i equal to 0 through $n - 1$ where n is the number of entries in the table. Forming the remaining minterms allows us to construct a table of minterms.

A	B	Minterm
0	0	$m_0 = \overline{A}\,\overline{B}$
0	1	$m_1 = \overline{A}B$
1	0	$m_2 = A\overline{B}$
1	1	$m_3 = AB$

Checking these minterms, we can see that if the values for the variables on the left are substituted into the minterm expression, the expression is true (1).

Now that we have the minterms, the function is easily formed by *oring* together the minterms for which the *function* is true (1). This form of the function is called the **Sum of Products (SOP)**. In our example the function is given by

$$F = \overline{A}\,\overline{B} + AB$$

Notice that there is a minterm for each of the entries in the truth table, but to form a function, we *or* only those minterms for which the function is true (1).

We can also write a function in terms of maxterms. To form the maxterms corresponding to the entries in the two-variable truth table, we complement the minterms. Thus the maxterm corresponding to m_0 is

$$M_0 = \overline{\overline{A}\,\overline{B}} = A + B$$

The maxterm expressions formed in this way are false (0) when the variable values from that row are substituted into the expression. If we complement all the minterms, we can form a table of maxterms.

A	B	Maxterm
0	0	$M_0 = A + B$
0	1	$M_1 = A + \overline{B}$
1	0	$M_2 = \overline{A} + B$
1	1	$M_3 = \overline{A} + \overline{B}$

The maxterms are labeled M_0, M_1, M_2, and M_3. Now, to see how to write a function in terms of maxterms, we first write the complement of the function in terms of minterms. For the foregoing example function F,

$$\overline{F} = \overline{A}B + A\overline{B}$$

If we complement this expression, we obtain

$$F = (A + \overline{B})(\overline{A} + B)$$

But this last expression shows us how to write the function directly in terms of maxterms. We simply *and* the maxterm expressions for which the function is 0. This form of the function is called the **Product of Sums (POS)**. Notice that, as with the minterms, there is a maxterm for each entry in the truth table, but *only* the maxterms for which the function is false (0) are used in writing the maxterm expression for the function.

Minterms and maxterms can be formed for any number of variables. Following are tables of minterms and maxterms for three and four variables.

Three-variable minterms and maxterms

A	B	C	Minterm	Maxterm
0	0	0	$m_0 = \overline{A}\,\overline{B}\,\overline{C}$	$M_0 = A + B + C$
0	0	1	$m_1 = \overline{A}\,\overline{B}C$	$M_1 = A + B + \overline{C}$
0	1	0	$m_2 = \overline{A}B\overline{C}$	$M_2 = A + \overline{B} + C$
0	1	1	$m_3 = \overline{A}BC$	$M_3 = A + \overline{B} + \overline{C}$
1	0	0	$m_4 = A\overline{B}\,\overline{C}$	$M_4 = \overline{A} + B + C$
1	0	1	$m_5 = A\overline{B}C$	$M_5 = \overline{A} + B + \overline{C}$
1	1	0	$m_6 = AB\overline{C}$	$M_5 = \overline{A} + \overline{B} + C$
1	1	1	$m_7 = ABC$	$M_7 = \overline{A} + \overline{B} + \overline{C}$

Four-variable minterms and maxterms

A	B	C	D	Minterm	Maxterm
0	0	0	0	$m_0 = \overline{A}\overline{B}\overline{C}\overline{D}$	$M_0 = A + B + C + D$
0	0	0	1	$m_1 = \overline{A}\overline{B}\overline{C}D$	$M_1 = A + B + C + \overline{D}$
0	0	1	0	$m_2 = \overline{A}\overline{B}C\overline{D}$	$M_2 = A + B + \overline{C} + D$
0	0	1	1	$m_3 = \overline{A}\overline{B}CD$	$M_3 = A + B + \overline{C} + \overline{D}$
0	1	0	0	$m_4 = \overline{A}B\overline{C}\overline{D}$	$M_4 = A + \overline{B} + C + D$
0	1	0	1	$m_5 = \overline{A}B\overline{C}D$	$M_5 = A + \overline{B} + C + \overline{D}$
0	1	1	0	$m_6 = \overline{A}BC\overline{D}$	$M_6 = A + \overline{B} + \overline{C} + D$
0	1	1	1	$m_7 = \overline{A}BCD$	$M_7 = A + \overline{B} + \overline{C} + \overline{D}$
1	0	0	0	$m_8 = A\overline{B}\overline{C}\overline{D}$	$M_8 = \overline{A} + B + C + D$
1	0	0	1	$m_9 = A\overline{B}\overline{C}D$	$M_9 = \overline{A} + B + C + \overline{D}$
1	0	1	0	$m_{10} = A\overline{B}C\overline{D}$	$M_{10} = \overline{A} + B + \overline{C} + D$
1	0	1	1	$m_{11} = A\overline{B}CD$	$M_{11} = \overline{A} + B + \overline{C} + \overline{D}$
1	1	0	0	$m_{12} = AB\overline{C}\overline{D}$	$M_{12} = \overline{A} + \overline{B} + C + D$
1	1	0	1	$m_{13} = AB\overline{C}D$	$M_{13} = \overline{A} + \overline{B} + C + \overline{D}$
1	1	1	0	$m_{14} = ABC\overline{D}$	$M_{14} = \overline{A} + \overline{B} + \overline{C} + D$
1	1	1	1	$m_{15} = ABCD$	$M_{15} = \overline{A} + \overline{B} + \overline{C} + \overline{D}$

Remember that we form the *minterm* for any row by *anding* together all the variables that are true (1) in that row and the complements of all the variables that are false (0) in that row. We form the *maxterm* for any row by *oring* together all the variables that are false (0) in that row and the complements of all the variables that are true (1) in that row.

Example 3.4-1

Write the function defined by the truth table in terms of minterms (sum of products) and maxterms (product of sums).

A	B	C	F
0	0	0	1
0	0	1	1
0	1	0	0
0	1	1	0
1	0	0	1
1	0	1	1
1	1	0	0
1	1	1	0

This function can be expressed in minterms (sum of products) as

$$F = \overline{A}\overline{B}\overline{C} + \overline{A}\overline{B}C + A\overline{B}\overline{C} + A\overline{B}C$$

This expression can be simplified (using logic adjacency) to

$$F = \overline{A}\,\overline{B}\,\overline{C} + \overline{A}\,\overline{B}C + A\overline{B}\,\overline{C} + A\overline{B}C$$

$$F = \overline{A}\,\overline{B} + A\overline{B} = \overline{B}$$

This function can be expressed in maxterms (product of sums) as

$$F = (A + \overline{B} + C)(A + \overline{B} + \overline{C})(\overline{A} + \overline{B} + C)(\overline{A} + \overline{B} + \overline{C})$$

This expression can also be simplified (using logic adjacency) to

$$F = (A + \overline{B} + C)(A + \overline{B} + \overline{C})(\overline{A} + \overline{B} + C)(\overline{A} + \overline{B} + \overline{C})$$

$$F = (A + \overline{B})(\overline{A} + \overline{B}) = \overline{B}$$

In the preceding the sum-of-products and product-of-sums expressions reduce to the same expression. Sometimes, however, they reduce differently, and one is simpler than the other.

Example 3.4-2

Write the function defined by the truth table in terms of minterms (sum of products) and maxterms (product of sums).

A	B	C	D	F
0	0	0	0	1
0	0	0	1	1
0	0	1	0	1
0	0	1	1	1
0	1	0	0	0
0	1	0	1	0
0	1	1	0	1
0	1	1	1	1
1	0	0	0	0
1	0	0	1	0
1	0	1	0	0
1	0	1	1	0
1	1	0	0	1
1	1	0	1	1
1	1	1	0	1
1	1	1	1	1

This function can be expressed in minterms (sum of products) as

$$F = \overline{A}\,\overline{B}\,\overline{C}\,\overline{D} + \overline{A}\,\overline{B}\,\overline{C}D + \overline{A}\,\overline{B}C\overline{D} + \overline{A}\,\overline{B}CD + \overline{A}BC\overline{D} + \overline{A}BCD$$
$$+ AB\overline{C}\,\overline{D} + AB\overline{C}D + ABC\overline{D} + ABCD$$

This sum-of-products expression can be simplified (using logic adjacency) to

$$F = \overline{A}\,\overline{B}\,\overline{C}\,\overline{D} + \overline{A}\,\overline{B}\,\overline{C}D + \overline{A}\,\overline{B}C\overline{D} + \overline{A}\,\overline{B}CD + \overline{A}BC\overline{D} + \overline{A}BCD$$

$$+ AB\overline{C}\,\overline{D} + AB\overline{C}D + ABC\overline{D} + ABCD$$

$$= \overline{A}\,\overline{B}\,\overline{C} + \overline{A}\,\overline{B}C + \overline{A}BC + AB\overline{C} + ABC + ABC$$

$$= \overline{A}\,\overline{B} + BC + AB$$

This function can be expressed in maxterms (product of sums) as

$$F = (A + \overline{B} + C + D)(A + \overline{B} + C + \overline{D})(\overline{A} + B + C + D)$$
$$(\overline{A} + B + C + \overline{D})(\overline{A} + B + \overline{C} + D)(\overline{A} + B + \overline{C} + \overline{D})$$

The product-of-sums expression can also be simplified (using logic adjacency) to

$$F = (A + \overline{B} + C + D)(A + \overline{B} + C + \overline{D})$$

$$(\overline{A} + B + C + D)(\overline{A} + B + C + \overline{D})$$

$$(\overline{A} + B + \overline{C} + D)(\overline{A} + B + \overline{C} + \overline{D})$$

$$= (A + \overline{B} + C)(\overline{A} + B + C)(\overline{A} + B + \overline{C})$$

$$= (A + \overline{B} + C)(\overline{A} + B)$$

In the preceding example the simplified sum-of-products and product-of-sums expressions are different, although with enough work we could show that they're the same. Under most circumstances it is worthwhile to examine the two different simplified expressions, because one may result in a simpler circuit than the other.

3.5 FORMING A TRUTH TABLE FROM A BOOLEAN EXPRESSION

Usually a problem comes to us in the form of a specification, as in Section 3.3, but sometimes we have a **Boolean expression (function)** and we want to find its truth table. If the function is written in minterms (or maxterms), we reverse what we did in the last section. For example, if

$$F = \overline{A}BC + A\overline{B}C + ABC + AB\overline{C}$$

then the truth table is

A	B	C	F
0	0	0	0
0	0	1	0
0	1	0	0
0	1	1	1
1	0	0	0
1	0	1	1
1	1	0	1
1	1	1	1

where we've placed a 1 in each row that corresponds to a minterm in the function, and a 0 in all other rows.

To form a truth table for a function written in maxterms, the process is almost identical. For example, if

$$F = (\overline{A} + B + C)(A + \overline{B} + C)(A + B + C)(A + B + \overline{C})$$

then the truth table is

A	B	C	F
0	0	0	0
0	0	1	0
0	1	0	0
0	1	1	1
1	0	0	0
1	0	1	1
1	1	0	1
1	1	1	1

where we've placed a 0 in each row that corresponds to a maxterm in the function, and a 1 in all other rows.

Now suppose we have a function that is not written in minterms (or maxterms)—for instance, the sum of products expression:

$$F = AB + B\overline{C}$$

This three-variable function is obviously not expressed in minterms, because any minterm for a three-variable function has three variables. To see what to do here, expand each of the terms into minterms, using logic adjacency:

$$AB = ABC + AB\overline{C}$$

$$B\overline{C} = AB\overline{C} + \overline{A}B\overline{C}$$

so

$$F = ABC + AB\overline{C} + AB\overline{C} + \overline{A}B\overline{C}$$

and the truth table is

A	B	C	F
0	0	0	0
0	0	1	0
0	1	0	1
0	1	1	0
1	0	0	0
1	0	1	0
1	1	0	1
1	1	1	1

Here two of the minterms in the function are the same, but, because $AB\overline{C} + AB\overline{C} = AB\overline{C}$, this presents no problem and only results in one row of the truth table being set to 1.

Now we can explore an easier way to write the truth table. In the preceding example, the first term, AB, results in a 1 in every row of the truth table for which $A = 1$ and $B = 1$, independent of the value of C (rows 6 and 7, taking the first row as 0). The term $B\overline{C}$ results in a 1 in every row of the truth table for which $B = 1$ and $C = 0$, independent of the value of A (rows 2 and 6). The pattern is simple: for any term in the function, place a 1 in each row of the truth table for which that term is true.

To see how this works, suppose that

$$F = A + B\overline{C}$$

The A results in a 1 in each row of the truth table for which A = 1, and the $B\overline{C}$ results in a 1 in each row of the truth table for which B = 1 and C = 0. Thus the truth table is

A	B	C	F
0	0	0	0
0	0	1	0
0	1	0	1
0	1	1	0
1	0	0	1
1	0	1	1
1	1	0	1
1	1	1	1

If a function is written as a product of sums, a similar procedure—the dual of the preceding—can be applied to produce the truth table. For example, suppose that

$$F = (\overline{A} + B)C$$

Here we put a 0 in each row of the truth table in which a factor in the function is false (0). The first factor, $\overline{A} + B$, is false (0) when A = 1 and B = 0. The second factor is false (0) when C = 0. Thus the truth table is

A	B	C	F
0	0	0	0
0	0	1	1
0	1	0	0
0	1	1	1
1	0	0	0
1	0	1	0
1	1	0	0
1	1	1	1

Example 3.5-1

Form the truth table for the function

$$F = (A + B)(C\overline{D} + BC)$$

Here we have a mixed expression that is neither the sum of products nor the product of sums. Using the distribution law, we can expand it into a sum of products,

$$F = AC\overline{D} + ABC + BC\overline{D} + BC$$

and the truth table is

A	B	C	D	F
0	0	0	0	0
0	0	0	1	0
0	0	1	0	0
0	0	1	1	0
0	1	0	0	0
0	1	0	1	0
0	1	1	0	1
0	1	1	1	1
1	0	0	0	0
1	0	0	1	0
1	0	1	0	1
1	0	1	1	0
1	1	0	0	0
1	1	0	1	0
1	1	1	0	1
1	1	1	1	1

3.6 KARNAUGH MAP PRELIMINARIES

We've been simplifying Boolean functions using logic adjacency. This requires some ingenuity, and it's not always clear that the simplest expression has been found. In this and succeeding sections, we'll develop a graphical technique for simplifying Boolean expressions which always results in the simplest expression being found.

First consider a two-variable Boolean function. If we write all the minterms for the function in an array of the form

$$\begin{array}{cc} \overline{A}\,\overline{B} & A\overline{B} \\ \overline{A}B & AB \end{array}$$

then all the logic-adjacent terms are next to each other, or *next neighbor* (neighbors both side-to-side and up-and-down). $\overline{A}\,\overline{B}$ is logic-adjacent to $\overline{A}B$, $\overline{A}B$ is logic-adjacent to AB, AB is logic-adjacent to $A\overline{B}$, and finally $A\overline{B}$ is logic-adjacent to $\overline{A}\,\overline{B}$. Now suppose we have a function

$$F = \overline{A}B + AB$$

and we put these terms into the array:

$$\begin{array}{cc} 0 & 0 \\ \overline{A}B & AB \end{array}$$

We can immediately see the logic adjacency, and the function simplifies to

$$F = B$$

The graphical technique is rather trivial for two-variable functions; we can easily see the logic adjacencies in these simple functions without the array. But now consider a three-variable function. For such a function, if we arrange the minterms into the array,

$$\begin{array}{cccc} \overline{A}\,\overline{B}\,\overline{C} & \overline{A}B\overline{C} & AB\overline{C} & A\overline{B}\,\overline{C} \\ \overline{A}\,\overline{B}C & \overline{A}BC & ABC & A\overline{B}C \end{array}$$

we find that we've arranged all the minterms so that logic-adjacent terms are next neighbor, and in addition we have logic adjacency across the ends. Now if we have a function

$$F = \overline{A}B\overline{C} + AB\overline{C} + ABC$$

and if we put it in the array,

$$\begin{array}{cccc} 0 & \overline{A}B\overline{C} & AB\overline{C} & 0 \\ 0 & 0 & ABC & 0 \end{array}$$

we can immediately see the logic adjacencies (one side-to-side and one top-to-bottom), and the function simplifies to

$$F = B\overline{C} + AB$$

In this simplification we used the term $AB\overline{C}$ twice. Notice that in the simplification of the function, the common terms in the logic-adjacent pairs remain.

With the function

$$F = \overline{A}B\overline{C} + \overline{A}BC + AB\overline{C} + ABC$$

an even more interesting simplification results. If we put it in the array

$$\begin{array}{cccc} 0 & \overline{A}B\overline{C} & AB\overline{C} & 0 \\ 0 & \overline{A}BC & ABC & 0 \end{array}$$

four terms group together in a square. When we simplify these terms, we get

$$F = \overline{A}B\overline{C} + AB\overline{C} + \overline{A}BC + ABC = B\overline{C} + BC = B$$

Four terms grouped together in a square gives a double simplification. The only variable remaining in the simplified expression for the square is the one that is common throughout the square (B). Both the variables A and C appear uncomplemented and complemented.

The idea of placing logic-adjacent minterms next to each other can be extended to four-variable functions using the array

$$\begin{array}{llll}
\overline{A}\,\overline{B}\,\overline{C}\,\overline{D} & \overline{A}B\overline{C}\,\overline{D} & AB\overline{C}\,\overline{D} & A\overline{B}\,\overline{C}\,\overline{D} \\
\overline{A}\,\overline{B}\,\overline{C}D & \overline{A}B\overline{C}D & AB\overline{C}D & A\overline{B}\,\overline{C}D \\
\overline{A}\,\overline{B}CD & \overline{A}BCD & ABCD & A\overline{B}CD \\
\overline{A}\,\overline{B}C\overline{D} & \overline{A}BC\overline{D} & ABC\overline{D} & A\overline{B}C\overline{D}
\end{array}$$

In this array all the neighbor elements are logic-adjacent side-to-side and up-and-down. In addition, they're logic-adjacent around the ends and from the top to the bottom.

Suppose we have the four-variable function

$$F = \overline{A}\,\overline{B}\,\overline{C}\,\overline{D} + \overline{A}B\overline{C}\,\overline{D} + AB\overline{C}\,\overline{D} + A\overline{B}\,\overline{C}\,\overline{D} + AB\overline{C}D + A\overline{B}\,\overline{C}D$$

If we put it into the array

$$\begin{array}{llll}
\overline{A}\,\overline{B}\,\overline{C}\,\overline{D} & \overline{A}B\overline{C}\,\overline{D} & AB\overline{C}\,\overline{D} & A\overline{B}\,\overline{C}\,\overline{D} \\
0 & 0 & AB\overline{C}D & A\overline{B}\,\overline{C}D \\
0 & 0 & 0 & 0 \\
0 & 0 & 0 & 0
\end{array}$$

we find we have a group of four logic-adjacent terms in a square (upper right corner) and, because a group of four terms in a column or row are logic-adjacent just like a square, a group of four logic-adjacent terms in a row (top row). With these groupings, we can immediately simplify the function to

$$F = \overline{C}\,\overline{D} + A\overline{C}$$

Notice that for logic-adjacent groups, only the variables that are common throughout the group remain in the simplification. In a four-variable array it's possible to have a grouping of eight logic-adjacent terms in a rectangle that reduces to a single variable. In the array

$$\begin{array}{llll}
\overline{A}\,\overline{B}\,\overline{C}\,\overline{D} & \overline{A}B\overline{C}\,\overline{D} & AB\overline{C}\,\overline{D} & A\overline{B}\,\overline{C}\,\overline{D} \\
\overline{A}\,\overline{B}\,\overline{C}D & \overline{A}B\overline{C}D & AB\overline{C}D & A\overline{B}\,\overline{C}D \\
0 & 0 & 0 & 0 \\
0 & 0 & 0 & 0
\end{array}$$

the eight logic-adjacent terms in the rectangle reduce to $\overline{C}$ (the common variable).

$$\overline{A}\,\overline{B}\,\overline{C}\,\overline{D} + \overline{A}B\overline{C}\,\overline{D} + AB\overline{C}\,\overline{D} + A\overline{B}\,\overline{C}\,\overline{D} + \overline{A}\,\overline{B}\,\overline{C}D + \overline{A}B\overline{C}D$$

$$+ AB\overline{C}D + A\overline{B}\,\overline{C}D = \overline{A}\,\overline{C}\,\overline{D} + A\overline{C}\,\overline{D} + \overline{A}\,\overline{C}D + A\overline{C}D$$

$$= \overline{C}\,\overline{D} + \overline{C}D = \overline{C}$$

3.7 KARNAUGH MAPS

In his paper, "The Map Method for Synthesis of Combinational Logic Circuits" (1953), M. Karnaugh developed an easy graphical method of simplifying Boolean functions using the idea of an array of next-neighbor, logic-adjacent minterms described in the last section. Rather than write out all the minterms, he used what has become known as a **Karnaugh map** to show the minterm relations. The Karnaugh map for a two-variable function has the form shown in Figure 3.7-1. Each square, or **cell**, in the map represents a minterm in the variables A and B. The value of one of the variables associated with a cell is given above the cell, and the value of the other variable associated with the cell is given to the left of the cell. Variable names are shown above and below a diagonal line at the upper left corner of the map. Here the column variable is A and the row variable is B. The number of the minterm associated with a cell is given in the upper left corner of the cell. For example, in this map the minterm associated with the upper left cell is $m_0 = \overline{A}\,\overline{B}$, the term that is true if A = 0 and B = 0.

To show the minterm relationships for a function in a Karnaugh map, we put a 1 in each cell that is a minterm of the function and a 0 in each of the other cells. For example, the Karnaugh map for the function

$$F = \overline{A}\,\overline{B} + AB$$

is shown in Figure 3.7-2. This function has no simplification. There are no minterms with neighbors.

The Karnaugh map for a three-variable function has the form shown in Figure 3.7-3. In this map each column has two variables associated with it. The values for the two variables are shown at the top of the column, and the variable names (A and B) are shown above the diagonal in the upper left corner. Each row has only one variable (C) associated with it, as in the two-variable map.

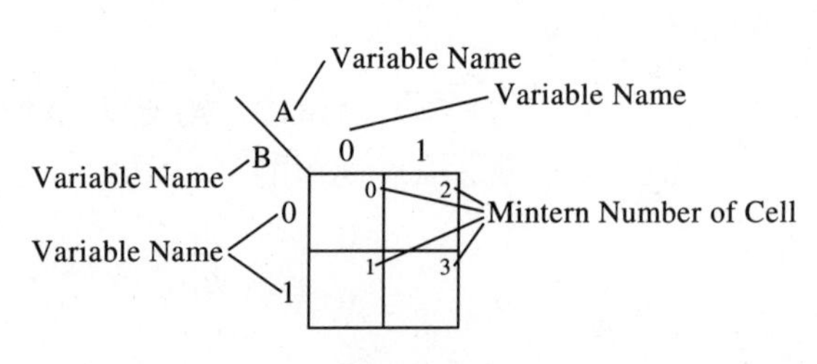

Figure 3.7-1

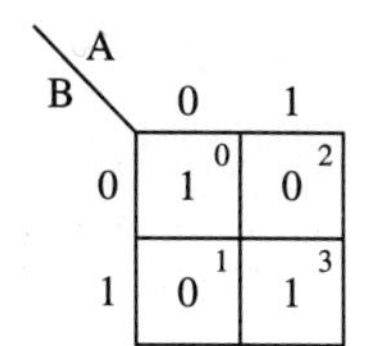

Figure 3.7-2

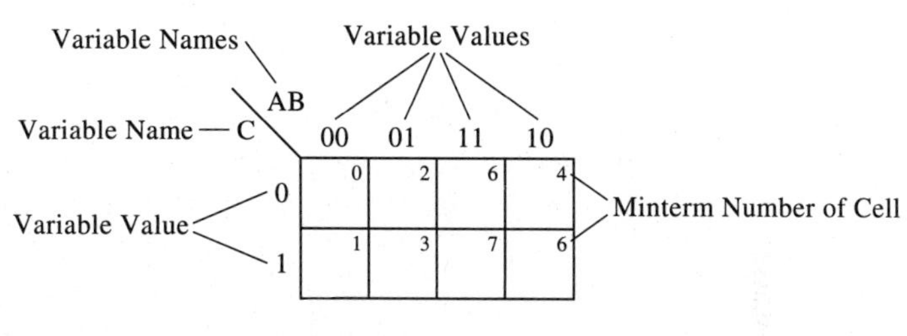

Figure 3.7-3

The minterm relationships for a function are shown in the three-variable map in the same way as in a two-variable map. For each minterm in the function, a 1 is entered in the cell corresponding to that minterm, and a 0 is entered in each of the other cells. For example, the Karnaugh map for the function

$$F = \overline{A}\,\overline{B}\,\overline{C} + A\overline{B}\,\overline{C} + \overline{A}\,\overline{B}C + A\overline{B}C$$

is shown in Figure 3.7-4. This map has a *grouping* of four minterms because of the logic adjacency across the ends of the map, so the simplified function is

$$F = \overline{B}$$

since $\overline{B}$ is the common variable (B = 0) in all the terms of the group.

The Karnaugh map for a four-variable function has the form shown in Figure 3.7-5. In this map both columns and rows have two variables associated with them. The minterm relationships of a function are again shown by placing a 1 in each cell representing a minterm in the function and a 0 in each of the other cells. For example, the Karnaugh map for the function

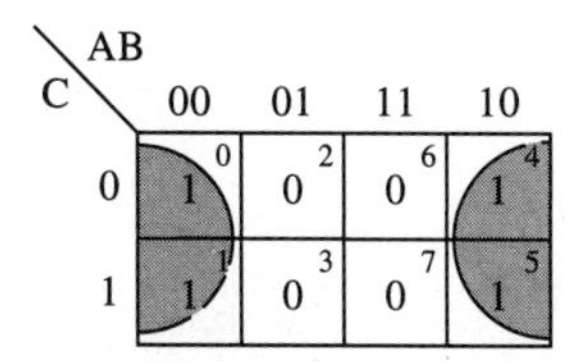

Figure 3.7-4

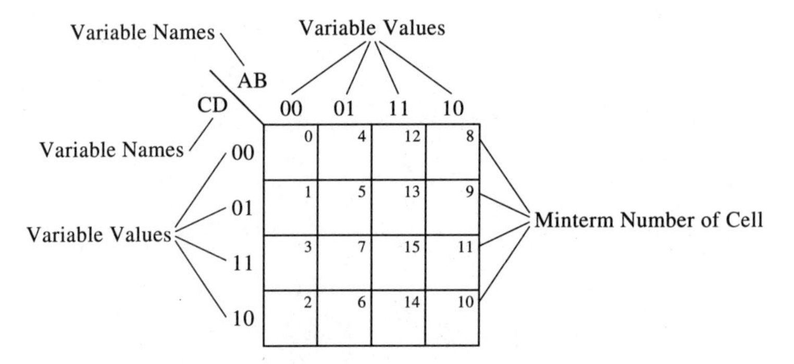

Figure 3.7-5

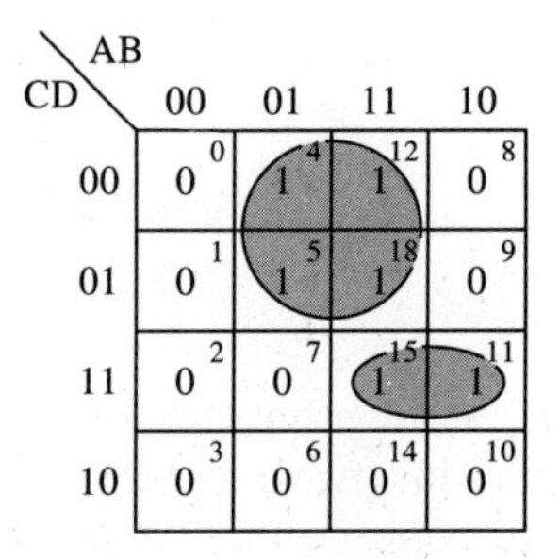

Figure 3.7-6

$$F = \overline{A}B\overline{C}\,\overline{D} + AB\overline{C}\,\overline{D} + \overline{A}B\overline{C}D + AB\overline{C}D + ABCD + A\overline{B}CD$$

is shown in Figure 3.7-6. The map has a grouping of four logic-adjacent minterms in a square and a pair of logic-adjacent minterms that simplify to

$$F = B\overline{C} + ACD$$

In each of the preceding examples, we found the Karnaugh map for a function expressed as a sum of products involving only minterms. We'll now examine some other methods of finding the map.

A	B	Minterm
0	0	m_0
0	1	m_1
1	0	m_2
1	1	m_3

B \ A	0	1
0	m_0 (0)	m_2 (2)
1	m_1 (1)	m_3 (3)

Figure 3.7-7

A	B	F
0	0	1
0	1	1
1	0	0
1	1	1

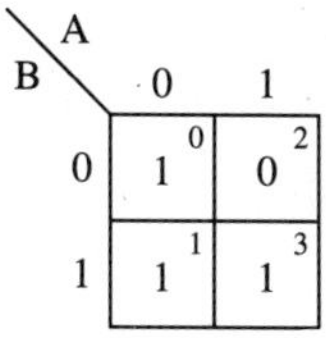

B \ A	0	1
0	1 (0)	0 (2)
1	1 (1)	1 (3)

Figure 3.7-8

A	B	C	Minterm
0	0	0	m_0
0	0	1	m_1
0	1	0	m_2
0	1	1	m_3
1	0	0	m_4
1	0	1	m_5
1	1	0	m_6
1	1	1	m_7

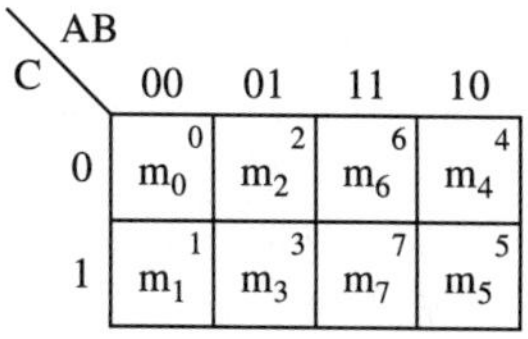

C \ AB	00	01	11	10
0	m_0 (0)	m_2 (2)	m_6 (6)	m_4 (4)
1	m_1 (1)	m_3 (3)	m_7 (7)	m_5 (5)

Figure 3.7-9

One of the simplest ways to find the Karnaugh map of a function is to use the truth table for that function. Each minterm in the truth table corresponds to a cell in the Karnaugh map. Figure 3.7-7 shows this relationship for a two-variable function. To form the Karnaugh map, we just transfer the function value in the truth table to the proper cell in the map. Note that the order is down the columns. A two-variable truth table and its map are shown in Figure 3.7-8.

Figure 3.7-9 shows the relationship between the truth table and the Karnaugh map for a three-variable function. The order for filling the map is again down the columns, but the fourth column is filled before the third. As in a two-variable map, the values from the truth table are transferred to the proper cell in the map. A three-variable truth table and its map are shown in Figure 3.7-10.

Figure 3.7-11 shows the relationship between the truth table and the Karnaugh map for a four-variable function. The order for filling the map is again down the columns, but the fourth row is filled before the third. As in the three-variable map the fourth column is filled before the third. A four-variable truth table and its map are shown in Figure 3.7-12.

Now that we've seen how to form a Karnaugh map from a truth table, we'll return to the idea of loading the map from a relation. We first developed Karnaugh maps for a sum-of-products function expressed in minterms. Suppose we have a function that is not expressed in minterms, such as

A	B	C	F
0	0	0	1
0	0	1	0
0	1	0	1
0	1	1	1
1	0	0	1
1	0	1	0
1	1	0	1
1	1	1	0

C \ AB	00	01	11	10
0	1 (0)	1 (2)	1 (6)	1 (4)
1	0 (1)	1 (3)	0 (7)	0 (5)

Figure 3.7-10

A	B	C	D	Minterm
0	0	0	0	m_0
0	0	0	1	m_1
0	0	1	0	m_2
0	0	1	1	m_3
0	1	0	0	m_4
0	1	0	1	m_5
0	1	1	0	m_6
0	1	1	1	m_7
1	0	0	0	m_8
1	0	0	1	m_9
1	0	1	0	m_{10}
1	0	1	1	m_{11}
1	1	0	0	m_{12}
1	1	0	1	m_{13}
1	1	1	0	m_{14}
1	1	1	1	m_{15}

CD \ AB	00	01	11	10
00	m_0 (0)	m_4 (4)	m_{12} (12)	m_8 (8)
01	m_1 (1)	m_5 (5)	m_{13} (13)	m_9 (9)
11	m_3 (3)	m_7 (7)	m_{15} (15)	m_{11} (11)
10	m_2 (2)	m_6 (6)	m_{14} (14)	m_{10} (10)

Figure 3.7-11

$$F = A + \overline{A}\,\overline{B}D$$

We can expand each term of the function into minterms, using logic adjacency,

$$\begin{aligned} F &= AB + A\overline{B} + \overline{A}\,\overline{B}CD + \overline{A}\,\overline{B}\,\overline{C}D \\ &= ABC + AB\overline{C} + A\overline{B}C + A\overline{B}\,\overline{C} + \overline{A}\,\overline{B}CD + \overline{A}\,\overline{B}\,\overline{C}D \\ &= ABCD + ABC\overline{D} + AB\overline{C}D + AB\overline{C}\,\overline{D} + A\overline{B}CD + A\overline{B}C\overline{D} \\ &\quad + A\overline{B}\,\overline{C}D + A\overline{B}\,\overline{C}\,\overline{D} + \overline{A}\,\overline{B}CD + \overline{A}\,\overline{B}\,\overline{C}D \end{aligned}$$

A	B	C	D	F
0	0	0	0	0
0	0	0	1	1
0	0	1	0	0
0	0	1	1	1
0	1	0	0	1
0	1	0	1	1
0	1	1	0	0
0	1	1	1	0
1	0	0	0	1
1	0	0	1	1
1	0	1	0	1
1	0	1	1	1
1	1	0	0	0
1	1	0	1	0
1	1	1	0	0
1	1	1	1	0

CD \ AB	00	01	11	10
00	0 (0)	1 (4)	0 (12)	1 (8)
01	1 (1)	1 (5)	0 (13)	1 (9)
11	1 (3)	0 (7)	0 (15)	1 (11)
10	0 (2)	0 (6)	0 (14)	1 (10)

Figure 3.7-12

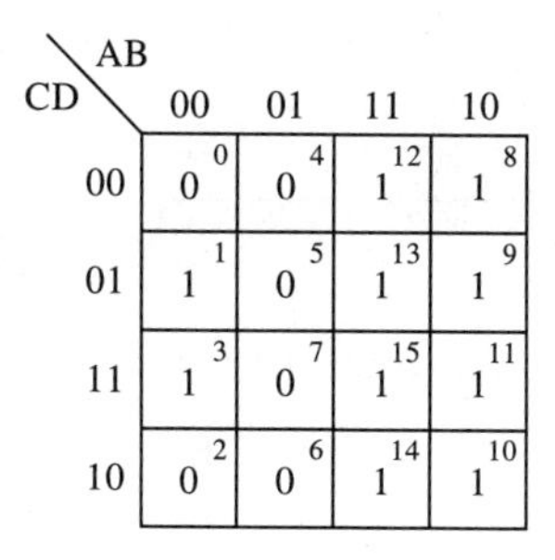

Figure 3.7-13

and then we can form the Karnaugh map just as we did in our first development. Figure 3.7-13 is the map for the preceding function.

There is a simpler way to form the map. The first term of our function (A) expands into eight minterms, all of which contain A, resulting in a 1 in each cell with A = 1. The second term of our function ($\overline{A}\,\overline{B}D$) expands into two minterms, both of which contain $\overline{A}\,\overline{B}D$, resulting in a 1 in all the cells with A = 0, B = 0, and D = 1. To fill the Karnaugh map for a function that is a sum of products, look at the terms one at a time and place a 1 in each cell for which the term is true. Complete the map by placing 0s in the remaining empty cells. Notice that the procedure for filling a Karnaugh map and the procedure we learned for forming a truth table are similar.

Example 3.7-1

Form a Karnaugh map for the function

$$F = AB + \overline{C}\,\overline{D}$$

From the first term of the function, we place 1s in all the cells of the map that have A = 1 and B = 1. From the second term, we place 1s in all the cells that have C = 0 and D = 0.

CD \ AB	00	01	11	10
00	1 (0)	1 (4)	1 (12)	1 (8)
01	0 (1)	0 (5)	1 (13)	0 (9)
11	0 (3)	0 (7)	1 (15)	0 (11)
10	0 (2)	0 (6)	1 (14)	0 (10)

Example 3.7-2

Form a Karnaugh map for the function

$$F = \overline{A}\,\overline{B}\,\overline{C} + AC$$

From the first term, we place 1s in all the cells of the map that have A = 0, B = 0, and C = 0. From the second term, we place 1s in all the cells that have A = 1 and C = 1.

C \ AB	00	01	11	10
0	1 (0)	0 (2)	0 (6)	0 (4)
1	0 (1)	0 (3)	1 (7)	1 (5)

CD \ AB	00	01	11	10
00	[0] 0	[4] 1	[12] 0	[8] 0
01	[1] 0	[5] 1	[13] 1	[9] 0
11	[3] 0	[7] 0	[15] 1	[11] 1
10	[2] 0	[6] 0	[14] 0	[10] 0

Figure 3.7-14

Now suppose we have a function that is expressed as a product of sums. We can form the Karnaugh map by using the dual of the preceding idea. Consider the example

$$F = (A + \overline{C})(\overline{A} + D)(B + C)$$

The Karnaugh map is shown in Figure 3.7-14. Here we consider one factor at a time, and we place 0s in the cells for which that factor is false. The first factor is false when $A = 0$ and $C = 1$, the second is false when $A = 1$ and $D = 0$, and the third is false when $B = 0$ and $C = 0$. We complete the map by placing 1s in all the unfilled cells.

Example 3.7-3

Form the Karnaugh map for the function

$$F = \overline{A}(A + B + \overline{C})$$

For the first factor, we place 0s in all the cells of the map for which $A = 1$. For the second factor, we place 0s in all the cells for which $A = 0$, $B = 0$, and $C = 1$. All the remaining unfilled cells are filled with 1s.

C \ AB	00	01	11	10
0	[0] 1	[2] 1	[6] 0	[4] 0
1	[1] 0	[3] 1	[7] 0	[5] 0

In this section we've seen how to form a Karnaugh map for a function defined by a truth table or by an expression. We also looked at some examples of how the

Karnaugh map helps us simplify functions. Now we need a systematic way to assure ourselves that we're getting the simplest expression from the map.

3.8 KARNAUGH MAP SIMPLIFICATION

Generally, to systematically find the simplest expression from a map, we choose groupings so that we *minimize the number of groupings required to cover all the entries in the map and at the same time maximize the size of the groupings.* We can do this by following a simple set of steps. We start with groupings of one term, then consider groupings of two terms, then four terms, and then eight. At each step, we look ahead and avoid forming groupings that can be covered by larger groupings.

1. Search the map, considering one entry at a time, and find all the single entries that cannot be grouped with any other entry. These are called *islands.*
2. Search the map, considering each entry not already grouped, and find all the entries that can be grouped in *only* one way with another single entry. Group these pairs of entries.
3. Search the map, considering each entry not already grouped, and find the entries that can be grouped in more than one way with one other entry. For these entries:
   (a) Group together the entries that will group with one other entry not already grouped.
   (b) Group together the entries that will group with one other entry already grouped.
   These steps can result in a situation where no clear choice is available. In that case, make one of the available choices and continue with this step.
4. Repeat steps 2 and 3 for groupings of four entries.
5. Repeat steps 2 and 3 for groupings of eight entries.

The process is complete when all map entries have been grouped—that is, covered. Let's look at some examples. The first has an island and some groupings in pairs.

Example 3.8-1

Simplify the function with the following Karnaugh map.

C \ AB	00	01	11	10
0	1 (0)	0 (2)	0 (6)	0 (4)
1	0 (1)	1 (3)	1 (7)	1 (5)

Using rule 1, we find that the entry in cell 0 cannot be grouped with any other entry. It's an island, as the following figure shows.

C \ AB	00	01	11	10
0	1 [0]	0 [2]	0 [6]	0 [4]
1	0 [1]	1 [3]	1 [7]	1 [5]

Using rule 2, we find that there are two single entries that can be grouped with another single entry in only one way. We can completely cover the map by grouping these entries (in cells 3 and 5) as shown in the following figure.

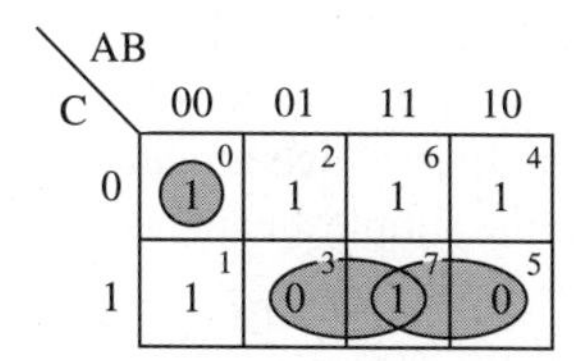

All that remains is to determine the simplified expression for each grouping. If we remember that variables common throughout the group remain in the simplification, this is easy. The island has no simplification, and it results in the term $\overline{A}\,\overline{B}\,\overline{C}$. The grouping of cells 3 and 7 has B = 1 and C = 1 common and simplifies to BC. The grouping of cells 7 and 5 has A = 1 and C = 1 common and simplifies to AC. The simplified function is thus

$$F = \overline{A}\,\overline{B}\,\overline{C} + BC + AC$$

Example 3.8-2

Simplify the function with the following Karnaugh map.

CD \ AB	00	01	11	10
00	1 [0]	1 [4]	1 [12]	1 [8]
01	0 [1]	1 [5]	0 [13]	1 [9]
11	0 [3]	0 [7]	0 [15]	1 [11]
10	0 [2]	0 [6]	1 [14]	0 [10]

Here we have no islands (rule 1). We do have three entries (cells 5, 11, and 14) that can be grouped with another entry in only one way (rule 2), as shown on the left (next page). Notice that the entry in cell 14 groups with the entry in cell 12 because of logic adjacency from the top to the bottom of the map. We're left with a group of four, which completes the covering, as shown on the right (rule 2).

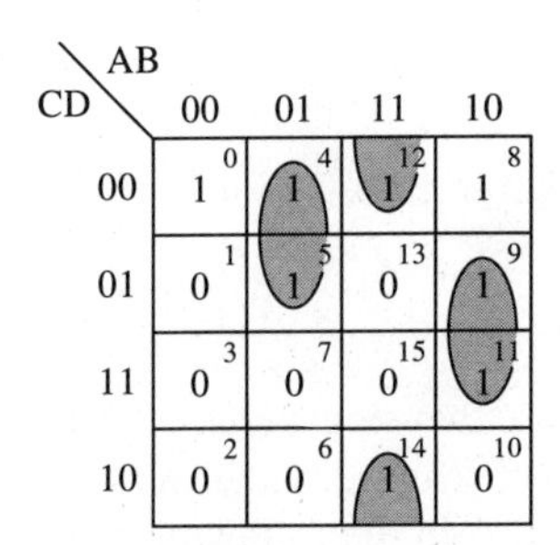

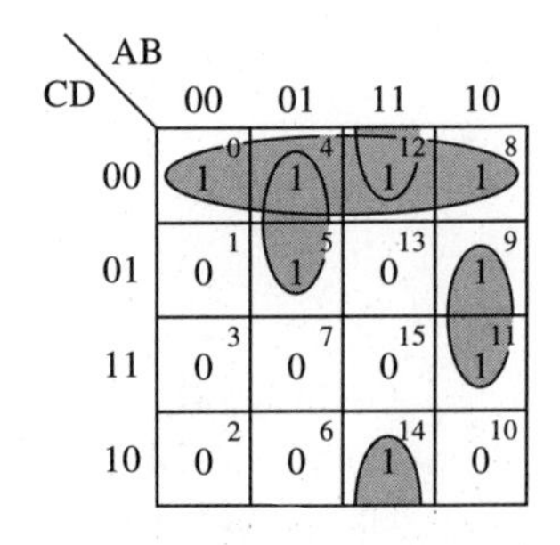

We should make a final check to see that we have not covered a grouping with a larger grouping. For example, we might have an unnecessary grouping of the entries in cells 8 and 12, which are already covered by the grouping of four entries.

Now, finding the simplest expression for each group (reading the map) gives us

$$F = \overline{A}B\overline{C} + AB\overline{D} + A\overline{B}D + \overline{C}\,\overline{D}$$

The terms in this expression are in the order in which we chose the groups. For the first groups (cells 4 and 5), A = 0, B = 1, and C = 0 are the common variables. For the second group (cells 12 and 14), A = 1, B = 1, and D = 0 are the common variables. For the third group (cells 9 and 11), A = 1, B = 0, and D = 1 are the common variables. For the fourth group (cells 0, 4, 12, and 8), C = 0 and D = 0 are the common variables.

Example 3.8-3

Simplify the function with the following Karnaugh map.

CD \ AB	00	01	11	10
00	[0] 1	[4] 0	[12] 1	[8] 0
01	[1] 0	[5] 0	[13] 0	[9] 0
11	[3] 0	[7] 1	[15] 0	[11] 1
10	[2] 1	[6] 1	[14] 1	[10] 1

The groupings of entries for the map follow (next page). There are no islands (rule 1). There are four entries (cells 0, 7, 11, and 12) that can only be grouped with another entry in only one way. These groups complete the cover. Even though the entries in cells 2, 6, 14, and 10 form a group of four, this group is not used because it's completely covered by necessary groupings of two entries.

CD \ AB	00	01	11	10
00	1	0	1	0
01	0	0	0	0
11	0	1	0	1
10	1	1	1	1

In reading the map, we get the simplified function:

$$F = \overline{A}\,\overline{B}\,\overline{D} + \overline{A}BC + AB\overline{D} + A\overline{B}C$$

Example 3.8-4

Simplify the function with the following Karnaugh map.

CD \ AB	00	01	11	10
00	0	0	0	0
01	1	1	0	1
11	0	1	1	1
10	0	0	0	0

The groupings of entries for the map follow. There are no islands (rule 1) and no single entries that can be grouped with another single entry in only one way (rule 2). In fact, all entries can be grouped in more than one way with another entry (rule 3), but there is no way to make a choice using rules 3a or 3b, so we make an initial arbitrary grouping. If we group the entries in cells 7 and 15, then all the other groupings follow from rule 3a. Groupings are chosen so that they cover only previously uncovered entries.

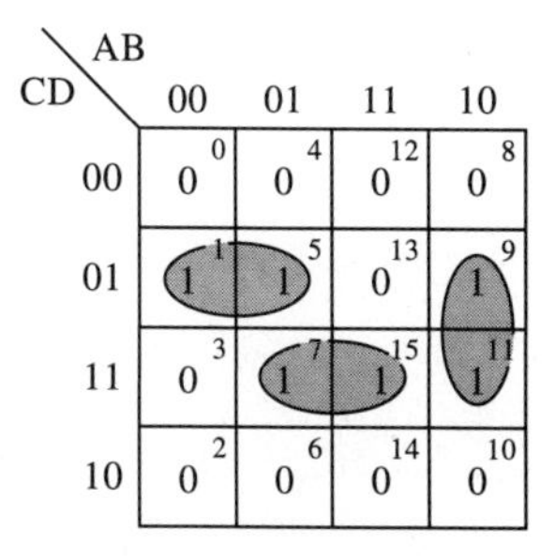

In reading the map, we find that the simplified function is

$$F = \overline{A}\,\overline{C}D + BCD + A\overline{B}D$$

Obviously, if we make a different initial choice, we'll get a different expression. It will be just as valid and of the same complexity as the one we have found.

If we change this example only slightly, we break the chain and the mapping proceeds quite differently. Consider the following map.

CD \ AB	00	01	11	10
00	0 (0)	0 (4)	0 (12)	0 (8)
01	1 (1)	1 (5)	0 (13)	1 (9)
11	0 (3)	1 (7)	1 (15)	1 (11)
10	0 (2)	0 (6)	1 (14)	0 (10)

Here, rule 2 applies to the entry in cell 14, and then rule 3a applies to the entries in cells 5 and 7 and cells 9 and 11 (see following map). This leaves only an application of rule 3b, which results in a choice between the entries in cells 1 and 5 and the entries in cells 1 and 9. We cannot choose both. Let's choose to group the entries in cells 1 and 9.

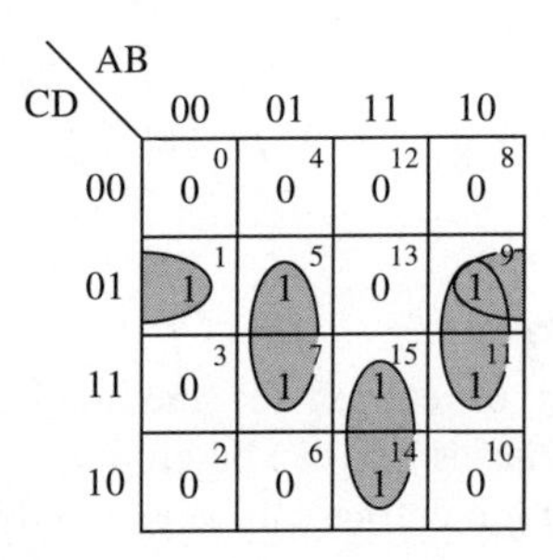

In reading the map, we find the simplified function:

$$F = ABC + \overline{A}BD + A\overline{B}D + \overline{B}\,\overline{C}D$$

Example 3.8-5

Simplify the function with the following Karnaugh map.

CD \ AB	00	01	11	10
00	1 (0)	1 (4)	0 (12)	1 (8)
01	1 (1)	0 (5)	0 (13)	1 (9)
11	0 (3)	0 (7)	0 (15)	0 (11)
10	1 (2)	0 (6)	0 (14)	1 (10)

The groupings of entries for this map follow. There are no islands (rule 1). There is one entry (cell 4) that can be grouped with another entry in only one way (rule 2). Notice that entries in cells 0, 1, 8, and 9 form a possible grouping of four entries, and entries in cells 0, 2, 8, and 10 form another. The corners are logic-adjacent. These groupings will cover all the remaining entries. Applying rule 2 to groupings of four entries, we find that both of the groupings of four must be used to cover all remaining entries.

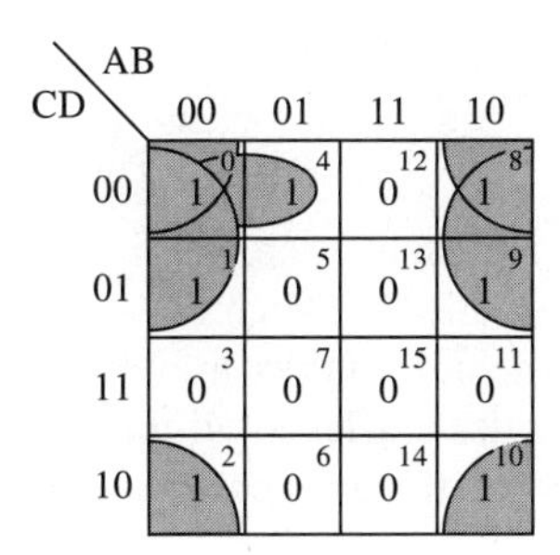

In reading the map, we find the simplified function:

$$F = \overline{A}\,\overline{C}\,\overline{D} + \overline{B}\,\overline{C} + \overline{B}\,\overline{D}$$

Example 3.8-6

Simplify the function

$$F = \overline{A}\,\overline{B} + \overline{A}BD + CD$$

In forming the Karnaugh map, we find that the first term in the function generates entries in all the cells for which A = 0 and B = 0; the second term generates entries in all the cells for which A = 0, B = 1, and D = 1; and the third term generates entries in all the cells for which C = 1 and D = 1.

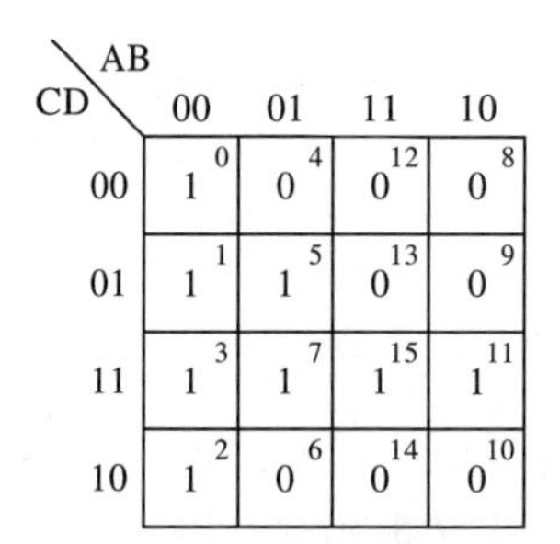

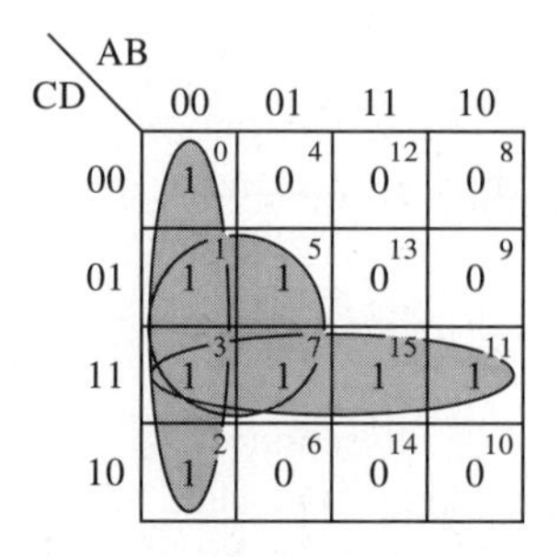

This map can be covered with three groups of four entries, as shown on the right (rule 2), and the function simplifies to

$$F = \overline{A}\,\overline{B} + \overline{A}D + CD$$

Notice that when we covered the map, we chose the largest possible groupings of entries. We never choose a grouping of two entries when we can cover it with a larger grouping of four entries.

Example 3.8-7

Simplify the function

$$F = AB\overline{C}\,\overline{D} + C\overline{D} + ABCD + \overline{B}\,\overline{C}\,\overline{D}$$

In forming the Karnaugh map, we find that the first term in the function generates entries in all the cells for which A = 1, B = 1, C = 0, and D = 0 (one cell); the second term generates entries in all the cells for which C = 1 and D = 0; the third term generates entries in all the cells for which A = 1, B = 1, C = 1, and D = 1; and the fourth term generates entries in all the cells for which B = 0, C = 0, and D = 0.

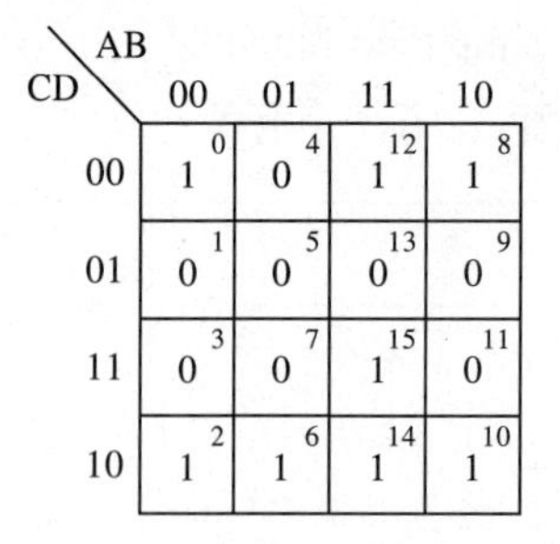

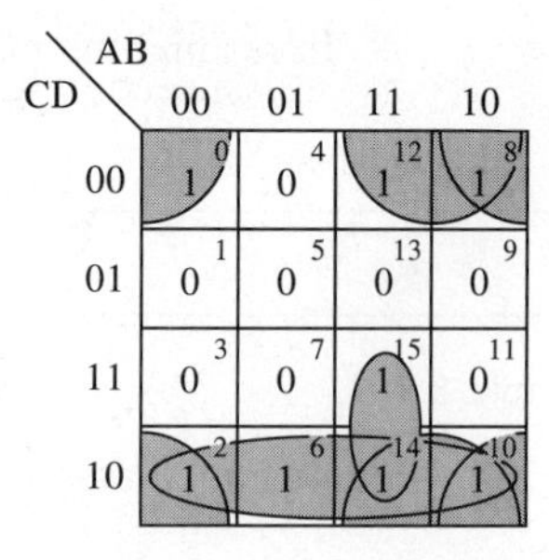

This map can be covered with one grouping of two entries (rule 2) and three groupings of four entries (rule 2), as shown on the right. The simplified function is

$$F = C\overline{D} + \overline{B}\,\overline{D} + A\overline{D} + ABC$$

The rules for covering a map with a minimal cover are useful for getting started. After some experience, patterns in the maps will begin to be recognizable and the rules will not be necessary.

3.9 FORMALIZING SIMPLIFICATION

In the last section we looked at methods of forming a minimal cover of a Karnaugh map. There is actually a formal structure associated with forming a minimal cover, which we ignored in that section, in which different types of groupings are given different names. In this section we'll fill in the formal structure.

Suppose a function has the Karnaugh map shown in Figure 3.9-1. On the map we've indicated all possible groupings of entries not completely covered by a larger grouping, called **prime implicants**. Not all of these groupings are required to cover the map.

Some of the prime implicants contain entries that can be grouped with other entries in only one way. The entries in the cells 0, 8, 1, and 9 make up such a group, because the entry in cell 9 can be grouped in only one way, with four other entries. The entries in cells 0, 4, 12, and 8 are another such group, because the entry in cell 4 can be grouped in only one way, with four other entries. Groupings of this type are called **essential implicants**.

Some of the prime implicants contain entries that can be grouped in only one way to cover two (or more) uncovered entries. The entries in cells 3 and 7 and in cells 15 and 14 are groupings of this type. Such groupings are called **necessary implicants**.

Some of the prime implicants contain entries that can be grouped in more than one way but are needed to cover otherwise uncovered entries in the map. Figure 3.9-1 has none of these, but when they exist they're called **optional implicants**.

Finally, some of the prime implicants are not needed to cover otherwise uncovered entries in the map. They cover entries already covered. The entries in cells 1 and 3, 7 and 15, and 12 and 14 are of this type; they are **redundant implicants**. Such entries should not be included in the **minimal cover**. The purpose of our rules in the last section was to exclude these redundant implicants.

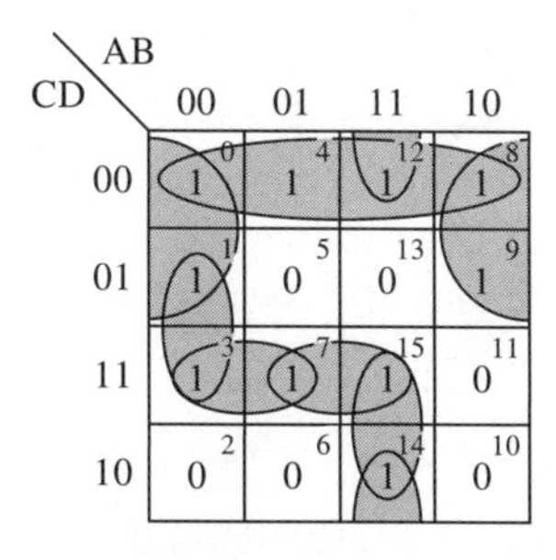

Figure 3.9-1

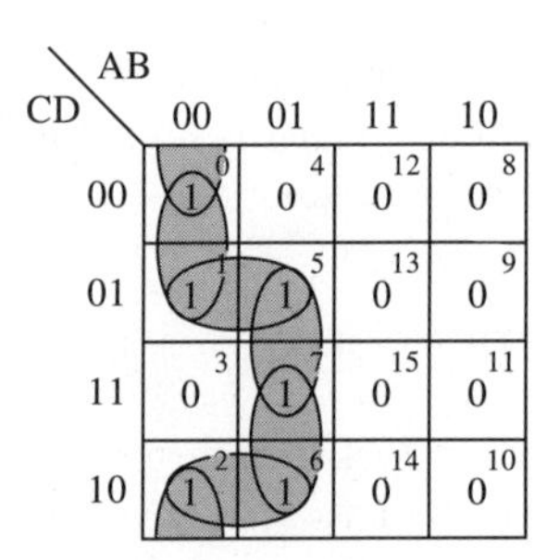

Figure 3.9-2

Now let's look at a Karnaugh map that contains optional implicants. Consider Figure 3.9-2. All the prime implicants are shown on the map, and all the prime implicants are also optional implicants. Each can be chosen in one of two ways, and we must make the choice. Once it is made, the other implicants are then either necessary implicants or redundant implicants. If we choose the grouping of the entries in cells 0 and 1 as an optional implicant, then the groupings in cells 5 and 7 and in cells 2 and 6 are necessary implicants, and this set of implicants completely covers the map.

To summarize, we see that all the groupings not completely covered by a larger grouping are called prime implicants. The set of all prime implicants consists of essential implicants, necessary implicants, optional implicants, and redundant implicants. Our purpose in finding a minimal covering for a map is to find all the essential, necessary, and optional implicants.

3.10 PRODUCT OF SUMS SIMPLIFICATION

In the preceding sections we have learned how to simplify a Boolean function using a Karnaugh map. The simplification we've found is the simplest sum-of-products (SOP) expression for the function. There is a dual of the process we've used; it results in a simplest product-of-sums (POS) expression for the function. In this simplification we work with the maxterms of the function, and because these maxterms map into the 0s, we work with the 0s of the map.

When we first developed the Karnaugh map, we arranged for the cells (which represent minterms) to be placed so that logic-adjacent terms were next to each other. In doing this, we also arranged the cells so that if they represented maxterms, the next-neighbor maxterms were logic-adjacent to each other, according to the dual logic-adjacent theorem:

$$(A + B)(A + \overline{B}) = A$$

To understand this, consider a three-variable function. If we arrange the maxterms for the function in the order corresponding to the Karnaugh map,

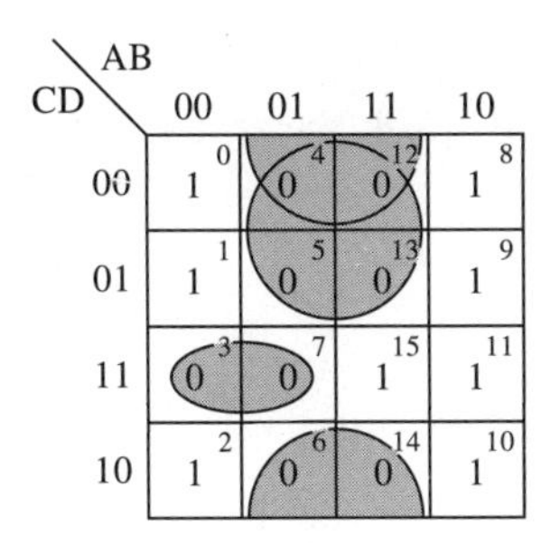

Figure 3.10-1

$$\begin{array}{cccc} A + B + C & A + \overline{B} + C & \overline{A} + \overline{B} + C & \overline{A} + B + C \\ A + B + \overline{C} & A + \overline{B} + \overline{C} & \overline{A} + \overline{B} + \overline{C} & \overline{A} + B + \overline{C} \end{array}$$

we have logic adjacency of near neighbors up-and-down, side-to-side, and around the ends, just as with the array of minterms.

Because we have logic adjacency of maxterms just as we had with minterms, we can form groupings and simplify as we did with minterms, except here we consider the 0 entries in the map, not the 1 entries. Consider Figure 3.10-1, the Karnaugh map for a function. Here we show a minimal covering of the 0s of the map. To read the map, we must remember that we want the simplified maxterm expression for the group. To get that expression, we again look for the common variables in the groups, but here we form the *or* of the complement of those variables. For the grouping of entries in cells 4, 5, 12, and 13, the common variables are B = 1 and C = 0. This grouping simplifies to the factor $(\overline{B} + C)$. Similarly, the grouping of entries in cells 4, 12, 6, and 14, which has common variables B = 1 and D = 0, simplifies to the factor $(\overline{B} + D)$. And the grouping of entries in cells 3 and 7, which has common variables A = 0, C = 1, and D = 1, simplifies to the factor $(A + \overline{C} + \overline{D})$. If we *and* these factors together, we get the complete simplified expression:

$$F = (\overline{B} + C)(\overline{B} + D)(A + \overline{C} + \overline{D})$$

There is another, almost mechanical, procedure for finding the simplest POS expression for a function. To understand it, reconsider the preceding example. After we've grouped the 0s into a minimal cover, we write the simplified SOP expression for this grouping just as if the 0s were 1s. This expression will be true (1) if any of its minterms are true (1), and otherwise it will be false (0). This is just the complement of the function we are seeking. We write the expression for the complemented function:

$$\overline{F} = B\overline{C} + B\overline{D} + \overline{A}DC$$

Now we can find F by complementing this expression (DeMorgan's theorem):

$$F = (\overline{B} + C)(\overline{B} + D)(A + \overline{C} + \overline{D})$$

This technique results in the same expression for F we found before. The first technique is more direct, the second more mechanical.

Example 3.10-1

Simplify the function with the Karnaugh map on the left as a POS.

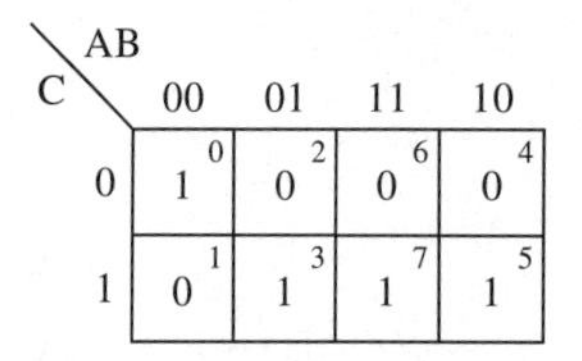

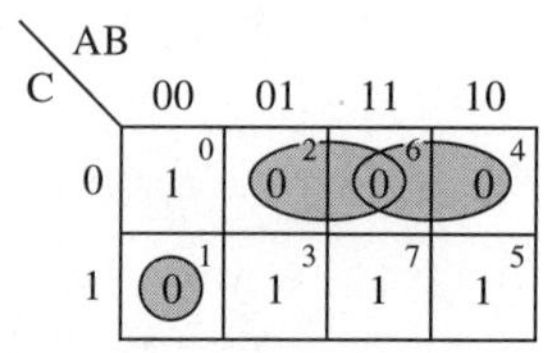

The minimal covering for the 0 entries in the map is shown on the right. All that remains is to determine the simplified expression for each grouping. This is easy if we remember that variables common throughout the group remain in the simplification but are complemented. The island has no simplification. It results in the factor $(A + B + \overline{C})$. The grouping of entries in cells 2 and 6 has $B = 1$ and $C = 0$ in common and simplifies to the factor $(\overline{B} + C)$. The grouping of entries in cells 6 and 4 has $A = 1$ and $C = 0$ in common and simplifies to the factor $(\overline{A} + C)$. The simplified function is thus

$$F = (A + B + \overline{C})(\overline{B} + C)(\overline{A} + C)$$

Example 3.10-2

Simplify the function with the Karnaugh map shown on the left as a POS.

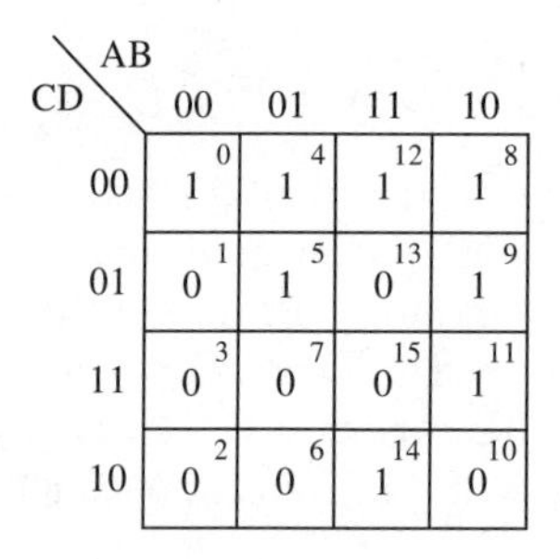

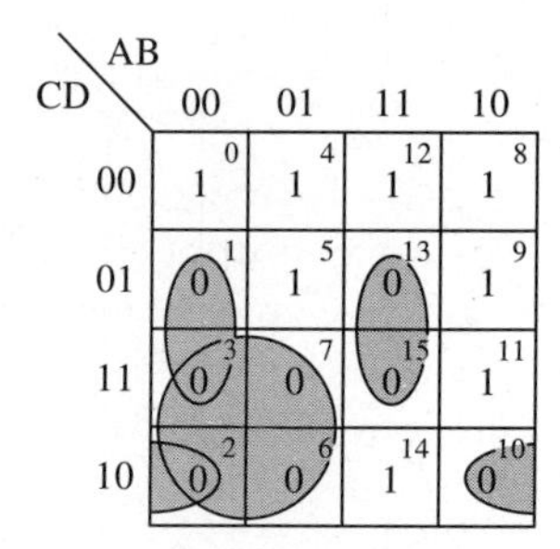

The minimal covering for the 0 entries in the map is shown on the right, so in reading the map we get the simplified expression:

$$F = (A + B + \overline{D})(\overline{A} + \overline{B} + \overline{D})(B + \overline{C} + D)(A + \overline{C})$$

The first factor in the expression is from the grouping of entries in cells 1 and 3 (A = 0, B = 0, and D = 1). The second factor is from the grouping of entries in cells 13 and 15 (A = 1, B = 1, and D = 1). The third factor is from the grouping of entries in cells 2 and 10 (B = 0, C = 1, and D = 0). The fourth factor is from the grouping of entries in cells 3, 7, 2, and 6 (A = 0 and C = 1).

Example 3.10-3

Simplify the function with the Karnaugh map on the left as a POS.

CD \ AB	00	01	11	10
00	0: 1	4: 0	12: 1	8: 0
01	1: 0	5: 0	13: 0	9: 0
11	3: 0	7: 1	15: 0	11: 1
10	2: 1	6: 1	14: 1	10: 1

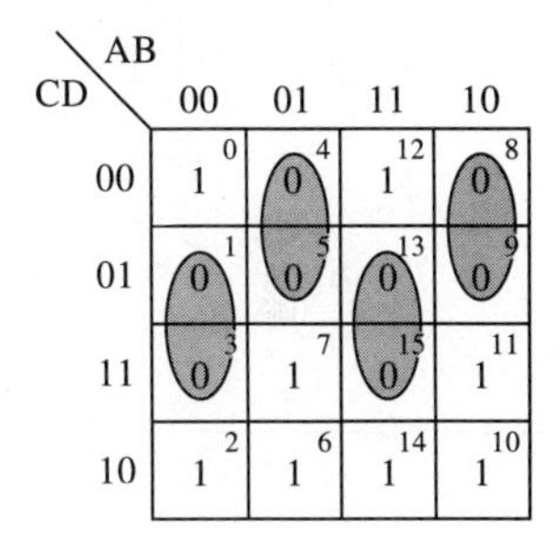

The minimal covering for 0 entries in the map is shown on the right, so in reading the map, we find that the simplified function is

$$F = (A + B + \overline{D})(A + \overline{B} + C)(\overline{A} + \overline{B} + \overline{D})(\overline{A} + B + C)$$

Example 3.10-4

Simplify the function with the Karnaugh map on the left as a POS.

CD \ AB	00	01	11	10
00	0: 0	4: 0	12: 0	8: 0
01	1: 1	5: 1	13: 0	9: 1
11	3: 0	7: 1	15: 1	11: 1
10	2: 0	6: 0	14: 0	10: 0

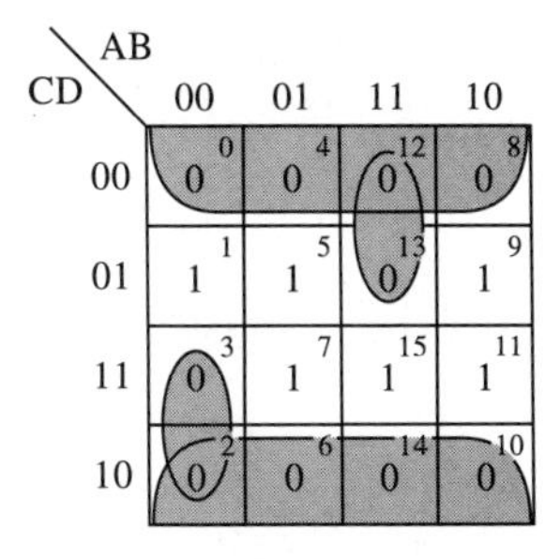

The minimal covering for the 0 entries in the map is shown on the right. Notice the grouping of eight entries. In reading the map, we find that the simplified function is

$$F = D(\overline{A} + \overline{B} + C)(A + B + \overline{C})$$

Example 3.10-5

Simplify the function with the Karnaugh map on the left as a POS.

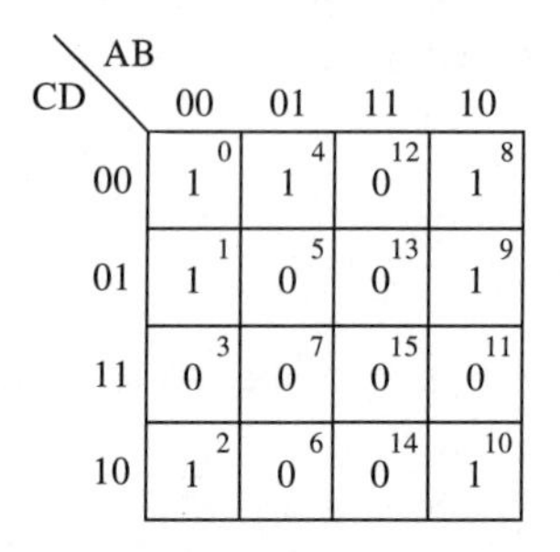

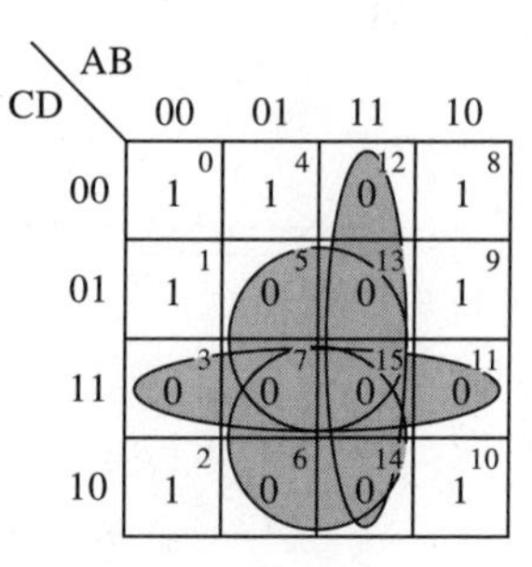

The minimal covering for the 0 entries in the map is shown on the right. Reading the map, we find that the simplified function is

$$F = (\overline{A} + \overline{B})(\overline{C} + \overline{D})(\overline{B} + \overline{D})(\overline{B} + \overline{C})$$

Example 3.10-6

Simplify the function

$$F + (A + B)(A + \overline{B} + \overline{D})(\overline{C} + \overline{D})$$

as a POS.

In forming the Karnaugh map that follows on the left, we notice that the first term generates 0 entries in all the cells for which A = 0 and B = 0 (the values that make this factor false); the second term generates 0 entries in all the cells for which A = 0, B = 1, and D = 1; and the third term generates 0 entries in all the cells for which C = 1 and D = 1.

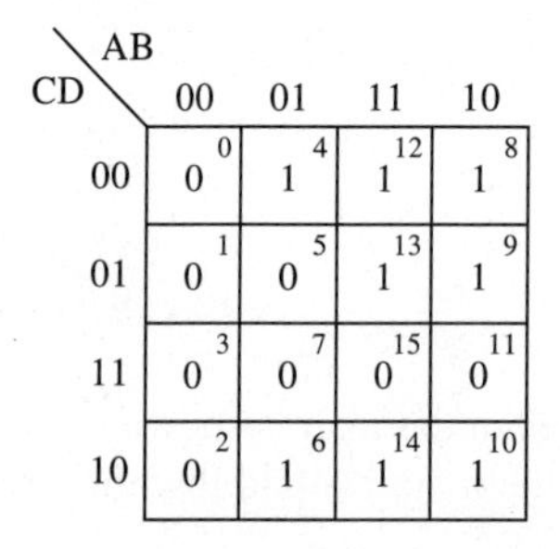

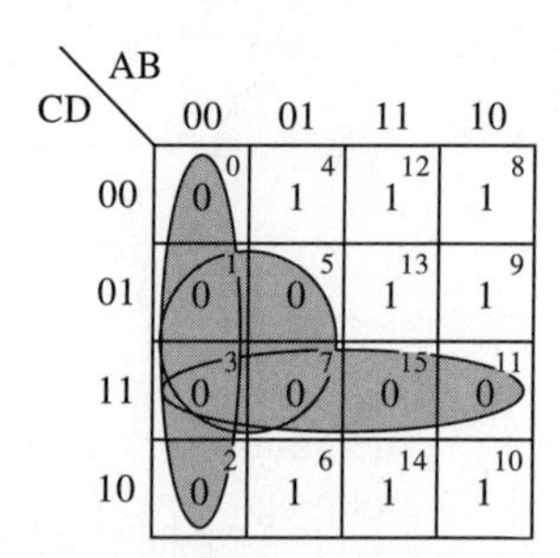

The 0 entries in this map can be covered with three groups of four entries, as shown on the right, and the function simplifies to

$$F = (A + B)(\overline{C} + \overline{D})(A + \overline{D})$$

Notice that when we covered the map, we chose the largest possible groupings of entries. We never choose a grouping of two entries when we can cover it with a larger grouping of four entries.

Example 3.10-7

Simplify the function

$$F = (A + B + \overline{C} + \overline{D})(C + \overline{D})(A + B + C + D)(A + \overline{B} + \overline{C} + \overline{D})$$

as a POS.

In forming the Karnaugh map that follows on the left, we notice that the first term generates 0 entries in all the cells for which A = 0, B = 0, C = 1, and D = 1 (one cell); the second term generates 0 entries in all the cells for which C = 0 and D = 1; the third term generates 0 entries in all the cells for which A = 0, B = 0, C = 0, and D = 0; and the fourth term generates 0 entries in all the cells for which A = 0, B = 1, C = 1, and D = 1.

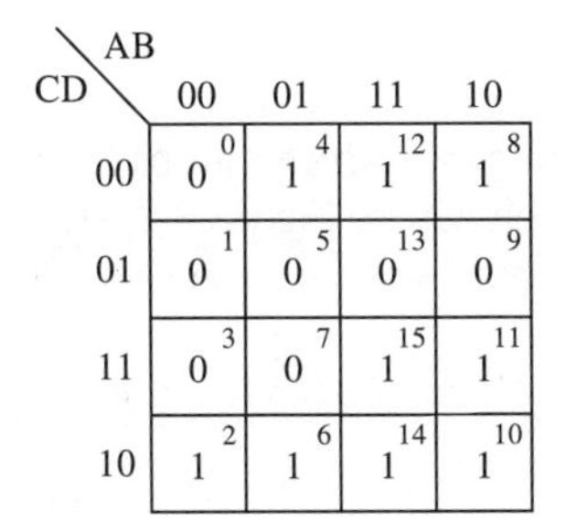

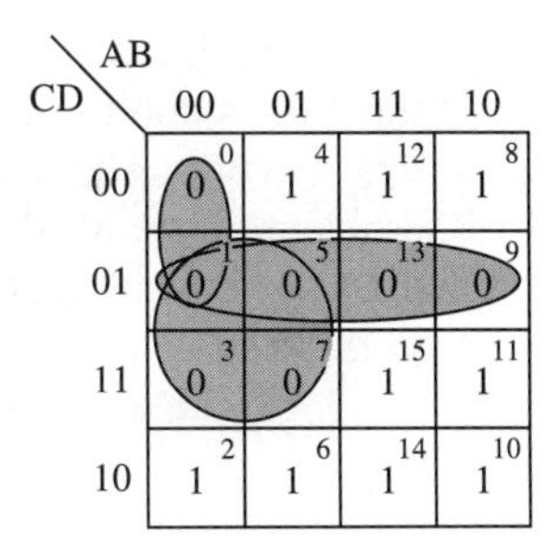

The 0 entries in this map can be covered with one grouping of two entries and two groupings of four entries, as shown on the right. The simplified function is

$$F = (A + B + C)(A + \overline{D})(C + \overline{D})$$

3.11 MAPS WITH *DON'T CARE* ENTRIES

Suppose we have a three-variable function which we want to be true when two, and only two, of the variables are true; false when one, and only one, of the variables is true; and either true or false (*don't care*) for any other combination of variables. The truth table for this function is

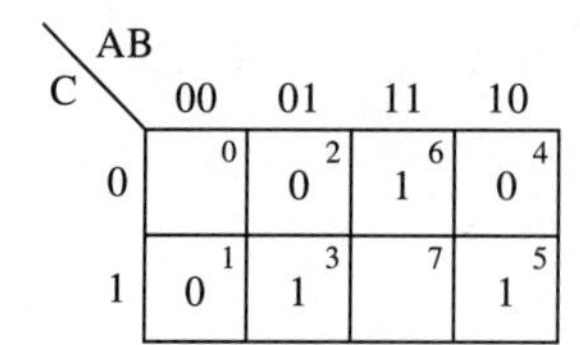

Figure 3.11-1

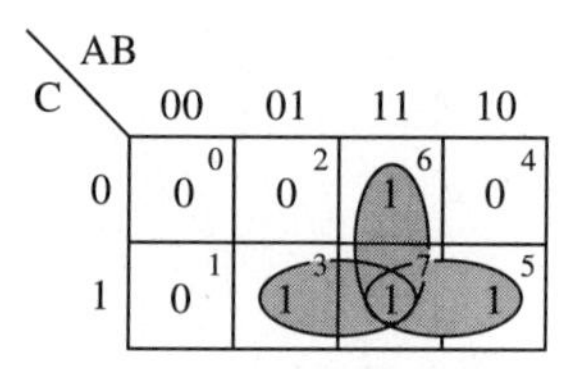

Figure 3.11-2

A	B	C	F
0	0	0	
0	0	1	0
0	1	0	0
0	1	1	1
1	0	0	0
1	0	1	1
1	1	0	1
1	1	1	

where we have not assigned values to combinations of variables for which the function is don't care (either true or false). Figure 3.11-1 is the Karnaugh map for this function. The map has two blank cells, which can be either 1s or 0s. In this problem, if we fill the blank cells as shown in Figure 3.11-2, we can get a better SOP simplification of the function because we can form larger groupings of entries in the map.

With the groupings shown, the simplified SOP function is

$$F = BC + AB + AC$$

For this map, the indicated choices for the blank cells also give the best POS simplification. The POS groupings are shown in Figure 3.11-3. The simplified POS expression for the function is

$$F = (A + C)(B + C)(A + B)$$

In this example we left the cells of the Karnaugh map blank to signify the don't

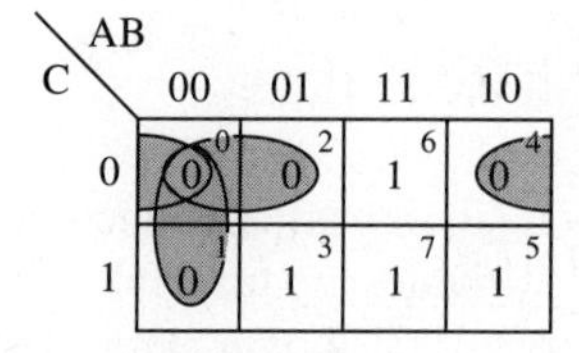

Figure 3.11-3

care function. A more common way of doing this is to use the symbol x in a cell that can be either 1 or 0. We use the same notation in the truth table. The truth table and map using this notation for the example problem are shown in Figure 3.11-4. The x entries are the don't care entries. They can be either 1 or 0; we choose the value that will make the simplest expression for the function.

A	B	C	F
0	0	0	x
0	0	1	0
0	1	0	0
0	1	1	1
1	0	0	0
1	0	1	1
1	1	0	1
1	1	1	x

C \ AB	00	01	11	10
0	x [0]	0 [2]	1 [6]	0 [4]
1	0 [1]	1 [3]	x [7]	1 [5]

Figure 3.11-4

Example 3.11-1

Simplify the function with the following Karnaugh map. Find both the SOP and POS expressions.

CD \ AB	00	01	11	10
00	1 [0]	x [4]	1 [12]	1 [8]
01	1 [1]	0 [5]	0 [13]	x [9]
11	0 [3]	1 [7]	0 [15]	0 [11]
10	1 [2]	0 [6]	0 [14]	x [10]

The minimal covering for the 1 entries in this map is on the right. Notice that we've chosen to use the don't cares in cells 4, 9, and 10 as 1s. In each case, this choice allows us to get a larger grouping of 1 entries.

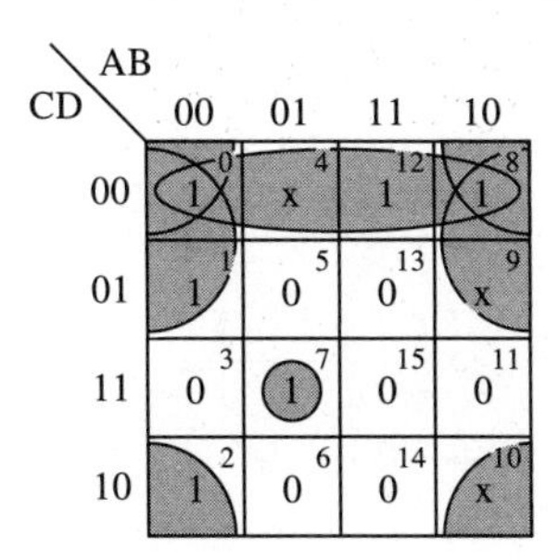

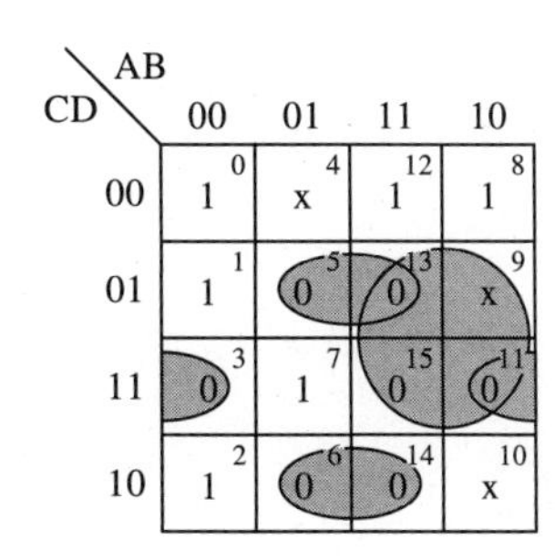

The simplified SOP expression for this function is

$$F_1 = \overline{C}\,\overline{D} + \overline{B}\,\overline{C} + \overline{B}\,\overline{D} + \overline{A}BCD$$

The minimal covering for the 0 entries in this map is shown on the right (see previous page). Notice that here we've chosen to use the don't cares in cells 4 and 10 as 1s and the don't care in cell 9 as a 0. The simplified POS expression for this function is

$$F_2 = (\overline{B} + C + \overline{D})(B + \overline{C} + \overline{D}) \cdot (\overline{B} + \overline{C} + D)(\overline{A} + \overline{D})$$

For maps with don't cares, the SOP and POS expressions may not be equivalent because, depending on the choice of the don't care entries, they may be different functions, as in this example—F_1 is 1 in cell 9, whereas F_2 is 0.

Example 3.11-2

Simplify the function with the Karnaugh map on the left. Find both the SOP and POS expressions.

CD \ AB	00	01	11	10
00	(0) 1	(4) x	(12) 0	(8) 1
01	(1) 1	(5) x	(13) 0	(9) x
11	(3) 0	(7) 1	(15) 0	(11) 0
10	(2) 1	(6) 1	(14) 0	(10) x

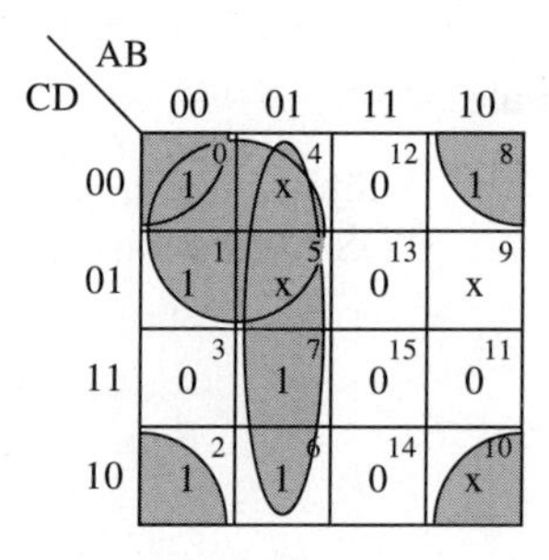

The minimal covering for the 1 entries in this map is shown on the right. We've chosen to use the don't cares in cells 4, 5, and 10 as 1s and the don't care in cell 9 as a 0. The simplified SOP expression for this function is

$$F_1 = \overline{A}\,\overline{C} + \overline{B}\,\overline{D} + \overline{A}B$$

The minimal covering for the 0 entries in this map follows. Here we've chosen to use all the don't cares as 1s.

CD \ AB	00	01	11	10
00	(0) 1	(4) x	(12) 0	(8) 1
01	(1) 1	(5) x	(13) 0	(9) x
11	(3) 0	(7) 1	(15) 0	(11) 0
10	(2) 1	(6) 1	(14) 0	(10) x

The simplified POS expression for this function is

$$F_2 = (\overline{A} + \overline{B})(B + \overline{C} + \overline{D})$$

The two functions are again different.

3.12 MAPS FOR FIVE AND SIX VARIABLES

Karnaugh maps can be extended to five and six variables by considering a three-dimensional array of cells set up so that all next-neighbor cells are logic-adjacent. This is a little hard to picture in a two-dimensional map, so we place side by side the sections of the map that are stacked up in the third dimension.

For a five-variable function we need two four-variable sections, which we consider stacked one on top of the other. The five-variable map is shown in Figure 3.12-1. At the top of each section of the map is the map heading variable. The section on the left is for A = 0 ($\overline{A}$), and the section on the right is for A = 1 (A). Notice that we've chosen the most significant variable as the map heading variable. We should visualize the left section of the map stacked on top of the right section. Cell 0 is then logic-adjacent to cell 16, cell 1 is logic-adjacent to cell 17, and so on. A five-variable map can be loaded from a five-variable truth table. In fact, the numbers shown in the cells correspond to the minterms of the truth table.

Suppose we have a function defined by Table 3.12-1, a five-variable truth table. Figure 3.12-2 is the Karnaugh map for this function.

In simplifying this function, we must consider groupings of entries in cells that lie one above the other as well as entries in cells that are neighbors side-to-side and top-to-bottom. The minimal cover of 1 entries for this map is shown in Figure 3.12-3. Notice that the entries in cells 0, 1, 4, 5, 16, 17, 20, and 21 form a grouping of eight entries in a cube, and the entries in cells 10 and 26 form a grouping of two

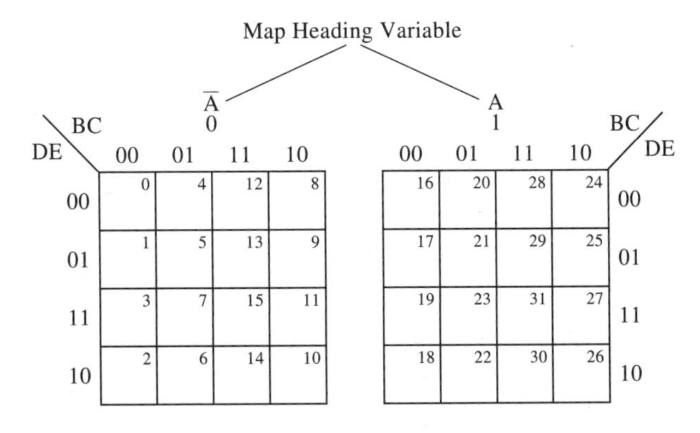

Figure 3.12-1

Table 3.12-1 Five-Variable Function Table

A	B	C	D	E	F
0	0	0	0	0	1
0	0	0	0	1	1
0	0	0	1	0	0
0	0	0	1	1	0
0	0	1	0	0	1
0	0	1	0	1	1
0	0	1	1	0	0
0	0	1	1	1	0
0	1	0	0	0	0
0	1	0	0	1	0
0	1	0	1	0	1
0	1	0	1	1	0
0	1	1	0	0	1
0	1	1	0	1	0
0	1	1	1	0	0
0	1	1	1	1	0
1	0	0	0	0	1
1	0	0	0	1	1
1	0	0	1	0	1
1	0	0	1	1	0
1	0	1	0	0	1
1	0	1	0	1	1
1	0	1	1	0	1
1	0	1	1	1	0
1	1	0	0	0	0
1	1	0	0	1	0
1	1	0	1	0	1
1	1	0	1	1	0
1	1	1	0	0	0
1	1	1	0	1	0
1	1	1	1	0	0
1	1	1	1	1	1

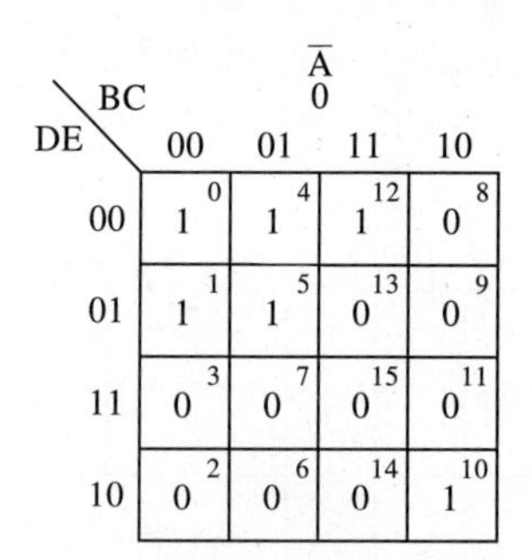

A = 1

DE \ BC	00	01	11	10
00	1 (16)	1 (20)	0 (28)	0 (24)
01	1 (17)	1 (21)	0 (29)	0 (25)
11	0 (19)	0 (23)	1 (31)	0 (27)
10	1 (18)	1 (22)	0 (30)	1 (26)

Figure 3.12-2

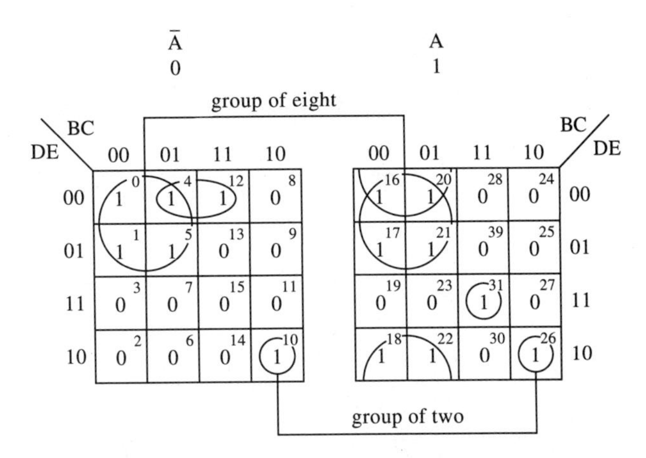

Figure 3.12-3

entries. To read the map, we remember that the variables which are common in a grouping are the variables which remain in the simplification. The simplified SOP function for this map is,

$$F = \overline{B}\,\overline{D} + \overline{A}C\overline{D}\,\overline{E} + A\overline{B}\,\overline{D} + B\overline{C}D\overline{E} + ABCDE$$

We can also do a POS simplification by grouping 0 entries. The minimal cover of 0 entries for the map is shown in Figure 3.12-4. Note the groupings of four entries

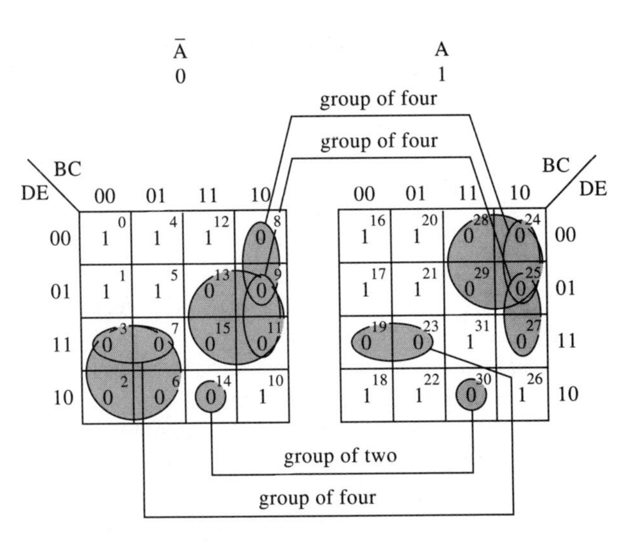

Figure 3.12-4

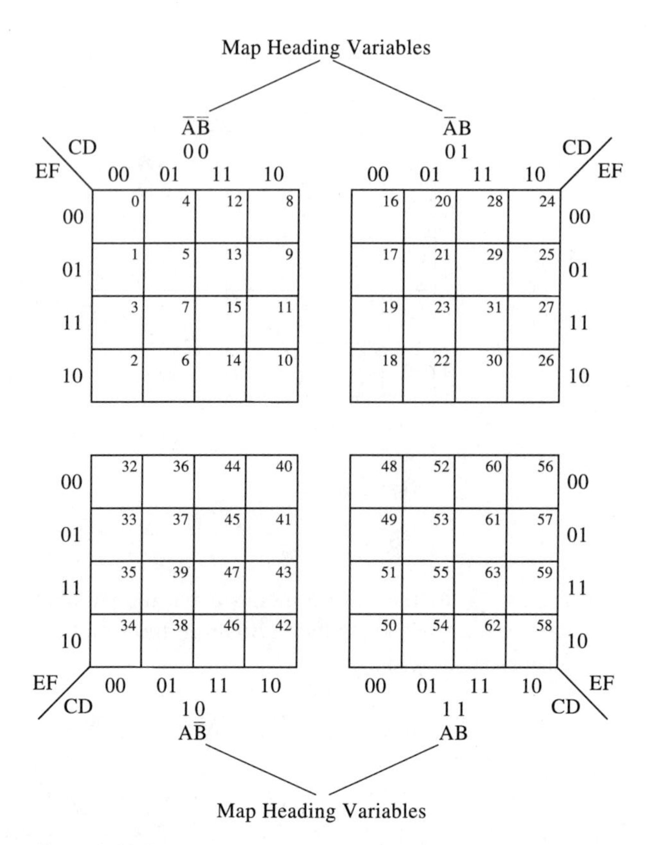

Figure 3.12-5

in cells 8, 9, 24, and 25; cells 9, 11, 25, and 27; and cells 3, 7, 19, and 23; and the grouping of two entries in cells 14 and 30. The simplified POS expression for this function is,

$$F = (A + \bar{B} + \bar{E})(A + B + \bar{D})(\bar{A} + \bar{B} + D)(\bar{B} + C + D)$$
$$(\bar{B} + C + \bar{E})(B + \bar{D} + \bar{E})(\bar{B} + \bar{C} + \bar{D} + E)$$

The method used above to map five-variable functions can be extended to six-variable functions using the six-variable map shown in Figure 3.12-5. In this map we use the two most significant variables as map headings for each of the four-variable sections of the map. We can think of this six-variable map as a three-dimensional map with the four four-variable sections stacked one on top of another, starting in the upper left corner with the 00 section and proceeding in a circle through the 01, 11, and 10 sections. We'll then have logic-adjacent cells next to each other in the stack. Notice that corresponding elements at the top and bottom of the stack are also logic-adjacent. We show this in our map by placing the four-variable sections of the map in a circle so that the 10 section is next to the 00 section. Actually, in this arrangement we have logic adjacency from each four-variable section of the map to the near-neighbor four-variable sections.

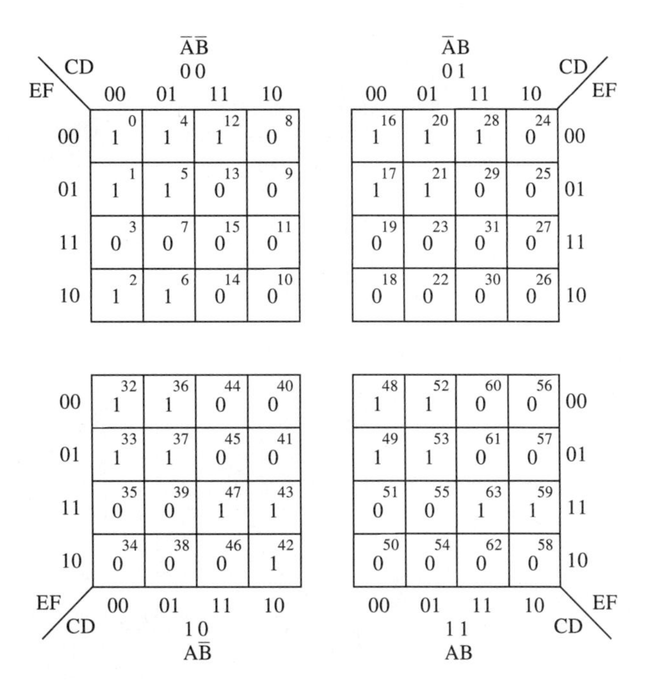

Figure 3.12-6

As an example of a function simplification using a six-variable map, consider the function with the map shown in Figure 3.12-6. The function could have been defined in a truth table (with sixty-four entries!) and entered directly into the map using the numbering in the cells for the minterms. The minimal covering of 1s in this map is shown in Figure 3.12-7. Note how the three-dimensional aspects of the map must be used to get a grouping of the sixteen entries in cells 0, 1, 4, 5, 16, 17, 20, 21, 48, 49, 52, 53, 32, 33, 36, and 37, and groupings of four entries in cells 4, 12, 20, and 28 and cells 47, 43, 63, and 59. Other groupings are in a single section of the map. The SOP simplified function (which we'll denote as G, since F is one of the logic variables in the problem) is

$$G = \overline{C}\,\overline{E} + \overline{A}D\overline{E}\,\overline{F} + ACEF + \overline{A}\,\overline{B}\,\overline{C}\,\overline{F} + A\overline{B}C\overline{D}E$$

We can also do a POS simplification by grouping the 0 entries in the map. A minimal covering (there are more than one) of 0s in this map is shown in Figure 3.12-8. Note how the three-dimensional aspects of the map must be used to get the various groupings of four and eight entries. No groupings are in a single section of the map. The simplified function (which we'll again denote as G, since F is a variable) is

$$G = (\overline{C} + D + E)(A + \overline{C} + \overline{F})(A + \overline{C} + \overline{E})(A + \overline{E} + \overline{F})(\overline{B} + \overline{E} + F)$$
$$(\overline{A} + \overline{C} + E)(\overline{A} + C + \overline{E})(\overline{A} + \overline{D} + \overline{E} + F)$$

As you can see, simplification of six-variable functions using three-dimensional maps can be messy and cumbersome. Rather than continue this technique, we'll

Figure 3.12-7

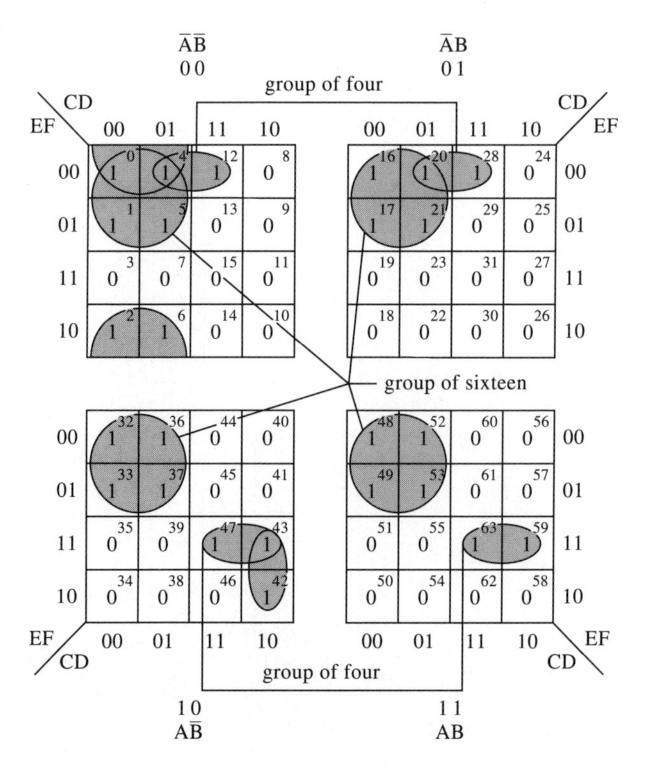

Figure 3.12-8

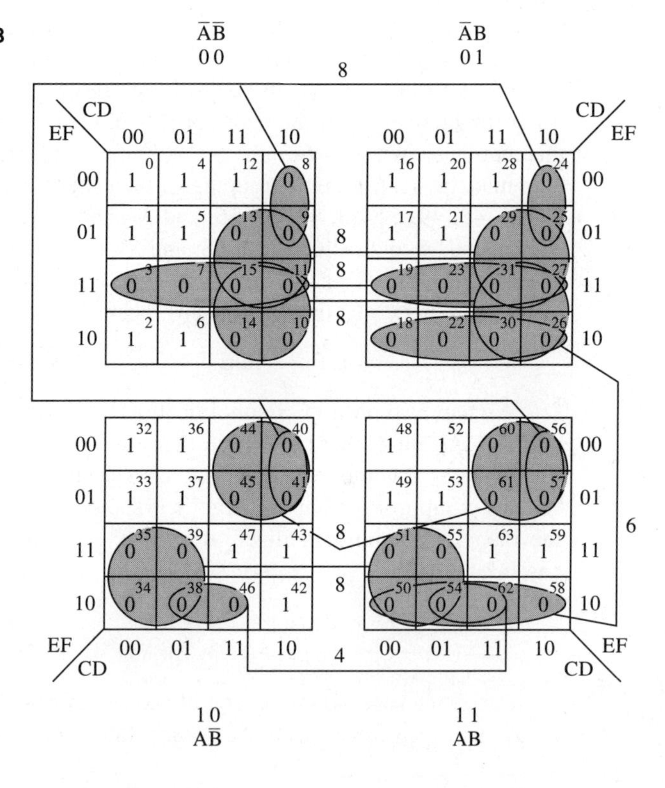

introduce a modified mapping technique that employs **variable entered maps**. It allows us to simplify five- and six- (and sometimes more) variable functions using four-variable maps.

3.13 VARIABLE ENTERED MAPS (VEMs)

In a Karnaugh map, each cell represents a minterm, and we enter a 1 or a 0 in the cell to specify whether the function is true or false for that minterm. Variable entered maps extend this idea. They allow us to enter variables in the cells of a map to represent minterms that have too many variables to be entered normally in the map.

This is most readily seen with an example. Consider the five-variable function

$$F = \overline{A}BCD + \overline{A}\,\overline{B}CDE + AB\overline{C}\,\overline{D}E + \overline{A}B\overline{C}D\overline{E}$$

We can represent this function in a four-variable map with *map variables* A, B, C, and D by showing explicitly, in the appropriate cells of the map, the values of E needed to make the function true (1) in each cell. Figure 3.13-1 is the VEM for this function. The map contains complete minterm information about the function. Part of the information is carried by the map variables, and part is explicitly entered in the cells as the appropriate value for E (the *map entered variable*). The first term in the function is true, independent of the value of E, so we enter a 1 in the cell corresponding to this term (A = 0, B = 1, C = 1, and D = 1). The second term in the function is true for A = 0, B = 0, C = 1, and D = 1, all of which correspond to a cell in the map, but for this term to be true, E must also be true. To show this relation, we enter E in the cell corresponding to the first four variables. We do the same kind of thing for the third and fourth terms.

To simplify the map, we use logic adjacency in the map variables (A, B, C, and D), first for terms with the entered variable and then for terms without the entered variable. We can think of this as a two-step process. First we cover all the map entered variables with maximum-sized groupings, then we cover the rest of the variables. In order to cover the map entered variables, we need to think of the 1s in the map in terms of the variable E. In these terms, the 1 really represents E + $\overline{E}$ because $\overline{A}BCD = \overline{A}BCDE + \overline{A}BCD\overline{E}$. So if we write out the 1s in terms of E, the map is as shown in Figure 3.13-2. Now, to cover the map entered variables, we

CD \ AB	00	01	11	10
00	0 (0)	0 (4)	E (12)	0 (8)
01	0 (1)	$\overline{E}$ (5)	0 (13)	0 (9)
11	E (3)	1 (7)	0 (15)	0 (11)
10	0 (2)	0 (6)	0 (14)	0 (10)

Figure 3.13-1

CD \ AB	00	01	11	10
00	0 (0)	0 (4)	E (12)	0 (8)
01	0 (1)	$\overline{E}$ (5)	0 (13)	0 (9)
11	E (3)	E+$\overline{E}$ (7)	0 (15)	0 (11)
10	0 (2)	0 (6)	0 (14)	0 (10)

Figure 3.13-2

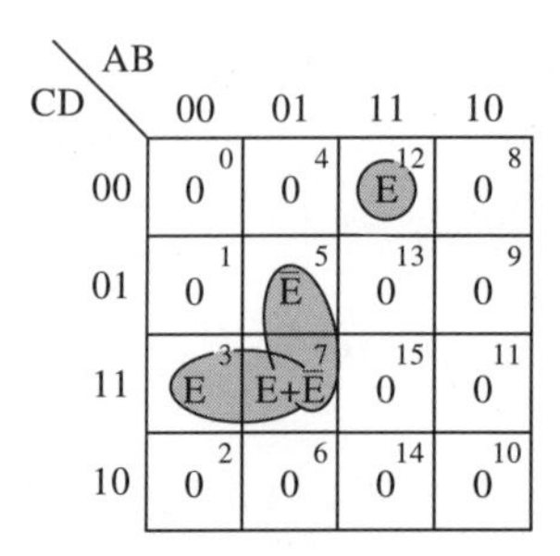

Figure 3.13-3

group entries in any near-neighbor cells that have the *same* entered variable, because they're logic-adjacent in the map variables (Fig. 3.13-3). Thus, we can group the $\overline{E}$ entries together to form a grouping of two ($\overline{A}B\overline{C}D\overline{E} + \overline{A}BCD\overline{E} = \overline{A}BD\overline{E}$), and the E entries together to form another grouping of two ($\overline{A}\overline{B}CDE + \overline{A}BCDE = \overline{A}CDE$). We need to emphasize that near-neighbor entries with the *same* entered variable are grouped together. The single E with no neighbors forms an island.

In this first step, we must cover all the single map entered variables (all the Es and $\overline{E}$s but not necessarily the E + $\overline{E}$s). In this example we don't need to take the second step and cover other variables, because we've covered all the entries in the map with the first step. Notice that the 1 (E + $\overline{E}$) is completely covered. The simplified function for this map is

$$F = AB\overline{C}\overline{D}E + \overline{A}BD\overline{E} + \overline{A}CDE$$

The map reading is the same as that for a normal map, except here we need to remember to include the common map entered variable in the simplified terms.

Suppose we change the example slightly so that we can see the two-step map reading process. If we delete the term in the function that contains $\overline{E}$, then the map is as shown in Figure 3.13-4. The covering for the map entered variables is shown in Figure 3.13-5. In this covering, the 1 (E + $\overline{E}$) is not completely covered ($\overline{E}$ is not grouped). In such a case we form a second map with all the map entered variables

CD \ AB	00	01	11	10
00	0^{0}	0^{4}	E^{12}	0^{8}
01	0^{1}	0^{5}	0^{13}	0^{9}
11	E^{3}	$E+\overline{E}^{\,7}$	0^{15}	0^{11}
10	0^{2}	0^{6}	0^{14}	0^{10}

Figure 3.13-4

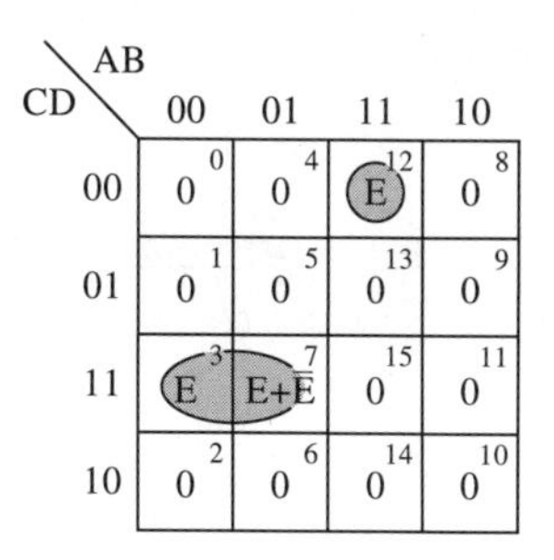

Figure 3.13-5

CD \ AB	00	01	11	10
00	0 (0)	0 (4)	0 (12)	0 (8)
01	0 (1)	0 (5)	0 (13)	0 (9)
11	0 (3)	1 (7)	0 (15)	0 (11)
10	0 (2)	0 (6)	0 (14)	0 (10)

Figure 3.13-6

that are completely covered entered as 0s and all the map entered variables that are not completely covered entered as 1s. All the *single* map entered variables must be covered in the first step and must not appear in the second step. For this example the second map is as shown in Figure 3.13-6. We now group the 1 entries in the second map. Here there is only an island.

The simplified function is found from reading both the first and second maps:

$$F = AB\overline{C}\,\overline{D}E + \overline{A}CDE + \overline{A}BCD$$

The first two terms are from reading the first map; the last term is from reading the second map. Later we'll formulate rules for grouping entries in a VEM to get a minimal covering, but first we'll consider some other examples of forming a VEM.

3.14 FORMING A VEM FROM A TRUTH TABLE

A VEM can be readily formed from a truth table for a function. Consider Table 3.14-1, which was used as an example in Section 3.12. If we want to form a VEM for this function, we can take the first four variables to be the map variables and the last variable to be the entered variable. Notice that if we do this, the map variables appear in groups of two in the truth table. The first two entries (rows) in the table have the same map variables, the second two have the same map variables, and so on. Thus, information from the first two entries in the table must be entered in cell 0 of the map, information from the second two entries must be entered in cell 1, and so on. To determine which value to enter in the cell, we look at the value of the

Table 3.14-1

A	B	C	D	E	F
0	0	0	0	0	1
0	0	0	0	1	1
0	0	0	1	0	0
0	0	0	1	1	0
0	0	1	0	0	1
0	0	1	0	1	1
0	0	1	1	0	0
0	0	1	1	1	0
0	1	0	0	0	0
0	1	0	0	1	0
0	1	0	1	0	1
0	1	0	1	1	0
0	1	1	0	0	1
0	1	1	0	1	0
0	1	1	1	0	0
0	1	1	1	1	0
1	0	0	0	0	1
1	0	0	0	1	1
1	0	0	1	0	1
1	0	0	1	1	0
1	0	1	0	0	1
1	0	1	0	1	1
1	0	1	1	0	1
1	0	1	1	1	0
1	1	0	0	0	0
1	1	0	0	1	0
1	1	0	1	0	1
1	1	0	1	1	0
1	1	1	0	0	0
1	1	1	0	1	0
1	1	1	1	0	0
1	1	1	1	1	1

function for the two entries associated with that cell ($\overline{E}$ and E). We enter the values of E for which the function is true.

Figure 3.14-1 is the map for this function. The first two entries in the truth table are true for both $\overline{E}$ and E, so $E + \overline{E}$ is entered in the 0 cell; the second two entries are not true for either $\overline{E}$ or E, so 0 is entered in the 1 cell; and so on.

The first-step covering for this map is shown in Figure 3.14-2. Notice that we have to cover all the single entered variables, but not the $E + \overline{E}$ entries. In fact, we should not group the $E + \overline{E}$ entries in the first step unless we need them to group

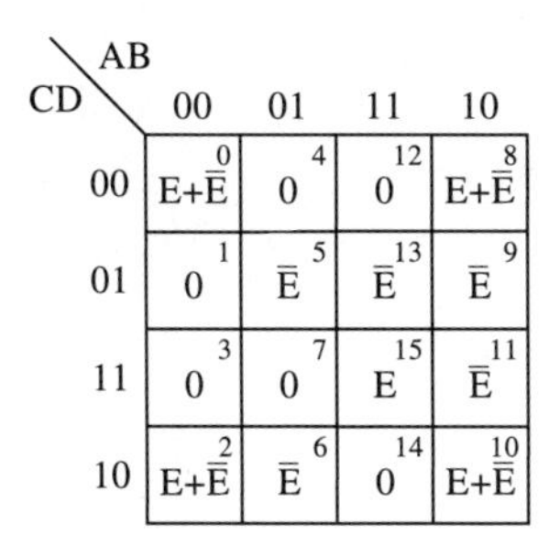

Figure 3.14-1

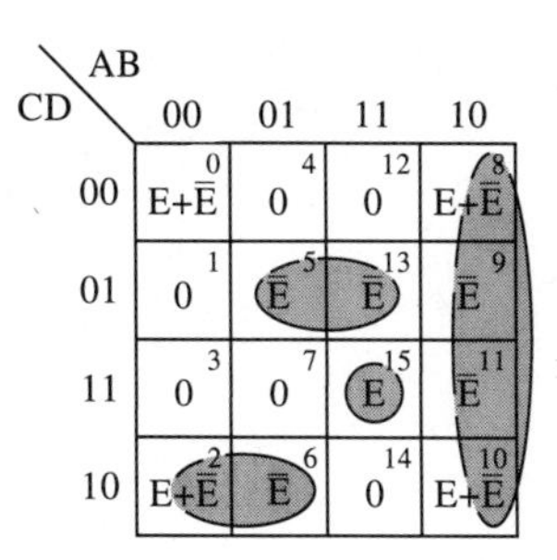

Figure 3.14-2

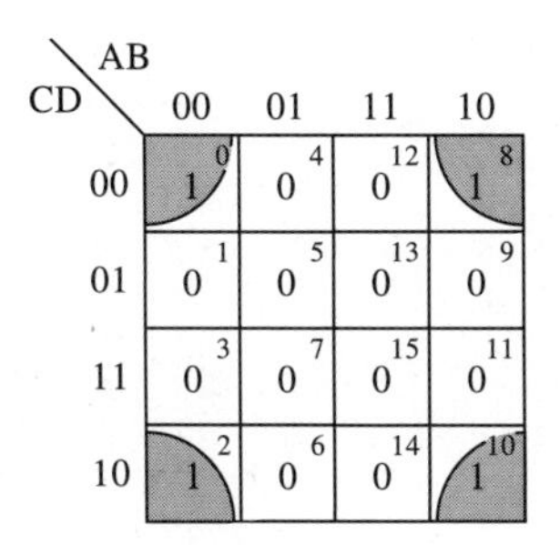

Figure 3.14-3

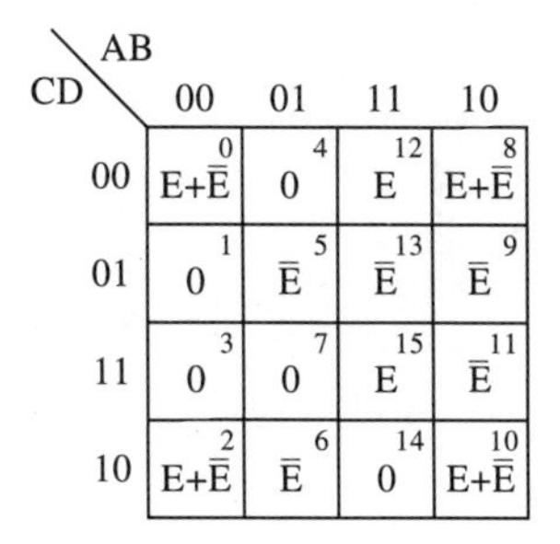

Figure 3.14-4

with one of the single entered variables. Again we emphasize that entries with a common entered variable are grouped. *Do not group E entries with $\overline{E}$ entries.*

Figure 3.14-3 is the second-step map for this function, with its minimal cover. The simplified function from these maps is

$$F = ABCDE + A\overline{B}\,\overline{E} + B\overline{C}D\overline{E} + \overline{A}C\overline{D}\,\overline{E} + \overline{B}\,\overline{D}$$

This is the same expression we found in Section 3.12. The first four terms are from the first (*variable entered*) map; the last term is from the second (*reduced*) map.

Now we want to modify this example to see how to handle a completely covered $E + \overline{E}$ term in the first step by using a don't care term in the second map. Suppose the truth table is changed so that the VEM for the function is as shown in Figure 3.14-4. (The function value in row 25 of the truth table is changed to 1.) When we group the entered variables in this map (Fig. 3.14-5), the $\overline{E} + E$ in the upper right corner is completely covered. This means that it need not appear in the reduced map. On the other hand, because we can group an entry in the map more than once, if this term is included in the reduced map it may be possible to get a better grouping in that map. We resolve this conflict by entering the term in the reduced map as a don't care (x) so that it need not be used unless it is required to simplify a grouping.

Figure 3.14-6 is the reduced map for this modified example. The modification does not change the grouping of entries in the reduced map, because in this example

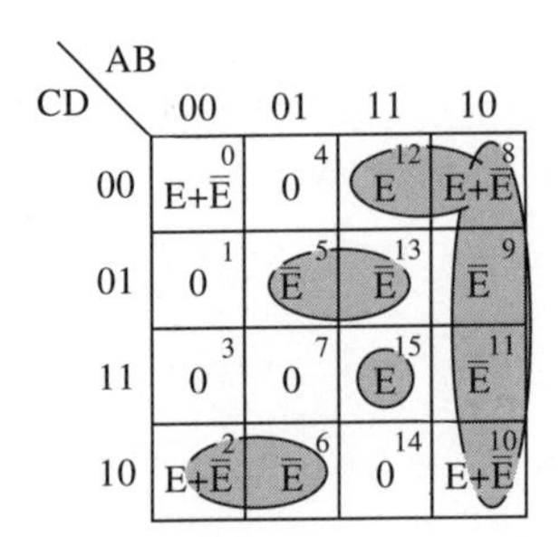

Figure 3.14-5

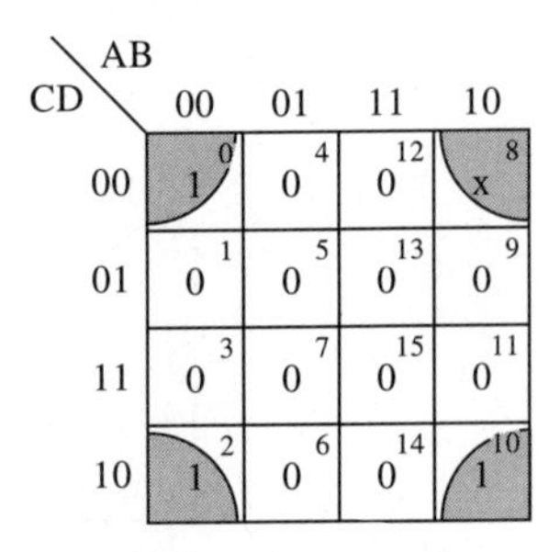

Figure 3.14-6

Table 3.14-2

A	B	C	D	E	F
0	0	0	0	0	x
0	0	0	0	1	x
0	0	0	1	0	0
0	0	0	1	1	0
0	0	1	0	0	1
0	0	1	0	1	x
0	0	1	1	0	0
0	0	1	1	1	0
0	1	0	0	0	0
0	1	0	0	1	0
0	1	0	1	0	1
0	1	0	1	1	0
0	1	1	0	0	1
0	1	1	0	1	0
0	1	1	1	0	0
0	1	1	1	1	0
1	0	0	0	0	1
1	0	0	0	1	1
1	0	0	1	0	x
1	0	0	1	1	0
1	0	1	0	0	1
1	0	1	0	1	1
1	0	1	1	0	1
1	0	1	1	1	0
1	1	0	0	0	0
1	1	0	0	1	0
1	1	0	1	0	1
1	1	0	1	1	0
1	1	1	0	0	0
1	1	1	0	1	0
1	1	1	1	0	0
1	1	1	1	1	1

CD \ AB	00	01	11	10
00	(0) $xE+x\bar{E}$	(4) 0	(12) 0	(8) $E+\bar{E}$
01	(1) 0	(5) $\bar{E}$	(13) $\bar{E}$	(9) $x\bar{E}$
11	(3) 0	(7) 0	(15) E	(11) $\bar{E}$
10	(2) $xE+\bar{E}$	(6) $\bar{E}$	(14) 0	(10) $E+\bar{E}$

Figure 3.14-7

CD \ AB	00	01	11	10
00	(0) x	(4) 0	(12) 0	(8) 1
01	(1) 0	(5) $\bar{E}$	(13) $\bar{E}$	(9) $x\bar{E}$
11	(3) 0	(7) 0	(15) E	(11) $\bar{E}$
10	(2) $xE+\bar{E}$	(6) $\bar{E}$	(14) 0	(10) 1

Figure 3.14-8

we include the don't care term in the grouping of the four corner entries. In some cases we may not want to use the don't care term. The simplified function for these maps is

$$F = ABCDE + A\bar{B}\bar{E} + B\bar{C}D\bar{E} + \bar{A}C\bar{D}\bar{E} + AC\bar{D}E + \bar{B}\bar{D}$$

In some cases we may want to map functions with don't cares in VEMs. Consider the function defined by Table 3.14-2. The VEM for this function is shown in Figure 3.14-7. In cell 0 of this map, the function value is don't care for both E and $\bar{E}$, so we enter this information in the cell. To show the don't care associated with the entered variable, we preface the variable with don't care (x). In cell 2, the function is 1 for $\bar{E}$ and don't care for E. In this case, both $\bar{E}$ and E are entered in the cell, but E is prefaced with a don't care (x). In cell 9, the function is don't care for $\bar{E}$ and 0 for E. In this case, only $\bar{E}$ is entered in the cell, again prefaced by x.

In the VEMs we have been forming, we've entered $\bar{E} + E$ for 1 and $x\bar{E} + xE$ for x so that it is easier to see how to group the entered variables. From now on we will enter 1s and xs. With this change, the VEM for this example is as shown in Figure 3.14-8. We must remember in making groupings that both the $\bar{E}$ and the E in the 1 can be grouped separately, and both the $x\bar{E}$ and the xE in the x can be grouped separately.

The minimal covering for the entered variables is shown in Figure 3.14-9, and the reduced map, with its minimal cover, is shown in Figure 3.14-10. Notice that the don't cares are all used in the reduced map to get the largest groupings possible. The simplified function for these maps is

$$F = ABCDE + B\bar{C}D\bar{E} + \bar{A}C\bar{D}\bar{E} + A\bar{B}\bar{E} + \bar{B}\bar{D}$$

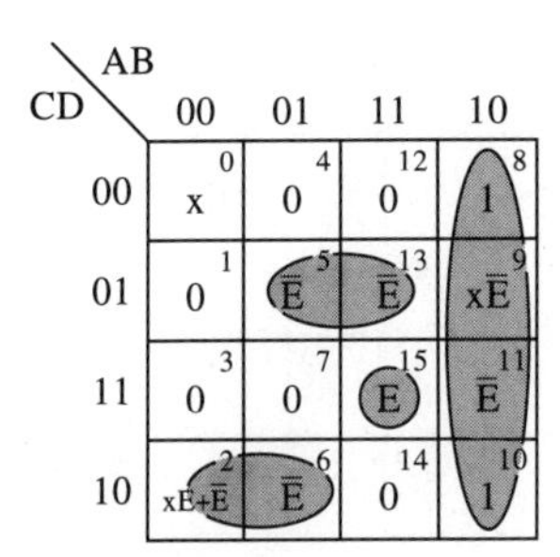

Figure 3.14-9

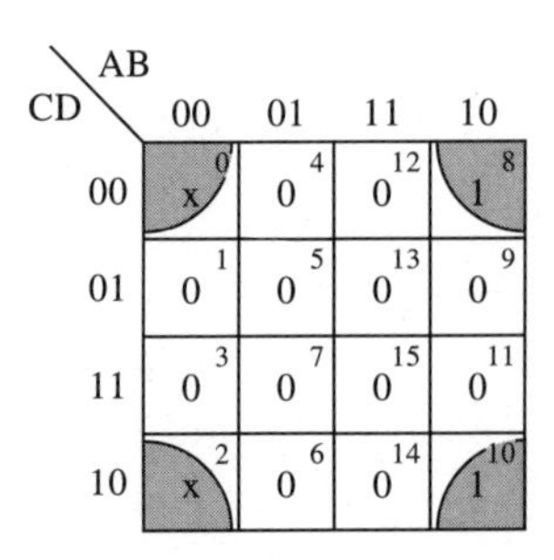

Figure 3.14-10

3.15 VEM MAP GROUPING RULES

In the last two sections we've formed groupings of entries in VEM maps, but we don't have a formal way of making groups for a minimal grouping. Forming groupings for a VEM is a two-step process. First we group all the single entered variables (EVs) that occur in the cells of the map. Then we form a reduced map with no entered variables, and we group the 1s in this map just as we did in a standard Karnaugh map. With the overall constraint that we never choose a grouping that is completely covered by a larger grouping, the rules for grouping the entered variables in the first step are as follows.

1. Examine all the single entered variables (E or $\overline{\text{E}}$) one at a time, and determine which variables cannot be grouped with another identical entered variable, a 1, a don't care for an identical variable, or a don't care. These are islands, as in the standard Karnaugh map.
2. Examine all the single entered variables one at a time, and determine which variables can be grouped in only one way with another identical entered variable, a 1, a don't care for an identical variable, or a don't care. Group these variables.
3. Examine all the remaining ungrouped single entered variables one at a time, and determine which variables can be grouped in more than one way. Try to group each of them first with another identical ungrouped entered variable and then with an uncovered or partially covered 1. Try to choose groupings that will completely cover the 1s, particularly 1s that will be islands in the reduced map. When grouping these variables with covered entered variables, covered 1s, don't care entered variables, and don't cares, just make an arbitrary choice.
4. Repeat steps 2 and 3, examining each ungrouped single entered variable and considering groupings of four entered variables, 1s, and don't cares.
5. Repeat steps 2 and 3, examining each ungrouped single entered variable and considering groupings of eight entered variables, 1s, and don't cares.

All the single entered variables must be covered in this first step. When that has been accomplished, stop and proceed to the second step.

The second step is to form the reduced map. Entries in the VEM are transferred to the reduced map using the following rules. EV means entered variable.

$0 \rightarrow 0$

$EV \rightarrow 0$

$\overline{EV} \rightarrow 0$

$1 \rightarrow x$, if it is completely covered

$1 \rightarrow 1$, if it is not completely covered

$x \rightarrow x$

$xEV \rightarrow 0$

$x\overline{EV} \rightarrow 0$

$EV + x\overline{EV} \rightarrow x$, if EV is covered

$EV + x\overline{EV} \rightarrow 1$, if EV is not covered

$xEV + \overline{EV} \rightarrow x$, if $\overline{EV}$ is covered

$xEV + \overline{EV} \rightarrow 1$, if $\overline{EV}$ is not covered

The reduced map is a normal map, and groupings are made in the normal way. Both maps must be read to determine the simplified function.

To see how these rules work, we'll apply them to some examples.

Example 3.15-1

Simplify the function represented in the following VEM.

CD \ AB	00	01	11	10
00	1 [0]	1 [4]	$\overline{E}$ [12]	$\overline{E}$ [8]
01	0 [1]	E [5]	0 [13]	$\overline{E}$ [9]
11	E [3]	0 [7]	E [15]	0 [11]
10	0 [2]	0 [6]	1 [14]	0 [10]

The minimal covering for entered variables follows on the left. The E in cell 3 is an island (rule 1). The E in cell 5 can be grouped only with the 1 in cell 4 (rule 2). The E in cell 15 can be grouped only with the 1 in cell 14 (rule 2). The $\overline{E}$ in cell 9 can be grouped only with the $\overline{E}$ in cell 8 (rule 2). The $\overline{E}$ in cell 12 can be grouped in one of three ways—with the 1 in cell 4, with the 1 in cell 14, or with the $\overline{E}$ in cell 8 (rule 3). The $\overline{E}$ in cell 8 is already covered, so it should not be grouped with the $\overline{E}$ in this cell if there are other choices. If we group the $\overline{E}$ with the 1 in cell 4 or cell 14, the 1 will be completely covered and may simplify the reduced map. By looking ahead, we can see that the best choice is to completely cover the 1 in cell 14, because it will be an island in the reduced map, whereas the 1 in cell 4 can be grouped with the 1 in cell 0.

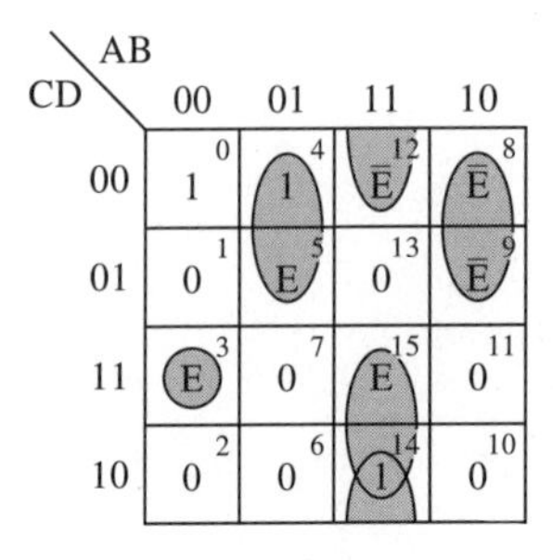

CD \ AB	00	01	11	10
00	1 (0)	1 (4)	0 (12)	0 (8)
01	0 (1)	0 (5)	0 (13)	0 (9)
11	0 (3)	0 (7)	0 (15)	0 (11)
10	0 (2)	0 (6)	x (14)	0 (10)

The reduced map for this example is shown on the right. The 1s in cells 0 and 4 are not completely covered, so they are transferred to the reduced map as 1s. The 1 in cell 14 is completely covered, so it is transferred as a don't care. The single entered variables and the 0s all transfer as 0s. The minimal covering for this map is as shown. The don't care in cell 14 is not covered, because its use will not result in a more minimal covering. The simplified function is

$$F = \bar{A}\bar{B}CDE + \bar{A}B\bar{C}E + AB\bar{D}\bar{E} + A\bar{B}\bar{C}\bar{E} + ABCE + \bar{A}\bar{C}\bar{D}$$

Example 3.15-2

Simplify the function represented in the following VEM.

CD \ AB	00	01	11	10
00	1 (0)	x (4)	$xE+\bar{E}$ (12)	$\bar{E}$ (8)
01	0 (1)	E (5)	$x\bar{E}$ (13)	$\bar{E}$ (9)
11	0 (3)	0 (7)	$\bar{E}$ (15)	$E+x\bar{E}$ (11)
10	0 (2)	0 (6)	1 (14)	0 (10)

The minimal covering for the entered variables follows on the left. There are no islands (rule 1). The E in cell 5 can be grouped only with the don't care in cell 4. The only other entered variable that cannot be grouped in a grouping of four is the E in cell 15. It can be grouped in more than one way (rule 3). It should be grouped with the E in cell 11 so that the entries in this cell are completely covered. Now, considering entered variables that can be grouped in groupings of four, the $\bar{E}$ in cell 9 can be grouped only with the entries in cells 12, 13, and 8. This completes the minimal covering for the entered variables. The possible grouping of four entries in the top row is not used because all the entered variables have been covered.

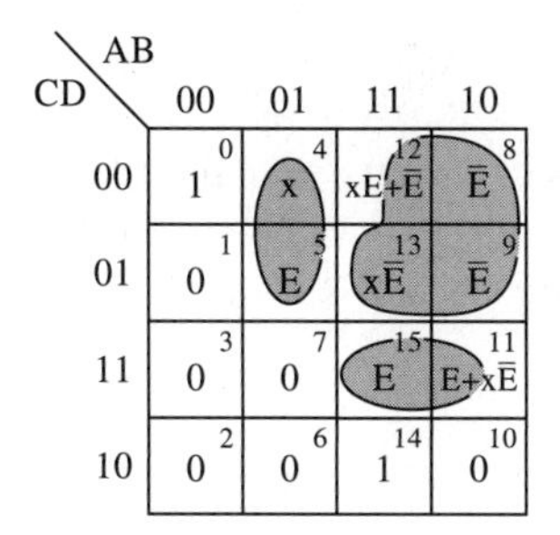

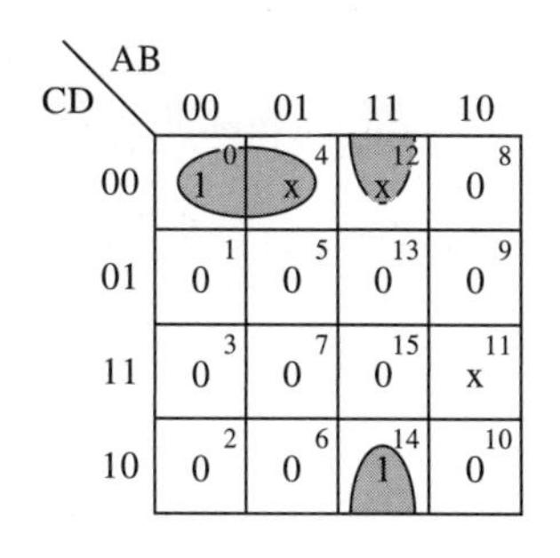

The reduced map for this example is shown on the right. The uncovered 1s in cells 0 and 14 transfer as 1s. The don't care in cell 4 transfers as a don't care. The $xE + \overline{E}$ in cell 12 and the $E + x\overline{E}$ in cell 11 transfer as don't cares because the essential variable in each is covered. The minimal cover for the reduced map is as shown on the map. The simplified function is

$$F = \overline{A}B\overline{C}E + ACDE + A\overline{C}\,\overline{E} + AB\overline{D} + A\overline{C}\,\overline{D}$$

Once we become familiar with VEM map simplifications, it's possible to find the map covering on one map without the use of the reduced map. To do so, proceed just as in the foregoing examples and cover all the entered variables. In this step, all the single entered variables must be covered. Once that is done, ignore the single entered variables and concentrate on the 1s that are not completely covered. They must be grouped with each other, with 1s that are completely covered, and with don't cares. To see how this is done, consider the following examples.

Example 3.15-3

Simplify the logic function defined by the following VEM.

CD \ AB	00	01	11	10
00	0 (0)	x (4)	$x\overline{E}$ (12)	xE (8)
01	1 (1)	1 (5)	$\overline{E}$ (13)	1 (9)
11	0 (3)	E (7)	xE (15)	E (11)
10	0 (2)	0 (6)	0 (14)	xE (10)

To cover this map, we first group all the single entered variables as shown on the left in the following figure. There are no islands. The E in cell 7 can form a grouping of two with the 1 in cell 5 or the xE in cell 15. It is grouped with the 1 (rule 3). The remaining entered variables can all be grouped in groupings of four. The E in cell 11 can be grouped in only one way, with a grouping of four (the xE in cell 8, the

1 in cell 9, and the xE in cell 10). The $\overline{E}$ in cell 13 can be grouped in two ways to form a grouping of four. The grouping with the 1s in cells 1, 5, and 9 is chosen over the grouping with the x$\overline{E}$ in cell 12, the x in cell 4, and the 1 in cell 5 (rule 3). Notice that this choice completely covers the 1s in both cells 5 and 9 and thus has more potential for simplifying the covering needed for the 1s in the next step of the simplification.

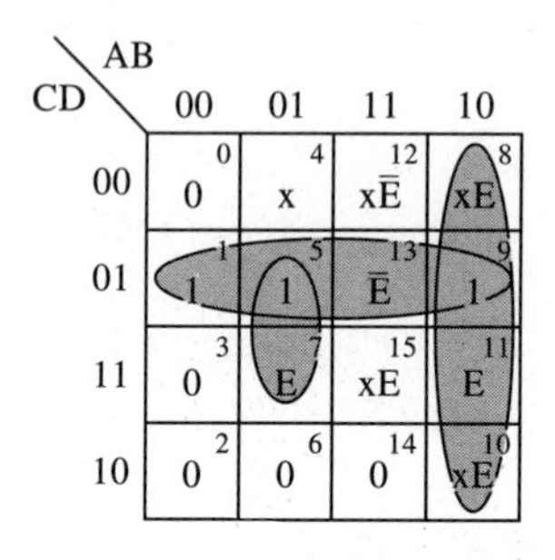

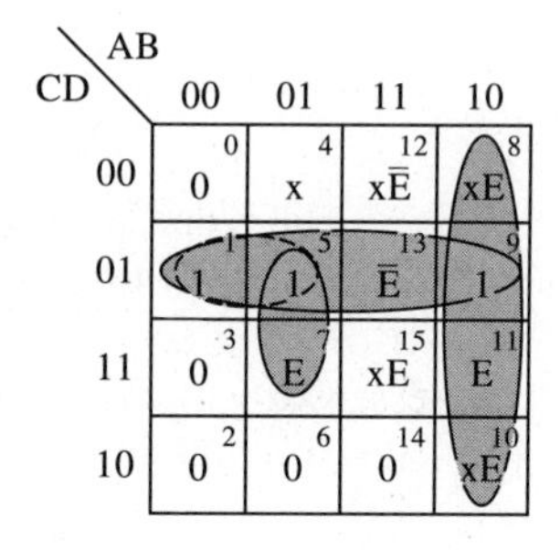

Now that all the entered variables are covered, we must cover all the 1s that are not yet completely covered. As it happens, the only one is in cell 1. It can be grouped with the 1 in cell 5 or with the 1 in cell 9. We choose to group it with the 1 in cell 5, as shown on the right (dashed line).

We've completed the covering of both the entered variables and the uncovered and partially covered 1s on a single map. To complete the simplification, we need to read the map, taking care not to forget the entered variables. The simplified function is

$$F = \overline{A}BDE + A\overline{B}E + \overline{C}D\overline{E} + \overline{A}\,\overline{C}D$$

The first three terms are from the groupings involving entered variables; the last term is from the grouping of 1s.

Example 3.15-4

Simplify the logic function defined by the following VEM.

CD \ AB	00	01	11	10
00	(0) E	(4) E	(12) 0	(8) 0
01	(1) 1	(5) 1	(13) 1	(9) 1
11	(3) 0	(7) 0	(15) 0	(11) xE
10	(2) E	(6) 0	(14) 0	(10) E

To cover this map, we first group all the single entered variables as shown on the left in the following figure. There are no islands. The E in cell 10 can be grouped in only one way, as a grouping of two with the E in cell 2 (rule 2). The other Es, in cells 0 and 4, can be grouped in only one way, as a grouping of four with the 1s in cells 1 and 5 (rule 2). The xE is not covered. We cover xs only when they help to give us a larger grouping with the entered variables.

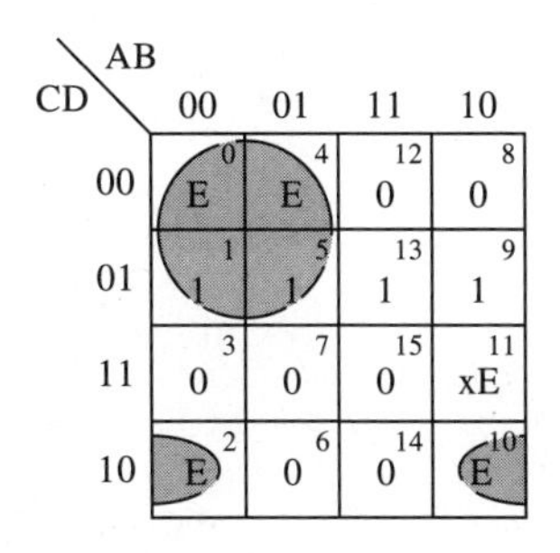

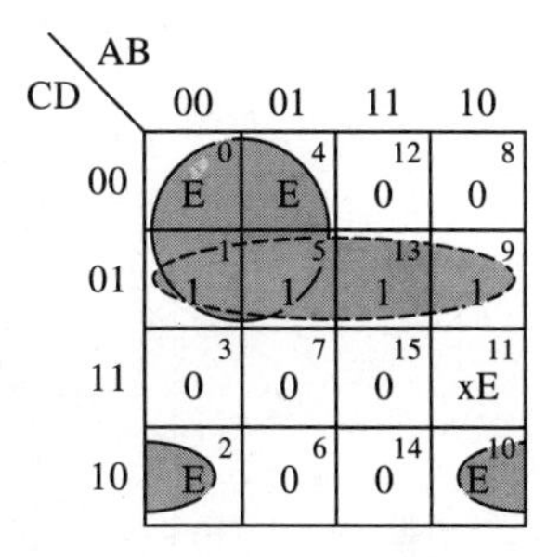

Now that all the entered variables are covered, we must cover all the 1s that are not yet completely covered. The 1s in cells 1, 5, 9, and 13 are not, although those in cells 1 and 5 are partially covered. These 1s can be grouped in only one way, to form a grouping of four, as shown on the right (dashed line).

We've completed the covering of both the entered variables and the uncovered and partially covered 1s on a single map. To complete the simplification, we need to read the map, taking care not to forget the entered variables. The simplified function is

$$F = \overline{B}C\overline{D}E + \overline{A}\,\overline{C}E + \overline{C}D$$

The first two terms are from the groupings involving entered variables; the last term is from the grouping of 1s.

Example 3.15-5

Simplify the logic function defined by the following VEM.

CD \ AB	00	01	11	10
00	^{0} 1	^{4} 1	^{12} 0	^{8} $\overline{E}$
01	^{1} 1	^{5} $x\overline{E}+E$	^{13} E	^{9} E
11	^{3} 0	^{7} $\overline{E}$	^{15} 0	^{11} 0
10	^{2} 1	^{6} $\overline{E}$	^{14} 0	^{10} 1

To cover this map, we first group all the single entered variables as shown on the left in the following figure. There are no islands. There are no entered variables that cannot be grouped in groupings of four. The $\overline{E}$ in cell 8 can be grouped in only one way, with the 1s in cells 0, 2, and 10—the corners (rule 2). The $\overline{E}$ in cell 7 can be grouped in only one way, with the $\overline{E}$ in cell 6, the $x\overline{E}$ in cell 5, and the 1 in cell 4 to form a grouping of four (rule 2). The Es in cells 9 and 13 can be grouped in only one way, with the E in cell 5 and the 1 in cell 1 to form a grouping of four.

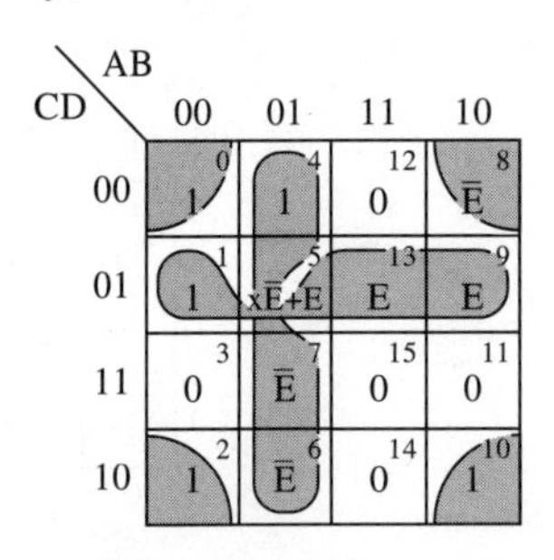

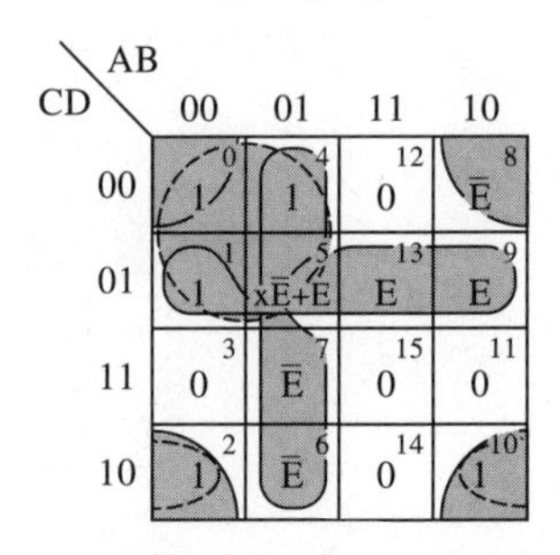

Now that all the entered variables are covered, we must cover all the 1s that are not yet completely covered. The 1s in cells 0, 1, 2, 4, and 10 are only partially covered. The covering for these 1s is shown on the right (dashed lines). Notice how we've used the $x\overline{E} + E$ in the grouping of four 1s in cells 0, 1, 4, and 5.

We've completed the covering of both the entered variables and the uncovered and partially covered 1s on a single map. To complete the simplification, we need to read the map, taking care not to forget the entered variables. The simplified function is

$$F = \overline{B}\,\overline{D}\,\overline{E} + \overline{A}B\overline{E} + \overline{C}DE + \overline{B}C\overline{D} + \overline{A}\,\overline{C}$$

The first three terms are from the groupings involving entered variables; the last two terms are from the groupings of 1s.

Example 3.15-6

Simplify the logic function defined by the following VEM.

CD \ AB	00	01	11	10
00	0 1	4 $\overline{E}$	12 0	8 $\overline{E}$
01	1 0	5 $x\overline{E}+E$	13 0	9 0
11	3 0	7 $\overline{E}$	15 0	11 0
10	2 1	6 $\overline{E}$	14 0	10 1

To cover this map, we first group all the entered variables as shown on the left in the following figure. This map is similar to that in the previous example. There are no islands. There are no entered variables that cannot be grouped in groupings of four. The $\overline{E}$ in cell 8 can be grouped in only one way, with the 1s in cells 0, 2, and 10—the corners (rule 2). The $\overline{E}$ in cell 7 can be grouped in only one way, with the $\overline{E}$ in cell 6, the x$\overline{E}$ in cell 5, and the 1 in cell 4 to form a grouping of four (rule 2). Note that the E in cell 5 does not need to be covered, because it occurs with x$\overline{E}$ and is thus not a single entered variable.

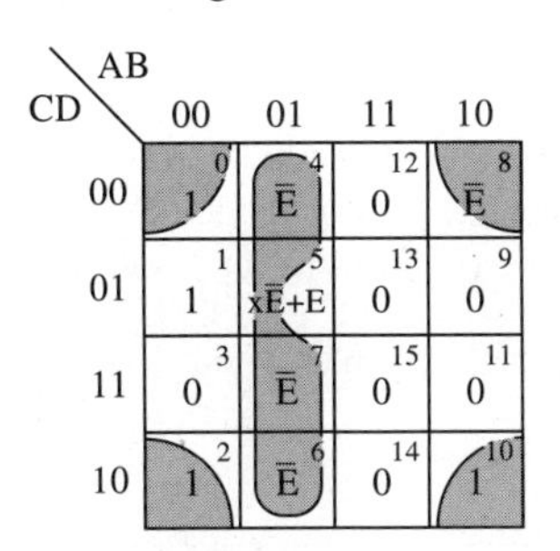

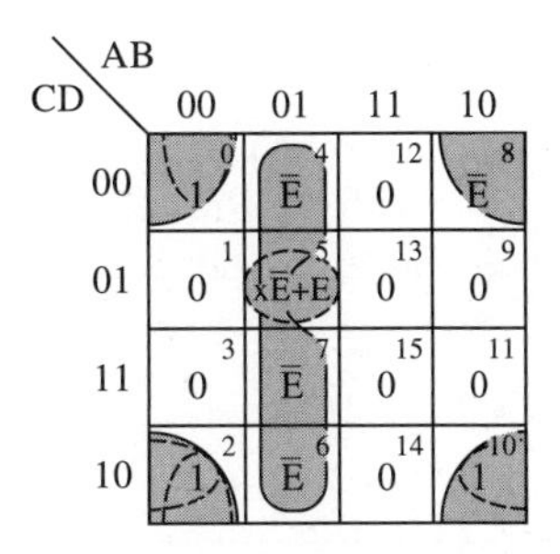

Now that all the entered variables are covered, we must cover all the 1s that are not yet completely covered. Because the E in cell 5 is not covered, the x$\overline{E}$ + E in that cell must be treated as a partially covered 1. Other partially covered 1s occur in cells 0, 2, and 10. The covering for the uncovered 1s is shown on the right (dashed lines). The x$\overline{E}$ + E in cell 5 is an island.

We've completed the covering of both the entered variables and the uncovered and partially covered 1s on a single map. To complete the simplification, we need to read the map, taking care not to forget the entered variables. The simplified function is

$$F = \overline{B}\,\overline{D}\,\overline{E} + \overline{A}B\overline{E} + \overline{A}\,\overline{B}\,\overline{D} + \overline{B}C\overline{D} + \overline{A}B\overline{C}D$$

The first two terms are from the groupings involving entered variables; the last three terms are from the groupings of 1s.

All of the foregoing examples were five-variable functions entered on a map with four map variables, so that there was one entered variable. In a later section on sequential logic design, we'll find that it is convenient to use VEMs with less than four map variables. For example, we might enter a four-variable function on a map with three map variables, or we might enter a three-variable function on a map with two map variables. We'll look at some of these smaller VEMs in the following examples, so you can become familiar with them.

Example 3.15-7

Simplify the three-variable function with the (two-map-variable) VEM on the left.

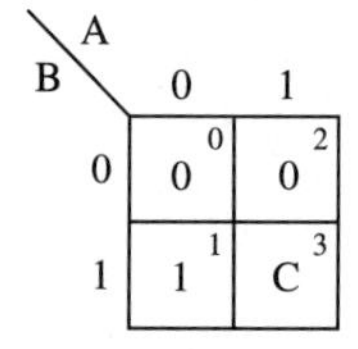

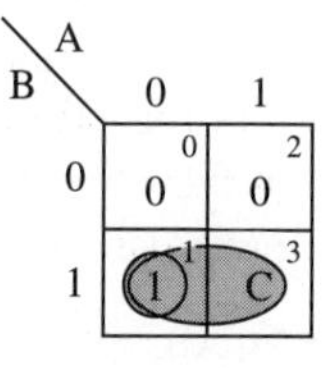

The one map cover for this map is shown on the right. The C in cell 3 can be grouped with the 1 in cell 1, but it does not completely cover the 1, so it must also be grouped as an island. Reading the map gives us the simplified function:

$$F = BC + \overline{A}B$$

Example 3.15-8

Simplify the three-variable function with the (two-map-variable) VEM on the left.

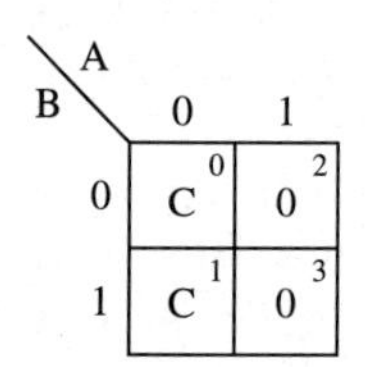

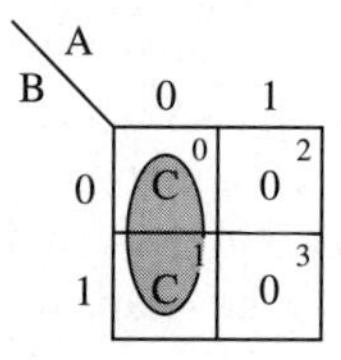

The one map cover for this map is shown on the right. Only the Cs need to be covered. Reading the map gives us the simplified function:

$$F = \overline{A}C$$

Example 3.15-9

Simplify the four-variable function with the (three-map-variable) VEM on the left.

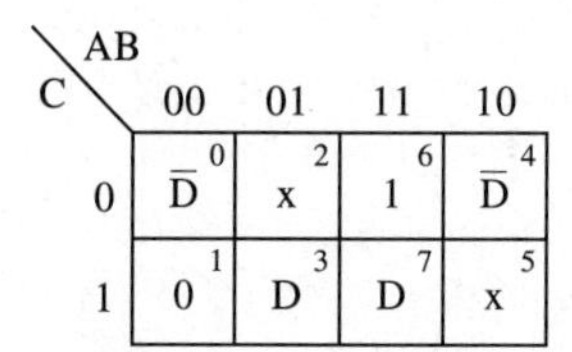

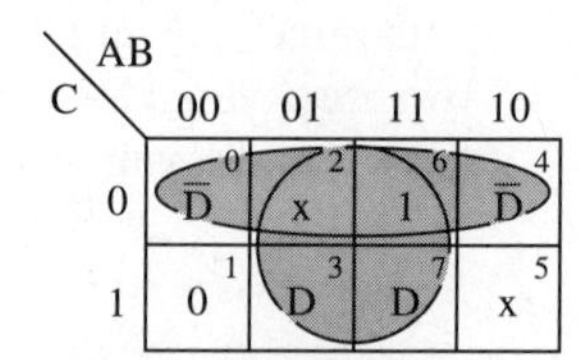

The one map cover for this map is shown on the right. The entered variables can be covered with the two groupings of four. These groupings completely cover the 1 in cell 6, so no other groupings are needed. Reading the map gives us the simplified function:

$$F = BD + \overline{C}\overline{D}$$

Example 3.15-10

Simplify the four-variable function with the (three-map-variable) VEM on the left.

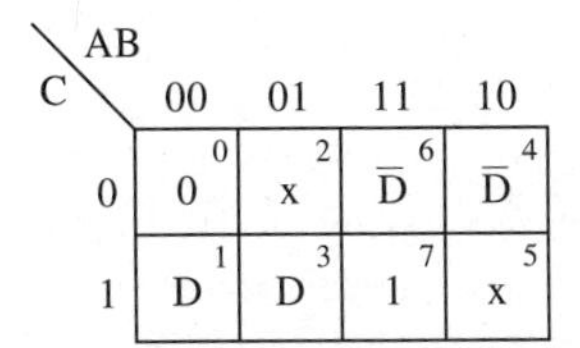

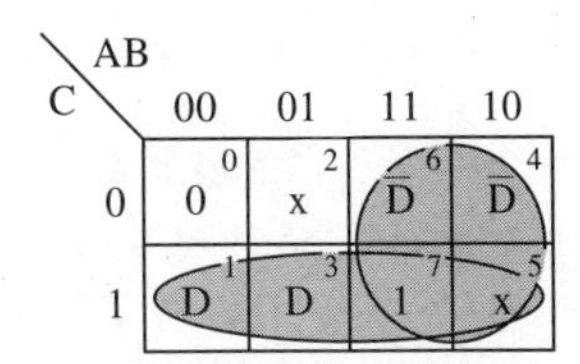

The one map cover for this map is shown on the right. The entered variables can be covered with the two groupings of four. These groupings completely cover the 1 in cell 7, so no other groupings are needed. Reading the map gives us the simplified function:

$$F = A\overline{D} + CD$$

Even though we've described how to cover the entries in a VEM without using a reduced map, the use of a reduced map will in many cases result in fewer errors. This is especially true when you are first learning how to do VEM coverings. If you have trouble with single map coverings, use reduced maps.

3.16 POS EXPRESSIONS FROM VEMs

Although it is possible to find a POS expression from a VEM, usually it is not worth the effort. However, in the case of designs for large production runs, we must assure ourselves that the simplest possible expression for the function has been found. Then it is necessary to look at the POS simplification. Remember that the POS expression is formed by the dual of the process used to form an SOP expression; thus we group 0s. This is most readily seen with an example.

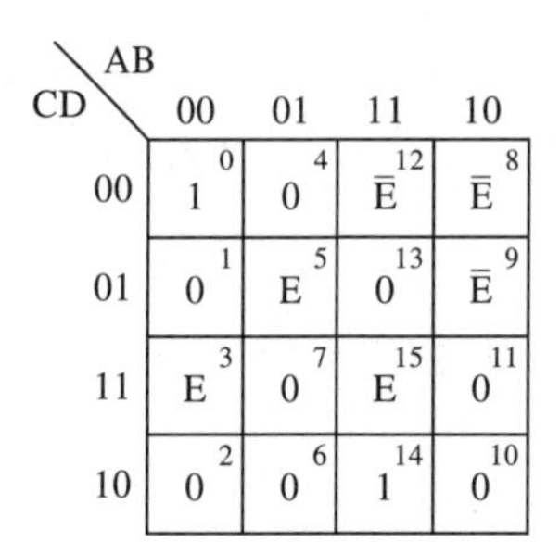

Figure 3.16-1

CD \ AB	00	01	11	10
00	1 (0)	0 (4)	0E (12)	0E (8)
01	0 (1)	$0\overline{E}$ (5)	0 (13)	0E (9)
11	$0\overline{E}$ (3)	0 (7)	$0\overline{E}$ (15)	0 (11)
10	0 (2)	0 (6)	1 (14)	0 (10)

Figure 3.16-2

Consider the VEM of Example 3.15-1, repeated in Figure 3.16-1. To form a covering of the 0 entered variables, we must think of $\overline{E}$ in a cell as actually meaning $\overline{E} + 0E$, and E as actually meaning $0\overline{E} + E$, where the 0 values are not explicitly given because they are not needed in a covering of 1s. In a covering of 0s, the 0 values are needed and the 1 values are not needed. Thus, $\overline{E}$ must be replaced with 0E and E must be replaced with $0\overline{E}$. We also have to think of the 0s as $0E + 0\overline{E}$s. Figure 3.16-2 is the resultant map.

The first step in covering this map is to cover the single 0 entered variables. This is similar to covering the single 1 entered variables in an SOP simplification. The minimal covering of 0 entered variables is shown in Figure 3.16-3. As is often true for a POS expression, the groupings are large. For this initial example, we will not attempt a complete cover on one map but rather will use a reduced map, shown in Figure 3.16-4. Notice that this is in some sense a dual of the usual reduced map. The single 0 entered variables, all of which must be covered, transfer as 1s; the 1s all transfer as 1s; the partially covered 0s transfer as 0s; and the completely covered 0s transfer as don't cares. The minimal covering of 0s for this map is as shown. The simplified POS expression for the function is

$$F = (\overline{A} + C + \overline{E})(\overline{B} + \overline{D} + E)(A + \overline{D} + E)(A + B + C + \overline{D})$$
$$(A + \overline{B} + \overline{C})(\overline{A} + B + \overline{C})(A + \overline{C} + D)$$

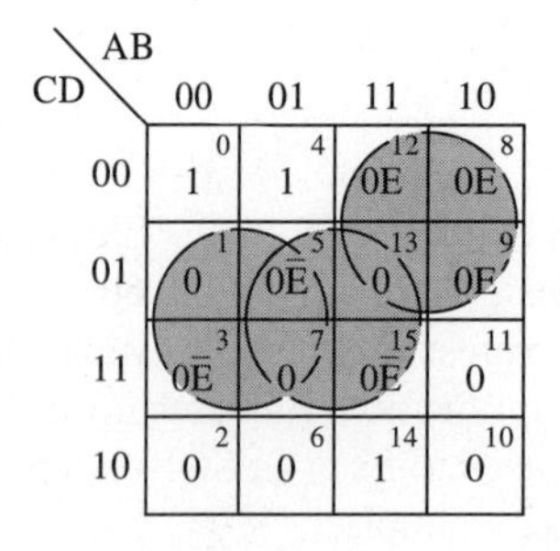

Figure 3.16-3

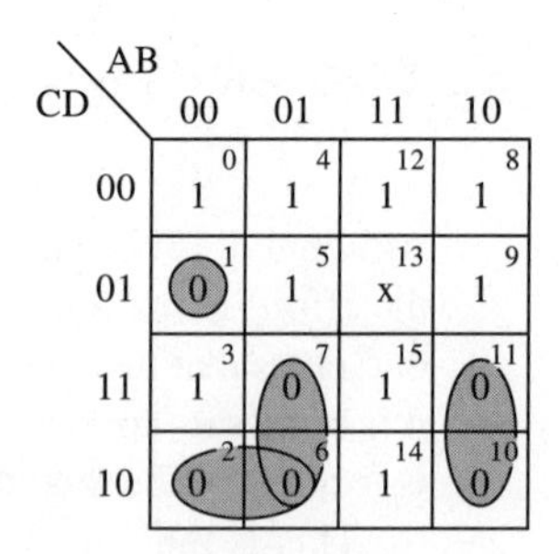

Figure 3.16-4

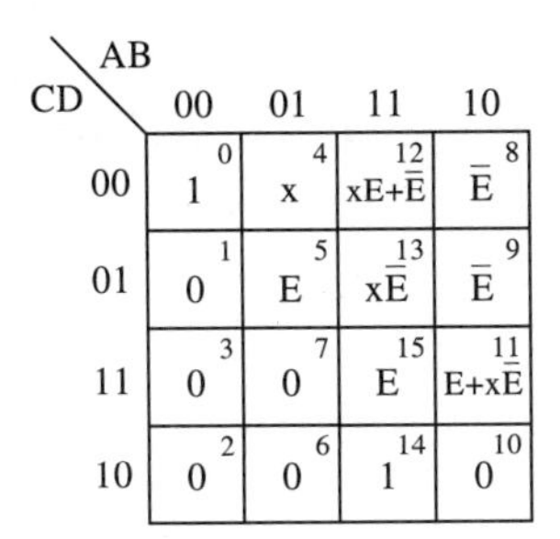

Figure 3.16-5

CD \ AB	00	01	11	10
00	1 (0)	x (4)	xE (12)	0E (8)
01	0 (1)	$0\overline{E}$ (5)	$0E+x\overline{E}$ (13)	0E (9)
11	0 (3)	0 (7)	$0\overline{E}$ (15)	$x\overline{E}$ (11)
10	0 (2)	0 (6)	1 (14)	0 (10)

Figure 3.16-6

Remember that to get this expression we form the not-true *or* expressions for each of the groups and then *and* them together. The only variables remaining in the simplified expressions are those that are common throughout the group, and these variables are complemented. Do not forget common entered variables in the grouping.

As another example, consider the VEM of Example 3.15-2, shown in Figure 3.16-5. Transforming all the entered variables to 0 entered variables gives the map in Figure 3.16-6. The don't cares and the don't care entered variables are unchanged, although the $x\overline{E}$ in cell 13 must be written as $0E + x\overline{E}$ with the 0E explicitly shown. The minimal covering for the 0 entered variables is shown in Figure 3.16-7. Figure 3.16-8 is the reduced map. Notice that xE and $x\overline{E}$ transfer as 1s, and $0E + x\overline{E}$ transfers as a don't care because the 0E is covered. The don't care transfers as a don't care. The minimal 0 covering for this reduced map is as shown. The simplified POS expression for the function is

$$F = (\overline{A} + C + \overline{E})(\overline{B} + \overline{D} + E)$$
$$(A + B + \overline{D})(B + \overline{C} + D)(A + \overline{C})$$

When the process becomes familiar, the POS simplification can be done on a

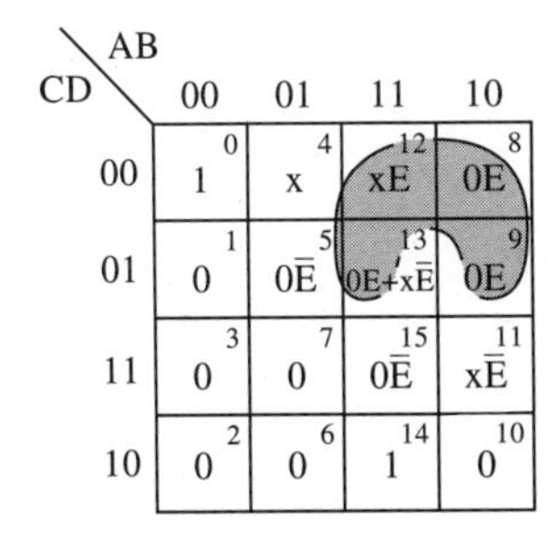

Figure 3.16-7

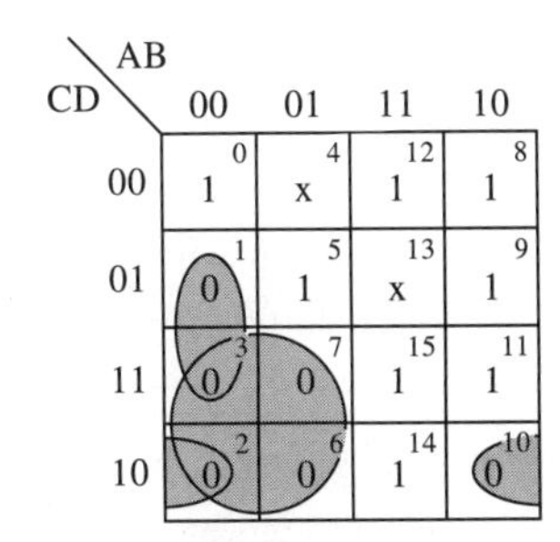

Figure 3.16-8

single map, without using a reduced map. It is still necessary to make a map showing 0 entered variables. The following example uses a single VEM to find a minimal 0 cover.

Example 3.16-1

Simplify the function with the VEM on the left as a POS expression.

CD \ AB	00	01	11	10
00	E (0)	E (4)	0 (12)	0 (8)
01	1 (1)	1 (5)	1 (13)	1 (9)
11	0 (3)	0 (7)	0 (15)	0 (11)
10	E (2)	0 (6)	0 (14)	xE (10)

CD \ AB	00	01	11	10
00	$0\overline{E}$ (0)	$0\overline{E}$ (4)	0 (12)	0 (8)
01	1 (1)	1 (5)	1 (13)	1 (9)
11	0 (3)	0 (7)	0 (15)	0 (11)
10	$0\overline{E}$ (2)	0 (6)	0 (14)	$xE+0\overline{E}$ (10)

The first step in the simplification is to form a map displaying the 0 entered variables, shown on the right. Then the single 0 entered variables are covered with a minimal covering, as follows on the left. This requires just one grouping of eight.

CD \ AB	00	01	11	10
00	$0\overline{E}$ (0)	$0\overline{E}$ (4)	0 (12)	0 (8)
01	1 (1)	1 (5)	1 (13)	1 (9)
11	0 (3)	0 (7)	0 (15)	0 (11)
10	$0\overline{E}$ (2)	0 (6)	0 (14)	$xE+0\overline{E}$ (10)

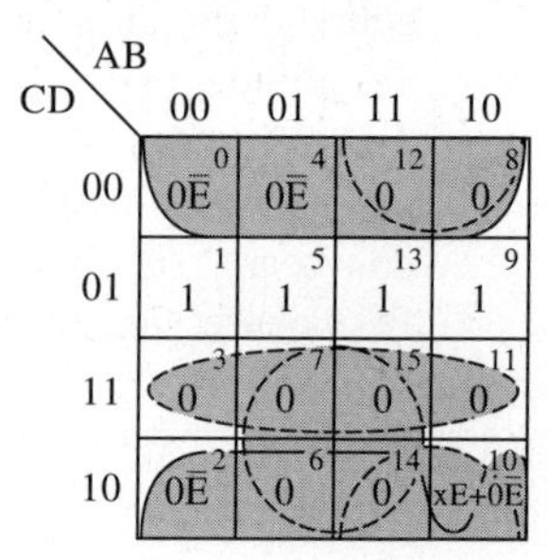

Finally, the 0s that are not completely covered are covered as shown (dashed) on the right. None of the 0s is completely covered, so they all must be covered in this step. The 0E + x$\overline{E}$ in cell 10 is completely covered (the 0E is covered), but it is used as a 0 since it will simplify the groupings of the 0s. We finish the simplification by reading the map and writing the simplified POS expression:

$$F = (D + E)(\overline{A} + D)(\overline{C} + \overline{D})(\overline{B} + \overline{C})$$

When you read a POS cover, remember to complement the variables, including the entered variables.

As further examples of POS expressions from VEMs, we'll find the POS simplifications for the small VEMs in the examples at the end of Section 3.15.

Example 3.16-2

Simplify the three-variable function with the (two-map-variable) VEM on the left as a POS.

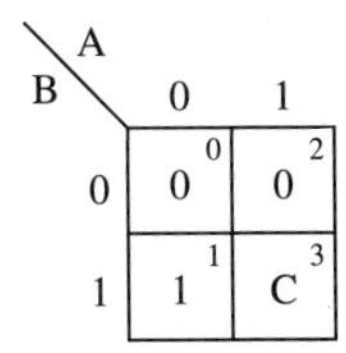
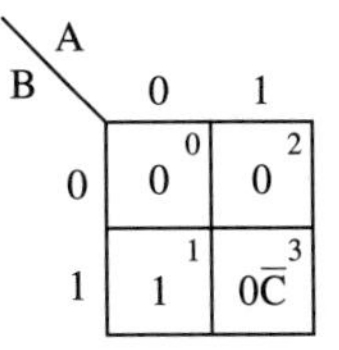

The map with 0 entered variables is shown on the right. The single map cover for the 0 entries of this map follows. The $0\overline{C}$ in cell 3 can be grouped with the 0 in cell 2. The 0 in cell 0 can be grouped with the partially covered 0 in cell 2.

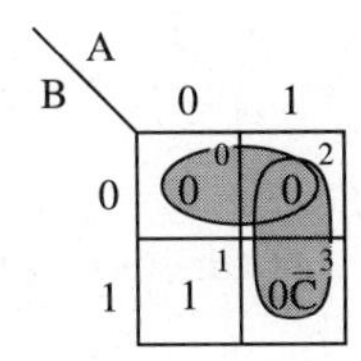

Reading the map gives us the simplified POS expression:

$$F = (\overline{A} + C)B$$

Example 3.16-3

Simplify the three-variable function with the (two-map-variable) VEM on the left as a POS.

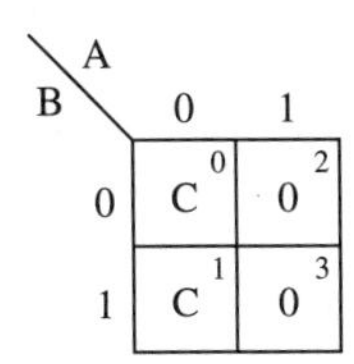
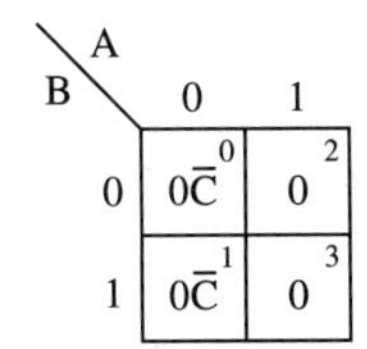

The map with 0 entered variables is shown on the right. The single map cover for the 0 entries of this map follows. The covering for the entered variables $0\overline{C}$ is the entire map. The 0s are only partially covered by the entered variables, so they must be grouped together.

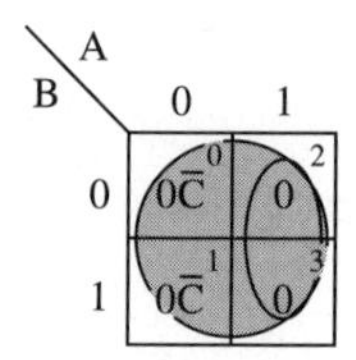

Reading the map gives us the simplified POS expression:

$$F = \overline{A}C$$

This is the same expression that was found for the SOP simplification in Section 3.15 (Example 3.15-8).

Example 3.16-4

Simplify the four-variable function with the (three-map-variable) VEM on the left as a POS.

C \ AB	00	01	11	10
0	$\overline{D}$ (0)	x (2)	1 (6)	$\overline{D}$ (4)
1	0 (1)	D (3)	D (7)	x (5)

C \ AB	00	01	11	10
0	0D (0)	x (2)	1 (6)	0D (4)
1	0 (1)	$0\overline{D}$ (3)	$0\overline{D}$ (7)	x (5)

The map with 0 entered variables is shown on the right. The single map cover for the 0 entries follows. Two groupings of four cover the entered variables and at the same time completely cover the 0 in cell 1.

C \ AB	00	01	11	10
0	0D (0)	x (2)	1 (6)	0D (4)
1	0 (1)	$0\overline{D}$ (3)	$0\overline{D}$ (7)	x (5)

Reading the map gives us the simplified POS expression:

$$F = (B + \overline{D})(\overline{C} + D)$$

Example 3.16-5

Simplify the four-variable function with the (three-map-variable) VEM on the left as a POS.

C \ AB	00	01	11	10
0	0 (0)	x (2)	$\overline{D}$ (6)	$\overline{D}$ (4)
1	D (1)	D (3)	1 (7)	x (5)

C \ AB	00	01	11	10
0	0 (0)	x (2)	0D (6)	0D (4)
1	$0\overline{D}$ (1)	$0\overline{D}$ (3)	1 (7)	x (5)

The map with 0 entered variables is shown on the right. The single map cover for the 0 entries of this map follows. Two groupings of four cover the entered variables and at the same time completely cover the 0 in cell 0.

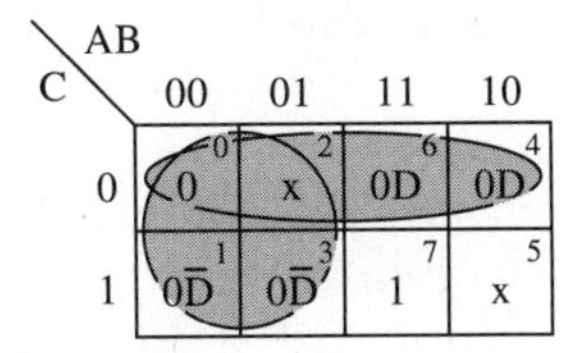

Reading the map gives us the simplified POS expression:

$$F = (A + D)(C + \overline{D})$$

3.17 VEMs WITH MORE THAN ONE ENTERED VARIABLE

VEMs can be formed for functions of six or more variables in much the same way as for functions of five variables. Instead of entering a variable in a cell of the map so that the combination of map variables and entered variables gives a true condition for that cell, an expression involving the mapped variables is entered in the cell so that the combination of map variables and the expression gives a true condition for that cell. This can be clarified with an example.

Suppose we wish to form a VEM for the function

$$G = ABCD\overline{E}F + ABCD\overline{E}\overline{F} + A\overline{B}CDE\overline{F} + ABC\overline{D}EF + A\overline{B}\overline{C}\overline{D}F + \overline{A}\overline{B}\overline{C}\overline{D}$$

The four-map-variable VEM is shown in Figure 3.17-1. The first and second terms go in cell 15 because the map variables for both of these terms are A, B, C, and D. The entered variable expression from the terms is $\overline{E}F + \overline{E}\overline{F}$, the expression that makes them true. The third term goes in cell 11 because the map variables are A, $\overline{B}$, C, and D, and the entered variable expression is $E\overline{F}$. The fourth term goes in cell 12 because the map variables are A, B, $\overline{C}$, $\overline{D}$, and the entered variable expression is

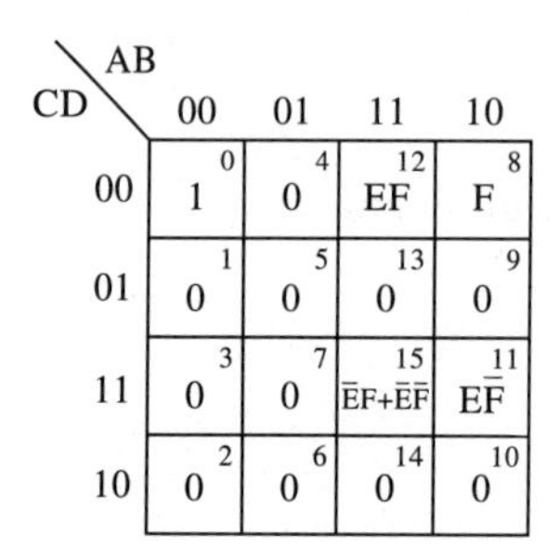

Figure 3.17-1

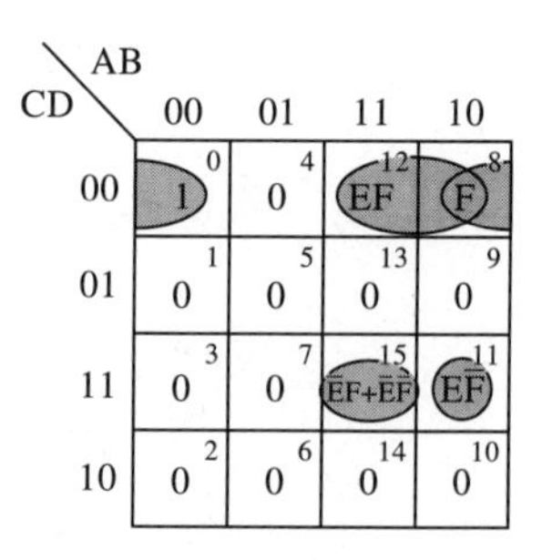

Figure 3.17-2

EF. The fifth term goes in cell 8, and the entered variable expression is F. This term is true, independent of the value of E. The sixth term goes in cell 0, and the entered variable expression is 1. This term is true, independent of the values of both E and F.

In this map it's tricky to cover the entered variables, but it can be done (Fig. 3.17-2). In grouping cells of the VEM, remember that the entered variable expressions in all the cells of the group must be identical. If we note that the F in cell 8 can be expanded to EF + $\overline{\text{E}}$F, then we can group EF in cell 12 with EF in cell 8. If we note that the 1 in cell 0 can be expanded to F + $\overline{\text{F}}$, then we can group the F in cell 8 with the F in cell 0. There is no way to group the entries in cells 15 and 11 because there are no common terms, although the two terms in cell 15 can be grouped together. These groupings give us the minimal covering for the entered variables shown.

Figure 3.17-3 is the reduced map. It is rather trivial—only the partially covered 1 is transferred to it. The simplified function for these maps is

$$G = A\overline{B}CDE\overline{F} + A\overline{C}\,\overline{D}EF + ABCD\overline{E} + \overline{B}\,\overline{C}\,\overline{D}F + \overline{A}\,\overline{B}\,\overline{C}\,\overline{D}$$

The common variables in each grouping remain in the simplified expression. We must be careful to include the entered variables.

Simplification of functions with six or more variables is much easier if the

CD \ AB	00	01	11	10
00	(0) 1	(4) 0	(12) 0	(8) 0
01	(1) 0	(5) 0	(13) 0	(9) 0
11	(3) 0	(7) 0	(15) 0	(11) 0
10	(2) 0	(6) 0	(14) 0	(10) 0

Figure 3.17-3

CD \ AB	00	01	11	10
00	$\bar{E}$ (0)	0 (4)	0 (12)	0 (8)
01	0 (1)	E (5)	1 (13)	0 (9)
11	0 (3)	0 (7)	F (15)	0 (11)
10	1 (2)	0 (6)	F (14)	F (10)

Figure 3.17-4

function is sparse in some of the variables. The following function does not have the variables E and F occurring together. It is a sparse function of these variables.

$$G = \bar{A}\bar{B}\bar{C}\bar{D}\bar{E} + \bar{A}B\bar{C}DE + AB\bar{C}D + ABCDF + \bar{A}\bar{B}C\bar{D} + ABC\bar{D}F + A\bar{B}C\bar{D}F$$

Figure 3.17-4 is the four-map-variable VEM for this function. Because there is no overlap between groupings of the entered variables E and F, the minimal covering of entered variables for this map is relatively easy to determine and is as shown in Figure 3.17-5. Figure 3.17-6 is the reduced map. Neither 1 is completely covered by the entered variables, so they transfer to the reduced map. The simplified expression for this function is

$$G = \bar{A}\bar{B}\bar{D}\bar{E} + B\bar{C}DE + ABDF + AC\bar{D}F + AB\bar{C}D + \bar{A}\bar{B}C\bar{D}$$

Our final examples of simplification of VEMs with more than one entered variable will be two- and three-map-variable VEMs with two entered variables, such as those we'll encounter later when we treat sequential logic circuit design.

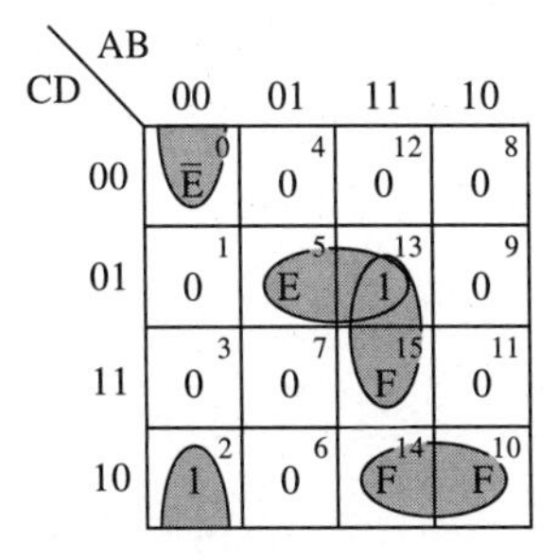

Figure 3.17-5

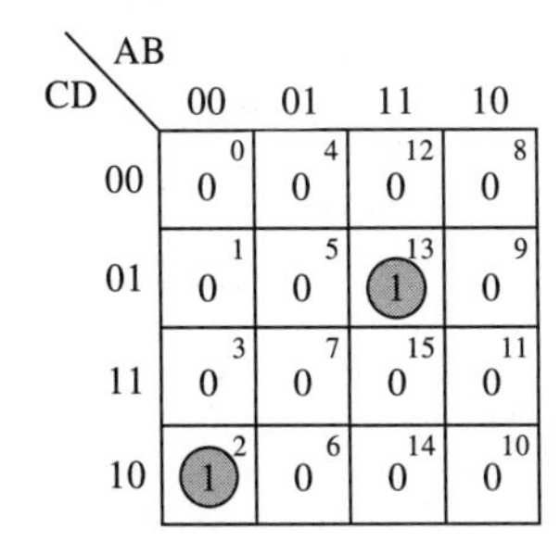

Figure 3.17-6

Example 3.17-1

Simplify the four-variable function with the (two-map-variable) VEM on the left.

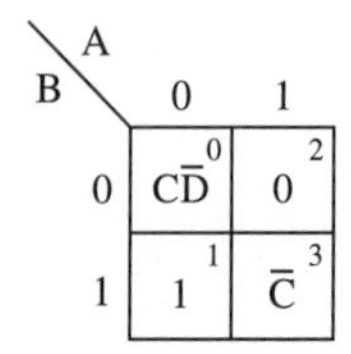

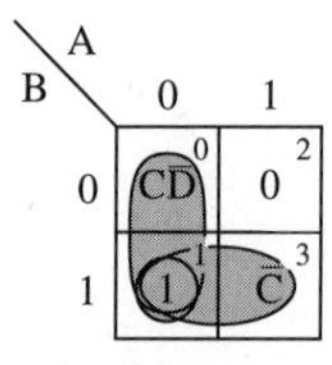

The single map cover for this map is shown on the right. The entered variables can be covered with the two groupings of two. Both $C\overline{D}$ and $\overline{C}$ can be grouped with the 1, but they do not cover it completely, so it must also form an island. Reading the map gives us the simplified expression:

$$F = B\overline{C} + \overline{A}B + \overline{A}C\overline{D}$$

Example 3.17-2

Simplify the four-variable function with the (two-map-variable) VEM on the left.

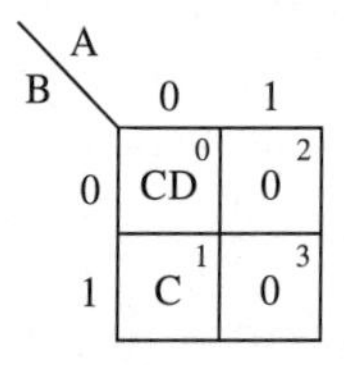

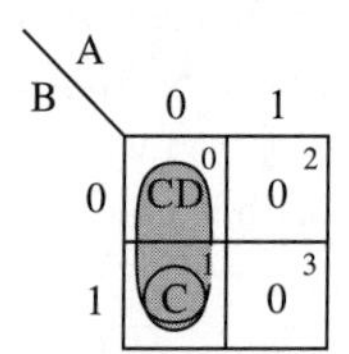

The single map cover for this map is shown on the right. The entered variable expression CD can be grouped with the entered variable C but does not cover it completely, so the C must also form an island. Reading the map gives us the simplified function:

$$F = \overline{A}CD + \overline{A}BC$$

Example 3.17-3

Simplify the five-variable function with the (three-map-variable) VEM on the left.

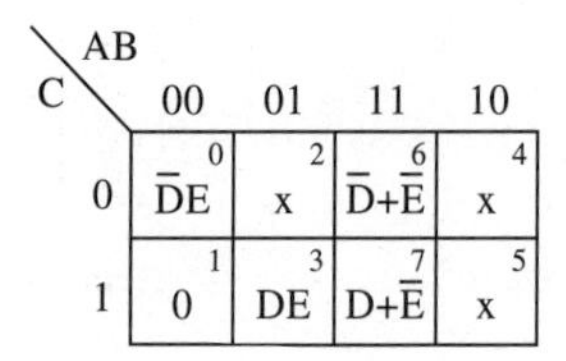

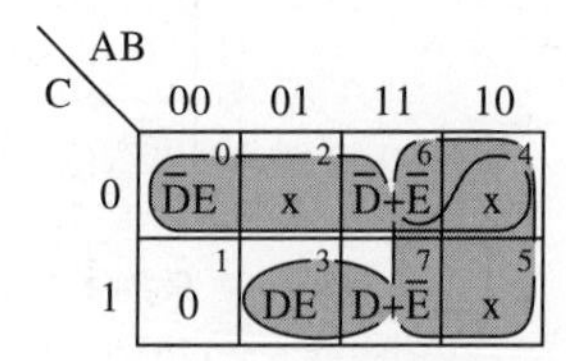

The single map cover for this map is shown on the right. The entered variable expression DE (cell 3) can be grouped with the D in cell 7. Notice that in an *and* expression such as DE, the entire expression DE is grouped as a single entity. When we cover the DE part of D in cell 7, we are left with only $\overline{E}$ since $D\overline{E} + \overline{E} = \overline{E}$. The other entered variable expressions can all be grouped as groupings of four. The $\overline{D}E$ (cell 0) can be grouped with the don't care in cell 3, the $\overline{D}$ in cell 6, and the don't care in cell 4. When we cover the $\overline{D}E$ part of $\overline{D}$ in cell 6, we are left with only $\overline{E}$ since $\overline{D}\overline{E} + \overline{E} = \overline{E}$. This leaves only the $\overline{E}$s in cells 6 and 7, which can be grouped with the don't cares in cells 4 and 5. With this covering, all the entered variables are completely covered. There are no 1s in this map, so there are no uncovered or partially covered 1s that need to be covered. Reading the map gives us the simplified expression:

$$F = \overline{C}\overline{D}E + A\overline{E} + BCDE$$

Example 3.17-4

Simplify the five-variable function with the (three-map-variable) VEM on the left.

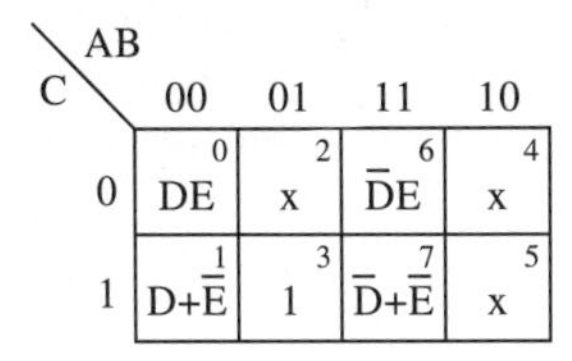

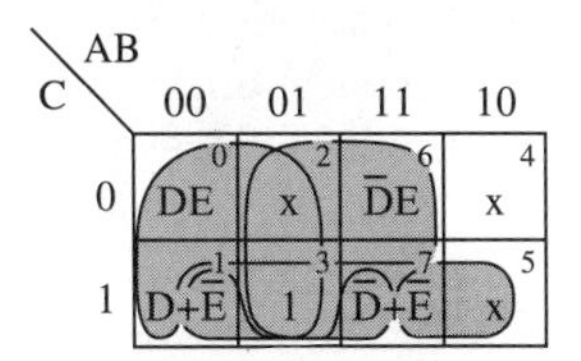

The single map cover for this map is shown on the right. All the entered variables can be grouped in groupings of four. The DE in cell 0 can be grouped with the D in cell 1, the 1 in cell 3, and the x in cell 2. The $\overline{D}E$ in cell 6 can be grouped with the $\overline{D}$ in cell 7, the 1 in cell 3, and the x in cell 2. Notice that all that remains in cells 1 and 7 is $\overline{E}$, since in cell 1 $D\overline{E} + \overline{E} = \overline{E}$ and in cell 7 $\overline{D}\overline{E} + \overline{E} = \overline{E}$. The $\overline{E}$s in cells 1 and 7 can be grouped with the 1 in cell 3 and the x in cell 5. This completely covers the 1 in cell 3, $DE + \overline{D}E + \overline{E} = 1$. Reading the map gives us the simplified expression:

$$F = \overline{A}DE + B\overline{D}E + C\overline{E}$$

Now that you have the mathematical tools, you'll find the design of logic circuits rather simple.

KEY TERMS

Absorption A simplification theorem which allows a term or factor in an expression to be absorbed into another term or factor as $AB + B = B$ or $(A + B)B = B$.

And A Boolean operation on two variables which indicated that the result of the operation is true if both of the variables are true. Denoted by a ·, but often just denoted by juxtaposition as $AB = A \cdot B$.

Boolean Algebra An algebra of two-valued variables with three operators—*not, and, or*.
Boolean Function A two-valued function of two-valued variables. The function can have only the values true (1) or false (0).
Boolean Variable A variable which can have only two values true (1) or false (0).
Cell A single square in a Karnaugh or Variable Entered Map.
Commutative Operations which produce the same result independent of the order of the variables are operated on, as $AB = BA$ or $A + B = B + A$.
Cover or Covering A grouping of all 1s or 0s in a Karnaugh map, or a grouping of all the entered variables in a VEM.
Covered Grouped with other near-neighbor entries or as an island.
Distributive Two different operations which produce the same result independent of which is performed first as $A(B + C) = AB + AC$ or $A + BC = (A + B)(A + C)$.
Don't Care or x An entry which can be either true (1) or false (0).
False, Not Asserted, or 0 One of the two possible values for a Boolean function or variable.
Karnaugh Map A graphical representation of a logic function used as an aid in simplication.
Logic Adjacency A simplification theorem which allows two binary terms or factors to be reduced to one as $AB + A\overline{B} = A$ or $(A + B)(A + \overline{B}) = A$.
Logic Adjacent Two terms or factors that can be simplified using logic adjacency. Also can be applied to 4 and 8 terms or factors.
Map Entered Variable or EV The variable or one of the variables used in the relation that is entered in the cell of a VEM to specify when the function is 1 (or 0) in that cell.
Map Variable The variables in a Karnaugh or VEM that are part of the map. A Karnaugh map has only map variables.
Maxterms N-variable *or* terms of the form $V_1 + V_2 + \cdots + V_n$ where the V_i are either the i-th variable or its complement. There are 2^n different n-variable maxterms—one for each binary number of length n. Each maxterm is false for only one combination of variable values and it is named for the binary value for which it is false. Thus, $M_3 = A + B + \overline{C} + \overline{D}$ is false if $A = 0$, $B = 0$, $C = 1$, and $D = 1$. Any function can be represented by *and*ing together all the maxterms for which the function is false.
Minimal Cover or Covering A covering of all the 1s, 0s, or entered variables with the smallest number of groupings and of the largest size of groupings.
Minterms N-variable *and* terms of the form $V_1V_2 \ldots V_n$ where the V_i are either the i-th variable or its complement. There are 2^n different n-variable minterms—one for each binary number of length n. Each minterm is true for only one combination of variable values and it is named for the binary value for which it is true. Thus, $m_3 = \overline{A}\overline{B}CD$ is true if $A = 0$, $B = 0$, $C = 1$, and $D = 1$. Any function can be represented by *or*ing together all the minterms for which the function is true.
Next-Neighbor Entries in a map which are next to each other and therefore logic adjacent.

Not A unitary Boolean operator which changes true (1) to false (0) and false (0) to true (0). Denoted by an overbar.
Or A Boolean operation on two variables which indicate the result of the operation is true if either of the variables are true. Denoted by a +.
Product of Sums (POS) A form of logic function in which *or* terms are *and*ed together.
Reduced Map The Karnaugh map which is obtained after all the entered variables have been covered in a VEM.
Sum of Products (SOP) A form of logic function in which *and* terms are *or*ed together.
True, Asserted, or 1 One of the two possible values for a Boolean function or variable.
Variable Entered Map (VEM) An extension of a Karnaugh map to allow for more variables.

EXERCISES

1. Prove the following identities using perfect induction.

$$1 + A = 1 \qquad 0 \cdot A = 0$$

$$0 + A = A \qquad 1 \cdot A = A$$

$$A + A = A \qquad A \cdot A = A$$

$$A + \overline{A} = 1 \qquad A \cdot \overline{A} = 0$$

2. Prove the commutative law

$$A + B = B + A$$

$$AB = BA$$

and the distributive law

$$A + (BC) = (A + B)(A + C)$$

$$A(B + C) = (AB) + (AC)$$

using perfect induction.

3. Prove the following.

(a) $AB + \overline{A}C + BC = AB + \overline{A}C$

(b) $(A + B)(\overline{A} + C)(B + C) = (A + B)(\overline{A} + C)$

4. Simplify the following logic expressions.

(a) $F = \overline{A}\,\overline{B}\,\overline{C}\,\overline{D} + \overline{A}B\overline{C}D + \overline{A}B\overline{C}\,\overline{D} + \overline{A}\,\overline{B}\,\overline{C}D$

(b) $F = \overline{A}\,\overline{B}\,\overline{C}\,\overline{D} + B\overline{C}\,\overline{D} + A\overline{C}\,\overline{D}$

(c) $F = \overline{A}\,\overline{B}\,\overline{C}\,\overline{D} + \overline{A}B\overline{C}\,\overline{D} + ACD + A\overline{B}C\overline{D}$

(d) $F = \overline{A}\,\overline{B}D + A\overline{B}D + \overline{B}CD$

(e) $F = \overline{A}\,\overline{B}CD + \overline{A}BD + AB\overline{C} + A\overline{B}\,\overline{C}\,\overline{D}$

(f) $F = \overline{A}\,\overline{B}CD + ABD + \overline{A}BD + A\overline{B}D$

(g) $F = \overline{A}\,\overline{B}D + B\overline{C}D + ABD + A\overline{B}D$

(h) $F = \overline{A}\,\overline{C}\,\overline{D} + \overline{A}\,\overline{C}D + B\overline{C}D$

(i) $F = \overline{A}\,\overline{B}CD + \overline{A}BC\overline{D} + ABCD + A\overline{B}C\overline{D} + A\overline{B}\,\overline{C}\,\overline{D}$

(j) $F = \overline{A}\,\overline{B}\,\overline{C}\,\overline{D} + ABCD + ABC\overline{D} + \overline{A}\,\overline{B}\,\overline{C}D + \overline{A}\,\overline{B}CD$

(k) $F = BD + ABD + CBD + \overline{A}\,\overline{B}\,\overline{C}$

(l) $F = \overline{A}\,\overline{B}\,\overline{C}\,\overline{D} + ABCD + \overline{A}\,\overline{B}CD + \overline{A}\,\overline{B}\,\overline{C}D$

(m) $F = ABCD + ABC\overline{D} + AB\overline{C}D + A\overline{B}CD + \overline{A}BCD$

(n) $F = AB + \overline{C}\,\overline{D} + \overline{A}B$

(o) $F = \overline{A}\,\overline{B}\,\overline{C} + \overline{A}\,\overline{B}\,\overline{C}D + A\overline{B}\,\overline{C} + ABCD$

(p) $F = \overline{A}\,\overline{B}\,\overline{C} + \overline{A}\,\overline{B}\,\overline{D} + ABC + ABD$

(q) $F = A + \overline{A}BCD + \overline{A}BC + \overline{A}\,\overline{B}\,\overline{C} + \overline{A}\,\overline{B}C$

5. Simplify the following logic expressions using DeMorgan's Theorem.

(a) $F = \overline{(A + B + C)(\overline{A} + B + \overline{C})}$

(b) $F = \overline{A + B\overline{C} + \overline{C}D}$

(c) $F = \overline{(A + B)CD}$

(d) $F = \overline{(A + B + \overline{C} + \overline{D})\overline{A}\,\overline{B}}$

(e) $F = \overline{\overline{ABC}\cdot\overline{\overline{B}C\overline{D}}}$

(f) $F = \overline{\overline{A + \overline{B} + \overline{C}}\cdot\overline{\overline{A} + \overline{B} + D}}$

6. Make truth tables for F_1 and F_2 in the following 2-bit, binary subtraction.

$$\begin{array}{r} A\ B \\ -C\ D \\ \hline F_2F_1 \end{array}$$

7. Make truth tables for F_1, F_2, F_3, and F_4 in the following 2-bit, binary multiplication.

$$\begin{array}{r} A\ B \\ \times C\ D \\ \hline F_4F_3F_2F_1 \end{array}$$

8. Make truth tables for each of the following 1-bit, binary expressions. (Notice that the operations are binary, not Boolean). Drop all carries and assume that a borrow can be made whenever it is needed.

(a) $F = A + B - C$

(b) $F = AB - C$ (AB is $A \times B$)

(c) $F = A + B + C + D$

(d) $F = A + B + C - D$

(e) $F = ABC - AD$

9. Express each of the following functions (defined by a truth table) as a sum of minterms. (Notice that several functions—F_a, F_b, and so on—are defined in a single table.)

A	B	C	F_a	F_b	F_c	F_d	F_e	F_f	F_g	F_h	F_i	F_j	F_k
0	0	0	0	1	0	1	0	1	0	1	0	1	0
0	0	1	1	1	1	0	0	1	1	0	1	0	1
0	1	0	0	0	1	0	1	1	1	0	0	0	1
0	1	1	0	1	0	1	0	1	0	1	1	0	1
1	0	0	0	1	1	1	0	0	0	1	1	0	1
1	0	1	1	0	1	0	0	0	1	1	1	1	0
1	1	0	1	1	0	0	1	1	1	0	0	0	1
1	1	1	0	0	0	0	0	1	1	1	1	1	1

10. Express each of the functions of exercise 9 as a product of maxterms.

11. Express each of the following functions (defined by a truth table) as a sum of minterms.

A	B	C	D	F_a	F_b	F_c	F_d	F_e	F_f	F_g	F_h	F_i	F_j	F_k
0	0	0	0	0	1	0	1	0	1	0	1	0	0	0
0	0	0	1	1	1	1	0	0	0	1	1	0	1	0
0	0	1	0	1	1	1	0	0	0	0	0	1	1	1
0	0	1	1	1	0	0	1	1	1	1	1	0	1	0
0	1	0	0	1	0	1	1	1	0	0	1	0	1	0
0	1	0	1	1	1	1	0	1	1	0	0	0	1	0
0	1	1	0	0	0	0	0	1	1	0	1	1	0	1
0	1	1	1	0	1	1	1	0	0	1	0	0	0	1
1	0	0	0	0	1	1	0	0	1	0	1	1	1	1
1	0	0	1	0	1	1	0	0	1	1	1	0	0	0
1	0	1	0	1	1	0	0	1	0	1	0	1	1	0
1	0	1	1	0	0	0	0	1	1	1	0	0	0	1
1	1	0	0	0	0	1	1	0	0	0	1	1	0	1
1	1	0	1	0	0	1	0	1	0	1	1	0	1	1
1	1	1	0	0	1	1	1	0	1	0	0	0	1	1
1	1	1	1	1	0	1	1	1	0	0	1	1	0	1

12. Express each of the functions of exercise 11 as a product of maxterms.

13. Make a truth table for each of the functions of exercise 4.

14. Make a truth table for each of the following functions.

(a) $F = (\overline{A} + \overline{B} + \overline{C} + \overline{D})(\overline{A} + B + \overline{C} + D)(\overline{A} + B + \overline{C} + \overline{D}) \cdot (\overline{A} + \overline{B} + \overline{C} + D)$

(b) $F = (\overline{A} + \overline{B} + \overline{C} + \overline{D})(B + \overline{C} + \overline{D})(A + \overline{C} + \overline{D})$

(c) $F = (\overline{A} + \overline{B} + \overline{C} + \overline{D})(\overline{A} + B + \overline{C} + \overline{D}) \cdot (A + C + D)(A + \overline{B} + C + \overline{D})$

(d) $F = (\overline{A} + \overline{B} + D)(A + \overline{B} + D)(\overline{B} + C + D)$

(e) $F = (\overline{A} + \overline{B} + C + D)(\overline{A} + B + D)(A + B + \overline{C})(A + \overline{B} + \overline{C} + \overline{D})$

(f) $F = (\overline{A} + \overline{B} + D)(A + B + D)(\overline{A} + B + D)(A + \overline{B} + D)$

(g) $F = (\overline{A} + \overline{B} + D)(B + \overline{C} + D)(A + B + D)(A + \overline{B} + D)$

(h) $F = (\overline{A} + \overline{C} + \overline{D})(\overline{A} + \overline{C} + D)(B + \overline{C} + D)$

(i) $F = (\overline{A} + \overline{B} + C + D)(\overline{A} + B + C + \overline{D})(A + B + C + D) \cdot (A + \overline{B} + C + \overline{D})(A + \overline{B} + \overline{C} + \overline{D})$

15. Make a Karnaugh map for each of the functions defined in the truth table of exercise 9. Using its map, express each function in its simplest SOP form.

16. Make a Karnaugh map for each of the functions defined in the truth table of exercise 11. Using its map, express each function in its simplest SOP form.

17. Make a Karnaugh map for each of the functions of exercise 4. Using its map, express each function in its simplest SOP form.

18. Make a Karnaugh map for each of the functions of exercise 14. Using its map, express each function in its simplest SOP form.

19. Find a minimum cover for each of the Karnaugh maps in Figure 3-E19, and express the function defined by the map in its simplest SOP form. Each map has some unusual feature.

20. Make a Karnaugh map for each of the functions defined in the truth table of exercise 9. Using its map, express each function in its simplest POS form.

21. Make a Karnaugh map for each of the functions defined in the truth table of exercise 11. Using its map, express each function in its simplest POS form.

22. Make a Karnaugh map for each of the functions of exercise 4. Using its map, express each function in its simplest POS form.

23. Make a Karnaugh map for each of the functions of exercise 14. Using its map, express each function in its simplest POS form.

24. Find a minimum 0s cover for each of the Karnaugh maps in Figure 3-E19, and express the function defined by the map in its simplest POS form. Each map has some unusual feature.

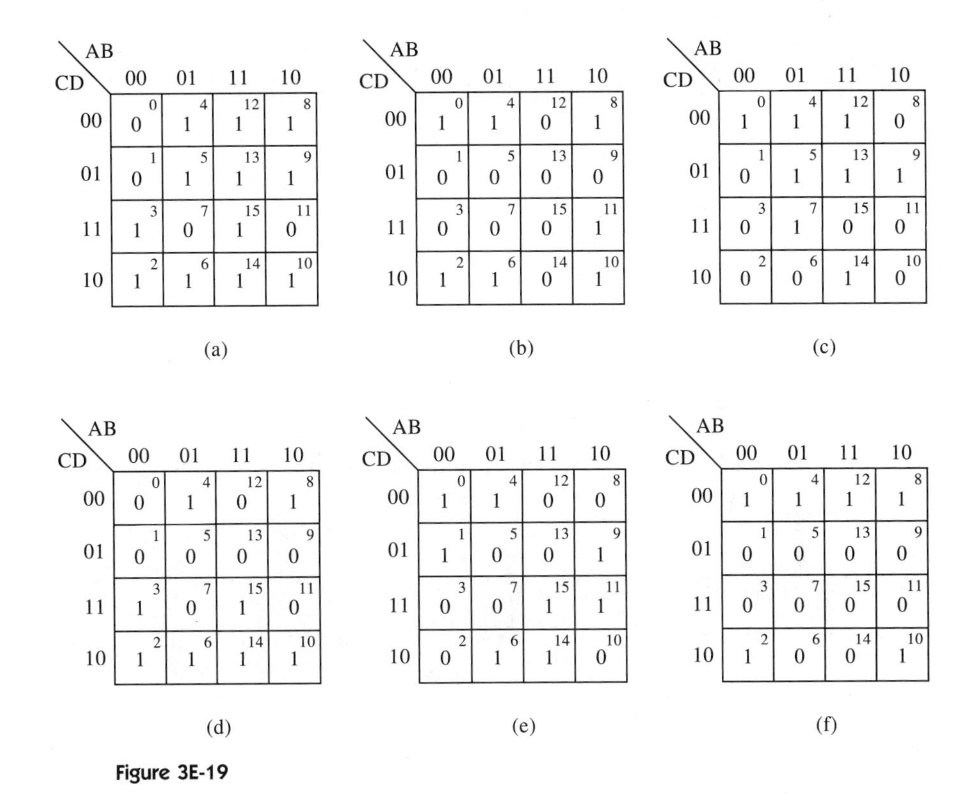

Figure 3E-19

25. Make a Karnaugh map for each of the functions defined in the truth table of exercise 9. Using its map, express each function in both its simplest SOP and simplest POS forms.

26. Make a Karnaugh map for each of the functions defined in the truth table of exercise 11. Using its map, express each function in both its simplest SOP and simplest POS forms.

27. Make a Karnaugh map for each of the functions of exercise 4. Using its map, express the function in both its simplest SOP and simplest POS forms.

28. Make a Karnaugh map for each of the functions of exercise 14. Using its map, express each function in both its simplest SOP and simplest POS forms.

29. Find a minimum 1s cover and minimum 0s cover for each of the Karnaugh maps in Figure 3-E19, and express the function defined by the map in both its simplest SOP and simplest POS forms. Each map has some unusual feature.

30. Find the minimum SOP and minimum POS expressions for each of the following functions (defined by a truth table).

A	B	C	F_a	F_b	F_c	F_d	F_e	F_f	F_g	F_h	F_i	F_j	F_k
0	0	0	0	x	0	1	0	1	x	1	0	1	0
0	0	1	x	x	1	0	0	x	x	x	1	0	1
0	1	0	x	x	x	0	1	x	1	0	0	x	1
0	1	1	0	1	0	1	x	1	x	1	1	x	x
1	0	0	0	1	1	1	x	0	0	x	x	0	1
1	0	1	1	0	1	x	0	0	1	1	1	1	0
1	1	0	1	1	x	0	1	1	1	0	0	0	x
1	1	1	0	0	x	0	0	1	1	1	1	1	x

31. Find the minimum SOP and minimum POS expressions for each of the following functions (defined by a truth table).

A	B	C	D	F_a	F_b	F_c	F_d	F_e	F_f	F_g	F_h	F_i	F_j	F_k
0	0	0	0	0	x	x	1	0	1	0	x	0	0	x
0	0	0	1	x	x	1	0	0	0	1	x	0	x	x
0	0	1	0	1	1	1	0	0	0	0	0	1	x	1
0	0	1	1	1	0	x	1	x	1	1	1	0	x	0
0	1	0	0	x	0	1	1	x	x	x	1	0	x	0
0	1	0	1	x	1	1	0	1	x	x	0	0	1	0
0	1	1	0	0	0	0	0	1	x	0	1	1	0	1
0	1	1	1	0	x	1	1	x	0	1	0	0	0	1
1	0	0	0	0	x	1	x	0	1	0	1	x	1	1
1	0	0	1	0	1	1	x	0	1	1	1	x	0	0
1	0	1	0	1	1	0	0	1	0	1	0	x	1	0
1	0	1	1	0	0	0	0	1	1	x	0	x	0	x
1	1	0	0	x	0	x	1	0	0	0	x	1	0	x
1	1	0	1	x	0	x	0	1	0	1	x	0	1	1
1	1	1	0	0	1	1	1	0	1	0	0	0	1	1
1	1	1	1	1	0	1	1	1	0	0	1	1	0	1

32. Find both the minimum SOP and minimum POS expressions for each of the following functions, using two-dimensional maps.

(a) $F = \overline{A}\,\overline{B}\,\overline{C}\,\overline{D}E + \overline{A}\,\overline{B}\,\overline{C}D + A\overline{B}\,\overline{C}D + \overline{A}\,\overline{C}DE + \overline{A}\,\overline{B}CDE + \overline{B}C\overline{D} + \overline{A}C\overline{D}E$

(b) $F = \overline{A}\,\overline{B}\,\overline{C}\,\overline{D}E + A\overline{B}\,\overline{C}\,\overline{D} + \overline{B}CE + \overline{B}CD + AB\overline{C}\,\overline{E} + A\overline{C}D\overline{E}$

(c) $F = \overline{A}\,\overline{B}\,\overline{C}\,\overline{D}E + A\overline{B}\,\overline{C}\,\overline{D} + A\overline{B}CE + CD + AB\overline{C}\,\overline{E} + A\overline{C}D\overline{E} + AC\overline{D}\,\overline{E}$

(d) $F = \overline{B}\,\overline{C}\,\overline{D}E + \overline{A}\,\overline{B}\,\overline{C}D + A\overline{B}\,\overline{C}D + \overline{A}\,\overline{C}DE + \overline{A}\,\overline{B}CDE + \overline{B}C\overline{D}$
$+ \overline{A}C\overline{D}E + A\overline{B}\,\overline{C}\,\overline{D}$

33. Find both the minimum SOP and minimum POS expressions for each of the functions in the following truth table, using two-dimensional maps.

A	B	C	D	E	F_a	F_b	F_c	F_d	F_e	F_f	F_g	F_i
0	0	0	0	0	1	1	0	0	1	1	0	0
0	0	0	0	1	1	1	1	x	0	1	x	x
0	0	0	1	0	0	0	0	0	0	0	0	0
0	0	0	1	1	0	0	0	x	1	0	0	x
0	0	1	0	0	0	1	1	0	0	1	1	0
0	0	1	0	1	0	1	x	0	0	1	1	0
0	0	1	1	0	0	0	1	0	0	1	1	0
0	0	1	1	1	0	0	1	0	0	1	1	0
0	1	0	0	0	1	1	0	0	x	1	0	0
0	1	0	0	1	1	1	x	x	1	1	1	x
0	1	0	1	0	0	0	0	0	0	0	0	0
0	1	0	1	1	1	1	1	x	1	1	0	0
0	1	1	0	0	1	0	1	0	1	0	0	0
0	1	1	0	1	0	0	1	0	0	0	1	1
0	1	1	1	0	0	0	0	0	0	1	0	1
0	1	1	1	1	0	0	0	0	0	1	0	0
1	0	0	0	0	x	0	0	1	1	0	0	1
1	0	0	0	1	1	1	1	x	1	1	x	1
1	0	0	1	0	0	1	0	1	0	0	0	1
1	0	0	1	1	1	1	0	0	1	0	0	0
1	0	1	0	0	0	0	0	1	0	0	0	1
1	0	1	0	1	0	0	1	1	0	0	x	1
1	0	1	1	0	1	0	0	1	1	0	0	1
1	0	1	1	1	0	0	0	1	0	0	0	1
1	1	0	0	0	1	1	1	0	1	1	1	0
1	1	0	0	1	1	1	1	1	1	1	1	1
1	1	0	1	0	0	1	0	0	0	1	0	0
1	1	0	1	1	0	0	0	1	0	1	0	1
1	1	1	0	0	0	0	0	0	1	1	1	1
1	1	1	0	1	0	1	0	0	1	0	1	1
1	1	1	1	0	0	0	0	0	1	0	1	0
1	1	1	1	1	0	0	0	1	1	0	1	0

34. Find both the minimum SOP and minimum POS expressions for each of the functions in exercise 32, using VEMs.

35. Find both the minimum SOP and minimum POS expressions for each of the functions given in the truth table of exercise 33, using VEMs.

36. Make a two-map-variable VEM for each of the functions of exercise 9, and use the VEM for the function to find its simplest SOP expression. Choose A and B as the map variables.

37. Make a two-map-variable VEM for each of the functions of exercise 30, and use the VEM for the function to find its simplest SOP expression. Choose A and B as the map variables.

38. Make a three-map-variable VEM for each of the functions of exercise 11, and use the VEM for the function to find its simplest SOP expression. Choose A, B, and C as the map variables.

39. Make a three-map-variable VEM for each of the functions of exercise 31, and use the VEM for the function to find its simplest SOP expression. Choose A, B, and C as the map variables.

40. Make a two-map-variable VEM for each of the functions of exercise 9, and use the VEM for the function to find its simplest POS expression. Choose A and B as the map variables.

41. Make a two-map-variable VEM for each of the functions of exercise 30, and use the VEM for the function to find its simplest POS expression. Choose A and B as the map variables.

42. Make a three-map-variable VEM for each of the functions of exercise 11, and use the VEM for the function to find its simplest POS expression. Choose A, B, and C as the map variables.

43. Make a three-map-variable VEM for each of the functions of exercise 31, and use the VEM for the function to find its simplest POS expression. Choose A, B, and C as the map variables.

44. Make a two-map-variable VEM for each of the functions of exercise 11, and use the VEM for the function to find its simplest SOP expression. Choose A and B as the map variables. Notice that these maps have two entered variables.

45. Make a two-map-variable VEM for each of the functions of exercise 31, and use the VEM for the function to find its simplest SOP expression. Choose A and B as the map variables. Notice that these maps have two entered variables.

46. Make a three-map-variable VEM for each of the functions of exercise 33, and use the VEM for the function to find its simplest SOP expression. Choose variables A, B, and C as the map variables. Notice that these maps have two entered variables.

4 Combinational Logic Circuits

4.1 INTRODUCTION

We've been concerned with developing techniques for manipulating and simplifying Boolean, or logic, functions. These techniques are the tools we need to design circuits that will perform logic operations—that is, evaluate logic expressions. These circuits are called logic circuits. They're peculiar in that they allow only two possible interpretations of signal levels—high and low. The signals are usually voltages, and for a very common kind of logic circuits, TTL circuits, the high level is a nominal 5 volts and the low level is a nominal 0 volts. We say ''nominal'' because an output voltage above 2.4 volts is usually considered high, and an output voltage below 0.4 is usually considered low. We specify *output* voltage because high and low *input* voltages are slightly different. A high input is normally 2.0 volts, and a low input is normally 0.8 volts; thus, an output has a *margin* of 0.4 volts when driving an input. (We will look more carefully at margins later.) For another common kind of logic circuit, the CMOS circuit, nominal 12 volts and 0 volts are sometimes used as high and low signal levels.

A logic circuit consists of interconnections of logic elements. In combinational logic circuits—the circuits we will consider initially—the logic elements are almost exclusively **gates**. Logic circuits can be constructed by interconnecting *discrete* (single) logic elements, using conductors such as wires or traces, on a circuit board. Figure 4.1-1 shows several logic devices called chips. (The quarter is for scale.) Each chip consists of up to four discrete gates. In Figure 4.1-2, several logic devices are shown installed on a printed circuit board. This is a very simple, one-layer circuit board with the logic devices on one side (shown on the top) and the conductors or traces used to connect the logic elements of the devices on the other (shown on the bottom). Complicated printed circuit boards can have many layers of conductors or

Figure 4.1-1 Logic devices or chips.

traces sandwiched together. Initially we'll describe the design of circuits using discrete gates, but this is by no means the only way, or even the most common way, to design logic circuits.

Logic circuits can also be constructed using **custom logic devices** or a number of different types of **programmable logic devices**. Each of these devices consists of many logic elements packaged together. In a custom logic device, interconnections between the logic elements are fixed at the time of manufacture. Custom logic devices are used for logic circuits that will have a large volume of use (that is, will be used at least tens of thousands and probably hundreds of thousands of times). The initial cost of setting up custom logic devices for production (design and tooling) is high, but the actual production cost per device is low. Custom logic is also used to produce specialized, very high speed circuits.

Some devices are programmable; that is, interconnections between their logic elements are programmed into them. Most new logic circuit designs use such devices. Programmable logic devices are more expensive than custom logic devices for high volume production, but they have no initial fixed cost, so they're used for initial designs and small and medium volume production. Some programmable logic devices are one-time programmable, and some can be reprogrammed.

We can divide programmable logic devices into two broad classes: those with fixed arrangements of logic elements (**fixed architectures**), such as **Programmable Read Only Memories (PROMs), Programmable Array Logic (PAL)** devices,

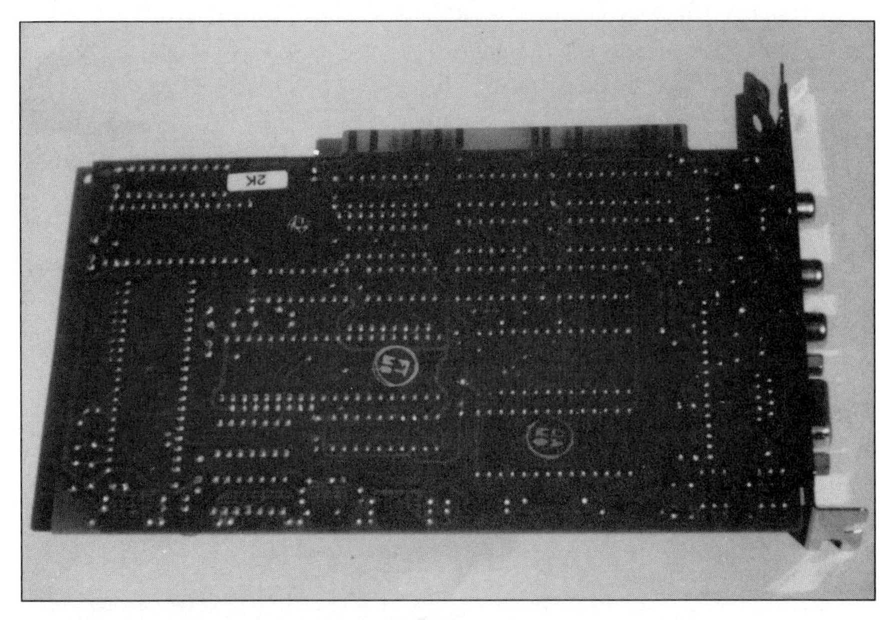

Figure 4.1-2 A simple one layer printed circuit board with logic devices (shown on the top) and traces (shown on the bottom).

and **Programmable Logic Array (PLA)** devices; and those with flexible arrangements of logic elements (**flexible architectures**)—the large **Field Programmable Gate Arrays (FPGAs)**. The PROMs, PALs, and PLAs used for combinational logic circuits all consist of gates very similar to the discrete gates we're about to discuss. However, the programming of devices with fixed architectures is rather specialized because of the software aids that are available. Custom logic and programmable gate arrays, on the other hand, can be designed using techniques like those we'll use to design circuits with discrete gates. Programming devices with flexible architectures does involve software, but the software is easy to use with the design methods we will consider for discrete logic.

Because custom logic and programmable gate array design closely parallels design with discrete gates, as we develop the design of combinational circuits using discrete logic we'll simply comment on the differences between it and the design of combinational logic circuits in custom logic and gate arrays. Later we'll discuss the special techniques needed to program fixed architecture devices as combinational logic circuits.

4.2 GATES

A gate is a many-input, 1-output device. If we initially consider only 2 inputs and 1 output, a general gate can be shown schematically as in Figure 4.2-1. We've labeled the inputs A and B and the output F. We can completely describe the function of this gate with a function table in which every possible combination of high and low input signal levels is assigned an output signal level.

Following is an example of a function table for a gate:

A	B	F
L	L	L
L	H	L
H	L	L
H	H	H

This table shows one of sixteen possible ways of assigning output signal levels for a 2-input gate, meaning that sixteen different kinds of 2-input gates are possible. Only a few of them are useful, however. This function table defines one of the useful gates.

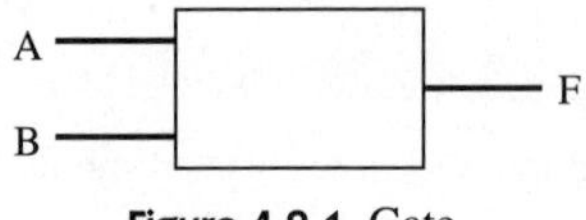

Figure 4.2-1 Gate.

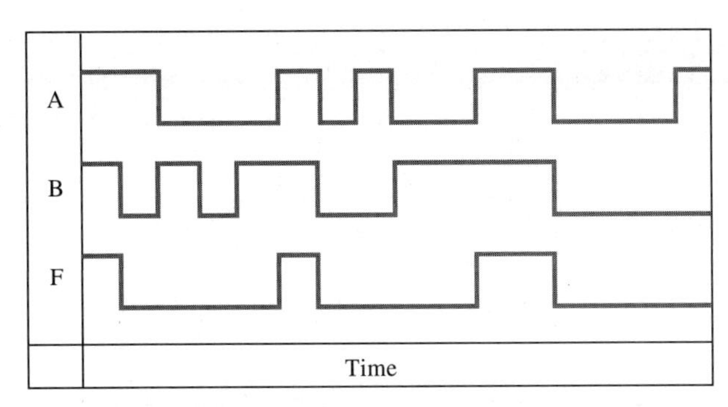

Figure 4.2-2 Timing diagram.

A function table is different from a truth table. A function table describes a physical device with input signals—voltages that can be either high or low—and an output signal that depends on the input signals. The output signal is also a voltage and can be either high or low. A truth table, in contrast, describes an abstract concept, a logic function of logic variables. The logic variables and logic function in a truth table can be either logical 1s (true) or logical 0s (false). We'll show how to convert a function table to a truth table when we use a gate to evaluate logic functions.

It is sometimes useful to supplement a truth table description of a gate with a plot showing how the output signal from the gate changes, as a function of time, as the input signals are changed. Such a plot is called a **timing diagram**. Figure 4.2-2 is an idealized timing diagram for the gate defined by the preceding function table. An actual gate has a delay between the time an input to the gate changes and the time the output changes, and the output signal is rounded as shown in Figure 4.2-3. The delay is called the **propagation delay**. There are actually two different delays: that for a low-to-high transition of the output, called TPLH, and that for a high-to-low transition of the output, called TPHL. The delays depend on the *technology (logic family)* and the type of gate, but they are usually on the order of nanoseconds for commonly used gates. Often the two delays are the same.

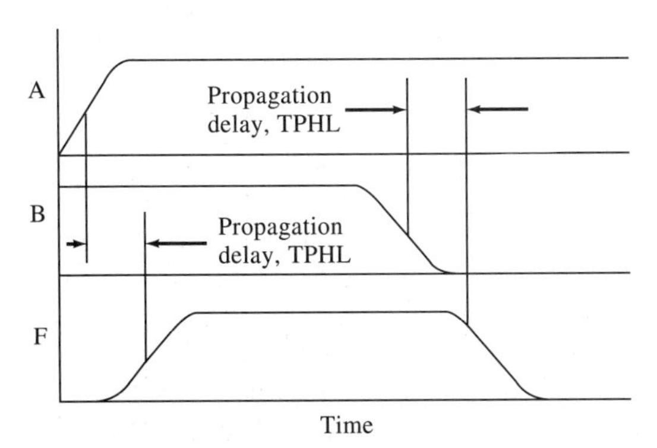

Figure 4.2-3 Actual signals.

Table 4.2-1 Typical Specification of Propagation Delays*

Switching Characteristics: $V_{CC} = 5$ V, $T_A = 25°C$							
Parameter	From (Input)	To (Output)	Test Conditions	Min	Typ	Max	Unit
t_{PLH}	A or B	Y	$R_L = 2$ k, $C_L = 15$ pF		9	15	ns
t_{PHL}					10	15	ns

*From *TTL Logic Data Book* (Texas Instruments, 1988).

Because of manufacturing tolerances it is impossible to make all gates of a given type identical. Usually, device manufacturers give three specifications for propagation delay: the maximum delay, which is the greatest delay any gate of a given type should have; the minimum delay for a gate of this type; and the typical delay for a gate of this type. The maximum delay is much used in design because it often represents the worst case. Table 4.2-1 shows how the propagation delay for a gate is specified in a manufacturer's data book. Notice that the minimum propagation delay isn't included. Manufacturers often omit the minimum for gates because it's not usually an important design parameter.

A discrete logic device (gate) is specified by a standardized part number in the form

Denotes commercial quality	Denotes technology	Denotes function
7 4	L S	0 0

The 74 indicates that the logic device is a commercial quality device (54 indicates military quality); the part numbers for all standard commercial quality logic devices start with 74. The designation LS, which stands for *low power Schottky,* designates the technology. A number of technologies are currently in use; they differ in speed (propagation delay), power consumption, and price. Table 4.2-2 lists some of the common technologies. The order of the table is generally from slowest to fastest. Faster devices normally consume more power, but CMOS devices use less power than other technologies. The 00 at the end of the part number designates the function. We will examine a variety of gate functions throughout this chapter. Devices with identical functions, and hence the same function number, generally exist in several technologies. Thus, the 74LS00 and the 74F00 have the same function. In the examples in this book, we will use devices with LS technology and give their part numbers. LS technology is a good choice where speed is not important, because it offers the advantage of both low power and low price.

Table 4.2-2 Device Technologies

Designation	Name	Power	Comments
None	Standard TTL	High	Not much used
LS	Low power Schottky	Low	Much used for slow circuits
ALS	Advanced low power Schottky	Low	Faster and lower power than LS
HC/HCT[1]	High speed CMOS	Low	Very low power
S	Schottky	Medium	Faster than LS
AS	Advanced Schottky	Medium	Faster and lower power than S
AC/ACT[1]	Advanced CMOS	Low	Faster than HC/HCT
F	Fast	High	Used where speed is important

[1]HCT and ACT denote CMOS devices that are TTL compatible.

Now that we've described a gate, let's see how we can use it to evaluate logic functions. We must relate the signals at the gate's inputs to the values of the logic variables in the function to be evaluated, and the signal at the gate's output to the value of the logic function. That isn't difficult, because the signals are two-valued (H and L), and so are the logic variables and function (0 and 1). There are actually two ways to make the correspondence. We can let a high signal (H) represent a logic 1 and a low signal (L) represent a logic 0, or vice versa. This idea can be clarified with a picture, but first we need to develop some notation. We'll use A(H) to denote the signal that represents the logic variable A with H = 1 (and L = 0), and A(L) to denote the signal that represents the logic variable A with L = 1 (and H = 0). Figure 4.2-4 shows how the signals A(H) and A(L) represent a logic variable A that is changing with time. The 1s and 0s in the figure are the values of the logic variable.

This notation is called *polarized mnemonics*. We give the signal the same label as the logic variable it represents, and we denote the level that corresponds to 1 by putting it in parentheses, as A(H). This is read to mean that if A is high, it corresponds to logic 1. A better way to say this is "A asserted high." Similarly, A(L) means "A

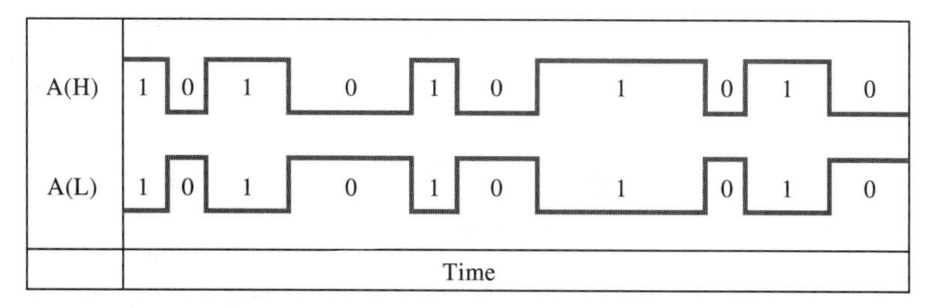

Figure 4.2-4 The signals A(H) and A(L), which represent a logic variable A which is changing with time. The 1s and 0s are the values of the logic variable.

asserted low,'' or A low corresponds to logic 1. The H or L is called the *polarity* of the signal.

Now a word about mnemonics. In our study of Boolean algebra, we labeled variables with letters such as A, B, and C and functions with letters such as F and F_1. We used the same labels for the signals that represented the variables and functions. This was just for convenience, however. In logic circuit design it's better to choose signal labels that have some meaning. Thus, a signal that is used for a reset function and is asserted high might be RST(H), whereas a signal that is used as a strobe and is asserted low might be STB(L). In many of our examples we'll continue to use simple labels for signals (and variables and functions) because we have no better labels. But remember that, for an actual circuit, signal labels that mean *something* are much more useful.

Now that we have a way of using gate signals to represent logic variables, let's see how to apply this idea to our sample gate. It's important to note that the polarity of each of the gate input signals (A and B) and the output signal (F) can be assigned independently. Suppose we assign the polarities A(H), B(H), and F(H). This means that

for signal A, H = 1 and L = 0
for signal B, H = 1 and L = 0
for signal F, H = 1 and L = 0

We can form a truth table from the function table by substituting 1 for H and 0 for L for all the signals:

A	B	F
0	0	0
0	1	0
1	0	0
1	1	1

This is the truth table for

$$F = AB$$

With the assignment of 1 to a high signal level for the signals A, B, and F, this gate evaluates an *and* of the logic variables A and B. It's important to understand what is meant by ''evaluates an *and*'' of A and B. The gate actually evaluates the expression AB and outputs a value for F that is the result of the evaluation. For example, if the logic variable A = 1, then gate input signal A is high, and if the logic variable B = 0, then gate input signal B is low. But a high input and a low input result in a low output, and a low output means that the logic function F = 0. All we need to do to evaluate AB is to apply the proper voltages to the inputs of the gate, and the output of the gate will give us the evaluation, or result.

The polarity assignments we've made—A(H), B(H), and F(H)—are but one of eight possible choices. Other assignments cause this gate to evaluate other logic functions. All possible assignments of polarities for the input and output signals, along with the resultant truth tables and the logic functions that are evaluated, follow. With polarity assignments A(L), B(L), and F(H),

A	B	F
1	1	0
1	0	0
0	1	0
0	0	1

$F = \overline{A}\,\overline{B}$

With polarity assignments A(H), B(L), and F(H),

A	B	F
0	1	0
0	0	0
1	1	0
1	0	1

$F = A\overline{B}$

With polarity assignments A(L), B(H), and F(H),

A	B	F
1	0	0
1	1	0
0	0	0
0	1	1

$F = \overline{A}B$

With polarity assignments A(L), B(L), and F(L),

A	B	F
1	1	1
1	0	1
0	1	1
0	0	0

$F = A + B$

With polarity assignments A(H), B(H), and F(L),

A	B	F
0	0	1
0	1	1
1	0	1
1	1	0

$F = \overline{A} + \overline{B}$

With polarity assignments A(L), B(H), and F(L),

A	B	F
1	0	1
1	1	1
0	0	1
0	1	0

$F = A + \overline{B}$

With polarity assignments A(H), B(L), and F(L),

A	B	F
0	1	1
0	0	1
1	1	1
1	0	0

$F = \overline{A} + B$

The gate described by the function table in the beginning of this section can evaluate eight different logic functions—four *and*s and four *or*s.

4.3 SYMBOLS

The analysis in Section 4.2 showed us how our sample gate can evaluate eight different logic functions, but the approach was rather indirect. If we're going to become proficient at designing logic circuits for evaluating logic functions, we need a more direct method of relating the function to be evaluated to the gate used in the evaluation. This more direct approach makes use of symbols to represent the operations a gate can perform and allows a determination of the logic function directly from the symbol.

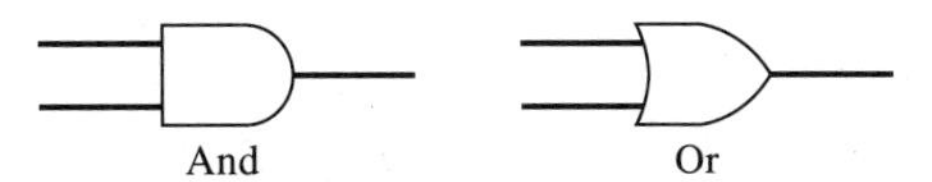

Figure 4.3-1 Symbols used to show the operation performed by a gate.

We said in Section 4.2 that our gate can perform two operations—the *and* operation and the *or* operation. The conventional circuit symbols for these two operations are shown in Figure 4.3-1. Their meanings are very specific. The basic *and* symbol means that a high input signal on the first input *and* a high input signal on the second input result in a high output signal. The basic *or* symbol means that a high input signal on the first input *or* a high input signal on the second input results in a high output signal.

Now look at the function table for the gate (page 142). The *and* symbol of Figure 4.3-1 exactly represents this gate. A high input on A and a high input on B cause a high output on F, but how do we use the *and* symbol to determine the logic function the gate is evaluating? Suppose we make the assertion levels of both input signals A and B high [A(H) and B(H)], and the assertion level of the output signal F high also [F(H)], as shown in Figure 4.3-2; then, if the logic variable A is 1 (asserted), the input signal A will be high, and if the logic variable B is 1 (asserted), the input signal B will be high. The output signal F will thus be high, and the logic function F will be 1 (asserted). That is, the gate will evaluate the function

$$F = AB$$

The notation used in the figure to express the function evaluated by the gate may need some clarification. Actually, both the function and the assertion level of the signal that represents the function are given. We should interpret the equation to mean that F = AB is the logic function evaluated by the gate, and the output signal representing this function is asserted high. (The output signal F is asserted high, and the signal AB, which is the same thing, is also asserted high.) Notice, in case you are tempted, that an expression such as F(H) = A(H)B(H) is meaningless in our notation. The expression AB is a logic expression that does not depend on the assertion levels of the signals chosen to represent the variables making up the expression.

Now let's consider different signal assertion levels for the input signals shown in Figure 4.3-3. Here the input signal assertion levels are both low [A(L) and B(L)], but the output signal assertion level is still high [F(H)]. Let's use the basic meaning

Figure 4.3-2 **Figure 4.3-3**

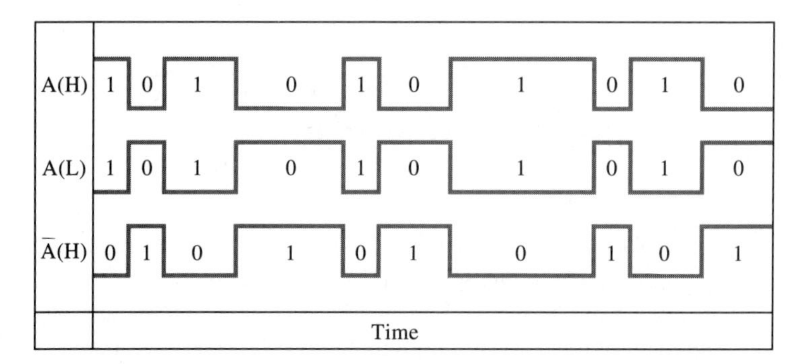

Figure 4.3-4 Timing diagram which shows that A(L) and A(H) are the same signals.

of the symbol to determine the logic function that is being evaluated. Remember that a high input on A and a high input on B cause a high output on F. But to get a high input on A, the logic variable A must be *not asserted* (0), and to get a high input on B, the logic variable B must be *not asserted* (0), which means that $\overline{A}$ and $\overline{B}$ cause F to be high, or asserted. That is, the function being evaluated is

$$F = \overline{A}\,\overline{B}$$

This example is an introduction to a very important principle in logic circuit design (as well as analysis). The basic gate symbol we've used means that if input signal A is high and input signal B is high, then output signal F will be high; but the assertion levels of signals A and B are both low. Their assertion levels are *incompatible* with, or do not match, the basic gate symbol. *When the assertion level of an input signal is incompatible with the basic gate symbol, the signal is complemented in the logic function the gate evaluates.* In this example, because both signals A and B have assertion levels that are incompatible with the gate symbol, they are both complemented in the logic function. At this point you may ask what happens if the output signal is incompatible with the gate symbol; but as you'll see, we never design circuits where this is true. Later, when we analyze circuits that may not have been designed in the way we're learning here, we'll see how to handle output incompatibilities.

Another way we could approach the gate with low input assertion levels is to see whether there is a way to make the input assertion levels compatible. Suppose we associate a signal with the logic variable $\overline{A}$ rather than A, and give it a high assertion level. This signal, $\overline{A}$(H), is the same as the signal A(L). We can see this in Figure 4.3-4, which shows A(H), A(L), and $\overline{A}$(H) for a logic variable A that changes in an arbitrary way with time. We can see in the figure that when A is 1, A(L) is low, but if A is 1 then $\overline{A}$ is 0, so $\overline{A}$(H) is low also. When A is 0, A(L) is high, but if A is 0 then $\overline{A}$ is 1, so $\overline{A}$(H) is high also. We can thus say that A(L) = $\overline{A}$(H) and that B(L) = $\overline{B}$(H). The inputs with low assertion levels in Figure 4.3-3 can thus be replaced by their complements with high input assertion levels so that there are no input incompatibilities and F = $\overline{A}\,\overline{B}$. Be careful with the fact that the signals A(L)

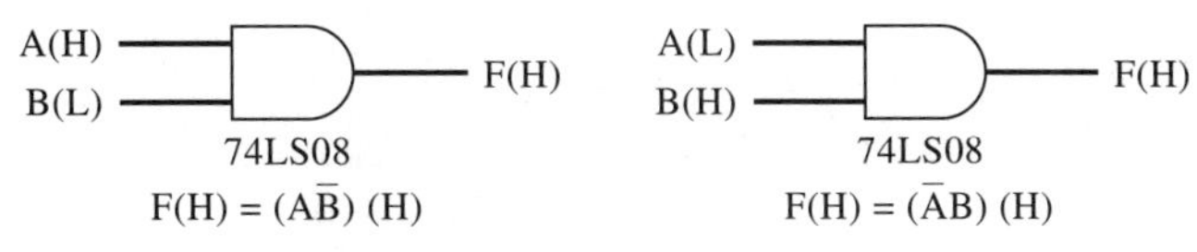

Figure 4.3-5

and $\overline{A}$(H) are equal. Although this is certainly true, it doesn't follow that the logic variables represented, A and $\overline{A}$, are equal, as can be seen in the figure. The idea that we can change the input variables to make them compatible was presented to clarify the notation we're using. We will use this idea not in our designs, but rather in our analysis of circuits later in the book. For now, we'll always use input signals that are uncomplemented and let incompatibilities at a gate input cause the logic variable to be complemented in the function being evaluated.

Our gate can evaluate two more *and* functions, as shown in Figure 4.3-5. Notice that in each case the basic symbol is the same—only the assertion levels of the input signals are changed—and in each case where the assertion level is incompatible with the gate symbol, the corresponding logic variable in the function being evaluated is complemented. To determine the function to be evaluated, we ask ourselves the question ''How can we make all the inputs high?'' and we answer, ''By making all the inputs that are asserted low (incompatible) not asserted (complemented).''

As seen in Section 4.2, our gate can also perform *or* operations, which the function table on page 142 shows. Note that a low input on A or a low input on B causes a low output on F. To have a useful circuit symbol for this function, we clearly need to use the symbol for the *or* operation; but the basic *or* symbol means that if input signal A is high or input signal B is high, output signal F will be high. To handle the preceding function with low assertion levels, we need a symbol for the operation—a low input signal A or a low input signal B causes a low output signal F. To form the proper symbol, we make use of an assertion level *changer* (or, equivalently, a signal inverter). This is a *bubble* placed on any of the inputs or the output of the basic gate symbol. Using this assertion level changer on all the inputs and the output of the basic *or* symbol, we arrive at the circuit symbol shown in Figure 4.3-6. This basic symbol with level changers means that a low input signal A or a low input signal B causes a low output signal F.

Now suppose we assume that the input signal assertion levels are both low [A(L) and B(L)] and the output signal assertion level is also low [F(L)], as shown in Figure 4.3-7. The logic function evaluated is clearly

$$F = A + B$$

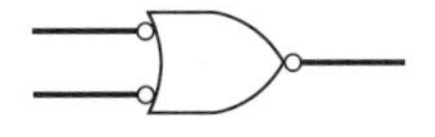

Figure 4.3-6

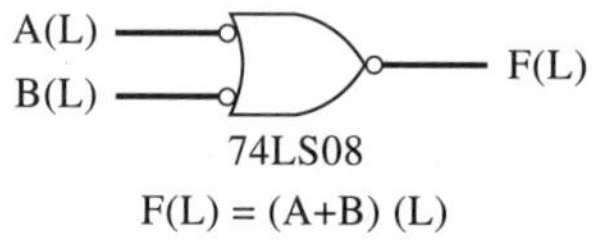

A(L)
B(L)
F(L)
74LS08
F(L) = (A+B) (L)

Figure 4.3-7

because if A is asserted (1), the input signal A will be low, or if B is asserted (1), the input signal B will be low; the output signal F will thus be low, that is, asserted (1). Notice that both the input signal assertion levels and the output signal assertion level are low and thus compatible with the basic gate symbol. There are no complemented variables in the logic function that is evaluated by the gate.

The other three *or* functions that our gate can evaluate have the circuit symbols shown in Figure 4.3-8. Notice that the output signal assertion level is always low to match the gate output. Only the input signal assertion levels are changed; and when the input signal assertion level is incompatible with the basic gate symbol, that variable is complemented in the logic function evaluated by the gate. Here the question we ask ourselves is "How can we make all the inputs low?" and we answer, "By making all the inputs that are asserted high (incompatible) not asserted (complemented)."

Now that we've found all the logic functions our basic gate can evaluate, we can give it a name—a "2-input, positive logic *and* gate." The name comes from the fact that, when signals A, B, and F are all asserted high (called positive logic), the gate performs an *and* of A and B. Even though this gate is called an *and* gate, we should always remember that it can perform both *and* and *or* operations and has both an *and* symbol and an *or* symbol, as shown in Figure 4.3-9. We should further note that, because we always choose the output signal assertion level to be compatible with the basic gate symbol, this gate always performs an *and* operation if the output signal assertion level is high and an *or* operation if the output signal assertion level is low. This gate is a standard discrete logic gate and is usually packaged four gates to a device (for example, the 74LS08).

A(H)
B(H)
F(L)
74LS08
F(L) = ($\bar{A}+\bar{B}$) (L)
A(L)
B(H)
F(L)
74LS08
F(L) = (A+$\bar{B}$) (L)
A(H)
B(L)
F(L)
74LS08
F(L) = ($\bar{A}$+B) (L)

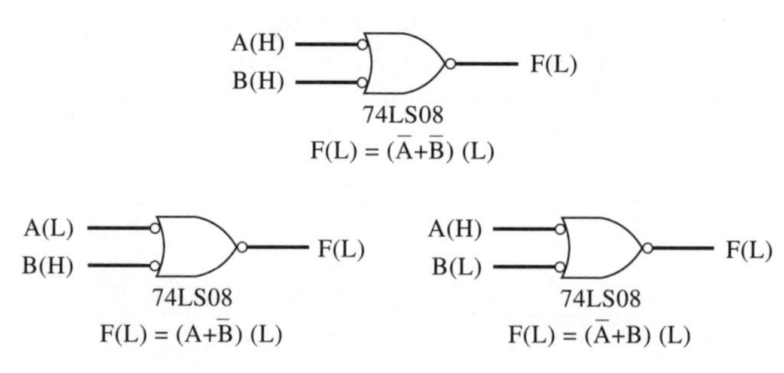

Figure 4.3-8

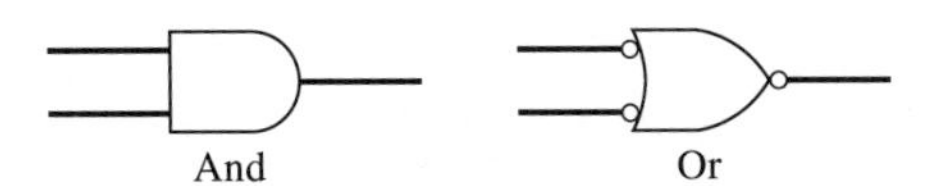

Figure 4.3-9 Two input, positive logic *and* gate.

Three other discrete logic gates are commonly used in logic circuit design. They are the positive logic *or* gate, the positive logic *nand* gate, and the positive logic *nor* gate. Our description of each gate will include a function table, a timing diagram, the two basic circuit symbols, and the eight logic functions the gate can evaluate.

The function table for the 2-input, positive logic *or* gate (e.g., the 74LS32) is

A	B	F
L	L	L
L	H	H
H	L	H
H	H	H

Figure 4.3-10 is a timing diagram. For this gate, if input signal A is high or input signal B is high, then output signal F is high. Figure 4.3-11 (left) shows the *or* symbol for this gate. Another way of describing the gate is to note that if input signal A is low and input signal B is low, then output signal F is low. Figure 4.3-11 (right) shows the *and* symbol for this gate. The gate can evaluate any of eight logic functions (four with *or* operations and four with *and* operations), depending on the assertion levels of signals A, B, and F. Figure 4.3-12 depicts the symbols, assertion levels, and logic functions that can be evaluated using this gate. Notice that the output assertion level determines the operation performed. When the output signal assertion level is high, the operation is always an *or,* and when the output assertion level is low, the operation is always an *and.*

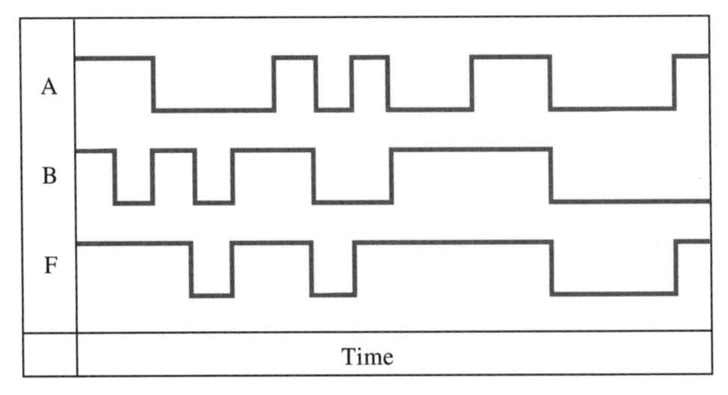

Figure 4.3-10 Timing diagram.

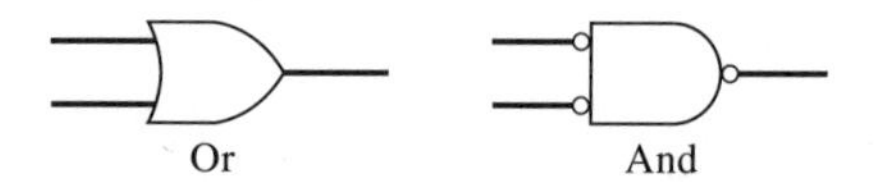

Figure 4.3-11 Two input, positive logic *or* gate.

The function table for the 2-input, positive logic *nand* gate (e.g., the 74LS00) is

A	B	F
L	L	H
L	H	H
H	L	H
H	H	L

Figure 4.3-13 is a timing diagram. For this gate, if input signal A is high and input signal B is high, then output signal F is low. This gate is a *not and* (abbreviated *nand*) gate—an *and* gate with the output complemented (*not*ed). Figure 4.3-14 (left) shows the *and* symbol for this gate. Another way of describing this gate is to note that if input signal A is low or input signal B is low, then output signal F is high.

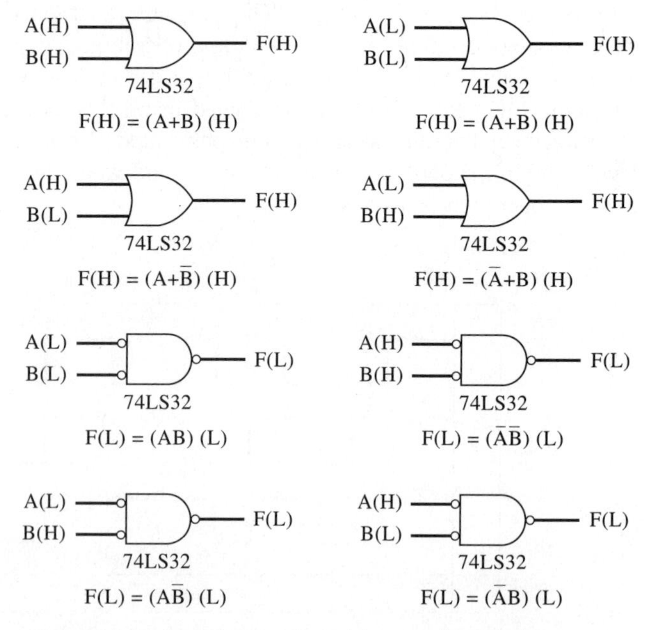

Figure 4.3-12 Functions evaluated by a positive logic *or* gate.

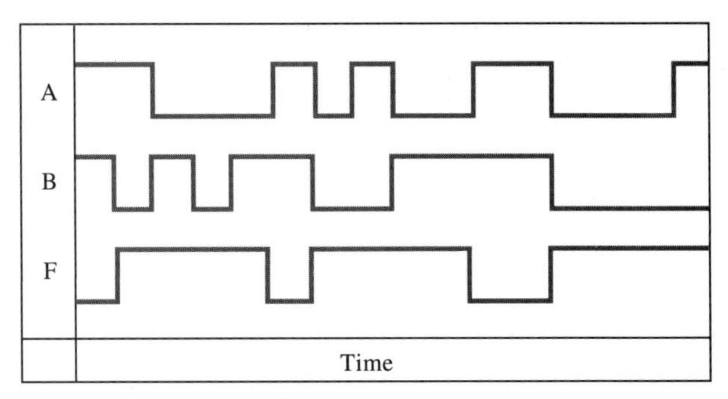

Figure 4.3-13 Timing diagram.

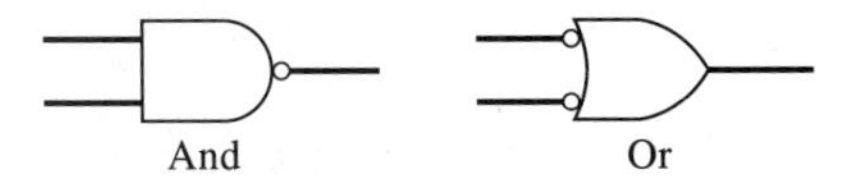

Figure 4.3-14 Two input, positive logic *nand* gate.

Figure 4.3-14 (right) shows the *or* symbol for this gate. The gate can evaluate any of eight logic functions, depending on the assertion levels of signals A, B, and F. Figure 4.3-15 depicts the symbols, assertion levels, and logic functions that can be evaluated using this gate. Notice that, as is true with the other gates, the output assertion level determines the operation performed. When the output signal assertion level is low, the operation is always an *and;* when the output assertion level is high, the operation is always an *or*.

The function table for the 2-input, positive logic *nor* gate (e.g., the 74LS02) is

A	B	F
L	L	H
L	H	L
H	L	L
H	H	L

Figure 4.3-16 is a timing diagram. For this gate, if input signal A is high or input signal B is high, then output signal F is low. This gate is a *not or* (abbreviated *nor*) gate—an *or* gate with the output complemented. Figure 4.3-17 (left) shows the *or* symbol for this gate. Another way of describing this gate is to note that if input signal A is low and input signal B is low, then output signal F is high. Figure 4.3-17 (right) shows the *and* symbol for this gate. The gate can evaluate any of eight logic functions, depending on the assertion levels of signals A, B, and F. Figure 4.3-18 depicts the symbols, assertion levels, and logic functions that can be evaluated

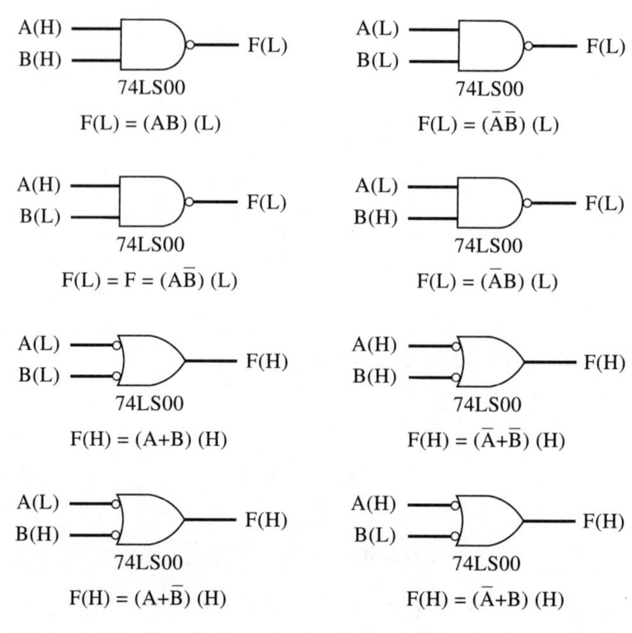

Figure 4.3-15 Functions evaluated by a positive logic *nand* gate.

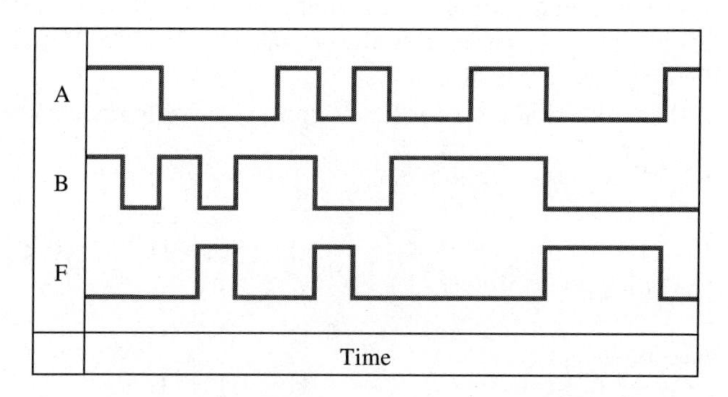

Figure 4.3-16 Timing diagram.

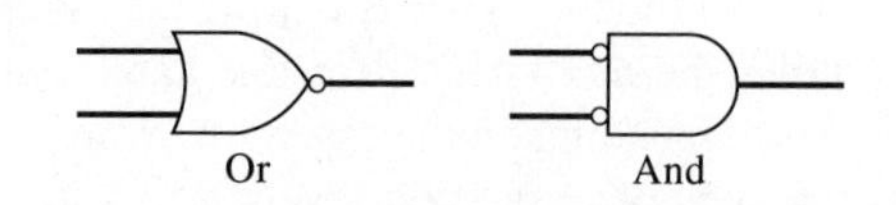

Figure 4.3-17 Two input, positive logic *nor* gate.

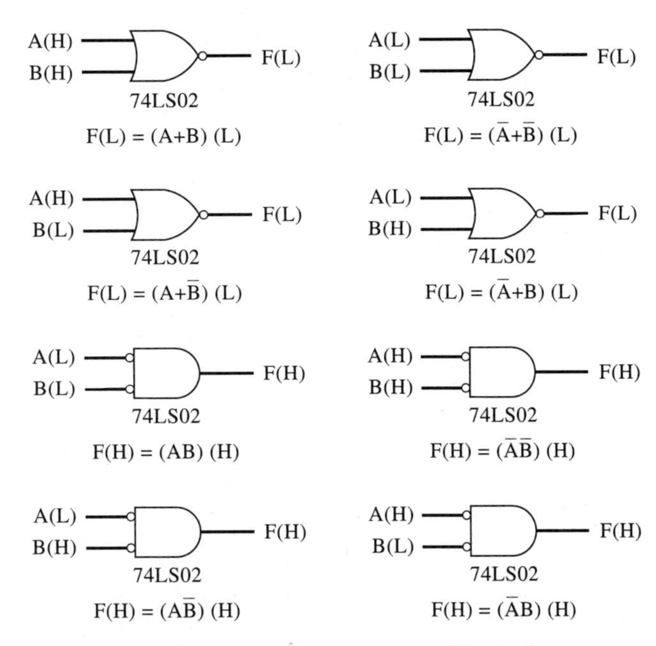

Figure 4.3-18 Functions evaluated by a positive logic *nor* gate.

by this gate. Notice that, as is true with the other gates, the output assertion level determines the operation performed. When the output signal assertion level is low, the operation is always an *or;* when the output assertion level is high, the operation is always an *and.*

Each of the logic functions in the foregoing figures can be verified by forming the truth table corresponding to the chosen assertion levels, but the function is immediately evident from the circuit symbol. In every case, if the assertion level for an input signal does *not* match, or is incompatible with, the circuit symbol, the logic variable corresponding to that input is complemented in the logic expression describing the function evaluated by the gate. Some designers flag the incompatibility. Figure 4.3-19 shows this flagging for a positive logic *and* gate performing both *and* and *or* operations. Notice that the flag on a gate performing an *and* operation is different from that on a gate performing an *or* operation. The flag on the gate performing an *and* operation is called an **inhibit**, and that on the gate performing an *or* operation is called an **enable**. The flagging doesn't change the circuit, but it does draw attention to the incompatibility and makes it easier to write the logic expression for the gate. We won't flag incompatibilities in this book.

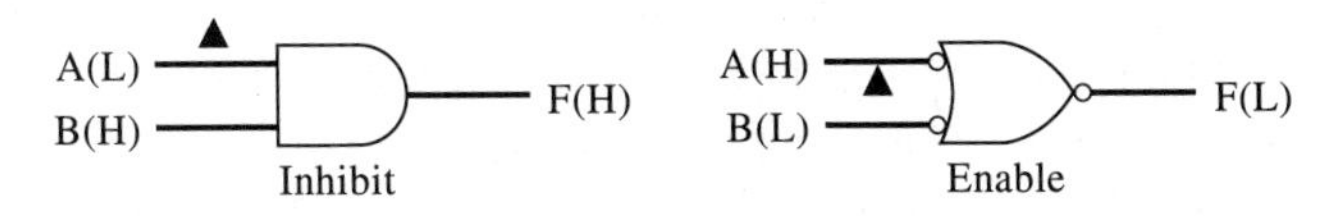

Figure 4.3-19 Flags which can be used to denote incompatibilities.

Gates are not limited to 2 inputs; they can have any number. Standard discrete gates usually have 2, 3, 4, or 8 inputs, although not all four types of gates are available with all these input configurations. *Nand* gates, which are much used, are even available in 12-input and 13-input configurations. Function tables and symbols for gates with more than 2 inputs are straightforward extensions of those for 2-input gates. Following is an example of a function table for a 4-input, positive logic *and* gate:

A	B	C	D	F
L	L	L	L	L
L	L	L	H	L
L	L	H	L	L
L	L	H	H	L
L	H	L	L	L
L	H	L	H	L
L	H	H	L	L
L	H	H	H	L
H	L	L	L	L
H	L	L	H	L
H	L	H	L	L
H	L	H	H	L
H	H	L	L	L
H	H	L	H	L
H	H	H	L	L
H	H	H	H	H

Figure 4.3-20 shows the symbols for a 4-input, positive logic *and* gate and an 8-input, positive logic *nand* gate. Notice that the symbol is expanded to allow more inputs. Extensions to other types of gates should be obvious. We haven't given a function table for the 8-input gate—it would be 256 lines long!

The inputs of the discrete gates we've described so far have input bubbles on either all or none of the inputs. This limitation exists only for the types of discrete logic gates that are manufactured. Custom logic and programmable gate arrays don't have such a limitation. Gates in custom logic can have mixed inputs, and gates in programmable gate arrays can be programmed to have mixed inputs; that is, some of the inputs can have bubbles and some can have no bubbles. In fact, any combination of inputs with and without bubbles is often available for a given gate.

Figure 4.3-21 depicts the types of gates available in one kind of field programmable gate array. This particular example shows 4-input gates; but all configurations of 2- and 3-input gates are also available. (The gate array shown has gates available

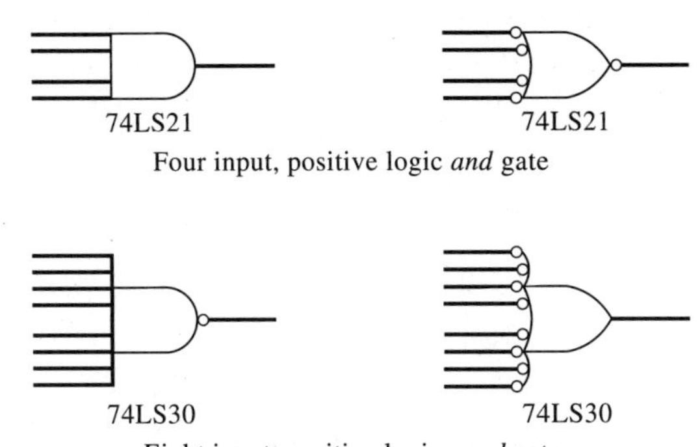

Figure 4.3-20 The symbols for a four input, positive logic *nand* gate and an eight input, positive logic *nand* gate.

with up to only 4 inputs. Some gate arrays have gates with more inputs.) The A, B, C, and D are input labels, and the Y is an output label. Each type of gate is labeled in the normal way as an *and,* a *nand,* a *nor,* or an *or* gate with a letter suffix to indicate the input configuration. Some of the other notations indicate limitations. The 2 in some of the gate symbols indicates that these gates are constructed with extra gate array modules. Such gates should be avoided in designs that use most of the modules in the array. Designs that fill an array are difficult to program. The triangle near the inputs of some of the gates in Figure 4.3-21 indicates that that input has extra propagation delay. Extra delay might be undesirable in some designs. Gates with mixed inputs, such as these found in gate arrays, allow for flexibility in the design of logic circuits, as we'll see in Example 4.3-4. Gates similar to those we have discussed so far are available both in custom logic and in other programmable gate arrays.

There are two special-purpose gates to cover before we leave our description of gates. They are the *exclusive or* (*xor*) and *exclusive nor* (*xnor*) gates. Both are 2-input gates. The function table for an *xor* gate follows.

A	B	F
L	L	L
L	H	H
H	L	H
H	H	L

Figure 4.3-22 shows the symbol for this gate and the logic function it evaluates when the inputs and output are asserted high. Notice that the *xor* function evaluated by the gate is given a special mathematical symbol.

The *xnor* gate is just an *xor* gate with the output asserted low. The function table for it is

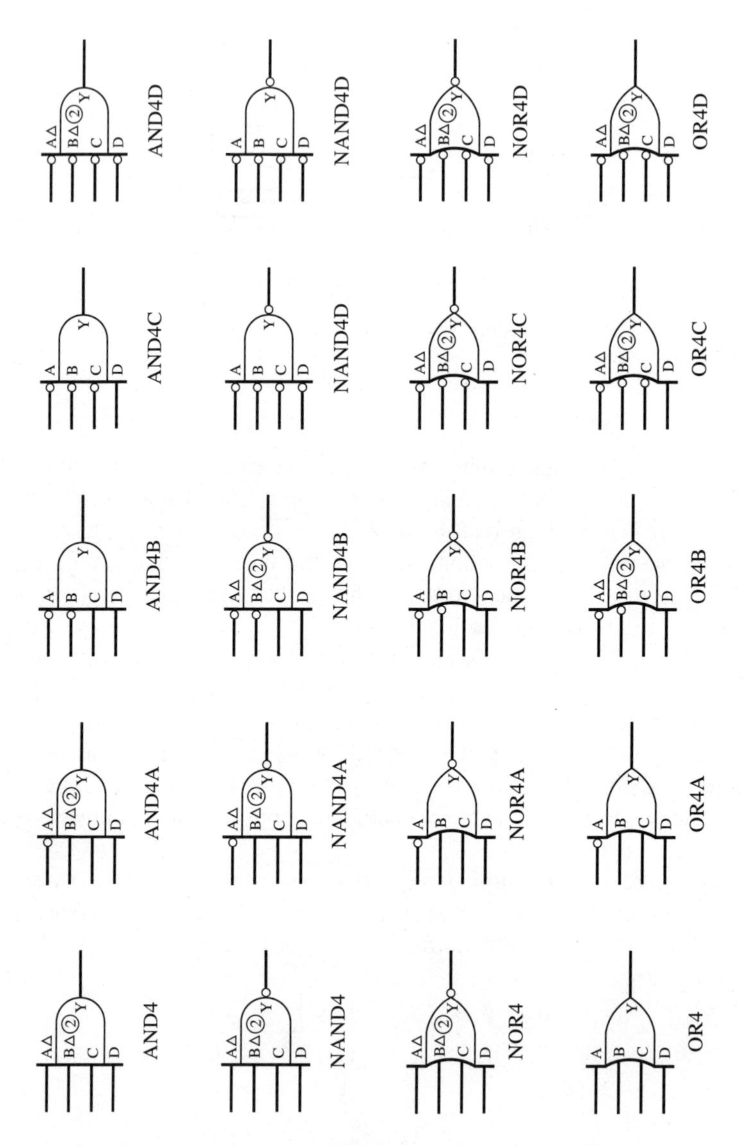

Figure 4.3-21 Example of mixed input gates available in gate arrays.

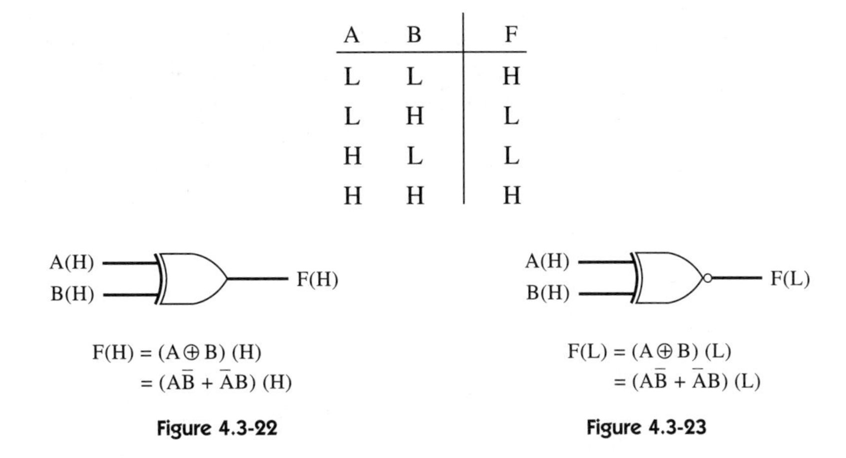

A	B	F
L	L	H
L	H	L
H	L	L
H	H	H

$$F(H) = (A \oplus B)\ (H) = (A\overline{B} + \overline{A}B)\ (H)$$

Figure 4.3-22

$$F(L) = (A \oplus B)\ (L) = (A\overline{B} + \overline{A}B)\ (L)$$

Figure 4.3-23

Figure 4.3-23 shows the symbol for this gate. The logic function evaluated is the same as that evaluated by the *xor* gate, but the output is asserted low.

In some special situations, *xor* and *xnor* gates can be used to advantage to simplify logic circuit design. We won't develop any procedure for doing this. Most modern designs use programmable logic devices, many of which don't use *xor* and *xnor* gates except for special purposes. The exception is gate arrays, which have both *xor* and *xnor* gates available.

Having looked at these special-purpose gates, let's return to logic circuit design with more conventional gates. Using single gates, we can design circuits to evaluate simple single-term logic functions, as the following examples show.

Example 4.3-1

Design a circuit to evaluate the logic function

$$F = \overline{A}B\overline{C}D$$

Evaluation of this function requires a 4-input gate that will perform an *and* operation. Because all four types of gates we've introduced perform *and* operations, we can potentially design four different circuits to evaluate this function. The different gates result in circuits with different assertion levels for the signals that represent the logic variables A, B, C, D, and F. We do have a practical limitation: some gates are not available (that is, are not manufactured commercially). If we limit the design to TTL gates, then no 4-input, positive logic *or* gate is available. With this limitation in mind, let's look at the other three designs.

We show the symbol for the *and* operation of a 4-input, positive logic *and* gate at the left in the following figure. To complete a design using this gate, we need to add the input and output signals (with their assertion levels), as shown on the right.

We always make the assertion level of the output signal compatible with the gate symbol. In this case, because there is no bubble on the output of the gate, the assertion level of the output signal is made high. The assertion levels of the input signals are chosen to give a proper function evaluation. The basic rule is that an input signal incompatibility produces a complement in the logic function being evaluated, so we make the A and C input assertion levels incompatible with the gate inputs.

To see what this circuit actually does and check that our design functions correctly, it's useful to computer simulate the operation of the circuit using logic circuit simulation software. More will be said about simulation in the next section. Following is a timing diagram for the circuit, generated with simulation software.

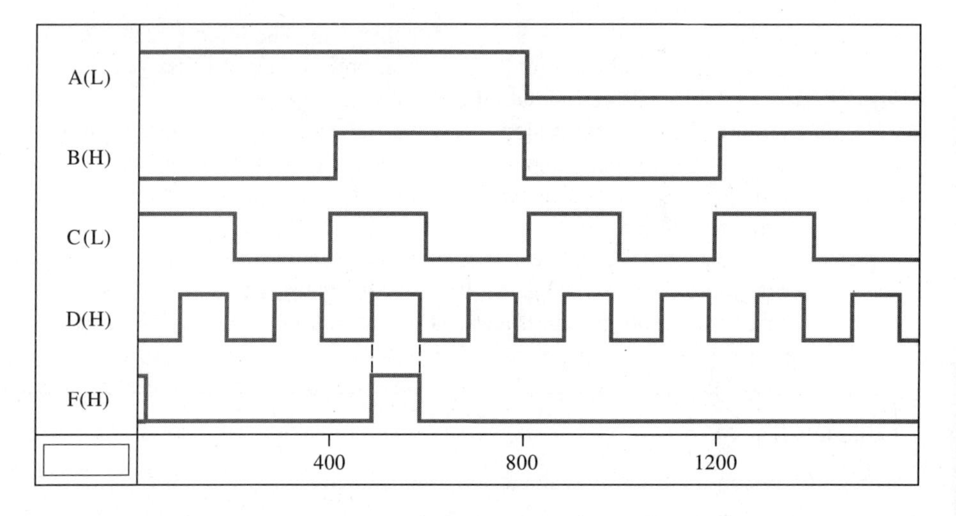

Notice that we must give the assertion levels of the signals on the timing diagram, because a high assertion level signal is different from a low assertion level signal even though they represent the same logic variable. The input signals for the simulation can be any time sequence. We've chosen them so that all possible combinations of inputs are tested. The input to the circuit is chosen to change every 100 nanoseconds (ns). This is an arbitrary choice. For the first 100 ns, logic variables A, B, C, and D are 0000. After 100 ns, D changes to 1 and the logic variables are 0001. For each subsequent 100 ns, the logic variables change in such a way as to count through the binary numbers 0000 to 1111 so that all combinations of inputs are tested.

We can see that the only time the output is high, which corresponds to F being asserted, or true, is when the input signals A(L), B(H), C(L), and D(H) are all high. Signal A(L) high means the logic variable A is not asserted, or 0; signal B(H) high means the logic variable B is asserted, or 1; signal C(L) high means the logic variable C is not asserted, or 0; and signal D(H) high means the logic variable D is asserted, or 1, just as required by the logic expression that is being evaluated.

The simulation actually shows the time delay between the change in the input signals and the change in the output signal (propagation delay), although it's difficult to see on this timing diagram because of the time scale—the propagation delay is 10 ns, and the total time is 1600 ns. Even so, using the dotted lines near the output signal change, which mark where the input signal changes, it is possible to see this delay.

The two additional circuits for evaluating the logic expression follow. The one on the left uses a 4-input, positive logic *nand* gate, and the one on the right uses a 5-input, positive logic *nor* gate. A suitable 4-input *nor* gate is not available in standard TTL. The fifth input of the *nor* gate is tied low so that it is always asserted.

This gate can be thought of as evaluating the function

$$F = \overline{A}B\overline{C}D1 = \overline{A}B\overline{C}D$$

Notice that the output assertion level is changed from circuit to circuit to match the outputs of the different gates. If the output assertion level were specified in the problem, then the selection of gates that could be used in the circuit design would be limited. For example, if specifications for the design required that the output be asserted low, then only the circuit using the positive logic *nand* gate could be used. Following are timing diagrams from software simulations of the operation of the two circuits.

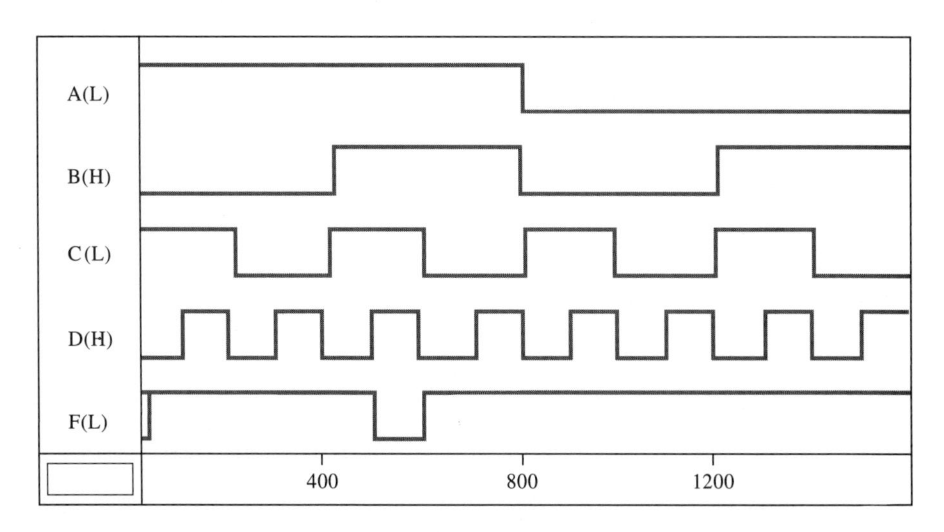

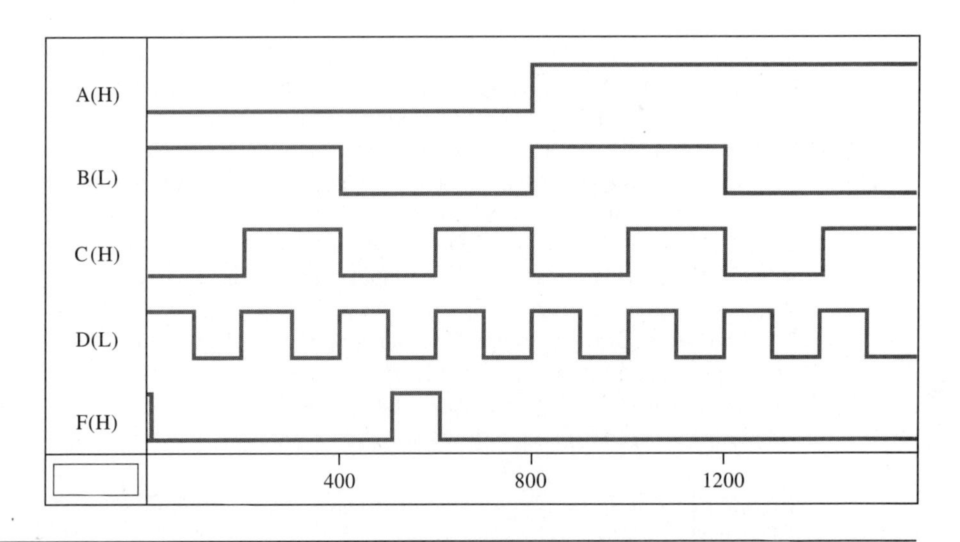

Example 4.3-2

Design a circuit to evaluate the logic function

$F = A + \overline{B} + \overline{C} + D$

Evaluation of this function requires a 4-input gate that will perform an *or* operation. Because all four types of gates we've introduced perform *or* operations, we can potentially design four different circuits to evaluate this function. The different gates will result in circuits with different assertion levels for signals A, B, C, D, and F. Again we have a practical limitation—some gates are unavailable. Perhaps the first choice that comes to mind for this evaluation is to use a positive logic *or* gate, but there is no 4-input, positive logic *or* gate if we again limit our choices to TTL devices. In this example, unlike the previous one, we won't let this limitation stop us. We can use three 2-input *or* gates to make a 4-input *or* gate. Suppose we group the function we're evaluating into two groups:

$F = (A + \overline{B}) + (\overline{C} + D)$

We can first evaluate $A + \overline{B}$ using a 2-input *or* gate, then evaluate $\overline{C} + D$ using another 2-input *or* gate, and finally evaluate $(A + \overline{B}) + (\overline{C} + D)$ using a third 2-input *or* gate. The following figure shows the circuit using three positive logic, 2-input *or* gates.

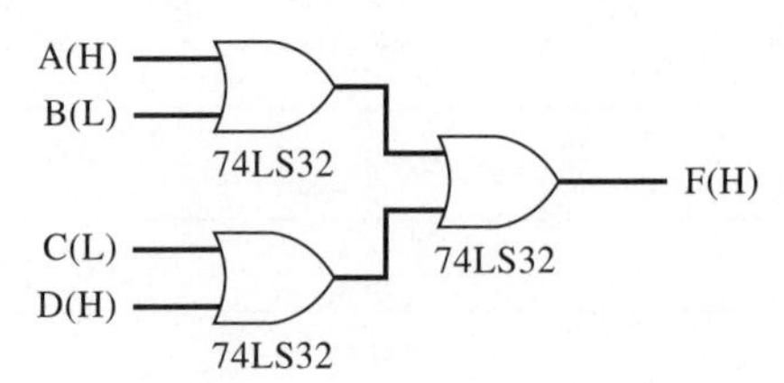

The gates can be cascaded because the output assertion levels of the two gates that evaluate $A + \overline{B}$ and $\overline{C} + D$ are high, so they match the inputs of the final gate. Using small gates like this to evaluate a function has the disadvantage of a propagation delay twice as long as with a single gate, since the signal must propagate through two gates. The technique of using small gates to evaluate a function with too many inputs for the available gates can also be used with positive logic *and* gates, but it doesn't work with positive logic *nand* or *nor* gates because the assertion levels between the gates don't match. (*Nand* gates can be used with a *nor* gate, and vice versa.) Following is a timing diagram that shows the results of a simulation of the operation of this circuit.

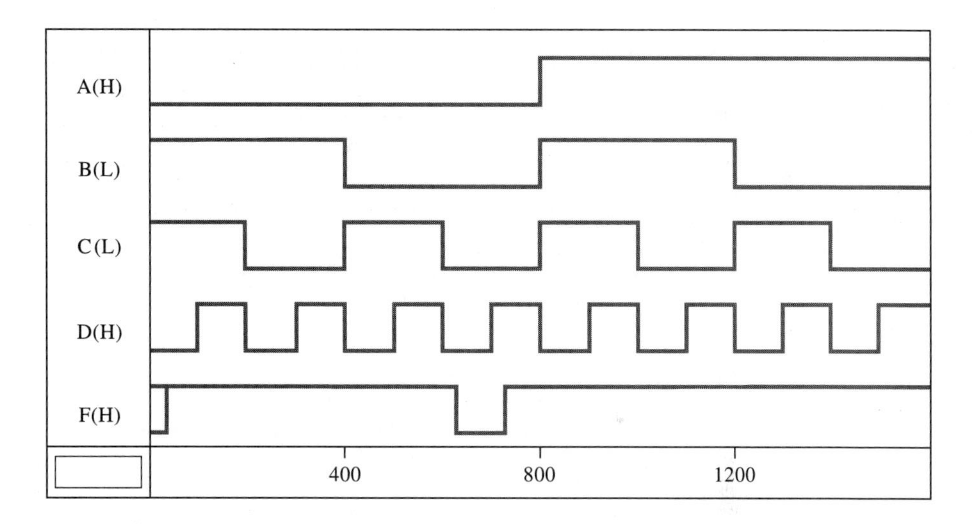

The following figures show the three other designs that are possible, all using single gates. Also shown are timing diagrams with the results from simulated operation. The first design, using a positive logic *nor* gate, actually uses a 5-input *nor* gate with one of the inputs tied to ground so that it is never high. This circuit can be thought of as evaluating the function

$$F = A + \overline{B} + \overline{C} + D + 0 = A + \overline{B} + \overline{C} + D$$

The other two circuits have no unusual features. Notice that in all the circuits, the output signal assertion level is chosen to match the gate output, and the input signal assertion level is chosen to match the gate input if the variable is not complemented, or chosen to not match the gate input if the variable is complemented.

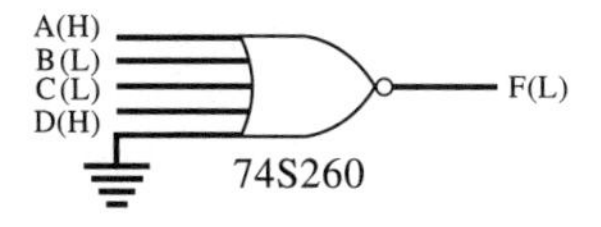

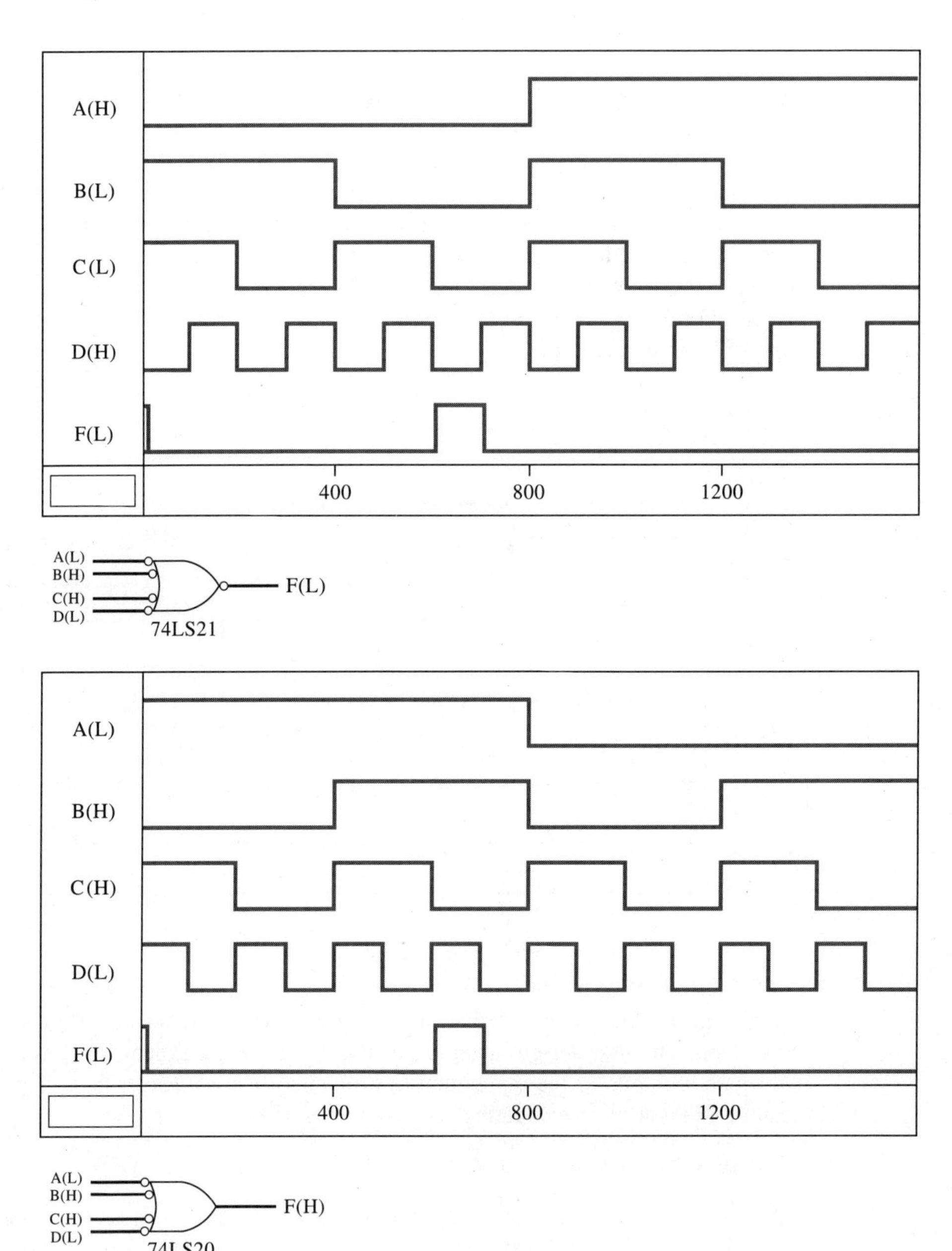
A(H)
B(L)
C(L)
D(H)
F(L)
400
800
1200
A(L)
B(H)
C(H)
D(L)
F(L)
74LS21
A(L)
B(H)
C(H)
D(L)
F(L)
400
800
1200
A(L)
B(H)
C(H)
D(L)
F(H)
74LS20

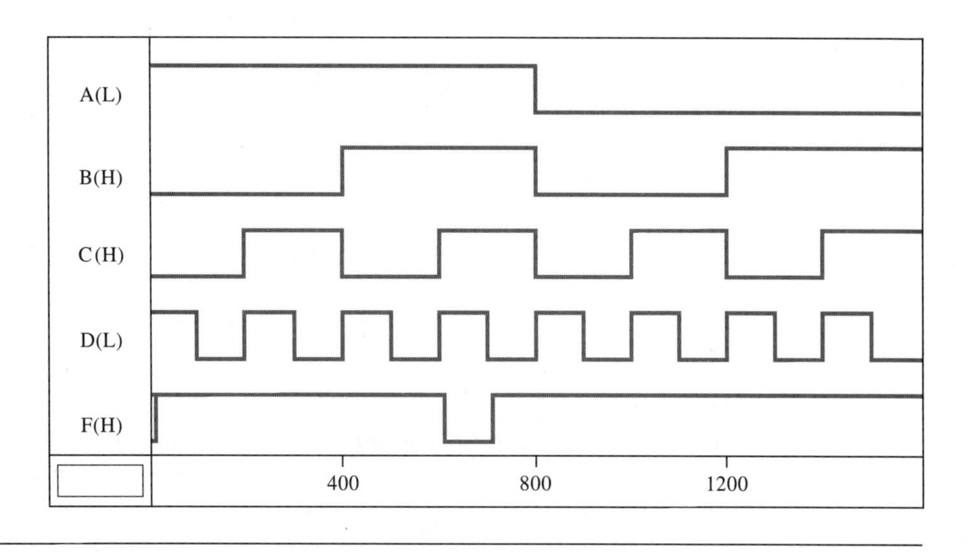

In Examples 4.3-1 and 4.3-2 we designed logic circuits using discrete TTL logic elements. This forced us to choose the input signal assertion levels in certain ways, depending on the logic element chosen for the design. In many cases, we are not free to choose the input signal assertion levels because they are fixed by constraints on the design. For example, they may be outputs of some other part of a larger logic circuit. In Section 4.5 we'll see how to overcome this limitation using *inverters*, but if we use gate arrays with their associated wide selection of gates, we don't need inverters, as can be seen in the following examples.

Example 4.3-3

Design a circuit to evaluate the logic function

$$F = A\overline{B}C\overline{D}$$

using gates found in a gate array like those in Figure 4.3-21. Assume the input signal assertion levels are all high, and make the output assertion level low.

Evaluation of this function requires a 4-input gate that will perform an *and* operation. To make the output assertion level low, we actually need an *and* with a bubble on the output—a *nand*. Now, how do we configure the input of this gate so that we can evaluate the required function with the input signal levels given? Signals A(H) and C(H) clearly need to be connected to inputs with no bubbles so that they will be compatible and will not be complemented. Signals B(H) and D(H), on the other hand, need to be connected to inputs with bubbles so that they will be incompatible and complemented. We thus need a gate having two inputs with no bubbles and two inputs with bubbles. Such a gate is readily available in a gate array; it is shown in the following figure with the input and output signals properly labeled. Also shown is the function that is evaluated.

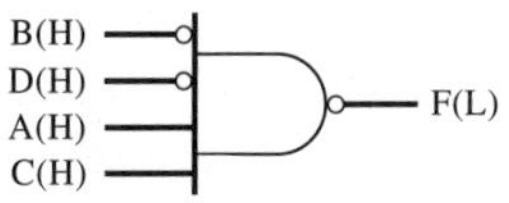

B(H)
D(H)
A(H)
C(H)
F(L)

NAND4B

$F(L) = (A\overline{B}C\overline{D})\ (L)$

We need to note especially two features of this design. First, the output of the gate is made compatible with the desired output signal assertion level, just as in the previous designs using discrete logic elements. Second, proper input signal compatibility or incompatibility at each input is determined by assigning a bubble or not assigning a bubble to the input, as needed. Thus, A(H) needed to be compatible, so the gate input was not assigned a bubble. B(H) needed to be incompatible, so the gate input was assigned a bubble. Similarly, C(H) was not assigned a bubble, and D(H) was. This procedure differs from that for a discrete gate, where compatibility and incompatibility are achieved by assigning a proper signal assertion level.

Example 4.3-4

Design a circuit to evaluate the logic function

$$F = A + \overline{B} + \overline{C} + D$$

Assume the input signal assertion levels are all low, and make the output assertion level high.

Evaluation of this function requires a 4-input gate that will perform an *or* operation. To make the output assertion level high, we need an *or* with no bubble on the output—an actual positive logic *or*. Now, how do we configure the input of this gate so that we can evaluate the required function with the input signal levels given? Signals A(L) and D(L) clearly need to be connected to inputs with bubbles so that they will be compatible and will not be complemented. Signals C(L) and B(L), on the other hand, need to be connected to inputs with no bubbles so that they will be incompatible and complemented. We thus need a gate having two inputs with no bubbles and two inputs with bubbles. Such a gate is readily available in a gate array; it is shown in the following figure with the input and output signals properly labeled, along with the function that is evaluated. As in Example 4.3-3, the output of the gates is made compatible with the output signal assertion level, and the input compatibilities are set, as needed, by placing bubbles or no bubbles on the inputs to the gate.

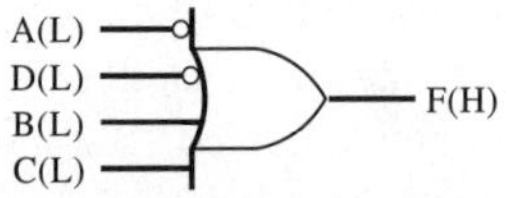

A(L)
D(L)
B(L)
C(L)
F(H)

OR4B

$F(H) = (A + \overline{B} + \overline{C} + D)\ (H)$

4.4 SIMULATION

In Section 4.3, simulation results were presented for the several circuits we designed. The use of software simulation is an important part of any modern design process. The normal steps in the design of a combinational (or sequential) logic circuit should be:

1. Paper design
2. Computer simulation
3. Redesign and resimulation as needed until a correct design is achieved
4. Prototype fabrication and testing

In fact, these should be the normal design steps for any electrical engineering design.

A number of software simulation packages for use with logic circuits are available. They vary in sophistication and price. Many can operate on a disk-based PC with extended memory. We'll illustrate the use of simulation software with a rather sophisticated system.

The first step in simulation using any software is to enter a description of the circuit into the computer. This is usually done using a schematic capture program, which allows the operator to draw the circuit directly on the computer monitor. This feature alone is a powerful aid to design. It frees the designer from the need to hand-draw circuit schematics. The circuit is displayed graphically on the computer display as it is drawn. The components that make up the circuit are stored in a library of components. For our purposes, the library consists of all possible gates. The gates can be retrieved from the library by name and placed anywhere on the circuit schematic, using a mouse. Once they're placed, the operator can connect their leads, again using a mouse to designate the beginning and the end of each connection and, if needed, the routing. Finally, the operator can label the inputs and outputs and use a mouse to position the labels. Labels are stored in the computer in such a way that they're associated with the actual inputs and outputs. They can therefore be used later, when simulations are run, as labels for signals. Schematic capture packages are versatile. Components can be moved either with or without their associated connections. Interconnections can be moved and removed. Labels can be moved, removed, and changed.

Once the circuit to be simulated is described to the simulation software, the input to the circuit must be described. Our sophisticated system allows the operator to enter the description of the input signals graphically. Consider the first circuit design of Example 4.3-1. Figure 4.4-1 is the simulation schematic for this circuit. It differs slightly from other schematics in the book because a different capture package has been used for it. This difference is evident in the style of the gates and the way

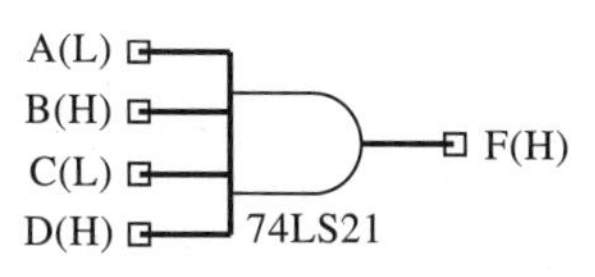

Figure 4.4-1

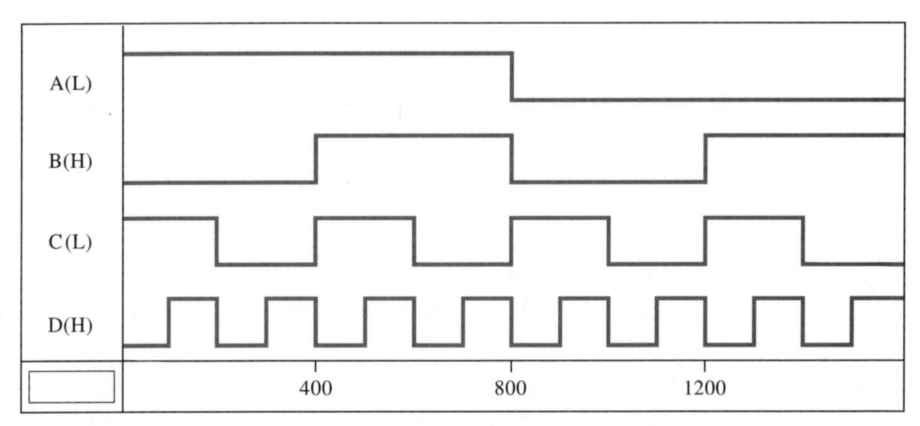

Figure 4.4-2 Simulation input signals.

pads are placed at each of the inputs and outputs. The software also uses a different way of devoting polarity [e.g., A-H rather than A(H)], but we have changed the notation to conform to that used throughout the text.

Graphical input signals for this circuit are shown in Figure 4.4-2. The operator can change the time scale and shape of the signals using the editing features of the software. The signals were produced so that the logic variables would count through the binary numbers 0000 to 1111, changing each 100 ns. The period for the fastest changing signal, D(H), was arbitrarily set to 200 ns, and the time axis was made 1600 ns long so that the variables could sequence through all of the binary numbers and thus all input conditions. The actual input to the simulation is specified in a command file. This command file is automatically generated from the graphical input wave forms.

Figures 4.4-3 through 4.4-7 are timing diagrams giving the results from a simulation, for 1600 ns, of the example circuit with the inputs of Figure 4.4-2. The

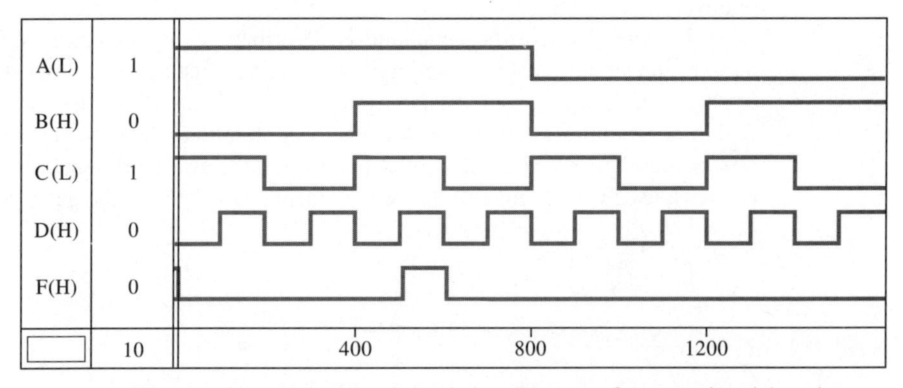

Figure 4.4-3 Timing of the beginning of simulation. Because of propagation delay, the output is undetermined before 10 ns.

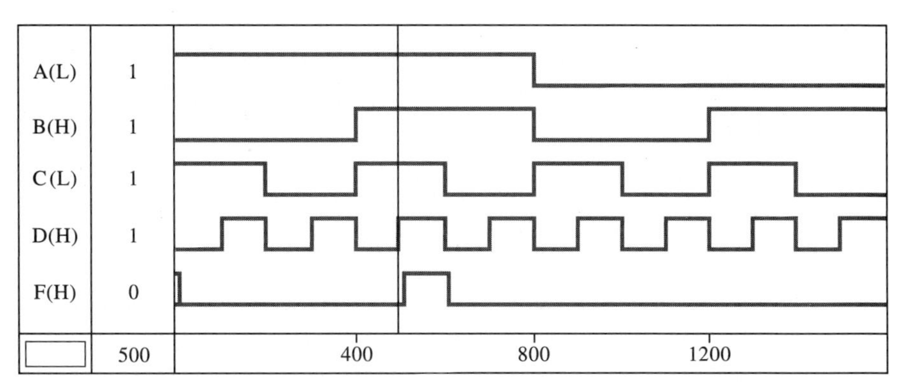

Figure 4.4-4 Timing of the input signal change which causes the start of the output pulse.

diagrams show how one of the interactive features of the simulation system can be used to measure the propagation delay through the circuit. The time of any feature of the timing diagram can be measured by positioning a marker at the feature to be measured. The time at the position of the marker is provided at the bottom of the second column in each diagram. (This column was deleted in the timing diagrams of Section 4.3.)

The first timing diagram (Fig. 4.4-3) shows how the output is delayed at the start of the simulation by the propagation delay of the circuit. For the first 10 ns the output from the simulation is undetermined. It is shown as both high and low. If the scale of the plot were large enough, the output would appear as cross-hatching to show that it is undetermined.

The second two timing diagrams (Figs. 4.4-4 and 4.4-5) show the time at which the input changes and the time at which the resulting output changes for the low-to-high transition of the output pulse. The difference is the propagation delay (TPLH = 8 ns). The final two timing diagrams (Figs. 4.4-6 and 4.4-7) show the

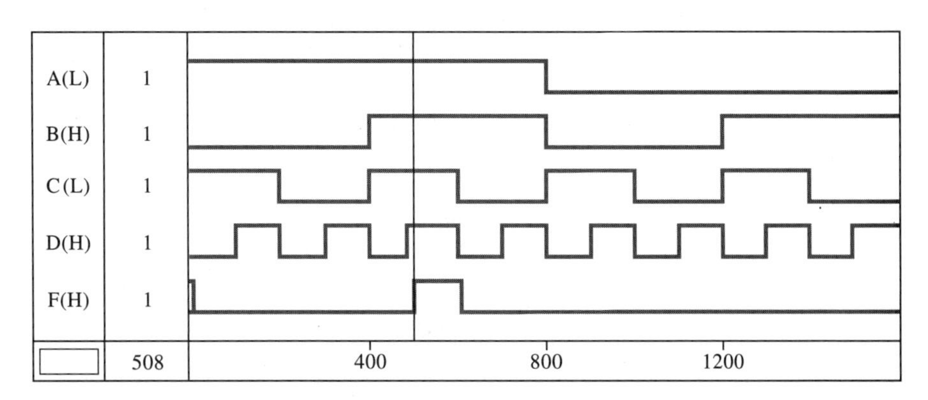

Figure 4.4-5 Timing of the start of the output pulse. It is delayed 8 ns for the change in the input which caused it (TPLH).

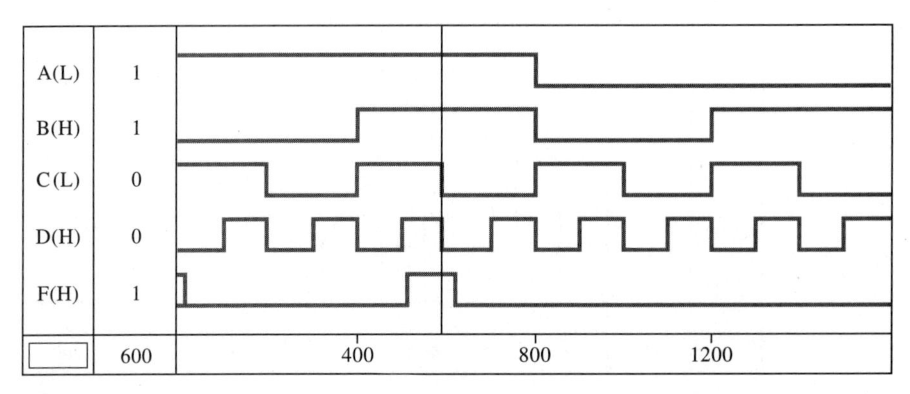

Figure 4.4-6 Timing of the input change which causes the end of the output pulse.

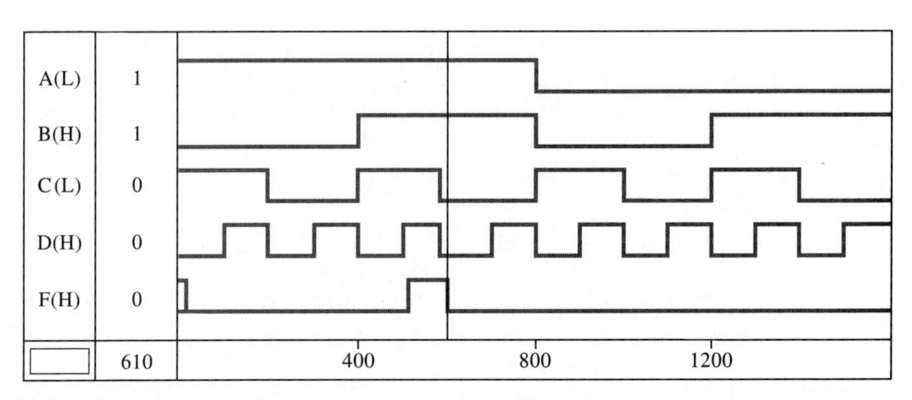

Figure 4.4-7 Timing of the end of the output pulse. It is delayed 10 ns for the change in the input which caused it (TPHL).

time at which the input changes and the time at which the resulting output changes for the high-to-low transition of the output pulse. The difference is the propagation delay (TPHL = 10 ns). Notice that TPLH is different from TPHL for this simulation.

We can see the propagation delay through two gates if we look at the first design of Example 4.3-2. Figure 4.4-8 is the simulation schematic for this design. Figures

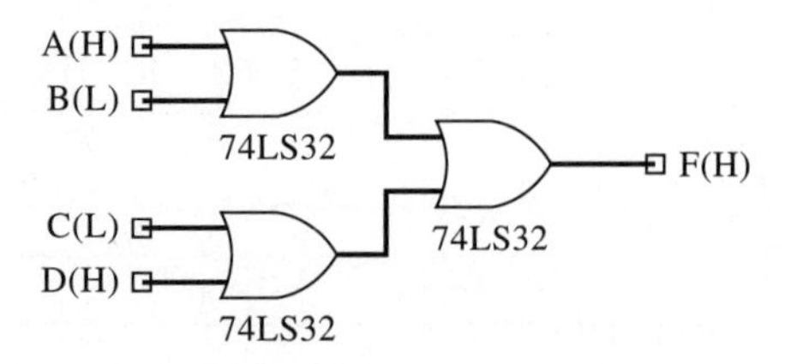

Figure 4.4-8

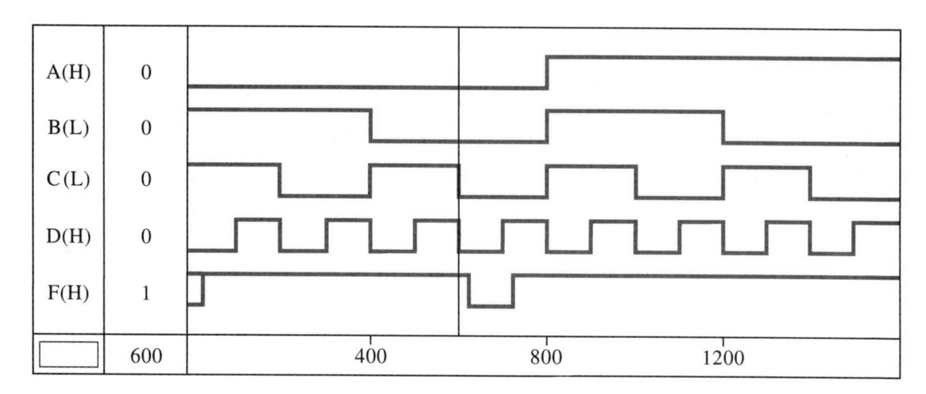

Figure 4.4-9 Timing for input change which causes a high to low transition in the output.

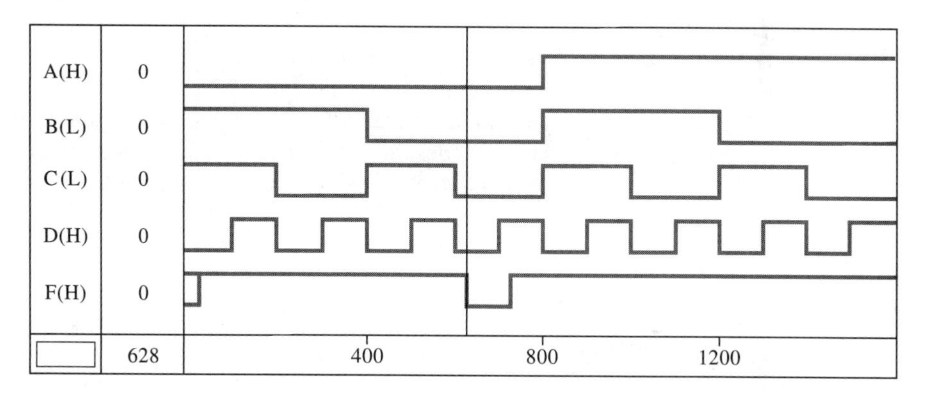

Figure 4.4-10 Timing of the high to low output transition. The propagation delay is 28 ns.

4.4-9 and 4.4-10 are timing diagrams showing the results of a simulation of the circuit. The time marked in Figure 4.4-9 is where the signal change that caused the high-to-low transition in the output occurred. The time marked in Figure 4.4-10 is where the output changed. The difference, which is the propagation delay through the circuit, is 28 ns. This is twice the delay for a single gate.

In this brief introduction to simulation we've discussed only a very few of the available features. It should be evident from the features presented, however, that simulation is a valuable part of design. We now look at the design of some more complicated circuits. With each circuit design we'll include a simulation.

4.5 INVERTERS

In our initial design examples with discrete logic elements, we were forced to use input signals with certain assertion levels. In some cases we may require an input

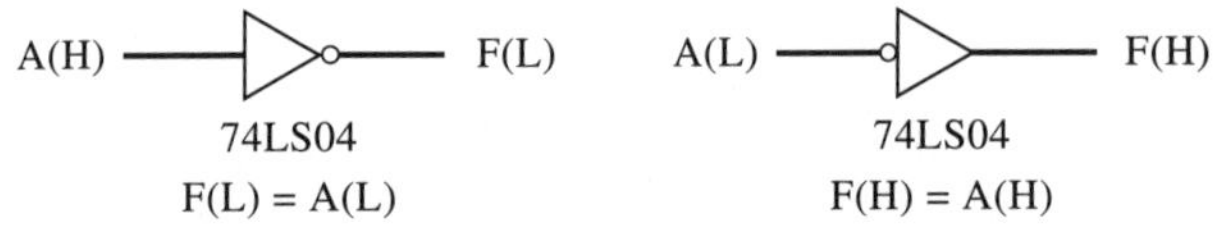

A(H) F(L)
74LS04
F(L) = A(L)

A(L) F(H)
74LS04
F(H) = A(H)

Figure 4.5-1 Inverter.

signal that is unavailable. We need a device to change the assertion level of a signal. Such a device is called an **inverter**. It's essentially a 1-input gate. A 1-input *and* or *or* gate would serve no useful logic purpose,[1] but a 1-input *nand* or *nor* gate serves as an inverter.

The circuit symbols for an inverter are shown in Figure 4.5-1 along with the logic function the inverter evaluates. The assertion levels are shown in the logic expressions as usual, but here the signal levels emphasize the assertion level change. The first symbol shows that if input signal A is asserted high, then output signal F is asserted low, but F is just A, so the output is A asserted low. The second symbol shows the opposite level change. We could use inverters to extend the designs of Examples 4.3-1 and 4.3-2 to the situation where the signal assertion levels we need are unavailable. Instead, we'll describe the design of circuits with more than single terms, deferring our introduction of the use of inverters until this description of more general designs.

4.6 LOGIC CIRCUIT DESIGN

Now that we've introduced gates and simple one-gate designs, we can easily extend the designs to logic expressions with more than single terms. A logic design almost always starts with a word description of some desired action. The action must then be formulated as a logic function. The formulation may require that a truth table be produced first, and the truth table may then be used to make a Karnaugh map.

Suppose the first part of a design has been done and we have found the Karnaugh map of Figure 4.6-1. The simplest logic circuit for evaluating a function is found by first reducing the function to its simplest form. For this map, the minimum SOP expression is

$$F = \overline{C}\,\overline{D} + \overline{A}\,\overline{B}\,\overline{C} + AB\overline{C} + A\overline{B}CD$$

We can write this expression as

$$F = W + X + Y + Z$$

[1]Actually, a 1-input noninverting amplifier performs the very useful function of a line driver where additional drive capacity is needed.

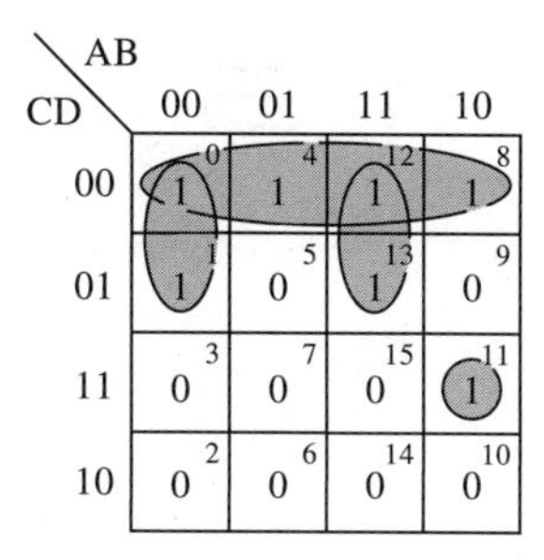

Figure 4.6-1

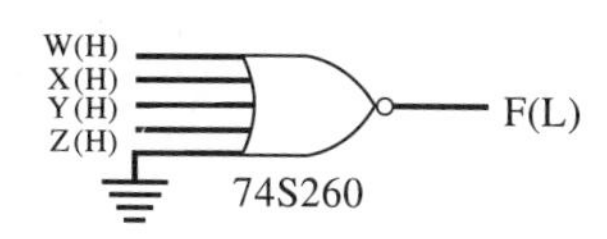

Figure 4.6-2 Output gate.

where

$$W = \overline{C}\,\overline{D}$$

$$X = \overline{A}\,\overline{B}\,\overline{C}$$

$$Y = AB\overline{C}$$

$$Z = A\overline{B}CD$$

The expression for F in terms of W, X, Y, and Z is a simple *or* of four terms. We saw in Section 4.3 how to design a circuit to evaluate this kind of function, but in order to complete the design we need to know the assertion level of the output, F. Normally this level is specified in the design requirements. Suppose the assertion level is specified as F(L). This level means that either a positive logic *nor* gate or a positive logic *and* gate can be used. Each will perform an *or* operation when the output is asserted low. Suppose we choose to use a 5-input *nor* gate (since a suitable 4-input *nor* gate isn't available). Figure 4.6-2 shows the design using this gate. Notice that all the input assertion levels are chosen high, compatible with the gate inputs, because none of the input variables is complemented. The fifth input is tied low, so it's never asserted.

To evaluate the W, X, Y, and Z functions, we require gates that perform *and* operations. For these gates, we'll choose the assertion levels of the input signals that represent A, B, C, and D to get a proper function evaluation, as we did in Section 4.3, but we'll always choose the assertion levels of the output signals to be compatible with the assertion levels for signals W, X, Y, and Z, already chosen. Positive logic *and* gates and positive logic *nor* gates have the proper output signal assertion levels (H) when they're used to perform their *and* operations. Figure 4.6-3 shows how the functions W, X, Y, and Z can be evaluated using four positive logic *and* gates. In these circuits, we've chosen the assertion levels of signals A, B, C, and D to be incompatible with the input gate symbol if the variable they represent is complemented in the logic expression to be evaluated. An incompatibility produces a complement.

Now, to finish our logic circuit we need only put the input and output circuits together as shown in Figure 4.6-4. In this circuit we need input signals A(H), A(L),

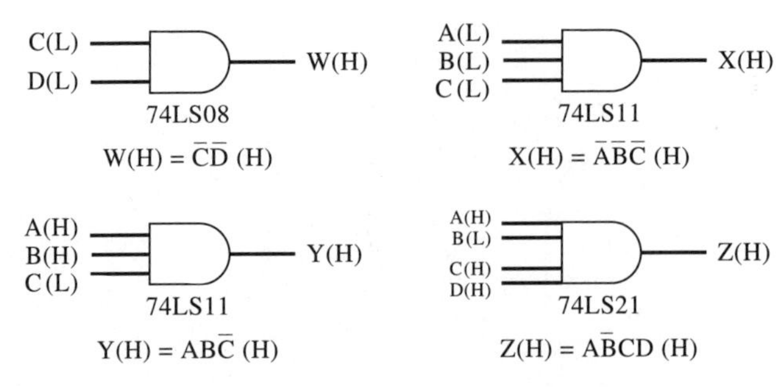

Figure 4.6-3 Input gates.

B(H), B(L), C(H), C(L), D(H), and D(L). Usually, both high and low assertion level signals for the variables are not available as input signals to a logic circuit. We must therefore use assertion level changers, or inverters, to produce the signals with the assertion levels we need from the input signals to the circuit. If we assume that the input signals available to the logic circuit are all asserted high and use inverters to produce signals with low assertion levels, the logic circuit becomes as shown in Figure 4.6-5.

In this circuit, we have not shown the connections between the inputs and inverted inputs on the left and the inputs to the gates. This is a common practice. When the connections are shown, they are often messy, and it's clearer to note that all the A(H) inputs to the gates are connected to A(H) at the inverter bank, all the A(L) inputs to the gates are connected to A(L) at the inverter bank, and so on. The set of all input signals and inverted input signals, shown on the left in this circuit, is called a *rail*. Some of these inputs may not be needed as inputs to the gates, but

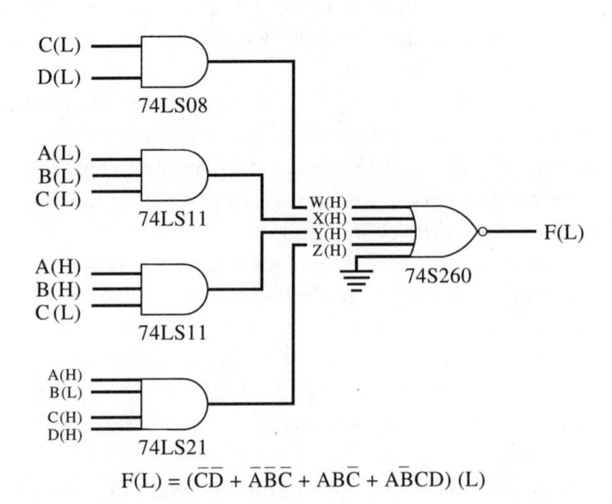

Figure 4.6-4 Complete design.

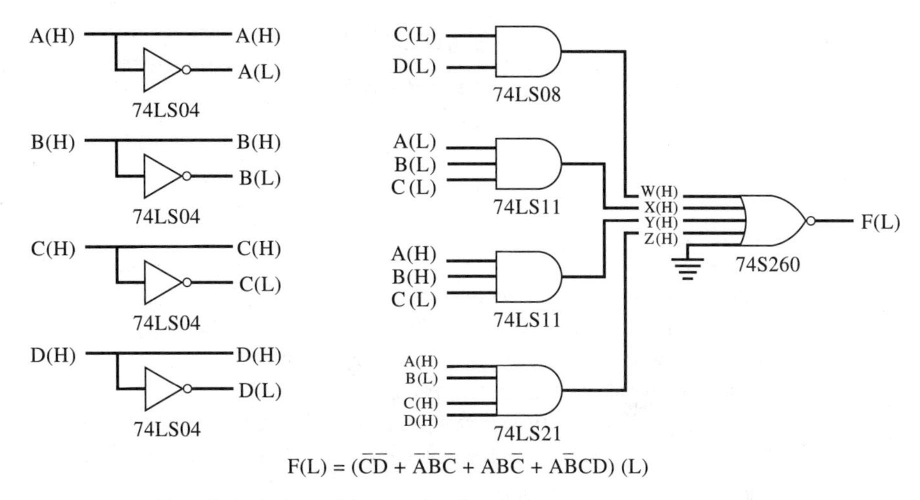

Figure 4.6-5 Complete design with assertion level changers.

inverters are packaged six to a chip, so it's convenient to produce signals with both assertion levels for each variable. This may be especially useful in developmental circuits, where modifications may be necessary.

To see just what this circuit does and to make sure it is performing correctly, it's useful to simulate the circuit's operation for a set of test inputs. Figure 4.6-6 is a timing diagram produced from a simulation of the circuit. The inputs were chosen so that all possible combinations of inputs were tested by letting the input logic variables A, B, C, and D count through the binary numbers 0000 through 1111.

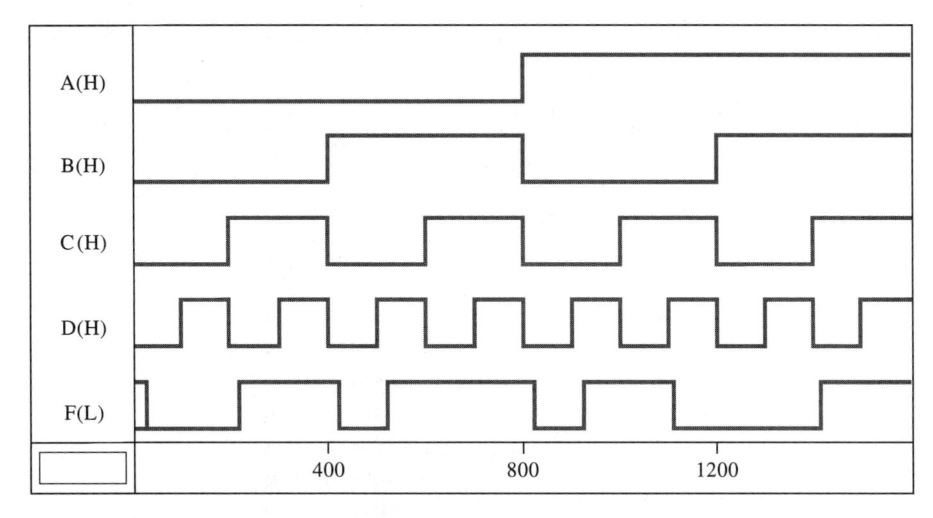

Figure 4.6-6 Timing diagram results for simulation.

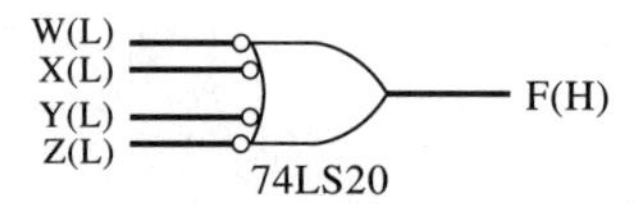

Figure 4.6-7 Output gate.

This circuit is one of many we can design using this technique, all of which evaluate the same logic function. Suppose the output assertion level needed is F(H), but we still have input signals with assertion levels A(H), B(H), C(H), and D(H). We can choose the output gate to be a positive logic *nand* gate or a positive logic *or* gate. These gates will perform an *or* operation when the output signal assertion level is high. Figure 4.6-7 shows the output circuit if we choose the *nand* gate. With this choice of output gate, the assertion levels for the intermediate signals must be W(L), X(L), Y(L), and Z(L).

Each of the input gates, which evaluate the functions W, X, Y, and Z, must perform an *and* operation and must be output compatible with the assertion levels already chosen for these signals (L). We can choose the input gates to be the same type as the output gate, positive logic *nand* gates, since this type of gate will perform an *and* operation when the output signal is asserted low. Once this choice is made, the circuit design is just the same as the preceding. Incompatibilities are chosen at the input gates to produce complements as needed. Figure 4.6-8 shows the design produced with this choice of gates. An interesting feature of this design is that it uses only one type of gate. Figure 4.6-9 is a timing diagram produced from a simulation of this circuit.

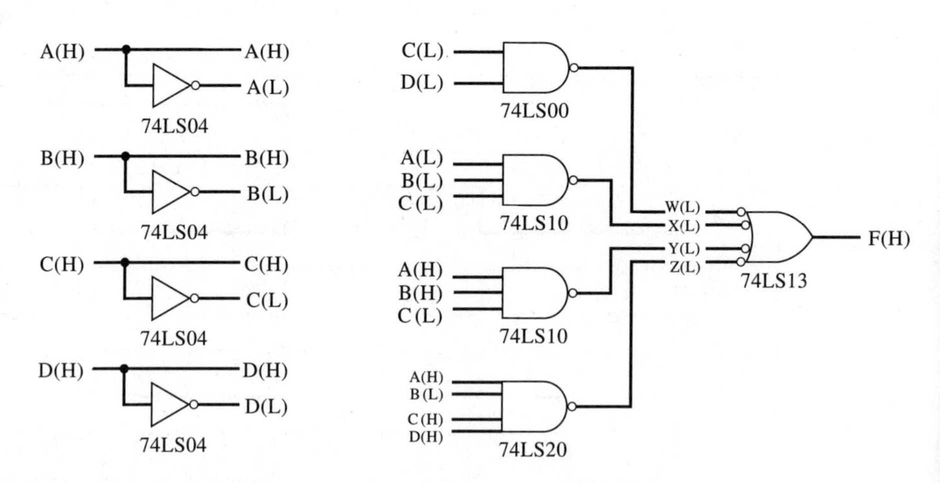

$$F(H) = (\overline{C}\overline{D} + \overline{A}\overline{B}\overline{C} + AB\overline{C} + A\overline{B}CD)\ (H)$$

Figure 4.6-8 An alternate design.

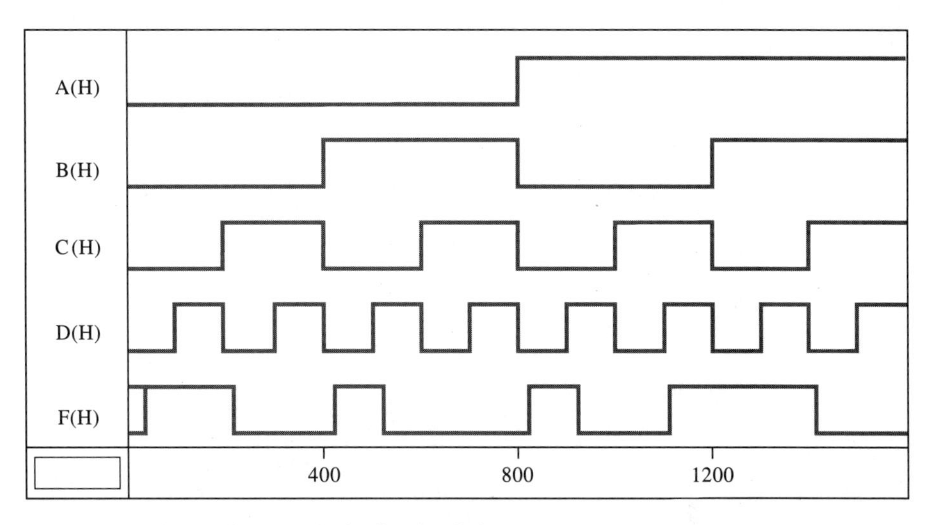

Figure 4.6-9 Timing diagram results for simulation.

Before we design any more logic circuits, let's summarize some of the features of our design. Keep in mind that the design is based on an SOP expression of the logic function. For this kind of design:

1. Two levels of gates are required—input gates and an output gate.
2. The output gate always performs an *or* operation.
3. The input gates always perform *and* operations.
4. We always choose the intermediate signal assertion levels to be compatible with the output gate.
5. Complements in the logic function to be evaluated are produced by incompatibilities at the inputs of the input gates.

Notice that feature 2 and the choice of output signal assertion level determine the type of gate we can use for the output. If the output signal assertion level is low (as in our first example), then the output gate must be either an *and* gate performing an *or* operation or a *nor* gate performing an *or* operation. If the output signal assertion level is high (as in our second example), then the output must be either an *or* gate performing an *or* operation or a *nand* gate performing an *or* operation. We'll find that features 2 and 3 are true only for SOP designs, but features 1, 4, and 5 are true for both SOP and POS designs.

Before considering other SOP design examples, remember that there are still a number of alternative SOP designs for the current example. If we assume that the input signals to the circuit are asserted high, as in the preceding examples, and the output signal is asserted low, then, using a positive logic *and* gate as the output gate and positive logic *nand* gates as the input gates, the design is as shown in Figure 4.6-10.

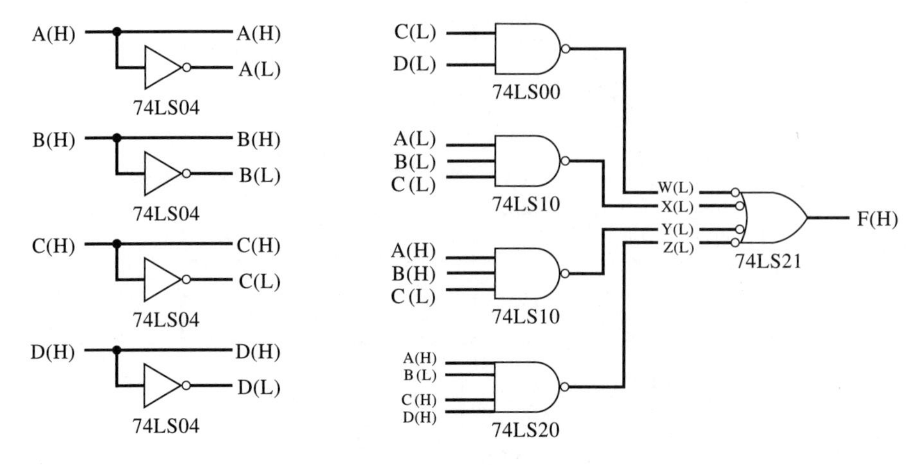

$$F(H) = (\overline{C}\,\overline{D} + \overline{A}\,\overline{B}\,\overline{C} + AB\overline{C} + A\overline{B}CD)\ (H)$$

Figure 4.6-10 Another alternate design.

If we assume the same input and output assertion levels, another design uses positive logic *nor* gates for both the output gate and the input gates, as Figure 4.6-11 shows. Notice that in this design all the signal assertion levels at the input gates are changed from the preceding design. This doesn't mean that the assertion levels of the signals input to the circuit are changed. It means only that the connections to the rail are changed. These last two designs will have the same timing diagrams as

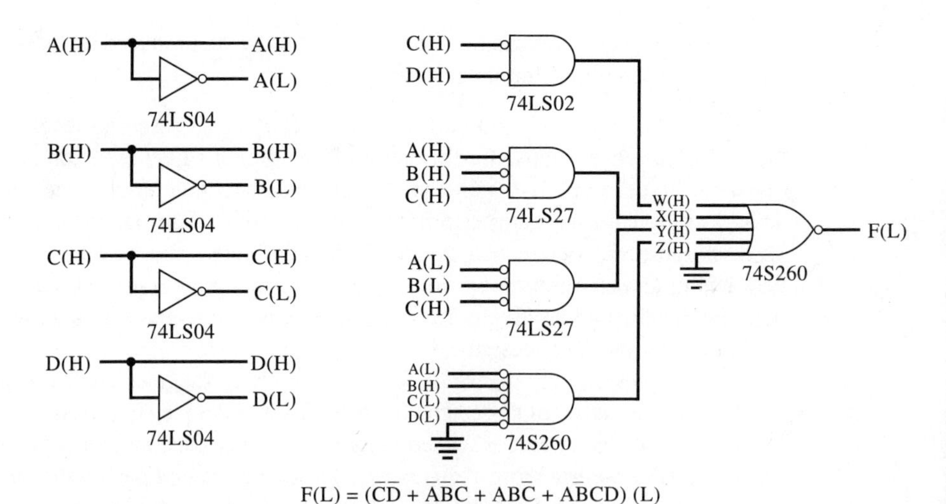

$$F(L) = (\overline{C}\,\overline{D} + \overline{A}\,\overline{B}\,\overline{C} + AB\overline{C} + A\overline{B}CD)\ (L)$$

Figure 4.6-11 A final alternate design.

the first design we considered in this section, because the input and output signal assertion levels are the same.

As we can see from the preceding example, the design process is not exclusive. Many different logic circuits will evaluate the required logic function. All of these circuits are correct. Some may be better than others, based on considerations in addition to the correct evaluation of the logic expression. Some of these additional considerations might be propagation delay through the circuit, the number of parts required, complexity of the circuit board (if the circuit is to be made on a circuit board), reliability, and ease of testing. The fact that many different possibilities exist is positive; it allows us to find the best possible design.

Example 4.6-1

Suppose we have completed the first part of a design, including the word description and the resultant truth table, and we have arrived at a logic function:

$$F = ABC + \overline{A}CD + A\overline{B}\,\overline{C}\,\overline{D}$$

Design a logic circuit to evaluate this logic function. Assume that the assertion levels of the inputs to the circuit are A(L), B(L), C(H), and D(L), and make the assertion level of the output from the circuit F(L). Use only *nand* or *nor* gates.

As the first step in the design, we make a Karnaugh map and simplify the function. Notice that this function doesn't simplify.

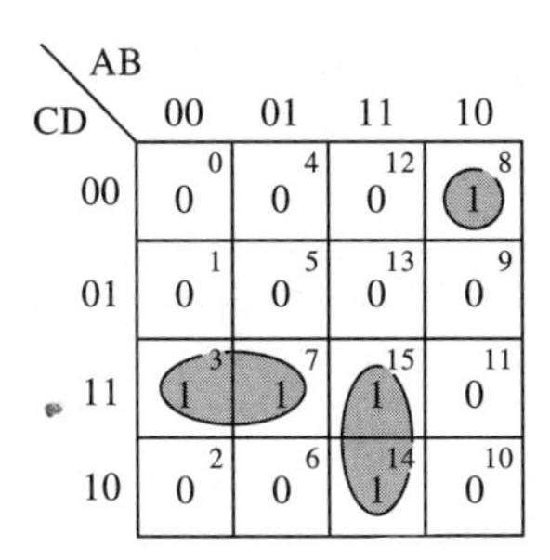

Now, because the output assertion level is low, the output gate must be either a positive logic *and* gate or a positive logic *nor* gate performing an *or* operation. But the specifications of the problem force us to use a positive logic *nor* gate. This means that the input gates must also be *nor*s (performing *and* operations) to make their outputs compatible with the inputs of the output gate. Because the assertion levels of the input signals to the circuit are specified, and both assertion levels for each of the signals are needed at the gate inputs, we must add a complete rail. The final design follows. Notice that the input is made incompatible when the variable in the logic expression being evaluated is complemented. A timing diagram showing the results from a simulation follows the circuit.

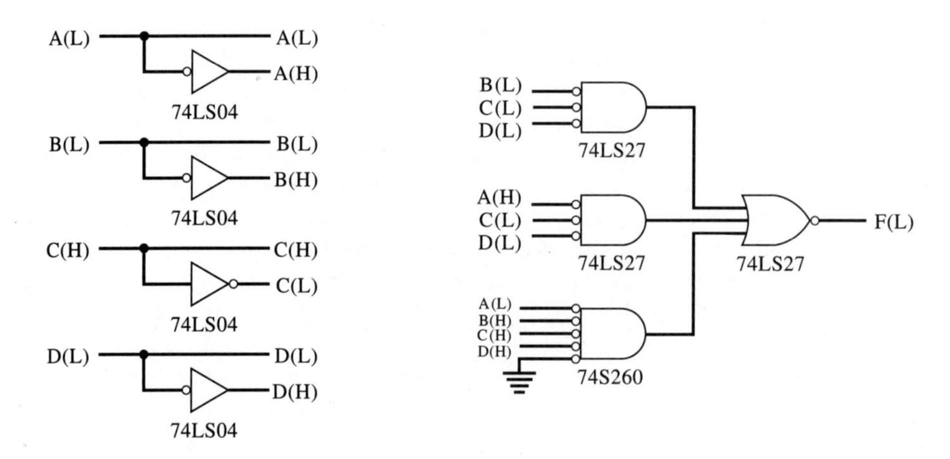

$$F(L) = (BCD + \bar{A}CD + A\bar{B}\bar{C}\bar{D})\,(L)$$

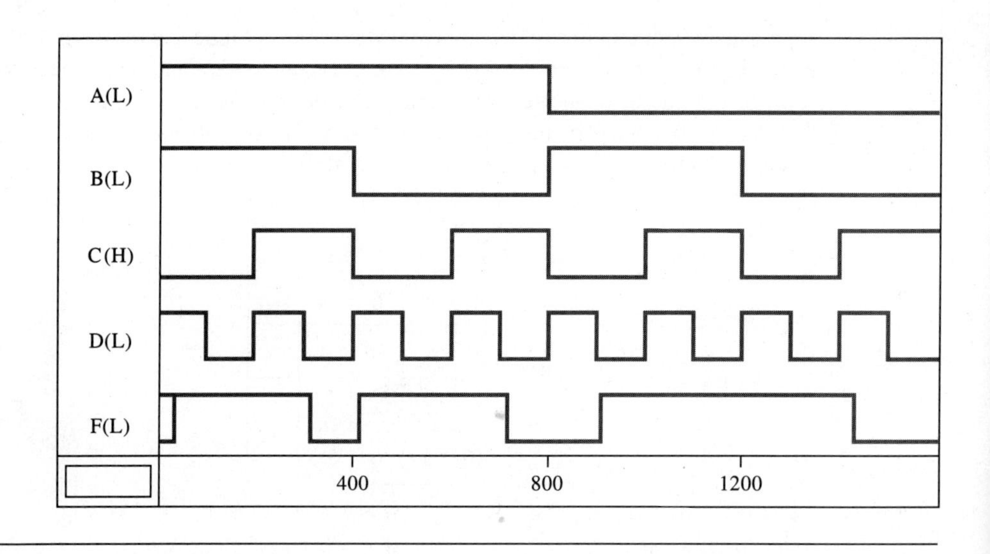

Example 4.6-2

Again assume that the first steps in the design of a logic circuit have been completed, and the following VEM results. Complete the design. Assume input assertion levels A(L), B(L), C(H), D(L), and E(H), and make the output assertion level F(H). Use only *nand* or *nor* gates.

CD \ AB	00	01	11	10
00	(0) E	(4) 1	(12) 1	(8) 1
01	(1) 0	(5) $\overline{E}$	(13) 0	(9) 0
11	(3) 0	(7) 1	(15) $x\overline{E}$	(11) 0
10	(2) 0	(6) 0	(14) $x\overline{E}$	(10) 0

We first simplify the function using the following VEM.

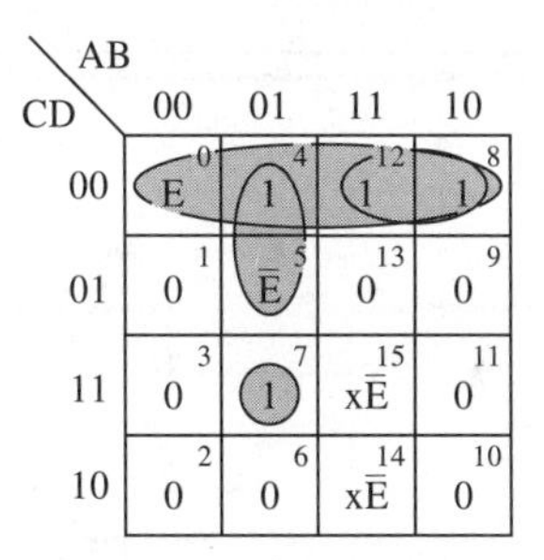

$$F = \overline{C}\,\overline{D}E + \overline{A}B\overline{C}\,\overline{E} + A\overline{C}\,\overline{D} + \overline{A}BCD$$

Now, because the output assertion level is high, the output gate must be either a positive logic *or* gate or a positive logic *nand* gate performing an *or* operation, but the specifications of the problem force us to use a positive logic *nand* gate. This means that the input gates must also be *nand*s (performing *and* operations) to make their outputs compatible with the inputs of the output gate. We add a full rail to produce both assertion levels of all the inputs. The final design follows. Notice how the inputs are made incompatible when the variable in the logic expression being evaluated is complemented. The only part of the rail we don't use is B(L). It's one of the given inputs, however, so we could not simplify the rail even if we wanted to. Following the circuit are the timing diagram results from a simulation.

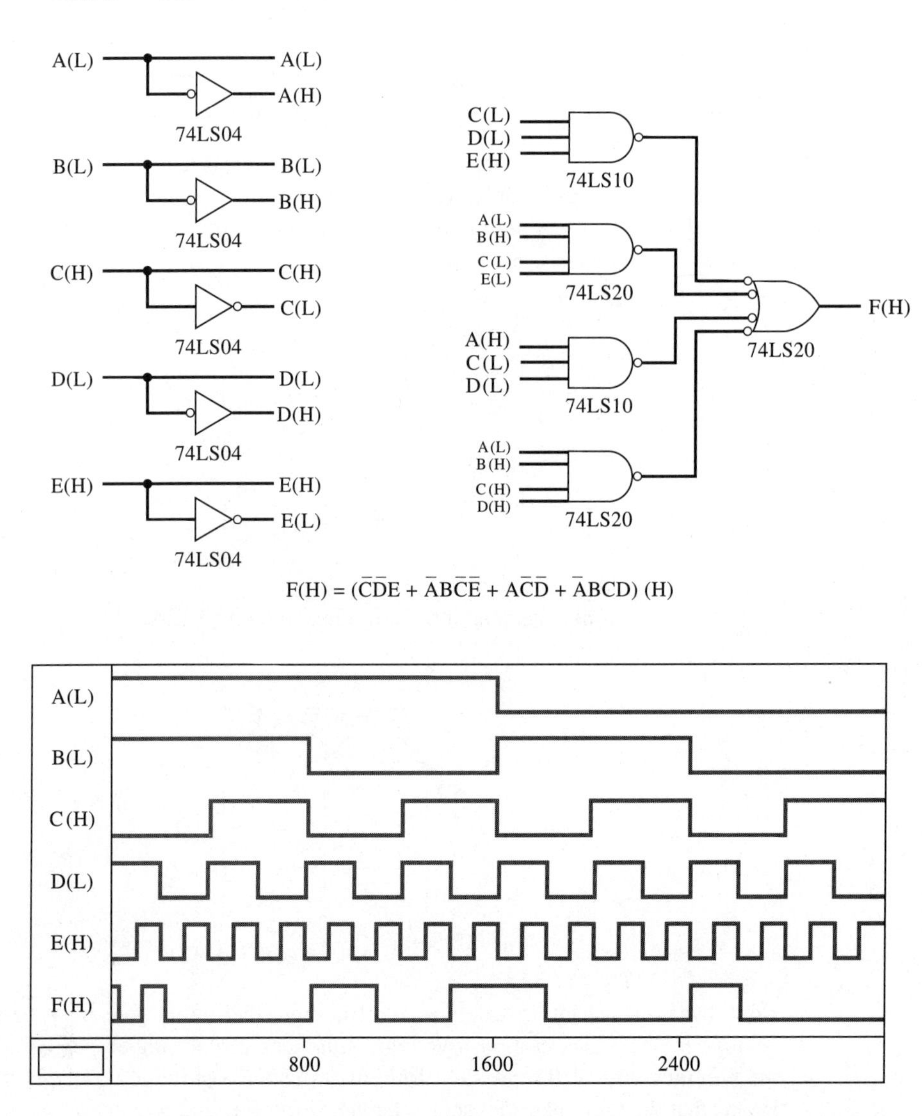

All of the foregoing circuit designs are based on the simplest SOP expression for the logic function. We can also design based on the POS expression; we'll return to our first example and do just that. Once the logic function to be evaluated has been formulated from the word description, the next step, in either an SOP or a POS design, is to simplify the function. We can do this most conveniently with a Karnaugh map, as shown in Figure 4.6-12. From the map, the simplest POS expression for the function is

$$F = (B + \overline{C})(\overline{A} + \overline{C})(\overline{C} + D)(\overline{A} + B + \overline{D})(A + \overline{B} + C + \overline{D})$$

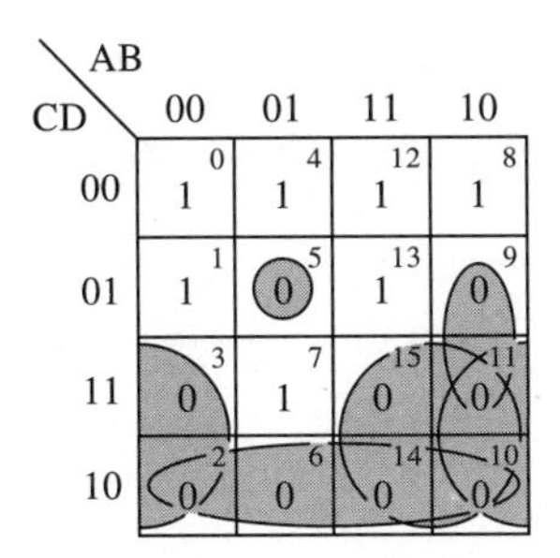

Figure 4.6-12

We can write this expression in the form

$$F = VWXYZ$$

where

$$V = B + \overline{C}$$

$$W = \overline{A} + \overline{C}$$

$$X = \overline{C} + D$$

$$Y = \overline{A} + B + \overline{D}$$

and

$$Z = A + \overline{B} + C + \overline{D}$$

In this form, we see that F is just the *and* of V, W, X, Y, and Z. If the assertion level for the output signal that represents F is low, F(L), then, as we saw in Section 4.3, this operation can be performed using either a positive logic *or* gate or a positive logic *nand* gate. The assertion level of the output is normally fixed in a design. There is a small problem: neither a 5-input, positive logic *or* gate nor a 5-input, positive logic *nand* gate is available as a standard TTL gate. Fortunately, we can use an 8-input positive logic *nand* gate. Figure 4.6-13 shows the output circuit design for this choice. Here we have 3 extra inputs, and we tie them to +5 volts so they will always be asserted.

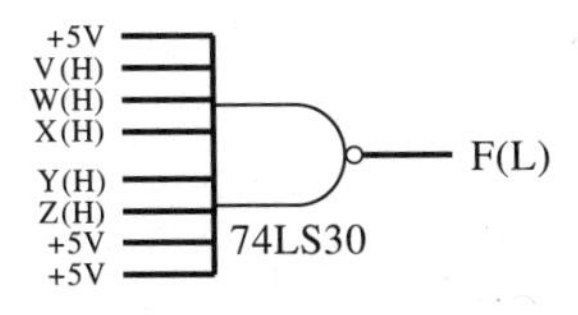

Figure 4.6-13 Output.

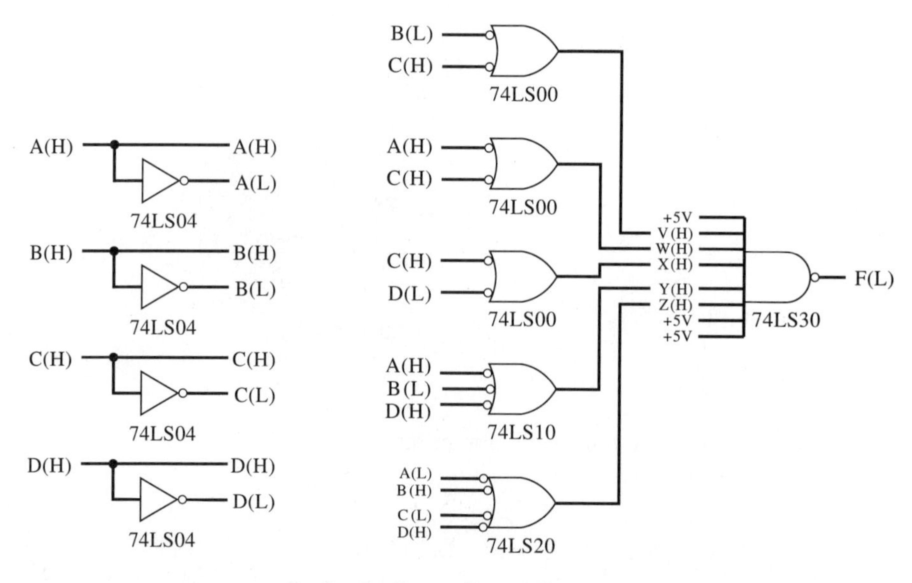

$$F(L) = ((B + \bar{C})\,(\bar{A} + \bar{C})\,(\bar{C} + D)\,(\bar{A} + B + \bar{D})\,(A + \bar{B} + C + \bar{D}))\,(L)$$

Figure 4.6-14 A POS based design.

The functions V, W, X, Y, and Z can be evaluated using positive logic *nand* gates, which perform *or* operations when their output signals are asserted high. We must choose either positive logic *or* gates or positive logic *nand* gates, and we choose positive logic *nand* gates because there are no standard TTL 4-input positive logic *or* gates. Figure 4.6-14 shows the complete logic circuit design. For this circuit we've assumed that the assertion levels for the signals furnished at the inputs are all high and produced a rail using inverters.

Following are the important features of a POS based design:

1. Two levels of gates are required—input gates and an output gate.
2. The output gate always performs an *and* operation.
3. The input gates always perform *or* operations.
4. We always choose the intermediate signal assertion levels to be compatible with the output gate.
5. Complements in the logic function to be evaluated are produced by incompatibilities at the inputs of the input gates.

The POS design is much like the SOP design except for the interchange of gate operations between the output and input gates. As with the SOP design, feature 2 and the output assertion determine the type of output gate that can be used. If the output assertion level is low (as in Example 4.6-1), then the output gate must be either a positive logic *or* gate performing an *and* operation or a positive logic *nand* gate performing an *and* operation. If the output assertion level is high (as in Example

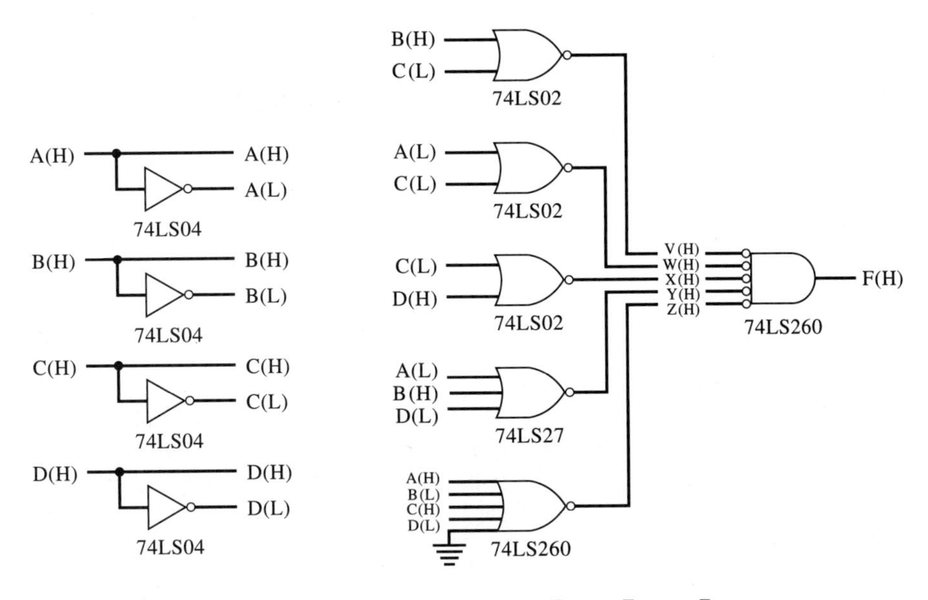

$$F(L) = ((B + \overline{C})\,(\overline{A} + \overline{C})\,(\overline{C} + D)\,(\overline{A} + B + \overline{D})\,(A + \overline{B} + C + \overline{D}))\,(H)$$

Figure 4.6-15 Another POS based design.

4.6-2), then the output must be either a positive logic *and* gate performing an *and* operation or a positive logic *nor* gate performing an *and* operation.

To finish our POS logic circuit design, we make a design with F(H) and all positive logic *nor* gates, shown in Figure 4.6-15. We again assume inputs A(H), B(H), C(H), and D(H). Simulation results are not shown for the POS designs because, except for small timing differences, they're the same as those for the SOP designs, with the same assertion levels. Now let's look at some other POS based circuit designs.

Example 4.6-3

Make a POS based logic circuit design to evaluate the logic function of Example 4.6-1,

$$F = ABC + \overline{A}CD + A\overline{B}\,\overline{C}\,\overline{D}$$

Input assertion levels are A(L), B(L), C(H), and D(L). The output assertion level must be F(L). Use only *nand* or *nor* gates.

As the first step in the design, we make a Karnaugh map and simplify the function as a POS. Because the output assertion level is low, the output gate must be either a positive logic *or* gate or a positive logic *nand* gate performing an *and* operation. In this design we're restricted to the *nand* gate.

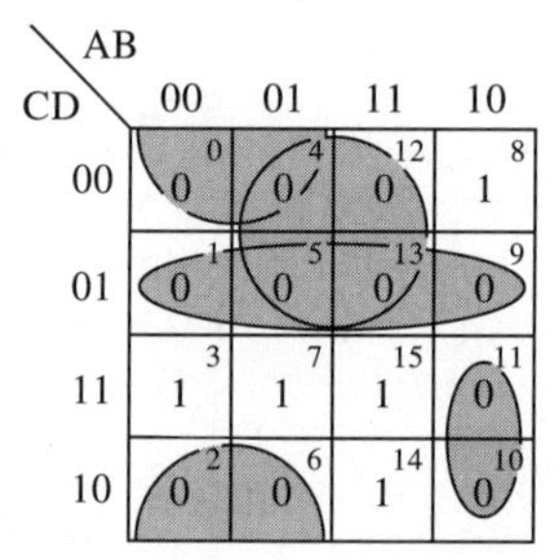

$F = (A + D)(\bar{B} + C)(C + \bar{D})(\bar{A} + B + \bar{C})$

With this output gate choice, the input gates must be *nand* gates performing an *or* operation to make the intermediate signal compatible and satisfy the constraint that the gates are either *nand* or *nor*. Following is the complete design with a full rail. If we examine the inputs to the input gates, we can verify that a full rail is actually needed. Simulation results are not given for the circuit because they are the same as those for the SOP design.

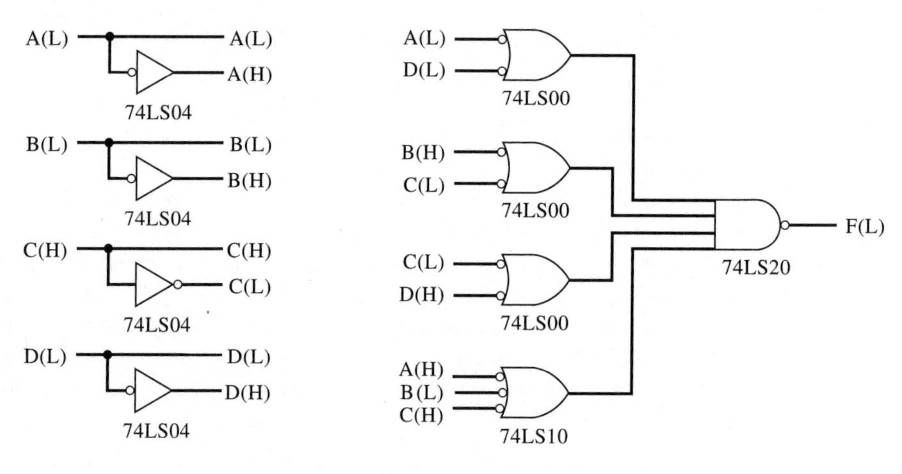

$F(L) = ((A + D)(\bar{B} + C)(C + \bar{D})(\bar{A} + B + \bar{C}))(L)$

Example 4.6-4

Make a POS based logic circuit design to evaluate the logic function of Example 4.6-2. Input assertion levels are A(L), B(L), C(H), D(L), and E(H). The output assertion level must be F(H). Use only *nand* or *nor* gates.

CD \ AB	00	01	11	10
00	E (0)	1 (4)	1 (12)	1 (8)
01	0 (1)	$\bar{E}$ (5)	$\bar{E}$ (13)	0 (9)
11	0 (3)	1 (7)	x$\bar{E}$ (15)	0 (11)
10	0 (2)	0 (6)	x$\bar{E}$ (14)	0 (10)

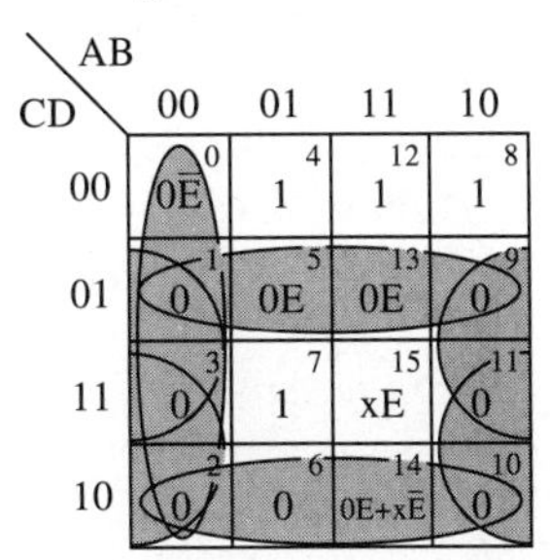

$F = (A+B+E)(C+\bar{D}+\bar{E})(B+\bar{D})(B+C)(\bar{C}+D)$

The logic function for this problem is defined in a VEM. To simplify the function as a POS, we need to make a VEM showing the zero entered variables. The VEM on the left is the original, and the one on the right shows the zero entered variables and the POS simplification of the logic function.

Because the output assertion level is high, the output gate can be either a positive logic *and* gate or a positive logic *nor* gate performing an *and* operation. The problem specifies a *nor*. With this choice of output gate, the input gates must be *nor* gates also, so that the intermediate signals will be compatible with the output gates. The final design, with a full rail, follows.

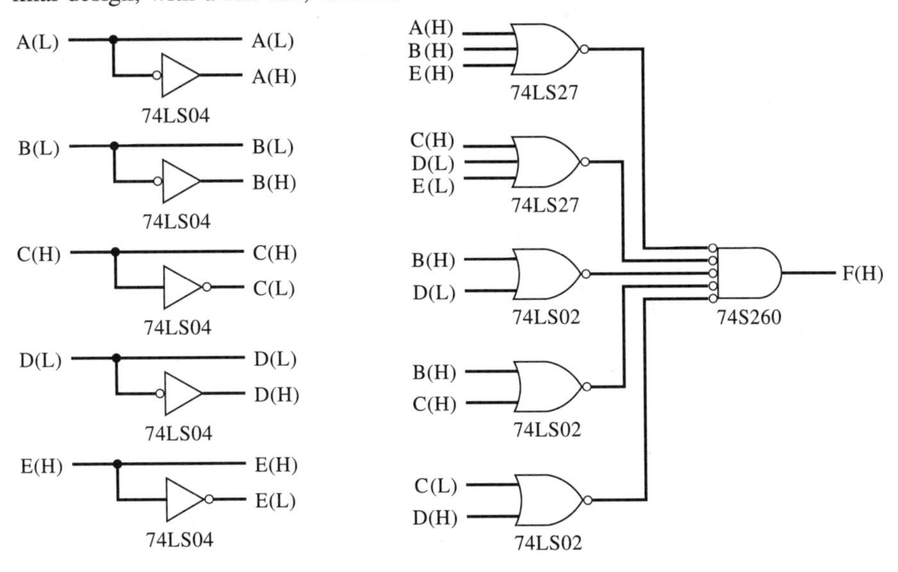

$$F(H) = ((A + B + E)(C + \bar{D} + \bar{E})(B + \bar{D})(B + C)(\bar{C} + D))(H)$$

In the designs for both SOP and POS based circuits to evaluate logic functions, we've often specified that the gates should be either positive logic *nand* gates or positive logic *nor* gates. These are good choices because a good variety of TTL *nand* and *nor* gates is available. This is especially true for *nand* gates. As we've seen in the examples, when we design with *nand* and *nor* gates, only one type of gate is needed for the entire circuit, and it is determined by the output assertion level and whether the circuit is based on an SOP simplification or a POS simplification. If the output assertion level is high, an SOP based circuit using only *nand* gates or a POS circuit using only *nor* gates can be designed. If the output assertion level is low, an SOP based circuit using only *nor* gates or a POS circuit using only *nand* gates can be designed. In some cases it's advantageous to use positive logic *and* gates in logic circuit design, but positive logic *or* gates should be avoided because only one TTL *or* gate is available, a 2-input *or* gate.

4.7 A REAL PROBLEM

Let's look at a problem that describes a physical situation—a very simple one. Suppose a loading dock for large trucks is shaped as shown in Figure 4.7-1. The dock is obviously blind; that is, the trucks entering cannot see into the dock area. We want a signal to warn trucks when not to enter the dock area. The constraints are that:

1. Only three trucks are allowed in the dock area at one time.
2. If slots 1 and 2 are filled in the dock area, other trucks may not enter.

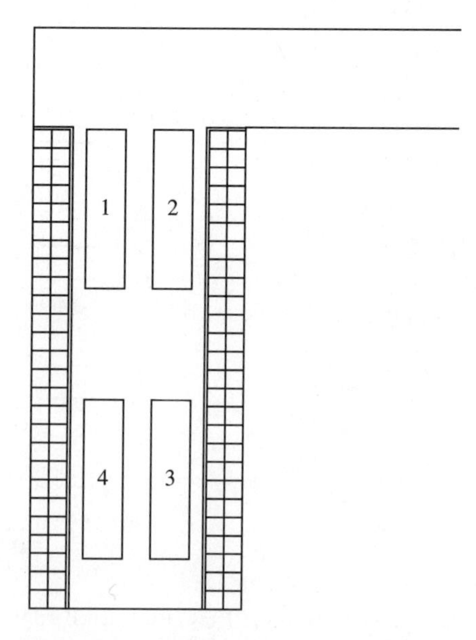

Figure 4.7-1 Loading dock.

When a truck is in one of the parking slots, a sensor in that slot produces a low output signal. A "no parking" light at the entrance to the lot should light when the dock area is full. A high output signal is required to light the "no parking" light.

The first step in solving a problem is to describe the problem in words. We have done that. The second step is to formulate the problem as a logic function and simplify the function. The third step is to design a logic circuit that will evaluate the logic function with the input and output assertion levels required. The fourth step is to simulate the logic circuit we have designed and determine whether it does indeed perform as expected.

To formulate the problem as a logic function, we use a truth table, which follows. The output asserted condition is "no parking." The input asserted condition is a truck in a slot.

T1	T2	T3	T4	NP	
0	0	0	0	0	
0	0	0	1	0	
0	0	1	0	0	
0	0	1	1	0	
0	1	0	0	0	
0	1	0	1	0	
0	1	1	0	0	
0	1	1	1	1	
1	0	0	0	0	
1	0	0	1	0	
1	0	1	0	0	
1	0	1	1	1	
1	1	0	0	1	
1	1	0	1	1	
1	1	1	0	1	
1	1	1	1	x	Not allowed by the specification

To simplify the function described by this truth table, we use a Karnaugh map (Fig. 4.7-2). With the covering shown, the simplified SOP function for this map is

$$NP = T2 \cdot T3 \cdot T4 + T1 \cdot T3 \cdot T4 + T1 \cdot T2$$

Many different logic circuits can be used to evaluate this function. Figure 4.7-3 shows the design for a circuit when we choose the input and output gates to be positive logic *nand* gates. We choose *nand* gates because the output is asserted low. If we could use positive logic *or* gates to perform the *and* for the input gates in this

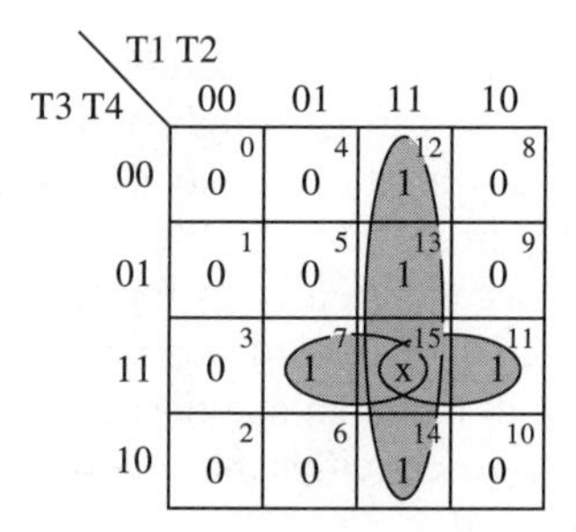

Figure 4.7-2

circuit, we would not need inverters. Unfortunately, if we use standard TTL devices, there are no 3-input positive logic *or* gates available.

To design a POS circuit that will evaluate this function, we first need to do a POS simplification of the function, using a Karnaugh map (Fig. 4.7-4). With the covering shown on the map, the simplified POS function is

$$NP = (T1 + T3)(T2 + T3)(T1 + T4)(T2 + T4)(T1 + T2)$$

and Figure 4.7-5 shows the logic circuit using positive logic *nor* gates for both input and output. The high output assertion level dictates *nor* gates. As we can see, this circuit can be designed so that no inverters are needed. A simulation of either of these circuits results in a timing diagram like that in Figure 4.7-6 except for small differences in propagation delays. Notice that the output is asserted only when three

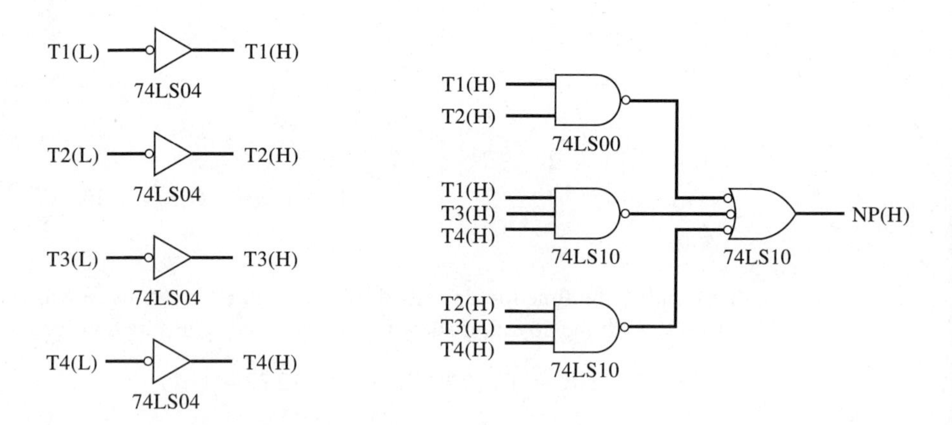

Figure 4.7-3 An SOP based design.

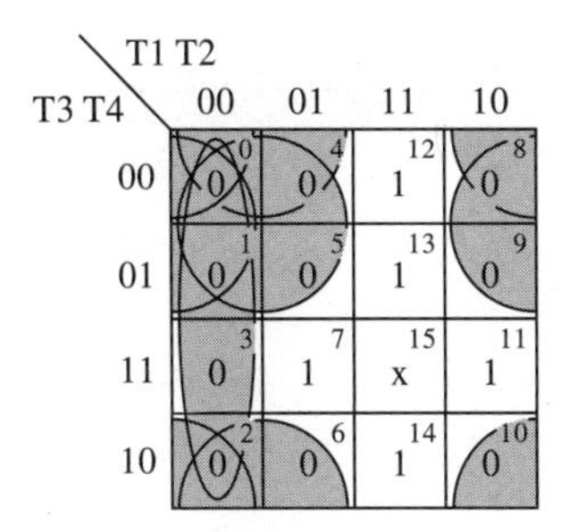

Figure 4.7-4

trucks are in the dock area or trucks 1 and 2 are in the dock area. It is also asserted when all four trucks are in the dock area, but this condition should not occur.

It is interesting to compare the two designs on the basis of the number of discrete logic devices used for each design. The SOP based design uses

3	3-input *nand* (74LS10) gates
1	2-input *nand* (74LS00) gate
4	inverters (74LS04)

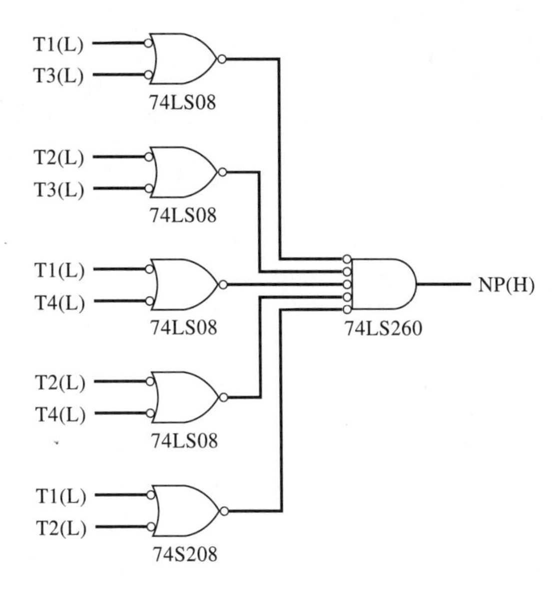

NP(H) = ((T1+T3) (T2+T3) (T1+T4) (T2+T4) (T1+T2)) (H)

Figure 4.7-5 A POS based design.

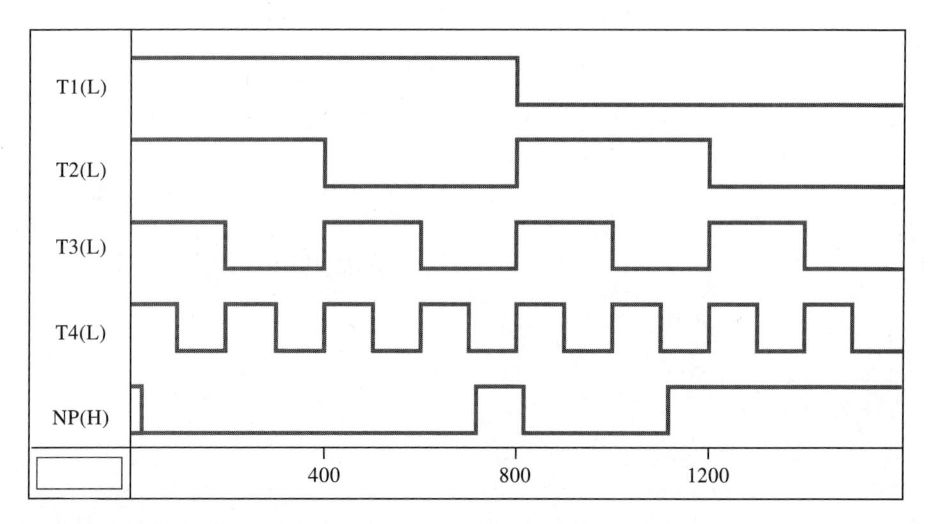

Figure 4.7-6 Simulation results.

There are three 3-input *nand* gates in a 74LS10, four 2-input *nand* gates in a 74LS00, and six inverters in a 74LS04. We therefore require

1	74LS10
1	74LS00
1	74LS04

Three of the 2-input *nand* gates in the 74LS00 and two of the inverters in the 74LS04 are not used. They are available for use in another circuit on the same board if they are needed.

The POS based design uses

5	2-input *and* (74LS08) gates
1	5-input *nor* (74S260) gate

There are four 2-input *and* gates in a 74LS08 and two 5-input *nor* gates in a 74S260, so we require

2	74LS08s
1	74S260

Three of the 2-input *and* gates in the 74LS08s are not used, and one of the 5-input *nor* gates in the 74S260 is not used. In this design the device count is the same for the SOP and POS based circuits. The POS based circuit is potentially more complex, but it does not require inverters, which reduces the device count.

Now let's look at some other examples of real problems.

Example 4.7-1

Design a 1-bit binary **adder** circuit. Include a provision for a 1-bit carry both in and out of the adder. A 1-bit adder with carry-in and carry-out is called a **full adder**. (An adder with no carry-in is called a half adder.)

First we need to see how to formulate this problem. Suppose we call the 2 bits we are going to add A and B, the sum S, the carry-in CI, and the carry-out CO. We can therefore set up the addition as follows:

$$\begin{array}{r} \text{CI} \\ \text{A} \\ +\text{B} \\ \hline \text{CO S} \end{array}$$

Notice that CI, A, B, CO, and S are all binary bits, and we are doing a binary addition of CI, A, and B to get a binary sum S and a binary carry CO. These latter two are the least significant and most significant bits in the sum of the binary bits CI, A, and B. Now, in order to evaluate S and CO using logic circuits, we need to formulate logic expressions that tell us how to find S and CO in terms of CI, A, and B, so we need to treat CI, A, and B as logic variables and S and CO as logic functions. To accomplish this second step in the design, we make truth tables for S and CO:

A	B	CI	CO	S
0	0	0	0	0
0	0	1	0	1
0	1	0	0	1
0	1	1	1	0
1	0	0	0	1
1	0	1	1	0
1	1	0	1	0
1	1	1	1	1

Then we use Karnaugh maps to simplify the functions CO and S. First we consider an SOP based design. The following maps show the minimum cover and SOP simplification of the two functions.

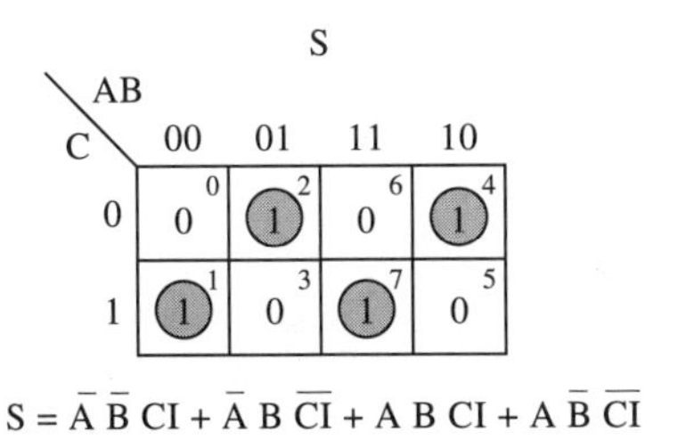

$S = \overline{A}\,\overline{B}\,CI + \overline{A}\,B\,\overline{CI} + A\,B\,CI + A\,\overline{B}\,\overline{CI}$

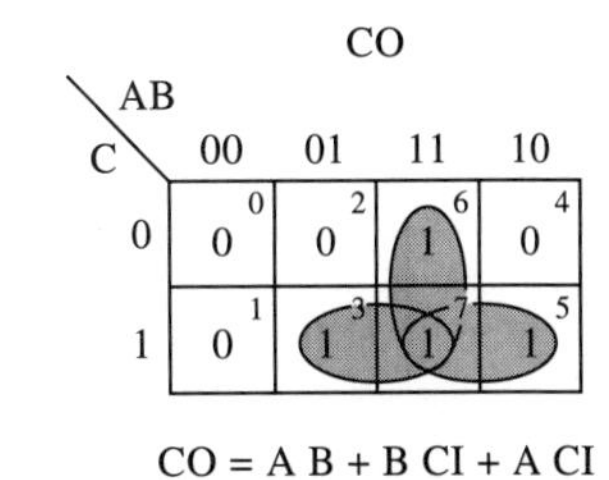

$CO = A\,B + B\,CI + A\,CI$

Up to this point we have said nothing about assertion levels. To continue the design, we need to specify assertion levels for the signals that represent the logic variable. Based on the assumption that all the input variable and output functions are asserted low, a logic circuit design using positive logic *nor* gates follows, along with the simulation results for the circuit. Notice that the carry-out (CO) is 1 (L) when both A and B are 1 (L); when either A or B is 1 (L) and the carry-in (CI) is 1 (L); and when A, B, and CI are all 1 (L)—just as it should be. Also, S is 1 (L) when only A or only B or only CI is 1 (L), or when A and B and CI are all 1 (L)—just as it should be.

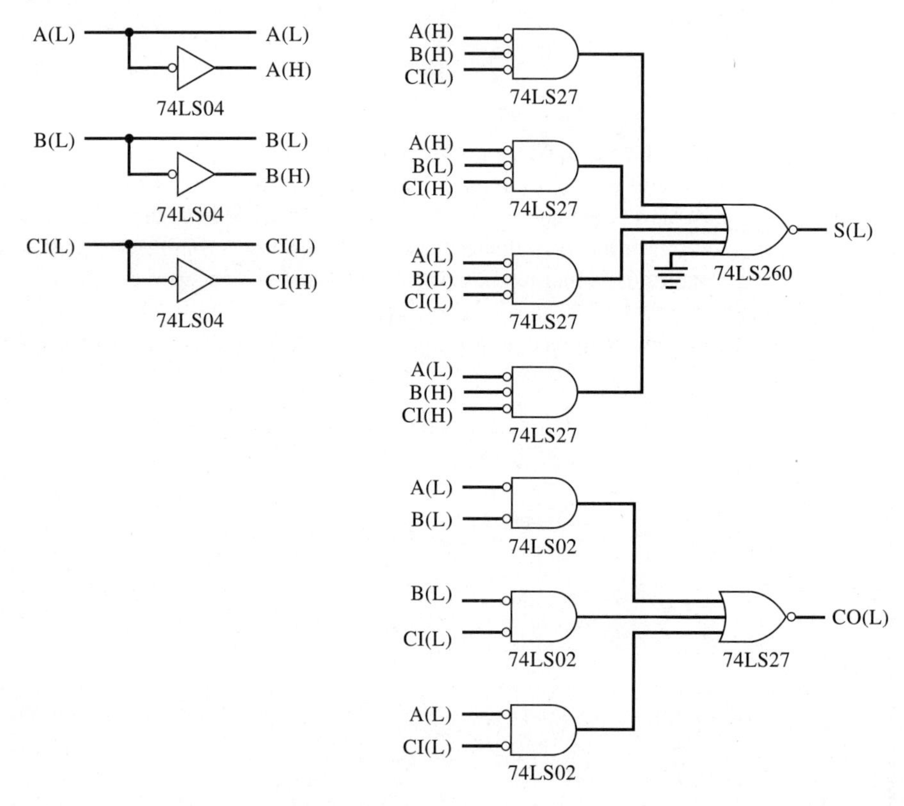

$$S(L) = (\bar{A}\,\bar{B}\,CI + \bar{A}\,B\,\overline{CI} + A\,B\,CI + A\,\bar{B}\,\overline{CI})\ (L)$$

$$CO(L) = (A\,B + B\,CI + A\,CI)\ (L)$$

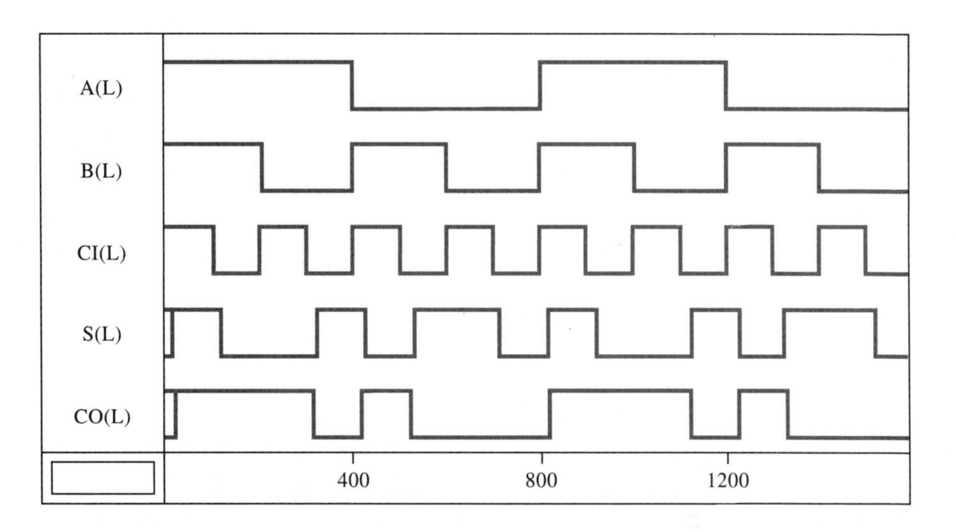

The adder we have designed is for adding binary bits. We had to consider the binary bits as logic variables and make the design accordingly. We can assign logic variables assertion levels just as we have been doing in all of our designs; but when we return to considering the binary bits as binary bits, it seems incongruous to talk about binary bits as being asserted or not asserted and to assign them assertion levels. Binary bits can clearly be either 0 or 1, and when we represent signals, we can assign the 0 to a low signal and the 1 to a high signal or vice versa, just as we do with logic variables. In fact, we can even use a similar notation. We just need to modify what the notation means. Thus, we'll use the notation S(L) to mean that, when the signal S(L), which represents the binary bit S, is low, the binary bit is 1; conversely, we'll use the notation S(H) to mean that, when the signal S(H), which represents the binary bit S, is high, the bit is 1. Now, how do we read S(L) (or S(H))? Although it's not entirely satisfactory, let's say that S(L) means "1 is low," and S(H) means "1 is high."

We should also do a POS based design; it may produce a simpler circuit. The Karnaugh maps and the simplification for this design follow. These logic functions are no simpler than the SOP logic functions—in fact, they are exact duals. The POS based design with the same assumption as before—that 1 is low for all the signals—follows the maps. The circuit, however, like the logic functions, is no simpler than the SOP design. Simulation results are not given; they would be similar to those for the SOP based design.

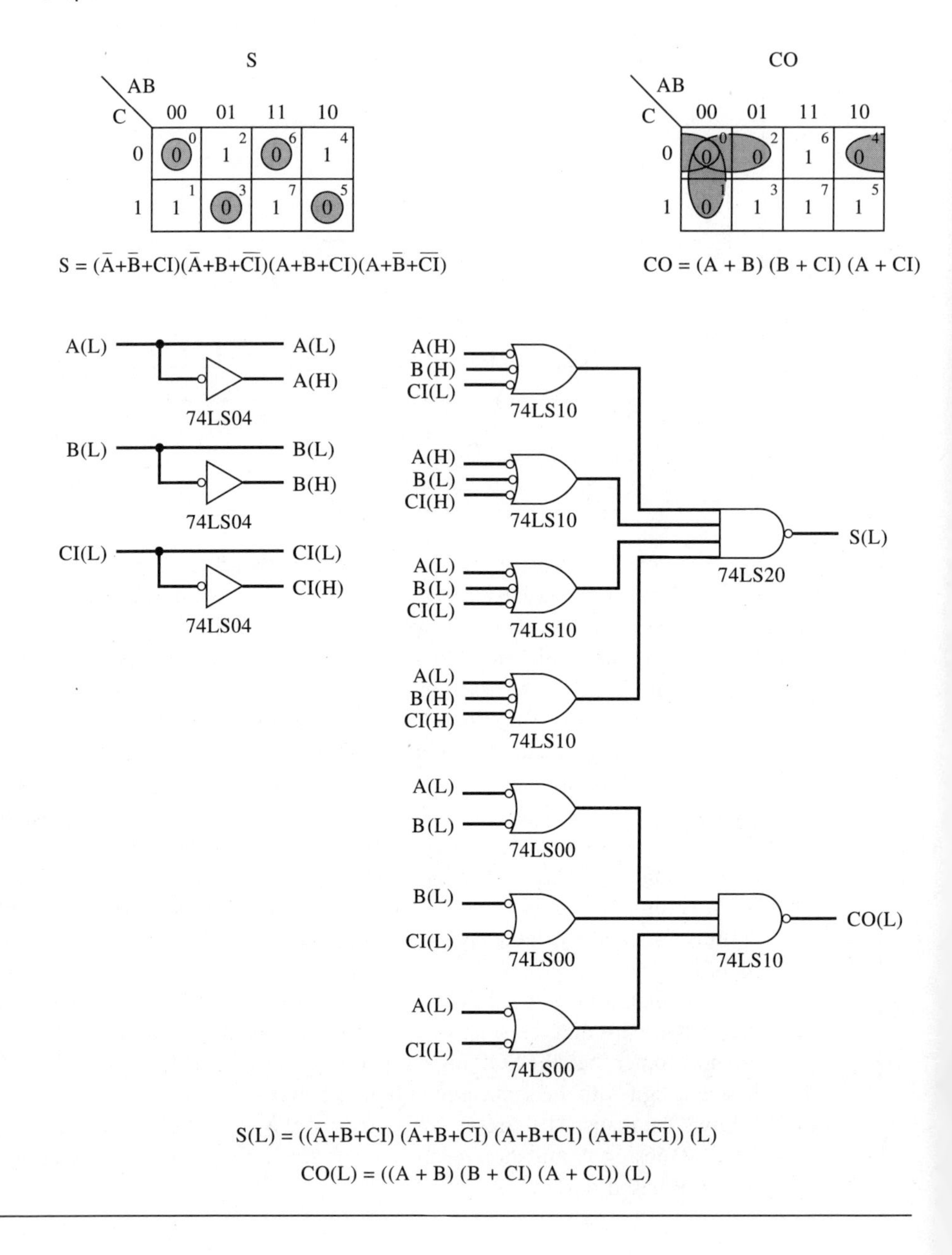

S
AB
C
00 01 11 10
CO
S = (Ā+B̄+CI)(Ā+B+C̄Ī)(A+B+CI)(A+B̄+C̄Ī)
CO = (A + B) (B + CI) (A + CI)
A(L)
A(H)
B(L)
B(H)
CI(L)
CI(H)
74LS04
74LS10
74LS20
74LS00
S(L)
CO(L)
S(L) = ((Ā+B̄+CI) (Ā+B+C̄Ī) (A+B+CI) (A+B̄+C̄Ī)) (L)
CO(L) = ((A + B) (B + CI) (A + CI)) (L)

Example 4.7-2

Design a 1-bit binary *subtracter* circuit. Include a provision for a 1-bit borrow both in and out of the subtracter.

First we need to see how to formulate this problem. Suppose we call the 2 bits we're going to subtract A and B, the difference D, the borrow-in BI, and the borrow-out BO. We can then set up the subtraction as follows:

$$\begin{array}{lr} -\text{BO}) & \text{A} \\ & -\text{B} \\ & \underline{-\text{BI}} \\ & \text{D} \end{array}$$

For this subtracter we actually need to evaluate two logic functions, to find the binary bits BO and D. As a first step in the design, we make truth tables for D and BO:

A	B	BI	BO	D
0	0	0	0	0
0	0	1	1	1
0	1	0	1	1
0	1	1	1	0
1	0	0	0	1
1	0	1	0	0
1	1	0	0	0
1	1	1	1	1

Then we use Karnaugh maps to simplify the functions BO and D. First we consider an SOP based design. The following maps show the minimum cover and SOP simplification of the two functions.

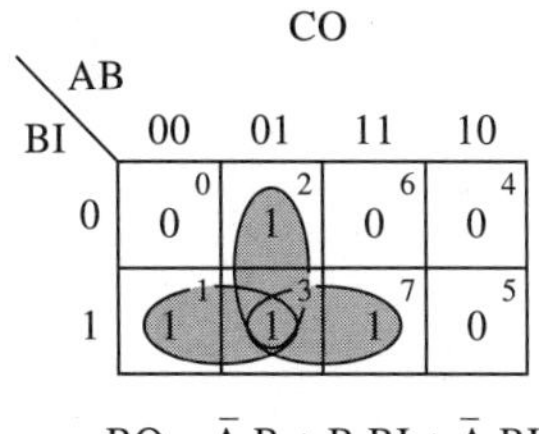

$BO = \bar{A}\,B + B\,BI + \bar{A}\,BI$

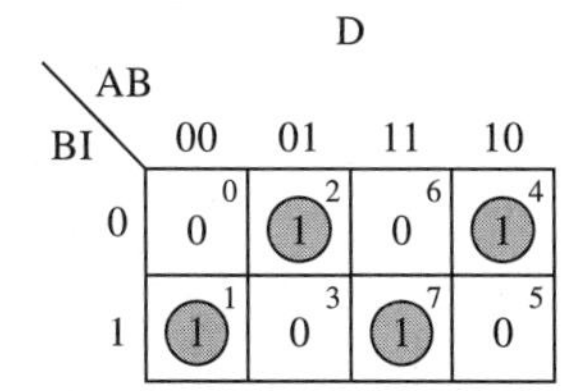

$D = \bar{A}\,\bar{B}\,BI + \bar{A}\,B\,\overline{BI} + A\,B\,BI + A\,\bar{B}\,\overline{BI}$

The circuit, assuming 1 is low for all the signals and using positive logic *nor* gates, follows, along with the simulation results for the circuit. Notice that a borrow-out (BO) is asserted when A is 0 and either B or BI is 1, or when B and BI are both 1 independent of A—just as it should be. Also notice that D is 1 when only A or only B or only BI is 1, or when A and B and CI are all 1—just as it should be.

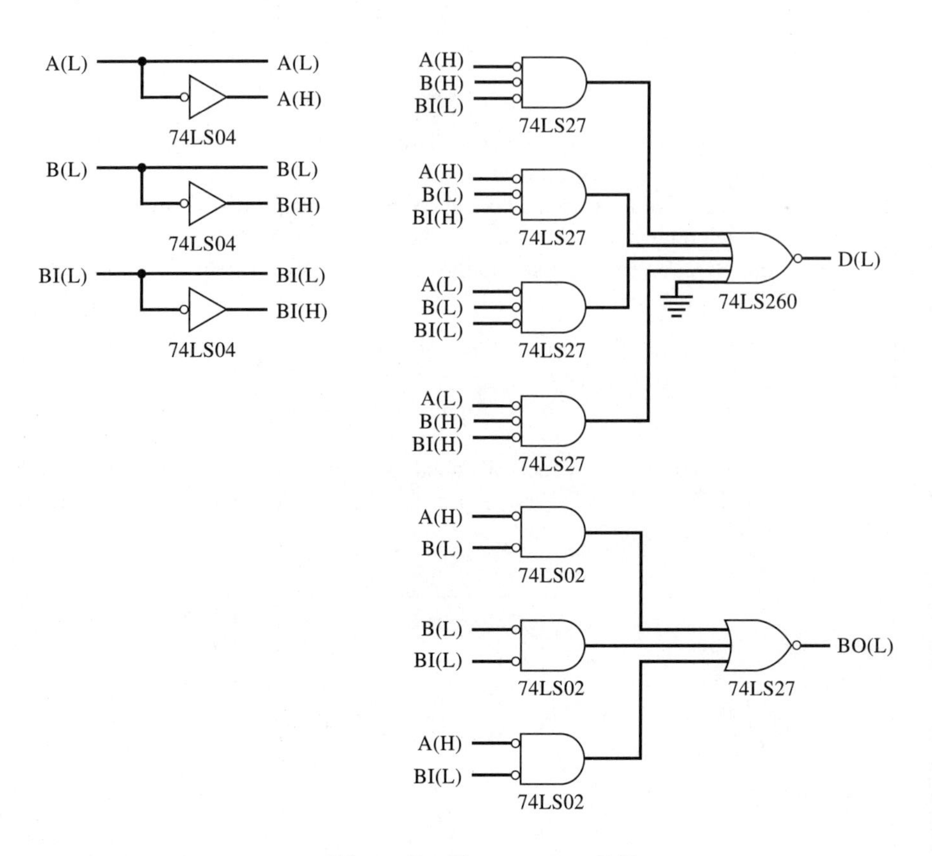

$$D(L) = (\bar{A}\ \bar{B}\ BI + \bar{A}\ B\ \overline{BI} + A\ B\ BI + A\ \bar{B}\ \overline{BI})\ (L)$$

$$BO(L) = (\bar{A}\ B + B\ BI + \bar{A}\ BI)\ (L)$$

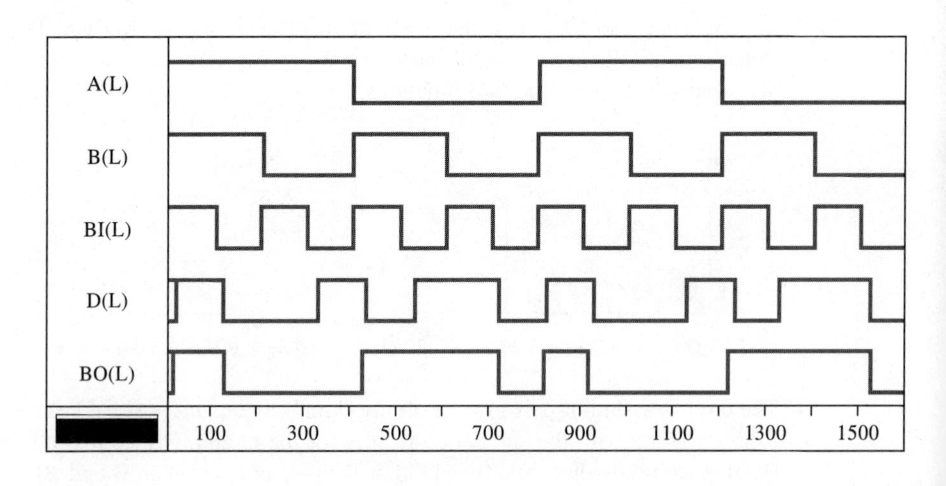

We should also do a POS based design; it may produce a simpler circuit. The Karnaugh maps and simplification follow. These logic functions, however, are no simpler than the SOP logic functions—in fact, they are exact duals. The POS based design follows the maps. Simulation results are not given; they would be similar to those for the SOP based design.

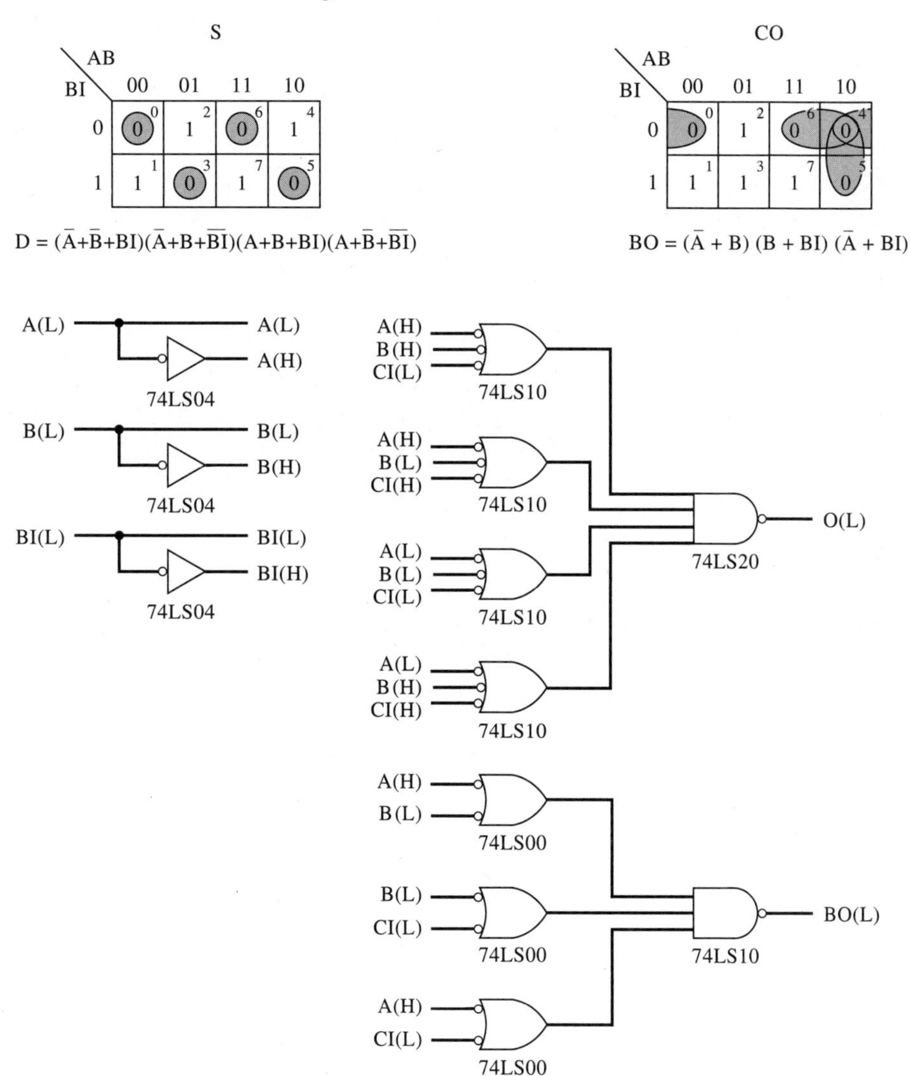

$$D(L) = ((\bar{A}+\bar{B}+BI)\ (\bar{A}+B+\overline{BI})\ (A+B+BI)\ (A+\bar{B}+\overline{BI}))\ (L)$$

$$BO(L) = ((\bar{A} + B)\ (B + BI)\ (\bar{A} + BI))\ (L)$$

Example 4.7-3

Design a circuit to compare two 2-bit binary numbers to determine whether they are the same. Call the numbers A1 A0 and B1 B0. For the bits of the numbers, let 1 be high. If the numbers are the same, assert an output Y(L).

The statement pretty much completes the formulation of the problem. Now we need to make a truth table that shows what values the function Y should have for all combinations of the inputs A1 A0 and B1 B0:

A1	A0	B1	B0	Y
0	0	0	0	1
0	0	0	1	0
0	0	1	0	0
0	0	1	1	0
0	1	0	0	0
0	1	0	1	1
0	1	1	0	0
0	1	1	1	0
1	0	0	0	0
1	0	0	1	0
1	0	1	0	1
1	0	1	1	0
1	1	0	0	0
1	1	0	1	0
1	1	1	0	0
1	1	1	1	1

The following Karnaugh map gives the function Y.

B1 B0 \ A1 A0	00	01	11	10
00	(0) 1	(4) 0	(12) 0	(8) 0
01	(1) 0	(5) 1	(13) 0	(9) 0
11	(3) 0	(7) 0	(15) 1	(11) 0
10	(2) 0	(6) 0	(14) 0	(10) 1

$$Y = \overline{A1}\ \overline{A0}\ \overline{B1}\ \overline{B0} + \overline{A1}\ A0\ \overline{B1}\ B0 + A1\ A0\ B1\ B0 + A1\ \overline{A0}\ B1\ \overline{B0}$$

As you can see, Y cannot be simplified. The circuit for evaluating Y follows, along with a simulation output.

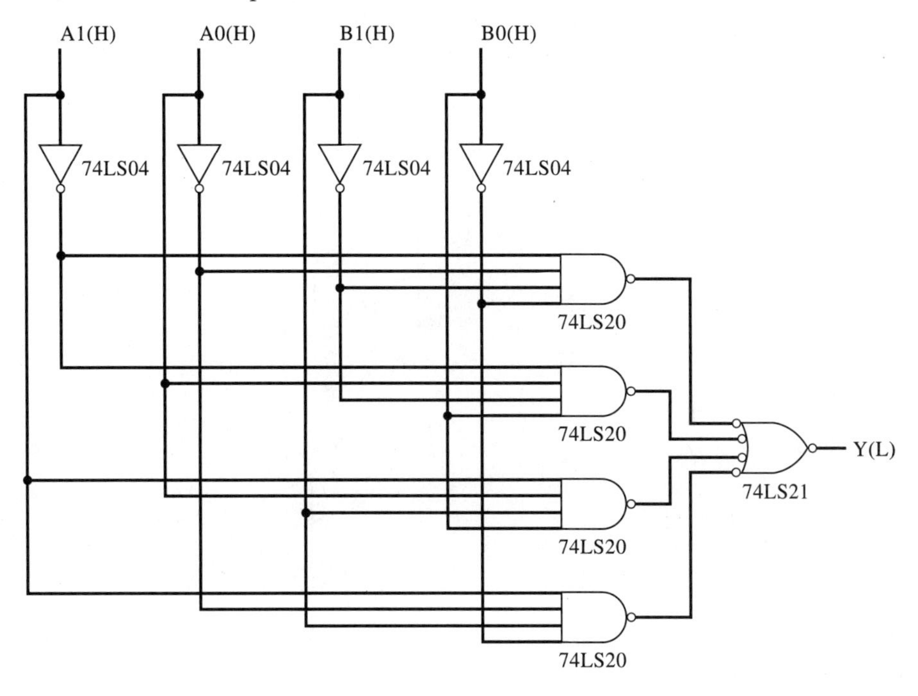

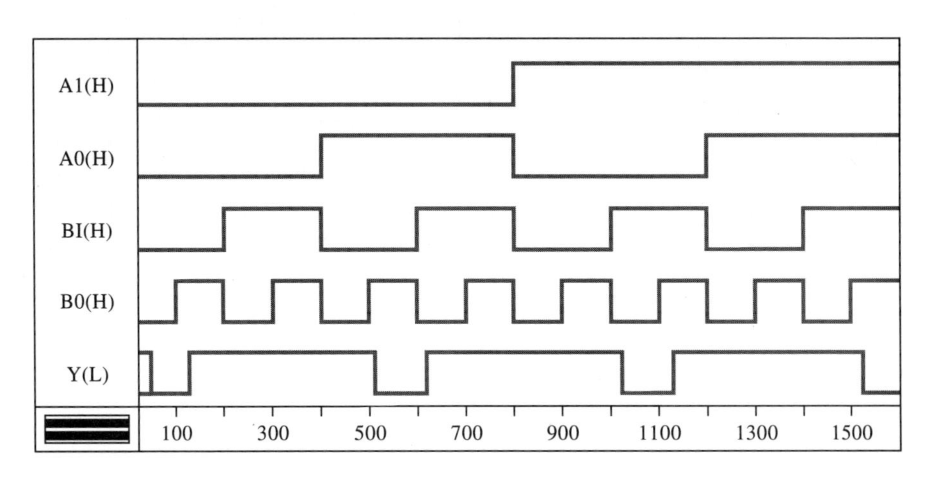

Now we should look at a POS based circuit to see if it is simpler. The POS Karnaugh map and simplified logic equation follow, along with the logic circuit. The circuit is indeed simpler.

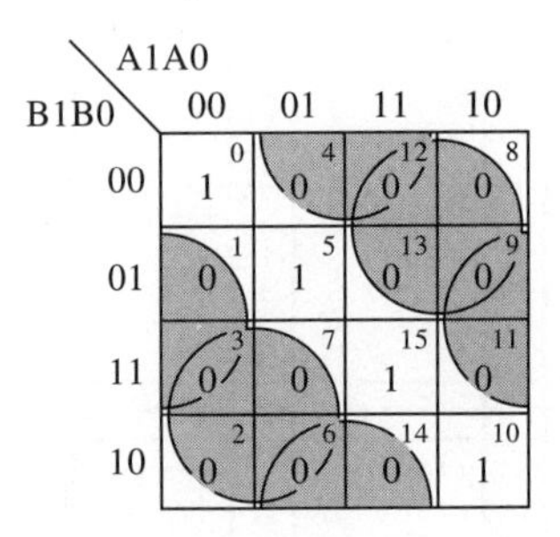

$$Y = (\overline{A0} + B0)(\overline{A1} + B1)(A0 + \overline{B0})(A1 + \overline{B1})$$

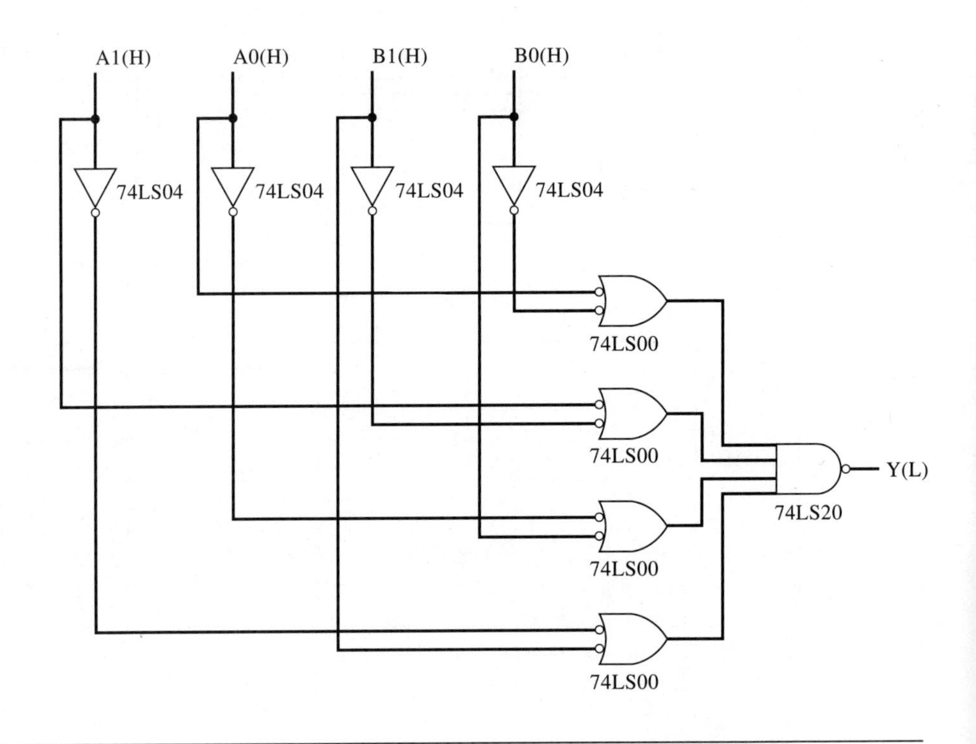

4.8 TIMING CONSIDERATIONS

In the examples of Section 4.7 we noted that there are time delays associated with the gates in the logic circuits, but in those examples time delays were not a critical

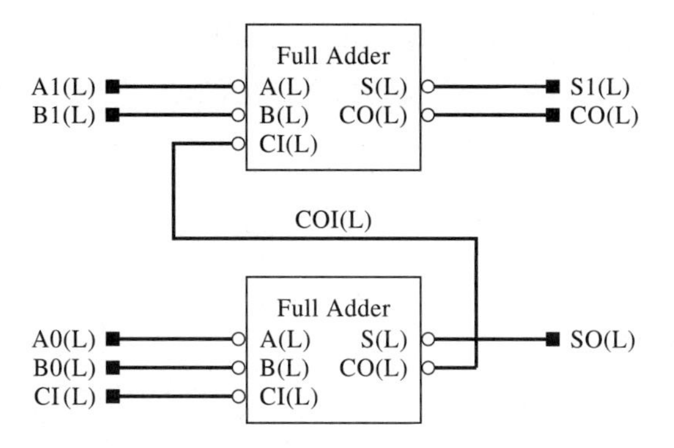

Figure 4.8-1 Two-bit adder using two 1-bit adders.

consideration. Now we'll present a case where timing is important. Suppose we want to make an adder that adds two 2-bit binary numbers as follows:

$$
\begin{array}{rrr}
 & & \text{CI} \\
 & \text{A1} & \text{A0} \\
 & \text{B1} & \text{B0} \\
\hline
\text{CO} & \text{S1} & \text{S0}
\end{array}
$$

where A1 A0 and B1 B0 are the numbers to be added, CI is the carry-in, S1 S0 is the sum, and CO is the carry-out. If this adder is to be used in a fast computer, its speed is important.

An easy way to design a 2-bit adder is to cascade two 1-bit adders like the one we designed in Section 4.7. Figure 4.8-1 shows such a 2-bit adder, in which we use a single symbol to represent the SOP based adder we designed in Example 4.7-1. Notice that all the inputs and outputs have bubbles to indicate that they are asserted low. Now, using an input signal that changes every 50 ns, let's simulate this circuit to see if it can produce a correct output. If we simulate for *typical* propagation delays (we will look at *maximum* propagation delays later), we get the simulation output shown in Figure 4.8-2. In this simulation, the inputs have been chosen so that all possible combinations of inputs occur. A first look at the output from the adder shows some strange results. For example, it appears that CO is asserted for very small input numbers. To see if the circuit is functioning properly, let's expand the first part and examine the results in some detail. Figure 4.8-3 shows the first part of the simulation timing diagram with vertical dashed lines placed every 50 ns where the input changes. Along each vertical line are the values of all the logic variables just before the change in the input.

We see on the first vertical line (first add) that

$$\text{A1 A0} = 00$$

$$\text{B1 B0} = 00$$

$$\text{CI} = 0$$

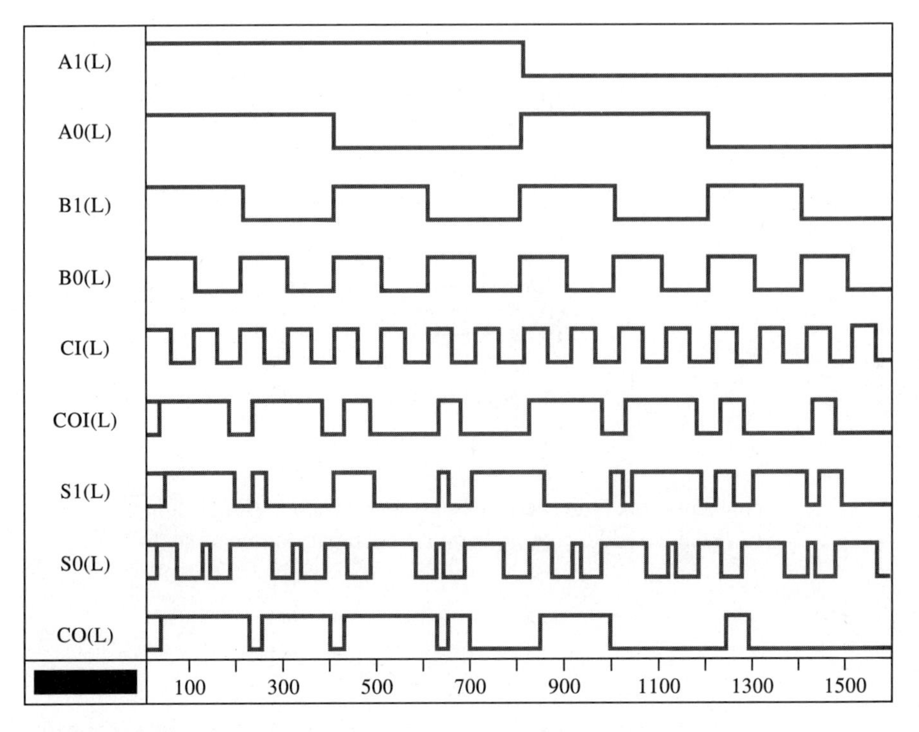

Figure 4.8-2 Simulation results for 2-bit adder.

and the output

$$\text{S1} \quad \text{S0} = 00$$

with no carry-out (CO = 0). This is correct. We see on the second vertical line (second add) that

$$\text{A1} \quad \text{A0} = 00$$

$$\text{B1} \quad \text{B0} = 00$$

$$\text{CI} = 1$$

and the output

$$\text{S1} \quad \text{S0} = 01$$

with no carry-out (CO = 0). This is also correct, but notice that S0 did not change until 16 ns after the input CI changed. Whether the result is correct depends on when we look at the result. If we look just before the input changes to a new value, we get a correct result for the previous value of the input. If we look during the 16 ns just after the input change, we get an incorrect value.

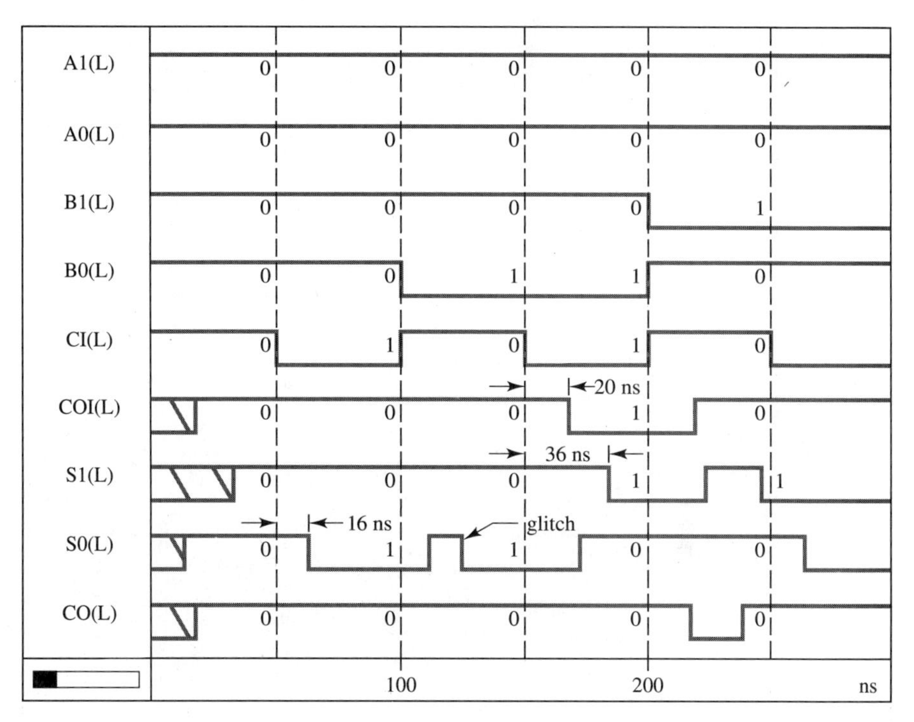

Figure 4.8-3 Details of the timing for the first several input changes.

Now look at the third vertical line (third add). Again the results are correct if we wait until just before the input changes to a new value, but we see an even stranger behavior in S0. It goes to an incorrect value for a short time and then returns to a correct value. This is called a *glitch* and shortly we'll see why glitches occur.

Look at the values at the fourth vertical line (fourth add). Again the circuit gives the correct results just before the input changes to a new value, but here the timing on S1 is more critical than the timing we found for S0. S1 does not reach a correct value until 36 ns after the input changes—this is 14 ns before the input changes again. The long delay results from the dependence of the change in S1 on a change in the carry between the low order adder and the high order adder (COI). S1 suffers a delay (20 ns) in the low order adder due to the intermediate carry and then an additional delay in the high order adder.

The kind of adder we have constructed is one with ripple carry. That is, the carry must ripple through the low order adder before it goes to the high order. There is even more delay in S1 at the next change in the input—the fifth vertical line (fifth add). There the delay is 46 ns, and the margin between the time when the correct value for the sum is reached and the input changes again is just 4 ns. If we continue our analysis through all the input changes we have in the simulation results, we find that we get correct results for any combination of inputs.

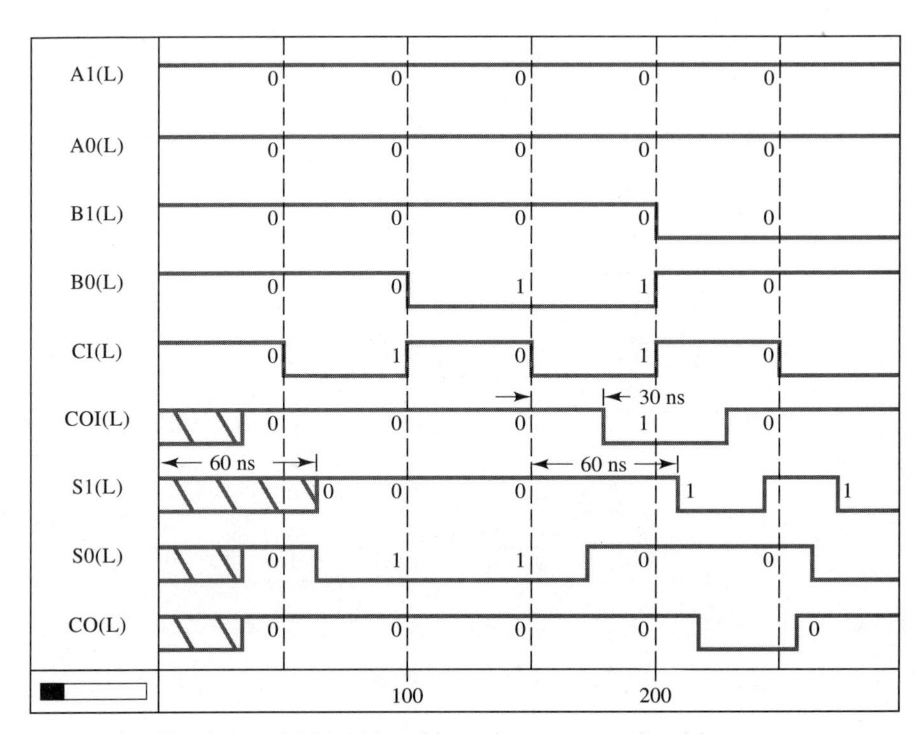

Figure 4.8-4 Simulation of 2-bit adder with maximum propagation delays.

We have constructed a circuit that functions correctly as long as we consider typical propagation delays and wait to look at the results of an addition until just before the input changes again. This means the addition takes almost 50 ns, and it also means that we need some way to catch and store the result of the addition. We will find out how to do this when we discuss sequential logic.

The simulation we have considered used typical propagation delays for the gates. For a more critical analysis we must use worst-case, or maximum, propagation delays. If a simulation with typical delays shows that our circuit will function correctly, and if we build a large number of circuits based on this assumption, most of them will function correctly. If a simulation with maximum propagation delays shows that our circuit will function correctly and if we build a large number of circuits based on this assumption, *all of them* will function correctly.

Figure 4.8-4 shows the results of a simulation with maximum propagation delays. We can see immediately that S1 does not reach a stable value until 60 ns after the simulation is started. This is 10 ns after the input value changes, and so the circuit is not fast enough to give correct results. We can see this problem again at the fourth add in the timing diagram. The correct result for S1 is delayed 60 ns from the input value change—10 ns after it again changes. This is due once more to a combination of a delay in the intermediate carry (COI) and a delay through the second order adder. The delay at the fifth add is even greater—75 ns. To make a useful adder, we normally need the correct results to be output before the input value

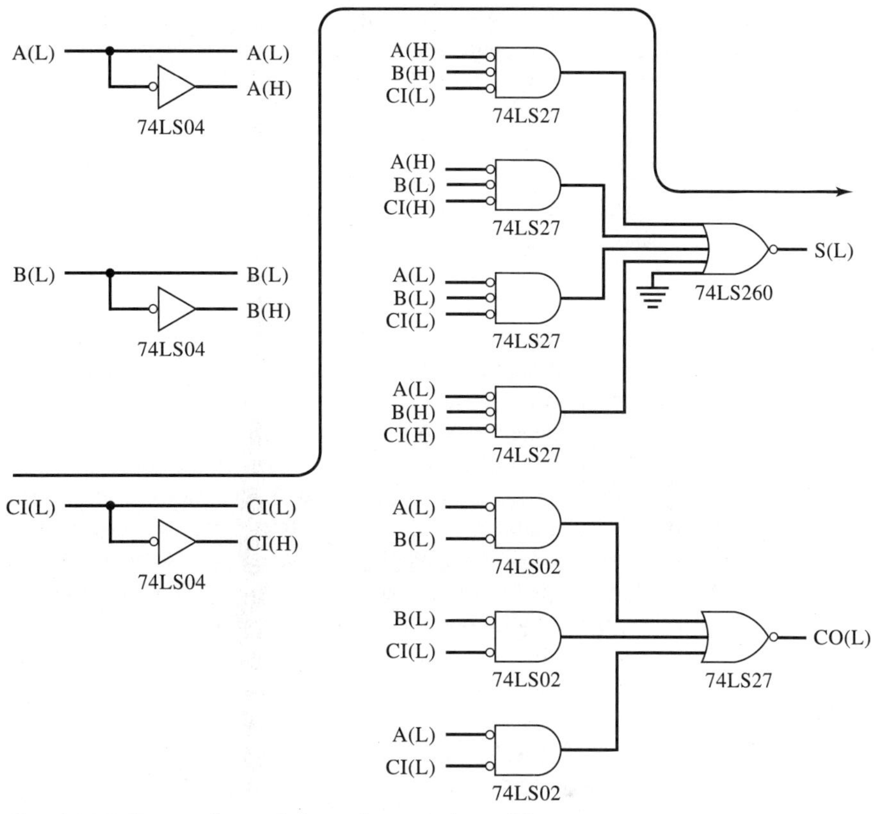

Figure 4.8-5 Propagation path for a change in input CI.

changes. The adder we have designed could not function faster than 75 ns per addition.

There are two ways in which we can improve the speed of the adder we have designed. The first is to change to a faster technology, such as 74F or 74ACT logic. We might also become clever and design a circuit that does not use ripple carry, but uses what is called look-ahead carry—a fast adder. We will not do this because look-ahead carry is not very useful for a device as simple as a 2-bit adder.

Our analysis of the timing in the 2-bit adder made use of a modern approach—simulation, a very quick and satisfying way to analyze timing. We can also do a timing analysis by hand (that is, without using simulation). In a hand analysis, we add the gate delays through all of the gates for each change at the input. For large circuits, hand analysis is a slow and tedious process; however, we will examine a partial hand analysis of the 2-bit adder, both because we should understand how to do a hand analysis and because it will give us insight into how the timing actually works. First, consider the delay from the change in the CI input to the change in S0 at the second add in the simulation of Figure 4.8-3. Figure 4.8-5 shows how this change propagates through the circuit. It propagates first through a 3-input *nor* gate (74LS27) and then through a 5-input *nor* gate (74LS260). The output of the first gate changes from low to high. The output of the second gate changes from high to low. If we consider typical delays, TPLH = 10 ns for the first gate and TPHL = 6 ns for the second gate, so the total delay through the two gates is 16 ns, just as the simulation shows (Fig. 4.8-3). If we consider maximum delays, TPLH for the first gate is 15 ns and TPHL for the second gate is also 15 ns, so the total delay through the two gates is 30 ns, just as the simulation shows (Fig. 4.8-4).

Now consider the timing for the fourth add in the simulation. Figure 4.8-6 shows how the change in CI propagates through the circuit. It first propagates through a 2-input *nor* gate (74LS02) and a 3-input *nor* gate (74LS270) in the low order adder; then it propagates through a 3-input *nor* gate (74LS27) and a 5-input *nor* gate in the high order adder. For typical propagation delays, TPHL and TPLH are both 10 ns for all gates except the 74LS260. For the 74LS260, the output changes from high to low and TPHL is 6 ns. The total delay through the low order adder is 10 + 10 = 20 ns, as shown in the simulation (Fig. 4.8-3); and the total delay through both the low and high order adders is 20 + 10 + 6 = 36 ns, as shown in the simulation (Fig. 4.8-3). If maximum propagation delays are considered, the delays TPHL and TPLH are both 15 ns for all the gates. Thus, the delay through the first adder is 30 ns, and the delay through both adders is 60 ns (Fig. 4.8-4).

Now consider the timing for the fifth add in the simulation. Figure 4.8-7 shows the propagation path through the circuit for the change in CI and the change in B0. Both of these changes suffer the same gate delays. Typical delays are summarized as follows:

TPHL (74LS02)	=	10 ns
TPLH (74LS27)	=	10 ns
TPHL (74LS04)	=	10 ns
TPLH (74LS27)	=	10 ns
TPHL (74LS260)	=	6 ns
		46 ns

The total delay is 46 ns, just as in the simulation (Fig. 4.8-3). If maximum delays are considered, then both TPLH and TPHL are 15 ns for all the gates, and the total delay is 75 ns (Fig. 4.8-4). For the fifth add of the simulation, CI and B0 are not the only inputs that change. The input B1 also changes, but the effect of a change in B1 propagates only through the gates of the high order adder; it is therefore delayed less than a change in either CI or B0. The path for CI or B0 is the worst-case, or *critical*, path. Before simulation became available for timing analysis, such analysis was performed by finding the critical path and adding the delays on that path. Finding the critical path could involve calculating the total delay for a number of paths to find the one with the largest delay.

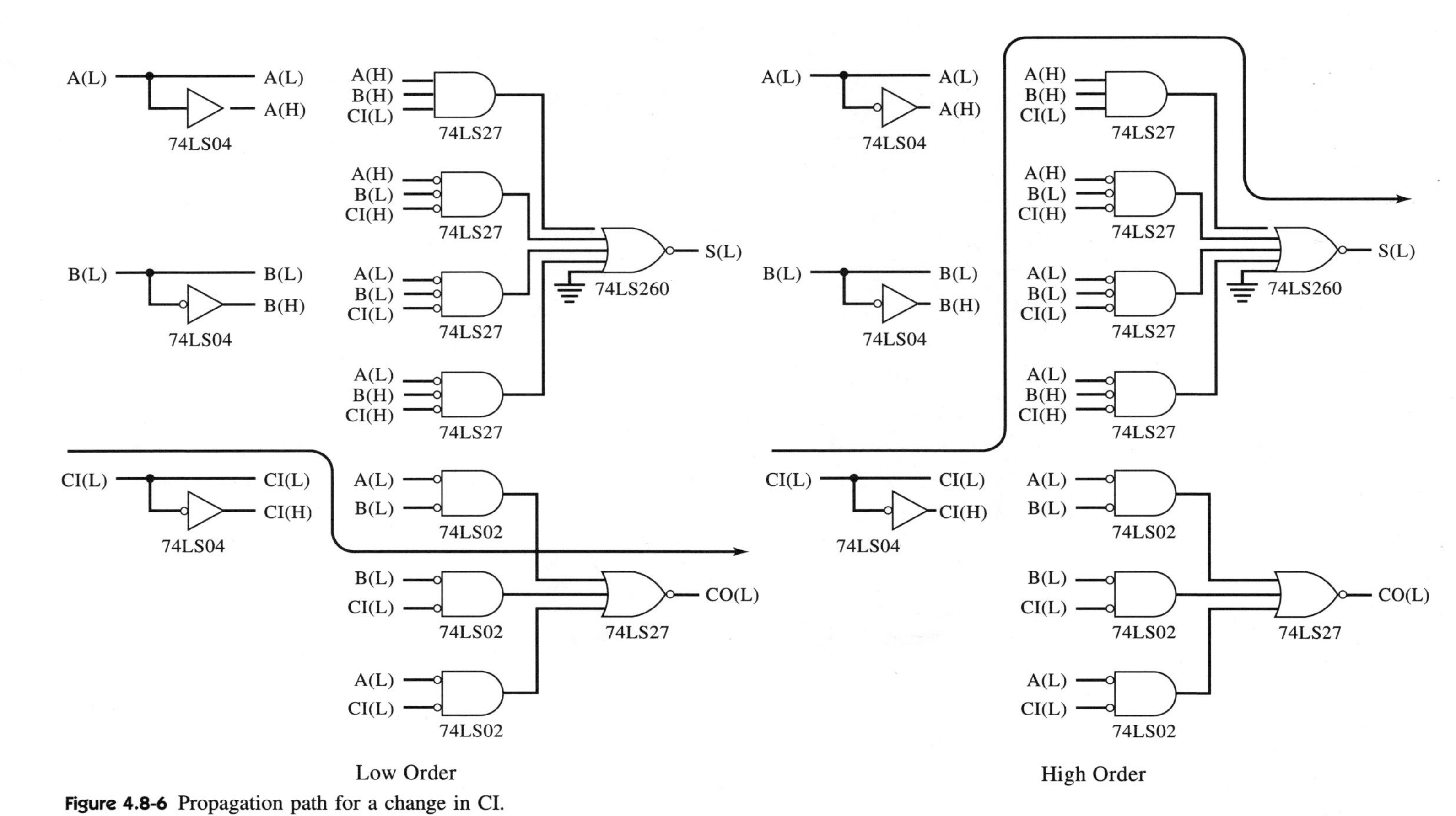

Figure 4.8-6 Propagation path for a change in CI.

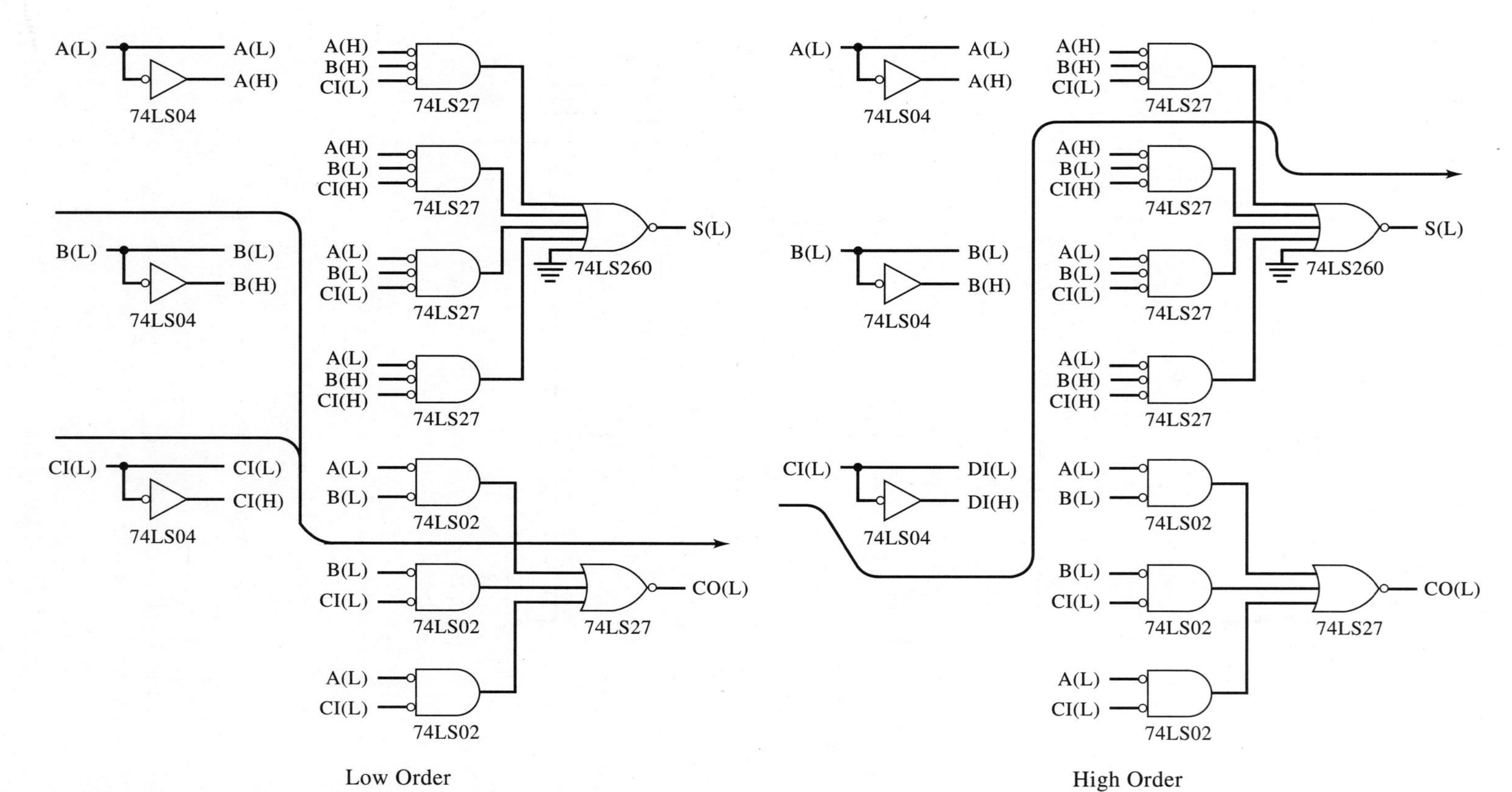

Figure 4.8-7 Propagation path for a change in B or CI.

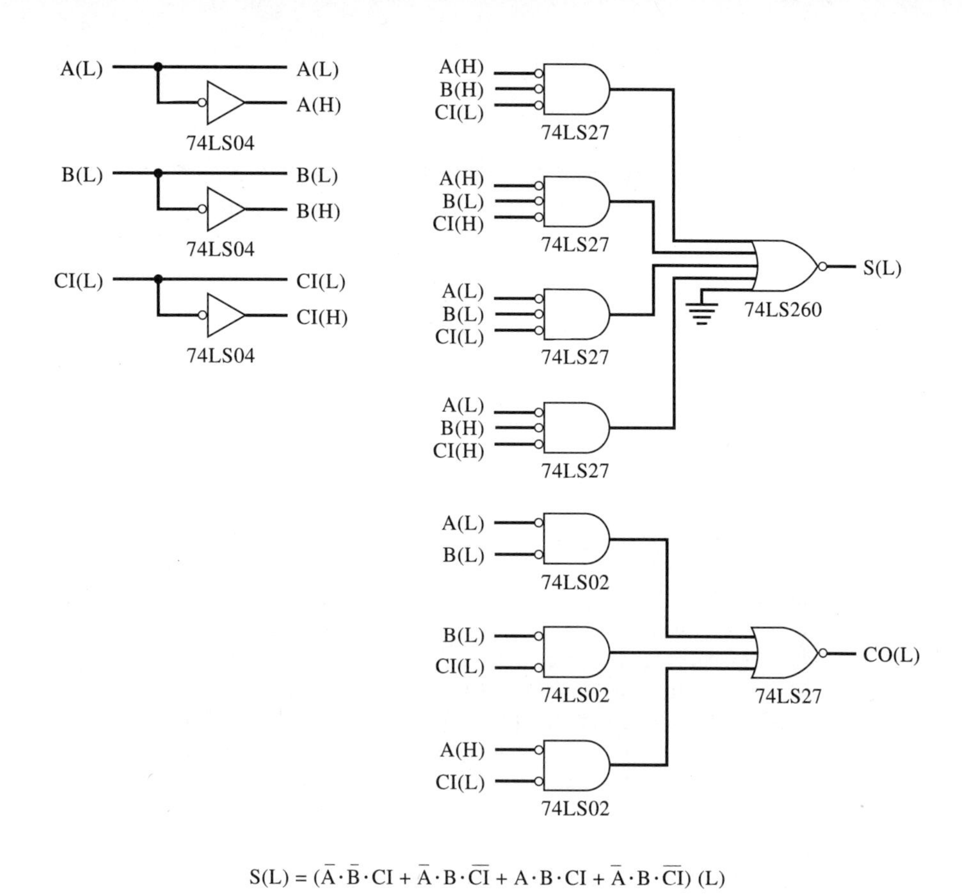

$$S(L) = (\bar{A}\cdot\bar{B}\cdot CI + \bar{A}\cdot B\cdot\overline{CI} + A\cdot B\cdot CI + \bar{A}\cdot B\cdot\overline{CI})\ (L)$$

$$CO(L) = (A\cdot B + B\cdot CI + A\cdot CI)\ (L)$$

Figure 4.9-1 Circuit from Example 4.6-1.

4.9 DRIVE CONSIDERATIONS

In Section 4.8 we considered one of the practical limitations of logic circuits—propagation delay through the circuit. There we saw that this circuit delay is caused by delays through the gates involved in the circuit. In fact, we saw that the propagation delay through a gate is one of the gate's basic properties, specified in its technical description. In this section we will consider another important limitation on gates, which is also specified in technical descriptions. It has to do with the drive capability of the gates involved in a logic circuit design.

To see the problem more clearly, consider the adder circuit in Figure 4.9-1, which we designed in Example 4.6-1. In a larger digital logic circuit of which this might be a part, the inputs to the input *nor* gates of the adder would also be driven by gates (or other logic devices). One of these input signals, A(L), connects to the third, fourth, fifth, and bottom gates and an inverter, so the device that produces the signal must be capable of enough drive to supply these five inputs.

To determine whether the drive is adequate, we need to examine how both the output and the input of a logic device (gate) function. First, let's consider the normal

Table 4.9-1 Loading and Fan-Out Table for the 74LS32

PIN	Description	54/74	54/74S	54/74LS
A, B	Inputs	1 ul	1 Sul	1 LSul
Y	Output	10 ul	10 Sul	10 LSul

From *TTL Data Manual* (Signetics, 1984).

input and output voltage specification for the common TTL families of logic devices. (We have been using devices from one of these families, the LS [Low Power Schottky] family, in all our examples.) The minimum input voltage required to force the output of a TTL device to either a high state or a low state, as appropriate for the device, is specified as 2.0 volts; it is called V_{IH}. The maximum voltage that will drive the output into the opposite state is specified as 0.8 volts; it is called V_{IL}. As we noted at the beginning of this chapter, an output specified for one of these families of devices will be more than adequate to drive an input. The output high drive specification is 2.4 volts, and the output low drive specification, 0.4 volts. These voltages are called V_{OH} and V_{OL}. Notice that V_{OH} is 0.4 volts greater than the 2.0 input volts needed to drive a device high. Similarly, the V_{OL} is 0.4 volts lower than the 0.8 volts needed to drive a device low. The differences $V_{OH} - V_{IH}$ and $V_{IL} - V_{OL}$ are called the noise margin.

A TTL device normally has a 5 volt supply, so we could expect the output to be near this 5 volt value when driven high. We can also expect the output to be near ground when it's driven low. As the device is made to drive more and more inputs, V_{OH} will be pulled down and V_{OL} will be pulled up. The loading that will pull V_{OH} down to 2.4 volts or pull V_{LO} up to 0.4 volts is the maximum loading under which the device is specified to operate correctly. We'll address this loading in some detail in the following two sections, but first we'll explore a simplistic but useful way of specifying drive capability.

This method of specification is useful when we employ only logic devices in a single logic family, such as LS. We use a unit called a unit load (ul) to specify both the loading caused by inputs and the drive capability of outputs. Table 4.9-1 shows the specification for a 74LS32 in terms of unit loads. The table is from an older data book for standard TTL, S, and LS logic families in both standard (74) and military (54) grades. (Some newer data books do not specify fanout in unit loads.) A and B are inputs to the 2-input *or* gate, and Y is the output. Notice how this manufacturer uses ''Sul'' to specify S unit loads and ''LSul'' to specify LS unit loads. Although this is usually not done, it does emphasize that a unit load applies only to a single family of devices. Table 4.9-2 tells us that a 74LS32 requires one unit load to drive either of its inputs and that its output is capable of driving a ten-unit loads. A device that is capable of driving a ten-unit loads is said to have a fanout of 10. A fanout of 10 is common for standard TTL, S, and LS families of logic.

Table 4.9-2 Loading and Fan-Out Table for the 74F32

PIN	Description	74F (U.L.) High/Low	Load Value High/Low
A, B	Inputs	1.0/1.0	20 uA/0.6 mA
Y	Output	50/33	1 mA/20 mA

From *Fast Data Manual* (Signetics, 1987).

We are now in a position to answer the question about drive we asked with respect to the circuit of Figure 4.9-1. We can expect any LS logic family device to easily drive 5 LS inputs. For some families of devices, the unit load drive is different for driving the output low and driving it high. (We will see why in Section 4.10.) Table 4.9-2 shows the drive specifications for the 74F32, which has this kind of asymmetrical drive capability. The table also shows load and drive currents, which we will discuss in the next two sections.

4.10 LOGIC DEVICE OUTPUT CIRCUITS

Figure 4.10-1 is a simple block diagram of a TTL logic device output. It consists of control components, which control the output based on some logic input, and pull-up and pull-down drivers connected to the output pin. It may also have an *output enable* input control, shown (to be discussed later). When the control components determine that the output should be driven high, the pull-up driver pulls the output up to V_{CC} ($+5$ volts). When the control components determine that the output should be driven low, the pull-down driver pulls the output down to ground. The pull-up and pull-down drivers are normally transistors.

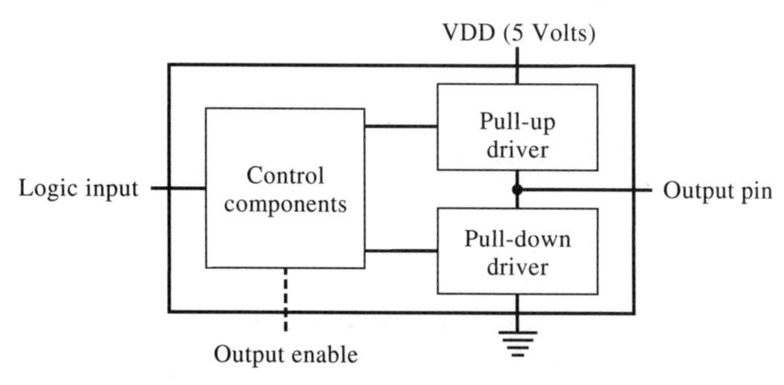

Figure 4.10-1 Block diagram of a logic circuit output.

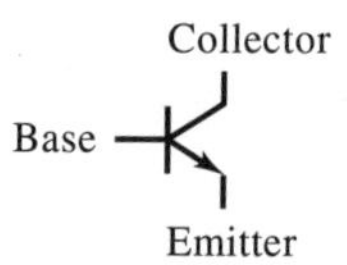

Figure 4.10-2 Transistor symbol.

The circuit symbol for a transistor is shown in Figure 4.10-2. A transistor has three terminals called the emitter, base, and collector, which are labeled in the figure. For a simple understanding of logic devices, we're concerned only with the saturated operation of the transistor. In saturated operation, the transistor functions like a switch that is controlled by the voltage on the base. Like a switch, it can be either on or off. In the *on* state, the transistor can pass a relatively high current from the collector to the emitter with near-zero voltage between them. In the *off* state, the transistor passes almost no current between the collector and emitter, even though the voltage between them is relatively large.

When we replace the pull-up and pull-down drivers in Figure 4.10-1 with transistors, we get the classic totem pole output circuit shown in Figure 4.10-3. Notice that there is a resistor in the collector of the upper transistor. If the control components wish to force the output pin to a high state, the upper transistor is turned on and the lower transistor is turned off. This switches the output to near V_{DD} (+5 volts). If the control components wish to force the output pin into the low state, the lower transistor is turned on and the upper transistor is turned off. This switches the output to ground. The limitations on the output circuit are caused by limitations in the output transistors.

When the pull-up transistor is turned on, it must source current to the load it's driving, as shown in Figure 4.10-4, and it can source only a limited current without developing a significant voltage from the collector to the emitter of the transistor.

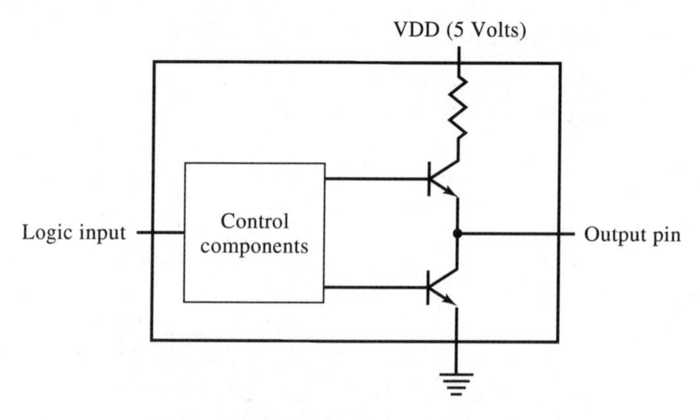

Figure 4.10-3 Logic device totem pole output.

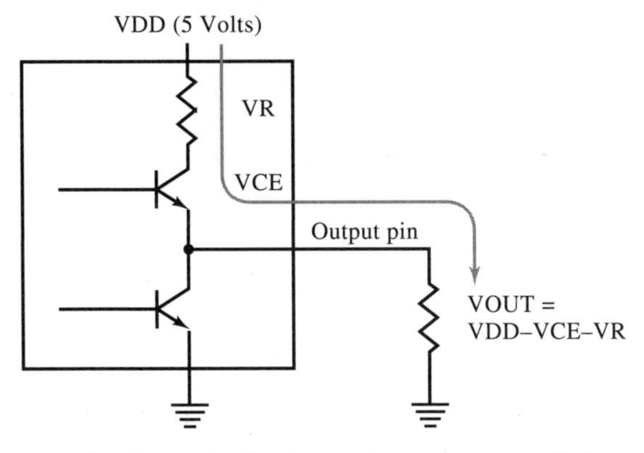

Figure 4.10-4 Logic output circuit sourcing current to pull the output high.

The source current must also flow through the current-limiting resistor in the collector of the pull-up transistor; this current flow also develops a voltage across the resistor. The collector-to-emitter and resistor voltages lower the voltage on the output pin to less than V_{DD}. Since the voltage must remain above 2.4 volts in order to drive subsequent devices high, the current that the device can source at 2.4 volts is usually given in the specifications as I_{OH}.

When the pull-down transistor is turned on, it must sink current to pull the output low, as shown in Figure 4.10-5. Again the transistor limits the current. As the transistor sinks more current, the collector-to-emitter voltage, which is the output voltage, increases from its ideal 0 volts. Since the output voltage must remain below 0.4 volts in order to drive subsequent devices low, the current that the device can sink at 0.4 volts is usually given in the specifications for the device as I_{OL}.

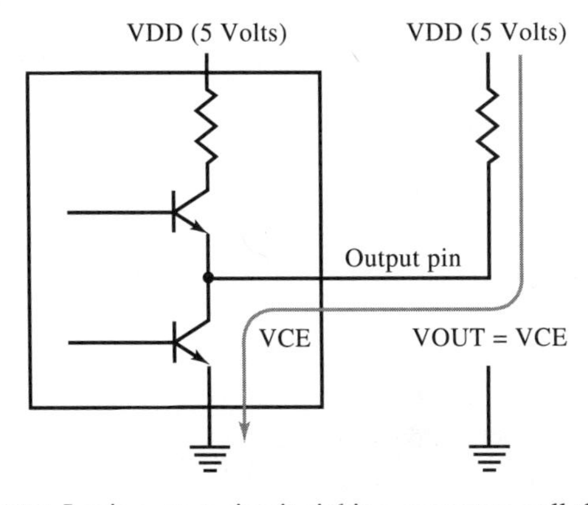

Figure 4.10-5 Logic output circuit sinking current to pull the output voltage low.

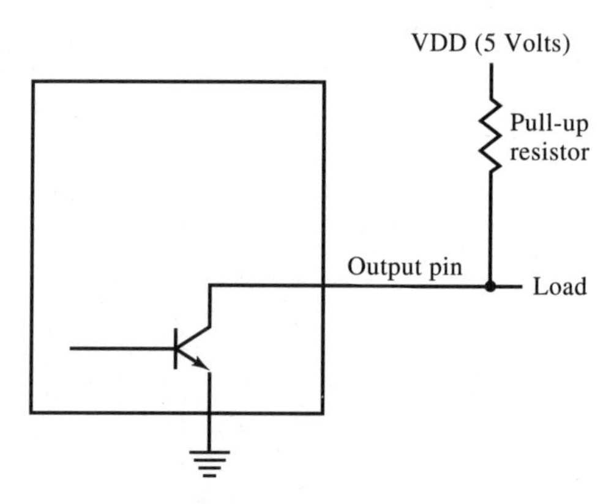

Figure 4.10-6 Open collector output circuit.

To complete our examination of drive capability, we also need to consider the inputs of the devices being driven. We will defer this examination until the next section and first look at two common variations of logic device outputs. In one variation, the device has no pull-up driver. It is called an *open collector* device, and the pull-up driver must be supplied external to it. Normally, the pull-up driver is simply a resistor externally connected to the device, as shown in Figure 4.10-6. When the pull-down transistor is turned off, the external resistor pulls the output pin high. When the pull-down transistor is turned on, the transistor pulls the output pin low. Notice that in the low output state the pull-down transistor must sink the current that flows through the pull-up resistor as well as any current from the loads being driven. This is a disadvantage of the open collector device. There are two possible advantages. First, the output source current does not flow through the device, so it does not heat it. Second, outputs of several open collector devices can be connected together to produce a hard-wired *or,* as shown in Figure 4.10-7. In the hard-wired *or,* the resistor pulls the output of all the logic circuits high unless any one of the logic circuits pulls the output low. This is an *or* operation (asserted low). Unlike the normal totem pole outputs, which contend when they are connected, open collector devices have no contention.

The other output variation is called a tristate output. It involves a modification of the totem pole output. In this variation, the control components are controlled not only by some logic input but also by the output enable, which was shown dashed in Figure 4.10-1. When the output enable is asserted, both the pull-up and pull-down transistors are turned off. This allows the output pin to float. It can be driven to any value, but only by sources outside the tristate device. The tristate output is useful in cases where it is desired to drive a single logic circuit from different sources at different times. The outputs of several logic devices can be connected together, and they will function correctly as long as all but one of the devices are tristated. If more than one of the device outputs are enabled at one time, then one of the devices may try to drive the common output low while another tries to drive it high. The outputs of the two devices will contend and result in an output that is neither high nor low

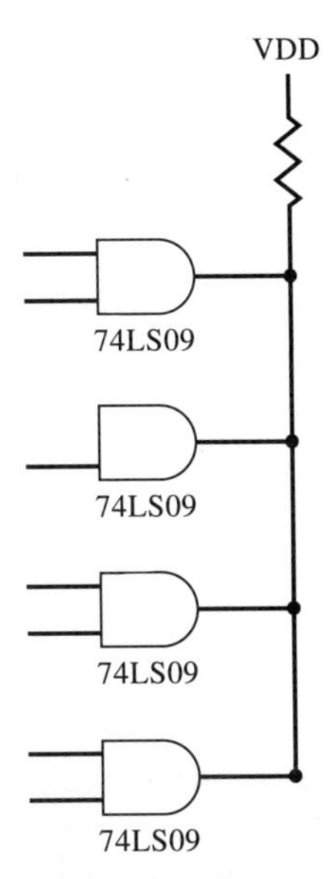

Figure 4.10-7 Wired *or* at the output of four open collector *and* gates.

but somewhere in between. The common output circuit is often called a bus, so this condition is called bus contention.

Another class of logic devices, commonly called CMOS devices, uses a different type of transistor. CMOS devices typically have different input and output specifications from the TTL families. We will not discuss their specifications in detail here; they can be found in data books. We will note, however, that some families of CMOS devices have been made TTL compatible. When this is true for a family, the family designation usually contains a "T" at the end, such as the 74ACT08.

4.11 INPUT LOADING

In Section 4.10 we were able to discuss the output of logic devices in terms of a simple transistor circuit, a model that is very close to the actual output configuration. The actual input configurations for logic devices, on the other hand, do not model easily. That is partly because the inputs vary from device to device, even in a single family, and partly because the input is more complicated. Consequently, we will not attempt to model the actual device inputs, but we'll present a simple equivalent model that is adequate for all inputs, shown in Figure 4.11-1. It consists of a pull-up resistor from the input to V_{CC}, a pull-down resistor to ground, and a capacitor to ground.

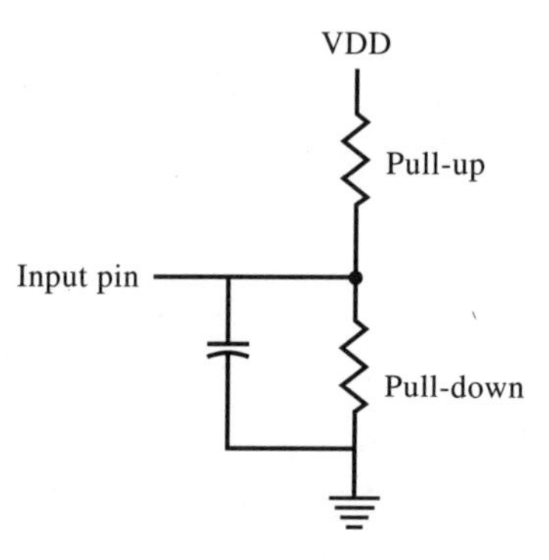

Figure 4.11-1 Input model for logic devices.

For TTL families, the resistors in the input normally dominate the effect of the capacitor, and the capacitor can be neglected. If we consider the input circuit being driven by an output circuit from Section 4.10, as shown in Figure 4.11-2, we can see that when the output is driven low the input loading causes current to flow as shown in part (a) of the figure—from V_{CC} through the pull-up resistor in the input circuit and the pull-down transistor in the output circuit. For this condition the input is a current source. If the output is now made to drive the input high, current flows from V_{CC} through the pull-up transistor of the output and the pull-down resistor of the input, as shown in part (b) of the figure. For this condition the input is a current sink. The maximum current the input will source when it's driven low is called I_{IL}. The maximum current the input will sink when it's driven high is called I_{IH}. These currents are normally something like 0.4 mA and 20 uA. Because the current is specified as the current that the input will sink, and the input sources this current, I_{IL} is negative. Table 4.11-1 lists the specifications of input and output characteristics for a 74LS08 just as they appear in a data book. Notice that some of the specifications are slightly different from nominal specifications we have discussed. In particular, V_{OH} is 2.7 volts rather than 2.4 volts. In general, specifications for the LS family of devices are slightly different from the nominal specifications for the TTL families. Table 4.11-1 contains some specifications we have not yet discussed. Some, such as V_{DD}, are obvious, but others need explanation. V_{IK} is the negative clamping voltage for the input. I_I is the input current when the input voltage is driven above the supply voltage as specified in the test conditions. I_{OS} is the output short-circuit current. I_{CCH} and I_{CCL} are supply currents for outputs high and low, respectively.

How are the specifications used? Suppose we have a 74LS08 output driving a number of other 74LS08 inputs; how many inputs can we drive? If the inputs are driven high, the driving 74LS08 has a recommended maximum source current of 0.4 mA ($I_{OH} = -0.4$ mA sink current). For this same high condition, each input sinks 20 uA ($I_{IH} = 20$ uA), so in this condition the 74LS08 will drive

$$0.4/0.02 = 20$$

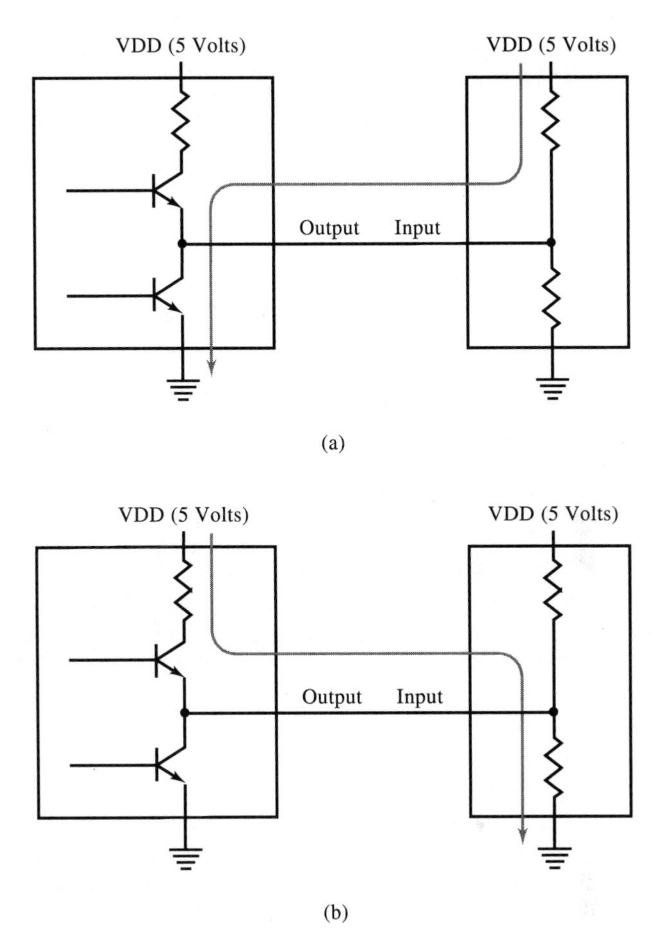

Figure 4.11-2 Logic device output circuit driving logic device input circuit.

inputs. Similarly, if the inputs are driven low, the driving 74LS08 has a maximum recommended sink current of 8 mA ($I_{OL} = 8$ mA). For this same low condition, each input sources 0.4 mA ($I_{IL} = -0.4$ mA sink current), so in this case the 74LS08 will drive

$$8/0.4 = 20$$

inputs again.

Now a note about CMOS device inputs. CMOS devices have very high impedance inputs. Usually, only the capacitor in the input circuit (such as that in Fig. 4.11-2) need be considered. The effect of this transistor is to slow the speed of any transition, because it takes time to charge and discharge.

Table 4.11-1 Electrical Specification for the 74LS08

Recommended Operating Conditions

	74LS08 Min	Nom	Max	Units
V_{CC} Supply voltage	4.75	5	5.75	V
V_{IH} High-level input voltage	2			V
V_{IL} Low-level input voltage			0.8	V
I_{OH} High-level output current			−0.4	mA
I_{OL} Low-level output current			8	mA
T_A Operating free-air temperature	0		70	°C

Electrical Characteristics of Recommended Operating Free-Air Temperature Range

Parameter	Test Conditions	74LS08 Min	Nom	Max	Units
V_{IK}	V_{CC} = MIN $I_I = -18$ A			−1.5	V
V_{OH}	V_{CC} = MIN V_{IH} = 2 V $I_{OH} = -0.4$ mA	2.7	3.4		V
V_{OL}	V_{CC} = MIN V_{IL} = MAX I_{OL} = 4 mA		0.25	0.4	V
	V_{CC} = MIN V_{IL} = MAX I_{OL} = 8 mA		0.35	0.5	V
I_I	V_{CC} = MAX V_I = 7 V			0.1	mA
I_{IH}	V_{CC} = MAX V_I = 2.7 V			20	uA
I_{IL}	V_{CC} = MAX V_I = 0.4 V			−0.4	mA
I_{OS}	V_{CC} = MAX	−20		−100	mA
I_{CCH}	V_{CC} = MAX V_I = 4.5 V		2.4	4.8	mA
I_{CCL}	V_{CC} = MAX V_I = 0 V		4.4	8.8	mA

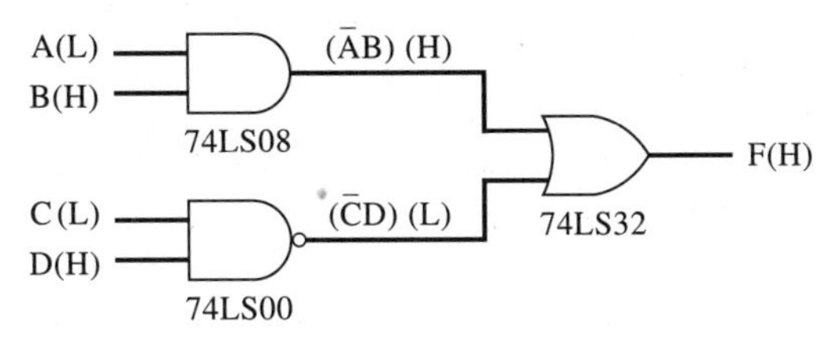

$$F(H) = (\bar{A}B + \overline{\bar{C}D})\ (H) = (\bar{A}B + C + \bar{D})\ (H)$$

Figure 4.12-1 Example logic circuit to be analyzed.

4.12 ANALYSIS

In our development of logic circuits we've neglected analysis because, with the techniques we're using, analysis is as simple as looking at the circuit. Now, however, we need to address analysis more carefully because we may encounter circuits designed by other, more tortuous techniques. Let's first investigate the analysis of a general circuit and then see why the circuits we have designed are easy to analyze.

To analyze a logic circuit, all we need to remember is that an assertion level incompatibility at the input of a gate causes that input to be complemented, and the output assertion level of a gate should be chosen to match the gate output. Consider the circuit in Figure 4.12-1. To analyze it, we first find the output of each input gate. These expressions are shown on the circuit. Notice that an incompatibility at the gate input causes the logic variable associated with that input to be complemented in the output expression. Notice also that the output expression at each gate is assigned an assertion level that is compatible with the gate output. If there is no output bubble, the assertion level is high. If there is an output bubble, the assertion level is low.

The second step of the analysis is to find the output of the output gate, using as inputs the output expressions from the two input gates. This final expression is also shown on the circuit. You can see that the second input is incompatible with the gate and must be complemented. The output expression contains complements of complements and can be simplified using DeMorgan's theorem (shown in the figure).

As already stated, our design of logic circuits makes analysis easy. We can write the logic expression for the circuit by inspection. Consider the SOP circuit from Section 3.8, which is repeated in Figure 4.12-2. The steps in the analysis of this circuit are shown. In this circuit there are no incompatibilities at the inputs of the output gate, so there is no need to use DeMorgan's theorem. This will always be the case with circuits designed using the concepts developed in this text.

We can easily obscure the inherent simplicity of the design by redrawing the circuit as shown in Figure 4.12-3. (Some perverse engineers actually do this.) The analysis for the redrawn circuit is shown. When the circuit is drawn like this, the output expression is not immediately evident and in fact requires the use of DeMorgan's theorem for simplification. The fact that we have lost the simplicity of the circuit by redrawing it suggests that we might try to recover the original circuit.

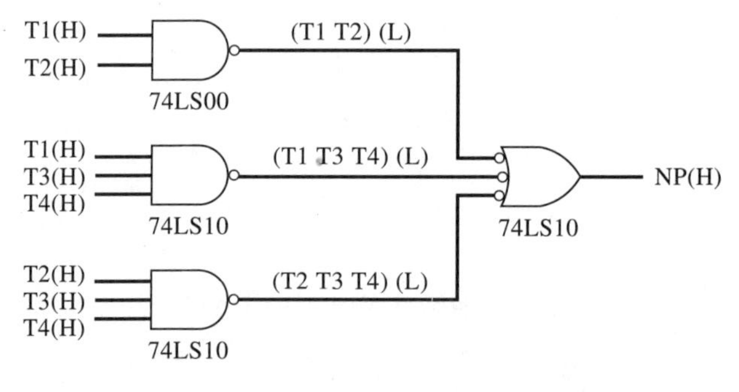

$$NP(H) = (T1\ T2 + T1\ T3\ T4 + T2\ T3\ T4)\ (H)$$

Figure 4.12-2 SOP based circuit to be analyzed.

Fletcher's text, *An Engineering Approach to Digital Design*, suggests a recovery method that can considerably reduce the work of the analysis and hence the possibility of error. Using Fletcher's technique, we start at the output of the circuit and try to make the inputs of all the gates compatible by changing their logic symbols. In the circuit of Figure 4.12-3 we would change the output gate to its equivalent *or* symbol (complement the gate) to make it compatible with the output. When we do this, we find that we have compatibility throughout the circuit. In fact, we have recovered the circuit we first analyzed.

Redrawing the circuit does not always result in compatibility throughout the circuit, however. Consider the logic circuit in Figure 4.12-4. The usual analysis of this circuit is shown. To simplify the output expression, we must use DeMorgan's

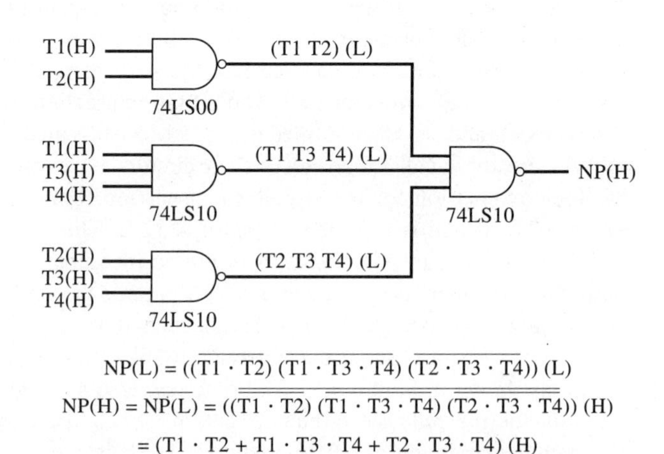

$$NP(L) = ((\overline{T1 \cdot T2})\ (\overline{T1 \cdot T3 \cdot T4})\ (\overline{T2 \cdot T3 \cdot T4}))\ (L)$$

$$NP(H) = \overline{NP(L)} = (\overline{\overline{(T1 \cdot T2)}\ \overline{(T1 \cdot T3 \cdot T4)}\ \overline{(T2 \cdot T3 \cdot T4)}})\ (H)$$

$$= (T1 \cdot T2 + T1 \cdot T3 \cdot T4 + T2 \cdot T3 \cdot T4)\ (H)$$

Figure 4.12-3 The circuit of Figure 4.12-2 redrawn to obscure its inherent simplicity.

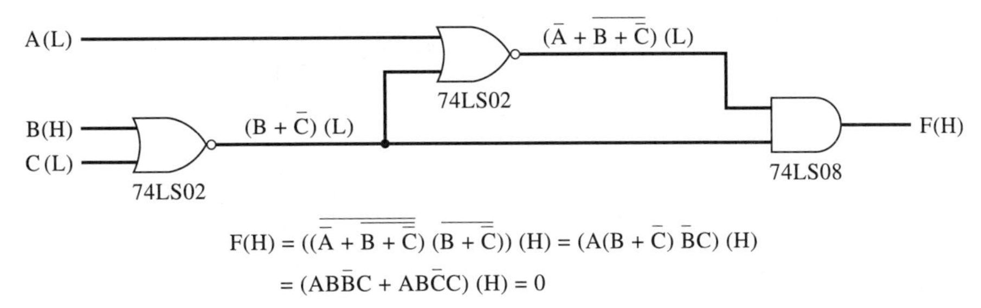

$$F(H) = ((\overline{\bar{A} + \overline{B + \bar{C}}})\ (\overline{B + \bar{C}}))\ (H) = (A(B + \bar{C})\ \bar{B}C)\ (H)$$
$$= (AB\bar{B}C + AB\bar{C}C)\ (H) = 0$$

Figure 4.12-4 Logic circuit for analysis.

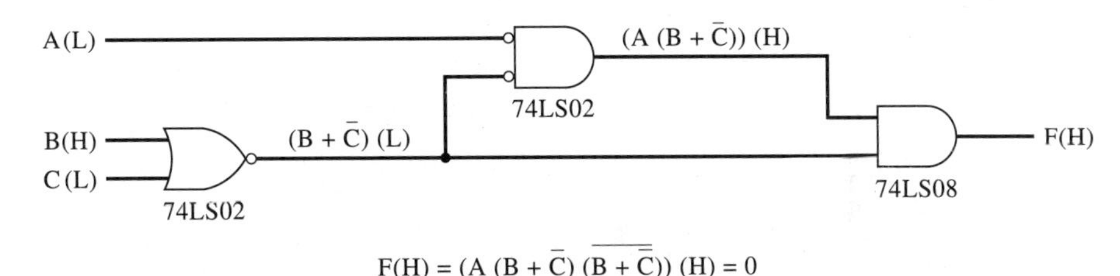

$$F(H) = (A\ (B + \bar{C})\ (\overline{B + \bar{C}}))\ (H) = 0$$

Figure 4.12-5 Logic circuit of Figure 4.12-4 redrawn to simplify analysis.

theorem twice. (We would obviously never build this circuit since it produces a trivial result.)

Now let's try to simplify the circuit before we analyze it. We start at the output and redraw the circuit to try to make all the assertion levels compatible, by changing (complementing) the operational symbols for the gates. The output of the output gate is already compatible with the gate, so we do not change the output gate operation symbol. The upper gate in the circuit has an output that is incompatible with the input to the output gate, so we change its operational symbol. The other input gate is not changed. Its output is incompatible with the input of the output gate, but if we change it, we will produce an incompatibility at the input to the upper gate. Here we cannot produce compatibility at both gate inputs, so we leave one incompatibility. The changes described produce the circuit in Figure 4.12-5. The analysis of this redrawn circuit is shown. By redrawing the circuit we have reduced the use of DeMorgan's theorem in the simplification of the output expression for the circuit.

4.13 POSITIVE LOGIC, NEGATIVE LOGIC

Polarized signals (mnemonics) give us a very general technique for the design, documentation, and analysis of logic circuits. They are not universally used, however. Let's examine some of the other notations that are used. One (the most common) assumes that all assertion levels are positive; it is called positive logic. We can

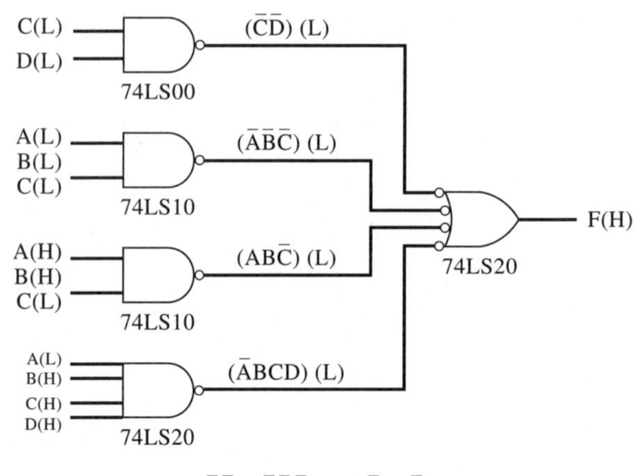

Figure 4.13-1 Example logic circuit.

change from polarized mnemonics to positive logic notation, and vice versa, by noting that

$$A(H) = A$$

$$A(L) = \overline{A}(H) = \overline{A}$$

With positive logic, the assertion level need not be given with the variable, because it is always high. Suppose we change one of the circuits we designed in Section 3.6 to positive logic notation. The circuit is first shown in Figure 4.13-1 with polarized mnemonics. It is redrawn using positive logic notation in Figure 4.13-2. The analysis for the redrawn circuit is also shown. With positive logic notation, a bubble at either an input or an output of a gate produces a complement.

You should now have an idea of how analysis is done using positive logic notation, but you needn't become proficient in these techniques. If you want to analyze a circuit that uses positive logic notation, change the notation to polarized mnemonics and then do the analysis.

Some designers draw this circuit in a form that makes the analysis more difficult (shown in Figure 4.13-3). This form changes the logic symbol for the output gate. The analysis of this circuit using positive logic notation is also shown. Notice the use of DeMorgan's theorem.

We have looked at positive logic notation where all assertion levels are assumed high. Another notation (much less common) assumes that all assertion levels are low; it is called negative logic. We can change from polarized mnemonics to negative logic notation, and vice versa, by noting that

$$A(L) = A$$

$$A(H) = \overline{A}(L) = \overline{A}$$

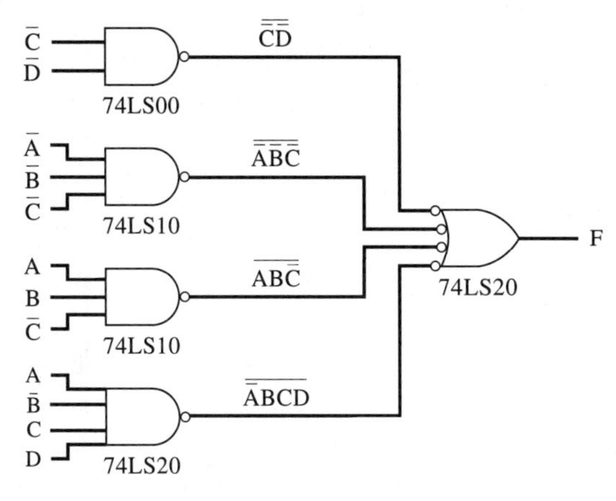

$$F = \bar{C}\bar{D} + \bar{A}\bar{B}\bar{C} + AB\bar{C} + \bar{A}BCD$$

Figure 4.13-2 Example logic circuit with positive logic notation.

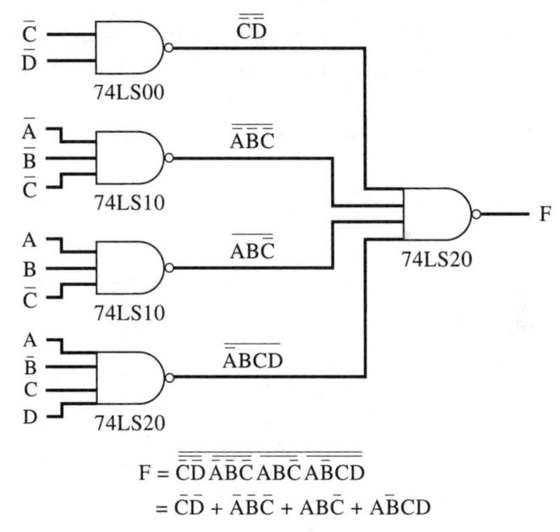

$$F = \overline{\overline{\bar{C}\bar{D}}\;\overline{\bar{A}\bar{B}\bar{C}}\;\overline{AB\bar{C}}\;\overline{\bar{A}BCD}}$$
$$= \bar{C}\bar{D} + \bar{A}\bar{B}\bar{C} + AB\bar{C} + \bar{A}BCD$$

Figure 4.13-3 The circuit of Figure 4.12-2 redrawn in a form that makes the analysis more difficult.

In negative logic notation, the assertion level need not be given with the variable, because it is always low. To change from positive logic to negative logic, and vice versa, all the variables are complemented. An example of negative logic notation applied to the circuit of Figure 4.12-1 is given in Figure 4.13-4, along with the negative logic analysis of the circuit. In a negative logic analysis, the variables are complemented when there are no bubbles. In this circuit, the inputs to all the gates are complemented (no bubbles), and the outputs are not complemented (bubbles).

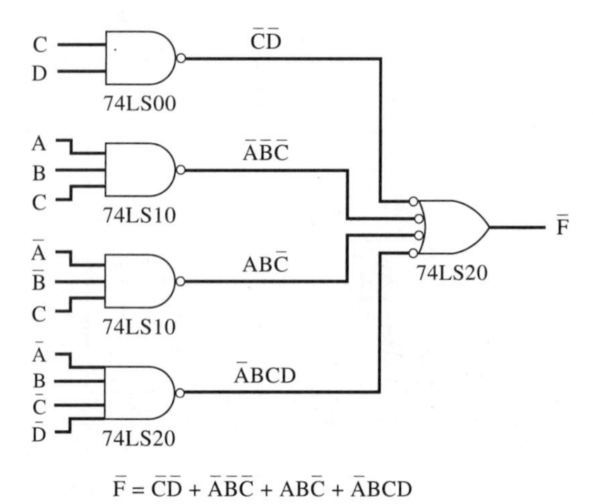

Figure 4.13-4 Circuit of Figure 4.12-1 with negative logic notation.

It is important to become proficient in changing back and forth between polarized mnemonics and positive logic notation and, to a lesser extent, between polarized mnemonics and negative logic notation. Don't worry about becoming proficient in analysis using these notations. You can always change to polarized notation and then analyze.

KEY TERMS

Gates A one-output, many-input logic element in which the output is a simple logic function of the input.

Custom Logic Device Logic circuit fabricated on a single chip.

Programmable Logic Device A logic device in which a large number of logic elements are fabricated on a single chip with programmable interconnections.

Fixed Architecture Programmable devices with a relatively fixed arrangement of logic elements.

Programmable Read Only Memory (PROM) Programmable memories which are useful as fixed architecture logic devices. They have fixed *and* gate followed by programmable *or* gates.

Programmable Array Logic (PAL) Fixed architecture logic devices with programmable *and* gates followed by fixed *or* gates.

Programmable Logic Array (PLA) Fixed architecture logic devices with programmable *and* gates followed by programmable *or* gates.

Flexible Architecture Programmable logic device in which the interconnection of the logic elements is almost unrestricted.

Field Programmable Gate Arrays (FPGAs) A flexible architecture programmable logic device.

Timing Diagram A diagram which shows the relation between the inputs and outputs of a logic device as a function of time.
Propagation Delay The time delay between a change in the input and the corresponding change in the output for a logic element.
Polarized Mnemonics A scheme used for labeling signals in a logic expression in which the polarity of the signal is shown explicitly as asserted high (H) or asserted low (L). The mnemonic part of the name implies that the name is chosen as a mnemonic for the signal.
Inhibit An input signal condition which stops an output of a logic element from being asserted.
Enable An input signal condition which allows an output of a logic element to be asserted.
Positive Logic *And* Gate A gate in which the output is high if all the inputs are high.
Positive Logic *Or* Gate A gate in which the output is high if one or more of the inputs is high.
Positive Logic *Nand* Gate A gate in which the output is low if all the inputs are high.
Positive Logic *Nor* Gate A gate in which the output is low if one or more of the inputs is high.
Exclusive *Or* (xor) Gate A 2-input gate in which the output is high if either input is high but not both.
Exclusive *Nor* (xnor) Gate A 2-input gate in which the output is low if either input is high but not both.
Full Adder A binary adder with carry-in and carry-out.
Half Adder A binary adder with no carry-in.
Glitch A (usually short duration) unwanted output in a logic circuit.
Output Enable A logic device input which allows the output to occur.

EXERCISES

1. Fill in the table that precedes the circuits. The columns labeled A(H), B(H), and F(H) should be filled in with H or L. Use the meaning of the basic gate symbol to fill in the F(H) column. F is a logic function, so it should have value 1 or 0.

A	B	A(H)	B(H)	F(H)	F
0	0				
0	1				
1	0				
1	1				

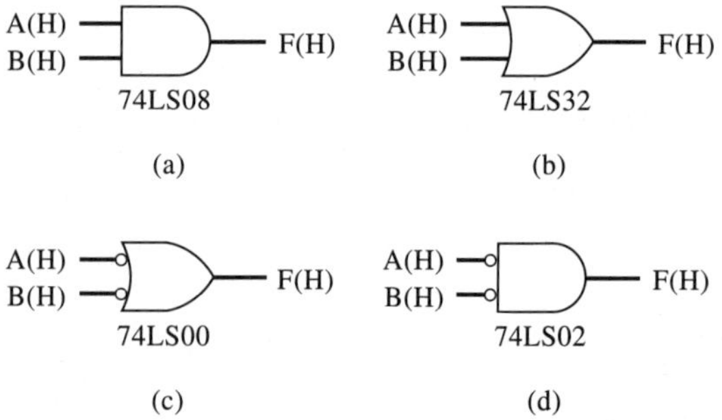

A(H)
B(H)
F(H)
74LS08
(a)

A(H)
B(H)
F(H)
74LS32
(b)

A(H)
B(H)
F(H)
74LS00
(c)

A(H)
B(H)
F(H)
74LS02
(d)

2. Fill in the table that precedes the circuits. The columns labeled A(H), B(L), and F(H) should be filled in with H or L. Use the meaning of the basic gate symbol to fill in the F(H) column. F is a logic function, so it should have value 1 or 0.

A	B	A(H)	B(L)	F(H)	F
0	0				
0	1				
1	0				
1	1				

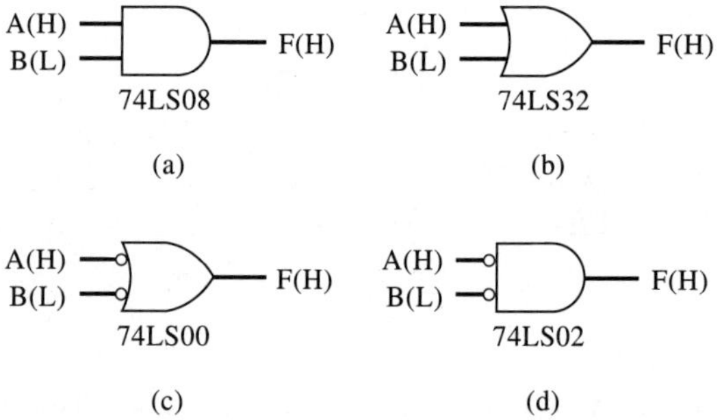

A(H)
B(L)
F(H)
74LS08
(a)

A(H)
B(L)
F(H)
74LS32
(b)

A(H)
B(L)
F(H)
74LS00
(c)

A(H)
B(L)
F(H)
74LS02
(d)

3. Fill in the table that precedes the circuits. The columns labeled A(L), B(H), and F(H) should be filled in with H or L. Use the meaning of the basic gate symbol to fill in the F(H) column. F is a logic function, so it should have value 1 or 0.

A	B	A(L)	B(H)	F(H)	F
0	0				
0	1				
1	0				
1	1				

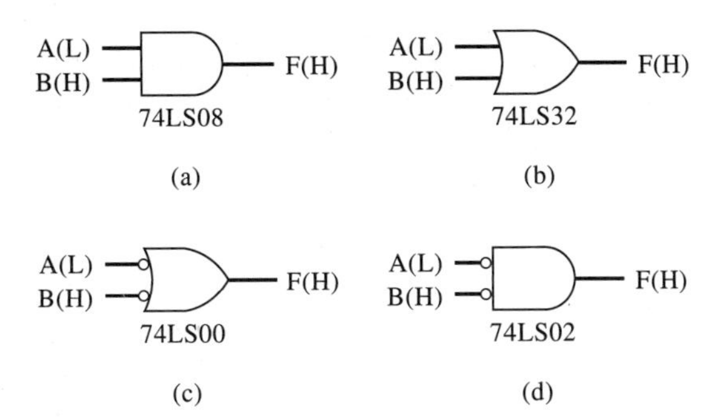

4. Fill in the table that precedes the circuits. The columns labeled A(L), B(L), and F(H) should be filled in with H or L. Use the meaning of the basic gate symbol to fill in the F(H) column. F is a logic function, so it should have value 1 or 0.

A	B	A(L)	B(L)	F(H)	F
0	0				
0	1				
1	0				
1	1				

A(L) B(L) F(H) 74LS08 (a)

A(L) B(L) F(H) 74LS32 (b)

A(L) B(L) F(H) 74LS00 (c)

A(L) B(L) F(H) 74LS02 (d)

5. Fill in the table that precedes the circuits. The columns labeled A(H), B(H), and F(L) should be filled in with H or L. Use the meaning of the basic gate symbol to fill in the F(L) column. F is a logic function, so it should have value 1 or 0.

A	B	A(H)	B(H)	F(L)	F
0	0				
0	1				
1	0				
1	1				

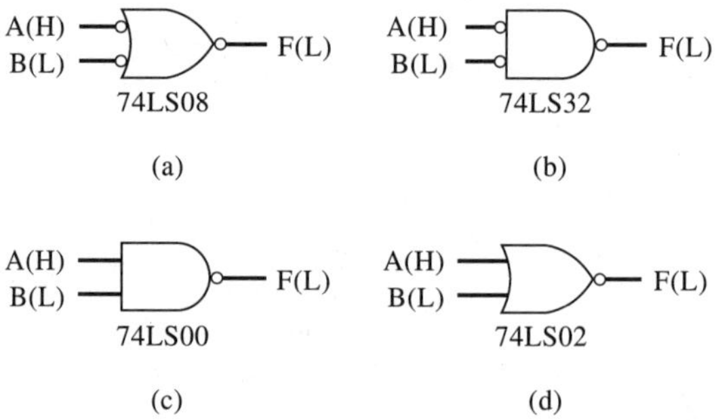

A(H) B(L) F(L) 74LS08

(a)

A(H) B(L) F(L) 74LS32

(b)

A(H) B(L) F(L) 74LS00

(c)

A(H) B(L) F(L) 74LS02

(d)

6. Fill in the table that precedes the circuits. The columns labeled A(H), B(L), and F(L) should be filled in with H or L. Use the meaning of the basic gate symbol to fill in the F(L) column. F is a logic function, so it should have value 1 or 0.

A	B	A(H)	B(L)	F(L)	F
0	0				
0	1				
1	0				
1	1				

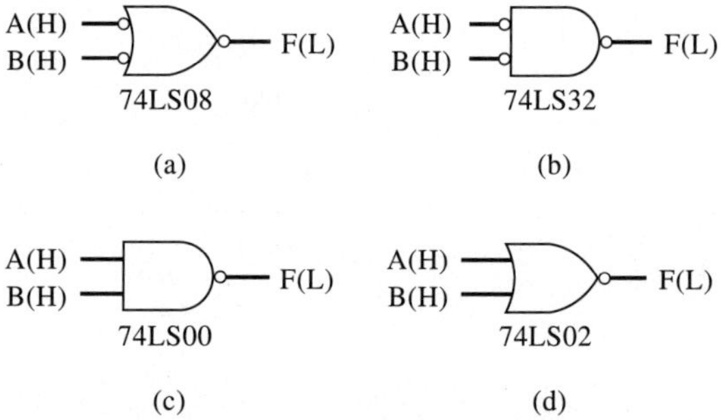

A(H) B(H) F(L) 74LS08

(a)

A(H) B(H) F(L) 74LS32

(b)

A(H) B(H) F(L) 74LS00

(c)

A(H) B(H) F(L) 74LS02

(d)

7. Fill in the table that precedes the circuits. The columns labeled A(L), B(H), and F(L) should be filled in with H or L. Use the meaning of the basic gate symbol to fill in the F(L) column. F is a logic function, so it should have value 1 or 0.

A	B	A(L)	B(H)	F(L)	F
0	0				
0	1				
1	0				
1	1				

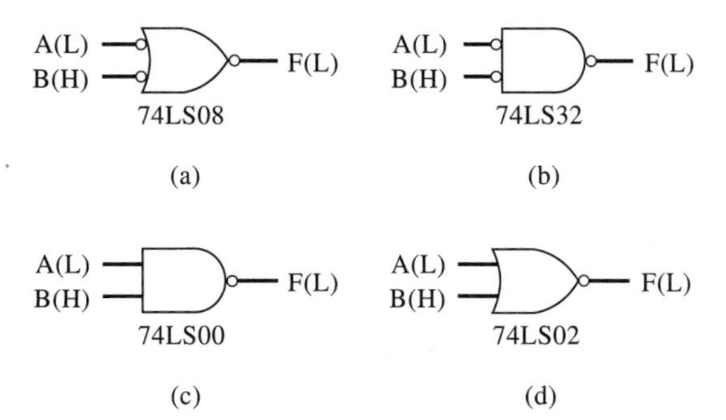

8. Fill in the table that precedes the circuits. The columns labeled A(L), B(H), and F(L) should be filled in with H or L. Use the meaning of the basic gate symbol to fill in the F(L) column. F is a logic function, so it should have value 1 or 0.

A	B	A(L)	B(H)	F(L)	F
0	0				
0	1				
1	0				
1	1				

A(L) B(H) F(L) 74LS08 (a)

A(L) B(H) F(L) 74LS32 (b)

A(L) B(H) F(L) 74LS00 (c)

A(L) B(H) F(L) 74LS02 (d)

9. Design logic circuits using discrete gates to evaluate the following single-term logic functions with the output signal assertion levels given. There are two correct circuits for each expression.

(a) $F = \overline{A}\,\overline{B}C$ F(L)
(b) $F = \overline{A}BC$ F(L)
(c) $F = \overline{A}\,\overline{B}C$ F(H)
(d) $F = \overline{A}BC$ F(H)
(e) $F = \overline{A}\,\overline{B}\,\overline{C}\,\overline{D}$ F(H)
(f) $F = \overline{A}\,\overline{B}\,\overline{C}\,\overline{D}$ F(L)
(g) $F = A + B + \overline{C}$ F(L)

(h) $F = \overline{A} + \overline{B} + C$ F(L)

(i) $F = \overline{A} + B + C$ F(L)

(j) $F = \overline{A} + \overline{B} + C$ F(H)

(k) $F = \overline{A} + B + C$ F(H)

(l) $F = \overline{A} + \overline{B} + \overline{C} + \overline{D}$ F(H)

(m) $F = \overline{A} + \overline{B} + \overline{C} + \overline{D}$ F(L)

10. Design logic circuits using the kinds of gates found in programmable gate arrays to evaluate the following single-term logic expressions with the signal assertion levels given.

(a) $F = \overline{A}\,\overline{B}C$ A(L), B(H), C(H), F(L)

(b) $F = \overline{A}BC$ A(H), B(L), C(L), F(L)

(c) $F = \overline{A}\,\overline{B}C$ A(H), B(H), C(L), F(H)

(d) $F = \overline{A}BC$ A(L), B(L), C(H), F(H)

(e) $F = \overline{A}\,\overline{B}\,\overline{C}\,\overline{D}$ A(H), B(H), C(L), D(H), F(H)

(f) $F = \overline{A}\,\overline{B}\,\overline{C}\,\overline{D}$ A(L), B(L), C(L), D(H), F(L)

(g) $F = A + B + \overline{C}$ A(L), B(H), C(H), F(L)

(h) $F = \overline{A} + \overline{B} + C$ A(H), B(L), C(L), F(L)

(i) $F = \overline{A} + B + C$ A(H), B(H), C(L), F(H)

(j) $F = \overline{A} + \overline{B} + C$ A(L), B(L), C(H), F(H)

(k) $F = \overline{A} + B + C$ A(H), B(H), C(L), F(H)

(l) $F = \overline{A} + \overline{B} + \overline{C} + D$ A(L), B(L), C(L), D(H), F(L)

(m) $F = \overline{A} + \overline{B} + \overline{C} + \overline{D}$ A(H), B(H), C(H), D(L), F(H)

11. Design SOP based logic circuits to evaluate the following logic expressions with the signal assertion levels given. Be sure to simplify the expression before you do the design. Use only positive logic *nand* and/or *nor* gates and inverters.

(a) $F = AB\overline{C} + A\overline{B}\,\overline{C} + ABC$; A(L), B(H), C(L), F(L)

(b) $F = \overline{A}\,\overline{B}\,\overline{C} + A\overline{B}\,\overline{C} + ABC + A\overline{B}C$; A(L), B(L), C(L), F(L)

(c) $F = AB + A\overline{B} + \overline{A}BC + \overline{A}\,\overline{B}C$; A(L), B(H), C(H), F(H)

(d) $F = \overline{A}\,\overline{C} + B\overline{C} + BC + AC$; A(H), B(H), C(H), F(L)

(e) $F = \overline{A}\,\overline{B}\,\overline{C}\,\overline{D} + \overline{A}B\overline{C}\,\overline{D} + ACD + A\overline{B}C\overline{D}$; A(L), B(H), C(L), D(L), F(H)

(f) $F = \overline{A}\,\overline{B}D + A\overline{B}D + \overline{B}CD$; A(H), B(H), C(L), D(L), F(H)

(g) $F = \overline{A}\,\overline{B}CD + \overline{A}BD + AB\overline{C} + A\overline{B}\,\overline{C}\,\overline{D}$; A(L), B(H), C(H), D(L), F(L)

(h) $F = \overline{A}\,\overline{B}CD + ABD + \overline{A}BD + A\overline{B}D$; A(H), B(H), C(L), D(H), F(H)

(i) $F = \overline{A}\,\overline{B}D + B\overline{C}D + ABD + A\overline{B}D$; A(H), B(H), C(H), D(L), F(L)

(j) $F = \overline{A}\,\overline{C}\,\overline{D} + \overline{A}\,\overline{C}D + B\overline{C}D$; A(H), B(H), C(L), D(L), F(H)

(k) $F = \overline{A}\,\overline{B}CD + \overline{A}BC\overline{D} + ABCD + A\overline{B}C\overline{D} + A\overline{B}\,\overline{C}\,\overline{D}$; A(L), B(L), C(L), D(L), F(H)

(l) $F = \overline{A}\,\overline{B}\,\overline{C}\,\overline{D} + ABCD + ABC\overline{D} + \overline{A}\,\overline{B}\,\overline{C}D + \overline{A}\,\overline{B}CD$; A(H), B(H), C(L), D(H), F(L)

(m) $F = BD + ABD + CBD + \overline{A}\,\overline{B}\,\overline{C}$; A(L), B(L), C(L), D(L), F(L)

(n) $F = \overline{A}\,\overline{B}\,\overline{C}\,\overline{D} + ABCD + \overline{A}\,\overline{B}CD + \overline{A}\,\overline{B}\,\overline{C}D$; A(H), B(L), C(L), D(L), F(H)

(o) $F = ABCD + ABC\overline{D} + AB\overline{C}D + A\overline{B}CD + \overline{A}BCD$; A(H), B(H), D(L), D(L), F(H)

(p) $F = \overline{A}\,\overline{B}\,\overline{C} + \overline{A}\,\overline{B}\,\overline{C}D + A\overline{B}\,\overline{C} + ABCD$; A(L), B(L), C(L), D(H), F(H)

(q) $F = \overline{A}\,\overline{B}\,\overline{C} + \overline{A}\,\overline{B}\,\overline{D} + ABC + ABD$; A(H), B(H), C(L), D(H), F(L)

(r) $F = A + \overline{A}BCD + \overline{A}BC + \overline{A}\,\overline{B}\,\overline{C} + \overline{A}\,\overline{B}C$; A(H), B(L), C(L), D(H), F(L)

(s) $F = ABCD + \overline{A}\,\overline{B}\,\overline{C}\,\overline{D} + \overline{A}\,\overline{B}CD + AB\overline{C}\,\overline{D}$; A(L), B(L), C(L), D(L), F(L)

12. Repeat exercise 11 with POS based logic. Use only positive logic *nand* and/or *nor* gates and inverters.

13. Design both SOP and POS based logic circuits to evaluate the logic expressions in exercise 11, using the kinds of gates found in programmable gate arrays and the signal assertion levels given. Use no inverters.

14. Design both an SOP based logic circuit and a POS based logic circuit to evaluate each of the functions in the following table. Use only positive logic *nand* and/or *nor* gates. For (a) through (d), assume the signal assertion levels are A(H), B(L), C(L), and F(H). For (e) through (i), assume A(L), B(L), C(H), and F(L).

A	B	C	F_a	F_b	F_c	F_d	F_f	F_g	F_h	F_i
0	0	0	0	0	1	0	0	0	1	0
0	0	1	1	1	0	0	0	1	0	1
0	1	0	0	1	0	1	0	0	0	1
0	1	1	0	0	1	0	1	1	0	1
1	0	0	0	1	1	0	1	1	0	1
1	0	1	1	1	0	1	1	1	1	0
1	1	0	1	0	0	1	0	0	0	1
1	1	1	0	0	0	1	1	1	1	1

15. Design both SOP and POS based logic circuits to evaluate the logic functions in the table of exercise 14, using the kinds of gates found in programmable gate arrays and the signal assertion levels given. Use no inverters.

16. Design both an SOP based logic circuit and a POS based logic circuit to evaluate each of the functions in the following table. For (a) through (d), assume the signal assertion levels are A(H), B(L), C(L), D(H), and F(H). For (e) through (h), assume A(L), B(L), C(H), D(L), and F(L). For (i) through (k), assume A(L), B(H), C(L), D(L), and F(L).

A	B	C	D	F_a	F_b	F_c	F_d	F_e	F_f	F_g	F_h	F_i	F_j	F_k
0	0	0	0	0	1	0	1	0	0	0	1	0	0	0
0	0	0	1	1	1	1	0	0	0	1	1	0	1	0
0	0	1	0	0	1	1	0	0	0	0	0	1	0	1
0	0	1	1	1	0	0	1	1	1	1	1	0	1	0
0	1	0	0	1	0	1	1	1	0	0	1	0	1	0
0	1	0	1	1	1	1	0	1	1	0	0	0	1	0
0	1	1	0	0	0	0	0	1	1	0	1	1	0	1
0	1	1	1	0	1	1	1	0	0	1	0	0	0	1
1	0	0	0	0	1	1	0	0	1	0	1	1	1	1
1	0	0	1	0	1	1	0	0	1	1	1	0	0	0
1	0	1	0	1	1	0	0	0	0	1	0	1	1	0
1	0	1	1	0	0	0	0	1	1	1	0	0	0	1
1	1	0	0	0	0	1	0	0	0	0	1	1	0	1
1	1	0	1	0	0	1	0	1	0	1	0	0	1	1
1	1	1	0	0	1	1	1	0	1	0	0	0	1	1
1	1	1	1	1	0	1	1	1	0	0	1	1	0	1

17. Design both SOP and POS based logic circuits to evaluate the logic functions in the table of exercise 16, using the kinds of gates found in programmable gate arrays and the signal assertion levels given. Use no inverters.

18. Design both an SOP based logic circuit and a POS based logic circuit to evaluate each of the following logic functions. Use only positive logic *nand* and/or *nor* gates and inverters. Assume the signal assertion levels are A(L), B(L), C(L), D(H), and F(L).

(a) $F = (\overline{A}+\overline{B}+\overline{C}+\overline{D})(\overline{A}+B+\overline{C}+\overline{D})(A+C+D)(A+\overline{B}+C+\overline{D})$

(b) $F = (\overline{A}+\overline{B}+D)(A+\overline{B}+D)(\overline{B}+C+D)$

(c) $F = (\overline{A}+\overline{B}+C+D)(\overline{A}+B+D)(A+B+\overline{C})(A+\overline{B}+\overline{C}+\overline{D})$

(d) $F = (\overline{A}+\overline{B}+D)(B+\overline{C}+D)(A+B+D)(A+\overline{B}+D)$

(e) $F = (\overline{A}+\overline{C}+\overline{D})(\overline{A}+\overline{C}+D)(B+\overline{C}+D)$

(f) $F = (\overline{A}+\overline{B}+C+D)(\overline{A}+B+C+\overline{D})(A+B+C+D)(A+\overline{B}+C+\overline{D})$
$(A+\overline{B}+\overline{C}+\overline{D})$

19. Design both SOP and POS based logic circuits to evaluate the logic expressions in exercise 18, using the kinds of gates found in programmable gate arrays and the signal assertion levels given. Use no inverters.

20. Design logic circuits to evaluate each of the following 1-bit, *binary* (not Boolean) expressions. Drop all carries, and assume that a borrow can be made whenever it is needed. Use any discrete gate you wish. Make the device count as small as possible. Assume that signals A, B, C, and D are all asserted high, and make the output signal F asserted low.

(a) $F = A + B - C$

(b) $F = AB - C$ (AB is $A \times B$)

(c) $F = A + B + C + D$

(d) $F = A + B + C - D$

(e) $F = ABC - AD$

21. Design logic circuits to evaluate the product $F_1\ F_2\ F_3\ F_4$ in the following 2-bit, binary multiplication.

$$\begin{array}{r} A\,B \\ \times C\,D \\ \hline F_4F_3F_2F_1 \end{array}$$

Use any discrete gates you wish. Assume that the signals A, B, C, and D are asserted high, and make the signals representing the product bits ($F_1\ F_2\ F_3\ F_4$) all asserted high.

22. Design both an SOP based logic circuit and a POS based logic circuit to evaluate each of the functions in the following table. Use only positive logic *nand* and/ or *nor* gates. For (a) through (d), assume the signal assertion levels are A(H), B(L), C(L), and F(H). For (e) through (g), assume A(L), B(L), C(H), and F(L).

A	B	C	F_a	F_b	F_c	F_d	F_e	F_f	F_g
0	0	0	0	0	x	1	0	1	0
0	0	1	x	1	x	x	1	0	1
0	1	0	x	x	1	0	0	x	1
0	1	1	0	0	x	1	1	x	x
1	0	0	0	1	0	x	x	0	1
1	0	1	1	1	1	1	1	1	0
1	1	0	1	x	1	0	0	0	x
1	1	1	0	x	1	0	0	1	x

23. Design both SOP and POS based logic circuits to evaluate the logic functions in the table of exercise 22, using the kinds of gates found in programmable gate arrays and the signal assertion levels given. Use no inverters.

24. Design both an SOP based logic circuit and a POS based logic circuit to evaluate each of the functions in the table below. For (a) through (d) assume the signal assertion levels are A(H), B(L), C(L), D(H), and F(H). For (e) through (h), assume A(L), B(L), C(H), D(L), and F(L). For (i) through (k), assume A(L), B(H), C(L), D(L), and F(L).

A	B	C	D	F_a	F_b	F_c	F_d	F_e	F_f	F_g	F_h	F_i	F_j	F_k
0	0	0	0	0	x	x	1	0	1	0	x	0	0	x
0	0	0	1	x	x	1	0	0	0	1	x	0	x	x
0	0	1	0	1	1	1	0	0	0	0	0	1	x	1
0	0	1	1	1	0	x	1	x	1	1	1	0	x	0
0	1	0	0	x	0	1	1	x	x	x	1	0	x	0
0	1	0	1	x	1	1	0	1	x	x	0	0	1	0
0	1	1	0	0	0	0	0	1	x	0	1	1	0	1
0	1	1	1	0	x	1	1	x	0	1	0	0	0	1
1	0	0	0	0	x	1	x	0	1	0	1	x	1	1
1	0	0	1	0	1	1	x	0	1	1	1	x	0	0
1	0	1	0	1	1	0	0	1	0	1	0	x	1	0
1	0	1	1	0	0	0	0	1	1	x	0	x	0	x
1	1	0	0	x	0	x	1	0	0	0	x	1	0	x
1	1	0	1	x	0	x	0	1	0	1	x	0	1	1
1	1	1	0	0	1	1	1	0	1	0	0	0	1	1
1	1	1	1	1	0	1	1	1	0	0	1	1	0	1

25. **(a)** Change each of the following circuits from positive logic notation to polarized notation.

(b) Analyze each circuit without altering its form except to change from positive logic notation to polarized notation.

(c) Simplify each circuit by changing the form of the gates to make them compatible as far as possible and then analyze.

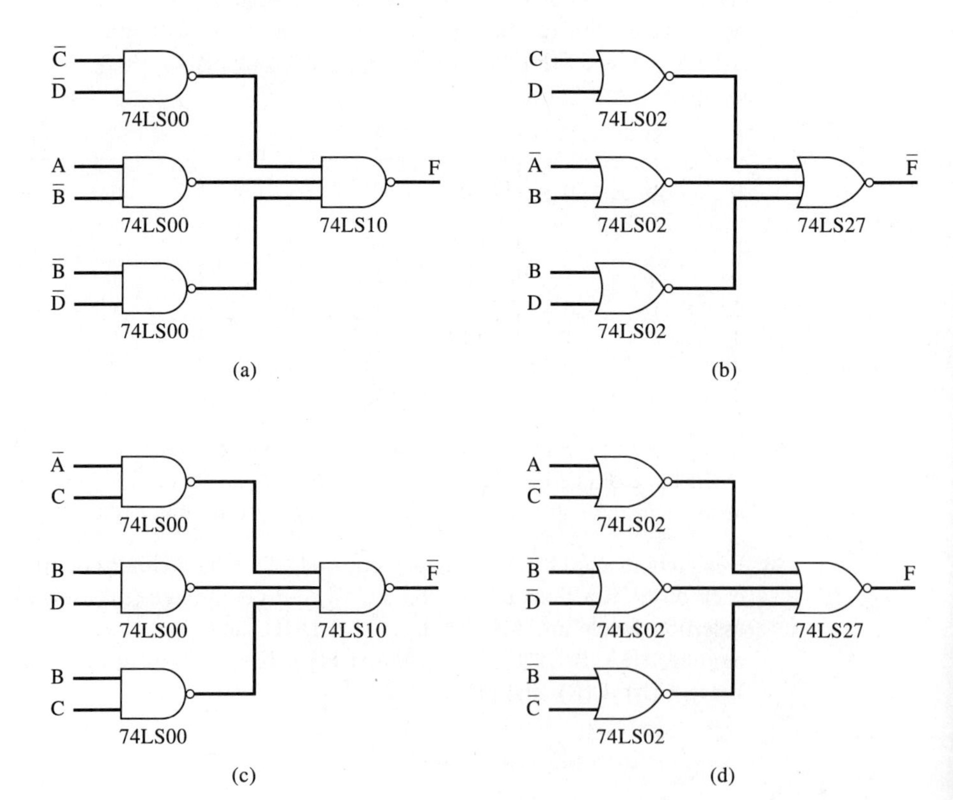

5 Examples of Combinational Logic Circuits

5.1 INTRODUCTION

Combinational logic circuits can be designed to perform an infinite number of logic operations. Some of these operations have proved so useful that the circuits to perform them are available commercially as discrete digital devices. Circuits to perform arithmetic operations on binary numbers are obvious candidates for these circuits, but also useful are others which perform electronic switching (multiplexing and demultiplexing), encoding and decoding of binary numbers, and parity generation. We'll describe the commercially available combinational logic circuits in this chapter.

These types of circuits are also available as predesigned subcircuits in FPGA designs. We'll also offer in this chapter some examples of these predesigned circuits.

5.2 MULTIPLEXERS

A **multiplexer** is an electronically controlled switch. Figure 5.2-1 shows the function of a 2-line to 1-line multiplexer—a multiplexer with two inputs and one output. The switch connects the output to one input or the other, depending on a select signal. Since there are only two possible ways to connect the input lines, only one select signal is needed. If the select line is high, the output will be switched to input 1, and if the select line is low, the output will be switched to input 0.

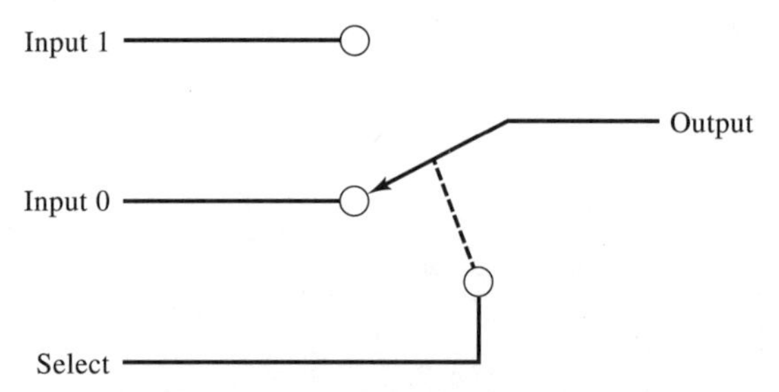

Figure 5.2-1 Function of a 2-line to 1-line multiplexer switched to input 0.

Figure 5.2-1 shows the actual function of a 2-line to 1-line analog multiplexer. In an analog multiplexer, the output is literally equal to the input for all values of input. We're more concerned with a digital multiplexer, in which the output will be high if the input is high, and the output will be low if the input is low. Following the normal combinational logic circuit design steps, we can design a circuit to perform the function of a 2-line to 1-line digital multiplexer.

To create a logic circuit design, we must first make a concise statement of the function of the multiplexer. The inputs to the multiplexer are the select signal (SEL), input 1 (IN1), and input 0 (IN0). There is only one output (OUT). If SEL is asserted, then OUT will be the same value as IN1, independent of the value of IN0. If SEL is not asserted, then OUT will be the same value as IN0, independent of the value of IN1. The following truth table shows the results of these conditions.

SEL	IN0	IN1	OUT
0	0	0	0
0	0	1	0
0	1	0	1
0	1	1	1
1	0	0	0
1	0	1	1
1	1	0	0
1	1	1	1

Figure 5.2-2 is the Karnaugh map and logic function for the multiplexer, formulated from the truth table. Figure 5.2-3 shows a design for the multiplexer circuit using positive logic *and* and *or* gates when the assertion levels for all input signals and the output signals are assumed to be high.

Normally a multiplexer would not be formed from discrete gates; rather, all its gates would be fabricated on a single integrated circuit. In fact, four 2-line to 1-line multiplexers are usually fabricated on a single integrated circuit and packaged

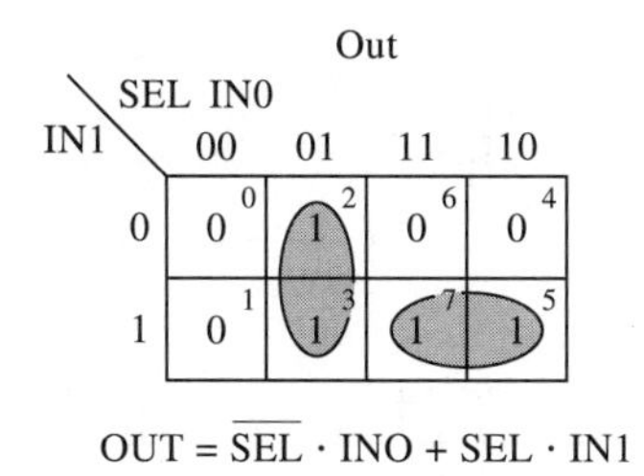

Figure 5.2-2 Two-line to 1-line multiplexer logic.

together. Figure 5.2-4 shows the logic circuits for a quadruple 2-line to 1-line multiplexer (74LS157) and an inverting quadruple 2-line to 1-line multiplexer (74LS158). Notice that these multiplexers have a strobe (STROBE $\overline{G}$ in the figure) as well as a select (SELECT $\overline{A}$/B) and that both the strobe and select are common to all four multiplexers. The select performs the function of selecting 1 of the 2 input lines of each of the multiplexers. The strobe turns the multiplexers on and off. If the strobe is asserted, the multiplexer is turned on and data can be passed from the selected input line of each multiplexer to the output line. If the strobe is not asserted, data cannot be passed and all the outputs are forced low. The function of the strobe can be readily seen in the circuits. If the strobe is not asserted (high), then the lower input to both of the *and* gates in each of the multiplexers will be low, both of the inputs to the *or* gates will be low, and the output will be low.

A multiplexer need not be limited to a two-way switching function. A multiplexer that performs a four-way switching function (a 4-line to 1-line multiplexer), one that performs an eight-way switching function (an 8-line to 1-line multiplexer), and one that performs a 16-way switching function (a 16-line to 1-line multiplexer) are all available. Figure 5.2-5 shows the logic circuit for a dual 4-line to 1-line multiplexer (74LS153). Notice that this multiplexer requires two select inputs (B and A). These select lines (asserted high) can be viewed as a binary input that selects, by number, the input line to be connected to the output for each multiplexer. Thus, BA = 00 selects the 0 input lines (1C0 and 2C0), BA = 01 selects the 1 input lines (1C1 and 2C1), and so on. This multiplexer, like the quadruple 2-line to 1-line multiplexer already described, has two strobes (STROBE $\overline{G}$) that function as

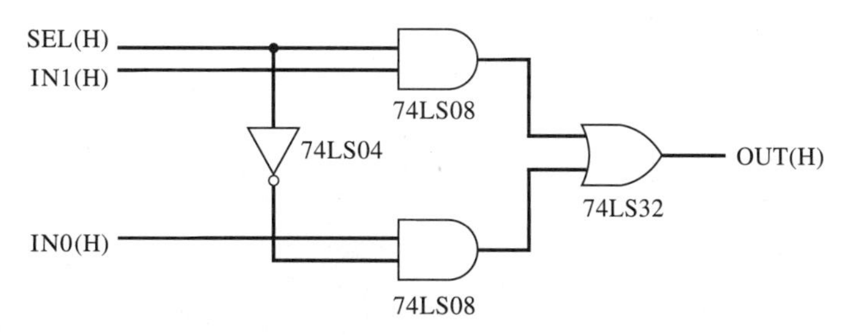

Figure 5.2-3 Circuit for a 2-line to 1-line multiplexer. Output signals are assumed to be high.

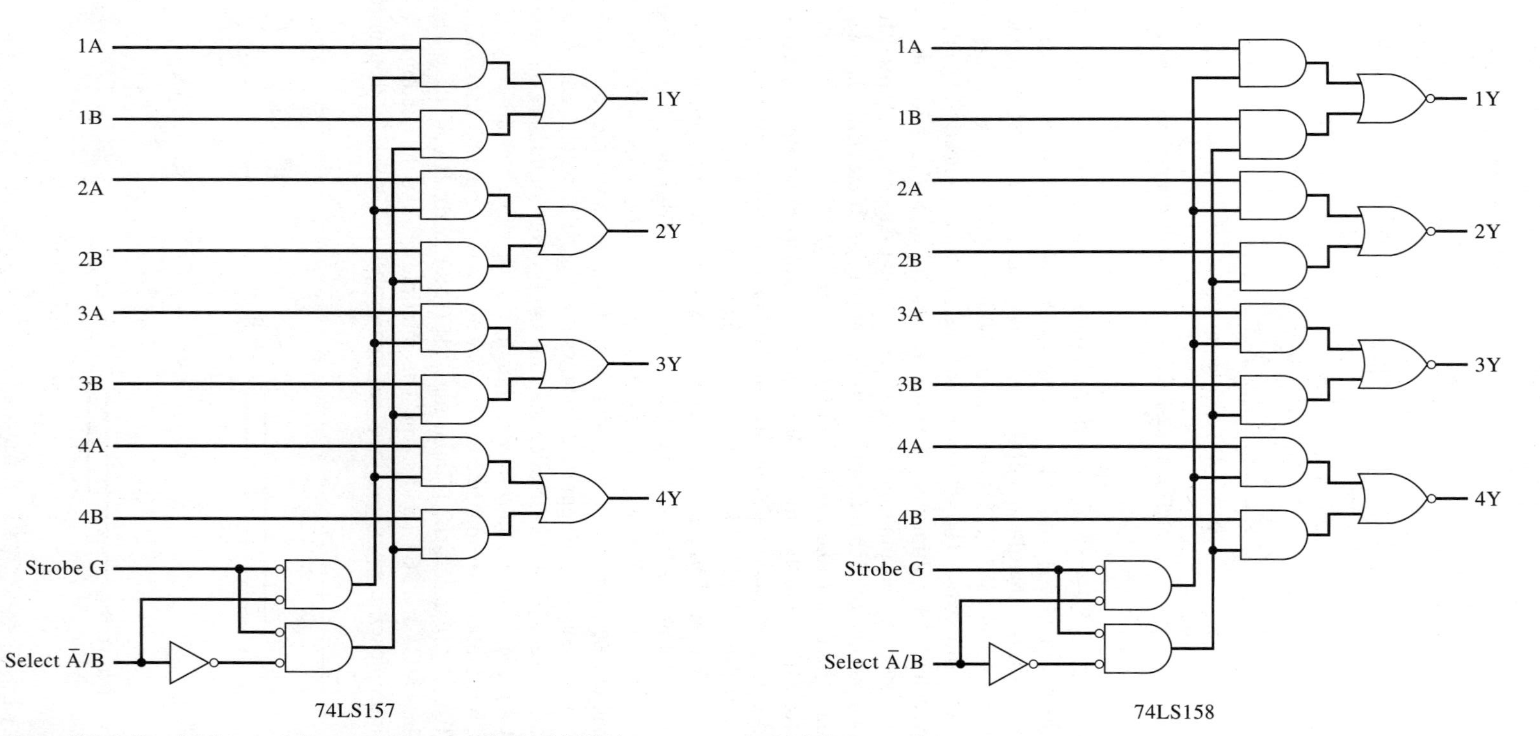

Figure 5.2-4 Quad 2-line to 1-line multiplexers. The 74LS157 has an output which is asserted high, while the 74LS158 has an output which is asserted low.

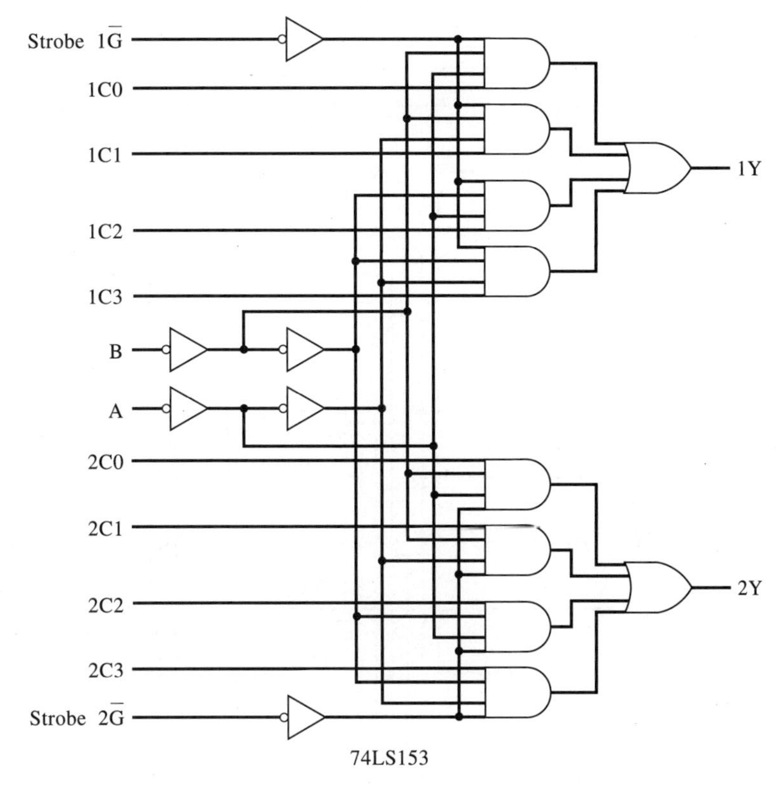

Figure 5.2-5 A 4-line to 1-line multiplexer.

switches to turn each of the multiplexers off and on. Table 5.2-1 is a function table for the 74LS153.

Table 5.2-1 Function Table for a 4-Line to 1-Line Multiplexer

Select Inputs		Data Inputs				Strobe	Output
B	A	C0	C1	C2	C3	$\overline{G}$	Y
x	x	x	x	x	x	H	L
L	L	L	x	x	x	L	L
L	L	H	x	x	x	L	H
L	H	x	L	x	x	L	L
L	H	x	H	x	x	L	H
H	L	x	x	L	x	L	L
H	L	x	x	H	x	L	H
H	H	x	x	x	L	L	L
H	H	x	x	x	H	L	H

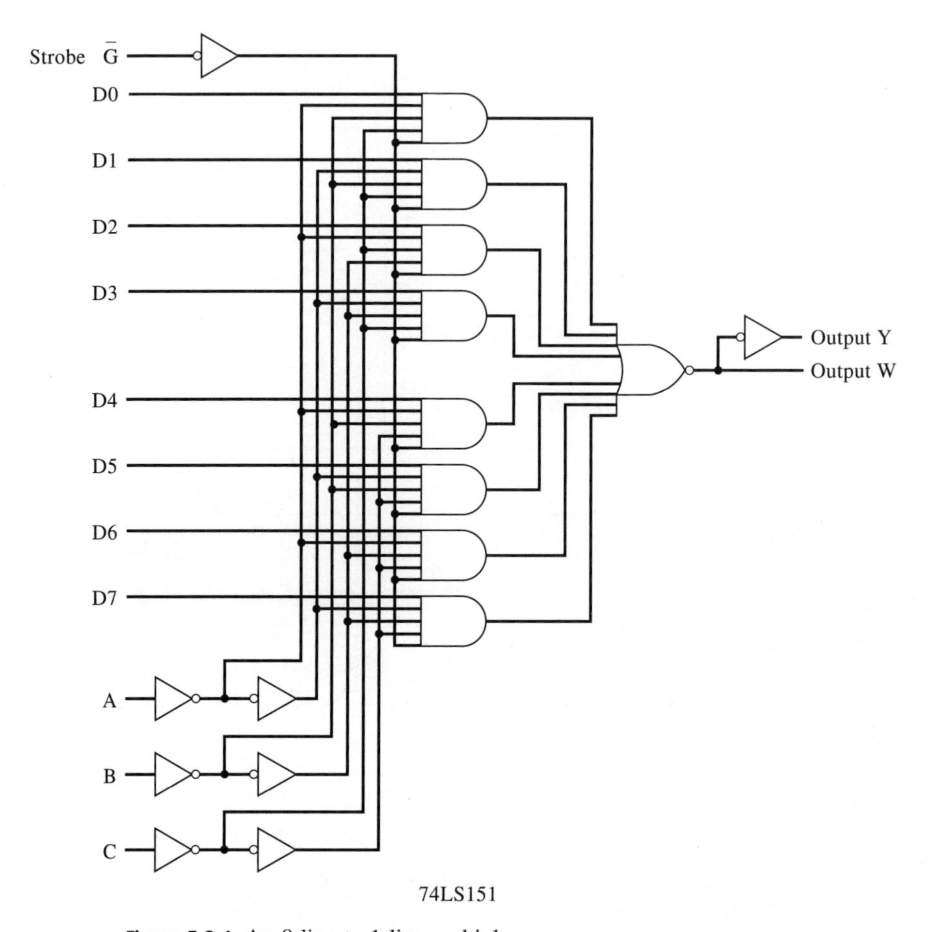

Figure 5.2-6 An 8-line to 1-line multiplexer.

Figure 5.2-6 shows the circuit for an 8-line to 1-line multiplexer (74LS151). This multiplexer requires three select inputs (C, B, and A), and these inputs (asserted high) can also be viewed as a binary input that selects, by number, the input line to be connected to the output. This multiplexer, like the others, has a strobe (STROBE $\overline{G}$) that functions as a switch to turn the multiplexer off and on. Table 5.2-2 is a function table for the 74LS151. The circuit and function table for a 16-line to 1-line multiplexer are not shown. This larger multiplexer is just an extension of the multiplexers we have already considered. It requires four select signals to select one of the 16 possible input lines.

Multiplexers are useful components in FPGA designs. One type of FPGA has a number of different types of multiplexers, which serve as basic modules. In fact, as we'll see in Chapter 6, this type of FPGA uses a certain configuration of multiplexers as its basic building block for gates and all other logic elements. Figure 5.2-7 shows the basic multiplexers available in this gate array. Notice the distinctive

Table 5.2-2 Function Table for 1-Line to 8-Line Multiplexer

Select Inputs			Strobe	Output	
C	B	A	$\overline{G}$	Y	W
x	x	x	H	L	H
L	L	L	L	D0	$\overline{D0}$
L	L	H	L	D1	$\overline{D1}$
L	H	L	L	D2	$\overline{D2}$
L	H	H	L	D3	$\overline{D3}$
H	L	L	L	D4	$\overline{D4}$
H	L	H	L	D5	$\overline{D5}$
H	H	L	L	D6	$\overline{D6}$
H	H	H	L	D7	$\overline{D7}$

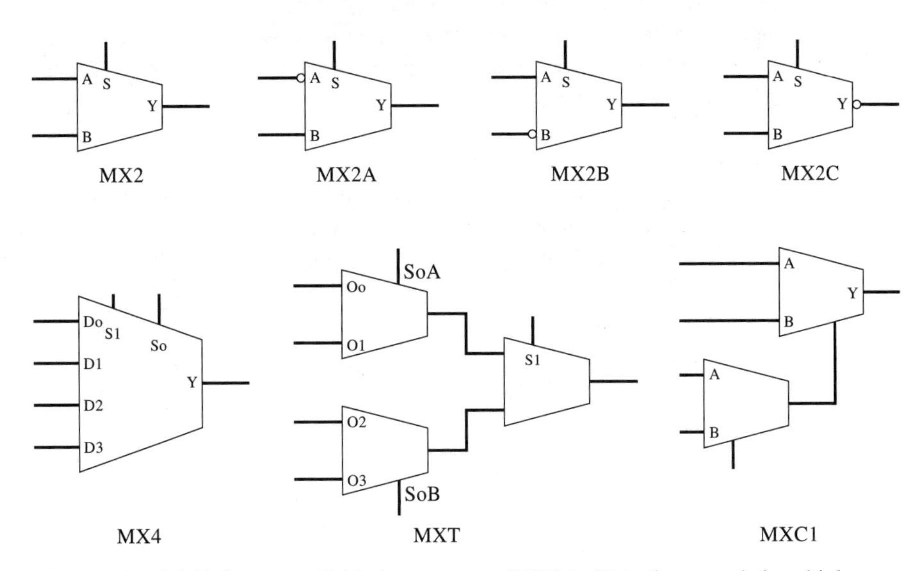

Figure 5.2-7 Multiplexers available in one type of FPGA. Note the cascaded multiplexers.

shape of the symbol that is commonly used to denote a multiplexer. In these devices, A, B, C, and D or D0, D1, D2, and D3 are inputs; S, S0, and S1 or S0A, S0B, and S are select lines; and Y is the output.

5.3 ENCODERS

An **encoder** is a combinational logic circuit that generates an output code. The value of the code depends on which of a number of input lines is selected. For example,

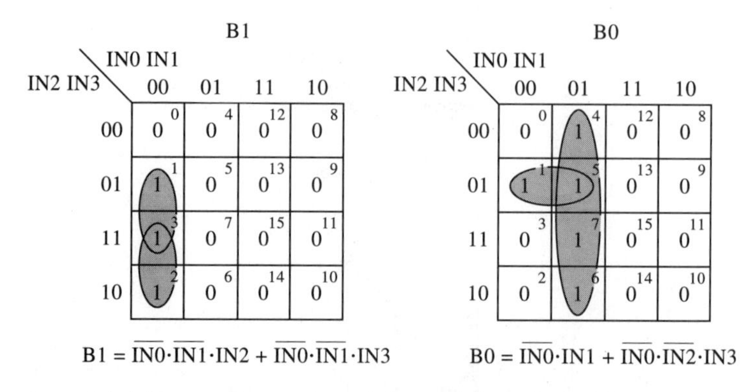

Figure 5.3-1 Karnaugh maps for a 4-line to 2-line priority binary encoder.

a 4-line to 2-line *binary* encoder generates a 2-bit binary output, which depends on which of the 4 input lines is selected. If the 0 line is selected, the binary output will be 00; if the 1 line is selected, the binary output will be 01; and so on. If two of the input (select) lines are asserted, the output is ambiguous. Most encoders are priority encoders, which resolve the ambiguity by assigning a priority to either the smallest or the largest asserted input.

We can design a 4-line to 2-line binary encoder using standard combinational logic design steps. The description of the encoder is as in the preceding paragraph except for a decision on the priority. Suppose we assign priority to the lower asserted input. Thus, for example, if the 0 input and the 2 input are asserted, the output should be 00, because the 0 input has priority. Let's call the inputs to the encoder IN0, IN1, IN2, and IN3 and the outputs B0 and B1, where B1 is the high order bit of the output code. The truth table for the encoder is as follows.

IN0	IN1	IN2	IN3	B1	B0
0	0	0	0	0	0
1	x	x	x	0	0
0	1	x	x	0	1
0	0	1	x	1	0
0	0	0	1	1	1

Because the output of the encoder is undefined if none of the inputs is asserted, we've arbitrarily assigned it to be 00.

The next step in the design is to make Karnaugh maps from the truth table. The truth table is actually for two different outputs, B1 and B0, so we need two Karnaugh maps. Figure 5.3-1 includes the maps and simplified logic expressions for B1 and B0. Figure 5.3-2 shows a circuit to evaluate those logic expressions, assuming that the inputs are asserted low and the outputs are asserted high. This design has a disadvantage: there is no way to distinguish between the output code for IN0 asserted

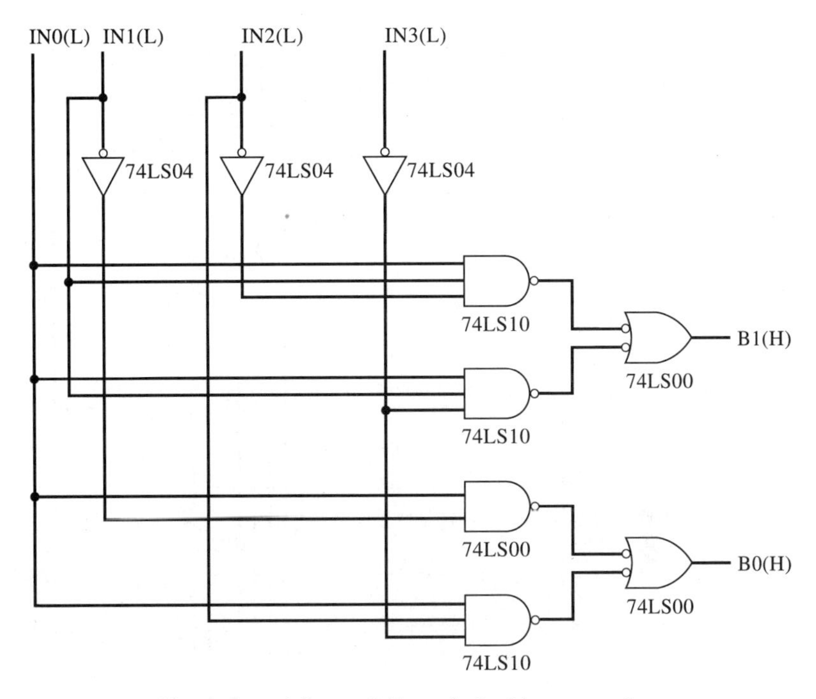

Figure 5.3-2 Circuit for a 4-line to 2-line priority binary encoder.

and the output code for no asserted inputs. We will discuss a circuit that overcomes this defect.

Encoders need not be 4-line to 2-line binary encoders, as in the design of Figure 5.3-2. They can also be 2-line to 1-line, 8-line to 3-line, or 16-line to 4-line binary encoders. There is also a 10-line to 4-line encoder that is actually a BCD encoder. A 2-line to 1-line binary encoder is fairly trivial, so we've chosen a 4-line to 2-line encoder as the design example. Larger binary encoders are simple extensions of the 4-line to 2-line binary encoder. Only 8-line to 3-line (74LS148) and 10-line to 4-line (74LS147) decoders are commercially available as single-chip devices.

Like the 4-line to 2-line binary encoder, the 8-line to 3-line binary encoder generates a 3-bit binary output that depends on which of the eight input lines is selected. If the 0 line is selected, the binary output will be 000; if the 1 line is selected, the binary output will be 001; and so on. The 8-line to 3-line encoder is a priority encoder, with priority assigned to the highest asserted output. Figure 5.3-3 shows the circuit for the encoder. It has an enable in (EI), an enable out (EO), and a group signal (GS) in addition to the usual input and output lines. The assertion levels are low for EI, EO, all of the inputs (0 through 7), and all of the outputs (A0 through A2). EI and EO are used for expanding the encoder; a data book should be consulted to see how this is done. The functions of the inputs to the encoder can be readily seen in the function table (Table 5.3-1). When EI is asserted, binary output, which depends on which input is asserted, is produced. When EI is not asserted, the output is driven to 000, independent of which input line is asserted. GS is an output which indicates that one of the inputs is active. It allows us to distinguish between

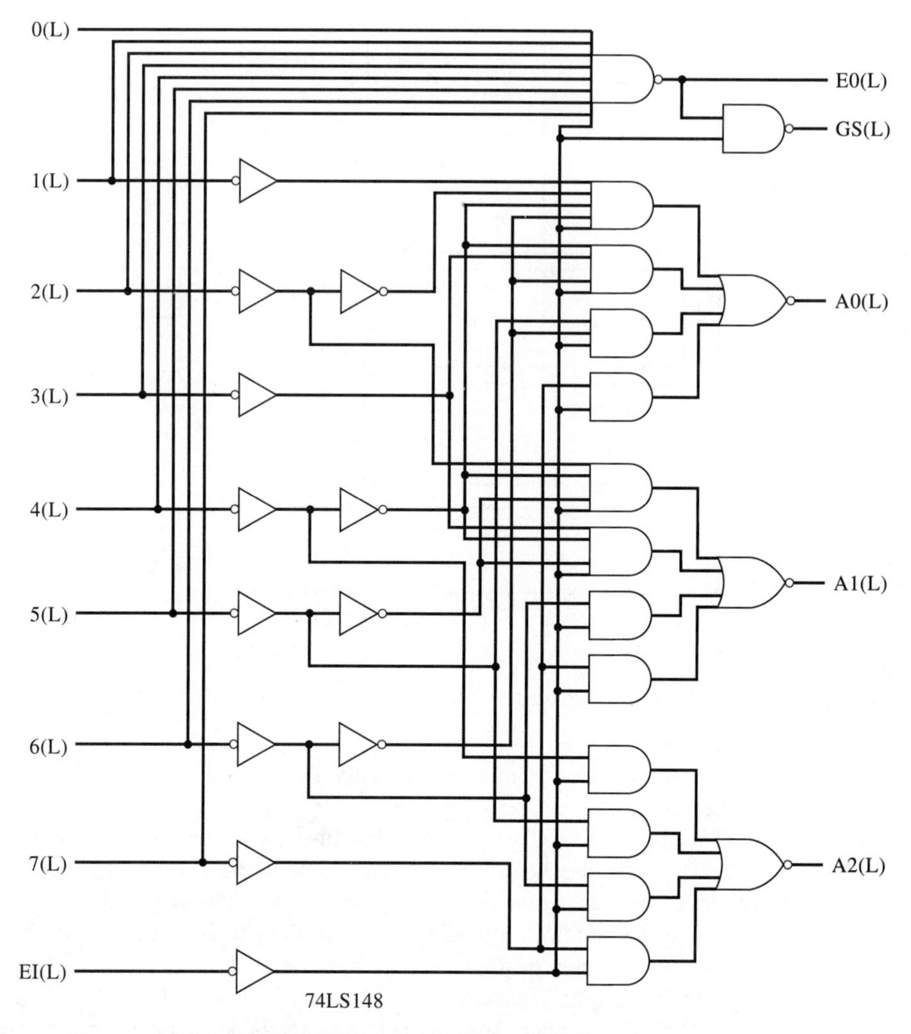

Figure 5.3-3 Eight-line to 3-line priority binary encoder.

the condition in which the 0 input is asserted and the condition in which none of the inputs is asserted.

The 10-line to 4-line encoder is rather specialized. Its function is similar to that of the 8-line to 3-line encoder, but it produces an output that extends from 0 (0000) to 9 (1001). It is used to produce BCD code.

5.4 DECODERS/DEMULTIPLEXERS

A **decoder/demultiplexer** is a device that performs both a decoding action, which is the inverse of encoding, and a demultiplexing action, which is the inverse of

Table 5.3-1 Function Table for a Priority 8-Input Binary Encoder (74LS148)

EI	0	1	2	3	4	5	6	7	A2	A1	A0	GS	EO
H	x	x	x	x	x	x	x	x	H	H	H	H	H
L	H	H	H	H	H	H	H	H	H	H	H	H	L
L	x	x	x	x	x	x	x	L	L	L	L	L	H
L	x	x	x	x	x	x	L	H	L	L	H	L	H
L	x	x	x	x	x	L	H	H	L	H	L	L	H
L	x	x	x	x	L	H	H	H	L	H	H	L	H
L	x	x	x	L	H	H	H	H	H	L	L	L	H
L	x	x	L	H	H	H	H	H	H	L	H	L	H
L	x	L	H	H	H	H	H	H	H	H	L	L	H
L	L	H	H	H	H	H	H	H	H	H	H	L	H

multiplexing. Let's first consider the decoding action. A decoder takes as input a binary code and asserts one of several output lines, depending on the input code. For example, a 2-line to 4-line decoder asserts one of four output lines, depending on the binary code on its input lines. If the input code is 00, output line 0 will be asserted; if the input code is 01, output line 1 will be asserted; and so on.

Now consider the demultiplexing action. A demultiplexer is an electronic switch. The action of a simple 1-line to 2-line demultiplexer can be seen in Figure 5.4-1. The input is switched between two output lines 0 and 1, depending on the assertion level of the select line. If the select line is asserted, the input is switched to line 1. If the select line is not asserted, the input is switched to line 0. If we extend the description of a 1-line to 2-line demultiplexer to a 1-line to 4-line demultiplexer, we'll have four output lines and two select lines. If the input to the select lines is coded 00, then the input is connected to line 0; if the input to the select lines is coded 01, then the input is connected to line 1; and so on.

We'll design a 1-line to 4-line demultiplexer and show how it will also function as a decoder. We've already given the description. Let's call the input IN, the select lines S1 and S0, and the output lines O0, O1, O2, and O3. The truth table is then as follows:

S1	S0	O0	O1	O2	O3
0	0	IN	0	0	0
0	1	0	IN	0	0
1	0	0	0	IN	0
1	1	0	0	0	IN

This is a variable-entered truth table. It shows that if select (S1 S0) is 00, then O0 is the same as IN (O0 = IN) and all the other outputs are 0; if select is 01, then O1

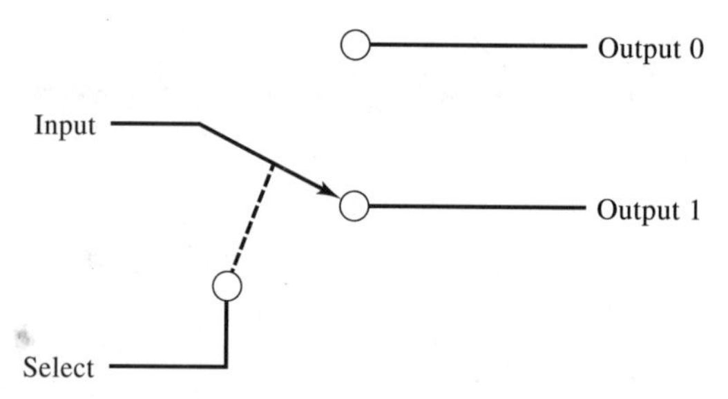

Figure 5.4-1 Function of a 1-line to 2-line demultiplexer.

is the same as IN and all the other inputs are 0; and so on. To see how to make this multiplexer into a decoder, consider what happens if IN is asserted (IN = 1). The truth table then becomes

S1	S0	O0	O1	O2	O3
0	0	1	0	0	0
0	1	0	1	0	0
1	0	0	0	1	0
1	1	0	0	0	1

which is simply the truth table for a 2-line to 4-line decoder, from the description at the beginning of this section.

Now, to continue the design, we need to make VEMs from the truth table for the multiplexer. We need four maps because there are four outputs. Actually, it should be clear from the truth table that maps will not give any simplification of the logic expressions. There is only one minterm in each output logic expression:

$$O0 = IN \cdot \overline{S1} \cdot \overline{S0}$$

$$O1 = IN \cdot \overline{S1} \cdot S0$$

$$O2 = IN \cdot S1 \cdot \overline{S0}$$

$$O3 = IN \cdot S1 \cdot S0$$

Figure 5.4-2 shows a circuit for this 1-line to 4-line demultiplexer, assuming that IN and the select lines (S1 and S0) are asserted high and the outputs (O0, O1, O2, and O3) are asserted high also.

Decoders/demultiplexers are normally specified by the decoder function. Dual 2-line to 4-line (74LS139A) and single 3-line to 8-line (74LS138A) and 4-line to 16-line (74LS154) decoders/demultiplexers are available commercially as discrete logic devices. Figure 5.4-3 shows the circuit for the dual 2-line to 4-line decoder/demultiplexer. It contains two independent decoder/demultiplexer circuits; Table 5.4-1 is a function table for one of them.

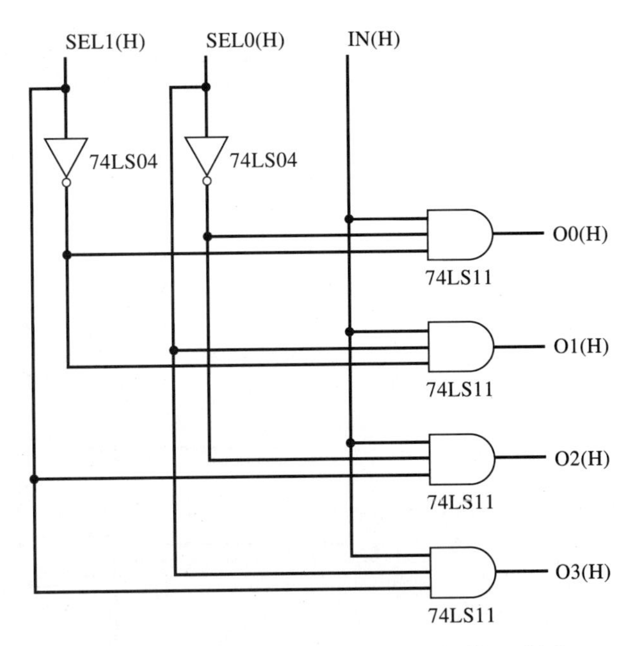

Figure 5.4-2 A circuit for our 2-line to 4-line decoder/demultiplexer.

When one of the decoder/demultiplexer circuits is used as a decoder, A and B are binary inputs that are asserted high; Y0, Y1, Y2, and Y3 are outputs that are asserted low; and $\overline{\text{G}}$ is an enable that is asserted low. If $\overline{\text{G}}$ is asserted (low), the decoder output functions normally and one of the outputs is asserted, depending on the binary code input on the inputs A and B. If $\overline{\text{G}}$ is not asserted, all the outputs are forced high (not asserted).

When one of the circuits is used as a demultiplexer, A and B are binary select inputs that are asserted high; $\overline{\text{G}}$ is the input; and Y0, Y1, Y2, and Y3 are the outputs. The output that is selected by the binary code on the select inputs (A and B) follows the input $\overline{\text{G}}$.

Figure 5.4-4 shows the circuit for the 3-line to 8-line decoder/demultiplexer, and Table 5.4-2 is its function table. This decoder/demultiplexer functions in much the same way as the 2-line to 4-line decoder, except it has three enables, G1, $\overline{\text{G2A}}$, and $\overline{\text{G2B}}$, which make it possible to easily expand the decoder without the use of extra logic. The enables also function as inputs when the device is used as a multiplexer. For example, G1 can be tied high, $\overline{\text{G2A}}$ tied low, and $\overline{\text{G2B}}$ used as an input to make a noninverting multiplexer, whereas $\overline{\text{G2A}}$ and $\overline{\text{G2B}}$ can be tied low and G1 used as an input to make an inverting multiplexer. The 4-line to 16-line decoder/multiplexer is an extension of the 3-line to 8-line decoder/demultiplexer. It has 2 enable inputs, G1 and $\overline{\text{G2}}$.

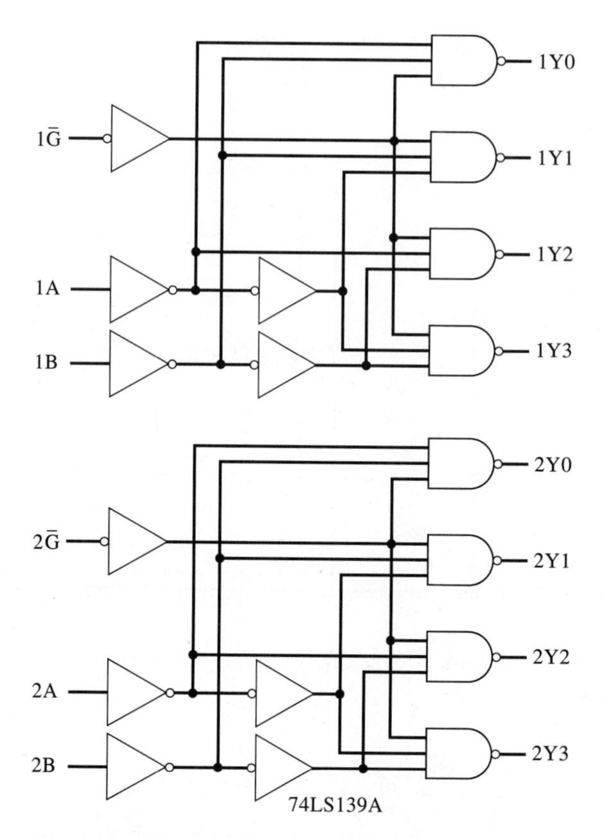

Figure 5.4-3 Dual 2-line to 4-line decoder/demultiplexer.

Table 5.4-1 Function Table for a 2-Line to 4-Line Decoder/Demultiplexer (74LS139A)

Inputs			Outputs			
Enable	Select					
$\overline{G}$	B	A	Y0	Y1	Y2	Y3
H	x	x	H	H	H	H
L	L	L	L	H	H	H
L	L	H	H	L	H	H
L	H	L	H	H	L	H
L	H	H	H	H	H	L

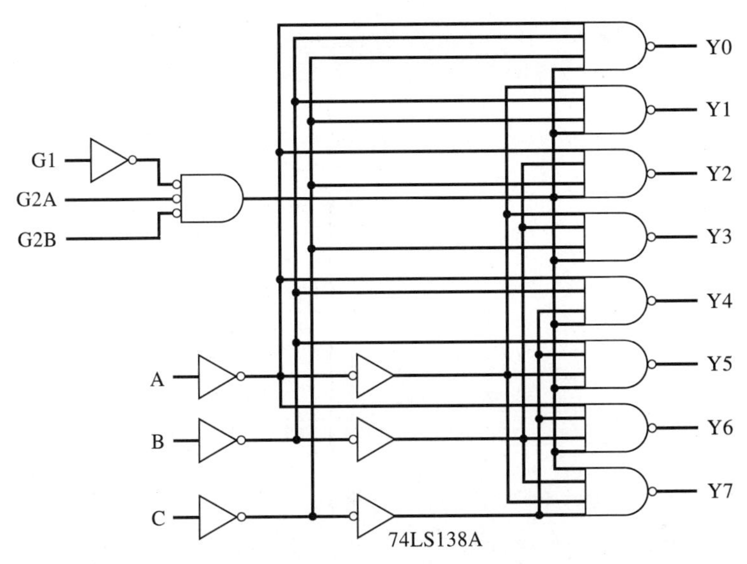

Figure 5.4-4 Three-line to 8-line decoder/demultiplexer.

Table 5.4-2 Function Table for a 3-Line to 8-Line Decoder/Demultiplexer (74LS138A)

Inputs						Outputs							
Enable			Select										
G1	G2A	G2B	C	B	A	Y0	Y1	Y2	Y3	Y4	Y5	Y6	Y7
L	x	x	x	x	x	H	H	H	H	H	H	H	H
x	H	x	x	x	x	H	H	H	H	H	H	H	H
x	x	H	x	x	x	H	H	H	H	H	H	H	H
H	L	L	L	L	L	L	H	H	H	H	H	H	H
H	L	L	L	L	H	H	L	H	H	H	H	H	H
H	L	L	L	H	L	H	H	L	H	H	H	H	H
H	L	L	L	H	H	H	H	H	L	H	H	H	H
H	L	L	H	L	L	H	H	H	H	L	H	H	H
H	L	L	H	L	H	H	H	H	H	H	L	H	H
H	L	L	H	H	L	H	H	H	H	H	H	L	H
H	L	L	H	H	H	H	H	H	H	H	H	H	L

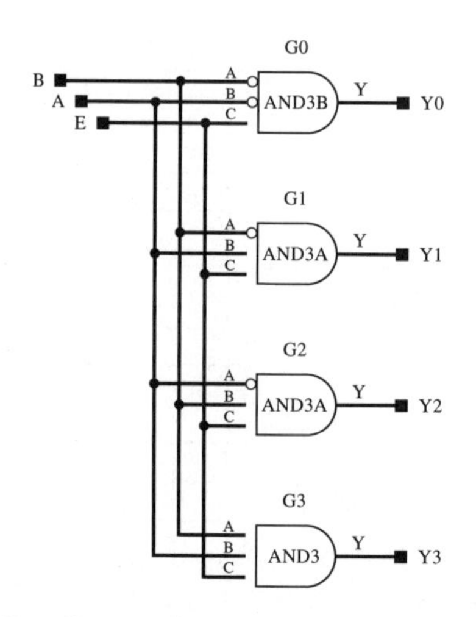

Figure 5.4-5 Two-line to 4-line decoder/demultiplexer found in an FPGA.

Decoders/demultiplexers are also available as predesigned modules in FPGAs. Figure 5.4-5 shows a 2-line to 4-line decoder, and Figure 5.4-6 shows a 3-line to 8-line decoder that is available in one type of FPGA.

5.5 ADDERS

In Sections 4.7 and 4.8 in the last chapter, we looked at the design of a simple 1-bit adder with carry-in and carry-out (called a full adder) and a simple 2-bit adder. The circuit for the 1-bit adder is repeated in Figure 5.5-1. It consists of two logic circuits, which we designed using our standard techniques. We can design a many-bit adder by cascading a number of 1-bit adders; we did this for the 2-bit adder design in Section 4.8. As we noted there, cascading 1-bit adders may not be a good solution to the design of many-bit adders if speed is a consideration. Figure 5.5-2 shows how a 4-bit adder can be made by cascading four 1-bit adders. In this design, the zero order bits must be processed in the zero order adder before the carry-in is available to the first order adder. The first order adder must then process its carry-in and input bits before its carry-out is available to the second order adder. Thus, the carry must ripple through each order of adder before the upper order bits of the sum become valid.

Normally, multibit adders are designed as fast adders. In a fast adder, we don't let the carry ripple through the adder; rather, we calculate each of the output bits independently of the result from the lower order bits. We could design a fast adder using our combinational logic design techniques, but the design of a 4-bit fast adder is rather complicated. The logic for the high order bits has 8 inputs. Rather than design a faster adder, then, let's investigate the design of a 4-bit fast adder that is available as a standard discrete device.

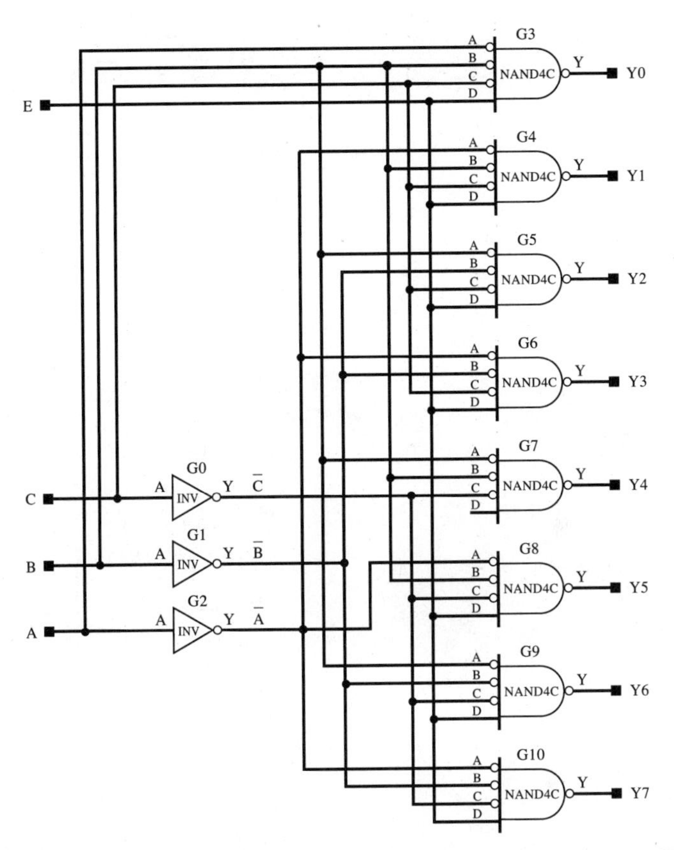

Figure 5.4-6 Three-line to 8-line decoder/demultiplexer found in one type of FPGA.

The circuit for this adder differs considerably from our smaller design, as can be seen in Figure 5.5-3. The adder is designed to both simplify the circuits for the upper bits and keep the speed of calculation as high as possible. The upper bit adders are complicated because they involve inputs from all the lower input bits. The design is actually a compromise between simplicity and speed. Part of the simplification is achieved by using *xor* gates in place of *and-or* gate combinations. Part is achieved by using common processing for the operations that are needed in more than one of the sum bits. Theoretically, each sum bit could be calculated in a two-level logic circuit with perhaps some inverters—that is, in a circuit with three gate delays. This would require many gates, and some of them would need a large number of inputs.

The actual design achieves considerable simplification with only slightly more propagation delay. Notice that there are four gate delays in the calculation of the 3 high order output bits and three gate delays in the calculation of the low order bit. Also, the carry-out bit is produced in a separate circuit with only three delays, making it faster than the 3 high order sum bits. This fast carry-out is useful when several

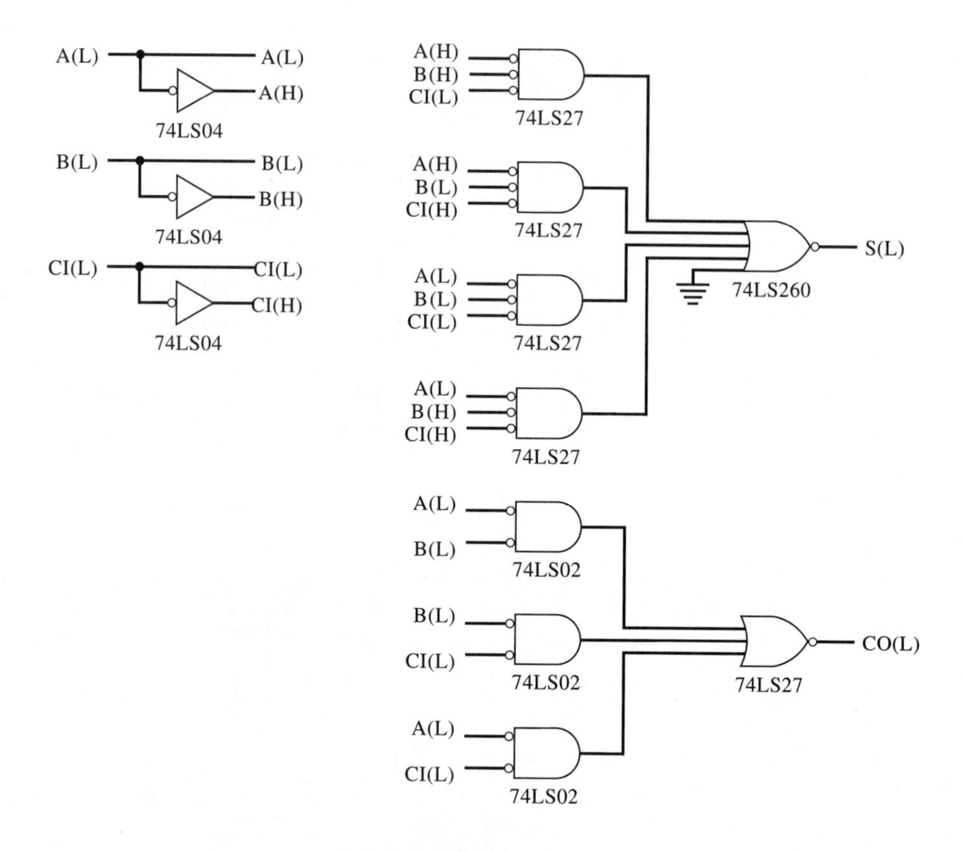

$$S(L) = (\bar{A}\ \bar{B}\ CI + \bar{A}\ B\ \overline{CI} + A\ B\ CI + A\ \bar{B}\ \overline{CI})\ (L)$$

$$CO(L) = (A\ B + B\ CI + A\ CI)\ (L)$$

Figure 5.5-1 A 1-bit adder designed using combinational logic design techniques.

adders are cascaded, as in Figure 5.5-4, in which two 4-bit fast adders are cascaded to make an 8-bit adder. The total delay in the cascaded adder is the delay for the carry-out in the first adder plus the delay from carry-in to sum-out in the second adder.

In order to see how the circuit of Figure 5.5-3 actually performs an addition, we can break the circuit into parts. Figure 5.5-5 shows the basic adder part of the circuit, which is common to the calculation of each of the sum bits. In this circuit we have labeled the input bits A and B for reference purposes. They could be any of the input bits, such as A2 and B2. The circuit is equivalent to two *xor* gates, as shown below the actual circuit. The first *xor* gate produces an output that is high if either input (A or B) is high ($1 + 0 = 0 + 1 = 1$, assuming 1 is high). Otherwise it produces a low. The second gate does the same thing for the carry-in (C0) and the result of $A + B$. This is just what is required to produce the proper 1-bit addition.

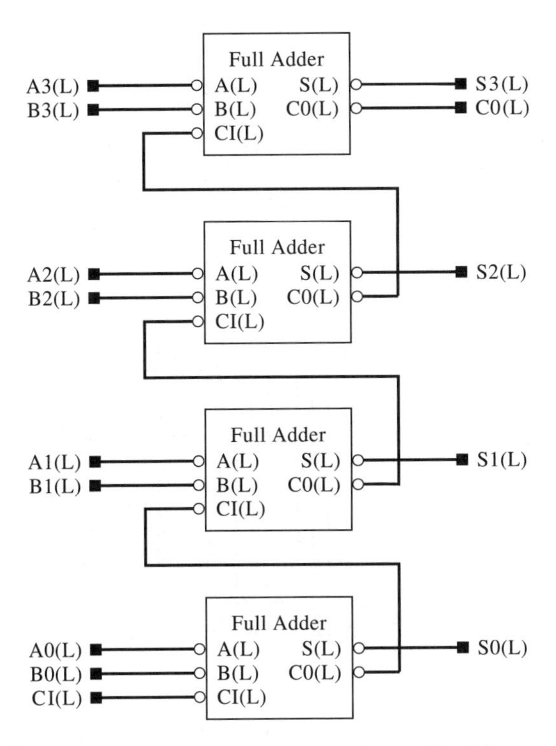

Figure 5.5-2 A 4-bit adder using four 1-bit adders.

Instead of using the first *xor* gate, the actual adders used the more complicated circuit shown, because the partial results $\overline{A} + \overline{B}$ and $\overline{A}\,\overline{B}$ are used repeatedly in the circuits that generate the carry-ins for each higher order sum bit, and we can simplify the circuit by calculating them only once. In addition to each of the basic adders, then, there is a circuit to generate the carry-in. This carry-in depends on all of the lower order inputs, so it becomes more complicated for the higher order adders. For example, the carry-in into the lowest order bit is simply the carry into the adder (C0). The carry-in to the second order bit depends on the 2 low order inputs (A1 and B1) and the carry-in to the adder (C0), and so on. We will not do a complete analysis of the carry generation circuits in this book. The carry-out of the overall adder is generated in the circuit at the top and depends on all the inputs and the carry into the overall adder. We will not do a complete analysis of this circuit, either.

FPGAs have predesigned modules for performing 8-bit, 12-bit, 16-bit, and even longer additions. The circuit for these additions can become complicated, but the idea is rather simple. The additions for the higher order bits in the adder are done in groups. For each group, the addition is performed for carry-ins of both 0 and 1. This addition is done while the carry bit is being calculated in the lower order part of the adder. When the carry has been calculated, the proper result for the upper order bits is selected based on the carry. Figure 5.5-6 shows the circuit for an 8-bit

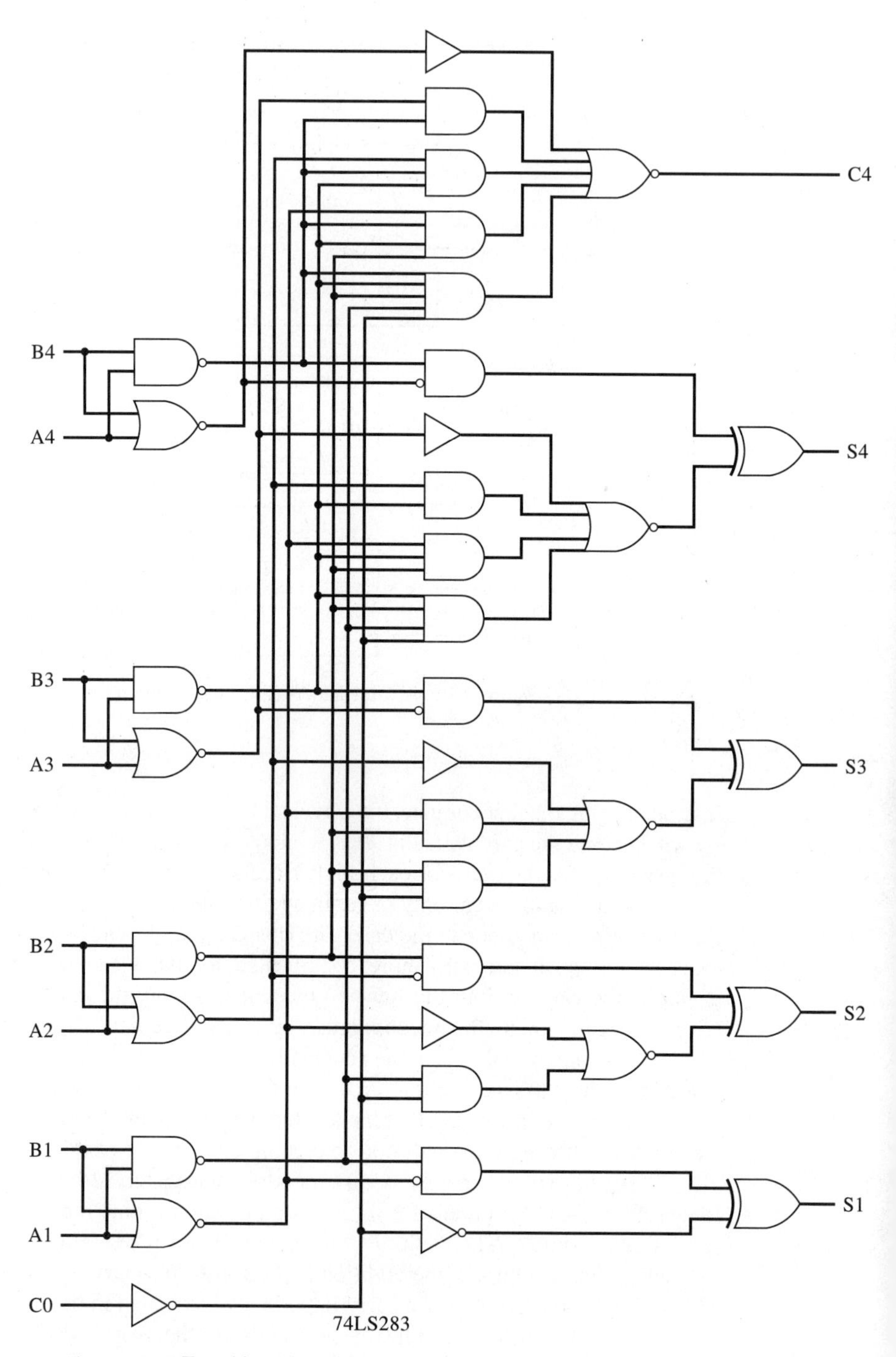

Figure 5.5-3 Four-bit adder with fast carry.

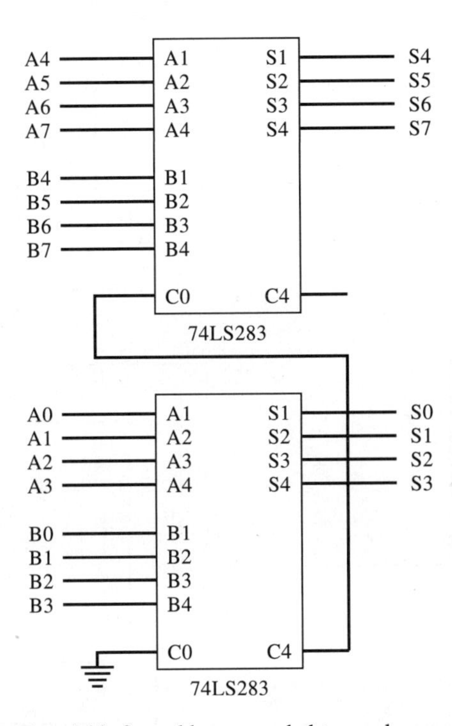

Figure 5.5-4 Two 4-bit fast adders cascaded to produce an 8-bit adder.

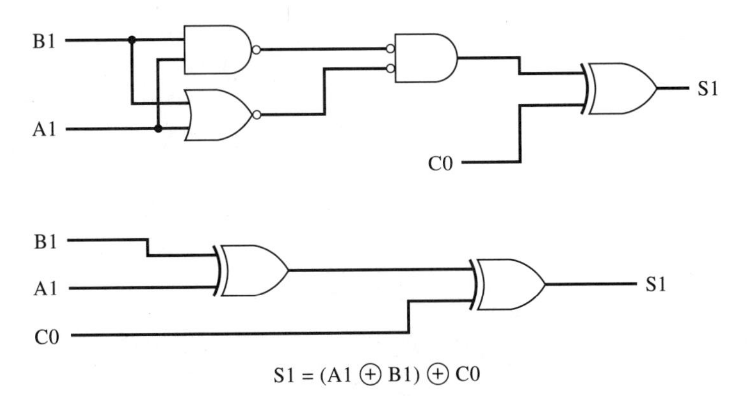

Figure 5.5-5 Detail of the 4-bit adder.

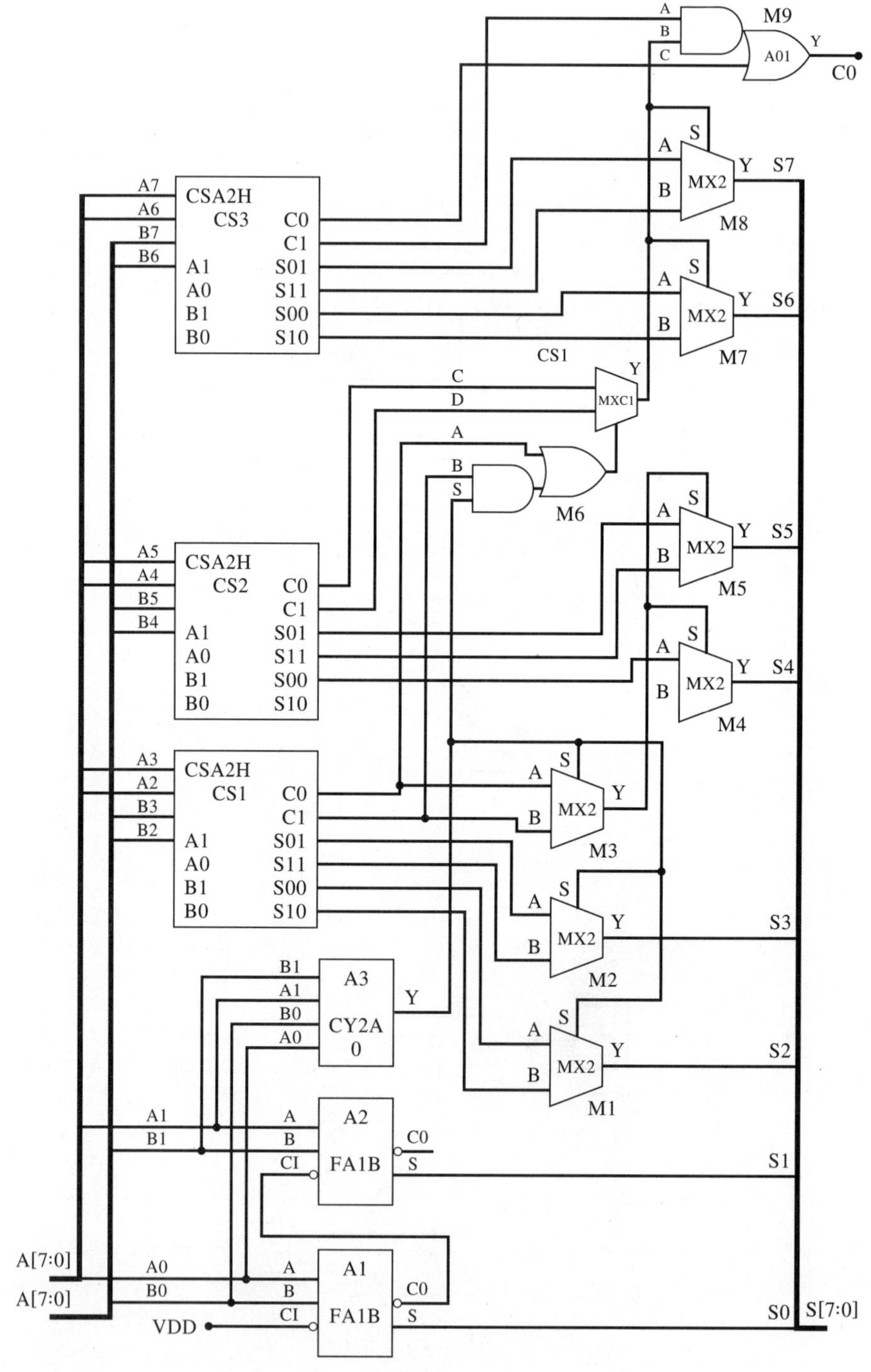

Figure 5.5-6 Eight-bit fast adder available in an FPGA.

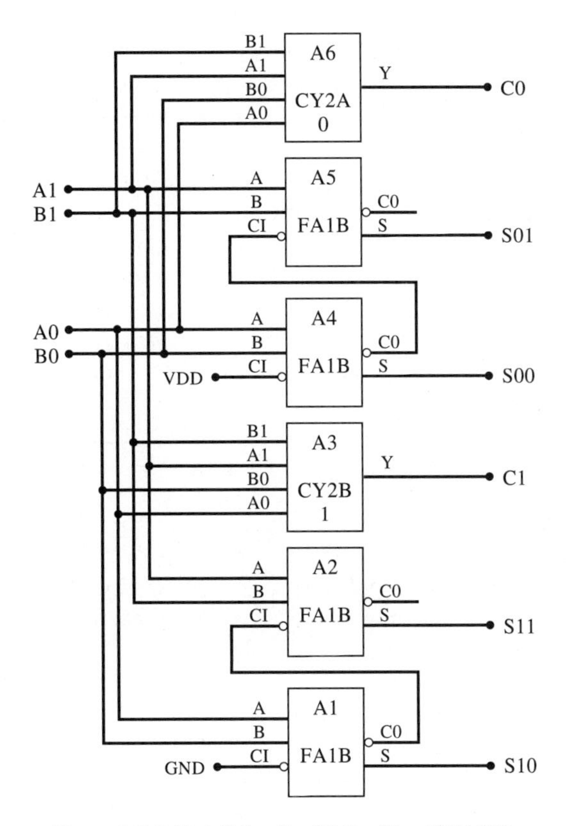

Figure 5.5-7 Detail for the 8-bit adder (CSA2H).

fast adder that is available in one type of FPGA. For this adder, the bits are added in twos. The sum for the 2 lower order bits (0 and 1) is simply calculated using two 1-bit adders with ripple carry. Simultaneously, the carry-out resulting from the addition of these same bits is calculated in a separate combinational circuit (in the box labeled CY2A in Fig. 5.5-6).

The sum of the two next higher order bits (2 and 3) is calculated using the circuitry in the box labeled CSA2H, CS1. Figure 5.5-7 shows this circuitry. It is just two 2-bit adders using simple 1-bit adders and ripple carry. One adder uses a carry-in of 0 and the other a carry-in of 1. Thus, two sums are generated: the first (S10 S00) is the sum for carry-in 0, and the second (S11 S01) is the sum for carry-in 1. At the same time the sum is calculated, the carry-outs are calculated in the combination circuits contained in box CY2A, which calculates the carry-out C0 for a carry-in of 0, and CY2B, which calculates the carry-out C1 for a carry-in of 1.

At this point (look back at Figure 5.5-4), we have sums and carry-outs for both a carry-in 0 and a carry-in 1, and we can select the proper sum and carry-out based on the carry we calculated in the lower 2 bits of the adder. The multiplexers on the output of CS1 select this proper sum and carry-out.

Bits 4 and 5 of the sum are calculated using the same circuitry as for bits 2 and 3. Notice how the proper look-ahead carry is used to multiply out the proper sum. The final bits, 6 and 7, use the same circuitry as bits 2 and 3. In addition, they have a final speed enhancement. The carry that is used for their output multiplexer is generated from the possible carry-outs out of bits 4 and 5 and bits 2 and 3, and the carry-out of bits 0 and 1. Again the circuit looks complicated, but the idea is simple. Speed is achieved at each stage by calculating the sums for both carry-in 0 and carry-in 1; then the proper sum is selected from the two possibilities. This means that we double the circuitry at each stage to achieve speed.

Longer adders use much the same idea as the 8-bit adder. It takes longer to generate the carry to the higher order bits, so adders for the higher order bits are combined into larger groups.

5.6 COMPARATORS

A **comparator** is a device that compares two binary numbers and produces some result. There are two different types of comparators. The simpler type, called an *identity comparator*, makes a bit-by-bit comparison of the two numbers and asserts an output if the numbers are bit-by-bit equal. The more complicated type, called a *magnitude comparator*, determines whether one of the inputs is larger, equal to, or smaller than the other input and asserts one of three inputs, depending on the results of the comparison.

In Example 4.6.3 we designed a 2-bit identity comparator using our standard logic circuit design techniques. To design a magnitude comparator, we can use the same techniques. Suppose we want to design a 2-bit magnitude comparator. Let the binary inputs to the comparator be A1 A0 and B1 B0. We need three outputs, $A < B$, $A = B$, and $A > B$, because we're computing three different results. The truth tables for the 3 outputs are

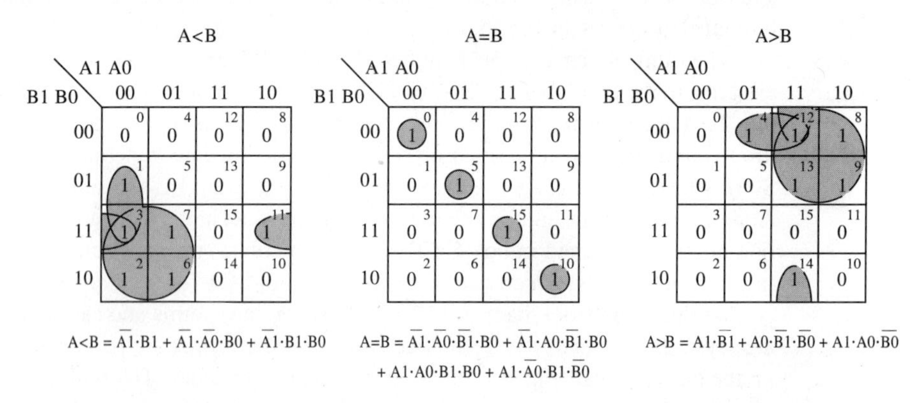

Figure 5.6-1 Karnaugh maps and logic expressions for 2-bit magnitude comparator.

A1	A0	B1	B0	A < B	A = B	A > B
0	0	0	0	0	1	0
0	0	0	1	1	0	0
0	0	1	0	1	0	0
0	0	1	1	1	0	0
0	1	0	0	0	0	1
0	1	0	1	0	1	0
0	1	1	0	1	0	0
0	1	1	1	1	0	0
1	0	0	0	0	0	1
1	0	0	1	0	0	1
1	0	1	0	0	1	0
1	0	1	1	1	0	0
1	1	0	0	0	0	1
1	1	0	1	0	0	1
1	1	1	0	0	0	1
1	1	1	1	0	1	0

Figure 5.6-1 shows the Karnaugh maps found from these truth tables, along with the simplified logic functions for the circuits we're designing. The A = B function does not simplify if we use only *and* and *or* gates. If we take a different approach to the simplification and allow *xnor* gates, we can achieve considerable simplification. Referring to Section 4.3, we see that an *xnor* gate has a high output when its inputs are equal ($F = \overline{A}\,\overline{B} + AB$, if both inputs and the output are asserted high). Thus, we can use one *xnor* gate to determine whether A1 and B1 are equal, and one to determine whether A0 and B0 are equal. Now, to determine if both are equal, we need to *and* the outputs of the two *xnor* gates. Figure 5.6-2 shows the circuit for the 2-bit magnitude comparator. This circuit used the simplification from the Karnaugh maps to evaluate A < B and A > B and used the simplification with *xnor* gates to evaluate A = B.

Comparators can be designed in any size. One 4-bit magnitude comparator (74LS85) is available, and some 8-bit identity comparators are available as standard devices. The circuit for the 4-bit comparator is shown in Figure 5.6-3; Table 5.6-1 is the function table. This 4-bit comparator has A < B, A = B, and A > B inputs that allow expansion to almost any size. To cascade comparators, the A < B output of the lower order comparator is connected to the A < B input of the higher order comparator, the A = B output is connected to the A = B input, and the A > B output is connected to the A > B input. Cascading results in a ripple carry, so the delays of all the comparators in the cascaded chain are added to arrive at the

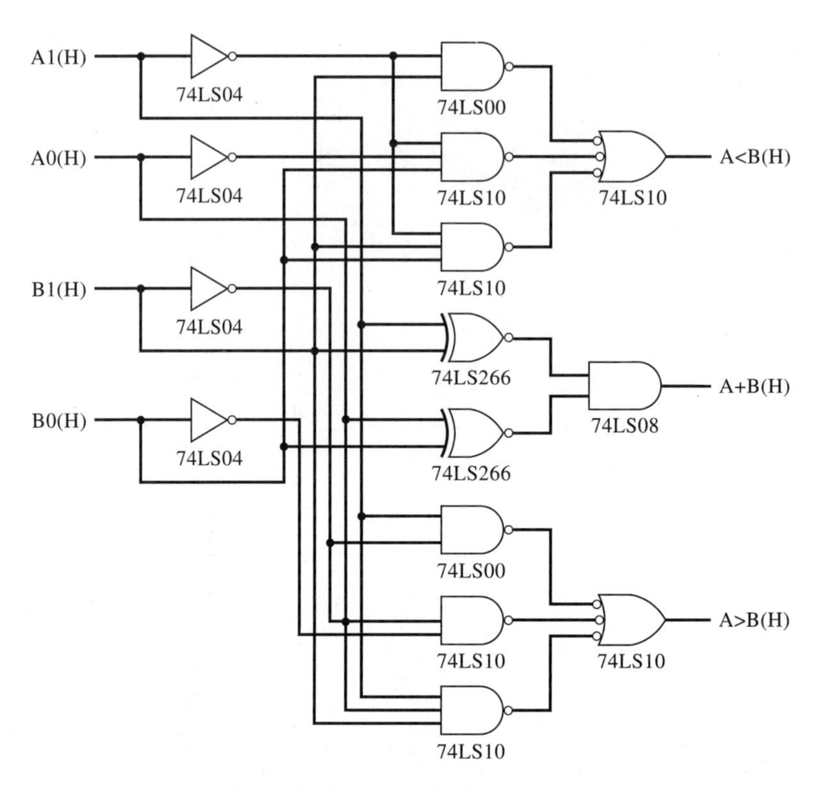

Figure 5.6-2 Circuit for 2-bit magnitude comparator.

delay in the total comparator. The comparators can also be expanded in a parallel arrangement, which decreases the delay. Consult a data book for the details of parallel expansion.

The 4-bit comparator has been designed to simplify the circuit by using common processing for the three different functions, $A < B$, $A = B$, and $A > B$. This increases the propagation delay, just as in the adder of the last section. It also makes the circuit difficult to analyze. We will reserve the analysis for an exercise at the end of this chapter.

Comparators, both identity and magnitude, are available as predesigned modules on FPGAs. One FPGA, for example, has available 4-, 8-, and 16-bit identity comparators and 2-, 4-, and 8-bit magnitude comparators. The circuit for the 2-bit magnitude comparator is shown in Figure 5.6-4. It has provisions for expansion; the 4-bit and 8-bit comparators are made by expanding the 2-bit comparator.

5.7 ARITHMETIC AND LOGIC UNIT (ALU)

An Arithmetic and Logic Unit (ALU) is a multipurpose calculation device. It performs a variety of arithmetic and logic operations on 2 binary multi-bit inputs,

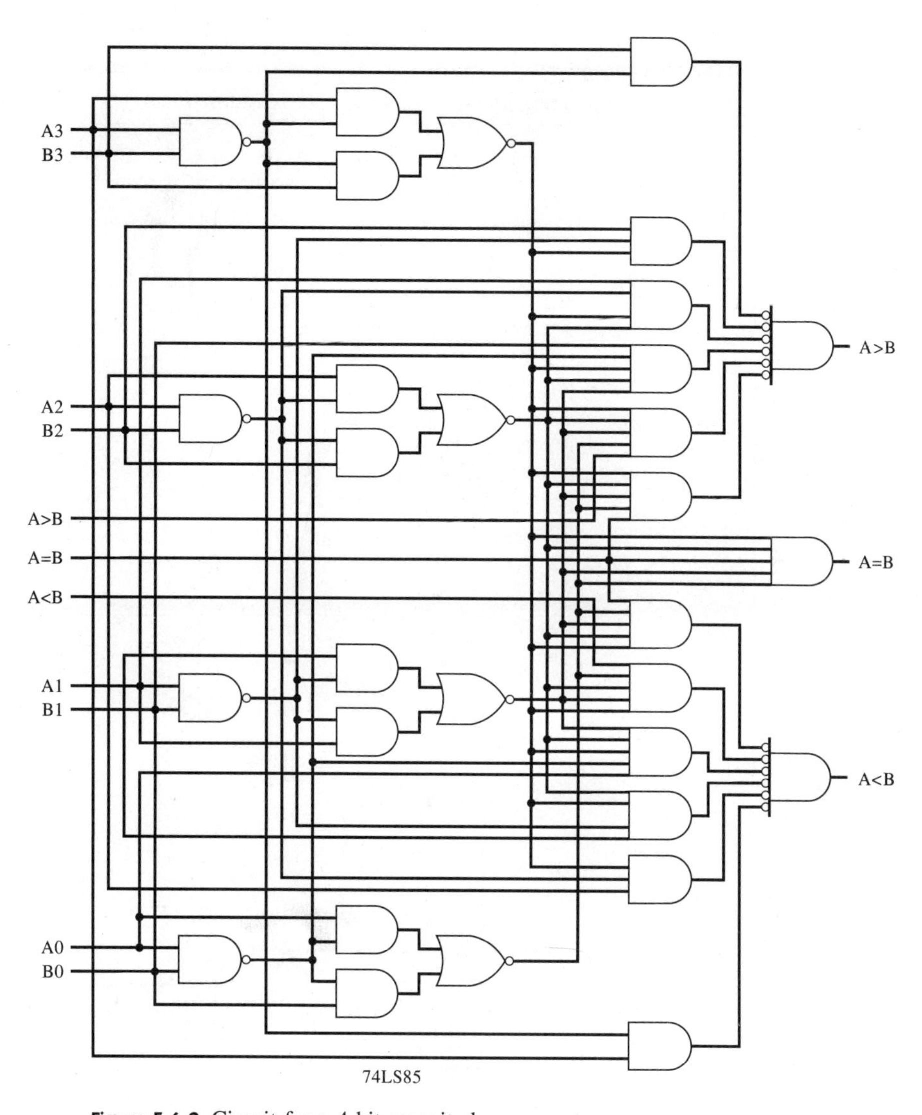

Figure 5.6-3 Circuit for a 4-bit magnitude comparator.

usually labeled A and B, and presents the results on the output, usually labeled F. Thus, an ALU can calculate A + B, A − B, and so on. The function to be performed is selected by setting select lines to the proper code for that function. Four-bit ALUs are available as standard devices (74LS181, 74F381, 74F382). Fast (20 ns) 16-bit ALUs are available from specialty manufacturers.

ALUs have provisions for carry-in and carry-out, so they can be expanded by cascading; therefore, two 4-bit ALUs can be cascaded to make an 8-bit ALU. Unfortunately, cascading means that the carry must ripple through all the ALUs in

Table 5.6-1 Function Table for 4-Bit Magnitude Comparator

Comparing Inputs				Cascading Inputs			Outputs		
A3, B3	A2, B2	A1, B1	A0, B0	A>B	A<B	A=B	A>B	A<B	A=B
A3 > B3	x	x	x	x	x	x	H	L	L
A3 < B3	x	x	x	x	x	x	L	H	L
A3 = B3	A2 > B2	x	x	x	x	x	H	L	L
A3 = B3	A2 < B2	x	x	x	x	x	L	H	L
A3 = B3	A2 = B2	A1 > B1	x	x	x	x	H	L	L
A3 = B3	A2 = B2	A1 < B1	x	x	x	x	L	H	L
A3 = B3	A2 = B2	A1 = B1	A0 > B0	x	x	x	H	L	L
A3 = B3	A2 = B2	A1 = B1	A0 < B0	x	x	x	L	H	L
A3 = B3	A2 = B2	A1 = B1	A0 = B0	H	L	L	H	L	L
A3 = B3	A2 = B2	A1 = B1	A0 = B0	L	H	L	L	H	L
A3 = B3	A2 = B2	A1 = B1	A0 = B0	x	x	H	L	L	H
A3 = B3	A2 = B2	A1 = B1	A0 = B0	H	H	L	L	L	L
A3 = B3	A2 = B2	A1 = B1	A0 = B0	L	L	L	H	H	L

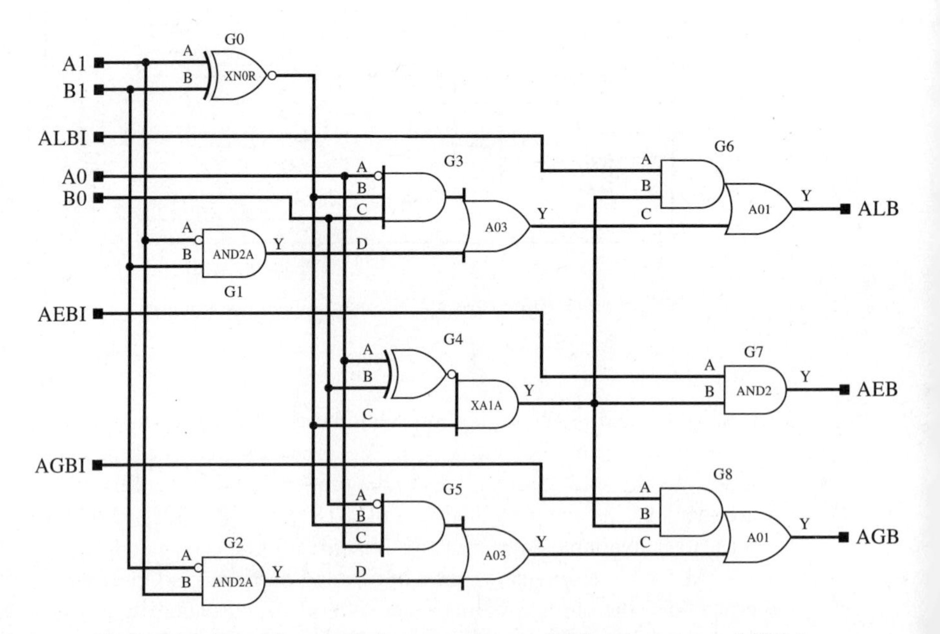

Figure 5.6-4 Two-bit magnitude comparator available in one type of FPGA.

the cascaded sequence, so propagation delays accumulate. To improve their speed, standard 4-bit ALUs can be combined with special look-ahead carry generators (74LS182, 74F182). These generators provide a fast carry to the higher order ALUs and should be investigated if high speed calculations with large operands are needed.

ALUs can be designed using standard logic design techniques, but because of their multiple functions, they are very complicated and are beyond the scope of this book. We will not attempt the design of even a small ALU, and we will not discuss the circuit for the standard devices. If an ALU is needed in the design of a logic system, data books, which describe ALUs in some detail, should be consulted.

KEY TERMS

Multiplexer A multiple input, single output electronically controlled switch.

Encoder A combinational logic circuit that generates output codes in response to different inputs.

Decoder/Demultiplexer A device with both a decoding action (inverse of encoding) and a demultiplexing action (inverse of multiplexing).

Comparator A device that compares two binary numbers and produces some result.

Identity Comparator A device that makes a bit-by-bit comparison of two numbers and asserts an output if the two numbers are bit-by-bit equal.

Magnitude Comparator A more complicated comparator that determines whether one of the inputs is larger, equal to, or smaller than the other input and asserts one of three outputs, depending on the results of the comparison.

Adder A device that adds two binary numbers.

Arithmetic and Logic Unit (ALU) A device that performs a variety of arithmetic and logic operations on two binary numbers.

EXERCISES

1. Show how a 4-line to 1-line multiplexer (74LS157) and a 1-line to 4-line demultiplexer (74LS139) can be used to transmit and recover signals from different sources over a single line. Assume that the different sources do not need to transmit at the same time.

2. Consult a data book and determine how to expand an 8-line to 3-line encoder (74LS148A) into a 16-line to 4-line encoder.

3. Consult a data book and determine how to expand a 3-line to 8-line decoder (74LS138) into a 6-line to 64-line decoder. Decoders such as the 74LS138A are used as address decoders in computers.

4. Based on the idea used in the fast adder that is found in an FPGA, design a 16-bit fast adder using discrete 4-bit fast adders (74LS283) and multiplexers (74LS157). Use the adders in 8-bit blocks by simply cascading two 4-bit adders. Calculate the upper 8 bits of the sum for carry-in 0 and carry-in 1; then, using the

carry-out from the lower 8-bit sum, select the proper upper 8-bit sum with the multiplexer. What is the maximum time required to do the 16-bit sum employing this technique? What is the maximum time required for four 4-bit cascaded adders? If simulation is available, simulate the fast adder and look at its timing delay.

5. Consult a data book and determine how to expand a 4-bit (74LS85) comparator to 24 bits so that the speed is not degraded as much as it would be for a simple cascading.

6. The part of the 4-bit comparator that performs the A = B function is not too difficult to analyze. The 5-input *and* gate that outputs A = B has four identical input circuits—one for each pair of input bits and an A = B input for cascading. As we might expect from the comparator design in the first part of Section 5.6, each of these input circuits should be equivalent to an *xnor* gate. To analyze the circuit, all we need to do is show the equivalence. Isolate one of the input circuits and show this equivalence—that is, show that $F = \overline{A}\,\overline{B} + AB$ if both inputs and the output are asserted high.

6 Programmable Logic Devices

6.1 LOGIC CIRCUIT DESIGN USING PROGRAMMABLE LOGIC

The circuit design we discussed in Chapter 4 uses discrete gates. Most discrete gates are arranged with several gates of the same type in a single package, but all leads to the gate are connected to pins on the package, and all circuit connections are made externally. Logic circuits can also be designed using devices that have in one package all the gates necessary for a logic circuit. Such devices have provisions for programming the interconnections of the gates internal to the device so that the logic can be customized.

There are two general types of programmable logic devices: those with flexible architecture and those with fixed architecture. Field Programmable Gate Arrays (FPGAs) have flexible architecture, and logic circuits using FPGAs can be designed in much the same way as logic circuits using discrete gates. In Chapter 4, along with our discussion of designs using discrete gates, we dealt with some of the considerations in designing with gate arrays. In this chapter we'll discuss some of the more general aspects of the architecture and programming of FPGAs.

Programmable Array Logic (PAL) devices and **Programmable Logic Array (PLA)** devices are programmable devices with fixed architecture. There is a third type of device with fixed architecture, a **Programmable Read Only Memory (PROM),** that is used primarily as a memory device but that can also be used as a programmable logic device. These three logic devices differ primarily in the way the internal gate connections can be made. The PROM is less flexible than either the PAL or the PLA. The PAL is more flexible than the PROM and is probably the most used of the fixed architecture devices. The PLA is the most flexible of the programmable logic devices, but in many logic circuit designs the additional flexibility is not needed. PROMs, PALs, and PLAs are collectively called **Programmable Logic Devices (PLDs).** We'll consider the architecture and programming of all of these devices after we look at FPGAs.

6.2 FIELD PROGRAMMABLE GATE ARRAYS

Field Programmable Gate Arrays can be designed in much the same way as discrete logic. All that's needed to program an FPGA is to capture a circuit diagram of the logic to be implemented in an appropriate schematic capture software, then process this circuit diagram using software available from the manufacturer of the FPGA. As we saw in Chapter 4, the logic elements used in the circuit must be available in the FPGA. These logic elements are normally furnished as a library of elements to be used in the schematic capture software. Chapter 4 contains some specific examples of these logic elements.

When using FPGAs, it is helpful to have at least a basic idea of how they're constructed. An FPGA is a *single* **Very Large Scale Integrated (VLSI) circuit**. It is constructed on a single piece of silicon and consists of identical, individually programmable modules laid out in a rectangular pattern, as shown in Figure 6.2-1. The modules are separated both horizontally and vertically by channels that contain horizontal and vertical metallic conductors. In addition, each module has conductors at its inputs and output that cross one or more of the channels. The figure shows these as vertical conductors, but they can be either vertical or horizontal. At each

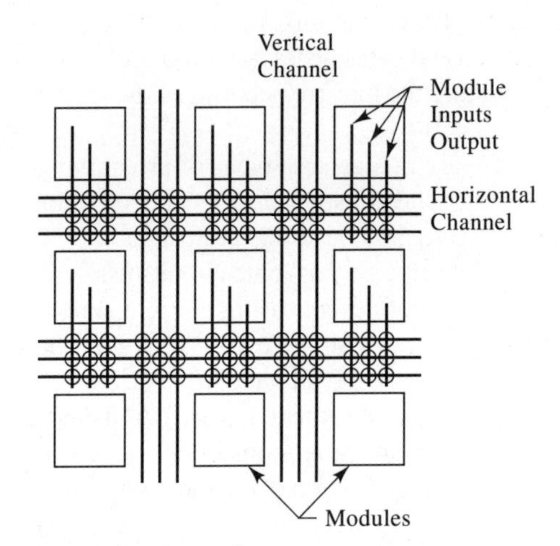

Figure 6.2-1 Architecture of a FPGA.

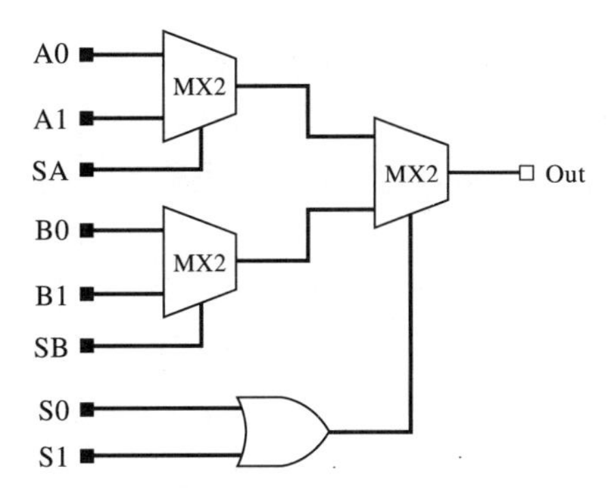

Figure 6.2-2 Circuit of one module used in a FPGA.

intersection between the horizontal and vertical conductors (symbolized in the figure by a small circle) is a programmable link—a provision for programming a connection—between the two conductors. The programmable links are used to interconnect the modules and also to program the individual modules.

The nature of the modules depends on the type of FPGA. For ease of use, the modules need to be programmable into the gates with which you're already familiar and into sequential elements, which we will cover in the last half of this book. In some cases a module may have both combinational and sequential components. In other cases there may be two different types of modules—combinational and sequential. Figure 6.2-2 is an example of the module used in one type of FPGA. It consists of three interconnected 2-line to 1-line multiplexers and an *or* gate. As we saw in Chapter 5, we can express the logic function of a multiplexer—for example, the upper left multiplexer—as

$$A = A0 \cdot \overline{SA} + A1 \cdot SA$$

where we have assumed all the inputs and the output are asserted high. Now, if we combine this expression with similar expressions for the other multiplexers and the proper logic expression for the *or* gate, the overall logic expression for the module is

$$\begin{aligned} OUT &= (A0 \cdot \overline{SA} + A1 \cdot SA)(\overline{S0 + S1}) + (B0 \cdot \overline{SB} + B1 \cdot SB)(S0 + S1) \\ &= A0 \cdot \overline{SA} \cdot \overline{S0} \cdot \overline{S1} + A1 \cdot SA \cdot \overline{S0} \cdot \overline{S1} + B0 \cdot \overline{SB} \cdot S0 + B0 \cdot SB \cdot S1 \\ &\quad + B1 \cdot \overline{SB} \cdot S0 + B1 \cdot SB \cdot S1 \end{aligned}$$

By tying the inputs of this module in different ways, the module can be configured as a variety of gates. In fact, it can be configured as any 2- or 3-input gate and most 4-input gates. If, for example, we tie A0, A1, B0, and S0 low,

$$OUT = B1 \cdot SB \cdot S1$$

and the module acts like a 3-input positive logic *and* gate. If, however, we tie A0 and A1 low and S0, S1, and SB high,

$$OUT = B1 + B0$$

the module acts like a 2-input *or* gate. These are just two examples of how the module can be configured. This module is from the example FPGA we used throughout Chapter 4, so the FPGA gates used in that chapter show the wide variety of gates into which this module can be programmed.

The sample module we've considered here is the module used by one manufacturer in one type of FPGA. Other FPGAs use different modules, but the idea of an array of programmable modules is fairly universal. So are the design steps.

As we noted, the first step in the design is to capture the logic circuit to be implemented with a suitable software package, using a library of logic elements available in the FPGA. Normally, the logic elements available in the library are the various configurations of basic modules in the array. Many FPGA libraries also contain more complicated entries, which one manufacturer calls ''soft macros.'' They are predesigned circuits for multiplexers, encoders, adders, and so on. We've given some examples of such circuits in Chapter 5. Predesigned circuits make design much easier. Even when they cannot be used directly, they can often be modified to perform a similar but slightly different function.

Once a logic circuit has been captured for the design, it's a good idea to simulate the circuit to determine whether it is functioning properly. This simulation is called a **functional simulation** because the delays for the various elements that make up the design are not known until the next step of the design.

The second step is to determine how to configure and interconnect the modules of the FPGA to produce the circuit for the circuit diagram. In modern design this is done by a computer using routing software (called a **router**). In some cases, the software is entirely automatic; it either achieves a complete routing for the circuit, or it fails. In other cases, the routing is partially automatic, with some software-assisted hand routing involved. In any case, the software must achieve a complete routing for the circuit to function properly. Sometimes it may be necessary to simplify the logic circuit in order to achieve complete routing. Sometimes the circuit is too large and must be split into two or more FPGAs. Once the routing is achieved, it is then possible to determine the actual circuit delays and to **back annotate** those delays into the simulation model.

Now an accurate simulation of the circuit can be made. For the FPGAs that are one-time programmable, this simulation is very important. With careful simulation, simple circuits should function on the first programming. For more complicated circuits, the first time the FPGA is programmed it may have timing problems, which can then be corrected and the device reprogrammed.

Programming is the step in the design in which the FPGA internal connections are made. It is completely computer automated. The routing of the devices determined in the previous step is now made into a fuse map, and the fuse map is used in conjunction with a device programmer to make the internal device connections. Once the device is programmed in this way, it must be tested. If it does not fulfill the design function during the test, it must be reprogrammed. With careful simulation, reprogramming can be kept to a minimum; this is very important for one-time reprogrammable devices. Simulation is also a good practice when using reprogrammable devices, because it allows timing checks on signals that may not be measurable in the actual device.

6.3 PROM LOGIC

The logic circuit diagram of what would be a very small PROM is shown in Figure 6.3-1. This PROM has 3 inputs, labeled I_0, I_1, and I_2. We'll assume initially that they are all asserted high. Each input is furnished with an inverter so that both high and low assertion levels of the input are available within the device. Unlike discrete logic circuits, PROMs have several outputs and can be used to evaluate several different

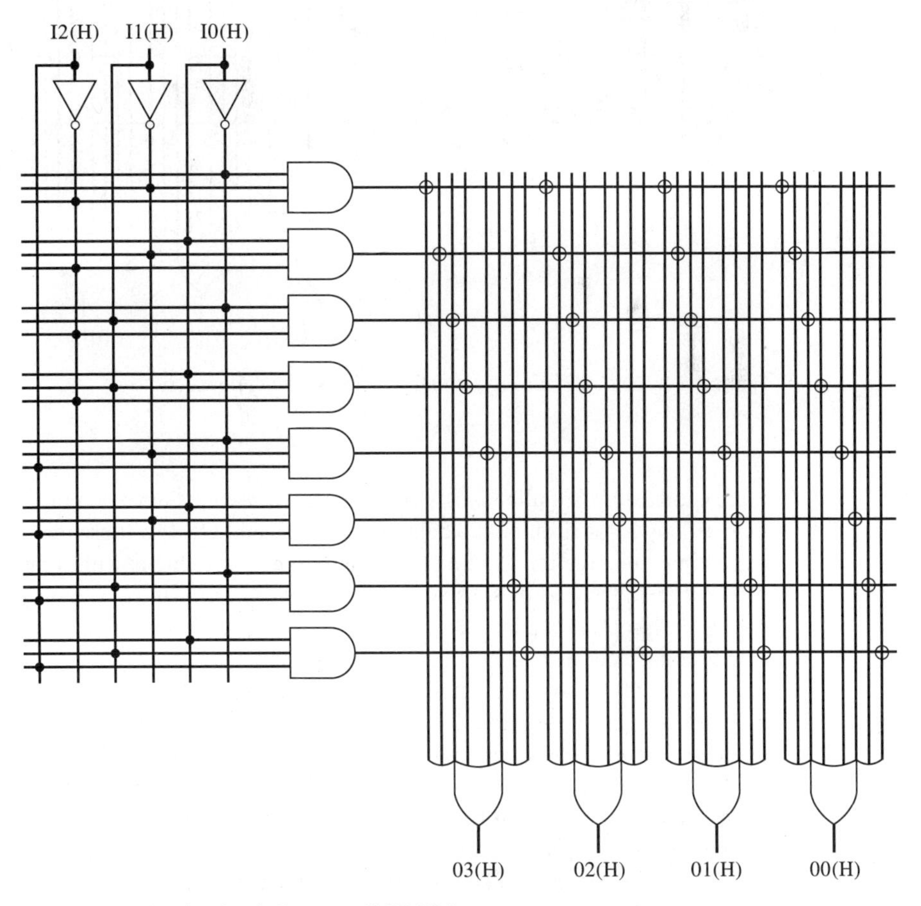

Figure 6.3-1 Logic circuit for a small PROM.

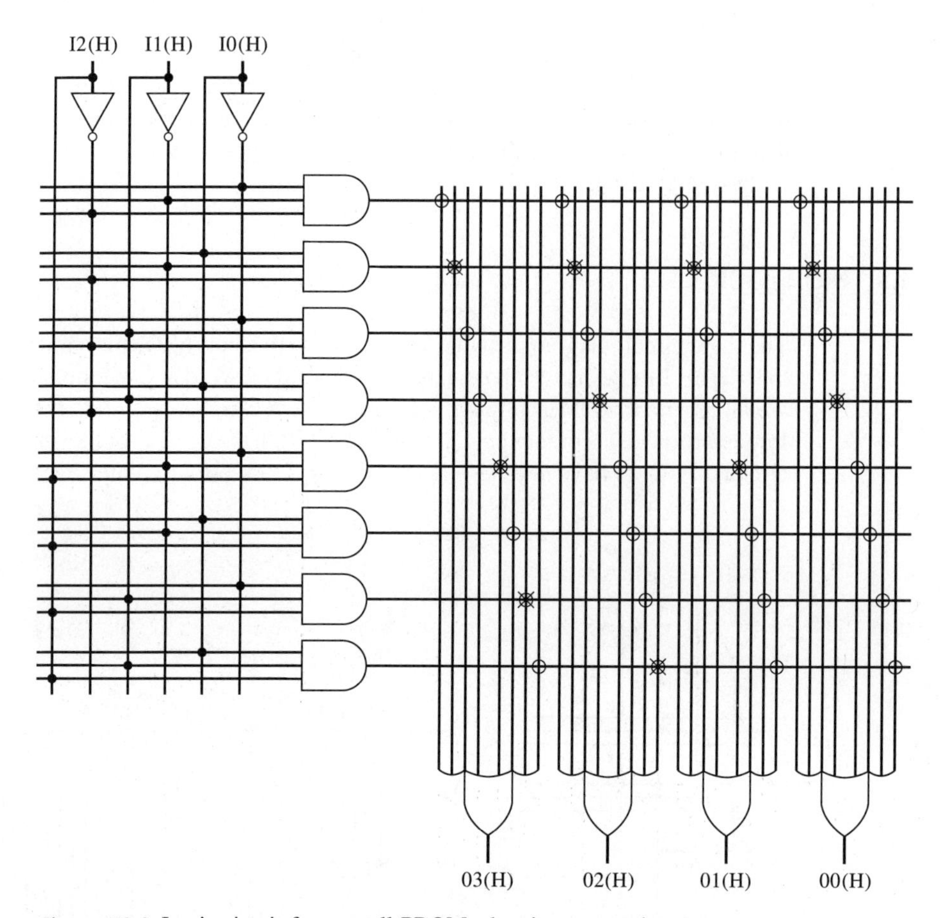

Figure 6.3-2 Logic circuit for a small PROM, showing connections.

logic expressions simultaneously if the logic expressions have common inputs. The device shown has 4 outputs labeled O_0, O_1, O_2, and O_3. We'll assume initially that the outputs, like the inputs, are all asserted high. Connections within the PROM are of two types: those shown as solid dots are permanent; those shown as unshaded circles are programmable. The programmable connections can be either connected or not connected, as needed in the circuit design.

In a PROM, the input gates are all permanently wired to evaluate *all* possible minterms of the input variables. There are three inputs in the figure, so there are eight *and* gates permanently wired to evaluate the eight minterms, $\bar{I}_2\bar{I}_1\bar{I}_0$ through $I_2I_1I_0$. The programmable parts of the PROM are the output gates. The wiring of any of the output *or* gates can be selected so that any combination of the minterms is *or*ed together. For example, if the connections shown by the $\times$s in Figure 6.3-2 are made, then the outputs O_3(H), O_2(H), O_1(H), and O_0(H) will be the values of the logic expressions:

$$O_3(H) = (\bar{I}_2\bar{I}_1I_0 + I_2\bar{I}_1\bar{I}_0 + I_2I_1\bar{I}_0)(H) = (\bar{I}_2\bar{I}_1I_0 + I_2\bar{I}_0)(H)$$

$$O_2(H) = (\bar{I}_2\bar{I}_1I_0 + \bar{I}_2I_1I_0 + I_2I_1I_0)(H) = (\bar{I}_2\bar{I}_1I_0 + I_1I_0)(H)$$

$$O_1(H) = (\bar{I}_2\bar{I}_1I_0 + I_2\bar{I}_1\bar{I}_0)(H)$$

$$O_0(H) = (I_2\bar{I}_1\bar{I}_0 + \bar{I}_2I_1I_0)(H)$$

Notice that inputs to the *or* gates that are not connected produce a zero in the *or* expression; they do not contribute to the expression.

When evaluating a logic expression using a PROM, the logic expression should not be simplified, because a PROM evaluates the unsimplified minterms. In the preceding example, for both O_3 and O_2, the expression before simplification shows exactly the connections to be made at the output *or* gate. The simplified expression does not.

PROMs are made only in certain sizes. The 3-input PROM of our example is much smaller than any actual available PROM. A PROM with eight inputs would be a small PROM. Because PROMs are made only in certain sizes and logic expressions come in all sizes, what happens when we cannot find a PROM with the number of inputs we need? In that case, we choose a larger PROM than we need and tie the unused inputs either high or low. For instance, we can evaluate the expression

$$O_3 = \bar{I}_2I_1 + I_2\bar{I}_1$$

using the 3-input PROM of our example if we tie the I_0 input low and program the connections as shown in Figure 6.3-3.

The logic circuit of a PROM (Fig. 6.3-1) has shown us how to use PROMs to evaluate logic expressions. The logic circuit is certainly a valid representation of the PROM, but it does not represent its normal use. A PROM is normally used as a memory device. A memory device can be thought of as an array of addressable storage locations (registers), as shown in Figure 6.3-4, which was drawn specifically for the 3-input, 4-output PROM used in the preceding explanation. Each of the addressable registers contains a 4-bit number. This number is placed on the output lines of the PROM when the register is addressed using the three input or address lines. For example, when the address lines are all low (000), the content of the zero register is placed on the output lines. In Figure 6.3-4, the input lines are labeled I and the output lines O, but normally in a memory device the input lines are labeled A for address and the output lines D for data. The normal labels of the lines are shown in parentheses. Besides the address and data lines, a memory device also has control lines such as a chip select, read, and/or write, which does not appear in the figure. A memory device is described by its size; that is, the number of registers in the memory and the size of the registers are given. A memory device with eight address lines and a 4-bit register size is a 256 × 4 memory, meaning it has 256 registers, and each register contains 4 bits. The number of registers is related to the number of address lines by the relation

$$\text{No. of registers} = 2^{\text{No. of address lines}}$$

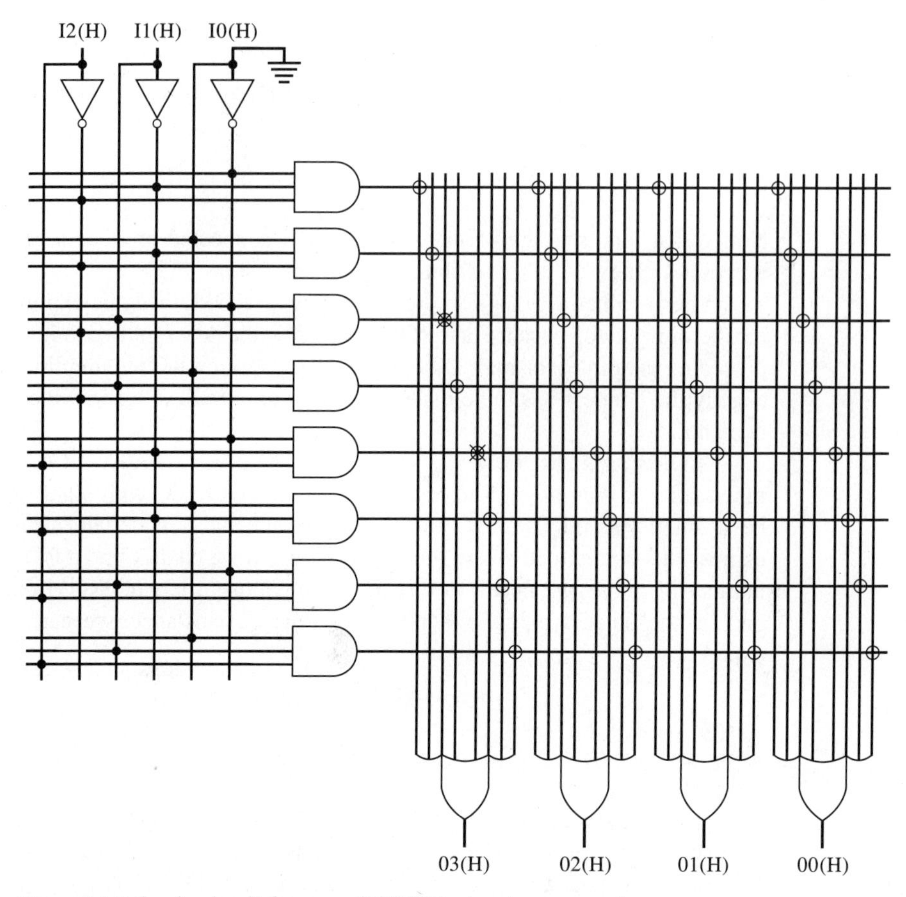

Figure 6.3-3 Logic circuit for a small PROM, showing connections.

The number of registers in a memory device is always a power of 2. For a memory device with eight address lines, the number of registers is

$$\text{No. of registers} = 2^8 = 256$$

There are many kinds of memory devices. The PROM is a Programmable Read Only Memory device. This means that the registers in the memory can be programmed to a binary number, but once they're programmed they cannot normally be changed. To program the PROM, we load each of the registers with a desired binary number. This is normally done using a PROM programmer or PROM burner.

Because we program a PROM by loading the registers with binary numbers, we need to see how to load the registers in order to evaluate a set of logic expressions. Consider the example of Figure 6.3-2. To determine what to load in address location 0, we look at the logic expressions for O_3, O_2, O_1, and O_0 and determine their values for $I_2 = 0$, $I_1 = 0$, and $I_0 = 0$. When we do this, we find that $O_3 = 0$, $O_2 = 0$, $O_1 = 0$, and $O_0 = 0$, so the register at address 000 is loaded with the value 0000.

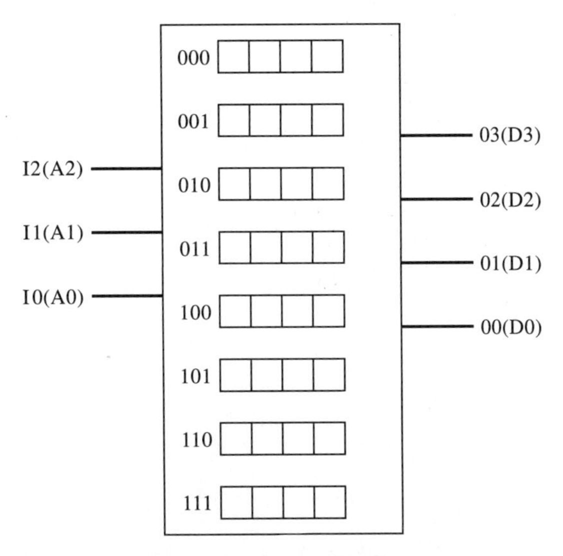

Figure 6.3-4 Array of registers.

If we look at each address location in turn and find the corresponding values of O_3, O_2, O_1, and O_0, we can form a table from the values to be loaded into the memory register for all the addresses (Table 6.3-1). Notice that the columns in the table of memory values are truth tables, one for each of the four logic expressions O_3, O_2, O_1, and O_0. This table is most easily formed a column at a time by filling in 1s for all the minterms of the logic expression corresponding to that column, then completing the column with 0s. For example, the minterms of O_3 are 001, 100, and 110, so in the left column the memory registers corresponding to those addresses are filled with 1s, and the memory registers corresponding to all other addresses are filled with 0s. The other columns can be filled in the same way by examining their logic expressions. We can think of the table of outputs versus addresses as a ''look-up table.'' For each address, we *look up* the value of the output logic functions.

Table 6.3-1

Address	Contents
000	0000
001	1110
010	0000
011	0101
100	1011
101	0000
110	1000
111	0100

Example 6.3-1

Use a 256 × 8 PROM to evaluate the logic expressions F_1 and F_2 defined by the following Karnaugh maps.

F_1

CD \ AB	00	01	11	10
00	0^{0}	1^{4}	0^{12}	1^{8}
01	1^{1}	0^{5}	1^{13}	0^{9}
11	0^{3}	1^{7}	0^{15}	1^{11}
10	1^{2}	0^{6}	1^{14}	0^{10}

F_2

CD \ AB	00	01	11	10
00	1^{0}	0^{4}	1^{12}	0^{8}
01	0^{1}	1^{5}	0^{13}	1^{9}
11	1^{3}	0^{7}	1^{15}	0^{11}
10	0^{2}	1^{6}	0^{14}	1^{10}

A 256 × 8 memory device has eight address (input) lines (A0 through A7) and 8-bit registers corresponding to outputs D0 through D7. We'll use only two bits in each register, since we're evaluating only two logic functions. In this example we'll use the high order bit (D7) for F_1 and the next bit (D6) for F_2.

Our logic functions have only four inputs, so we'll let A7 = A, A6 = B, A5 = C, and A4 = D, and tie A3, A2, A1, and A0 low. Now we can program the registers of the PROM using values from the Karnaugh maps. Only the two higher order bits of the registers need to be programmed. The other bits can be set to zero, one, or a combination of zeros and ones, because they are not used. We'll set them to zero. A 4-input logic function has 16 minterms, so we need a table showing the 16 addresses and the corresponding register values. Notice that in the table the four low order bits in all the addresses are zero, because the address lines corresponding to these bits are tied low. The four high order address bits are the actual minterms. For each minterm, the value for F_1 (from the Karnaugh map) is placed in the higher order register bit (D7) and the value for F_2 is placed in the next register bit (D6). The registers corresponding to all the other addresses can be loaded with any value, because they are not used. They correspond to addresses in which the four low order bits of the address are not all zeros. These registers cannot be addressed.

Address	Register	Address	Register	Address	Register
00000000	01000000	01100000	01000000	10110000	10000000
00010000	10000000	01110000	10000000	11000000	01000000
00100000	10000000	10000000	10000000	11010000	10000000
00110000	01000000	10010000	01000000	11100000	10000000
01000000	10000000	10100000	01000000	11110000	01000000
01010000	01000000				

Table 6.3-2

Address	Register	Address	Register	Address	Register
00000000	10111111	01100000	10111111	10110000	01111111
00010000	01111111	01110000	01111111	11000000	10111111
00100000	01111111	10000000	01111111	11010000	01111111
00110000	10111111	10010000	10111111	11100000	01111111
01000000	01111111	10100000	10111111	11110000	10111111
01010000	10111111				

In our logic design using PROMs, we've assumed that all the inputs are asserted high. Suppose now that some of the inputs are asserted low. Using the relation

$$A(L) = \overline{A}(H)$$

we can produce a situation where all the inputs are asserted high by using the complements of the inputs that are asserted low. The design can be done just as before except that, when we form the minterms (addresses) in the logic expression to be evaluated, we use the complements of the inputs that are asserted low. For example, if we want to evaluate the logic expression

$$F = ABC + \overline{A}\,\overline{B}C$$

with assertion levels A(H), B(L), and C(H), then we need to write this expression as

$$F = A\overline{B}C + \overline{A}BC$$

and load the registers of the PROM so that the values for the addresses 101 and 100 are both 1, and the values for the addresses corresponding to all the other minterms are 0.

It's no more difficult to change the output assertion level to low: we simply complement this output. For example, if we want to change all the output assertion levels to low in Example 6.3-1, we complement the outputs in the table and produce a new table, Table 6.3-2. Actually, we need complement only the two high order bits that correspond to the two actual outputs, but it's just as easy to complement all the bits.

Now consider an example with mixed assertion levels.

Example 6.3-2

Repeat Example 6.3-1 with the following mixed assertion levels: A(H), B(L), C(L), D(H), F_1(L), and F_2(H).

Just as in Example 6.3-1, we'll use the high order bit (D7) of the PROM for F_1 and the next bit (D6) for F_2. We'll also use the same input assignments—A7 = A,

A6 = B, A5 = C, and A4 = D—and tie A3, A2, A1, and A0 low. Now we can form a table of address and register contents just like the table in Example 6.3-1, but with the columns corresponding to the low assertion levels that are complemented.

Address	Register	Address	Register	Address	Register
01100000	11000000	00000000	11000000	11010000	00000000
01110000	00000000	00010000	00000000	10100000	11000000
01000000	00000000	11100000	00000000	10110000	00000000
01010000	11000000	11110000	11000000	10000000	00000000
00100000	00000000	11000000	11000000	10010000	11000000
00110000	11000000				

6.4 PAL LOGIC

Figure 6.4-1 is the logic diagram of what would be a very small PAL. This PAL has only four inputs and two outputs. The four inputs are labeled I_0, I_1, I_2, and I_3, and we'll initially assume they are all asserted high. Each input is furnished with an inverter so that both high and low assertion levels of the input are available in the device. Like PROM logic circuits, PALs have several outputs and can be used to evaluate several different logic expressions if they have common inputs. This device has two outputs, labeled O_0 and O_1, and we'll assume they're asserted high like the inputs. Connections within the PAL are of two types: those shown as solid dots are permanent, and those shown as unshaded circles are programmable.

The PAL differs from the PROM in that the input *and* gate connections are programmable in a PAL, whereas the output *or* gate connections are programmable in a PROM. This difference usually results in a reduction of the gate count in a PAL as compared to a PROM performing the same function. The PAL is naturally in the form of a standard two-level logic circuit for evaluating an SOP logic expression, so it is easy to use. For example, if we want to evaluate the logic expressions

$$O_1 = I_2\, I_1\, I_0 + \bar{I}_2\, \bar{I}_1\, \bar{I}_0 + I_3\, I_1 + I_3\, \bar{I}_2$$

$$O_0 = \bar{I}_1\, I_0 + I_3\, I_2 + \bar{I}_3\, \bar{I}_2$$

using our sample PAL, we program the connections shown by the Xs in Figure 6.4-2. Notice that when there is no connection to an input line of an *and* gate, the input is pulled high (asserted) by the PAL circuitry.

The first expression (O_1) has four terms and uses all the *and* gates connected to the first output gate. The second expression (O_0) has only three terms. The fourth *and* gate (the bottom gate in the array), which would represent a fourth term in the logic expression, is connected to all the input lines (both high and low assertion levels), so it is never asserted. This is easily seen because the product contains a number of pairs of terms of the form $I_i\, \bar{I}_i$, which is always zero. In this case we did

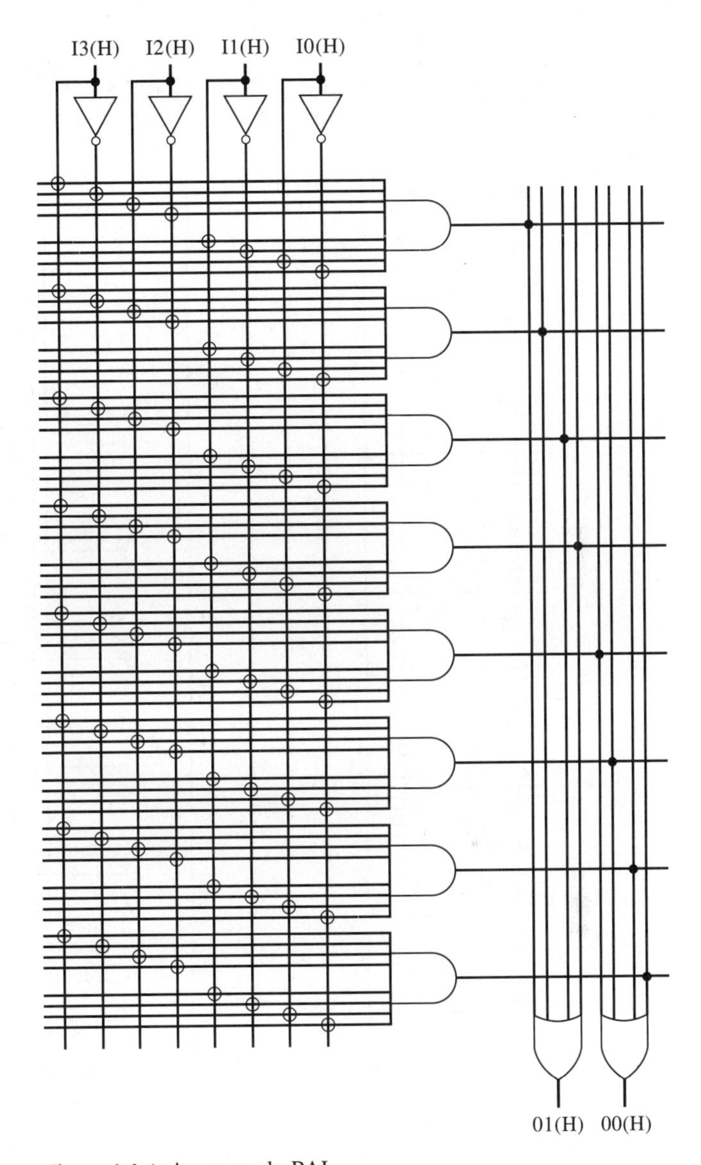

Figure 6.4-1 An example PAL.

not simplify the logic expressions, because they cannot be simplified. Normally we would simplify the expressions using Karnaugh maps. We should always try to simplify, because the simplified expressions are easier to implement with the PAL and in some cases result in a smaller PAL.

If we have input lines to the PAL that are asserted low, we can handle them much as we did for a PROM. Using the relation

$$A(L) = \overline{A}(H)$$

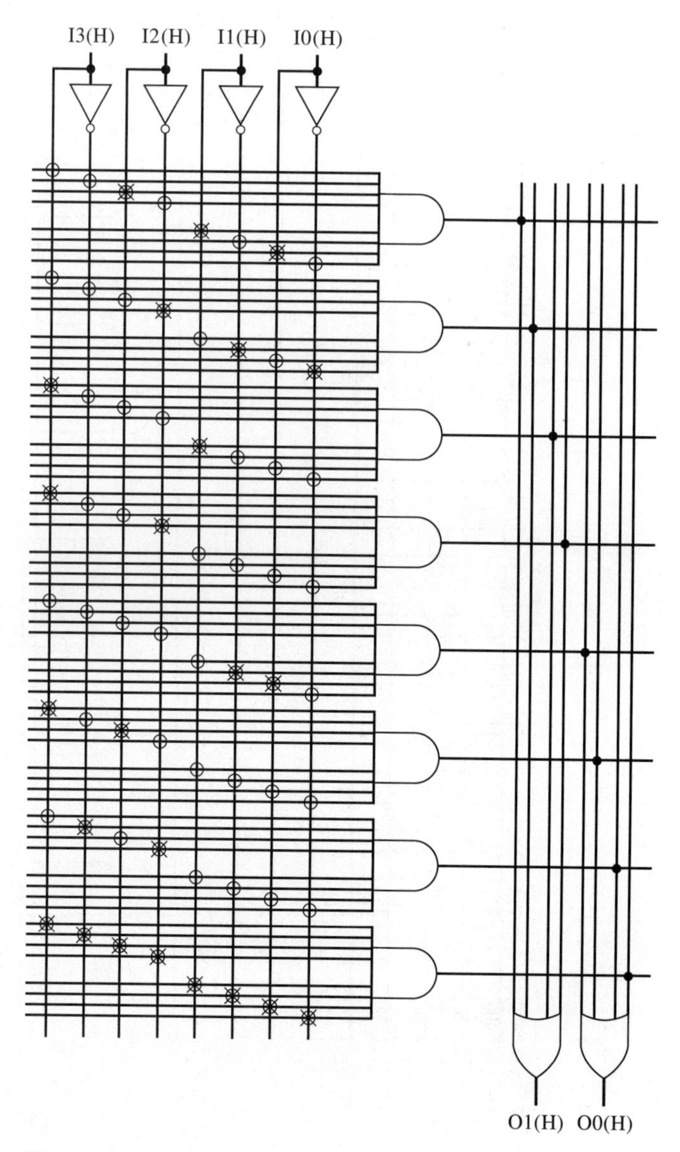

Figure 6.4-2

we can produce a situation where all the inputs are asserted high. The design can be done just as before except that, when we form the logic expression to be evaluated, we use the complement of the inputs that are asserted low. For example, if the input assertion levels in the preceding case are $I_3(L)$, $I_2(H)$, $I_1(H)$, and $I_0(L)$, then the logic expressions become (complementing I_3 and I_0)

$$O_1 = I_2\, I_1\, \bar{I}_0 + \bar{I}_2\, \bar{I}_1\, I_0 + \bar{I}_3\, I_1 + \bar{I}_3\, \bar{I}_2$$

$$O_0 = \bar{I}_1\, \bar{I}_0 + \bar{I}_3\, I_2 + I_3\, \bar{I}_2$$

and the PAL connections change to those shown in Figure 6.4-3. Again, as might be suspected, the expressions cannot be simplified.

To handle outputs with low assertion levels, we choose to use PALs with low output assertion levels if all the outputs have low assertion levels. PALs having low output assertion levels are readily available, as will be seen when we consider designs using some actual PALs. Mixed-output assertion levels are more difficult. One way to handle mixed assertion levels is to express the logic function that is to have the low assertion level in POS form, and then complement it to get a low assertion level

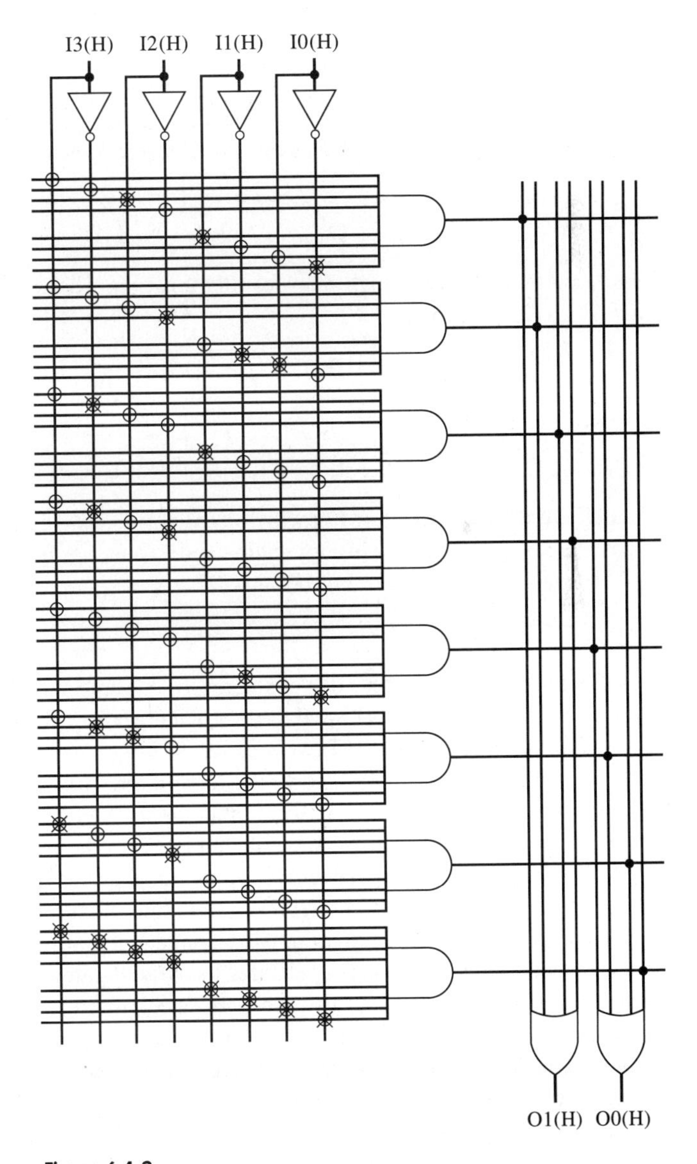

Figure 6.4-3

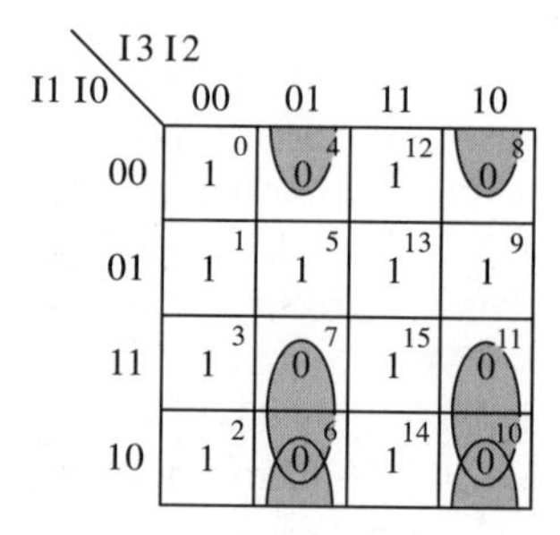

Figure 6.4-4

SOP expression—the form of expression we need for programming in a PAL. This process can be better seen with an example. Consider the original O_0 of the preceding case:

$$O_0 = \bar{I}_1 I_0 + I_3 I_2 + \bar{I}_3 \bar{I}_2$$

Figure 6.4-4 is the Karnaugh map for this function. The POS expression from the map is

$$O_0 = (I_3 + \bar{I}_2 + I_0)(I_3 + \bar{I}_2 + \bar{I}_1)(\bar{I}_3 + I_2 + I_0)(\bar{I}_3 + \bar{I}_2 + \bar{I}_1)$$

So the complement is

$$\overline{O_0} = \bar{I}_3 I_2 \bar{I}_0 + \bar{I}_3 I_2 I_1 + I_3 \bar{I}_2 \bar{I}_0 + I_3 I_2 I_1$$

Actually, the complement of O_0, which is what we want, can be read directly from a covering of the map 0s. This complemented expression is in the form we need for a PAL evaluation of $\overline{O_0}(H)$, but because

$$O_0(L) = \overline{O_0}(H)$$

it is also the expression we need to evaluate $O_0(L)$. The PAL connections that must be programmed to evaluate $O_0(L)$ and the original $O_1(H)$ are shown in Figure 6.4-5.

As already stated, the PAL we've used in the above examples is smaller than any actual PAL. Figure 6.4-6 shows the logic circuit for an actual PAL consisting of eight separate logic circuits with common inputs. Figure 6.4-7 shows one of these logic circuits, with the various parts labeled to aid in the following description. It's a two-level logic circuit with seven 32-input *and* gates followed by a 7-input *or* gate. (The top *and* gate is not used directly in the logic circuit; it has a different use, as we'll see presently.) The programmable connections are between the 32 vertical input lines and the horizontal input lines to the *and* gates. To save drawing space, a single horizontal input line to each *and* gate is used to represent the 32 actual inputs to the gate. (A single line that represents a number of connections is called a **bus.**) The seven *and* gates are permanently connected to an *or* gate.

External connections to the circuit are indicated by pin numbers on the circuit diagram. The single logic circuit shows only one of the dedicated inputs, pin 3, and its associated **driver**. The driver provides both high and low assertion level inputs,

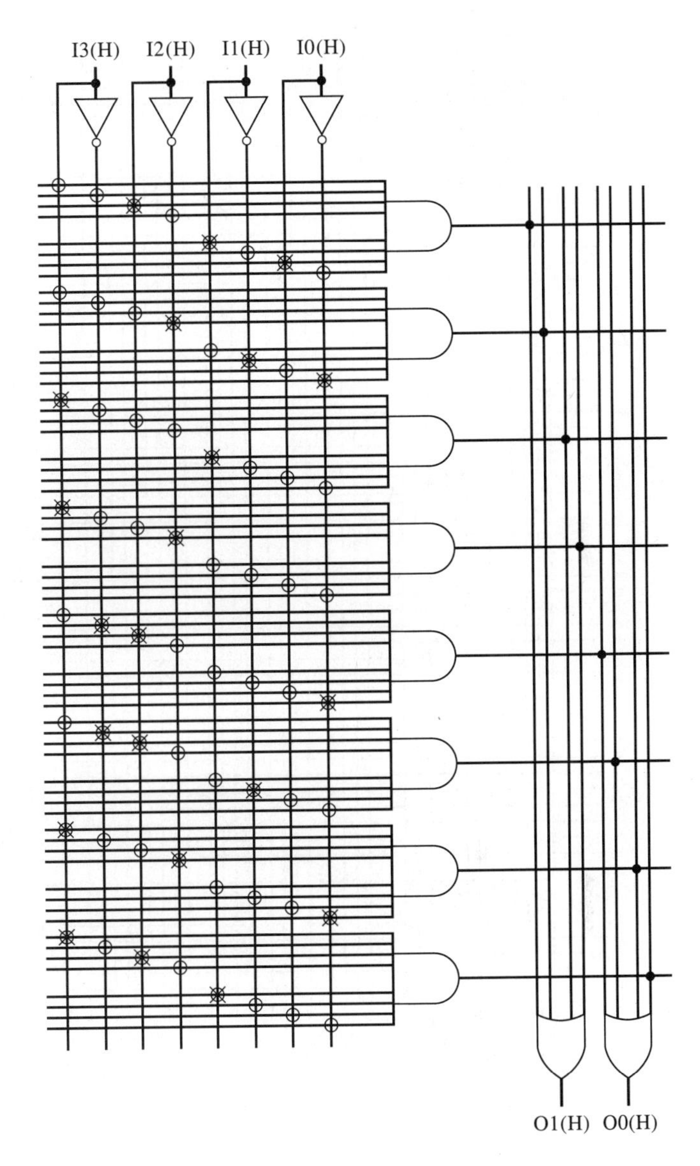

Figure 6.4-5

so both the input and its complement are available. Pin 18 of the single logic circuit can serve as either an input or an output because of the tristate output driver. This driver has an enable input, shown as a connection at the top of the driver. When the enable is asserted, the driver is "turned on" and passes the output of the logic circuit to pin 18. When the enable is not asserted, the output of the driver is placed in a high impedance state (essentially disconnected from the output line). As can be seen from the circuit, the enable for the driver is a programmable *and* function of the

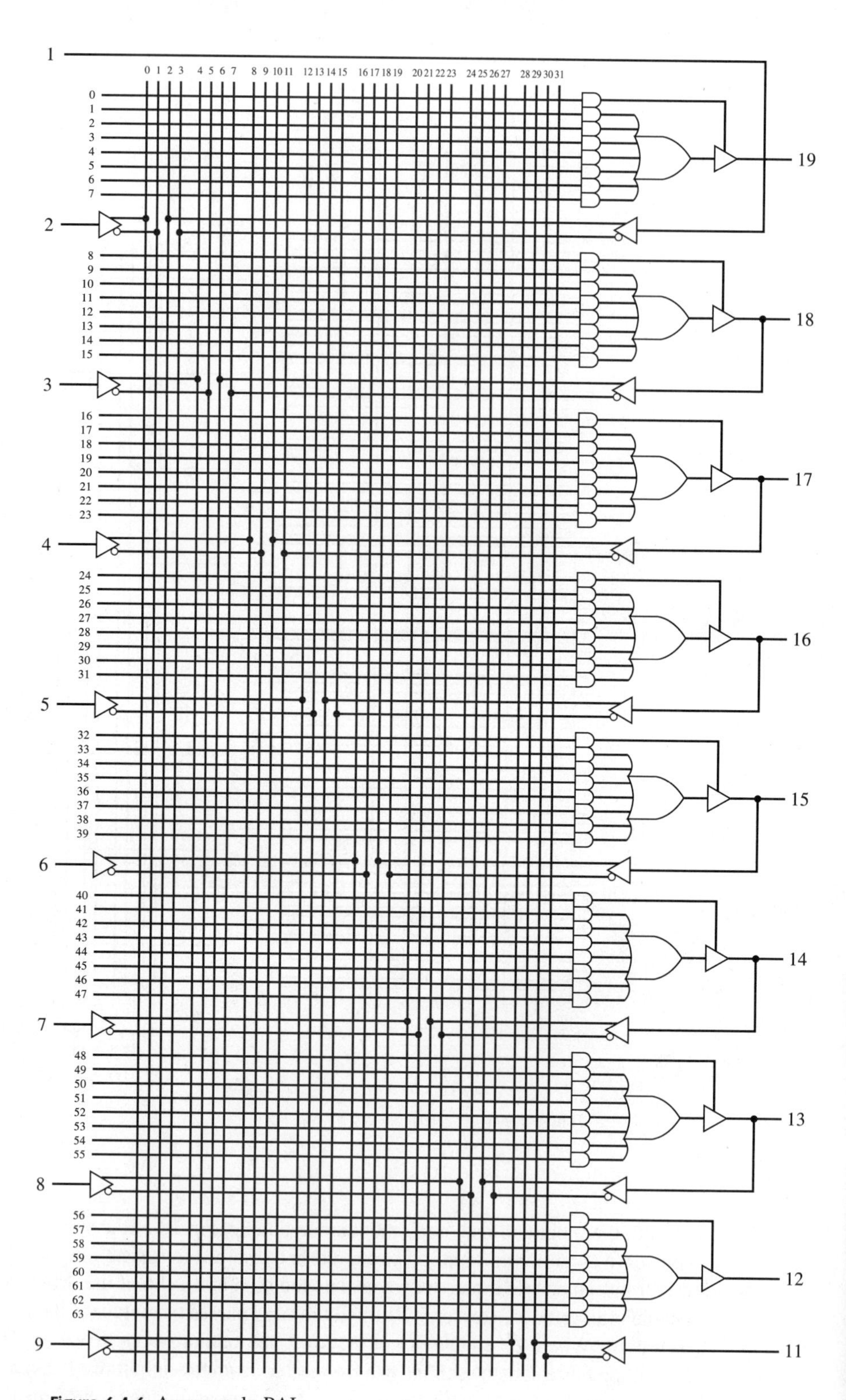

Figure 6.4-6 An example PAL.

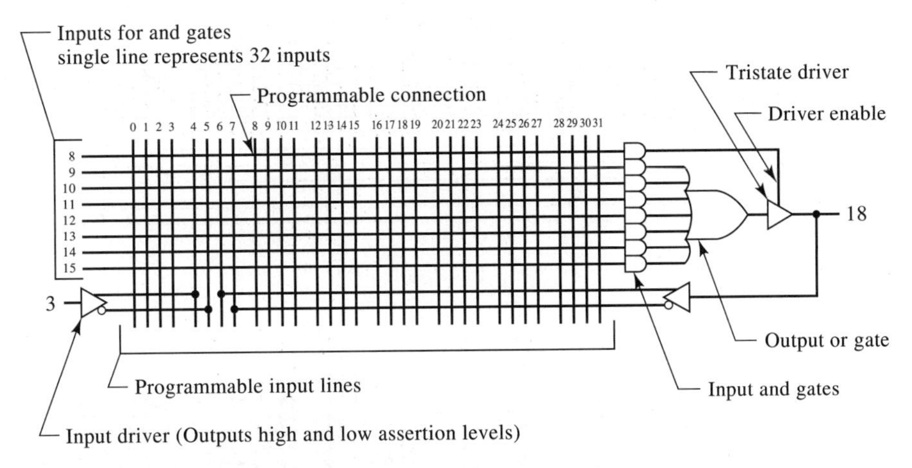

Figure 6.4-7 A single logic circuit from a PAL.

inputs to the PAL (using the top *and* gate). If the output driver is enabled, then pin 18 must function as an output. If the output driver is not enabled, then pin 18 can function as an input through its input driver. An additional feature of this single logic circuit is the internal connections available from its output to the inputs of each of the other logic circuits of the PAL. This allows cascading of logic circuits without external connections.

The single logic circuit we've described is repeated eight times in a PAL, except that the top and bottom logic circuits are slightly different. These circuits have dedicated outputs and an additional dedicated input. Thus, the top circuit has a dedicated input through pin 2, similar to the typical single logic circuit already described, but it also has a dedicated input through pin 1, and it has a dedicated output through pin 19. The bottom circuit has dedicated inputs through pins 9 and 11 and a dedicated output through pin 12. The PAL has a total of 10 dedicated inputs, eight external connections that function as either inputs or outputs, and two dedicated outputs. The cascading feature described for the typical single logic circuit is not available in the top or the bottom logic circuit without external connections.

The PAL we've described is one of a number of available PALs. PALs generally differ in the number and size of the individual logic circuits they contain and in their output configurations. Figures 6.4-8, 6.4-9, and 6.4-10 show some of the output configurations that are available. The first is identical to the circuit of Figure 6.4-6 except that it has a low rather than a high output assertion level. The other two configurations have eight rather than seven *and* gates available for logic, and they do not have tristate drivers; instead, they have dedicated outputs. Another output configuration, not shown, is a register output, which we'll discuss when we cover sequential logic. In most modern PALs the output configuration, as well as the logic connections, is programmable. This means that the output type—simple *or* gate, or register—and the output assertion level can be chosen by proper programming.

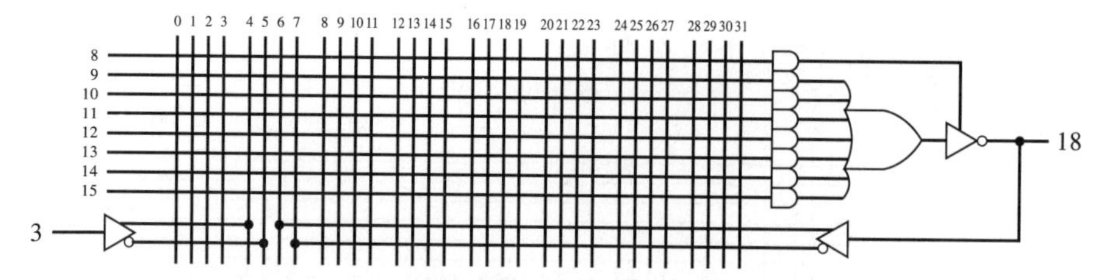

Figure 6.4-8 PAL with low output assertion level.

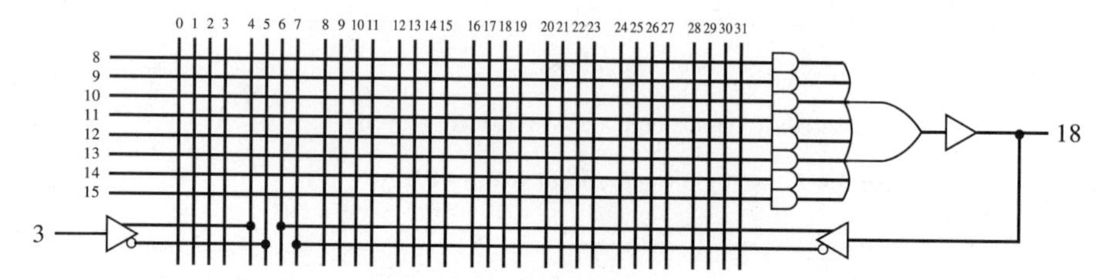

Figure 6.4-9 PAL with dedicated high assertion level output.

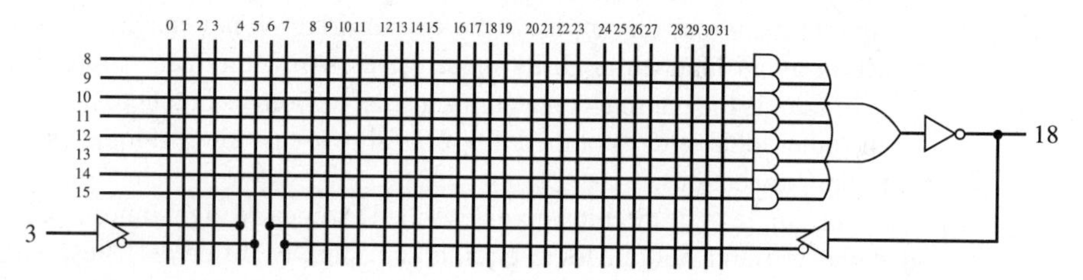

Figure 6.4-10 PAL with dedicated low assertion level output.

PALs are programmed with a PAL programmer that actually makes the desired connections in the circuit. In some PALs, all the programmable connections are initially connected with fusible links, and the programmer burns away the connections that are not desired. In this type of PAL the programming is permanent. Some PALs have provisions for erasable programming so that they can be used and changed a number of times.

In any case, software exists to help the designer program the PAL. The software differs depending on the manufacturer, but usually only the logic expressions to be evaluated and the pin connections for the input and output signals need be input to the software. The actual internal connections (called a mask or a fuse map) needed in the PAL are generated by the software, and most software simplifies the logic expressions. Sometimes the software is actually run on the programmer, which is a small computer; sometimes it's run separately and the mask or fuse map generated is input to the programmer. An example of a PAL design will help to clarify this process. The design is for a specific PAL manufactured by Advanced Micro Devices and a specific type of software (PALASM). Designs for PALs from other manufacturers and using different software would differ in details.

Example 6.4-1

A common device needed in computer design is an address decoder such as that discussed in Chapter 5. Using three address lines, it's possible to generate eight addresses (000 through 111). A 3-input, 8-output address decoder is a device that asserts one of the 8 output lines depending on the address on the address lines. Thus, if the address on the address lines is 000, the low order output line is asserted; if the address is 001, the next order output line is asserted. Design a 3-to-8 decoder. Assume in the design that the address lines (S_0, S_1, and S_2) are asserted high and the output lines (D_0 through D_7) are asserted low.

A function table for this device follows. It actually describes eight different functions, D_0 through D_7. Because the output assertion level is low for each output, we'll choose a PAL (PAL16L8) with a low output assertion level. (Figure 6.4–6 with a bubble on each output.) The PALs we've been discussing have a maximum of 8 outputs if we make all the external pins—which can be either outputs or inputs—*outputs*. Eight is just the number we need. This leaves ten input pins, and we need use only three because we have three inputs.

S2	S1	S0	D7	D6	D5	D4	D3	D2	D1	D0
L	L	L	H	H	H	H	H	H	H	L
L	L	H	H	H	H	H	H	H	L	H
L	H	L	H	H	H	H	H	L	H	L
L	H	H	H	H	H	H	L	H	H	H
H	L	L	H	H	H	L	H	H	H	H
H	L	H	H	H	L	H	H	H	H	H
H	H	L	H	L	H	H	H	H	H	H
H	H	H	L	H	H	H	H	H	H	H

Normally, the next step in the design would be to simplify the logic expressions using Karnaugh maps. For the logic functions described by the table with high address assertion levels and low output assertion levels, there is no simplification. Only one term in each column is 1, so the Karnaugh map has only one nonzero entry. We can thus read the logic expressions directly from the table. Actually, we do not need to simplify the logic expressions in any case, because the software will do it for us.

To program the PAL, we'll use software called PALASM. Programming is accomplished by presenting an input or source code that describes the logic circuit to be designed to PALASM for processing. The input code for this example is shown in the following list. It's divided into the *declaration*, *equations*, and *simulation* sections.

```
; PALASM Design Description
; ------------------------------ Declaration Segment ----------
TITLE     3 TO 8 DECODER
PATTERN   3-8DEC
REVISION  A
AUTHOR    ALAN W SHAW
COMPANY   UTAH STATE UNIVERSITY
DATE      04/07/92
CHIP   A38DEC   PAL16L8
; ------------------------------ PIN Declarations -------------
PIN  1     S2                                  ; INPUT
PIN  2     S1                                  ; INPUT
PIN  3     S0                                  ; INPUT
PIN  4     NC                                  ;
PIN  5     NC                                  ;
PIN  6     NC                                  ;
PIN  7     NC                                  ;
PIN  8     NC                                  ;
PIN  9     NC                                  ;
PIN  10    GND                                 ;
PIN  11    NC                                  ;
PIN  12    /D0                 COMBINATORIAL   ; OUTPUT
PIN  13    /D1                 COMBINATORIAL   ; OUTPUT
PIN  14    /D2                 COMBINATORIAL   ; OUTPUT
PIN  15    /D3                 COMBINATORIAL   ; OUTPUT
PIN  16    /D4                 COMBINATORIAL   ; OUTPUT
PIN  17    /D5                 COMBINATORIAL   ; OUTPUT
PIN  18    /D6                 COMBINATORIAL   ; OUTPUT
PIN  19    /D7                 COMBINATORIAL   ; OUTPUT
PIN  20    VCC                                 ;
; ------------------------------ Boolean Equation Segment ------
EQUATIONS
D0        = /S2*/S1*/S0
D1        = /S2*/S1* S0
D2        = /S2* S1*/S0
D3        = /S2* S1* S0
D4        =  S2*/S1*/S0
D5        =  S2*/S1* S0
D6        =  S2* S1*/S0
D7        =  S2* S1* S0
D0 . TRST = VCC
D1 . TRST = VCC
D2 . TRST = VCC
D3 . TRST = VCC
D4 . TRST = VCC
D5 . TRST = VCC
```

```
D6 . TRST  =  VCC
D7 . TRST  =  VCC
; ------------------------------  Simulation Segment  ----------
SIMULATION
TRACE_ON S2  S1  S0 /D0 /D1 /D2 /D3 /D4 /D5 /D6 /D7
SETF  /S2  /S1  /S0
SETF  /S2  /S1   S0
SETF  /S2   S1  /S0
SETF  /S2   S1   S0
SETF   S2  /S1  /S0
SETF   S2  /S1   S0
SETF   S2   S1  /S0
SETF   S2   S1   S0
SETF  /S2  /S1  /S0
TRACE_OFF
; ----------------------------------------------------------------
```

The declaration section contains information to document the design, the device identification, pin assignments, and string substitutions. The beginning of the declaration section of the example code is delineated by a comment line consisting of dashes and the words ''Declaration Segment.'' In any line, anything following a semicolon (;) is a comment and is ignored by PALASM. If a line begins with a semicolon, the entire line is a comment. The first six lines of the declaration are called the *design header*. As can be seen from the example, each line starts with a keyword (TITLE, PATTERN, and so on). The intended use of the line should be evident from the keyword, although a line can consist of any text or can be blank. The next line after the header, which starts with the keyword, CHIP, contains two entries. The first is a name the user can assign to the chip. The second is the manufacturer's designation for the device that is to be used in the design (for instance, PAL16R4). These two items are separated by at least one space.

After the line describing the chip is a listing of the signal names that are assigned to the pins of the device. Something must be listed for each pin. The polarity of the signal is specified by adding a / in front of the name if the assertion level is low. For our purpose all the output pins are assigned a storage type COMBINATORIAL (combinational), because the circuits we are describing are combinational logic circuits. The INPUT or OUTPUT designation for the pins is a comment (it follows a semicolon) but may be used by the program. In the example, the output signals D0 through D7 are all asserted low, matching the outputs of the device. It's not actually necessary to choose assertion levels that match, but it makes it easier to understand the logic produced. The pins that are not used are labeled NC (no connection). The power pin is labeled VCC and the ground GND. In this example program we've assigned the input signals to pins 1, 2, and 3 and the output signals to pins 12 through 19.

If we use the PALASM editor to produce the source program, the declaration section is just a matter of filling in blanks on the screen shown in the table below. As we can see from the table, all the keywords are supplied by the editor and many of the blanks are even filled by selecting from lists of appropriate entries for that

PDS Declaration Segment

Title
Pattern
Revision
Author
Company
Date 05/28/92

CHIP ChipName = ______ Device = ______

P/N	Number	Name	Paired with PIN	Storage	;comment

Enter PIN/NODE Data. [Press ⟨ESC⟩=abort, F1=help, F2=choices, F10=save & exit]

blank. Selections are available for DEVICE (a list of all the PALs which can be programmed by the software), P/N (PIN or NODE), storage (blank, COMBINATORIAL, REGISTERED, or LATCHED), and comment (blank, INPUT, OUTPUT, IO, CLOCK, or ENABLE). The selection list appears in response to the F2 key.

The remainder of the source code is entered using a text editor. (The normal PALASM editor is a Wordstar-like editor.) This remaining code includes the final lines of the declaration section, which can contain string substitution statements to allow substitution of simple names for complex statements. String substitutions are a convenience in writing the subsequent program but aren't necessary. The syntax is:

```
STRING name 'string'
```

The string is delimited by single quotes. An example of a string substitution statement is,

```
STRING   NAME1 ' INPUT1 * INPUT2 + RST '
```

There are no string statements in the example program.

The beginning of the equations section of the program is delineated by a comment line of dashes and the words ''Boolean Equation Segment.'' The comment line is followed by the keyword, EQUATIONS (see example input code listing). The equations section consists of a listing of the logic expressions to be programmed. The syntax for the logic expressions is

output logic variable = logic expression

The logic expression is a function of the input logic variables (signals) that we specified in the declaration segment. In it, we use * for *and,* + for *or,* and a slash (/) before a variable name to indicate *not asserted.* This use of a slash makes computer input easier, because most keyboards do not have an overbar key. Notice the difference in the meanings of / in the pin assignments and / in the logic expressions. In the pin assignments, / means asserted low. In the logic expressions, / means *not asserted.* If you keep this simple distinction in mind, you'll avoid a lot of confusion.

The equations section also contains expressions that describe the output tristate condition. Their syntax is

output logic variable.TRST = logic expression

One of these expressions is needed for each tristated output. The .TRST is the keyword that denotes a tristate condition. The logic expression can be a single *and* term logic expression. In the example, the tristate condition is VCC, which is always asserted.

The final section of the program is the simulation section. Its beginning is delineated by a comment line of dashes and the words ''Simulation Segment.'' Following the comment line is the keyword, SIMULATION. The simulation section describes a sequence of input signals that are to be applied to the machine. The syntax for all the simulation commands is

command   signal list

In the example, the first command is TRACE_ON. It is used to specify the signals that are to be traced and can be viewed separately. The signal list lists the signals to be traced. A signal with no / is asserted high, and one with / is asserted low. In the TRACE_ON command, we again use / to specify the assertion level of the signal, as we did when we named the pins. The TRACE_ON command has a companion command, TRACE_OFF, which turns off the signals to be traced. Tracing allows select signals to be viewed. All the signals are recorded and can be viewed, but unless the TRACE_ON and TRACE_OFF commands are used, the signals to be viewed cannot be limited.

The second command is SETF. It causes the signals listed to be asserted or not asserted (/ means not asserted). By changing the inputs to the PAL with SETF commands, we can simulate all combinations of input and determine whether the logic circuit is performing as desired.

Simulation allows loops and conditional branching, which we will not cover.

When our example program is processed using PALASM, a number of outputs are produced, including a JEDEC output for programming the actual programmable device, a fuse map, and a timing diagram. The time scale in the timing diagram is not truly proportional to time as we measure it, but it does show the relative positions of the transitions.

When the input of this example is used in the programming software, it generates the JEDEC file shown in the following table. Each row of the table represents 32 possible programmable connections (i.e., the input connections to one *and* gate). All the programmable connections are numbered starting with zero (0000) in the upper left corner of the PAL and working across the row, then across the next row. In the JEDEC file, the closed connections are shown as 1s and the open connections as 0s. The figure at the end of this example shows the logic circuit for this design. When an × is placed at an intersection of an input line and an input to an *and* gate, it indicates a connection. An × in an *and* gate indicates that all inputs to the gate are connected. The figure that directly follows the table shows the simulation results. Notice that when S2, S1, and S0 are all low (000), the output D0 is low (asserted); when S1 and S1 are low and S0 is high (001), the output D1 is low (asserted); and so on. The simulation is a PALASM simulation. It does not show the timing accurately.

```
PALASM4 PAL ASSEMBLER - MARKET Version 1.2 (5-31-91)
  (c) - COPYRIGHT ADVANCED MICRO DEVICES INC., 1991

TITLE     :3 TO 8 DECODER       AUTHOR   :ALAN W SHAW
PATTERN   :3-8DEC               COMPANY  :UTAH STATE UNIVERSITY
REVISION  :A                    DATE     :04/07/92

487
PAL16L8
A38DEC*
QP20*
QF2048*
G0*F0*
L0000 11111111111111111111111111111111*
L0032 01010111111111111111111111111111*
L0064 00000000000000000000000000000000*
L0096 00000000000000000000000000000000*
L0128 00000000000000000000000000000000*
L0160 00000000000000000000000000000000*
L0192 00000000000000000000000000000000*
L0224 00000000000000000000000000000000*
L0256 11111111111111111111111111111111*
L0288 01011011111111111111111111111111*
L0320 00000000000000000000000000000000*
L0352 00000000000000000000000000000000*
L0384 00000000000000000000000000000000*
L0416 00000000000000000000000000000000*
L0448 00000000000000000000000000000000*
L0480 00000000000000000000000000000000*
L0512 11111111111111111111111111111111*
```

```
L0544 10010111111111111111111111111111*
L0576 00000000000000000000000000000000*
L0608 00000000000000000000000000000000*
L0640 00000000000000000000000000000000*
L0672 00000000000000000000000000000000*
L0704 00000000000000000000000000000000*
L0736 00000000000000000000000000000000*
L0768 11111111111111111111111111111111*
L0800 10011011111111111111111111111111*
L0832 00000000000000000000000000000000*
L0864 00000000000000000000000000000000*
L0896 00000000000000000000000000000000*
L0928 00000000000000000000000000000000*
L0960 00000000000000000000000000000000*
L0992 00000000000000000000000000000000*
L1024 11111111111111111111111111111111*
L1056 01100111111111111111111111111111*
L1088 00000000000000000000000000000000*
L1120 00000000000000000000000000000000*
L1152 00000000000000000000000000000000*
L1184 00000000000000000000000000000000*
L1216 00000000000000000000000000000000*
L1248 00000000000000000000000000000000*
L1280 11111111111111111111111111111111*
L1312 01101011111111111111111111111111*
L1344 00000000000000000000000000000000*
L1376 00000000000000000000000000000000*
L1408 00000000000000000000000000000000*
L1440 00000000000000000000000000000000*
L1472 00000000000000000000000000000000*
L1504 00000000000000000000000000000000*
L1536 11111111111111111111111111111111*
L1568 10100111111111111111111111111111*
L1600 00000000000000000000000000000000*
L1632 00000000000000000000000000000000*
L1664 00000000000000000000000000000000*
L1696 00000000000000000000000000000000*
L1728 00000000000000000000000000000000*
L1760 00000000000000000000000000000000*
L1792 11111111111111111111111111111111*
L1824 10101011111111111111111111111111*
L1856 00000000000000000000000000000000*
L1888 00000000000000000000000000000000*
L1920 00000000000000000000000000000000*
L1952 00000000000000000000000000000000*
L1984 00000000000000000000000000000000*
L2016 00000000000000000000000000000000*
C3EC4*
^CE9E8
```

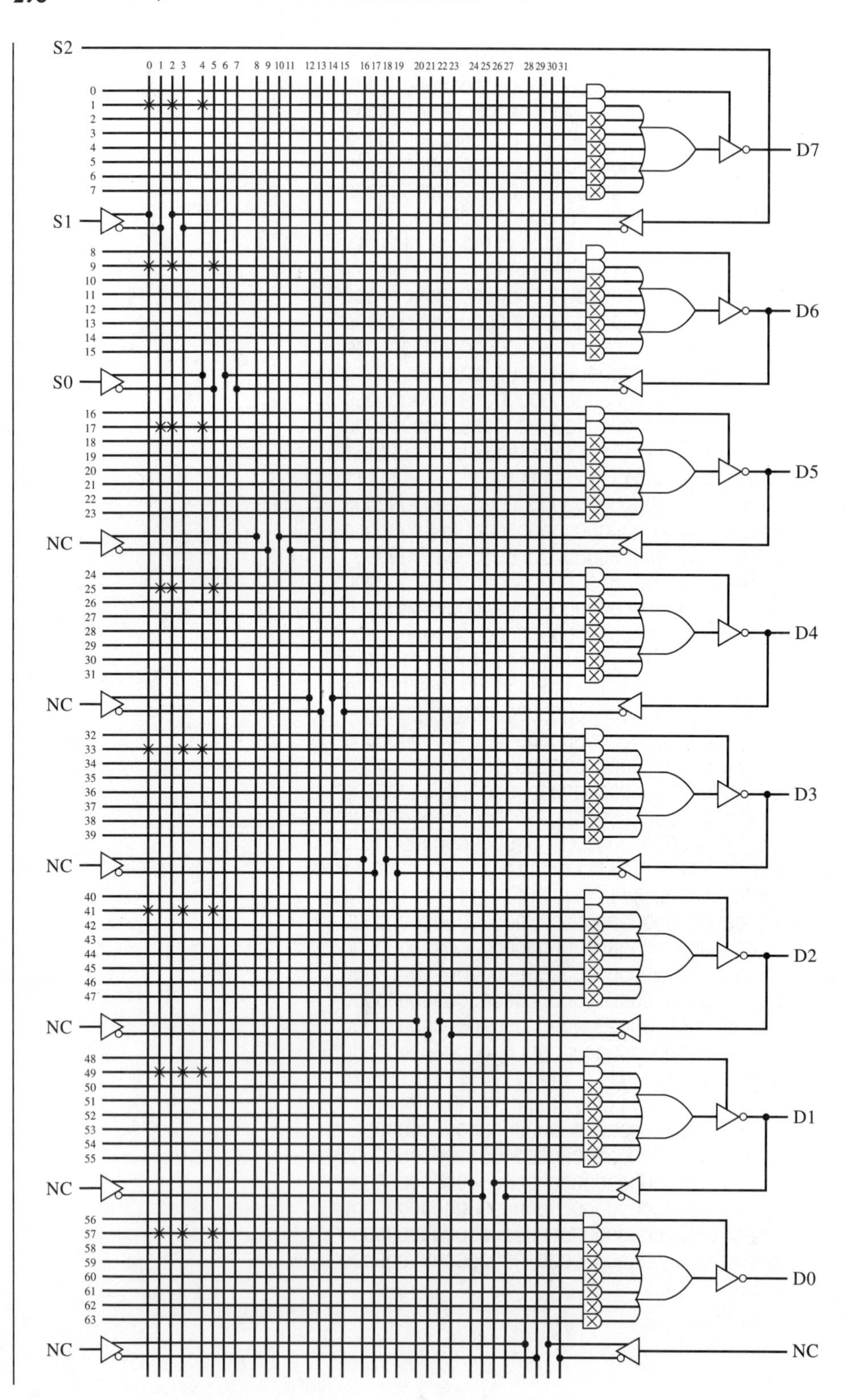
S2
S1
S0
NC
NC
NC
NC
NC
NC
D7
D6
D5
D4
D3
D2
D1
D0
NC

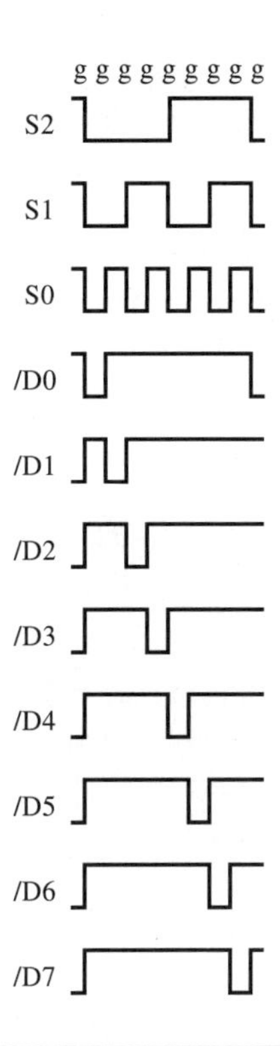

In the preceding example we showed the JEDEC file and the circuit of the PAL with the proper connections marked. This is a useful exercise because it illustrates what actually happens when a PAL is programmed, but it's not necessary in an actual design. The details of the problem of getting from logic expressions to programmed PAL are usually of interest only to the engineers who design the software and to the PAL programmers.

Example 6.4-2

Design a 4-bit identity comparator in a PAL similar to the PAL described in the text.

As we saw in Section 5.5 in the last chapter, an identity comparator is a device that compares two binary numbers and asserts an output if they're identical. That is, it does a bit-by-bit comparison of the two numbers. Suppose we let the bits of the numbers be denoted by A4, A3, A2, and A1 and B4, B3, B2, and B1. It does not matter whether we assume 1 is high or 1 is low, just so long as we make the same assumption for all the bits of both A and B. We'll denote the output with OUT, and we'll assert it low because we're going to design with a PAL that matches a low asserted output.

At this point in the design, we would normally form a truth table for the function we want to evaluate, form Karnaugh maps from the truth tables, and write a simplified logic expression from the Karnaugh maps. There is a problem, however, because the function has eight variables and a truth table would be impossibly large, whereas a map would have to be a VEM with four entered variables. Fortunately, we can write a logic expression for the comparator from its description, without the intermediate step of a truth table, and we do not need maps to simplify the expression because our PAL programming software will simplify it for us.

Suppose we consider only A1 and B1. They'll be identical if both A1 and B1 are 1 or if both A1 and B1 are 0—that is, if $A1 \cdot B1 + A1 \cdot B1$ is true. Now, if we make a similar expression for each of the other three pairs of bits, all of these expressions must be true. So

$$\begin{aligned} OUT = {} & (A1 \cdot B1 + \overline{A1} \cdot \overline{B1})(A2 \cdot B2 + \overline{A2} \cdot \overline{B2}) \\ & (A3 \cdot B3 + \overline{A3} \cdot \overline{B3})(A4 \cdot B4 + \overline{A4} \cdot \overline{B4}) \end{aligned}$$

We might be able to simplify this expression, but we won't even try. We'll instead go directly to a PALASM program, presented in the following table. We've chosen the same PAL as in the last example, the PAL16L8, which has outputs that match our required low output assertion level.

```
;PALASM Design Description

;---------------------------------------- Declaration Segment -------------------------------
TITLE      Four Bit Identity Comparator
PATTERN    4BCOMP
REVISION   A
AUTHOR     ALAN W SHAW
COMPANY    UTAH STATE UNIVERSITY
DATE       04/09/92
CHIP    A4BCOMP     PAL16L8

;----------------------------------------------- PIN Declarations --------------------------
PIN    1       A1                                          ; INPUT
PIN    2       A2                                          ; INPUT
PIN    3       A3                                          ; INPUT
PIN    4       A4                                          ; INPUT
PIN    5       B1                                          ; INPUT
PIN    6       B2                                          ; INPUT
PIN    7       B3                                          ; INPUT
PIN    8       B4                                          ; INPUT
PIN    9       NC                                          ;
PIN    10      GND                                         ;
PIN    11      NC                                          ;
PIN    12      NC                                          ;
PIN    13      NC                                          ;
PIN    14      NC                                          ;
PIN    15        /OUT                    COMBINATORIAL ; OUTPUT
PIN    16      NC                                          ;
PIN    17      NC                                          ;
```

```
PIN     18        NC                                          ;
PIN     19        NC                                          ;
PIN     20        VCC                                         ;
;---------------------------------- Boolean Equation Segment ---------------------------
EQUATIONS

OUT = (A1*B1+/A1*/B1)*(A2*B2+/A2*/B2)*(A3*B3+/A3*/B3)*(A4*B4+/A4*/B4)
OUT.TRST = VCC
;----------------------------------------- Simulation Segment -----------------------------
SIMULATION

;----------------------------------------------------------------------------------------------------
```

Unfortunately, when we try to process this program, we get a very fundamental error:

```
Equation being processed for output ==>>     OUT
| > ERROR X3120
Signal OUT cannot have more than 7 product term(s).
```

Looking back at the circuit diagram for the PAL16L8 (Figure 6.4–6 with output bubbles) in Example 6.4-1, we see that it indeed has only seven *and* gates in each section. To overcome this limitation, we need to choose a different PAL. The PAL22V10 (see Example 10.2–2 for circuit) is a 24-pin device with ten logic sections, two of which have 16 *and* gates apiece. This PAL also has the virtue of very high speed. If we use the PAL22V10, then we must rewrite the PALASM program as follows. The main changes are to move VCC and GND because of the 24-pin package, and to make sure that OUT is assigned to a pin with 16 *and* gates in its logic section. Pin 18 is such a pin. As already noted, the output of the PAL22V10 is programmable; all we need to do to program an output is to assign the assertion level when we enter the signal name associated with the output, set the storage type to COMBINATORIAL, and then specify that the pin is an output. The program for the PAL22V10 follows.

This program processes with no errors and produces a JEDEC file, which can be used to program the PAL. The simulation for some representative inputs is shown in the figure following the program. Note that OUT is low only when all the bits of A match the bits of B. The simulation does not test all possible combinations of the bits of A and B, because there are 16 × 16 different possibilities. The fuse map is not given.

```
;PALASM Design Description

;----------------------------------------- Declaration Segment ------------------------------
  TITLE        Four Bit Identity Comparator
  PATTERN      4BCOMP
  REVISION     A
  AUTHOR       ALAN W SHAW
  COMPANY      UTAH STATE UNIVERSITY
  DATE         04/09/92
```

```
CHIP     A4BCOMP     PAL22V10

;------------------------------------------------ PIN Declarations ----------------------
PIN     1        A1                                             ; INPUT
PIN     2        A2                                             ; INPUT
PIN     3        A3                                             ; INPUT
PIN     4        A4                                             ; INPUT
PIN     5        B1                                             ; INPUT
PIN     6        B2                                             ; INPUT
PIN     7        B3                                             ; INPUT
PIN     8        B4                                             ; INPUT
PIN     9        NC                                             ;
PIN     10       NC                                             ;
PIN     11       NC                                             ;
PIN     12       GND                                            ;
PIN     13       NC                                             ;
PIN     14       NC                                             ;
PIN     15       NC                                             ;
PIN     17       NC                                             ;
PIN     18         /OUT                          COMBINATORIAL ;
OUTPUT
PIN     19       NC                                             ;
PIN     20       NC                                             ;
PIN     21       NC                                             ;
PIN     22       NC                                             ;
PIN     23       NC                                             ;
PIN     24       VCC                                            ;
;---------------------------- Boolean Equation Segment -------------------------
EQUATIONS

OUT = (A1*B1+/A1*/B1)*(A2*B2+/A2*/B2)*(A3*B3+/A3*/B3)*(A4*B4+/A4*/B4)
OUT.TRST = VCC
;------------------------------------------- Simulation Segment ---------------------------
SIMULATION

SETF /A1 /A2 /A3 /A4 /B1 /B2 /B3 /B4
SETF A1 /A2 /A3 /A4 /B1 /B2 /B3 /B4
SETF /A1 A2 /A3 /A4 /B1 /B2 /B3 /B4
SETF /A1 /A2 /A3 /A4 B1 /B2 /B3 /B4
SETF A1 A2 A3 A4 B1 B2 B3 B4
SETF A1 A2 A3 A4 /B1 B2 B3 B4
SETF A1 /A2 A3 /A4 B1 /B2 B3 /B4
SETF /A1 /A2 A3 A4 /B1 /B2 B3 B4
SETF A1 /A2 /A3 A4 /B1 /B2 /B3 /B4
SETF A1 A2 A3 A4 /B1 /B2 /B3 /B4
```

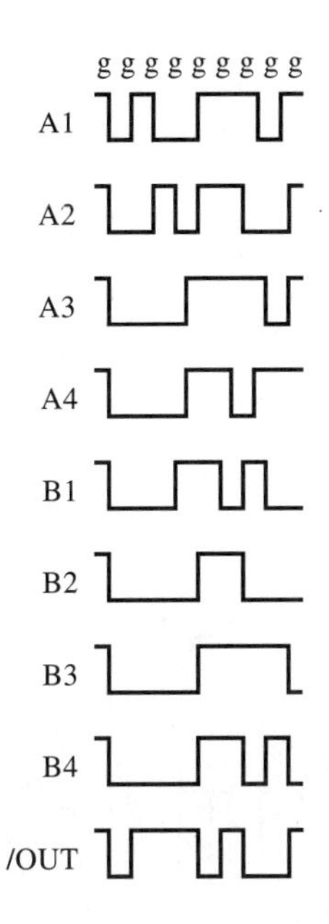


```
;-----------------------------------------------------------------------------------------------------------
```

6.5 PLA LOGIC

Figure 6.5-1 is the logic diagram of what would be a very small PLA. This PLA has four inputs labeled I_0, I_1, I_2, and I_3 and 4 outputs labeled O_0, O_1, O_2, and O_3. We'll initially assume that all the inputs and outputs are asserted high. Programmable connections in this logic circuit are denoted by open circles. Both the input connections to the *and* gate and the input connections to the *or* gates are programmable. This gives the PLA more flexibility than the PAL, but it also gives it more complexity. As a result, designers have tended to use the PAL more frequently than the PLA.

There is one situation in which a PLA is quite useful: where the logic expressions that we want to evaluate have common terms. As an example, consider the four logic functions O_3, O_2, O_1, and O_0, defined by the Karnaugh maps in Figure 6.5-2. The simplified SOP logic expressions for these functions are

$$O_3 = \bar{I}_1\,\bar{I}_0 + \bar{I}_3\,\bar{I}_2$$

$$O_2 = \bar{I}_3\,\bar{I}_1\,\bar{I}_0 + \bar{I}_3\,I_1\,I_0 + \bar{I}_3\,\bar{I}_2$$

$$O_1 = \bar{I}_3\,\bar{I}_1\,\bar{I}_0 + \bar{I}_3\,I_1\,I_0$$

$$O_0 = \bar{I}_1\,\bar{I}_0 + \bar{I}_3\,I_1\,I_0$$

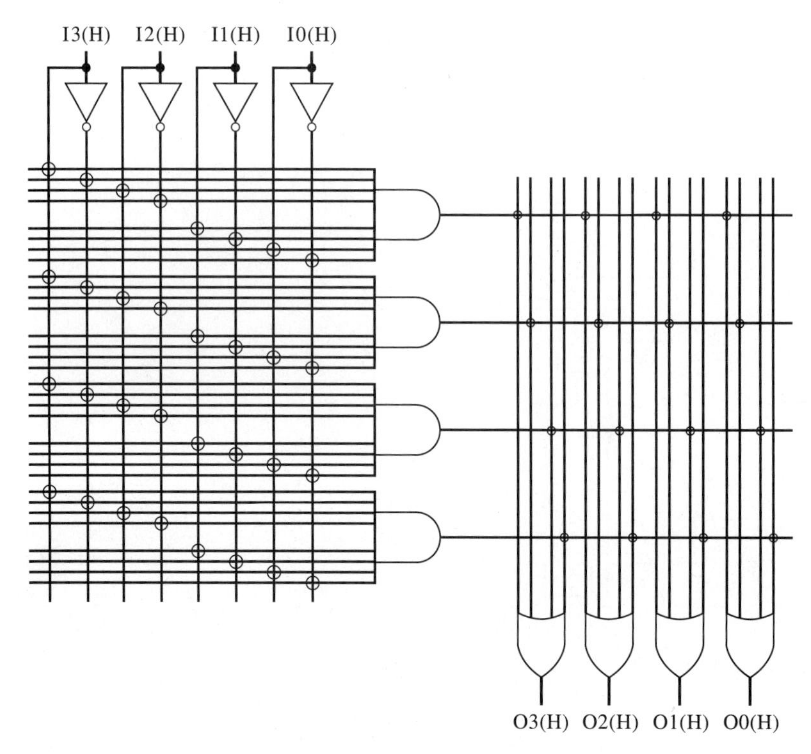

Figure 6.5-1 An example PLA.

Notice that these four functions have only four different *and* terms: $\bar{I}_1\,\bar{I}_0$, $\bar{I}_3\,\bar{I}_2$, $\bar{I}_3\,\bar{I}_1\,\bar{I}_0$, and $\bar{I}_3\,I_1\,I_0$. This means that with PLA logic we need only four *and* gates, because we can use each term several times by properly programming the *or* connections. Figure 6.5-3 shows the connections that must be programmed in our small sample PLA to evaluate the four logic functions. As you'll probably notice, the connections have been contrived to show the advantages of a PLA. Real problems for which a PLA can be used to advantage may not be so evident.

If we have low assertion level inputs in a PLA design, we can handle the inputs by complementing them, just as we did in our PAL designs. Low and mixed assertion level outputs are handled even more easily, as will be seen when we discuss an actual PLA.

Figure 6.5-4 is an example of an actual PLA. Notice that the orientation of the PLA in the figure is different from the orientation of the PALs in the figures of Section 6.4. This particular PLA device has 16 inputs (I0 through I15) and 8 inputs (F0 through F7). Each input has a driver that produces both high and low assertion levels, internal to the device. There are programmable connections between the horizontal input lines and the vertical input buses to the *and* gates. The *and* gates are 32-input *and* gates, so each bus is a 32-line bus. There are also programmable connections between the vertical output lines from the *and* gates and the horizontal input buses to the *or* gates. The *or* gates are 48-input *or* gates, so each bus is a 48-line bus.

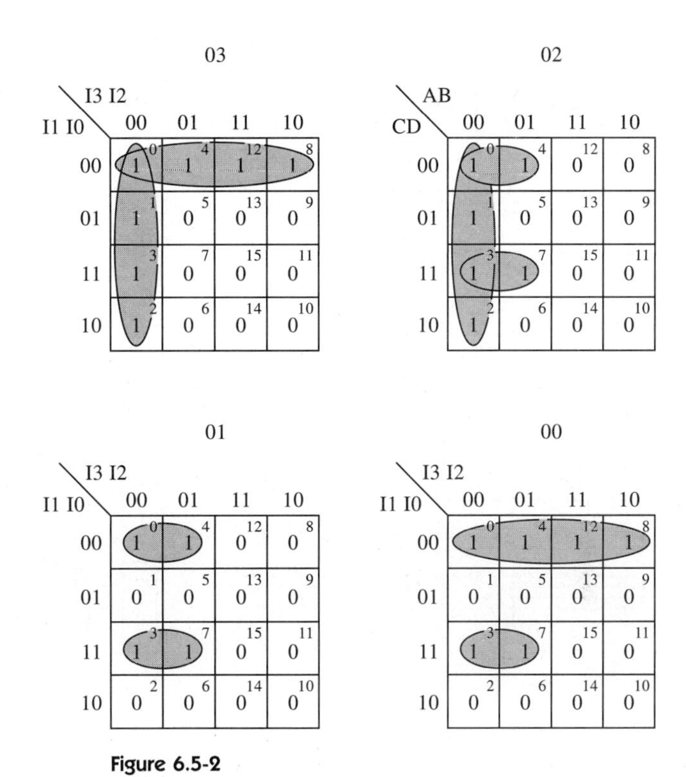

Figure 6.5-2

The output drivers are tristate drivers with a common external enable, labeled $\overline{CE}$. The enable in this circuit is drawn somewhat differently (as a dashed line connecting the drivers) from that in the PAL circuit, but it performs the same function. When $\overline{CE}$ is asserted (low), the outputs are all turned on. When it is not asserted (high), the outputs are all set to a high impedance. Each of the output circuits has a 2-input exclusive *or* (*xor*) gate between the output *or* gate and output driver. One of the inputs of this gate is programmable either high (open) or low (grounded). As we can see from the function table for an *xor* gate (Table 6.5-1), when the second input (IN2) is low (grounded), the output (OUT) is the same as the first input (IN1); but when the second input is high (open), the output is the complement of the first input. The *xor* is thus a programmable assertion level changer. With the programmable connection grounded, the output assertion level is high; with the programmable connection open, the output assertion level is low. (Some PALs have this same programmable output assertion level feature, although we did not show any in our discussion.) This is a representative PLA. Other PLAs differ in detail.

The actual programming of a PLA, done with a PLA programmer, is much the same as the programming of a PAL. PLAs with permanent programming (fusible links) and with erasable programming are both available. Software is available so that the mask for the PLA can be generated from the logic expressions for the functions that are to be evaluated. Most programmers can program both PALs and PLAs, and most software can generate masks for both PALs and PLAs. We will not

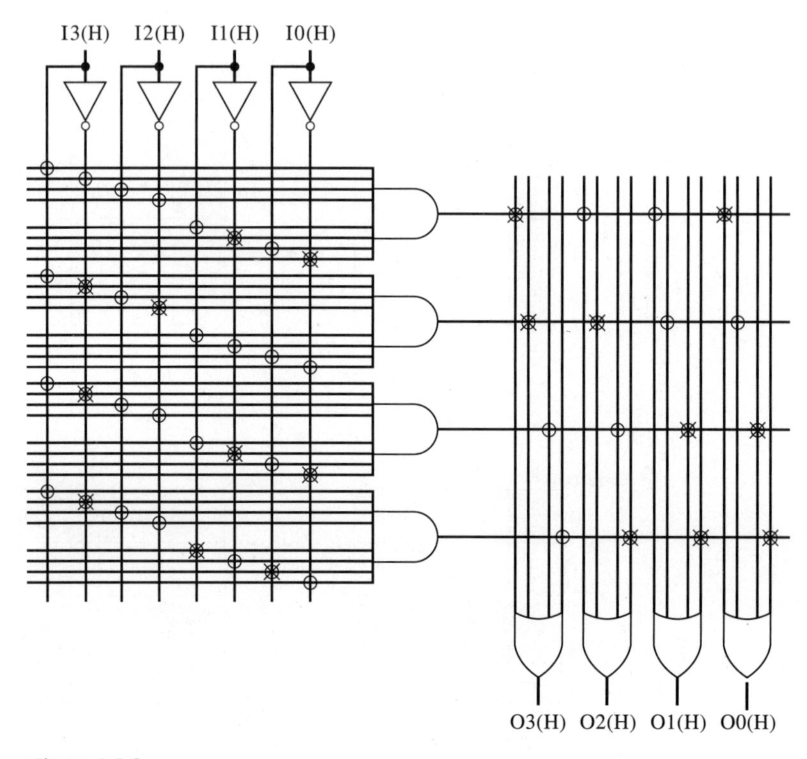

Figure 6.5-3

Table 6.5-1

IN1	IN2	OUT
L	L	L
L	H	H
H	L	H
H	H	L

give a design example for an actual PLA, but the software used in Example 6.4-2 (PALASM) can be used to program PLAs in the same way it's used to program PALs. Specific software documentation can be consulted to determine how to set up the logic equations for other mask generation programs.

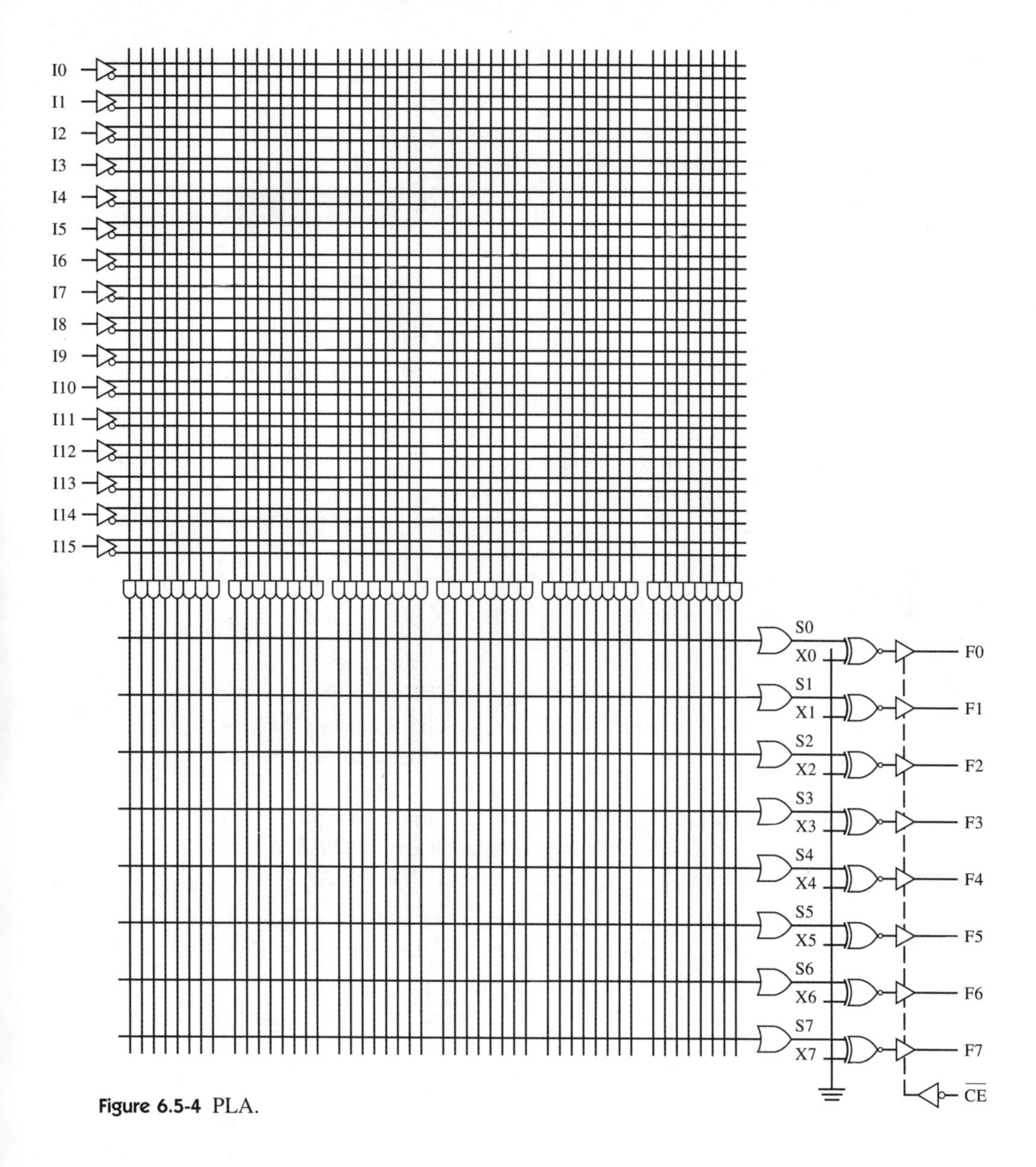

Figure 6.5-4 PLA.

KEY TERMS

Discrete Logic Logic devices in which the logic elements are interconnected by conductors external to the device.

Field Programmable Gate Array A device consisting of a large number of programmable modules with programmable interconnections.

Functional Simulation A simulation that determines whether a circuit is functioning properly but does not check timing.

Fuse Map Shows fuses to be programmed in a logic device.

JEDEC File The standard file needed to program a programmable logic device.
Programmable Array Logic (PAL) Fixed architecture programmable devices with an *and–or* input array and with only the *and* gates programmable.
Programmable Logic Array (PLA) Fixed architecture programmable devices with an *and–or* input array and with both *and* and *or* gates programmable.
Programmable Logic Device (PLD) A generic term applied to all programmable devices. Often means either PALs and PLAs.
Programmable Read Only Memory (PROM) Memory devices which can be used as programmable logic devices. They have a fixed architecture with an *and–or* array with the *or* gates programmable.
Programmer or Burner A device for programming programmable logic devices. More modern programmers will program PROMs, PALs and PLDs. FPGAs usually require special programmers.
Source Program The user-produced program needed to program a programmable logic device to perform a specific logic function.
Very Large Scale Integrated (VLSI) Circuit An integrated circuit with a very large number of elements.

EXERCISES

1. Design a dual 4-line to 1-line inverting multiplexer in a PAL16L8. (Multiplexers are described in Section 5.2.) Assume the inputs to the first multiplexer are I1A, I1B, I1C, and I1D and the output is Q1. Assume the inputs to the second multiplexer are I2A, I2B, I2C, and I2D and the output is Q2. Make the select lines A1 and A0. The select lines are to be asserted high. A1 is the high order bit of the select lines, and 00 should select I1A and I2A.
   (a) Write a PALASM sources program that will properly configure the PAL. In the program, assign signal names to the pins of the PAL and write logic expressions for all of the outputs.
   (b) Copy the diagram for a PAL16L8 and mark the connections that must be made by the programmer to program the PAL.
2. Design an 8-line to 1-line inverting multiplexer in a PAL16L8. (Multiplexers are described in Section 5.2.) Assume the inputs to the multiplexer are I0, I1, I2, I3, I4, I5, I6, and I7 and the output is Q. Make the select lines A2, A1, and A0. The select lines are to be asserted high. A2 is the high order select line, and 000 should select I0.
   (a) Write a PALASM sources program that will properly configure the PAL. In the program, assign signal names to the pins of the PAL and write logic expressions for all of the outputs.
   (b) Copy the diagram for a PAL16L8 and mark the connections that must be made by the programmer to program the PAL.
3. Design a dual 4-line to 2-line binary encoder in a PAL16L8. Make it a priority encoder, with the lowest asserted input having priority. (Encoders are described

in Section 5.3.) Assume the input lines for the first encoder are IA0, IA1, IA2, and IA3, all asserted high, and the output lines are CA1 and CA0. The levels for the other encoder should be the same but with the A changed to B. Make 1 high for the output code bits. C1 is the most significant bit.

(a) Write a PALASM sources program that will properly configure the PAL. In the program, assign signal names to the pins of the PAL and write logic expressions for all of the outputs.

(b) Copy the diagram for a PAL16L8 and mark the connections that must be made by the programmer to program the PAL.

4. Design an 8-line to 3-line binary encoder in a PAL16L8. Make it a priority encoder, with the lowest asserted input having priority. (Encoders are described in Section 5.3.) Assume the input lines are I0, I1, I2, I3, I4, I5, I6, and I7, all asserted high, and the output lines are C2, C1, and C0. Make 1 high for the output code bits. C2 is the most significant bit.

   Write a PALASM sources program that will properly configure the PAL. In the program, assign signal names to the pins of the PAL and write logic expressions for all of the outputs.

5. Design a dual 2-line to 4-line decoder in a PAL16L8. (Decoders are described in Section 5.4.) Assume the input select lines for the first decoder are A11 and A10, with high equal to 1, and the output lines are O10, O11, O12, and O13, all asserted low. Label the second decoder line the same, but change the 1 to a 2. A1 is the most significant bit, and 00 input should cause I0 to be asserted.

(a) Write a PALASM sources program that will properly configure the PAL. In the program, assign signal names to the pins of the PAL and write logic expressions for all of the outputs.

(b) Copy the diagram for a PAL16L8 and mark the connections that must be made by the programmer to program the PAL.

6. Design a 4-bit adder with carry-in and carry-out. This will probably require a PAL22V10. You will need to consult a data book to see exactly how the PAL22V10 is configured. (Adders are described in Section 5.5.) The carry-in should be labeled C0. The input bits should be labeled A4, A3, A2, and A1 and B4, B3, B2, and B1. The output bits should be labeled S4, S3, S2, and S1. The output carry should be labeled C4. Make high equal to 1 for all the bits and carries. Do not try to make a truth table. Approach the problem as in Example 6.4-2. Write a logic expression for each bit of the sum, starting with the low order bit. Write another expression for the carry-out of each bit.

   Write a PALASM sources program that will properly configure the PAL. In the program, assign signal names to the pins of the PAL and write logic expressions for all of the outputs. Try to make the delay through the PAL as short as possible. Process this program if you have access to software. You will find that some of the terms must be broken up into parts, and each part processed in a different logic segment of the PAL, or the design will not fit even on a PAL22V10.

7 Sequential Machines

7.1 INTRODUCTION

In the first part of this book we studied combinational logic circuits, which are circuits that evaluate logic expressions. The output of such circuits depends only on the immediate input. Furthermore, the output generated by a change in the input appears instantaneously after the change in the input (except for the time delays through the circuit gates).

In this chapter we will address a different kind of logic circuit, called a *sequential* logic circuit. In fact, because a sequential logic circuit is one way of building a sequential machine, we will take a general approach and discuss **sequential machines**. In a sequential machine, the output *may* depend on the immediate input to the machine, but it *always* depends on the *previous condition* of the machine. A sequential machine must have some kind of memory, because the previous condition of the machine must be saved to determine the present output.

Sequential machines can be thought of as machines that proceed through an orderly set of conditions. The conditions are called **states**, so sequential machines are also called **state machines**. The transitions of the machine from one state to the next are most often timed by a clock, in which case the machine is a **synchronous machine**. When the machine is not clocked, a transition from one state to the next is determined by a change in the input to the machine. Because an input change may occur at any time, a machine that is not clocked is an **asynchronous machine**. Sequential machines may cycle repetitively through a finite set of states, or they may step through an infinite set of states and never repeat. We are interested only in machines with a finite set of states, called **finite state machines (FSMs)**.

Finite state machines are useful in a very common type of control problem. Suppose some system must execute a certain repetitious sequence of operations that is initiated by a command (input). The system could be an elevator that moves to a given floor and stops when one of the control buttons is pushed. It could be an arithmetic processing unit in a computer that, when given an add instruction, undertakes a fixed sequence of operations resulting in an addition. Systems such as these can be conveniently controlled by finite state machines.

7.2 MODEL FOR A GENERAL SEQUENTIAL MACHINE

Figure 7.2-1 shows the model for a general sequential machine. The present state of the machine is stored in the memory element. For now, the memory can be any device capable of storing enough information to specify the states of the machine. For example, a state machine might have three states specified by the letters A, B, and C, so the memory could be any device that can store codes representing these three letters.

The next state of the machine is determined by the present state of the machine and by the input. The **next-state decoder** uses logic operations on the present state of the machine and the input to the machine to produce a code, which generates the next state of the machine in the memory. The code might, in the simplest case, be

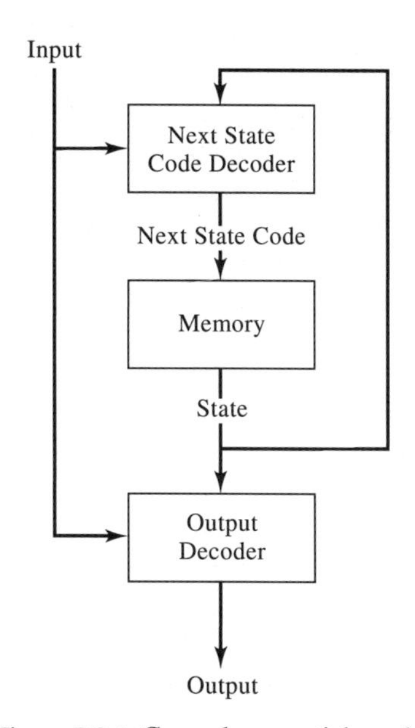

Figure 7.2-1 General sequential machine.

the next state of the machine. The sequential machine is a **feedback** system. The present state of the machine is *fed back* to the next-state decoder and used, along with the input, to determine the next state.

The output of the machine is determined by the present or previous state of the machine and possibly by the input to the machine. The **output decoder** uses logic operations on the state of the machine and the input to the machine to generate the output. The output of a synchronous machine may be clocked, just as the state transitions are clocked. In some cases, the output is simply the state of the machine, and no output decoder is needed; but often the state must be translated before it is output. In any case, when the output is a function of the input, it will certainly be translated differently, according to the input. Sequential machines have traditionally been classified as **Moore machines** if their output depends only on the state, and as **Mealy machines** if their output depends on both the state of the machine and the input to the machine. We'll find that this simple classification breaks down when we begin to design state machines using logic circuits. We'll then introduce an expanded classification.

7.3 STATE AND STATE TRANSITION DIAGRAMS

The concept of state is a very general one that is not limited to sequential machines. We've been using *state* in a rather intuitive way: the state of a machine is the condition of the machine at a given time. A more formal definition of state for a system has to do with the way the system evolves with time: the state of a system is a *description* of the condition of the system at a given time. The description must be thorough enough that the evolution of the system from that time forward can be completely determined from the description and the subsequent inputs to the system. Thus, if the state of a system is known at time t_0, and the inputs are known from time t_0 to time t_1, then the state of the system can be found at t_1.

A convenient way to describe a finite state machine is by using a state transition diagram. This is a graphical way of depicting how the machine changes from one state to another. It also depicts the output generated by the machine. A graphical description of a finite machine is possible because the machine is finite and all states and transitions can be shown.

Figure 7.3-1 is an example of a state transition diagram for a synchronous machine—a machine in which the transitions are controlled by a clock. The ellipses represent the states of the system (machine). The state identification is written within each ellipse. In this diagram there are three states, which we've called A, B, and C. The transitions between the states are represented by lines, with arrows to indicate their directions. Some of the transitions do not change the states, such as the ones above state A and above state B. All of the transitions in this synchronous machine are controlled by the clock. If a transition is also conditional on an input condition, the condition is written next to the line representing the transition. The self transition at the top depends on the input condition

$$\overline{\mathrm{START}} + \mathrm{RESET}$$

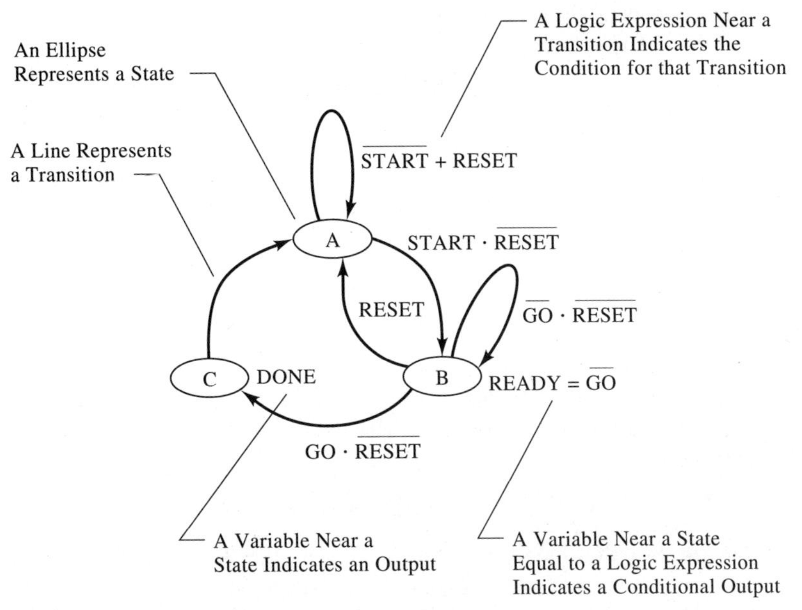

Figure 7.3-1 Example of a state transition diagram.

If no input conditions are written near the transition, it occurs unconditionally on the next clock cycle of the machine.

For a synchronous machine, the transitions out of any given state are not complete unless they cover all possible input conditions. The machine must be able to determine what to do each time it's clocked. As the figure shows, if the machine is in state A, it makes a self transition on the clock (stays in state A) for the input conditions

$$\overline{\text{START}} + \text{RESET} = \overline{\text{START}} \cdot \overline{\text{RESET}} + \overline{\text{START}} \cdot \text{RESET} + \text{START} \cdot \text{RESET}$$

independent of the condition of the input GO, and makes a transition to state B on the clock for the input conditions

$$\text{START} \cdot \overline{\text{RESET}}$$

again independent of the condition of the input GO. Notice that transitions for all four possible input combinations of START and RESET are given explicitly, whereas the input condition of GO is implied by its absence in the expressions. The transitions are all independent of GO. In state B, the machine makes a self transition on the clock if the input conditions are

$$\overline{\text{GO}} \cdot \overline{\text{RESET}}$$

independent of the condition of the input START; makes a transition to state C on the clock if the input conditions are

$$\text{GO} \cdot \overline{\text{RESET}}$$

independent of the condition of the input START; and makes a transition to state A on the clock if the input conditions are

$$\text{RESET} = \overline{\text{GO}} \cdot \text{RESET} + \text{GO} \cdot \text{RESET}$$

independent of the condition of the input START. Finally, in state C, the machine makes a transition to state A on the clock, independent of any input conditions.

Outputs are indicated by writing the name of the output variable near the state in which the output is asserted. State C produces an output DONE. If the output depends on an input (a Mealy machine), the input condition that produces the output is given. There is such a conditional output for state B:

$$\text{READY} = \overline{\text{GO}}$$

This notation indicates that the output READY is asserted when the machine is in state B and GO is not asserted. The conditions on the READY output could be written more fully as

$$\text{READY} = (\text{state B}) \cdot \overline{\text{GO}}$$

but the fact that the output is placed near state B indicates that state B must be asserted, so it need not be explicitly indicated.

The state diagram just described is for a synchronous machine. For an asynchronous machine, every transition must have an input condition because the transitions are caused by changes in the inputs. In the preceding case, if the transition from state C to state A were controlled by an input condition, then the clock could be dispensed with and the machine operated as an asynchronous machine. The asynchronous machine would, however, perform quite differently from the synchronous machine. An asynchronous machine changes state only when an input changes, so, unlike the case of a synchronous machine, transitions for all possible input conditions need not be shown for each state; only those transitions caused by a change in an input condition need be shown.

The notational scheme we used includes a slight variation on the notation introduced by Fletcher in his textbook: the notation for input conditions comes directly from Fletcher, but the notation for outputs differs slightly from Fletcher's. Overall, this scheme differs markedly from the standard notation used on state diagrams, but it's much easier to remember and use.

7.4 PRESENT STATE–NEXT STATE AND OUTPUT TABLES

For the purposes of analysis and design, it is useful to describe the state transitions of a state machine in a present state–next state table. Such a table lists all possible combinations of inputs and present states, and gives the corresponding next state for each combination. It contains the same information about the transitions of a state machine as a state transition diagram, but it does not necessarily contain output information. Table 7.4-1 is the present state–next state table for the state machine of Section 7.3. It was taken directly from the state diagram by considering the inputs

Table 7.4-1 Present State–Next State Table of the Machine of Section 7.3

RESET	START	GO	Present State	Next State
0	0	0	A	A
0	0	0	B	B
0	0	0	C	A
0	0	1	A	A
0	0	1	B	C
0	0	1	C	A
0	1	0	A	B
0	1	0	B	B
0	1	0	C	A
0	1	1	A	B
0	1	1	B	C
0	1	1	C	A
1	0	0	A	A
1	0	0	B	A
1	0	0	C	A
1	0	1	A	A
1	0	1	B	A
1	0	1	C	A
1	1	0	A	A
1	1	0	B	A
1	1	0	C	A
1	1	1	A	A
1	1	1	B	A
1	1	1	C	A

and state for each row of the table and consulting the state diagram to determine the next state.

The present state–next state table can be rather large if there are many inputs and/or states. However, many of the state transitions do not depend on all the inputs (for example, the transition from state C to state A does not depend on any input). We can use a don't care entry (x) for these inputs in the table. With this notation, Table 7.4-1 can be simplified to Table 7.4-2.

To complete the description of the state machine, we need an output table to show what outputs occur for the various states and inputs. Table 7.4-3 is a reduced output table for the state machine of Section 7.3. Notice that the outputs are for the *present* state. They depend only on the present state and possibly on the input. Table 7.4-4 shows how to combine the present state–next state table and the output table. In this combined table it's important to remember that the outputs are outputs for the present state. They depend only on the present state and possibly the input and are *independent of the next state*.

Table 7.4-2 Reduced Present State–Next State Table for the Machine of Section 7.3

RESET	START	GO	Present State	Next State
0	0	x	A	A
0	1	x	A	B
0	x	0	B	B
0	x	1	B	C
1	x	x	A	A
1	x	0	B	A
1	x	1	B	A
x	x	x	C	A

Table 7.4-3 Reduced Output Table for the Machine of Section 7.3

RESET	START	GO	Present State	READY	DONE
x	x	x	A	0	0
x	x	0	B	1	0
x	x	1	B	0	0
x	x	x	C	0	1

Table 7.4-4 Combined Present State–Next State and Output Table for the Machine of Section 7.3

RESET	START	GO	Present State	Next State	READY	DONE
0	0	x	A	A	0	0
0	1	x	A	B	0	0
0	x	0	B	B	1	0
0	x	1	B	C	0	0
1	x	x	A	A	0	0
1	x	0	B	A	1	0
1	x	1	B	A	0	0
x	x	x	C	A	0	1

7.5 TIMING DIAGRAMS

State machines are systems that change with time. One way of showing this change is by means of a timing diagram, which shows pictorially how the various quantities associated with the state machine change with time. We'll show a timing diagram for the state machine of Section 7.3, but first we need to consider some clocking details.

A synchronous machine, such as the one described in Section 7.3, has a clock associated with it. As we've noted, the clock causes the transitions of the state machine. The clocking can be done in several ways. The cleanest way is to use either the rising edge or the trailing edge of the clock signal. This **edge triggering** of the transitions is particularly desirable because it gives a definite time at which the inputs to the system (which affect the transitions) are sampled.

Another way to clock the transitions is to use either the high or low portion of the clock signal. This kind of **level triggering** creates problems when the input signals change during the active clocking level: the state of the machine may change in an unexpected way. We'll look at some level triggered devices presently, but for now, let's assume that the machine described in Section 7.3 is **positive edge triggered**; that is, the transitions occur at the rising edges of the clock signal.

Figure 7.5-1 shows the timing diagram for this machine. The clock signal is at the top. The clock does not have an assertion level associated with it in the normal sense. An up arrow indicates that the clocking is at the positive-going edge of the clock. The figure has vertical lines at the clock edges where transitions are clocked. The three input signals (RESET, START, and GO), which we assume to be asserted high, are below the clock. The inputs are arbitrary, and these were chosen

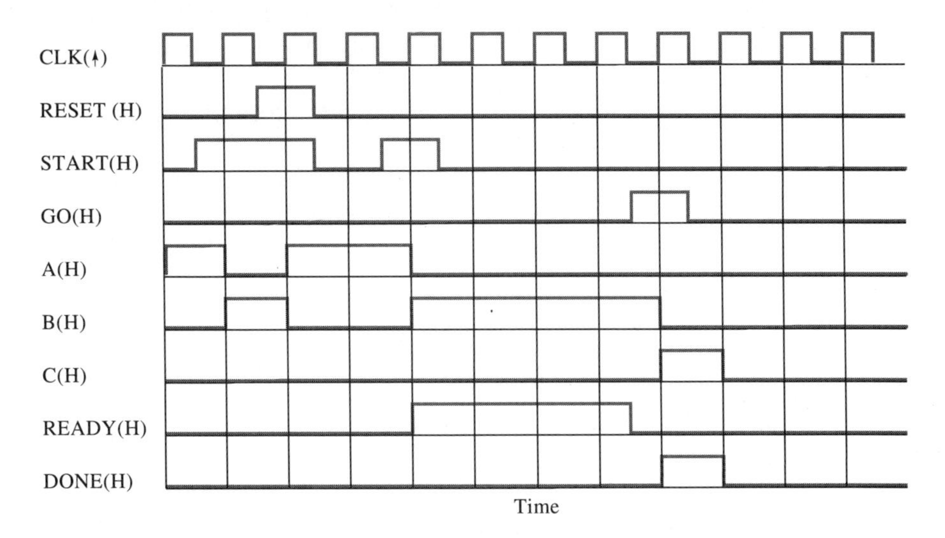

Figure 7.5-1 Timing diagram for the state machine of Section 7.3.

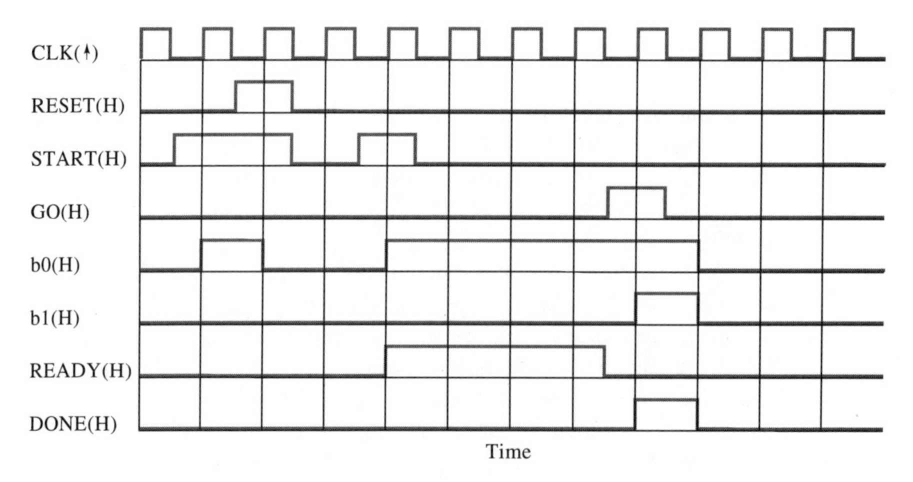

Figure 7.5-2 Timing diagram for the state machine of Section 7.3, showing state encoding.

to show some of the state transitions that can occur. Notice that the chosen inputs do not change at the rising clock edge. Input signals that change on the clocking edge can give ambiguous results because they will be sensed if they occur just before the clocking edge, and they will not be sensed if they occur just after the edge. The timing for the states is shown below the inputs. The states are assumed to be asserted high, and an asserted state indicates that the machine is in that state. The timing for the outputs (READY and DONE), which are also assumed to be asserted high, are shown below the states. Notice that the output READY is conditional on the input GO. It occurs for only part of the time the machine is in state B (when GO is not asserted).

In this timing diagram we've shown the states as if they were signals, and in a certain type of sequential logic circuit, this is true. In more traditional sequential logic circuits, the state is stored as a binary code. For example, state A might have the code 00, state B the code 01, and state C the code 11. In this case, the logic circuit signals that represent the states are the values of the binary bits of the code. If we call the binary bits b1 and b0, then the part of the preceding diagram that shows the states is replaced with b1 and b0, the signals that encode the state. Figure 7.5-2 shows the new timing diagram with b1 and b0 assumed to be asserted high.

Now let's look at a simple example of an asynchronous machine timing diagram. Figure 7.5-3 shows the state transition diagram for a simple SET-RESET memory cell, which is a two-state machine with no clock and two inputs (SET and RESET) that should not be asserted at the same time. The two states are encoded in a 1-bit memory that can have the two values 1 (SET) and 0 (RESET). Transitions occur only when the input changes. Figure 7.5-3 also shows a timing diagram for this two-state machine, with representative inputs. The assertion levels for all the signals are assumed to be high.

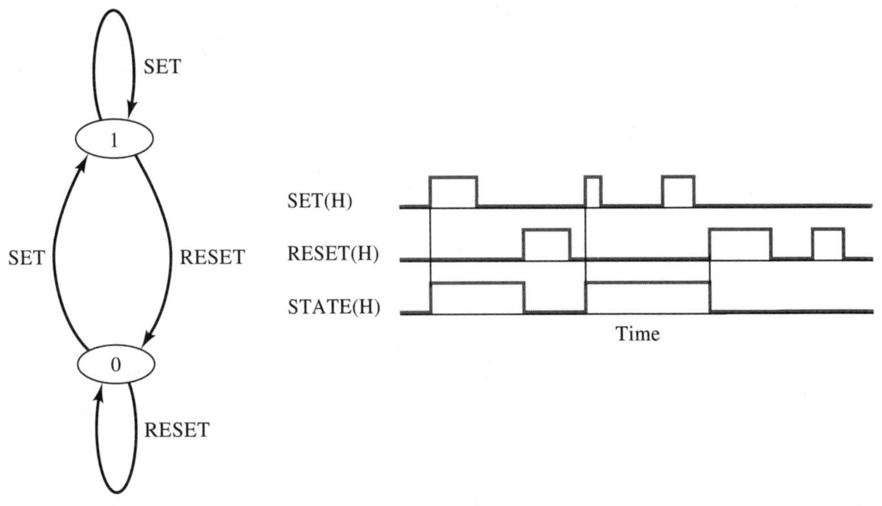

Figure 7.5-3 Asynchronous state machine and representative timing diagram.

7.6 INITIAL DESIGN STEPS FOR A STATE MACHINE

The last three sections introduced three ways of describing a state machine. The formulation of these descriptions actually constitutes three important steps in the design of a state machine, but these three steps must be preceded by an extremely important first step: to *make a clear word description of the state machine* we wish to design. Thus, the initial steps in the design of a state machine are as follows:

1. Make a clear word description of the state machine. Include a listing of all the inputs and the effects of those inputs—that is, the outputs caused by the inputs and when the outputs should occur with relation to the inputs causing them. A formal description of a machine is called a *specification*.
2. Make a state transition diagram from the specification. Usually, this is the most difficult part of a state machine design.
3. Make a timing diagram from the state transition diagram to check the timing of the inputs and outputs. For simple state machines, this step may not be necessary—the timing will be obvious.
4. Make a combined present state–next state table and output table. When you become familiar with the design process, you'll find this step is not necessary.

The timing diagram of step 3 becomes more important as the state machine becomes more complicated; but the tables of step 4 will eventually seem cumbersome and we will stop using them. It's very important that the timing diagram be constructed to reflect the timing of the state transition diagram and then checked to determine if it fits the description of the machine to be designed. Remember that the timing diagram is a check to see if the state transition diagram is correct.

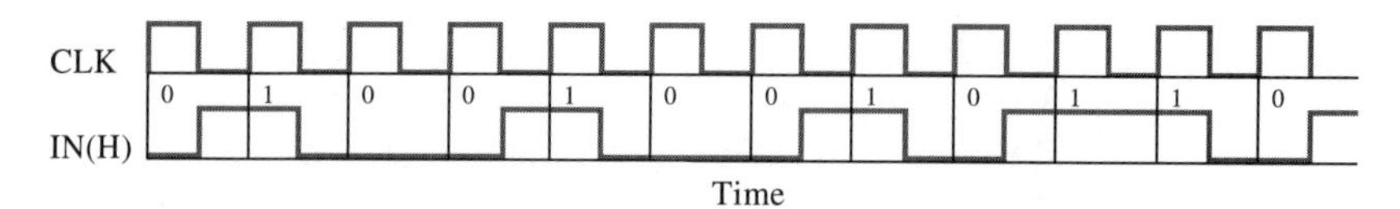

Figure 7.6-1 Input signal timing.

Let's consider an example. Suppose we want to design a state machine to analyze a stream of incoming bits in a signal, IN(H), and determine when the incoming bit pattern is 010. While we're searching for the first bit of this pattern, we want to assert an output signal, SRCH(H); each time we find the first bit of the pattern, we want to assert an output signal, START(H); we will continue to assert the output signal START(H) when the second bit of the pattern is found; and each time we find all the bits of the pattern, we want to assert an output signal, FOUND(H). We do not want to miss any of the bits in the input stream, and we want to avoid using any bit twice. We'll assume that the incoming bits are synchronous with a clock signal (CLK) so that they change only on the negative clock edge, as shown in Figure 7.6-1.

The first step in the design is a clear word description of the problem, or a specification, which we have in the preceding paragraph. The second step is a state transition diagram, the third step is a timing diagram, and the fourth step is a present state–next state and output table.

Figure 7.6-2 is a tight state transition diagram—a diagram with no unneeded states—for our machine. (We can actually make a machine with only three states, but it has no advantage over a machine with four states.) In the description of the problem, we gave the assertion levels of the input signal IN and the output signals SRCH, START, and FOUND. These assertion levels are not needed in the state transition diagram, which is concerned only with logic.

Because the state machine has tight timing constraints on missed bits and bits

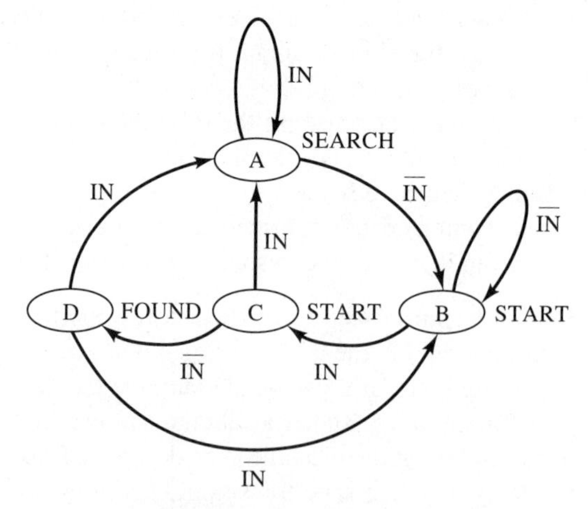

Figure 7.6-2 State transition diagram.

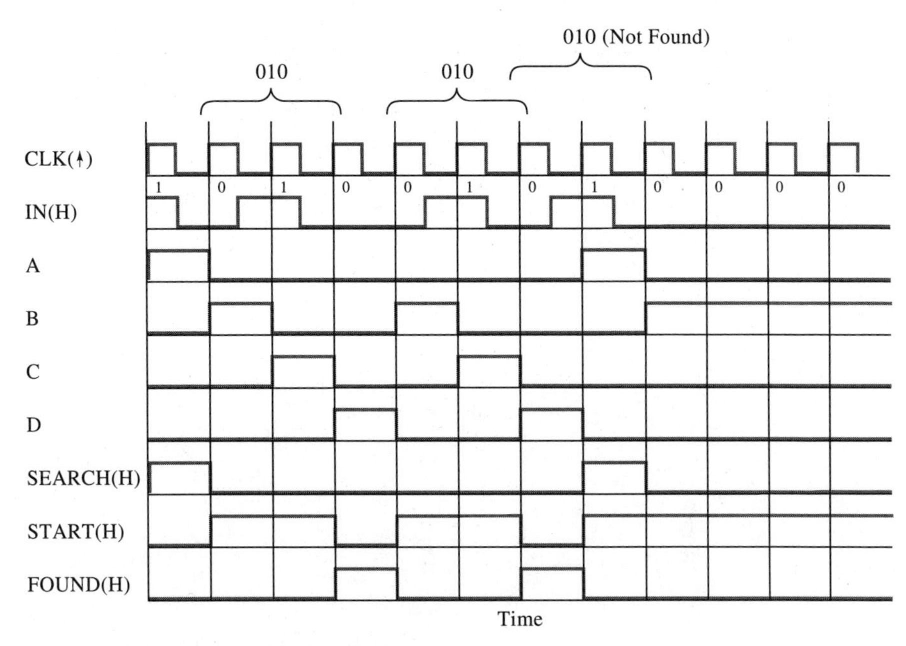

Figure 7.6-3 Timing diagram.

counted twice, we need to make a timing diagram from the state transition diagram to ensure that we've satisfied these constraints. Figure 7.6-3 is a timing diagram for the proposed system. Notice that the timing diagram, unlike the state transition diagram, needs assertion levels for the inputs and outputs. The diagram shows signal levels, not logic variables. It has two 010 sequences one of which follows directly after the other and two sequences that overlap. The machine described by the state transition diagram correctly finds both of the sequences that follow one after the other, and it does not find the overlapping sequence. It appears to satisfy the specifications.

Table 7.6-1 Present State–Next State and Output Table

IN	Present State	Next State	SRCH	START	FOUND
0	A	B	1	0	0
1	A	A	1	0	0
0	B	B	0	1	0
1	B	C	0	1	0
0	C	D	0	1	0
1	C	A	0	1	0
0	D	B	0	0	1
1	D	A	0	0	1

Example 7.6-1

Design a machine to find the code sequence 010 in an input stream just as in the machine we already designed, but use only three states.

Following is a state transition diagram for a state machine that will find the sequence 010 using only three states.

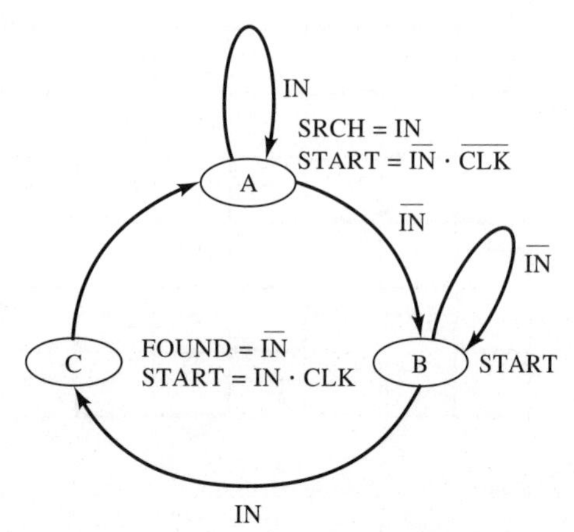

It accomplishes this by making the outputs conditional on the input (A Mealy Machine). The following timing diagram for this state machine demonstrates that the machine meets the constraints on missed bits and overlapping sequences. It shows that two sequences that follow one after another are correctly found, and an overlapping sequence is not found.

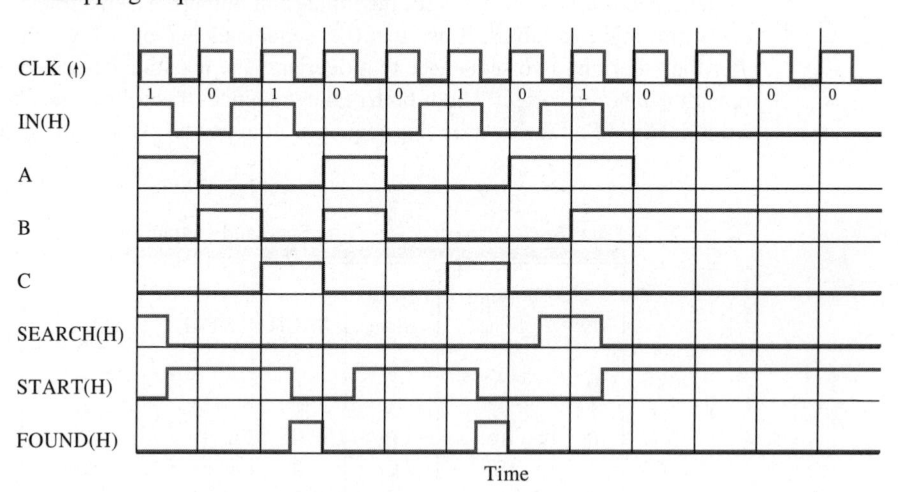

Some of the outputs are conditional on the clock. By making START conditional on the clock, we can assert it in state A as soon as we detect the first 0 in the sequence 010. If we do not make it conditional on the clock, it will be asserted for half of a clock cycle after the machine has found a 010 sequence and enters state A. By making START conditional on the clock in state C, we can assert START until the last 0 of the sequence 010 is found. Again, if we do not use the clock, we'll continue to assert START even when we've failed to find the sequence 010 and are starting over.

Just a word of caution on using the clock to select output signals. The procedure works fine for ideal state machines with synchronous inputs, but in actual state machines that have time delays, or in situations where the input is not synchronous, it can cause significant problems. We'll look at the problems caused by time delays when we discuss how to complete the design of state machines in Chapters 8, 9, and 10.

Table 7.6-1 is the combined present state–next state and output table formed from the state transition diagram.

It's interesting to compare the timing of this state machine with the timing in Figure 7.6-3 for a Moore machine that is designed to do the same thing. SRCH and START are much the same in the two machines, but they both occur one half of a clock cycle earlier in the machine of Example 7.6-1. In the latter machine, FOUND is asserted for one half of a clock cycle at the end of the sequence 010. In Figure 7.6-3, FOUND is asserted for a full clock cycle after the sequence 010 is found. In the design of Example 7.6-1 the timing of the input signal, IN, affects the lengths of SRCH, START, and FOUND. The reason we can use the clock to select SRCH in state A and START in state C is that IN changes synchronously with the negative clock edge. The reason FOUND lasts for one half of a clock cycle is also that IN changes on the negative clock edge. If IN were asynchronous, the machine of Example 7.6-1 would not function very well. In particular, FOUND could be any length from a full clock cycle to very short, depending on when IN changes. For many applications this would be undesirable. We'll say more about conditional output that depends on asynchronous input in Section 9.3.1.

7.7 FORMING THE STATE TRANSITION DIAGRAM

As you know, the first step in the design of a state machine is to make a clear word description, which should include a listing of all the inputs and the actions (outputs) they cause, and a clear statement of any critical timing constraints on the inputs and outputs. The state machine design of Section 7.6 had one input, IN, and three outputs, SRCH, START, and FOUND. Timing constraints on IN were described in a timing diagram. The effect of the IN was that SRCH should be asserted until the first 0 of the sequence 010 was found; then START should be asserted until the last 0 was found or the sequence failed; and finally FOUND should be asserted when the entire sequence was found. Critical timing constraints were no missed bits and no bit counted in two different sequences.

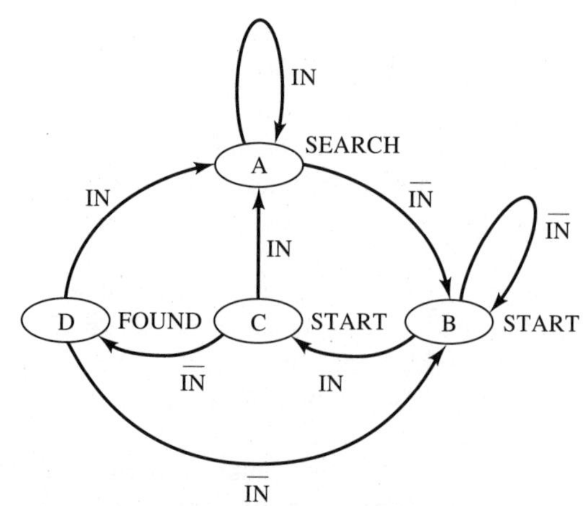

Figure 7.7-1 State transition diagram.

The second step in the design is to use the description and produce a state transition diagram for the machine to perform the functions described. In this first state transition diagram, do not try to assign state codes. Call the states A, B, C, and so on, or 1, 2, 3, and so on. Make the state transition diagram with a minimum number of states, by eliminating equivalent states. *Two states are equivalent if they have the same outputs and if they have the same next-state transitions.* When two states are equivalent, they can be combined. The best way to eliminate equivalent states is to *not* create them as you draw the diagram. Each time you look at the next-state transitions out of a state, examine all the states you already have and determine whether you can use one of these already-existing states as the next state. Remember that you can use an existing state as the next state for a transition if it has the *outputs* you need and the *next-state transitions* you need.

Let's look again at the four-state state transition diagram from Section 7.6. Figure 7.7-1 repeats this diagram. The first equivalent state is state A. There are two transitions out of A, one for IN and one for $\overline{\text{IN}}$. If we did not recognize that when the input is IN, the next state for A is A itself, we would produce the partial state transition diagram in Figure 7.7-2. In this diagram, the next state for A is E if the input is IN, but E is equivalent to A because it has the same outputs (none) and the

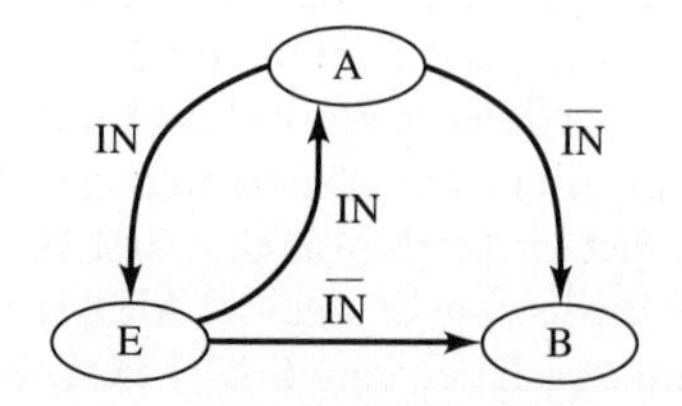

Figure 7.7-2 Partial state transition diagram.

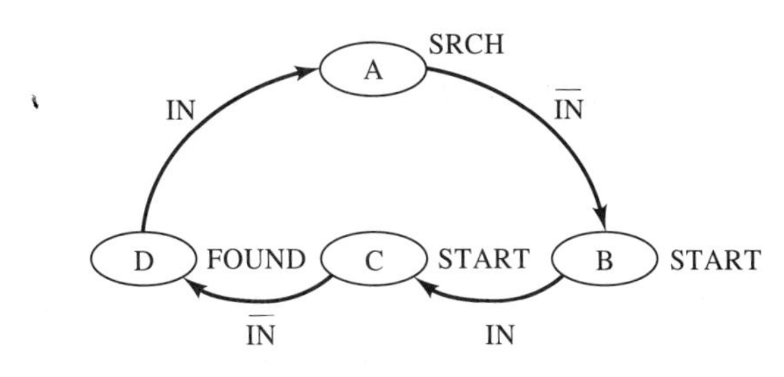

Figure 7.7-3

same next-state transitions. This is a little tricky to see because the next-state transitions for A are E if the input is IN and B if the input is $\overline{\text{IN}}$, whereas the next state transitions for E are A if the input is IN and B if the input is $\overline{\text{IN}}$. These transitions are equivalent only if we indeed consider E and A to be the same. Because E is equivalent to A, E should be eliminated by taking the transition that goes to E back to A and removing E. If we're careful in the first place, we'll recognize that if the input is IN, the next state for A is A itself, because A has the required outputs and next-state transitions.

It's very important to eliminate equivalent states by recognizing and using next states that already exist (as in Fig. 7.7-1, where we recognized that the next state for A, if the input is IN, is A itself). If we do not do this, we may produce state transition diagrams that are incredibly complicated and that can have an infinite number of states. All the states in the state diagram of Figure 7.7-1 eventually have transitions to states that already exist, as must be the case for the machine to be finite.

At the beginning of this section, we discussed how to avoid equivalent states in a state transition diagram. Now we need some idea of how to start drawing a state transition diagram. A useful technique is to start with the sequence of actions you want the machine to normally follow (or one of the sequences, if there are more than one) and draw the states and transitions that are needed to follow this sequence. The machine in Section 7.6 is to find the input sequence 010, so we first follow through the states and transitions needed to find this sequence and to produce the partial state transition diagram of Figure 7.7-3.

Once we have a partial state transition diagram for the normal sequence of the machine, then we consider, one at a time, other sequences through which the machine might go. This amounts to considering inputs other than those we considered in our initial partial diagram. In our example, suppose the input in state A is IN rather than $\overline{\text{IN}}$. In this case we need a self transition back to A. In state B, suppose the input is $\overline{\text{IN}}$ rather than IN. In this case we need a self transition back to B. Continuing this process through all the states, we complete the diagram. When completing the diagram, we must make sure that we consider all the input conditions that can occur for a state and indicate the transition that will occur for each of these input conditions. In the example there is one input, so there are *two* input conditions, IN and $\overline{\text{IN}}$, and we must indicate transitions for each of these conditions at each state. If there were two inputs (IN1 and IN2), there would be four input conditions ($\overline{\text{IN1}} \cdot \overline{\text{IN2}}$,

$\overline{IN1} \cdot IN2$, $IN1 \cdot \overline{IN2}$, and $IN1 \cdot IN2$) and four transitions for each state. For three inputs, there are nine input conditions for each state, so multi-input machines can rapidly become very messy.

Once we have a state transition diagram that we believe will meet the specifications contained in our description, we should draw a timing diagram for the state transition diagram. The timing diagram should consider the critical input conditions. For example, in Section 7.6 we drew a timing diagram for two input sequences, 010, that followed one after the other to determine whether both sequences were found, and for an overlapping 010 sequence to determine whether it was incorrectly found. The timing diagram should also consider the output timing, especially for the outputs that are conditional on an input. The durations of outputs conditional on asynchronous inputs can vary greatly and can be very short. This condition might be unacceptable.

Example 7.7-1

Design a state machine (first four steps) that will sample an input SIG(H) in 2-bit slices and determine whether the two sampled bits are 11. The machine is to start sampling when an input SYNC(L) is asserted and continue sampling as long as SYNC remains asserted. The machine will output a signal COUNT(H) each time the two bits in a slice are 11. Both SIG and SYNC change synchronously with the negative-going edge of the clock. The timing diagram of SIG, SYNC, and COUNT follows. Notice the 2-bit time slices.

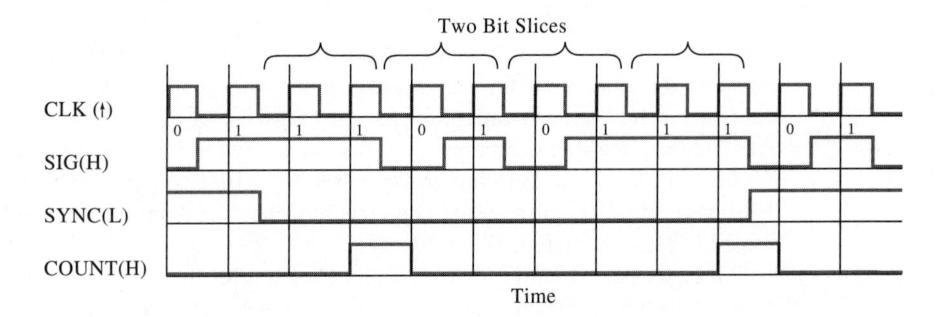

The first step in the design, the description of the problem, has already been accomplished. The second step is to make a state transition diagram. To draw it, we'll first consider the *normal* action of the machine—finding the sequence 11 if SYNC is asserted. Consider the following diagram on the left.

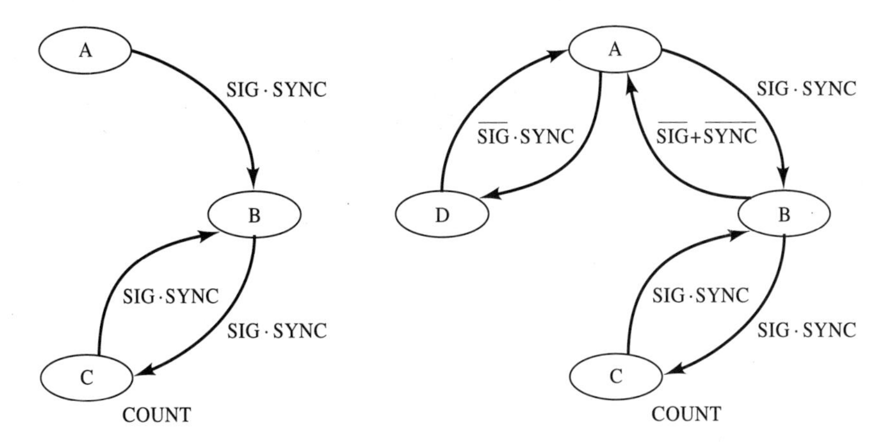

If we're in state A and both SIG and SYNC are asserted, we've found the first 1 of the sequence 11, so we make a transition to state B. If we're in state B and both SIG and SYNC are asserted, we've found the second 1 in the sequence 11, so we make a transition to state C. At this point we've found the sequence 11, so we output COUNT in state C. Now suppose we're in state C and SIG and SYNC are again asserted. We've found the first 1 of the second 11 sequence, so we should make a transition back to B. We do not go back to A because we've found the first 1 of the sequence 11, and if we went back to A we would miss one bit.

Now consider other sequences that can occur. Suppose we're in state A and SIG is not asserted but SYNC is asserted. We need to start a slice, but it will not be a slice with the sequence 11, so we need another state, D, as shown in the figure above to the right.

The state D is just for counting a 2-bit slice. To keep the counting properly synchronized, we need to count two bits even when they are not 11, and return to state A so we can start again. The transition from state D back to state A is an unconditional transition to keep the count correct. There is one other sequence we need to consider when we start in state A. Suppose we have a 10 sequence. We'll make the transition to state B, but then we need to go back to state A on the condition $\overline{\text{SIG}} \cdot \text{SYNC}$. If we look ahead a little, we can also see that we need to make transitions back to state A from any state if SYNC is not asserted. This means that the transition for state B to state A should be conditional on

$$\overline{\text{SIG}} \cdot \text{SYNC} + \overline{\text{SYNC}} = \overline{\text{SIG}} + \overline{\text{SYNC}}$$

as shown in the diagram.

Now consider the following diagram.

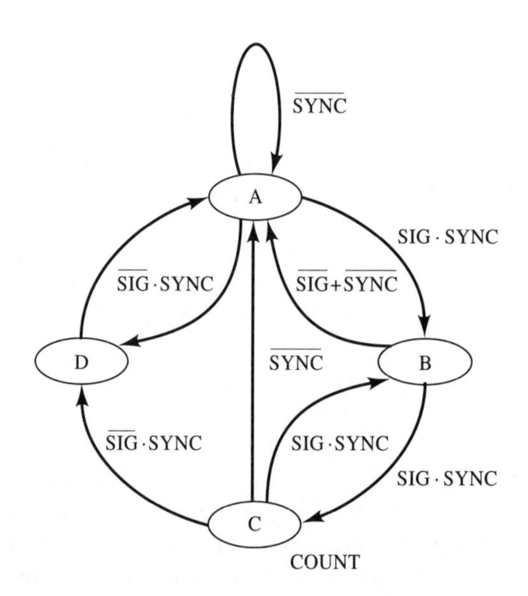

Suppose we're in state C and SIG is not asserted but SYNC is asserted. Here we have the start of a sequence that is not a 11 sequence, so again we need only count by making a transition to state D. Finally, let's consider what happens when SYNC is not asserted. In that case, we want the machine to return to state A no matter what state it's in. We already did this for state B. We also did this for state D because state D always makes a transition to A, independent of the input conditions. We need to add transitions from state C to A, conditional on $\overline{SYNC}$, and from A back to A, also conditional on $\overline{SYNC}$. We can make a final check on our diagram by ensuring that all the possible transitions have been considered for each state. Because there are two inputs, there are four conditions: $\overline{SIG} \cdot \overline{SYNC}$, $\overline{SIG} \cdot SYNC$, $SIG \cdot \overline{SYNC}$, and SIG · SYNC. For example, the conditions on the transitions out of state A are $\overline{SIG} \cdot SYNC$, SIG · SYNC, $\overline{SYNC} = \overline{SIG} \cdot \overline{SYNC} + SIG \cdot \overline{SYNC}$. We should make this check for all four states.

The third step in the design is a timing diagram, formed from the state transition diagram, to check that the machine performs as it should. Forget the specifications of the machine; the question is: What will a machine, with the diagram we've made, *do*? The timing diagram follows.

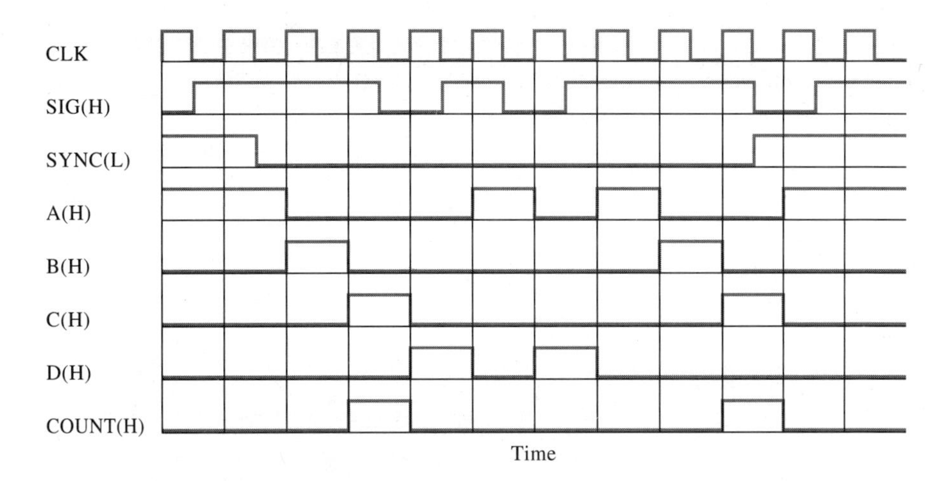

We must make sure we can see how it is produced from the state transition diagram. In this timing diagram, the SIG and SYNC are arbitrary. We should choose them so that they represent a good test. That is, we should have some sequences that are 11 and some that are not 11. This timing diagram does not test the transition from state C to state B, and it does not test the transitions associated with $\overline{\text{SYNC}}$.

In this timing diagram we've treated the states as signals. They may or may not be actual signals; but in any case, they occur only one at a time. This allows us to make the following timing diagram, which designates the states the machine is in at any given time by its name. This kind of diagram is much more compact, and we'll use it in the remainder of the book.

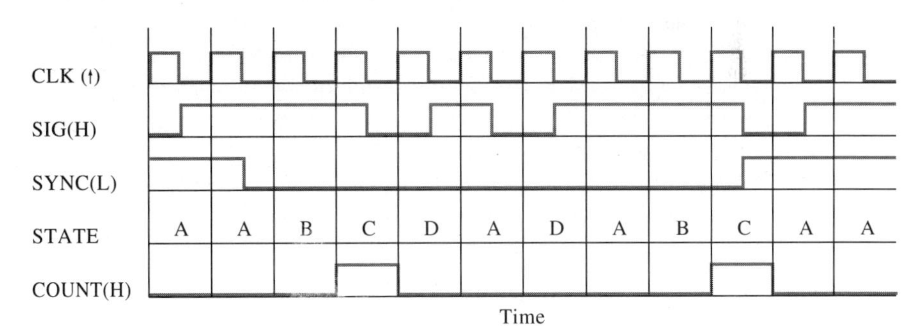

The final step in the design is a present state–next state and output table. It is formed directly from the state transition diagram on page 328.

SYNC	SIG	Present State	Next State	COUNT
0	x	x	A	0
1	0	A	D	0
1	1	A	B	0
1	0	B	A	0
1	1	B	C	0
1	0	C	D	1
1	1	C	B	1
1	x	D	A	0

7.8 USING COMPUTERS AS STATE MACHINES

Later in this text we'll concentrate on designing logic circuits to implement state machines, but state machines can be made using any system that has memory and the means of making controlled transitions from one memory state to another. One system that meets these requirements is a general purpose computer. Memory is inherent in a computer, and because the transitions of the memory from one state to another are controlled by a computer program (software) or code, it's very easy to change the state machine.

There are two reasons why we may want to make a state machine using a computer. One reason is to test the design of a state machine that will later be made using a logic circuit. Writing a program to test a state machine is much less common than it once was, because easy-to-use, general purpose computer software exists that will test (simulate) the logic design of any state machine. We've already looked at simulation in conjunction with combinational logic circuit design, and that kind of simulation can be extended to logic circuits used in state machine design.

A second reason for using a computer to build a state machine is that it's often the easiest way. Computer implementations of state machines are more common than logic circuit implementations. They are easier and faster to design, easier to debug, and easier to change. Normally, the only state machines that are made using logic circuits are those that are too simple to warrant the use of a computer and those that must operate too fast to use a computer.

In a computer used as a state machine, a computer program (software) must be written to cause the computer to perform three functions: storing the state of the machine, determining the actions performed within the state (the outputs), and determining the next state—that is, implementing the next-state logic of the machine. The first function is easy for a computer: the state of the machine can be coded in a variable.

There are two different methods for performing the second and third functions. The first method is to use in-line code. A state machine implemented with in-line code has a separate section of code for each state of the machine. Each code performs the actions required by the state (outputs), determines the next state of the machine, and branches to that state at the appropriate time.

The second method is to use look-up tables. A table-driven machine has a common section of control code that is executed for any state of the machine. For every state of the machine and every set of inputs, both the actions performed in that state and the next-state transition are found in the look-up tables. The look-up tables are formed from the present state–next state and output table we discussed in Section 7.4. This table is put in a form that is convenient to store and access on a computer. Each time a state transition is to be made, the control code accesses the present state–next state table to determine the next state. After the transition it accesses the output table to determine the actions (outputs) in the new state. For a Mealy machine only, the output table must also be accessed each time an input changes to determine whether the outputs change. A table-driven machine implementation uses less code and, for complicated machines, is easier to write. An in-line code implementation usually takes more code but runs faster.

A software implementation of a finite state machine can be written in any language that is available for the target machine. The language can be an assembly language or a high level language such as FORTRAN, PASCAL, or C.

For software implementation of state machines, we must consider the problem of timing (clocking) the machine. If we desire only a simulation of the state machine, timing is usually controlled by the input. The software is designed to request input once in each state, and execution of the software is suspended until the input is supplied (usually from a terminal). The other parts of the software (table look-ups, outputs, and state changes) take some time to execute, but this time is usually very short compared to the time required for input. Thus, the input essentially controls the clocking.

For a software implementation of a state machine that is to be used as an actual synchronous state machine, some form of timing (clock) must be supplied. On most computers a timing function is available that can be program controlled and used to supply a clocking function to the software. The details of the clocking function depend on the actual computer used. In some cases the clocking can be performed with a simple WAIT(TIME) or similar command, which gives a fixed time delay specified by TIME. Software-implemented state machines also need a way of sampling real time inputs at the clocking time to determine the next-state transition and the outputs. For Mealy machines, which can change output during a state if an input changes, the inputs have to be sampled during the state as well as at the state transition. Software implementations have a basic timing limitation. The code takes some time to execute from one state change to the next, and this time sets a limit on the clocking time—the clock period can be no briefer than this time.

We'll look at two examples of implementations of state machines in computers. Don't worry if you do not understand all the details. Just try to get the general idea.

Example 7.8-1

Implement a state machine, with the following state transition diagram, in software. Use in-line code. The inputs to the machine are A and B, and the outputs are C and D. The SANITY input is to show that the machine always starts in state 0. Design the machine so that it's clocked by the internal clock of the computer.

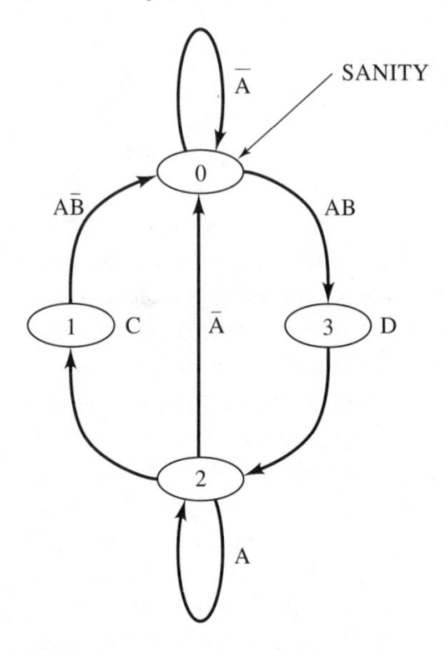

Even though an in-line code is being generated, it's convenient to generate it using a present state–next state table and an output table. In making these tables, we'll modify them to make them more convenient for use in a table-driven implementation of the machine (which we will do in the next example). First notice that the only combinations of inputs that cause transitions are $\bar{A}$, $A\bar{B}$, and AB, so only these inputs need be considered in the present state–next state table. We'll call these inputs 0, 1, and 2. The following present state–next state table has the form we'll use—a matrix of next states, with present state as rows and input as columns. It's a very convenient form for programming because it can be put in a 4 × 3 array.

	Input		
Present State	0 $\bar{A}$	1 $A\bar{B}$	2 AB
0	0	1	3
1	2	2	2
2	0	2	2
3	2	2	2

The output table for this machine, which follows, is very simple because the machine is a Moore machine and the outputs depend only on the state. The output actions are called 0, 1, and 2 and are described in the ''Description'' column of the table. Notice that the output table can be stored in a 1 × 4 array.

State	Output	Description
0	0	No output
1	1	Assert C
2	0	No output
3	2	Assert D

When implementing state machines with a language such as FORTRAN, PASCAL, or C, it's convenient to organize the code so that the actions performed in a state are coded in the first lines of code for that state, and the next-state logic is coded in the following lines.

The C source code for the in-line implementation of the state machine follows. The code contains three procedures (subroutines) that handle the clocking function (check_clock()), the inputs (get_input()), and the outputs (send_output()), and a main program that executes the in-line code. The variable *state,* which stores the machine state, is initialized to 0 (SANITY). The main program is an infinite loop (FOREVER). In the FOREVER loop, the first line of code is a check to determine whether the next clock cycle has occurred, using the procedure check_clock. The cycle occurs each second. When a new cycle is encountered, the remainder of the loop is executed. First, the inputs are read from a parallel port (0 × 200) using the procedure get_input. Second, the outputs are sent to the parallel port using the procedure send_output. Third, the state is changed using a case statement, which depends on the present state. For example, if the present state is 0, then the case is 0, and the next state is set to 1 if A is asserted and B is not asserted (A && !B, && means *and* and ! means *not*) or set to 3 if both A and B are asserted (A && B).

C Source Code for In-Line Driven Machine

```
/*
   file:              inline.c
   purpose:           implement a state machine with in-line code.
   implementation:    The inputs A and B correspond to bits 0 and 1 of the byte
                      read in from the parallel port.
                      The outputs C and D correspond to bits 2 and 3 of the byte
                      sent to the parallel port.
                      We assume that there's a parallel port on the computer with
                      bits 0 and 1 programmed as inputs and bits 2 and 3
                      programmed as outputs.
```

```
*/

#include <stdio.h>
#include <time.h>

#define FOREVER for(;;)

/* Define input and output ports to be a parallel port with port address 200h
*/

#define IN 0×200
#define OUT 0×200

long int SECONDS;
int A, B;
main()
{
      int state, input;
      void get_input(), snd_output();
      int check_clock();

      state = 0;
      SECONDS = 0;

      FOREVER
           {
                if (check_clock())
                   {
                        get_input();
                        snd_output(state);
                        switch(state)
                           {
                                case 0: if (A && !B)
                                           state = 1;
                                        if (A && B)
                                           state = 3;
                                        break;

                                case 1: state = 2;
                                        break;

                                case 2: if (!A)
                                           state = 0;
                                        break;

                                case 3: state = 2;
                                        break;
                           }
                   }
           }
}
```

```
void check_clock()

/* Use Turbo C function clock(), which counts the number of clock
   "ticks" since the program began running. The constant CLK TCK
   gives the number of "ticks" per second.                              */


{
   long int T;
   if ((T = (long int)(clock()/CLK TCK)) ! = SECONDS)
    {
    SECONDS = T;
    return(1);
    }
   else
    return(0);
}

get_input()
{
     int byte_in;

     byte_in = inportb(IN);
     /* AND the byte read to determine if bits 0 and 1 are set */
     A = byte_in & 0×0001;
     B = byte_in & 0×0002;
}

snd_output (state)
int state;
{
     int byte_out;
     switch (state)
        {
            case 1: outportb(OUT, 0×0004);                  /* bit 2 is set */
                    break;
            case 3: outportb(OUT, 0×0008);                  /* bit 3 is set */
                    break;
        }
}
```

Example 7.8-2

Implement the machine of Example 7.8-1 with table-driven code.

To implement this machine using look-up tables, we need the present state–next state table and output table that we constructed in Example 7.8-1. They're repeated here for convenience.

Present State	Input		
	0 $\overline{A}$	1 $A\overline{B}$	2 AB
0	0	1	3
1	2	2	2
2	0	2	2
3	2	2	2

State	Output	Description
0	0	No output
1	1	Assert C
2	0	No output
3	2	Assert D

A C source program for the implementation of the state machine follows. In this program, the present state–next state transition array is *nxt_state_tbl* and the output action array is *output_tbl.* In these arrays we use the indices to denote state numbers and input numbers. The first index (row) denotes the state number, and the second index (column) denotes the input number.

The procedures used in this program to check the time, to input, and to output are the same as those in Example 7.8-1. The variable *state* that stores the machine state is initialized to 0 (SANITY). The main program is an infinite loop (FOREVER). In the FOREVER loop, the first line of code is a check to determine whether the next clock cycle has occurred, using the procedure check_clock. The cycle occurs each second. When a cycle is encountered, the remainder of the loop is executed. The software first updates the inputs, then gets the output action for the state from the output table and executes that action. It then gets the next state from the present state–next state table and causes a transition to the next state (updates *state*). The process is repeated until it's stopped by some external means.

C Source Code for Table-Driven Machine

```
/*
   file name:         table.c
   purpose:           implement a state machine using state and output tables.
   implementation:    The inputs A and B correspond to bits 0 and 1 of the byte
                      read in from the parallel port.
                      The outputs C and D correspond to bits 2 and 3 of the byte
```

```
                    sent to the parallel port.
                    We assume that there's a parallel port on the computer with
                    bits 0 and 1 programmed as inputs and bits 2 and 3
                    programmed as outputs.
*/

#include <stdio.h>
#include <time.h>

#define FOREVER for(;;)

/* Define input and output ports to be a parallel port with port address 200b */

#define IN 0×200
#define OUT 0×200

long int SECONDS;
int A, B;

main( )
{
    int state, input;
    static int nxt_state_tbl[4][3] =
     {
        {0, 1, 3},
        {2, 2, 2},
        {0, 2, 2},
        {2, 2, 2}
     };
    static int output_tbl[4] = {0, 1, 0, 2};
    void get_input( ), snd_output( );
    int check_clock( );

    state = 0;

    FOREVER
      {
      if (check_clock( ) )
         {
         get_input( );
         /* Translate A and B into input values:
           !A = 1, A && !B = 2, A && B = 3        */
         if (!A)
           input = 0;
         else
           if (!B)
             input = 1;
           else
             input = 2;
         snd_output(output_tbl[state]);
         state = nxt_state_tbl[state][input];
         }
      }
}
```

```
int check_clock( )

/*Use Turbo C function clock( ), which counts the number of clock "ticks" since the
  program began running. The constant CLK TCK gives the number of "ticks" per
  second.                                                                        */

{
  long int T;
  if ((T = (long int)(clock( )/CLK_TCK) != SECOND)
    {
    SECONDS = T;
    return(1);
    }
  else
    return(0);
}

void get_input( )
{
  int byte_in;

  byte_in = inportb(IN);
  /* AND the byte read to determine if bits 0 and 1 are set */
  A = byte_in & 0x0001;
  B = byte_in & 0x0002;
}

void snd_output(output)
int output;
{
  int byte_out;

  switch(output)
    {
        case 1: outportb(OUT, 0x0004);                    /* bit 2 is set */
                break;
        case 2: outportb(OUT, 0x0008);                    /* bit 3 is set */
                break;
    }
}
```

KEY TERMS

Sequential Machines Machines in which the output *may* depend on the immediate input to the machines but *always* depends on the *previous* condition of the machine. Sequential machines step through a set of conditions.

Feedback System A system in which the output is *fed back* to the input and is used along with the external input to determine the output.

State The condition of a sequential machine.

State Machine Another name for a sequential machine.

Output Decoder The part of a state machine that uses logic operations on the state of machine and its input to generate an output.

Synchronous Machine A state (or sequential) machine timed by a clock.

Asynchronous Machine A machine that is not clocked.

Finite State Machine (FSM) A state (or sequential) machine that cycles repetitively through a finite set of states.

Next-State Decoder The part of a state machine that uses logic operations on the present state of a machine and its input to produce a code, which in turn generates the next state of the machine in the memory.

Moore Machine A machine in which the output depends only on the state.

Mealy Machine A machine in which the output depends on both the state of the machine and its input.

Edge Triggered A description for a device or a state machine that uses either the rising edge or trailing edge of the clock signal to cause a transition to the next state. Most devices and state machines are edge triggered.

Level Triggered A description for a device or a state machine that uses either a high or low portion of the clock signal to cause a transition to the next state. Level triggered device and state machines are potentially unstable.

Positive Edge Triggered Edge triggering in which the transition occurs at the rising edge of the clock signal.

Negative Edge Triggered Edge triggering in which the transition occurs at the falling edge of the clock signal.

EXERCISES

1. Make present state–next state and output tables for state machines with the following state transition diagrams.

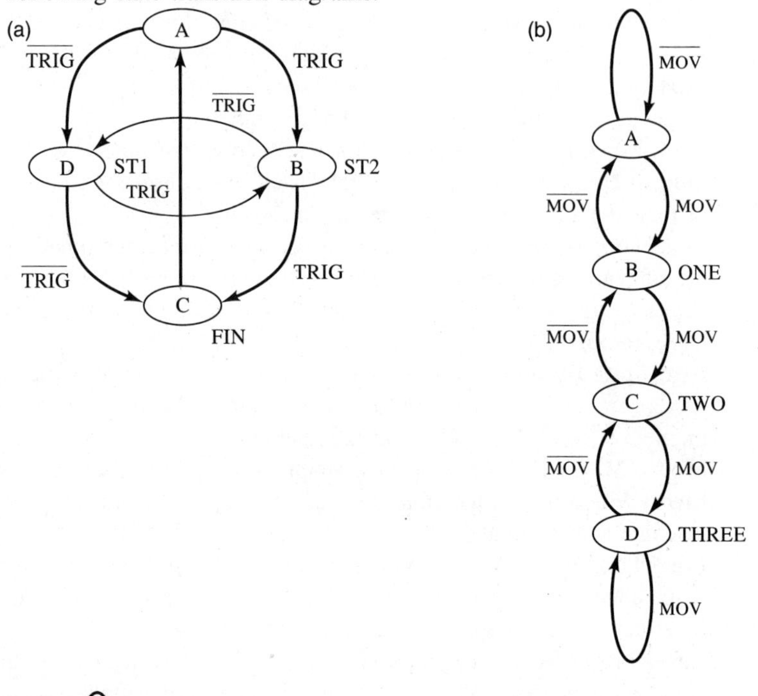

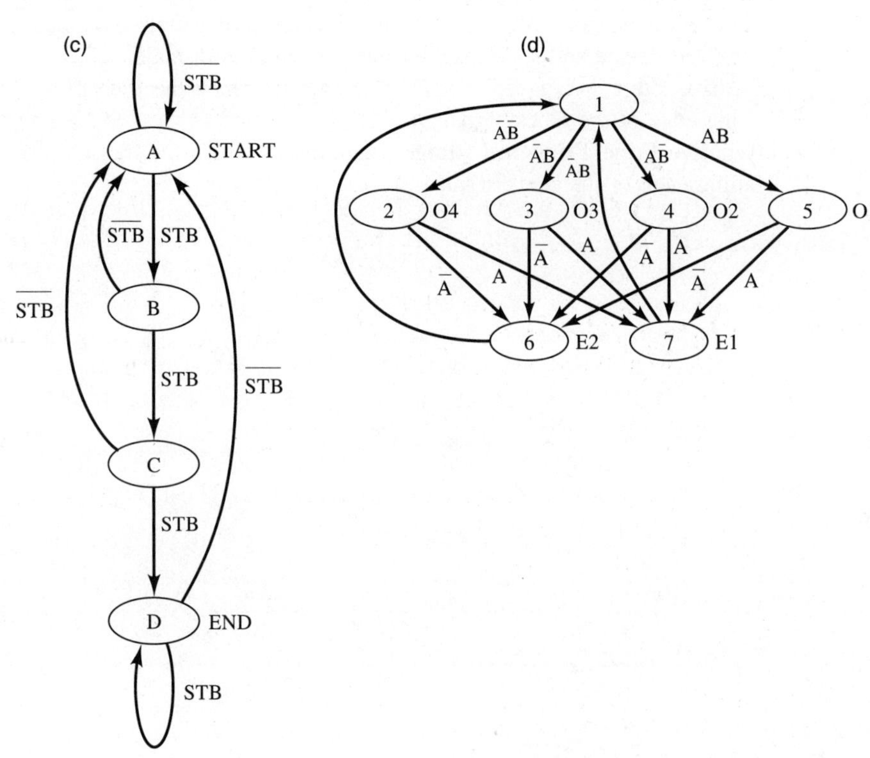

2. Draw timing diagrams for the four state machines of exercise 1. Assume the machines are synchronous machines that are positive-edge triggered. Use the following inputs.

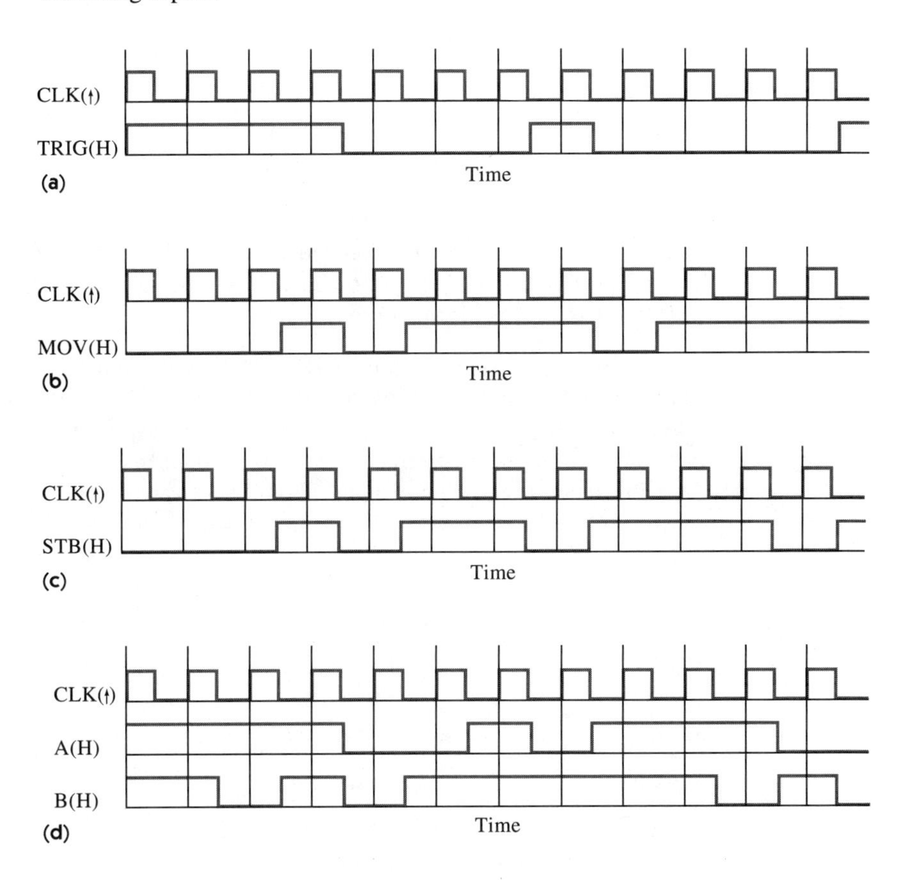

3. Design a state machine (first four steps) to find one of the following sequences in an input signal, SIG(H). Each time the sequence is found, an output, FND, should be asserted. When a complete sequence is found, the machine should start looking for the sequence again without missing a bit. When the sequence is not found, the machine should immediately start looking again. That is, if the sequence to be found is 001, and a 1 or a 01, is found, the machine should start looking for a new correct sequence on the next bit. Changes in SIG are synchronous with the negative-going edge of the clock, as shown in the following timing diagram. Make the designs Moore machine designs. The timing diagram should include at the least the desired sequence and another sequence.

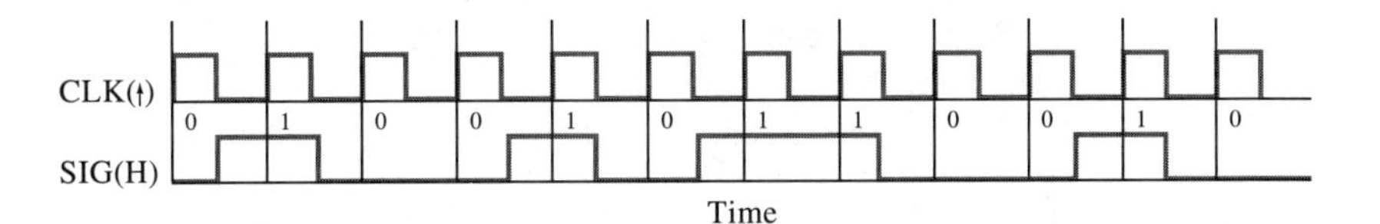

(a) 000
(b) 001
(c) 011
(d) 100
(e) 101
(f) 110
(g) 111
(h) 0000
(i) 0001
(j) 0010
(k) 0011
(l) 0100
(m) 0101
(n) 0110
(o) 0111
(p) 1000
(q) 1001
(r) 1010
(s) 1011
(t) 1100
(u) 1101

4. Make a Mealy machine design for each of the machines of exercise 3. In some cases, the output will need to be dependent on the clock in order for the Mealy machine to function properly. You should be able to make a Mealy machine with fewer states than a Moore machine.

5. Design a state machine (state transition and timing diagram but no present state–next state table) that will sample an incoming signal SIG(H) in 3-bit slices and determine whether the three bits are as specified in each of the following cases. Each time the bits are as specified, the machine should assert an output, CNT(H), for a full clock cycle. The machine should start sampling when an incoming signal, SYNC(L), is asserted. The machine should stop sampling when it has completed a 3-bit slice *and* SYNC is no longer asserted. Both SIG and SYNC are synchronized to change on the negative-going clock edge. The timing diagram should include at least one desired sequence and one other sequence with SYNC asserted.

(a) 000
(b) 001
(c) 010
(d) 011
(e) 100
(f) 101
(g) 110
(h) 111

6. Repeat exercise 5, but have the machine look for two different sequences as follows. Each time the first sequence is found, the machine should assert output CNT1(H). Each time the second is found, it should assert output CNT2(H).

(a) 000 001
(b) 000 010
(c) 000 011
(d) 000 100
(e) 000 101
(f) 000 110
(g) 000 111
(h) 001 010
(i) 001 011
(j) 001 100
(k) 001 101
(l) 001 110
(m) 001 111
(n) 010 011
(o) 010 100
(p) 010 101
(q) 010 110
(r) 010 111
(s) 011 100
(t) 011 101
(u) 011 110

7. Design a synchronous state machine (state transition diagram and timing diagram only) to find the first positive-going edge of an incoming signal SIG(H) that occurs after the machine is *armed*. The machine is to have three inputs: FDEDGE(H), a signal that arms the machine and causes it to start looking for an

edge; SIG, as specified before; and RESET(L), a signal that causes the machine to get ready to find another edge. The machine should assert an output signal, FND(H), when the edge is found. FND should remain asserted until the input RESET is asserted. Be sure to consider both the case in which SIG is high when FDEDGE is asserted and the case in which SIG is low when FDEDGE is asserted.

8. Repeat exercise 7, but design an asynchronous machine.

9. Design a state machine (state transition diagram and timing diagram only) to determine whether an input signal SIG(L) is asserted one, two, or three clock cycles after an input START(L) is asserted. Assert an output ONE, TWO, or THR depending on which condition is found. Repeat each time START is asserted. Assume that changes in SIG and START are synchronous with the negative-going edge of the clock. Show all three conditions in the timing diagram.

10. Design a state machine (state transition diagram and timing diagram only) to control processes that require one of the following sets of outputs with the timing shown. In each case, choose a suitable clock frequency, but make it as low as possible.

    (a) (Start the sequence when RESET is asserted.)
    START—for 10 ms
    (Wait for 3 ms.)
    STOP—for 1 ms
    (Wait for 5 ms.)
    INC—until GO is asserted
    (Repeat when RESET is asserted.)

    (b) (Start the sequence when START is asserted.)
    FWD—for 6 ms
    (Wait for 1 ms.)
    RVS—until STP is asserted
    (Wait for 2 ms.)
    INC—until NEW is asserted
    (Repeat sequence until START is deasserted.)

    (c) (Start the sequence when NCT is asserted.)
    DOWN—until BTM is asserted
    (Wait for 3 ms.)
    STOP—for 2 ms
    (Wait for 5 ms.)
    UP—for 5 ms
    (Repeat if NCT is still asserted or wait until NCT is asserted and repeat.)

    (d) (Start the sequence when RESET is asserted.)
    START—for 7 ms
    Starting at the same time as START, LEVEL—for 3 ms
    (Wait for 3 ms after the end of START.)
    STOP—for 1 ms

(Wait for 3 ms.)
STOP—for 1 ms
(Wait for 3 ms.)
(Repeat the sequence endlessly.)

(e) (Start the sequence when GO is asserted.)
START—until STOP is asserted
Starting at the same time as START, NEWCT—for 2 ms
FNL—for 1 ms after STOP is asserted
(Wait for 5 ms.)
NEW—for 3 ms
(Repeat until GO is deasserted.)

11. Using the programming language of your choice, implement one of the foregoing machines in software. Design the machine so that it clocks once each time it receives an input if it's a synchronous machine. Each time the machine is clocked, indicate the condition of every output (that is, asserted or not asserted). You can do this by outputting a 0 or 1 for each output. Run the program with a combination of inputs that will demonstrate that it functions correctly.

8 Two-State State Machines (Flip-Flops)

8.1 INTRODUCTION

In Chapter 7 we introduced the idea of a state machine and showed how such a machine could be implemented on a general purpose computer. In this chapter we'll see how to implement a state machine with a logic circuit, using gates and memory elements. There are two reasons why we might implement a state machine in this way rather than on a general purpose computer. Either the machine might be so simple that it would not warrant the use of a computer, or it might run at faster rates than a computer implementation can achieve.

The simplest kind of state machine has only two states. Figure 8.1-1 is a state diagram for a general two-state machine. Such a machine can be thought of as a memory cell that is capable of storing 1 bit of information. It can be in either an *asserted* state (1) or a *not asserted* state (0). Two-state machines are called flip-flops because they flip from one state to another and then flop back. There are a number of kinds of flip-flops that differ in how their inputs and clocks cause transitions between the two states. The conditions that cause transitions are represented in Figure 8.1-1 as COND1, COND2, COND3, and COND4 and can be thought of as logic conditions involving inputs and the clock. We will examine the design and analysis of flip-flops as our first venture into state machine design techniques.

Looking back at Figure 5.2-1, which shows a general state machine, we see that the three components of a state machine are a next-state code decoder, a memory, and an output decoder. Often the output decoder is not needed because the state is the output. The design of a state machine amounts to designing the next-state decoder so that it produces the proper next-state code inputs to the memory to cause the desired state transitions. This means that we need a memory, and we need to know what inputs are required to change the state of the memory.

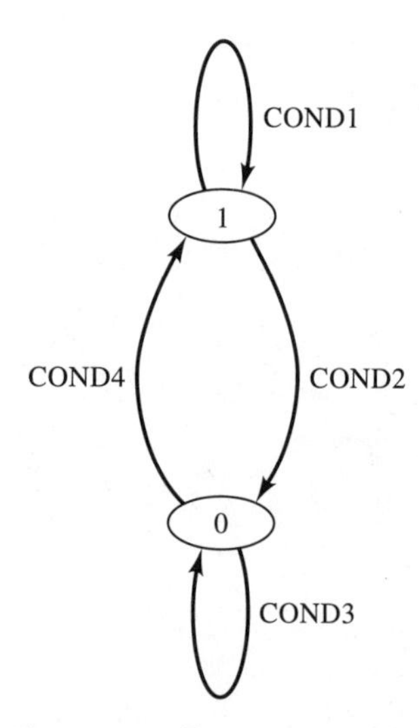

Figure 8.1-1 Two-state machine.

The simplest possible memory is needed for the design of flip-flops. It needs to store 1 bit of information, and it needs inputs that will allow the bit to be changed from *asserted* to *not asserted* and back. In Section 8.2 we'll discuss such a SET-RESET memory cell, which is itself an asynchronous flip-flop.

8.2 SET-RESET MEMORY CELL

To store 1 bit of information in a logic circuit, we need a circuit which can be placed either in a state with a high output that does not change or in a state with a low output that does not change. We can build a circuit with these characteristics using two inverters. If we cascade two inverters and feed the output of the second back to the input of the first, as shown in Figure 8.2-1, the output, Q(H) in the figure, will be stable (unchanging) and be either *asserted* (high) or *not asserted* (low). This is easy to show by analyzing the circuit. If Q is asserted, then Q(L), the input to the second inverter, will be low. This will make the output of the second inverter, Q(H), high, so the input to the first inverter will be high. Now a high input to the first inverter will make the output, Q(L), low, which is just what we started with. This is positive feedback: the feedback into the first inverter reinforces the input into the

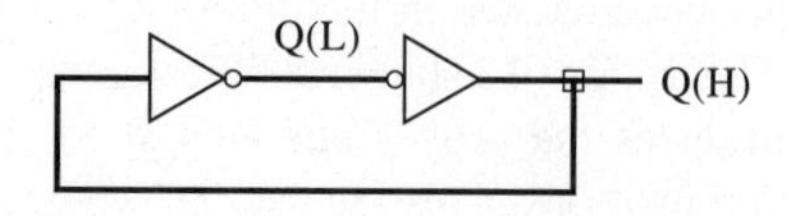

Figure 8.2-1 Basic memory circuit.

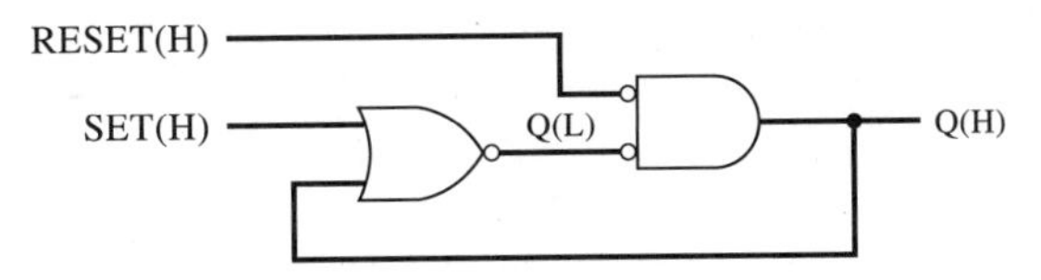

Figure 8.2-2 Basic memory circuit (cell) with inputs.

second inverter. If the situation is reversed and Q is not asserted, so that Q(L) is high, then the output of the second inverter will be low. This will feed a low input into the first inverter and cause a high output, reinforcing the input into the second inverter. In this circuit, Q will either be asserted and remain asserted or be not asserted and remain not asserted. This memory circuit has no provisions for changing its state; it remains in its initial condition until it's turned off. In order for it to be useful, we need a provision for changing its state.

If we change the second inverter in the circuit into a 2-input *nor* gate performing an *and* operation, and the first inverter into a 2-input *nor* gate performing an *or* operation, as shown in Figure 8.2-2, we still have the same basic memory circuit with positive feedback; but when the output of the second gate is high (Q asserted), it can be changed to low by asserting the RESET(H) input to that gate; and when the output of the first gate is high (Q not asserted), it can be changed to low by asserting the SET(H) input to that gate. In either of these cases, we can see, by tracing around the feedback path, that the change that is made produces a new stable set of signals in the circuit, and the circuit remains in this new stable state even when the input (SET or RESET) is no longer asserted.

Now suppose we assert RESET(H) when the output of the second gate is low (Q not asserted). The output will remain low, and the state of the circuit will not change. Similarly, if SET(H) is asserted when the output of the first gate is low (Q asserted), the output will remain low and the state of the circuit will not change. If both SET(H) and RESET(H) are asserted at the same time, both Q(H) and Q(L) will be forced low. This is an ambiguous state: the system is neither SET nor RESET, and if both the inputs are removed simultaneously, the system may enter either a SET or a RESET state.

It's interesting to simulate the SET-RESET cell using a logic simulator to check the analysis. Figure 8.2-3 shows the results of such a check. In the figure, the inputs SET(H) and RESET(H) are arbitrary. Notice that when SET changes to an asserted level (high), Q changes to an asserted level Q(H) (high) unless it's already asserted, and when RESET changes to an asserted level (high), Q changes to a not asserted level Q(H) (low) unless it's already not asserted. This is a simulation for actual gates, so the responses in Q(H) and Q(L) that are due to the changes in SET(H) and RESET(H) are delayed slightly (10 ns), owing to gate delays in the two gates. Normally Q(H) and Q(L) have opposite values. At the end of the simulation, we see what happens if both SET and RESET are asserted (high): both Q(H) and Q(L) are forced low, as would be expected.

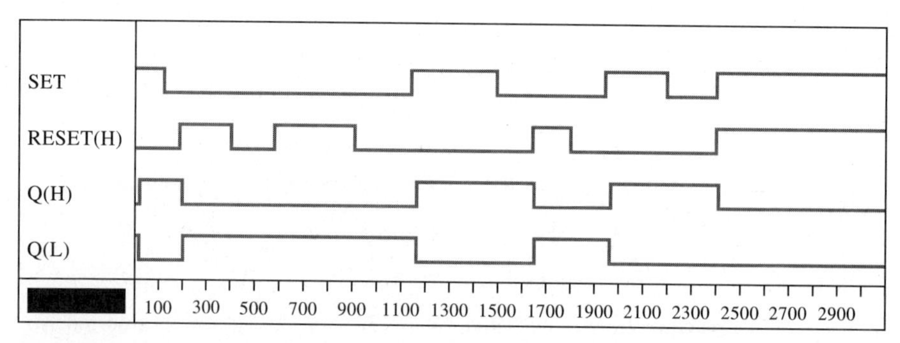

Figure 8.2-3 Simulation results for a SET-RESET cell.

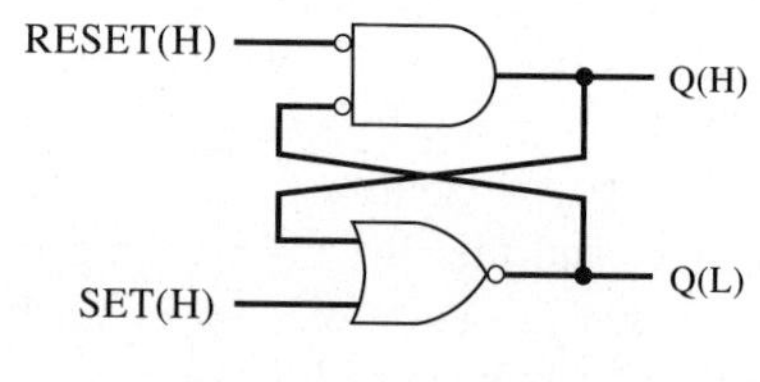

Figure 8.2-4 Circuit for a memory cell using *nor* gates as it is normally drawn.

We've described a memory cell using *nor* gates, and we've drawn it in a particular way to make the analysis easier. Figure 8.2-4 shows a more common way to draw a cell and illustrates why a memory cell is said to be constructed from cross-coupled gates. We can make a memory cell, using *nand* gates, that functions in the same way as the one just described except that the assertion levels for all the signals are reversed. Figure 8.2-5 shows this *nand* memory cell.

As already stated, the memory cell we've described is an asynchronous two-state machine—a SET-RESET flip-flop. Figure 8.2-6 is a state transition diagram for this machine. Table 8.2-1 is a present state–next state table in which the present state is called Q_n and the next state Q_{n+1}; this is standard notation for flip-flops.

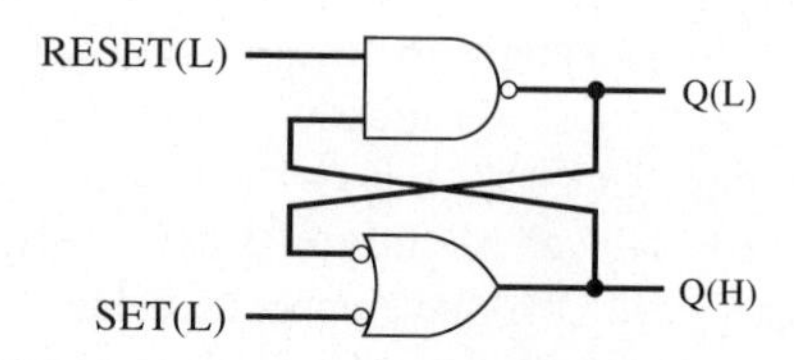

Figure 8.2-5 Circuit for a memory cell using *nand* gates.

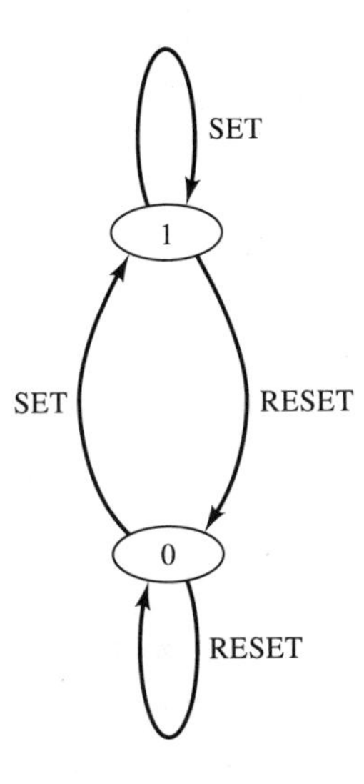

Figure 8.2-6 State transition diagram for asynchronous SET-RESET flip-flop.

Table 8.2-1 SET-RESET Flip-Flop Present State–Next State Table*

SET	RESET	Q_n	Q_{n+1}
0	0	0	0
0	0	1	1
0	1	0	0
0	1	1	0
1	0	0	1
1	0	1	1
1	1	0	x
1	1	1	x

Q_n	Q_{n+1}	SET	RESET
0	0	0	x
0	1	1	0
1	0	0	1
1	1	x	0

*This table is shown in two forms. The one on the right, called the excitation table, will be more useful for later designs.

8.3 DESIGN OF A LEVEL CLOCKED D FLIP-FLOP

Our first design will be a level clocked D flip-flop using the basic SET-RESET memory cell. We've already found that level clocked sequential machines can have problems, but a level clocked D flip-flop, which is used in a transparent D latch, is

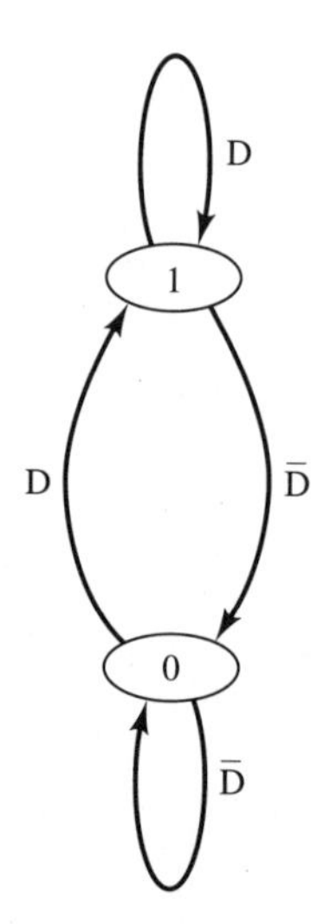

Figure 8.3-1 D flip-flop.

rather useful. In any case, a level clocked machine is easier for an initial design effort. A clocked D flip-flop is a flip-flop with two inputs—a clock input and an input called D. When the clock is asserted, the flip-flop changes state to whatever is on the D input. When the clock is not asserted, the flip-flop does not change state.

The first step in the design of a state machine is to put the word description into a precise form. The ultimate form needed for the design is a present state–next state table. In some cases it is useful to progress from a word description to a state transition diagram and then to a present state–next state table. A timing diagram is usually not needed for these simple machines. Figure 8.3-1 is the state transition diagram for the D flip-flop. The clock input is not shown explicitly because transitions occur only when the clock is asserted; Ck would therefore be present in all the logic expressions describing the conditions for a transition. Table 8.3-1 is the present state–next state table, which we could actually develop directly from the word description. Notice that the present state–next state table is a precise statement of what we want the machine being designed to *do*.

Sometimes it's useful to have a logic expression to describe the operation of a flip-flop, although it is not required for the design. We can find the logic expression by making a Karnaugh map for the next state (from the present state–next state table), as shown in Figure 8.3-2(a). The expression for Q_{n+1} from this Karnaugh map is

$$Q_{n+1} = Ck \cdot D + \overline{Ck} \cdot Q_n$$

The second term tells us that the state does not change unless the clock is asserted. We're usually interested only in the transitions that occur when the clock is asserted. Figure 8.3-2(b) is the Karnaugh map if the clock is asserted and the logic expression for these transitions is

$$Q_{n+1} = D.$$

Table 8.3-1 Present State–Next State Table for D Flip-Flop

Ck	D	Q_n	Q_{n+1}
0	0	0	0
0	0	1	1
0	1	0	0
0	1	1	1
1	0	0	0
1	0	1	0
1	1	0	1
1	1	1	1

Once we have a precise description of what we want our machine to do, we must make the machine do it. For each set of inputs and each present state, we need to find the inputs to the memory element that will cause the memory element to make the transition into the proper next state. These inputs are the next-state code. They are found by searching the excitation table. The **excitation table** is the present state–next state table for the memory element, with the columns rearranged so that the present state and the next state are the first two columns, and the inputs are the remaining columns. This arrangement makes it easy to determine the next-state code, if we know the present state and the next state. In the course of the design, we usually just add the next-state code, found from the excitation table, to the present state–next state table of the flip-flop to be designed. The next-state code for this design will be the memory cell inputs SET and RESET that are needed for each transition in the present state–next state table.

Table 8.3-2 is the present state–next state table with the next-state code added. Included is the excitation table for the SET-RESET memory element. To fill out the table with the next-state code, we consider each desired present state–next state transition and match it with the same present state–next state transition in the excitation table for the memory element. For example, the desired present state–next

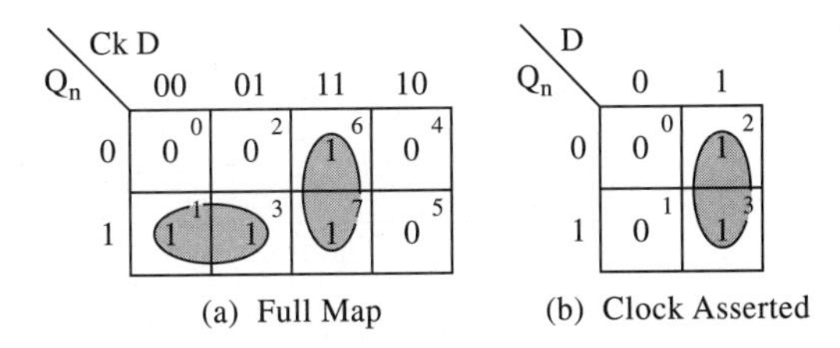

(a) Full Map (b) Clock Asserted

Figure 8.3-2 Next-state maps for D flip-flop.

Table 8.3-2 Present State–Next State Table with Next-State Code and the Excitation Table for the Memory Cell

Ck	D	Q_n	Q_{n+1}	SET	RESET
0	0	0	0	0	x
0	0	1	1	x	0
0	1	0	0	0	x
0	1	1	1	x	0
1	0	0	0	0	x
1	0	1	0	0	1
1	1	0	1	1	0
1	1	1	1	x	0

Memory Cell Excitation Table

Q_n	Q_{n+1}	SET	RESET
0	0	0	x
0	1	1	0
1	0	0	1
1	1	x	0

state transition for the first row of the table is $Q_n = 0$, $Q_{n+1} = 0$. This matches the first row of the excitation table for the memory element, so the next-state code is SET = 0, RESET = x. For the second row, $Q_n = 1$, $Q_{n+1} = 1$ matches the fourth row of the excitation table for the memory cell, so the next-state code is SET = x, RESET = 0.

Once the next-state code is determined, we can design the next-state decoder. This is a combinational logic circuit design. We know the inputs (Ck and D) and the present state (Q_n), and we want to design combinational logic circuits to find SET and RESET. We'll use Karnaugh maps to simplify the logic expressions in the design. We need two maps—one for SET and one for RESET. Figure 8.3-3 shows the maps and the resulting simplified decoder logic expressions. These expressions can be implemented using any gate that has an *and* operation with high assertion level output. Figure 8.3-4 shows the level clocked D flip-flop design using *and* gates in the next-state decoder, with inputs Ck and D asserted high. There are many possible variations of this design that use the other memory cell (with *nand* gates) having low assertion level inputs and/or different next-state decoder gates.

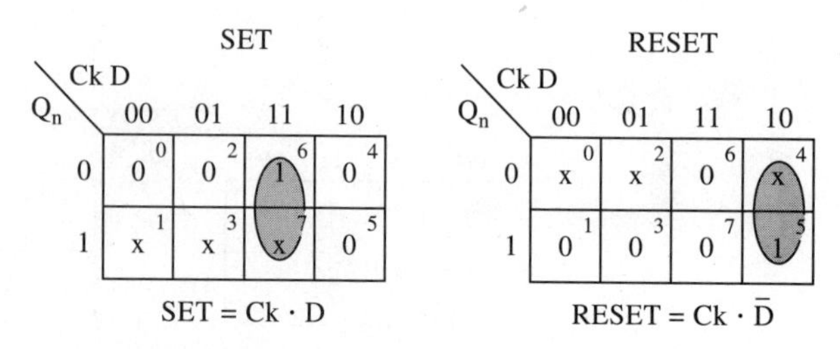

Figure 8.3-3 Karnaugh maps for next-state decoder.

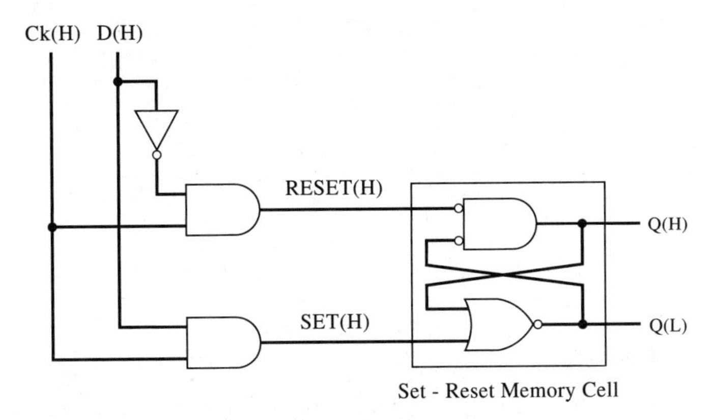

Figure 8.3-4 Level clocked D flip-flop.

The flip-flop in Figure 8.3-4 is a level clocked device. We've noted that level clocked devices can have problems if the input changes while the clock is asserted. In this flip-flop, a change in D while the clock is asserted is not necessarily troublesome. In fact, a change while the clock is asserted results in a useful feature. When we analyze the circuit of Figure 8.3-4 to see how a change in D affects Q when the clock is asserted, we get the results shown in Table 8.3-3. It's clear from this table that Q follows D as long as the clock is asserted. When the clock is asserted, we say the flip-flop is *transparent*, because changes in D propagate directly through to Q. When the clock is not asserted, Q does not change; rather, it's latched (held) in the last state that existed before the clock was changed to not asserted. These results are also evident from the logic expression we found for Q_{n+1}. A device consisting of several level clocked D flip-flops that are packaged together and use a common clock input is called a **transparent latch**.

A level clocked D flip-flop is only partially synchronous even though it has a clock. When the clock is not asserted, the device's output does not change, and the device acts like a synchronous device. However, when the clock is asserted, the device's output changes with the input, and the device acts like an asynchronous device. We can clearly see this partially synchronous, partially asynchronous behavior in the timing diagram of Figure 8.3-5, which was found by simulating the circuit of Figure 8.3-4. It shows the intermediate signals SET(H) and RESET(H) as well as

Table 8.3-3 Clock Asserted

D	SET	RESET	Q
0	0	1	0
1	1	0	1

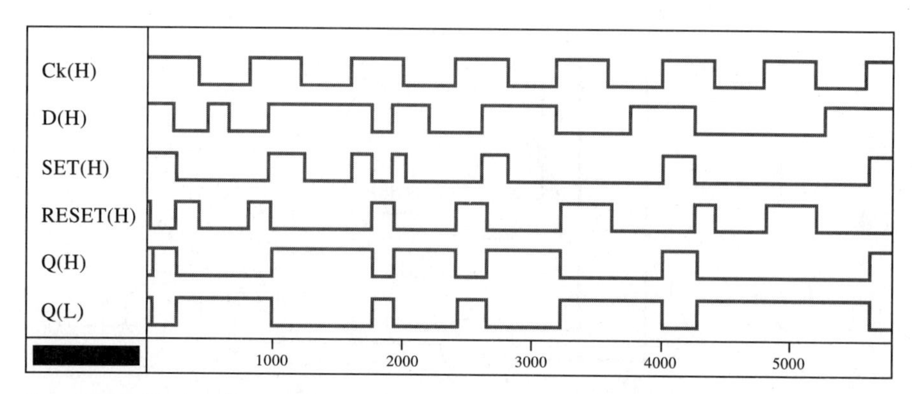

Figure 8.3-5 Timing diagram from a simulation of the level clocked D flip-flop of Figure 8.3-4.

the output signals Q(H) and Q(L). The inputs Ck(H) and D(H) cause SET(H) and RESET(H), which in turn cause Q(H) and Q(L).

8.4 LEVEL CLOCKED T (TOGGLE) FLIP-FLOP

In this section we'll design a level clocked T flip-flop using a basic SET-RESET memory cell. A clocked T (toggle) flip-flop has 2 inputs—a clock input and an input called T. When the clock is asserted, the flip-flop changes state (toggles) if T is asserted but stays in the same state if T is not asserted.

The steps in the design of this flip-flop are the same steps we took in designing the D flip-flop of Section 8.3. Having gone through the design once, we can now summarize these steps.

1. Make a clear word description of the desired flip-flop.
2. Make a present state–next state table for the desired flip-flop. This may be facilitated by drawing a state transition diagram.
3. Using the excitation table for the chosen memory cell, add the next-state code to the present state–next state table.
4. Design a next-state decoder, using Karnaugh maps to simplify the logic required.
5. Draw a circuit, including the decoder logic and the memory cell.

These steps are a little different from the first four steps in the design of a state machine that were outlined in Chapter 7. Step 2 (state transition diagram) and step 4 (present state–next state table) have been combined, and step 3 (timing diagram) has been omitted. Usually the timing diagram is not needed in a simple two-state design.

We completed the first step for a T flip-flop at the beginning of this section. Now, from the word description we develop the state transition diagram and present

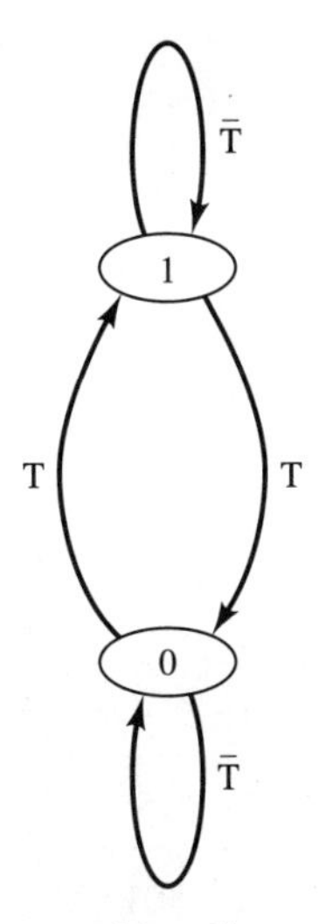

Figure 8.4-1 State transition diagram.

Table 8.4-1 Present State–Next State Table

Ck	T	Q_n	Q_{n+1}
0	0	0	0
0	0	1	1
0	1	0	0
0	1	1	1
1	0	0	0
1	0	1	1
1	1	0	1
1	1	1	0

state–next state table (Fig. 8.4-1 and Table 8.4-1). At this point, the Karnaugh map for next-state transitions (clock asserted) (Fig. 8.4-2) allows us to find a useful logic expression that describes the transitions of the T flip-flop:

$$Q_{n+1} = \overline{T} \cdot Q_n + T \cdot \overline{Q}_n,$$

although we do not need it for our design.

The next step in the design is to add the next-state code to the present state–next state table. We do this, as in the D flip-flop design, by considering each present state–next state transition along with the excitation table for the memory cell, and determining the memory cell inputs (SET and RESET) that cause the desired

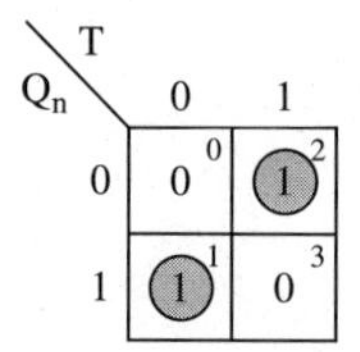

Figure 8.4-2 Next-state map.

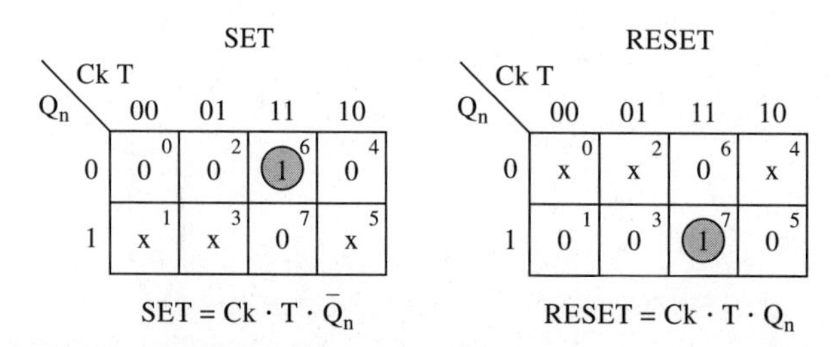

Figure 8.4-3 Karnaugh maps for the inputs to the memory cell.

transition. Table 8.4-2 is the present state–next state table with the next-state code added. Included is the excitation table for the SET-RESET memory cell that is used to determine the next-state code.

Now that we've found the next-state code, we make Karnaugh maps for the memory inputs and design the logic circuits for the next-state decoder. Figure 8.4-3 shows the maps and simplified logic expressions for the memory inputs. Actually, the logic expressions cannot be simplified.

The last step is to put the next-state decoder logic and the memory cell together in the final design. Figure 8.4-4 shows a final design that uses positive logic *and*

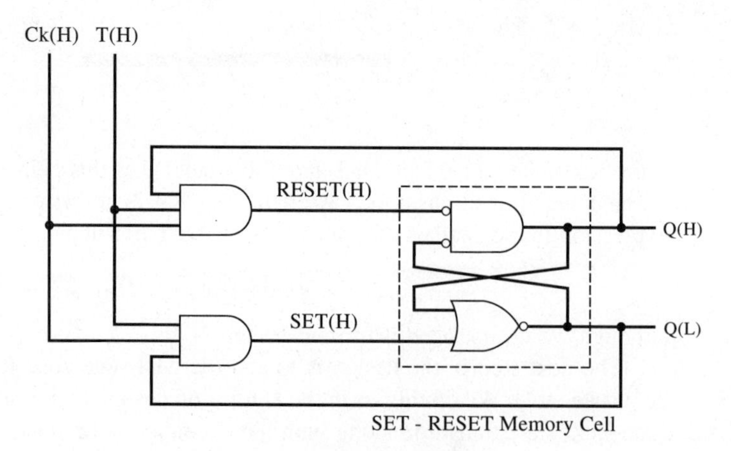

Figure 8.4-4 Level clocked T flip-flop design.

Table 8.4-2 Present State–Next State Table and Next-State Code with Excitation Table for SET-RESET Memory Cell

Ck	T	Q_n	Q_{n+1}	SET	RESET
0	0	0	0	0	x
0	0	1	1	x	0
0	1	0	0	0	x
0	1	1	1	x	0
1	0	0	0	0	x
1	0	1	1	x	0
1	1	0	1	1	0
1	1	1	0	0	1

Memory Cell Excitation Table

Q_n	Q_{n+1}	SET	RESET
0	0	0	x
0	1	1	0
1	0	0	1
1	1	x	0

gates for the decoder logic and *nor* gates for the SET-RESET memory cell. Both Ck and T are asserted high.

The level clocked T flip-flop is not a useful device, because it toggles back and forth between its two states as long as the clock and T are asserted. To see how this happens, consider Table 8.4-3, which shows how Q_n changes when both T and the clock are asserted. We can see from the table that the flip-flop toggles from Q_n, the present state, to the opposite state; but as soon as this happens, the new state becomes the present state, so the flip-flop toggles again. The time it takes for the state to change is determined by the time delay through the gates of the next-state decoder and the memory cell. The device will be useful only if it toggles no more than once (unless we want to make an oscillator). One way to make it toggle only once is to make the time period during which the clock is asserted shorter than the time delay through the next-state decoder and memory cell. Then the clock will no longer be asserted after the state changes, and only one transition will occur.

Figure 8.4-5 shows the timing for the T flip-flop, found by simulating the circuit of Figure 8.4-4. When T is not asserted, the flip-flop behaves as expected—it does not change state. When T is asserted and the clock is not asserted, the flip-flop still behaves as expected—it does not change state. But if both T and the clock are

Table 8.4-3 Clock Asserted

T	Q_n	SET	RESET	Q_{n+1}
1	0	1	0	1
1	1	0	1	0

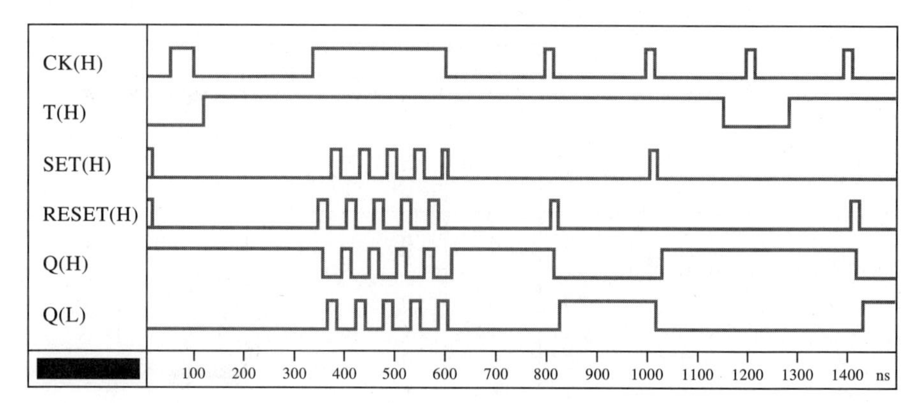

Figure 8.4-5 Timing for T flip-flop showing toggling problem.

asserted for a long period, as shown next, the flip-flop toggles back and forth between its two possible states. Figure 8.4-6 is an enlargement of the region where the toggling occurs. Notice that the first transition of Q(H) is delayed from the clock edge by 18 ns—the propagation delay through the decoder *and* gate (8 ns) and the delay through the memory cell *nor* gate (10 ns). The transition of Q(L) is further delayed by the propagation delay through the second *nor* gate in the memory (10

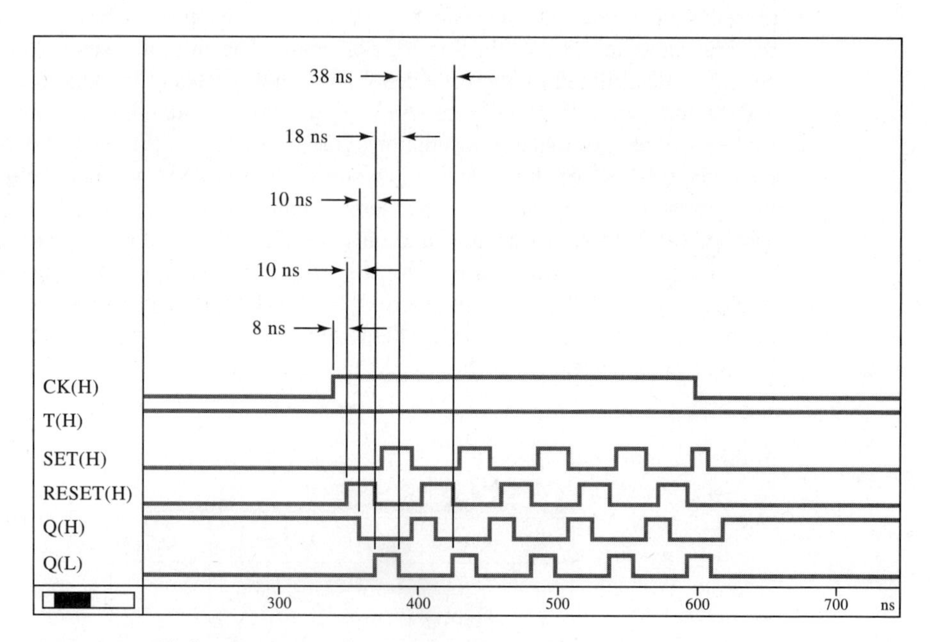

Figure 8.4-6 Timing for the toggling in Figure 8.4-5.

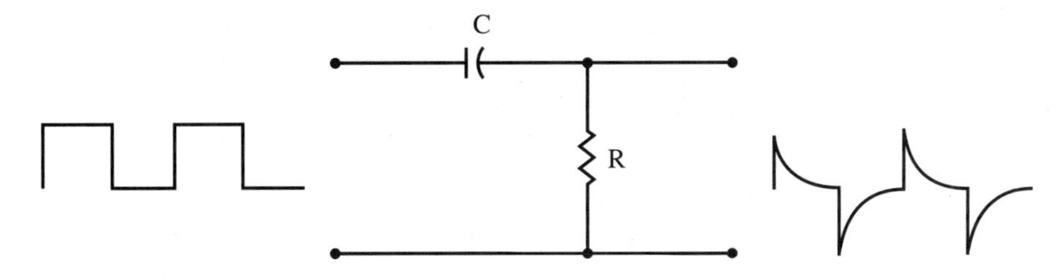

Figure 8.4-7 Circuit for producing short spikes at clock edges.

ns). The transition of Q(L) causes SET(H) to be asserted, which in turn causes a second transition of Q(L). If the clock is no longer asserted when the first transition of Q(L) occurs, no further transitions occur.

The latter part of the simulation results in Figure 8.4-5 shows that a short clock assertion period (10 ns) will make the flip-flop function as expected. When T is asserted, the flip-flop toggles on the clock. When T is not asserted, the flip-flop does not toggle. When the clock assertion period is this short, the clock is deasserted before the state changes, and only one transition occurs for each clock pulse. A problem with this type of operation is that the clock must have a short asserted period. Usually the clock period is not short, but circuitry to form short pulses at either the leading edge or the trailing edge of the clock could be included in the flip-flop, making it an edge clocked (or *edge triggered*) device. Figure 8.4-7 shows a possible circuit for producing a short spike of voltage at each leading edge of a clock to make a *positive edge triggered* device. (The circuit also produces a short negative spike at each trailing edge of the clock and, with an inverter, can be used for a *negative edge triggered* device.)

Edge triggered flip-flops are the most common type of flip-flop. However, they are usually designed not with short clock pulses, but as asynchronous state machines with more than two states, and are made to appear externally as if they had only two states. Asynchronous state machines are discussed later in the book.

8.5 LEVEL CLOCKED JK FLIP-FLOP

In this section we'll design a level clocked JK flip-flop using a basic SET-RESET memory cell. A level clocked JK flip-flop has 3 inputs—a clock input and 2 inputs called J and K. When the clock is asserted, the flip-flop changes state, depending on J if the present state is *not asserted* and depending on K if the present state is *asserted.* When the present state is *not asserted*, the next state is the same as J. When the present state is *asserted*, the next state is the complement of K.

The steps in the design of this flip-flop are the same steps we took in designing

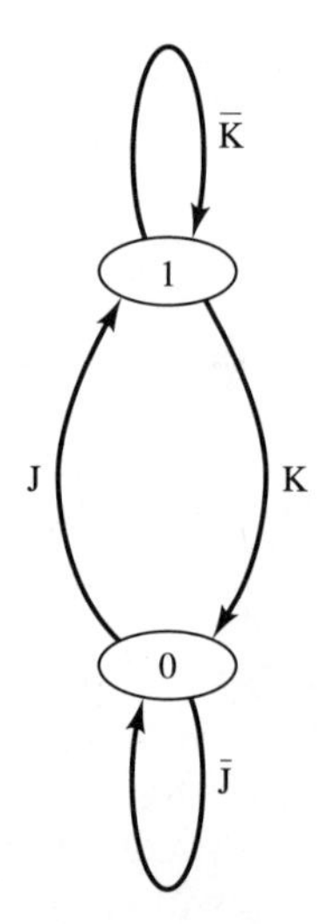

Figure 8.5-1 State transition diagram.

the D and T flip-flops. We've already done the first step—a clear word description. Now, from this word description, we develop the state transition diagram (Fig. 8.5-1) and present state–next state table (Table 8.5-1). At this point, the Karnaugh map for next-state transitions (clock asserted) (Fig. 8.5-2) allows us to find a useful logic expression that describes the transition of the JK flip-flop:

$$Q_{n+1} = J\overline{Q}_n + \overline{K}Q_n$$

although we do not need it for our design.

The next step in the design is to add the next-state code to the present state–next state table. To do this, we consider each present state–next state transition and search the excitation table to determine the memory cell inputs (SET and RESET) that cause the desired transition. Table 8.5-2 is the present state–next state table with the next-state code added. Included is the excitation table for the SET-RESET memory cell we used to determine the next-state code.

Now that we've found the next-state code, we make Karnaugh maps for the memory inputs and design the logic circuits for the next-state decoder.

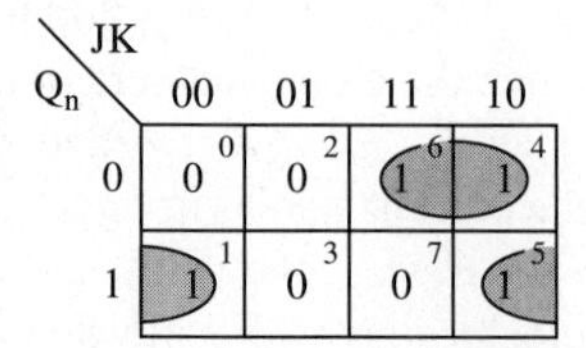

Figure 8.5-2 Next-state map.

Table 8.5-1 Present State–Next State Table

Ck	J	K	Q_n	Q_{n+1}
0	0	0	0	0
0	0	0	1	1
0	0	1	0	0
0	0	1	1	1
0	1	0	0	0
0	1	0	1	1
0	1	1	0	0
0	1	1	1	1
1	0	0	0	0
1	0	0	1	1
1	0	1	0	0
1	0	1	1	0
1	1	0	0	1
1	1	0	1	1
1	1	1	0	1
1	1	1	1	0

Table 8.5-2 Present State–Next State Table and Next-State Code with the Excitation Table for the SET-RESET Memory Cell

Ck	J	K	Q_n	Q_{n+1}	SET	RESET
0	0	0	0	0	0	x
0	0	0	1	1	x	0
0	0	1	0	0	0	x
0	0	1	1	1	x	0
0	1	0	0	0	0	x
0	1	0	1	1	x	0
0	1	1	0	0	0	x
0	1	1	1	1	x	0
1	0	0	0	0	0	x
1	0	0	1	1	x	0
1	0	1	0	0	0	x
1	0	1	1	0	0	1
1	1	0	0	1	1	0
1	1	0	1	1	x	0
1	1	1	0	1	1	0
1	1	1	1	0	0	1

Memory Cell Excitation Table

Q_n	Q_{n+1}	SET	RESET
0	0	0	x
0	1	1	0
1	0	0	1
1	1	x	0

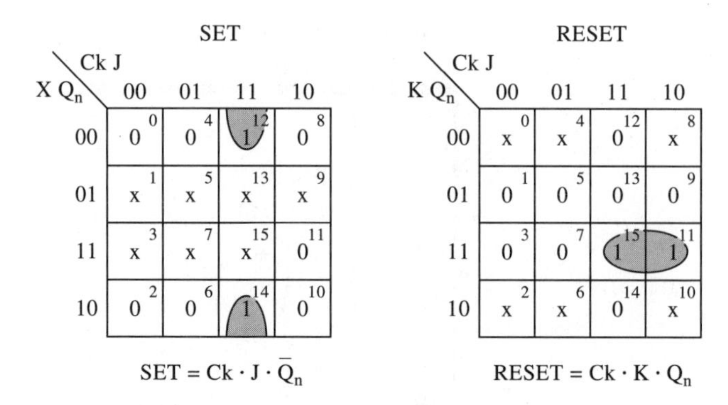

Figure 8.5-3 Karnaugh maps for the inputs to the memory cell.

Figure 8.5-3 shows the maps and simplified logic expressions for the memory inputs.

The last step is to put the next-state decoder logic circuit and the memory cell together in the final design. Figure 8.5-4 shows a final design that uses positive logic *and* gates for the next-state decoder logic circuits and *nor* gates for the SET-RESET memory cell.

We can check our design by simulating the circuit of Figure 8.5-4. Figure 8.5-5 shows the results of such a simulation. Notice that the flip-flop behaves according to the design requirements except that it toggles repeatedly when J and K are both asserted. Because of this toggling, the level clocked JK flip-flop, like the level clocked T flip-flop, is not a useful device. The reason for the toggling can easily be seen from the logic expression for the next-state transition with J = 1 and K = 1:

$$Q_{n+1} = \overline{Q}_n$$

This means that if the present state is *asserted*, it will change to *not asserted*; but then the present state will be *not asserted*, so it will change to *asserted*. This unfortunate toggling will continue until one of the inputs changes. The only way we

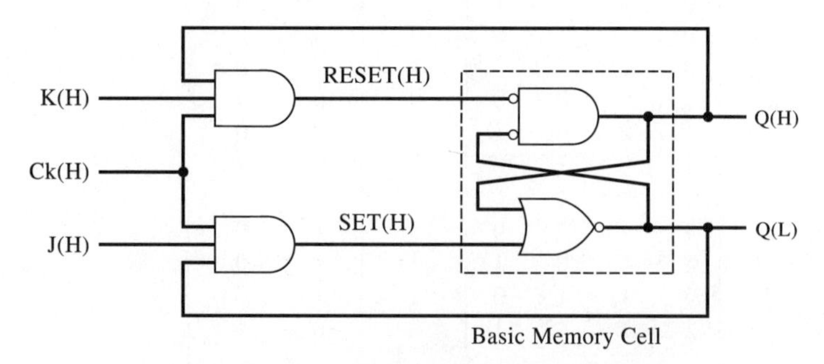

Figure 8.5-4 Level clocked JK flip-flop design.

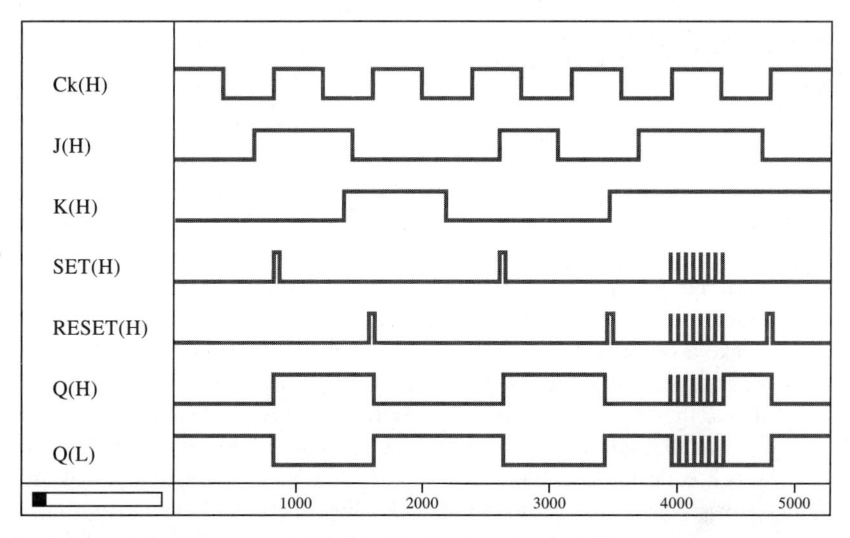

Figure 8.5-5 Simulation results for the JK flip-flop circuit in Figure 8.5-4.

can make the JK flip-flop useful is to make it an edge triggered device. We can do that by making the clock short, as discussed in the section on T flip-flops, or by making an asynchronous state machine with more than two states. Usually, JK flip-flops are asynchronous machines with more than two states. Figure 8.5-6 shows simulation results for a short clock. Notice that this JK flip-flop, which is essentially edge triggered, behaves quite differently from the one in Figure 8.5-5.

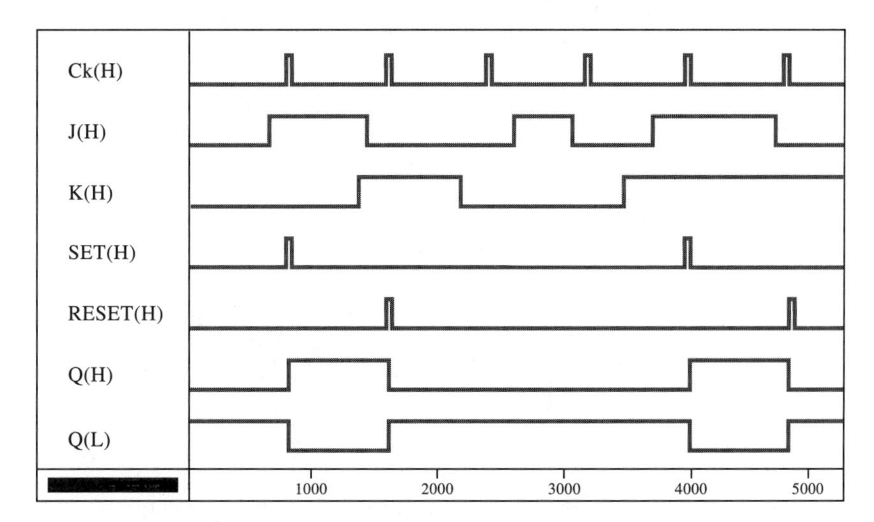

Figure 8.5-6 Simulation for a short clock. This is essentially edge triggered.

8.6 COMMERCIAL FLIP-FLOPS

Edge triggered D and JK flip-flops are both available in a number of different commercial versions. T flip-flops are not available commercially, but a JK flip-flop can easily be changed to a T (see Section 8.9). Commercial flip-flops are usually asynchronous state machines that appear externally to have two states, which change synchronously with the clock edge, but internally have more than two states. We have not yet provided the tools to complete the design of this type of state machine, but it's easy to see how to make a state transition diagram for it. The objective is to make a machine that will change state as soon as the clock is asserted but will not change state again until the clock is first deasserted. For both the D and JK flip-flops, this actually requires four states, as shown in Figures 8.6-1 and 8.6-2. The idea is that after each transition, the machine is placed in a holding state until the clock is deasserted. It's then placed in an armed state and can make a transition as soon as the clock is again asserted. Thus, there are actually two internal machine states for the "0" external state of the machine, and two internal states for the "1" external state of the machine. In each case, one state is the holding state and the other is the armed state, as shown in the diagrams. We'll show how to complete the design of these machines later in the text.

Figure 8.6-3 shows circuit symbols for some examples of commercial TTL flip-flops. The symbol for a flip-flop is a rectangle with the inputs and outputs labeled. Since the device is edge triggered, the clock input is given a special triangle symbol. The bubble and triangle indicate a negative edge trigger. All three example flip-flops

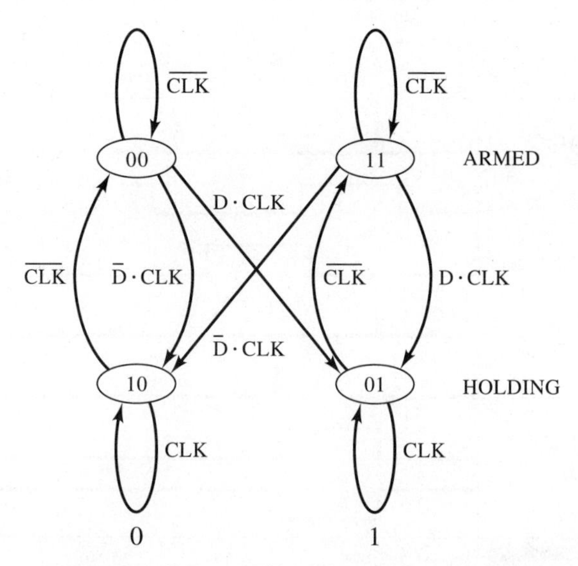

Figure 8.6-1 State transition diagram for a possible edge triggered D flip-flop.

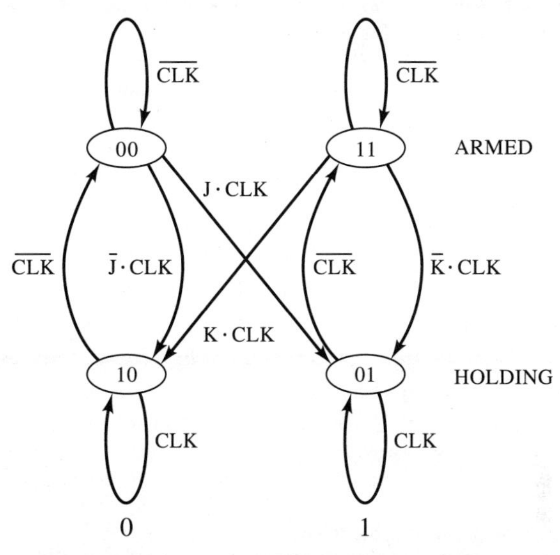

Figure 8.6-2 State transition diagram for a possible edge triggered JK flip-flop.

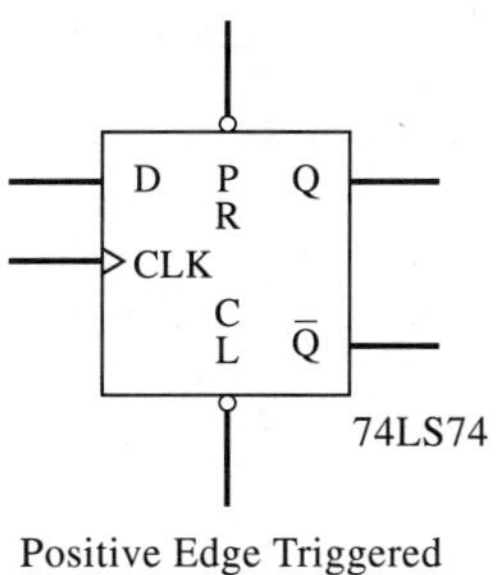

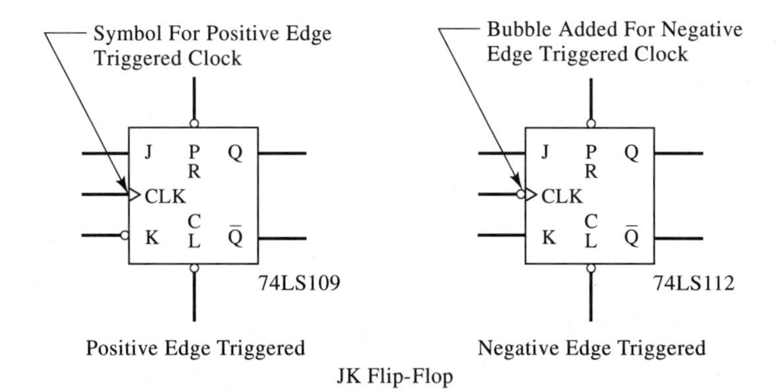

Figure 8.6-3 Example commercial flip-flops.

Table 8.6-1 D Flip-Flop Function Table

Mode Select-Function Table

Operating Mode	Inputs				Outputs	
	PR	CL	CLK	D	Q	$\overline{Q}$
Asynchronous preset	L	H	x	x	H	L
Asynchronous clear	H	L	x	x	L	H
Undetermined*	L	L	x	x	H	H
Load "1" (set)	H	H	↑	h	H	L
Load "0" (reset)	H	H	↑	l	L	H

H = HIGH voltage level steady state.
h = HIGH voltage level one setup time prior to the LOW-to-HIGH clock transition.
L = LOW voltage level steady state.
l = LOW voltage level one setup time prior to the LOW-to-HIGH clock transition.
x = Don't care.
↑ = LOW-to-HIGH clock transition.
**Note:* Both outputs will be HIGH while both PR and CL are LOW, but the output states are unpredictable if PR and CL go HIGH simultaneously.

Table 8.6-2 JK Flip-Flop Function Table

Mode Select-Function Table

Operating Mode	Inputs					Outputs	
	PR	CL	CLK	J	K	Q	$\overline{Q}$
Asynchronous preset	L	H	x	x	x	H	L
Asynchronous clear	H	L	x	x	x	L	H
Undetermined*	L	L	x	x	x	H	H
Toggle	H	H	↓	h	h	$\overline{q}$	q
Load "1" (set)	H	H	↓	h	l	H	L
Load "0" (reset)	H	H	↓	l	h	L	H
Hold "no change"	H	H	↓	l	l	q	$\overline{q}$

H = HIGH voltage level steady state.
h = HIGH voltage level one setup time prior to the HIGH-to-LOW clock transition.
L = LOW voltage level steady state.
l = LOW voltage level one setup time prior to the HIGH-to-LOW clock transition.
x = Don't care.
q = Lowercase letters indicate the state of the reference output prior to the HIGH-to-LOW clock transition.
↓ = HIGH-to-LOW clock transition.
**Note:* Both outputs will be HIGH while both PR and CL are LOW, but the output states are unpredictable if PR and CL go HIGH simultaneously.

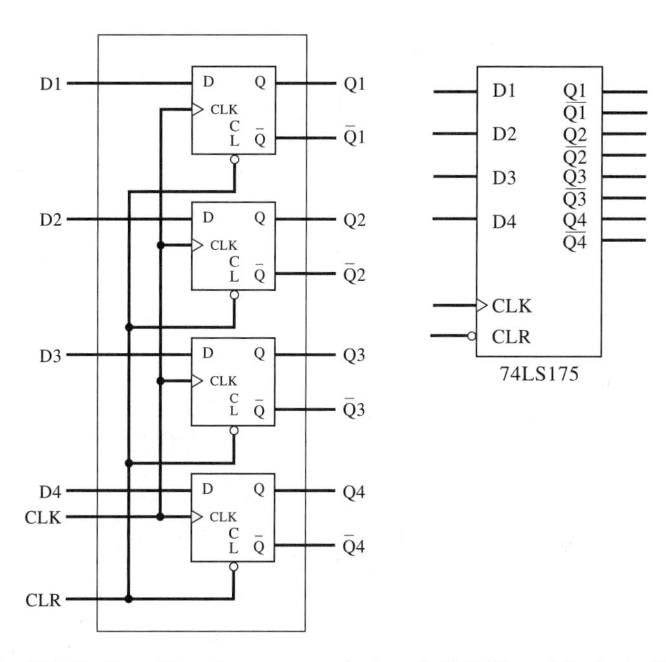

Figure 8.6-4 Quad D flip-flop. Note the common clock and CLEAR and the lack of PRESET.

have asynchronous presets and clears. The preset (PR) sets the state of the flip-flop to *asserted*, independent of the clock. The clear (CL) sets the state of the flip-flop to *not asserted*, independent of the clock. Both PR and CL are asserted low for the sample flip-flops. PR and CL perform the same functions as SET and RESET in a basic memory cell, and they are often called SET and RESET.

Tables 8.6-1 and 8.6-2 are function tables for a positive edge triggered D flip-flop and a negative edge triggered JK flip-flop. They are similar to the tables found in a data book. Except for the added PR and CL and edge triggered clock, these flip-flops perform the same functions as the level clocked D and JK flip-flops described in the last two sections. The function table for a JK flip-flop is actually the same as the foregoing description of the operation of a JK flip-flop, although that may not be evident. Here we have a toggle, a set, a reset, and a hold function. In the foregoing description we said that the clock causes the flip-flop to change state, depending on J if the present state is *not asserted* and depending on K if the present state is *asserted*. When the present state is *not asserted*, the next state is the same as J. When the present state is *asserted*, the next state is the complement of K. You'll find that this is consistent with the function table. The setup time referred to in the table footnotes is a timing constraint on the flip-flops. We'll look at timing in Section 8.7.

The sample D and JK flip-flops are packaged two flip-flops to a chip. A number of other dual JK flip-flops are available in TTL logic; some do not have presets, and some do not have clears. Other D flip-flops packaged with four, six, and eight flip-flops to a chip are also available. Figure 8.6-4 shows an example of a quad D flip-

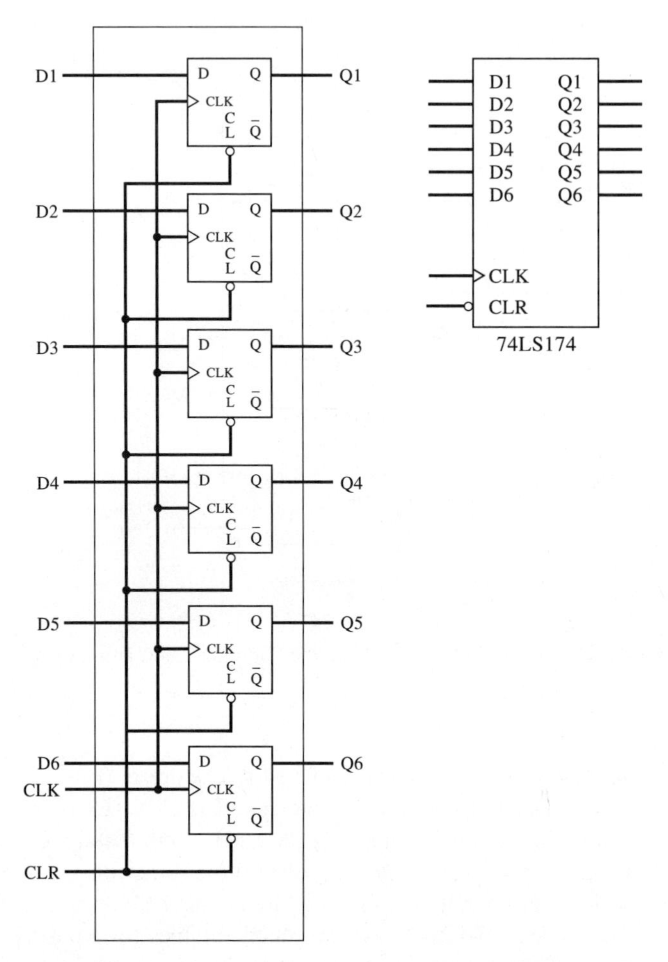

Figure 8.6-5 Hex D flip-flop. Note the common clock and CLEAR and the lack of PRESET. Note also that only the Q is output.

flop; Figure 8.6-5 shows an example of a hex D flip-flop; and Figure 8.6-6 shows an example of an octal D flip-flop with tristate output. In each figure, the internal logic of the chip is shown along with the circuit symbol. These multiple D flip-flops do not have some of the features that are present in the dual D flip-flop. All of the multiple flip-flops have common clocks. The quad and hex flip-flop have common clears, but no presets. The octal flip-flop has neither clears nor presets. Both the hex and octal flip-flops have only Q outputs (no $\overline{Q}$). The octal flip-flop shown has a tristate driver for each output (see Fig. 8.6-6). The output drivers are enabled from a common enable called the output control ($\overline{OC}$).

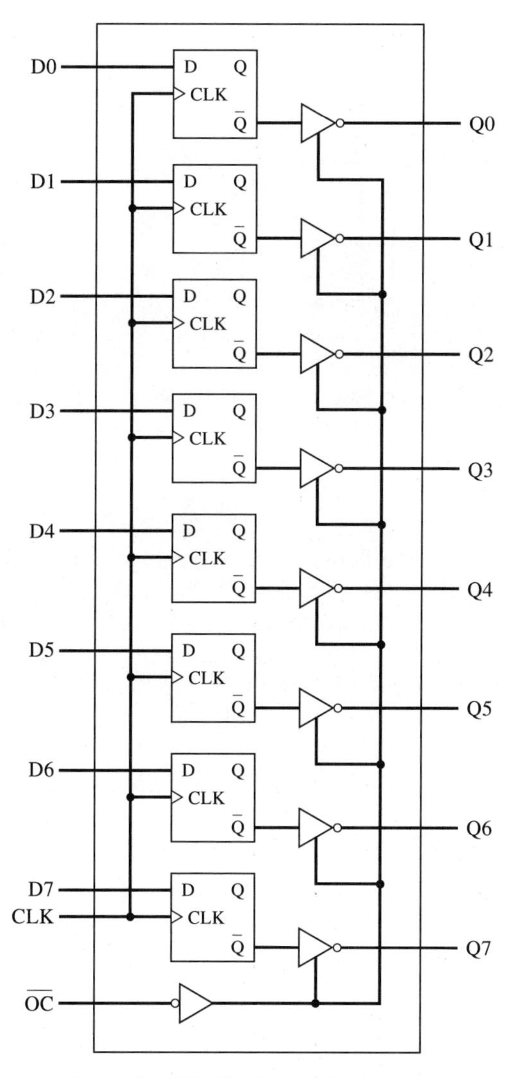

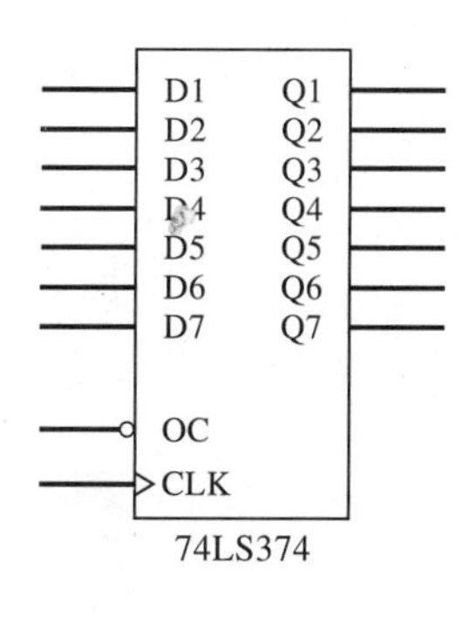

Figure 8.6-6 Octal D flip-flop with tristate output. Note the common clock and output control (OC) and the lack of both PRESET and CLEAR. Note also that only Q is output.

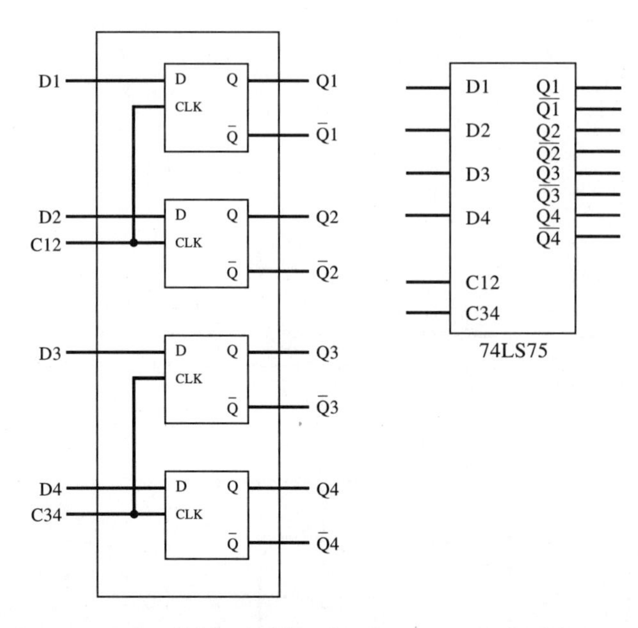

Figure 8.6-7 Quad D latch. Note that there are two clock inputs, one for the first two flip-flops and one for the last two.

Multiple, level clocked D flip-flops are also available commercially. Level clocked D flip-flops are called *latches*. Figure 8.6-7 shows an example of a quad D latch, and Figure 8.6-8 shows an example of an octal D latch with tristate output. The quad latch has two clock inputs, one for each pair of flip-flops (C12 and C34). The octal latch has a common clock (G) for all the flip-flops. Some latches have clears, but the examples shown do not. The octal latch shown has tristate output drivers just like the octal D flip-flop.

In addition to the representative samples illustrated here, a number of other quad, hex, and octal flip-flops and latches are available, with and without tristate output drivers and with different combinations of clocks, presets, clears, and outputs. Data books should be consulted to determine the exact devices available.

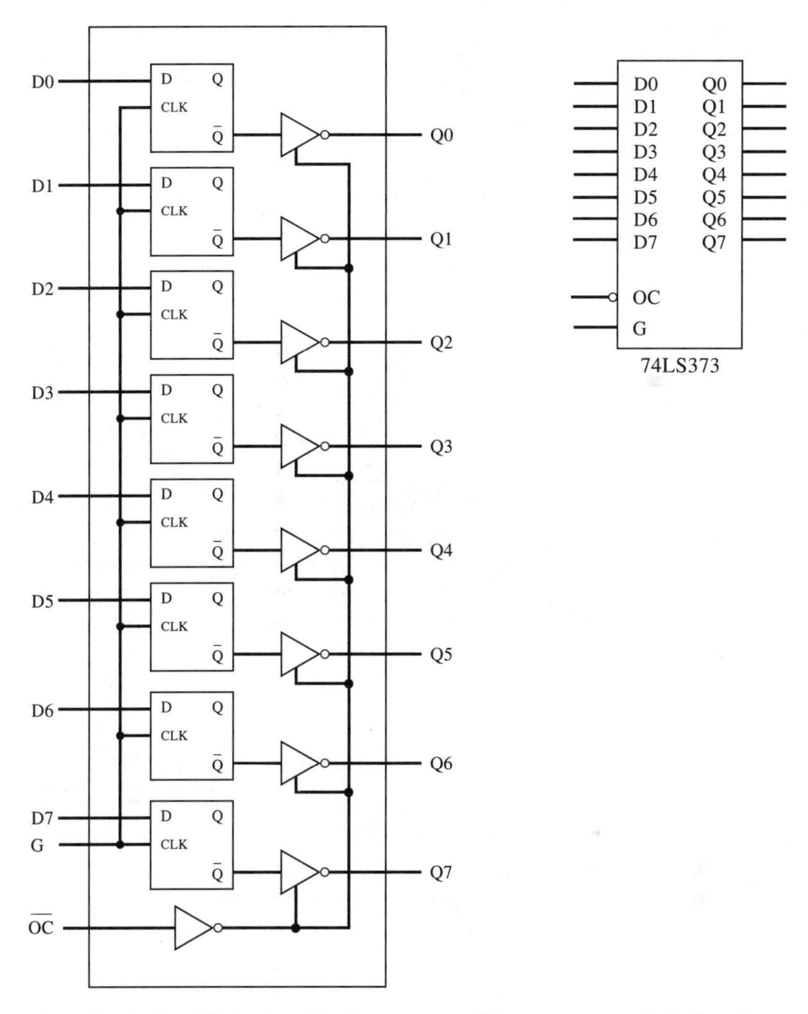

Figure 8.6-8 Octal D latch with tristate output. Note the common clock and output control. Note also that only Q is output.

8.7 TIMING CONSTRAINTS ON EDGE TRIGGERED FLIP-FLOPS

Figure 8.7-1 shows the important timing for an edge triggered flip-flop. The critical times are the setup times [$t_s(L)$ and $t_s(H)$] and the hold times [$t_h(L)$ and $t_h(H)$]. Two setup times and two hold times are shown in the figure because two different transitions of D, J, or K (low to high and high to low) can occur, and the setup and hold times may be different depending on the transition. The setup time is necessary because the internal logic of the flip-flop has propagation delays associated with gates and other logic elements. If D, J, or K changes at a time less than the hold time before the clock edge, the transition is unpredictable. For normal TTL logic,

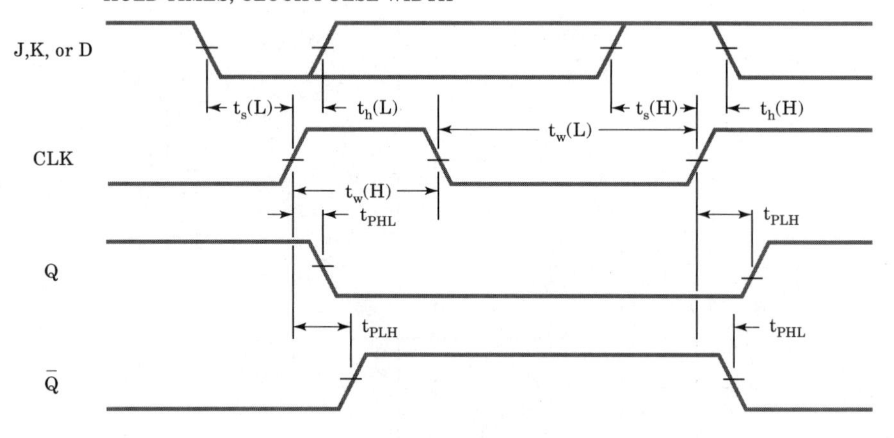

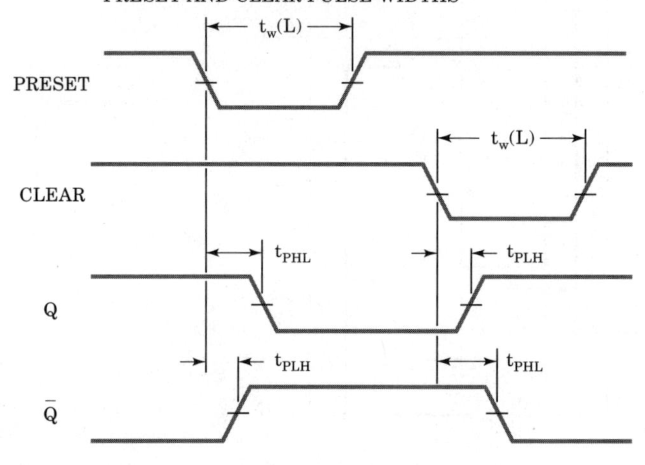

Figure 8.7-1 Timing diagram for an edge triggered flip-flop.

setup times are typically 10s of nanoseconds. D, J, or K also must not change for a certain time—the hold time—after the clock edge, or the transition is unpredictable. The hold time is usually zero or near zero.

In any actual flip-flop, there is a time delay between the clock edge and the outputs Q and $\overline{Q}$, called a propagation delay. There are two different delays, depending on whether the output changes from high to low (t_{PHL}) or from low to high (t_{PLH}). Propagation delays are typically 10s of nanoseconds for normal logic devices. Flip-flops have speed limitations that are specified by mininum constraints on the high clock pulse width [$t_w(H)$] and the low clock pulse width [$t_w(L)$].

The timing for PR and CL is shown in the same figure. The preset and clear must have a minimum pulse width [$t_w(L)$], and there are propagation delays to the output changes (t_{PHL} and t_{PLH}).

8.8 PROGRAMMABLE GATE ARRAY FLIP-FLOPS AND LATCHES

In our discussion of combinational logic circuits we've looked at the large variety of gates available in a programmable gate array. Equally large varieties of flip-flops and latches are available. Figure 8.8-1 shows the flip-flops available in one type of

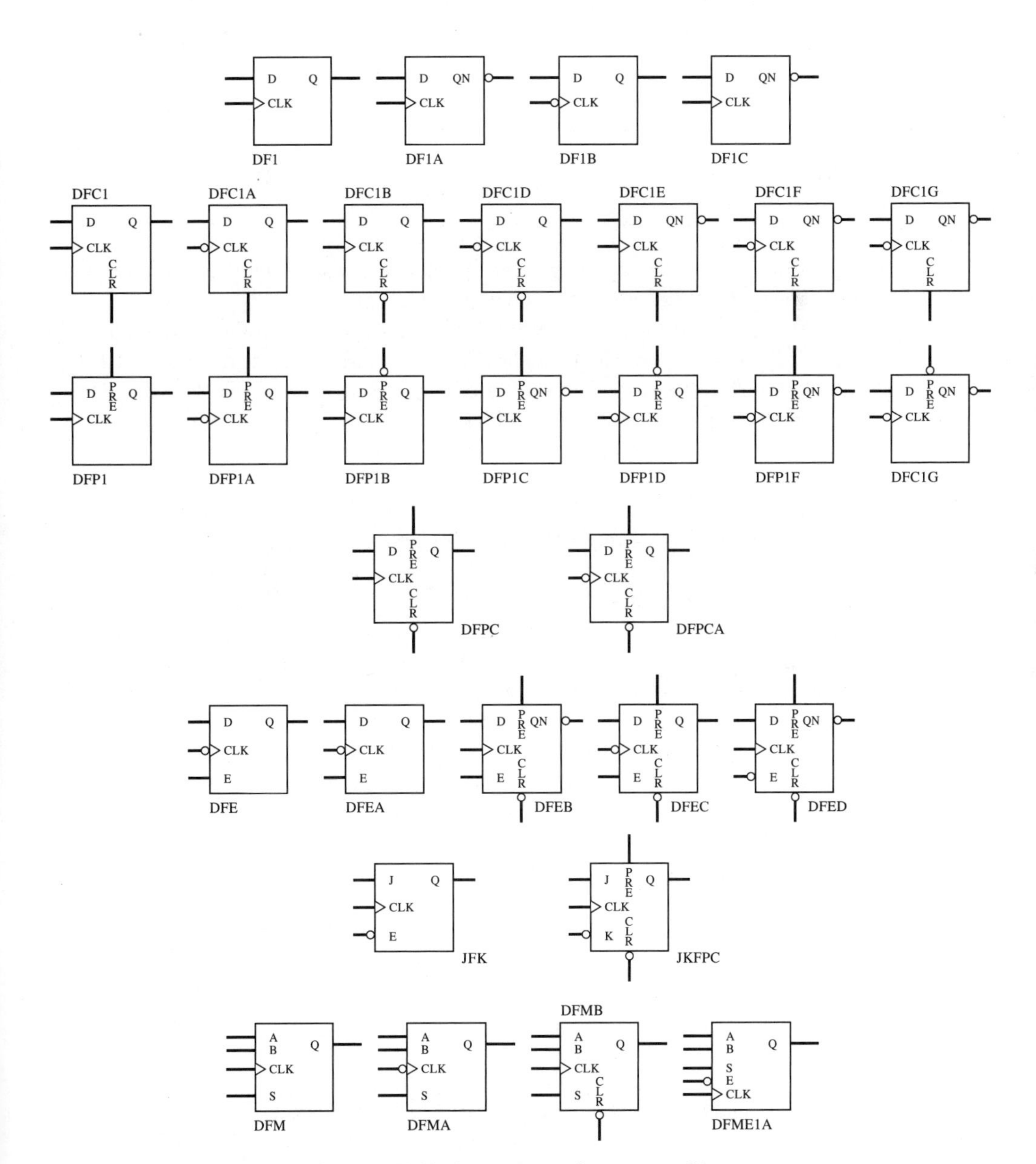

Figure 8.8-1 Flip-flops available in one type of programmable gate array.

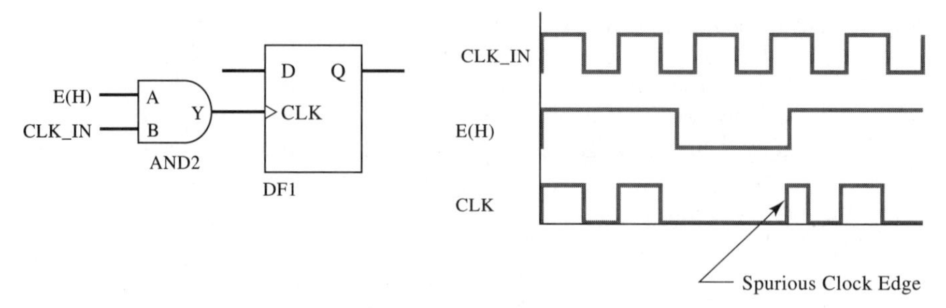

Figure 8.8-2 A gated clock can cause a spurious clock edge.

gate array. A flip-flop in this gate array requires more of the gate array resources than a gate does: two modules as compared to one.

All but two of the available flip-flops are D types. With the variety of D flip-flops and gates available in a gate array, there is little need for the flexibility of the JK flip-flop. Flip-flops exist with almost all possible combinations of assertion levels for the clock, preset, and clear. Although each flip-flop has only one output, this output is available in either assertion level. Because of this, and because gates with all combinations of mixed inputs are also available, we'll find that the single output does not place any limitation on the use of these flip-flops in designs.

Two enhancements that are available in these flip-flops, but that are not available in discrete flip-flops, are an enable (E) and multiplexed inputs. A flip-flop with an E input is an enabled flip-flop. The enable is a clock enable. If the enable is asserted, the clock will be active and the flip-flop will change state on the appropriate clock edge. If the enable is not asserted, the flip-flop will not change state on the clock edge, although it can be made to change state using the preset or clear. The enable is more than a gate that turns the clock on. If the clock input to a flip-flop is gated as shown in Figure 8.8-2, changes in E can cause the flip-flop to be clocked as shown in the timing diagram in the figure. This kind of spurious clocking will not occur in the enabled flip-flop—only clock edges will clock the flip-flop.

The flip-flops in the last row of Figure 8.8-1 have multiplexed inputs. Multiplexed input allows the input to be switched between A and B, using the select input (S). If S is asserted, input B is selected. If S is not asserted, input A is selected. Both the enable and multiplexed inputs increase the flexibility of the flip-flop without adding to the number of programmed modules in the array. Perhaps the most common use of multiplexed inputs is in adding a gate to the input of the flip-flop, as shown in Figure 8.8-3.

Figure 8.8-4 displays the variety of latches available in one type of programmable array. These latches, like the flip-flops, have only one output, although latches with either assertion level output are available. Latches are available with both clock assertion levels, with clears (only asserted low), with both assertion levels of enables (E), and with multiplexed inputs. The enable in a latch, like the enable in a flip-flop,

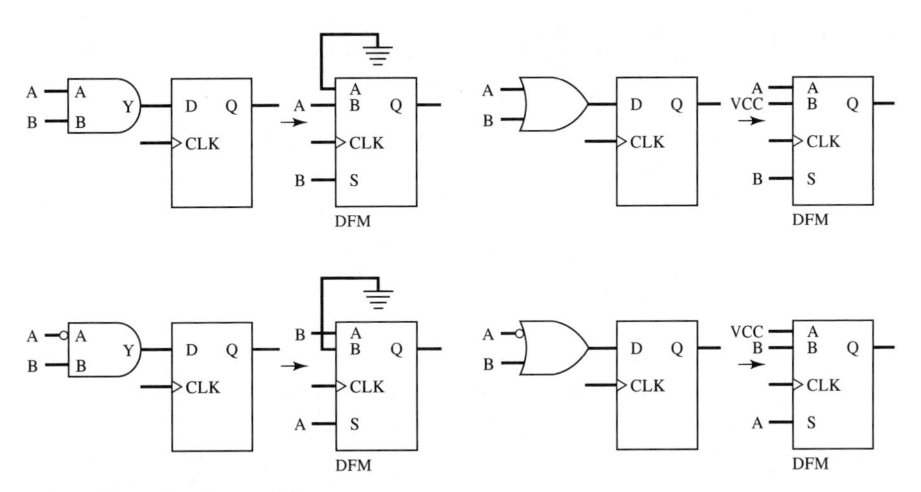

Figure 8.8-3 Flip-flops with input gates.

enables the clock. The multiplexed input to the latch functions in the same way as the multiplexed input to the flip-flop. For ease of design, both programmable gate array flip-flops and latches can be grouped together into software packages (called soft modules) much like the arrays found in discrete logic.

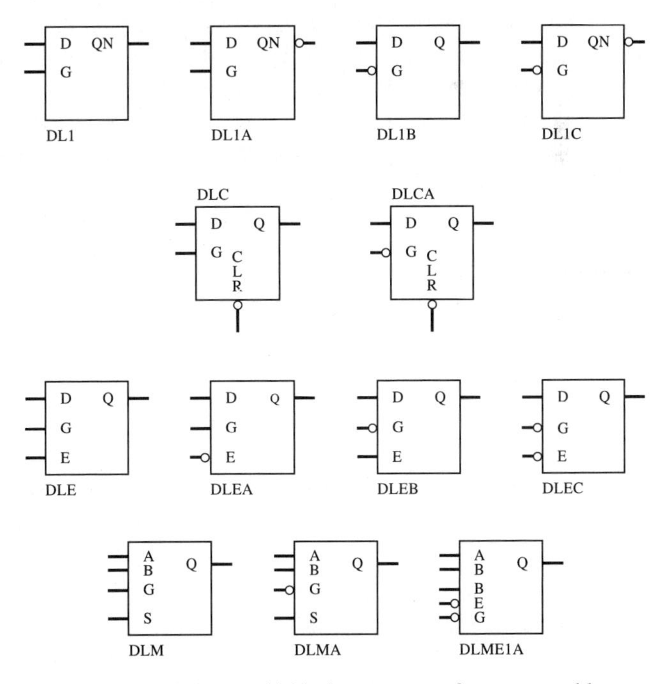

Figure 8.8-4 Latches available in one type of programmable gate array.

8.9 DESIGNING AN EDGE TRIGGERED T FLIP-FLOP

An edge triggered T flip-flop can easily be made from an edge triggered D or JK flip-flop. The making of a T flip-flop from a D or JK flip-flop is a design problem similar to those we've just solved. In the designs of flip-flops that appeared earlier in this chapter, we used the basic SET-RESET memory cell for the memory element. Now we'll use an edge triggered D or JK flip-flop as the memory element, which actually makes the design easier. The clock function is already taken care of in the memory, so it does not need to be considered as an input for the next-state code decoder logic design.

First consider the design of a T flip-flop using a D flip-flop as the memory element. The first step in the design is to describe the flip-flop. We've already described a T flip-flop: it changes state (toggles) on the clock (edge) if T is asserted, but does not change state if T is not asserted. Table 8.9-1 is the present state–next state table for a T flip-flop. This table does not include a clock, because the clock function is handled by the memory element of the state machine (edge triggered D flip-flop). *If the clock is included as an input to the next-state code decoder logic, the design will not function because of timing constraints.*

Now that we have the present state–next state table, we need to add the next-state code. Remember that to do this we need the excitation table for the memory. The excitation table for a D flip-flop is very simple; see Table 8.9-2. D is always equal to the next state, independent of the present state. This means that if we have a state machine with D flip-flop memory, we do not have to add a next-state code to the present state–next state table, because the next-state code is the same as the next state. For our design D, the next-state code is simple Q_{n+1}.

The next step is to design the logic circuits to produce the next-state code from the inputs and present state. In this design, the input is T and the present state is Q_n. Figure 8.9-1 is the Karnaugh map for the next-state decoder logic design. Remember that the next-state code is the same as the next state. The figure also includes the next-state decoder logic expression, which allows us to design the next-state decoder circuit. Figure 8.9-2 shows the final design using positive logic *nand* gates for the

Table 8.9-1 Present State–Next State Table

T	Q_n	Q_{n+1}
0	0	0
0	1	1
1	0	1
1	1	0

Table 8.9-2 Excitation Table

Q_n	Q_{n+1}	D
x	0	0
x	1	1

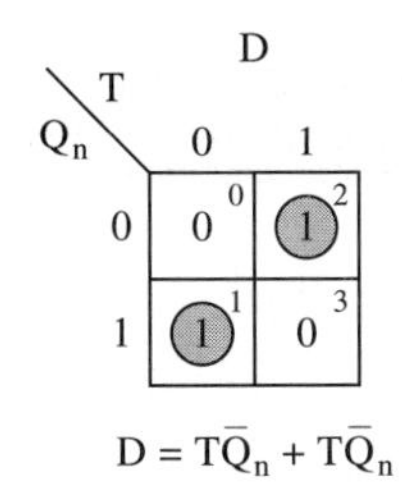

Figure 8.9-1 Decoder map.

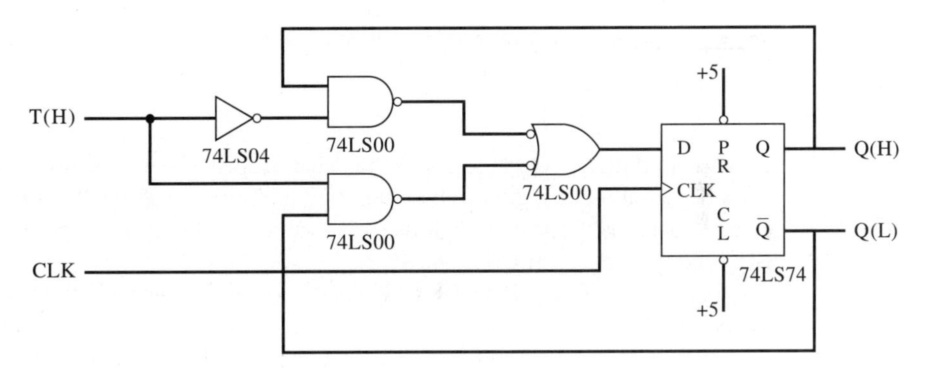

Figure 8.9-2 T flip-flop with D flip-flop memory.

next-state decoder circuits. Notice that we can use the PR and CL for the D flip-flop that functions as the state machine memory, as PP and CL for the T flip-flop, or we can tie them high as shown in the figure. Tying them high disables them, so the D flip-flop is never preset or cleared.

To test our design, we simulate the circuit of Figure 8.9-2; the resulting output waveforms are shown in Figure 8.9-3. The input we've chosen for the simulation is

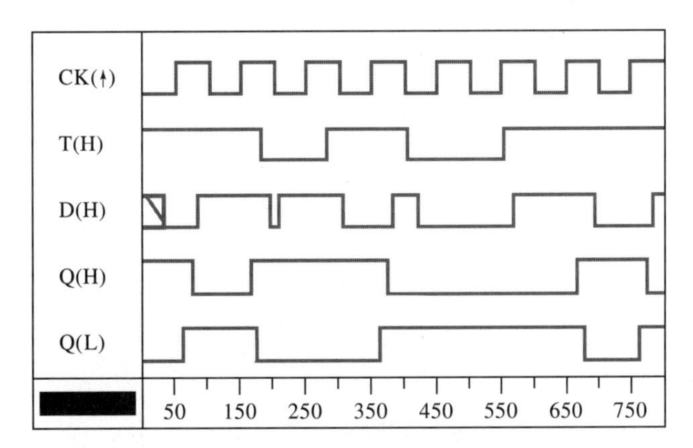

Figure 8.9-3 Simulation results for the circuit of Figure 8.9-2.

Table 8.9-3 Present State–Next State Table for T Flip-Flop, with the Excitation Table for the JK Flip-Flop

T	Q_n	Q_{n+1}	J	K
0	0	0	0	x
0	1	1	x	0
1	0	1	1	x
1	1	0	x	1

Excitation Table

Q_n	Q_{n+1}	J	K
0	0	0	x
0	1	1	x
1	0	x	1
1	1	x	0

somewhat arbitrary; we chose it to show what happens when T is either asserted or not asserted at a clock edge. The time scale of the figure is small enough and the clock transitions are fast enough that it's easy to see the delay between the clock edge and the change in Q(H) and Q(L).

To design a T flip-flop with a JK flip-flop memory, we follow the same steps as for the D flip-flop memory until the present state–next state table is produced. Now, unlike the design with D memory we must add the JK next-state code. Table 8.9-3 is the present state–next state table with the next-state code for JK flip-flop memory added. Also included is the JK excitation table used in finding the next-state code. Figure 8.9-4 is the Karnaugh maps for J and K, which we produce from the present state–next state table and use for designing the next-state logic decoders. As we can see, the logic expressions for J and K are particularly simple. No logic gates are needed, only a wire to tie the J and K inputs together. Figure 8.9-5 shows the design of the T flip-flop using JK flip-flop memory. In this design we've again tied both PR and CL high so that they are disabled. The clock is inverted to make the flip-flop positive edge triggererd. Notice that the design using the JK flip-flop for memory is simpler than the design using D flip-flop for memory. This is usually true in any state machine design.

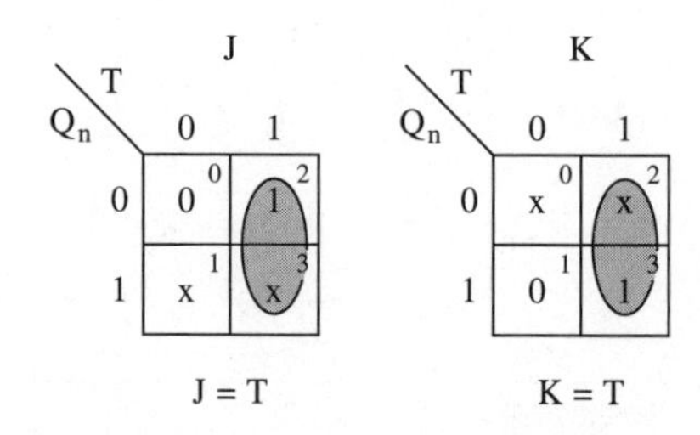

Figure 8.9-4 Karnaugh maps for next-state decoder.

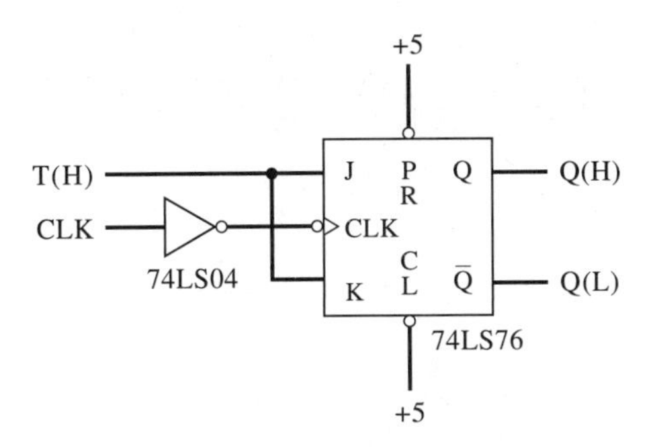

Figure 8.9-5 T flip-flop with JK flip-flop memory.

Now we should address the timing constraints for our T flip-flop designs. If we add logic before the memory flip-flop, we always increase the setup time and decrease the hold time compared to the setup time and hold time of the memory flip-flop. This is because the logic has a propagation delay. To find the new setup time, we must add the *maximum* propagation delay through the logic gates of the decoder logic circuit to the setup time of the memory flip-flop; and to find the new hold time, we must subtract the *minimum* propagation delay from the hold time of the memory flip-flop. If we assume the setup time for the D flip-flop used as memory in the first design to be 20 ns, the gate delay for the *nand* gates to be 10 ns, and the delay for the inverter to be 10 ns, the setup time for the T flip-flop is then

$$\begin{aligned} t_s &= 2(10) + 10 + 20 \\ &= 50 \text{ ns} \end{aligned}$$

where we have two gate delays and an inverter delay in the decoder logic, and the setup time of the D flip-flop. If we assume the hold time of the D flip-flop is 0 ns, then the hold time for the T flip-flop is

$$t_h = 0 - 2(10) = -20 \text{ ns}$$

Here we use only the gate delays (the minimum propagation delay) and subtract. The negative sign indicates that the input can change 20 ns before the clock edge. In the design using JK flip-flop memory, there are no logic gates in the decoder logic. The setup and hold times are therefore the same for the T flip-flop as for the JK flip-flop used as memory except for the delay in the clock from the inverter. This delay decreases the setup time and increases the hold time.

As a check on the design of the T flip-flops with JK flip-flop memory, we simulate the circuit of Figure 8.9-5; the resulting waveforms arc shown in Figure 8.9-6. The input is the same as was used in the simulation of the T flip-flop with D flip-flop memory.

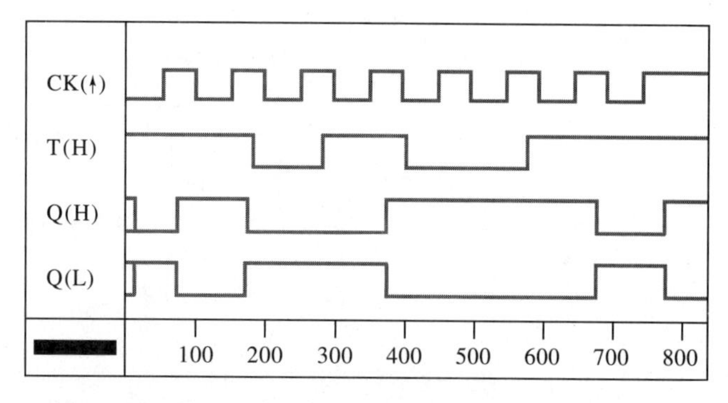

Figure 8.9-6 Simulation results for the circuit of Figure 8.9-5.

8.10 CHANGING ONE TYPE OF FLIP-FLOP TO ANOTHER TYPE

The problem of changing from one type of edge triggered flip-flop to another type is really the design of the second type of flip-flop using the first type as memory. This is an easy design, because the clocking problem is already taken care of in the memory flip-flop. All we have to do is to follow the same steps we followed in Section 8.9 to design a T flip-flop. We can, in fact, design any kind of edge triggered flip-flop we can dream up (and describe concisely) using the design technique we used for designing the edge triggered T flip-flop.

Example 8.10-1

Design two synchronous, positive edge triggered SET-RESET flip-flops with priority SET—one of them with a D flip-flop memory and one of them with a JK flip-flop memory. That is, design flip-flops that function in the same way as the SET-RESET cells we've been using as memory cells, but design them so that they're synchronous flip-flops that change state on the positive clock edge. Also, the SET-RESET cell has an ambiguity when both SET and RESET are asserted; resolve this ambiguity by making the machine do a SET function when both SET and RESET are asserted (priority SET).

The first step in the design is to describe the flip-flop. This step is the same for both designs, and we have already done it. Remember that a normal asynchronous SET-RESET flip-flop either makes a transition into or stays in the 1 state when SET is asserted, and either makes a transition into or stays in the 0 state when RESET is asserted. If we add the priority SET to these conditions, we get the following present state–next state table—the second design step, which is also the same for both designs. This table does not include a clock, because the clock function is handled by the memory element of the state machine (the edge triggered D or JK flip-flop). If the clock is included as an input to the next-state code decoder logic, the design will not function because of timing constraints.

SET	RESET	Q_n	Q_{n+1}
0	0	0	0
0	0	1	1
0	1	0	0
0	1	1	0
1	0	0	1
1	0	1	1
1	1	0	1
1	1	1	1

Now consider the design using a D flip-flop as the memory element. We have the present state–next state table, so it seems we should add the next-state code. In Section 8.9, however, we learned that the next-state code for D flip-flop memory is the same as the next state; therefore, we do not need to add the next-state code to the table, because we already have it ($D = Q_{n+1}$). The next step, then, is to design the logic circuits to produce the next-state code from the inputs and present state. In this design, the inputs are SET and RESET and the present state is Q_n. The Karnaugh map for the next-state decoder logic design follows. The next-state decoder logic expression appears below the map. This expression allows us to design the next-state decoder circuit.

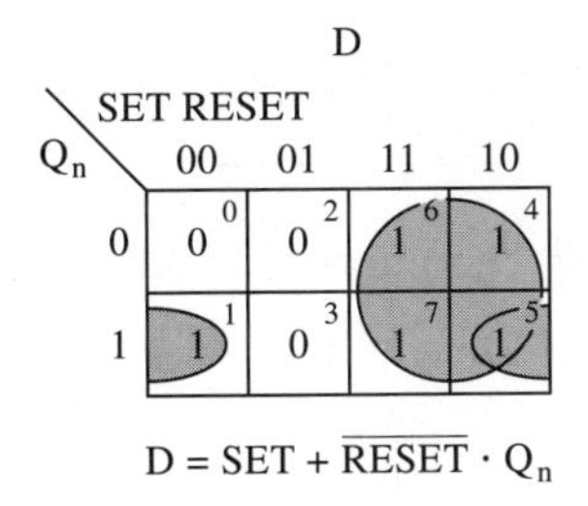

$$D = SET + \overline{RESET} \cdot Q_n$$

Following is the final design, using positive logic *nand* gates for the next-state decoder circuits. Notice that we can use PR and CL for the D flip-flop that functions as the state machine memory for the SET-RESET flip-flop, or we can tie them high as shown in the figure. Tying them high disables them, so the D flip-flop is never preset or cleared.

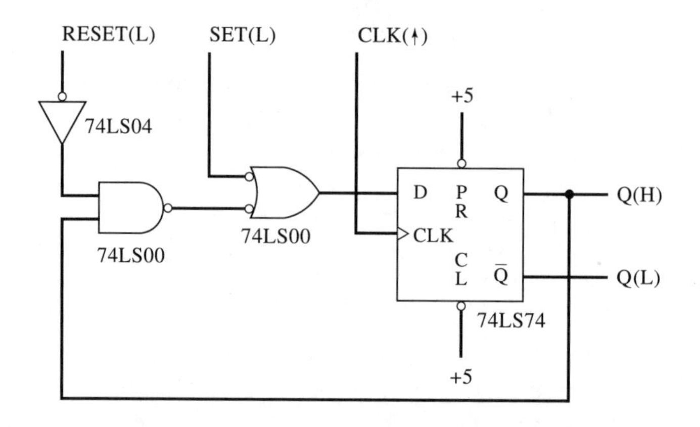

To check our design, we simulate the circuit, with the following results. The timing scale is such that we can readily see the propagation delays from the positive-going clock edges to the change in Q. Notice in particular the priority feature when both SET and RESET are asserted.

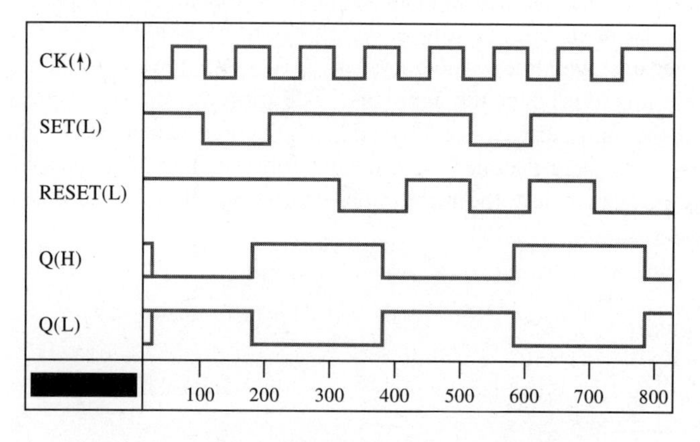

To design the SET-RESET flip-flop with a JK flip-flop memory, we need to add the JK next-state code to the present state–next state table, as follows. Also shown is the JK excitation table used in finding the next-state code.

SET	RESET	Q_n	Q_{n+1}	J	K
0	0	0	0	0	x
0	0	1	1	x	0
0	1	0	0	0	x
0	1	1	0	x	1
1	0	0	1	1	x
1	0	1	1	x	0
1	1	0	1	1	x
1	1	1	1	x	0

Excitation Table

Q_n	Q_{n+1}	J	K
0	0	0	x
0	1	1	x
1	0	x	1
1	1	x	0

Following are the Karnaugh maps for J and K. They are produced from the present state–next state and next-state code table, and we use them for designing the next-state decoders. Remember that the input variables are SET, RESET, and Q_n. The next-state code decoder logic for J and K appears below the maps.

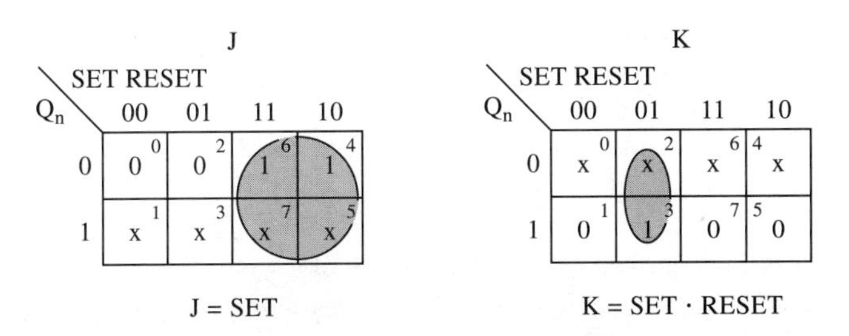

All that remains in the design is to draw the circuits that will implement the logic. The following figure shows a circuit using a positive logic *nand* gate.

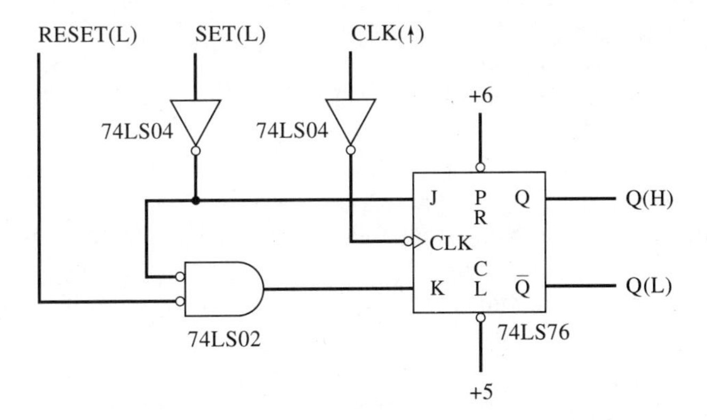

We can check the design by simulating the circuit, with the following output results.

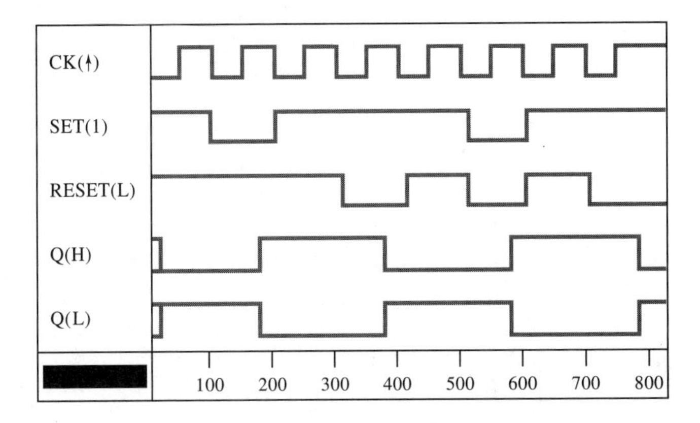

KEY TERMS

Asserted In the true, active, or on state.
Not Asserted In the false, inactive, or off state.
D Flip-flop A flip-flop in which the state is set equal to the input D, on the clock.
Enable An input to a flip-flop which must be asserted for the flip-flop to clock.
Excitation Table A form of the present state–next state table in which the transition is listed first in the table and then the inputs which cause this transition are listed.
Hold Time The time after the clock an input to flip-flop must be stable in order to cause a proper change in the output.
JK Flip-flop A flip-flop which changes state to J if it is in the not asserted (0) state and to $\overline{K}$ if it is in the asserted (1) state on the clock.
Level Clocked Device A device which changes when the clock is high or low.
Negative Edge-Triggered Device A device which changes on the high to low clock transition.
Positive Edge-Triggered Device A device which changes on the low to high clock transition.
SET Same as asserted.
RESET Same as not-asserted.
SET-RESET Memory Cell or Flip-flop The basic memory cell. SET places the cell in an asserted state and RESET places it in a not-asserted state.
Setup Time The time before the clock an input to a flip-flop must be stable in order to cause a proper change in the output.
Transparent The output immediately following the input except for a propagation delay.
T Flip-flop A flip-flop in which the state toggles on the clock if T is asserted.

EXERCISES

1. The SET-RESET memory cell (flip-flop) has an ambiguity when both SET and RESET are asserted. The ambiguity can be resolved by giving a priority to either SET or RESET. If SET is given precedence, then the resulting flip-flop is called a SET-RESET flip-flop with priority SET. If RESET is given precedence, then the resulting flip-flop is called a SET-RESET flip-flop with priority RESET.
   (a) Design an asynchronous, priority SET, SET-RESET flip-flop using a SET-RESET memory cell. To avoid confusion, use S and R for the input of the flip-flop you are designing, and reserve SET and RESET for the inputs to the memory cell.
   (b) Design an asynchronous, priority RESET, SET-RESET flip-flop using a SET-RESET memory cell. To avoid confusion, use S and R for the input of the flip-flop you are designing, and reserve SET and RESET for the inputs to the memory cell.
2. Exercise 1 describes priority SET and priority RESET SET-RESET flip-flops and asks for the design of an asynchronous flip-flop of each priority. We can also design a level clock flip-flop of each priority.

(a) Design a level clocked, priority SET, SET-RESET flip-flop using a SET-RESET memory cell. To avoid confusion, use S and R for the input of the flip-flop you are designing, and reserve SET and RESET for the inputs to the memory cell. Is the flip-flop you have designed stable?

(b) Design a level clocked, priority RESET, SET-RESET flip-flop using a SET-RESET memory cell. To avoid confusion, use S and R for the input of the flip-flop you are designing, and reserve SET and RESET for the inputs to the memory cell. Is the flip-flop you have designed stable?

3. Design a level clocked LM flip-flop using a SET-RESET memory cell. An LM flip-flop functions as follows. If it's in the RESET (0) state and the inputs are the same, it changes state when the clock is asserted, but if the inputs are different it stays in the RESET state. If it's in the SET (1) state and input L is asserted, it changes state when the clock is asserted; if L is not asserted, it stays in the SET state. The behavior in the SET state is independent of M.

4. Design a level clocked SN flip-flop using a SET-RESET memory cell. An SN flip-flop functions as follows. The next state is set equal to $\overline{N}$ when the clock is asserted, except that when the state is SET (1) and S is asserted, the state always remains SET.

5. Change a positive-edge clocked D flip-flop to a positive-edge clocked JK flip-flop.

6. Change a negative-edge clocked JK flip-flop to a negative-edge clocked D flip-flop.

7. Change a positive-edge clocked T flip-flop to a positive-edge clocked JK flip-flop.

8. Change a negative-edge clocked T flip-flop to a negative-edge clocked D flip-flop.

9. The SET-RESET memory cell (flip-flop) has an ambiguity when both SET and RESET are asserted. The ambiguity can be resolved in two ways. If SET is given precedence, then the resulting flip-flop is called a SET-RESET flip-flop with priority SET. If RESET is given precedence, then the resulting flip-flop is called a SET-RESET flip-flop with priority RESET.

   (a) Design two negative-edge clocked, priority RESET, SET-RESET flip-flops, using a D flip-flop memory cell for the first and a JK flip-flop memory cell for the second.

   (b) Determine the setup and hold times for the resulting flip-flops, assuming that the D and JK flip-flops have setup and hold times of 20 ns and 0 ns for both the high and low transitions, that any gates have gate delays of 15 ns for both transitions, and that any inverters have delays of 10 ns for both transitions.

10. (a) Design two positive-edge clocked LM flip-flops, using a D flip-flop memory cell for the first and a JK flip-flop memory cell for the second. An LM flip-flop functions as follows. If it's in the RESET (0) state and the inputs are the same, it changes state on the clock, but if the inputs are different it stays in the RESET state. If it's in the SET (1) state and input L is asserted, it changes state on the clock; if L is not asserted, it stays in the SET state. The behavior in the SET state is independent of M.

(b) Determine the setup and hold times for the resulting flip-flops, assuming that the D and JK flip-flops have setup and hold times of 20 ns and 0 ns for both the high and low transitions, that any gates have gate delays of 15 ns for both transitions, and that any inverters have delays of 10 ns for both transitions.

11. (a) Design two negative-edge clocked SN flip-flops, using a D flip-flop memory cell for the first and a JK flip-flop memory cell for the second. An SN flip-flop functions as follows. The next state is set equal to $\overline{N}$ on the clock except that when the state is SET (1) and S is asserted, the state always remains SET.

    (b) Determine the setup and hold times for the resulting flip-flops, assuming that the D and JK flip-flops have setup and hold times of 20 ns and 0 ns for both the high and low transitions, that any gates have gate delays of 15 ns for both transitions, and that any inverters have delays of 10 ns for both transitions.

9 Many-State Synchronous State Machines

9.1 INTRODUCTION

In Chapter 8 we learned how to design two-state state machines. In this chapter we extend the design procedure to state machines with more than two states. Initially we will limit the designs to synchronous machines, but at the end of Chapter 10 we'll use the same procedure to extend the design to some simple asynchronous machines. The model for a general sequential machine, shown in Figure 9.1-1, is the same for a twenty-state state machine as it is for a two-state state machine. The principal parts of the machine are the next-state decoder, the memory, and the output decoder. The memory and the next-state decoder for a many-state state machine are designed in much the same way as those for a two-state state machine. The principal differences are that (1) the memory must be larger than one bit in order to store state codes for more than two states, and (2) the next-state decoder must have logic circuits for each bit. The two-state state machines we designed didn't have output decoders; most many-state state machines do have them.

State machines can be designed using discrete logic gates and flip-flops; using programmable logic devices such as PROMs, PALs, and PLAs; or using logic elements available in large, possibly programmable, gate arrays. The most efficient arrangement of the memory elements and the decoders in a state machine depends on the type of hardware available. In a conventional state machine, characteristically the memory is made as small as possible, thus minimizing the number of flip-flops (memory elements) in the machine; however, this can complicate the next-state decoder. It is not surprising that conventional designs are well suited to discrete logic and PROMs, PALs, and PLAs, given that PROMs, PALs, and PLAs were essentially customized for such designs. Conventional designs are less suited to large gate arrays, however. After discussing conventional state machine design, we'll

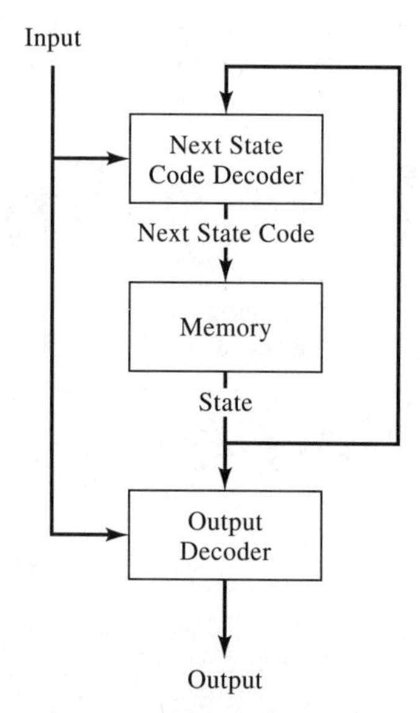

Figure 9.1-1 State machine model.

suggest an alternative design using one flip-flop per state. In many cases this alternative is better for large gate arrays, and in some cases it may be attractive for fast machines using discrete elements.

Initially, the many-state state machines we'll design will be synchronous and will all use edge triggered flip-flops for memory. Because the clocking is handled by the memory flip-flop, we will not have to use the clock as an input in the next-state decoder logic. Some of the machines we design will be large and complicated, so we'll introduce simplifications that allow us to progress directly from the state transition diagram to the Karnaugh maps or to the VEMs that are used to simplify the next-state decoder logic.

9.2 DESIGN OF A CONVENTIONAL MANY-STATE STATE MACHINE

The design of a state machine with more than two states follows the same steps as the design of a two-state state machine with some additions. The additions handle the added complexity arising from more than a single flip-flop in the memory.

Suppose we want to complete the design of the state machine we started in Chapter 7. First let's repeat the description. The machine is to analyze a stream of incoming bits, IN(H), in order to determine when the incoming bit pattern is 010. While the machine is searching for the first bit of this pattern, we want it to assert the output signal SRCH(H); each time it finds the first bit of the pattern, we want it to assert the output signal START(H); the machine should continue to assert the output signal START(H) when the second bit of the pattern is found; and each time

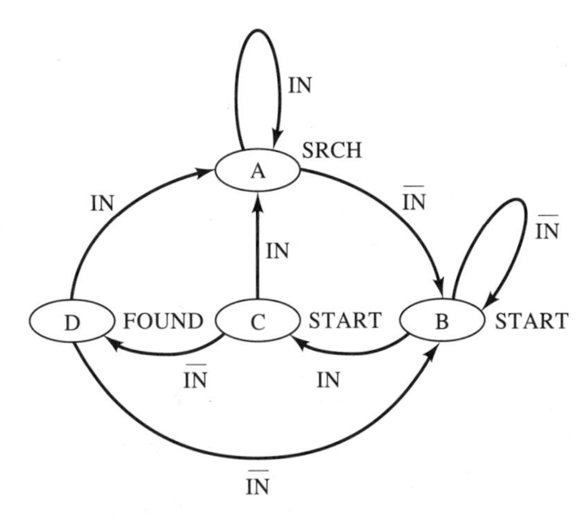

Figure 9.2-1 Primitive state transition diagram.

it finds all the bits of the pattern, we want it to assert the output signal FOUND(H). We do not want to miss any of the bits in the input stream, and we do not want to use any bit twice. We'll assume that the incoming bits are synchronous with a clock signal (CLK) so that they change only on the negative clock edge.

The initial steps of the design were completed in Chapter 7, but we'll go over them again. The first step is a clear verbal description of the problem, and the second is a state transition diagram. Remember that the state transition diagram should be tight—that is, it should have the minimum number of states. In the design of two-state state machines, we found that the state transition diagram could be omitted, but in the design of many-state state machines it should always be included. The third step is a timing diagram to check the critical timing constraints of the problem and make sure the state transition diagram is correct. If the machine is simple, the timing diagram may not be needed. The fourth step is a present state–next state table.

Figure 9.2-1 is the tight state transition diagram, and Table 9.2-1 is the present state–next state table we found for the machine in Chapter 7. We do not need to repeat the timing check we did in Chapter 7 to complete the design. The state transition diagram and present state–next state table are referred to as **primitive** because they do not have *binary state codes* assigned to the states.

To continue the design, we need to assign each state a binary code rather than the letters we used in the primitive diagram and table. We assign binary codes because the memory will consist of flip-flops that can store only binary bits. With a two-state state machine, the assignment of state codes was automatic. One state had binary code 1, and the other had binary code 0. Only a 1-bit memory was needed. With a many-state state machine, the assignment is no longer automatic. For our present machine we'll need two binary bits (two memory flip-flops) to allow us to have distinct codes for the four different states. Two bits will allow us to form the

Table 9.2-1 Present State–Next State Table

IN	Present State	Next State
0	A	B
1	A	A
0	B	B
1	B	C
0	C	D
1	C	A
0	D	B
1	D	A

four codes—00, 01, 10, and 11. We'll assign one of these codes to each state. Sometimes there are reasons for assigning particular codes to particular states. In this problem, however, we'll assign the codes arbitrarily to all states except the initial state, to which we'll assign the code 00. Figure 9.2-2 shows the state diagram with state code assignments. Once we've made the assignments, we no longer need the state names A, B, C, and D; the codes can be used to name the states.

Table 9.2-2 is the present state–next state table with state codes assigned. We've reversed the positions of the present-state code and the input, placing the present state first. This change makes it easier to form some of the maps we will use later to simplify our design procedure. Notice that both the present state and the next state have 2-bit binary codes. We've called the bits B1 and B0. For an edge triggered flip-

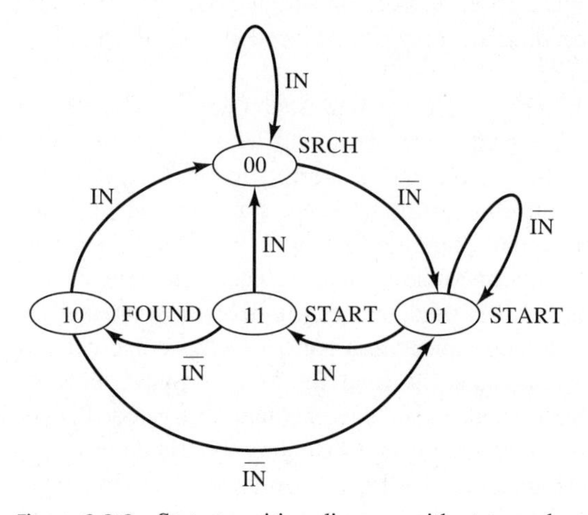

Figure 9.2-2 State transition diagram with state codes.

Table 9.2-2 Present State–Next State Table with State Code

Present State		IN	Next State	
B1	B0		B1	B0
0	0	0	0	1
0	0	1	0	0
0	1	0	0	1
0	1	1	1	1
1	0	0	0	1
1	0	1	0	0
1	1	0	1	0
1	1	1	0	0

flop memory, we don't need to include the clock as an input. This state machine has only one input and a 2-bit state code, so the present state–next state table is not very large.

Now that we have the present state–next state table, we should add the next-state code to it. We'll make three different designs: one with D flip-flop memory, using discrete logic elements; one with D flip-flop memory, using the logic elements available in a gate array; and one with JK flip-flop memory, using discrete logic elements. In the two designs using D memory, we don't need to add the next-state code to the present state–next state table because it's the same as the next state. This means we can go directly to the Karnaugh maps for the next-state code decoder design. The memory has two flip-flops, so we'll need logic circuits to produce the next-state code to input to each of these flip-flops. We'll call the flip-flop inputs D0 and D1. The next-state code decoder inputs are the bits in the present-state code and the inputs to the state machine (B0, B1, and IN). The present state–next state table can be thought of as two truth tables: one for D1 (next state B1) and one for D0 (next state B0). In both tables the inputs are B1, B0, and IN. Figure 9.2-3 shows the Karnaugh maps that are used to simplify the logic expressions for D0 and D1, along with the simplified expressions. The maps come directly from the truth tables for D1 and D0. The logic expressions enable us to design logic circuits to determine the next-state code from the present-state code and the input to the state machine. But let's leave this logic circuit until we've completed some other aspects of the design.

To finish the state machine design, we need to make three output tables and then design the logic to produce the outputs, which will be the same for D and JK memories. Before designing the output decoder, we'll design the next-state decoder for JK memory. Table 9.2-3 is the present state–next state table with the JK next-

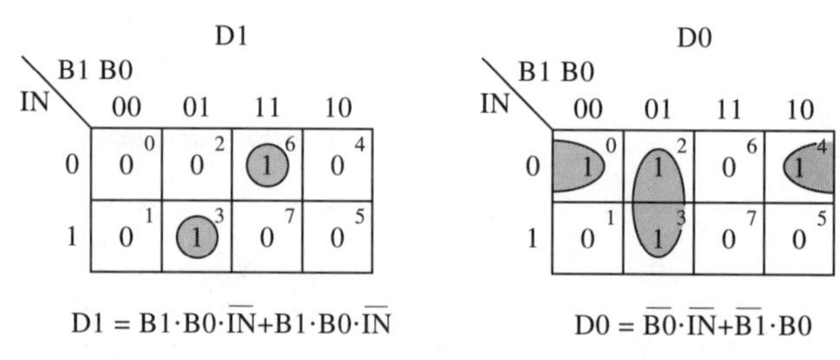

Figure 9.2-3 Karnaugh map for next-state code (D memory).

state code added. We find the next-state code by using the excitation table for a JK flip-flop, just as we did in the design of two-state state machines with JK memory. Here there are four columns of code—J0, K0, J1, and K1—because there are a J input and a K input for each of the two memory flip-flops. Remember that we determine J0 and K0 using present- and next-state B0 bits, whereas we determine J1 and K1 using present- and next-state B1 bits. For example, in the first row the present-state bit for determining J1 and K1 is B1 = 0 and the next state bit is B1 = 0, so J1 = 0 and K1 = x. In that same row, the present-state bit for determining J0 and K0 is B0 = 0 and the next state bit is B0 = 1, so J0 = 1 and K0 = x. We can find the next-state code for JK memory most easily if we remember that when the present state is 0, J is equal to the next-state bit and K is x; when the present state is 1, J is x and K is equal to the complement of the next-state bit. Figure 9.2-4 shows the Karnaugh maps that are used to design the next-state code decoder and the simplified logic expressions obtained. These maps come directly from the

Table 9.2-3 Present State–Next State Table with Next-State Code

Present State		IN	Next State		Next-State Code			
B1	B0		B1	B0	J1	K1	J0	K0
0	0	0	0	1	0	x	1	x
0	0	1	0	0	0	x	0	x
0	1	0	0	1	0	x	x	0
0	1	1	1	1	1	x	x	0
1	0	0	0	1	x	1	1	x
1	0	1	0	0	x	1	0	x
1	1	0	1	0	x	0	x	1
1	1	1	0	0	x	1	x	1

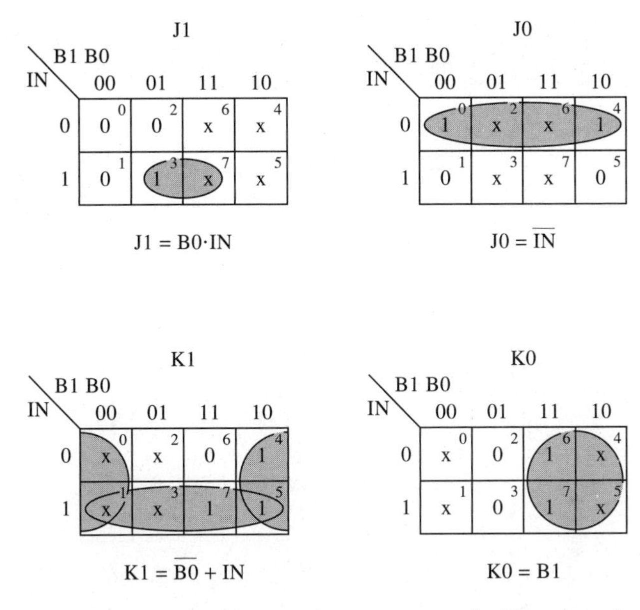

Figure 9.2-4 Karnaugh maps for next-state code (JK memory).

table, which can be thought of as a number of truth tables, one for each bit in the next-state code.

A many-state state machine usually requires an output decoder because the outputs are different from the states. We did not need an output decoder in a two-state state machine, because the output *was* the state. To design an output decoder, we need to make an output table. Table 9.2-4 is the output table for our state machine,

Table 9.2-4 Output Table

Present State		IN			
B1	B0		SRCH	START	FOUND
0	0	0	1	0	0
0	0	1	1	0	0
0	1	0	0	1	0
0	1	1	0	1	0
1	0	0	0	0	1
1	0	1	0	0	1
1	1	0	0	1	0
1	1	1	0	1	0

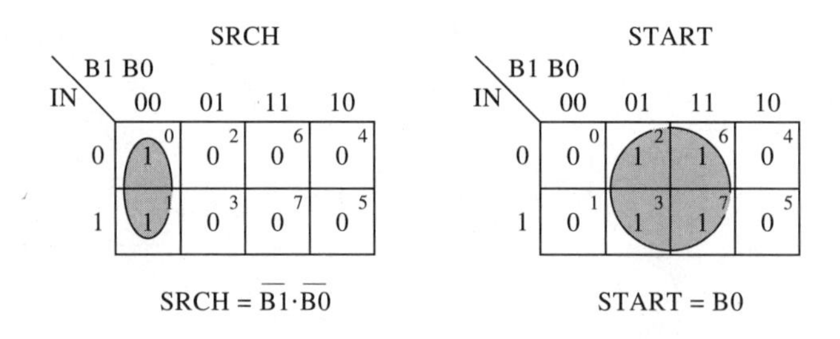

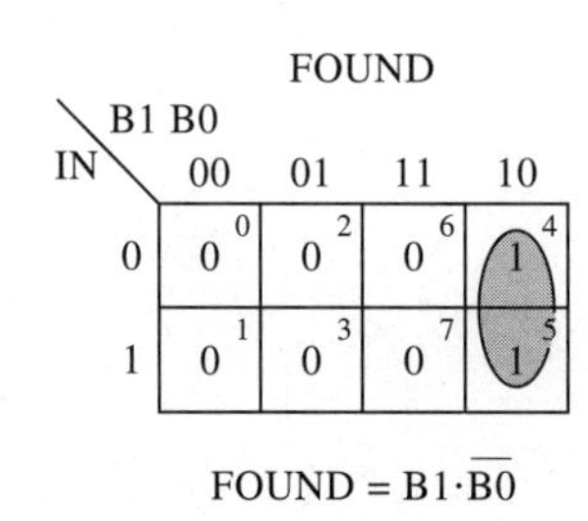

Figure 9.2-5 Karnaugh maps for outputs.

and Figure 9.2-5 shows the Karnaugh maps that are used to simplify the output decoder logic and the simplified logic expressions obtained. Notice that we can again think of the output table as several truth tables, one for each of the outputs, and form maps directly from these tables.

The only task that remains in this design is to turn the next-state decoder logic expressions and the output decoder logic expression into logic circuits and put them together with the memory flip-flops. We've done this in Figure 9.2-6 for D memory using discrete logic elements, in Figure 9.2-7 for D memory using gate array logic elements, and in Figure 9.2-8 for JK memory using discrete logic elements.

Notice that the design using gate array elements contains no inverters, even though the D flip-flops don't have a low assertion level output. This is made possible by using the wide variety of gates available. Thus, for example, the upper *nand* gate in the next-state decoder, which needs to produce an $\overline{\text{IN}}$ in the logic expression, uses a bubble instead of an inverter, and the second *nand* gate uses B1(H) and a bubble instead of B0(L) to produce $\overline{\text{B0}}$. An additional minor simplification is the choice of D flip-flops with no preset (PRE). Such flip-flops do not exist as discrete logic elements.

The clock in the JK design has to be inverted because the JK flip-flop used is negative-edge triggered. We need a positive-edge triggered memory element because the incoming signal is synchronous with the negative clock edge. In all three circuits an asynchronous RESET has been added to put the state machine into the first state. This is easily done using the clear (CL or CLR) function of the flip-flops, because we made the code for the first state 00. The choice of discrete gates in the logic

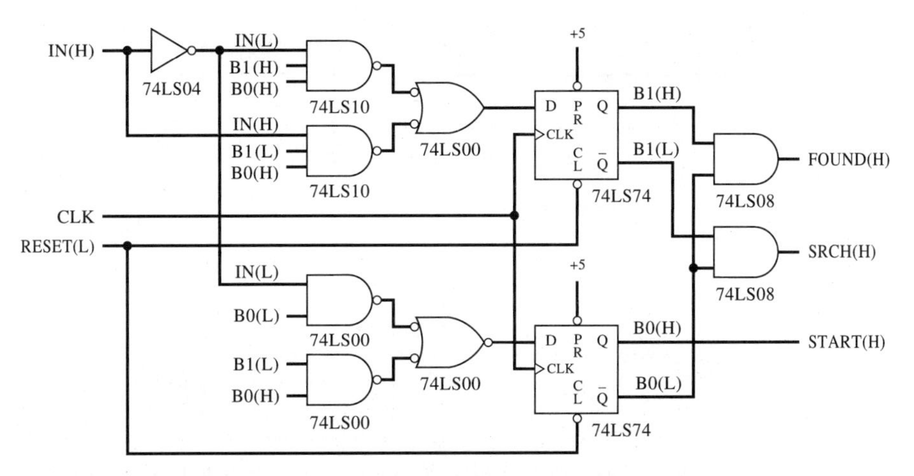

Figure 9.2-6 State machine with D memory using discrete logic elements.

circuits using discrete logic elements was partly dictated by available gates (only commercially available gates were used), but there were also some free choices. Other designs that use different gates are clearly possible. Some of the connects from the outputs of the flip-flops back to the inputs of the logic are not shown but are referenced. The label B1(H) on an input means the input is connected to the B1(H) output from a flip-flop. The state machine with JK memory requires simpler next-state code decoder logic (which is usually true).

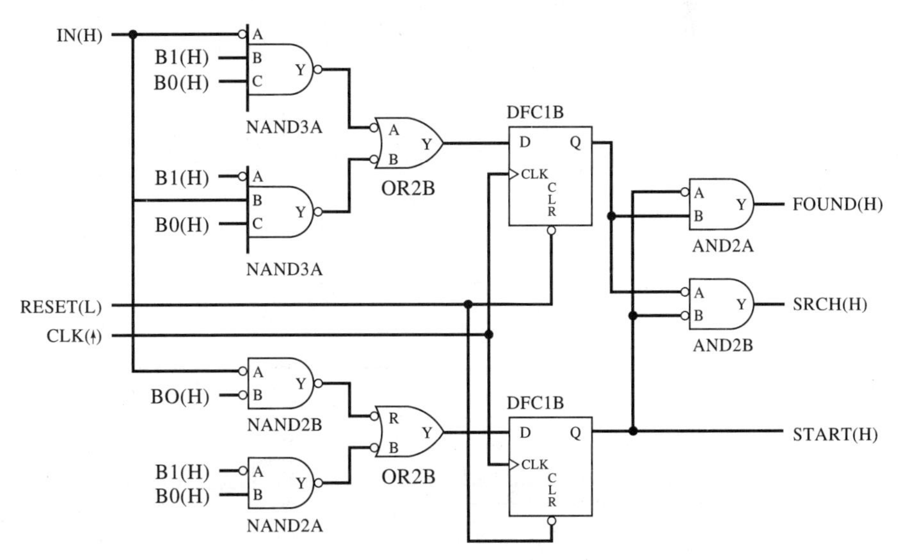

Figure 9.2-7 State machine with D memory using gate array modules.

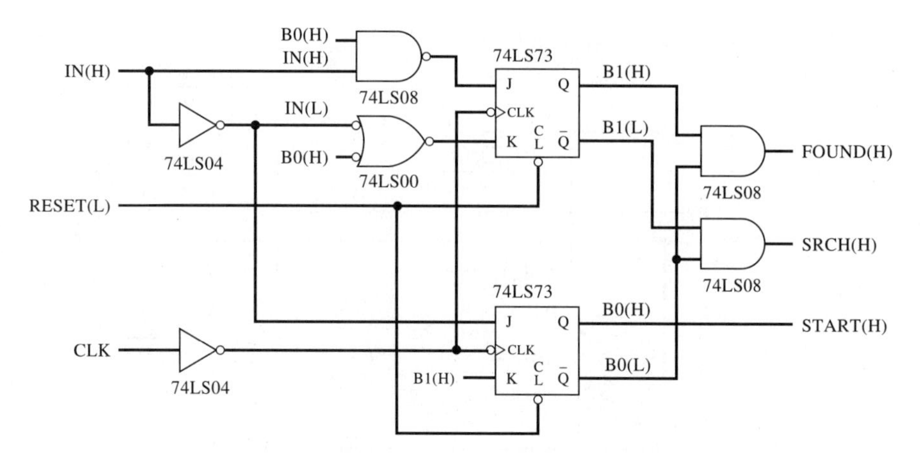

Figure 9.2-8 State machine with JK memory using discrete logic elements.

Any logic circuit design should be simulated to test its operation before it is built. Figure 9.2-9 shows the output signals found in a simulation of the circuit of Figure 9.2-6, and Figure 9.2-10 shows the output signals found in a simulation of the circuit of Figure 9.2-8. The design using gate array elements was not simulated. To get a good simulation with delays for the gate array design, not only the state machine but the complete circuit to be placed in the gate array must be designed and routed, because the delays depend on how all the modules in the array are routed.

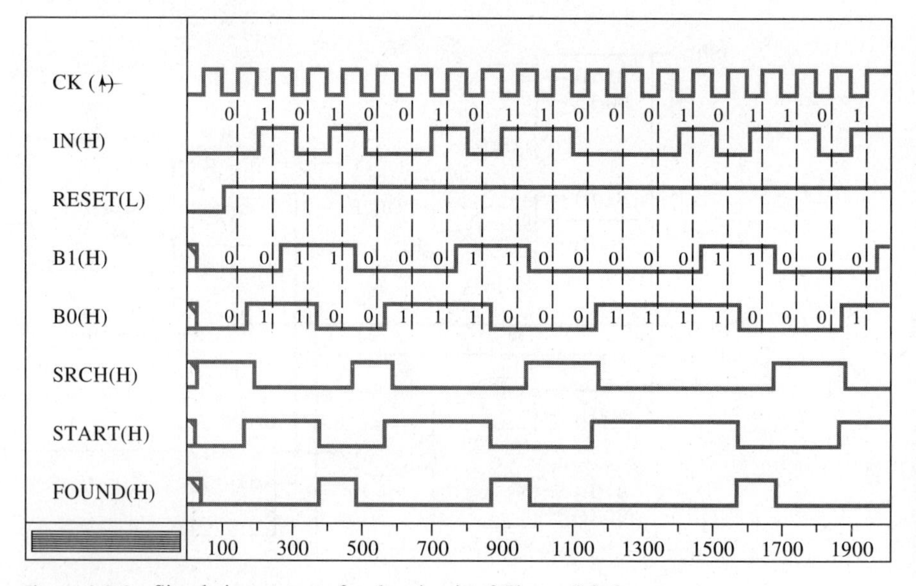

Figure 9.2-9 Simulation outputs for the circuit of Figure 9.2-6.

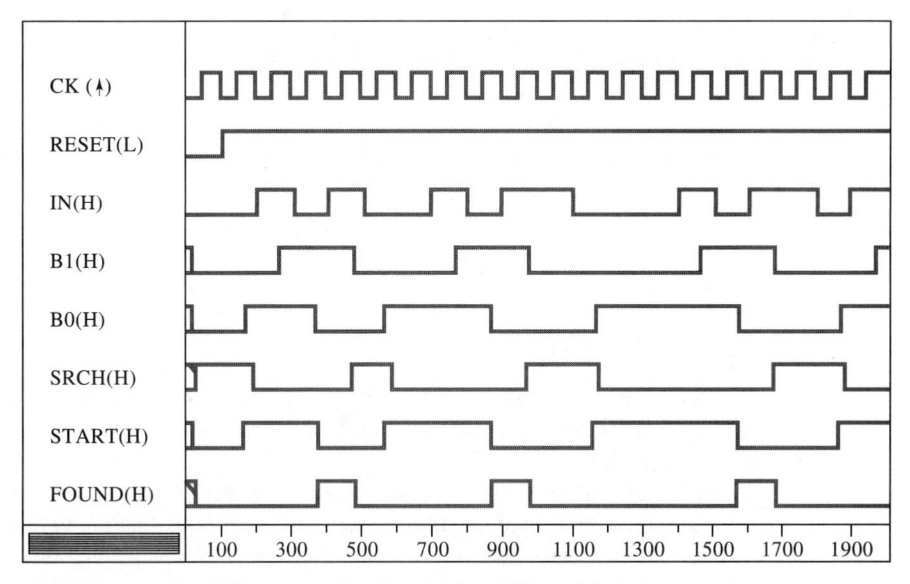

Figure 9.2-10 Simulation output for the circuit of Figure 9.2-8.

The simulations for the two different circuits are almost identical; they differ slightly in timing. To help us interpret the simulation results, we have added some notation on the simulation for the first circuit so that, as the machine makes transitions from state to state, the current state can be easily seen. Because the state is clocked by a clock edge and then takes some time to stabilize, a good place to look at the state is just before the clock edge that causes a new transition. If the machine is properly designed, the state will be stable at this time. In Figure 9.2-9, a vertical dashed line has been drawn at each clock edge and a 1 or 0 placed on the state code bits, just before the line, indicating the state code. The 1 or 0 on the input (IN) indicates its value just before the clock edge. The input at the clock edge determines the next transition.

To interpret the simulation, start at the left. Notice that we have first asserted RESET, which asynchronously resets the state code to 00 (B1 = 0, B0 = 0). On the first clock edge after the reset, IN = 0, so the state changes to 01; on the second clock edge, IN = 1, so the state changes to 11; and on the third clock edge, IN = 0, so the state changes to 10. The input for the simulation has been chosen so that it will simulate a 010 sequence, which is the sequence to be found, and the simulation shows the state transitions we expect for this sequence. After the 010 input sequence, the input is IN = 1. Because 1 is not the start of a new 010 sequence, the state changes to 00 and should remain 00 until an input IN = 0 is received. This happens on the next clock edge, changing the state to 01. The simulation continues for a variety of other input sequences, and in every case the simulation shows that the state machine acts in the way we expect it to.

The input chosen for the simulation should cause all the transitions of the state machine to be tested. We have not chosen our input sequence very well—the transition from the 11 to the 00 state and the transition from the 10 to the 01 state are not tested.

In this simulation we have been concerned primarily with the proper transitions through the states. We can also verify the outputs, SRCH, START, and FOUND. Notice that SRCH is asserted when the machine is in the 00 state; START is asserted when the machine is in the 01 or 11 state; and FOUND is asserted when the machine is in the 10 state. This is just as it should be.

Now that we've designed a many-state state machine, we'll summarize the design steps, which are the same steps we used in the design of the two-state state machines, with some additions.

1. Make a clear verbal description of the desired state machine. Take particular care to describe the characteristics of all inputs and the desired characteristics of all outputs.
2. Make a tight (smallest possible number of states) state transition diagram for the desired state machine. If necessary, draw a timing diagram to make sure that the state transition diagram describes a machine that will function according to the verbal description. The timing diagram is especially important when there are tight timing constraints on the design.
3. Make state code assignments, and change the state transition diagram to reflect these codes.
4. Make a present state–next state table for the desired state machine.
5. Referring to the excitation table for the memory cells to be used, add the next-state code to the present state–next state table. For D memory this step is unnecessary.
6. Design a next-state decoder, using Karnaugh maps to simplify the logic required.
7. Make an output table.
8. Design an output decoder, using Karnaugh maps to simplify the logic required.
9. Draw a circuit, including the next-state code decoder logic, the memory cell, and the output decoder logic.
10. Simulate the circuit (for as many different input conditions as you can).

Before discussing these steps, let's examine another design—a variation of the one we just completed.

Example 9.2-1

Complete the design that was started in Example 7.6-1, for a machine to find the input sequence 010 in an input stream, using only three states.

The figure on the left shows the state transition diagram for this state machine that was found in Chapter 7. The machine finds the input sequence 010, using three states, by making the outputs—FOUND, SRCH, and START—conditional on the input.

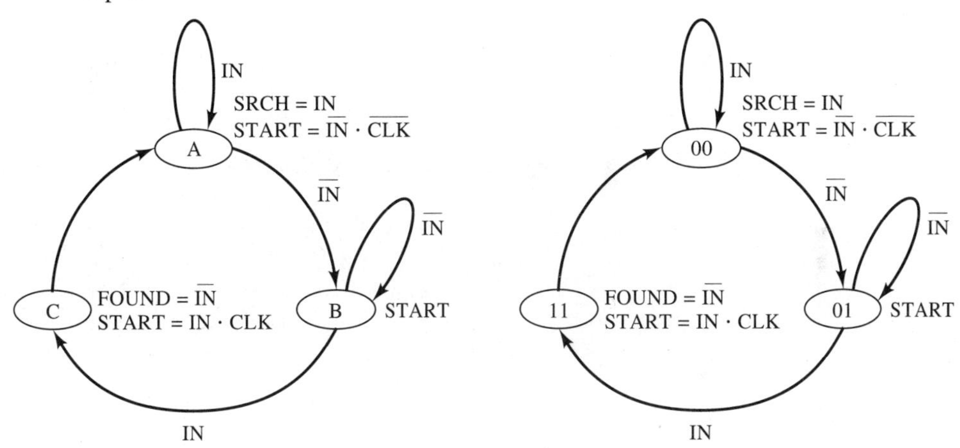

The timing check was done in Chapter 7, and we will not repeat it here. Once we have a state transition diagram and have checked the timing constraints of the problem description, we should assign state codes. For three states, we need a 2-bit code (two memory flip-flops). The figure on the right shows the state transition diagram with state code assignments. The assignments are arbitrary except for the first one; we assigned the code 00 to the first state.

The next step in our design is to make the present state–next state table, which follows. Notice that in this state machine the 10 state is unused. Because the machine should never get into the unused state, we can let the next state for this unused state be anything (*don't care*). Actually, we may get into an unused state because of some kind of error, and it is important to design the machine so that it will recover. We'll discuss this important problem later.

The present state–next state table allows us to immediately make a D memory design of the next-state code decoder, because the next-state code is the same as the next state. The figure following the table shows the Karnaugh maps for simplifying the next-state code logic and the simplified logic expressions.

Present State		IN	Next State	
B1	B0		B1	B0
0	0	0	0	1
0	0	1	0	0
0	1	0	0	1
0	1	1	1	1
1	0	0	x	x
1	0	1	x	x
1	1	0	0	0
1	1	1	0	0

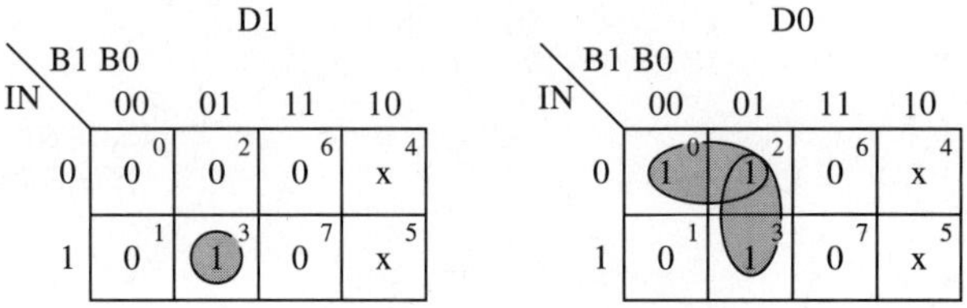

D1

IN \ B1 B0	00	01	11	10
0	0	0	0	x
1	0	1	0	x

$D1 = \overline{B1} \cdot B0 \cdot IN$

D0

IN \ B1 B0	00	01	11	10
0	1	1	0	x
1	0	1	0	x

$D0 = \overline{B1} \cdot B0 + \overline{B1} \cdot \overline{IN}$

D Next-State Decoder Maps

To make a JK memory design, we need to add the next-state code to the present state–next state table, as follows. This enables us to form Karnaugh maps for the next-state code logic, which follow the table, and write the simplified logic expressions.

Present State–Next State Table with JK Next-State Code

Present State		IN	Next State		Next-State Code			
B1	B0		B1	B0	J1	K1	J0	K0
0	0	0	0	1	0	x	1	x
0	0	1	0	0	0	x	0	x
0	1	0	0	1	0	x	x	0
0	1	1	1	1	1	x	x	0
1	0	0	x	x	x	x	x	x
1	0	1	x	x	x	x	x	x
1	1	0	0	0	x	1	x	1
1	1	1	0	0	x	1	x	1

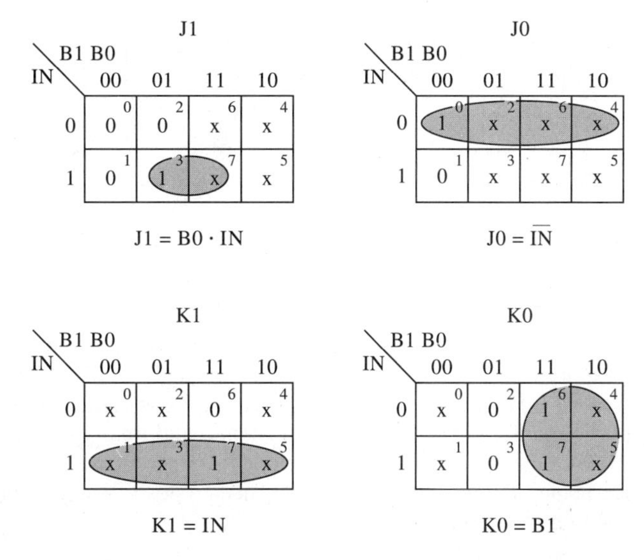

JK Next State Code Decoder Maps

Now that we have the next-state code logic for both the D and JK memory designs, we'll design the output decoder. We form the following output table directly from the state transition diagram. Notice the entered variable CLK in the zeroth and seventh row of the START column. We've used an entered variable rather than adding CLK as an input in the table, which would make it twice as long. The entered variable gives an additional condition on the output START. Following the output table are the Karnaugh maps for simplifying the next state decoder logic and the simplified logic expression.

Output Table

State		IN	SRCH	START	FOUND
B1	B0				
0	0	0	0	$\overline{\text{CLK}}$	0
0	0	1	1	0	0
0	1	0	0	1	0
0	1	1	0	1	0
1	0	0	x	x	x
1	0	1	x	x	x
1	1	0	0	0	1
1	1	1	0	CLK	0

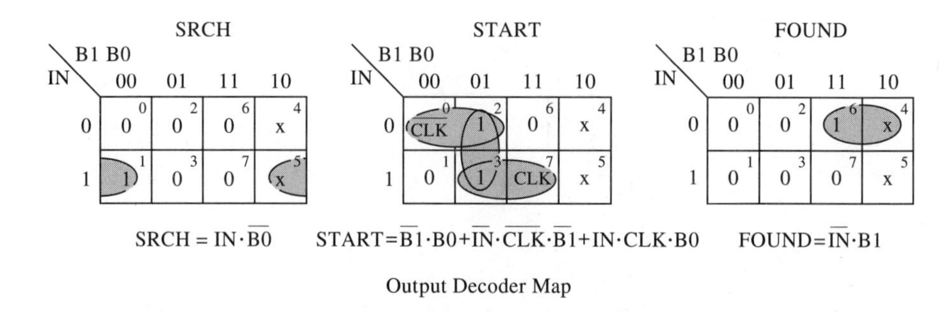

Output Decoder Map

All that remains to be done in the design is to put the decoder logic circuits, memory flip-flops, and output decoder circuits together. This is done in the following figures, first for D memory using discrete logic elements, then for D memory using programmable gate array logic elements, and finally for JK memory using discrete logic elements. Notice the inverter in the clock for the JK memory. The JK flip-flops are negative-edge triggered, and we need a positive-edge triggered memory flip-flop because of the input timing.

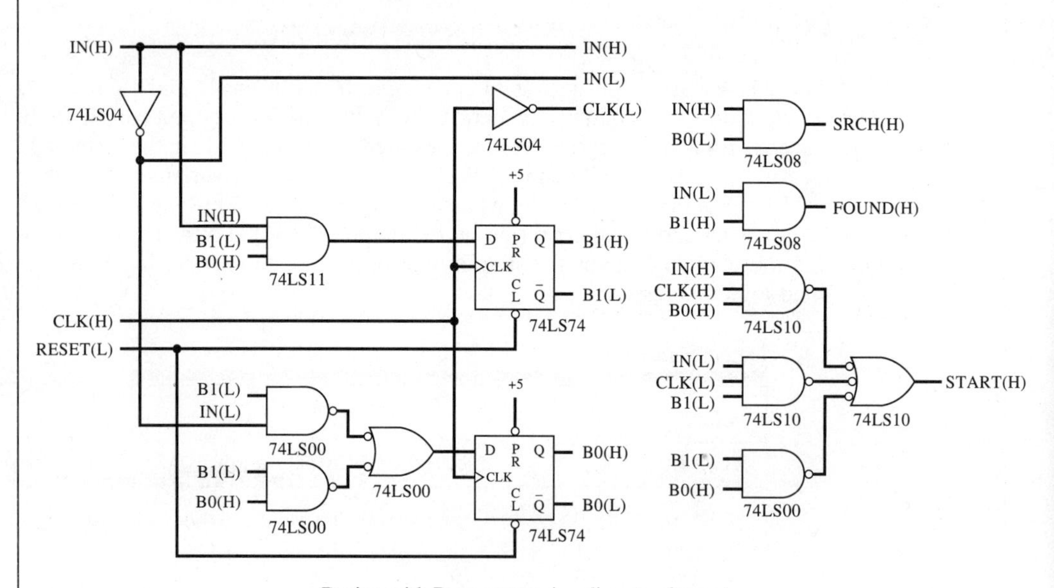

Design with D memory using discrete elements

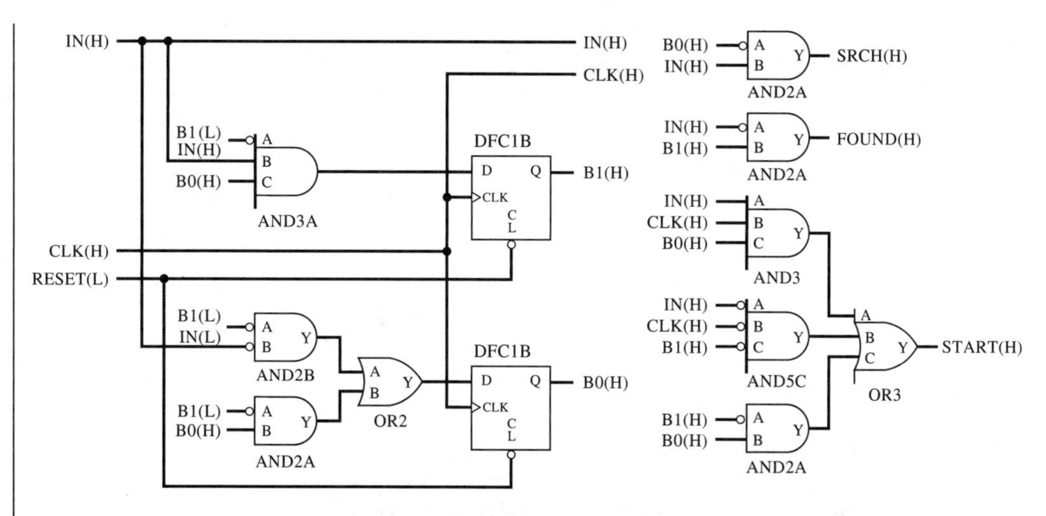

Design with D memory using programmable gate array logic

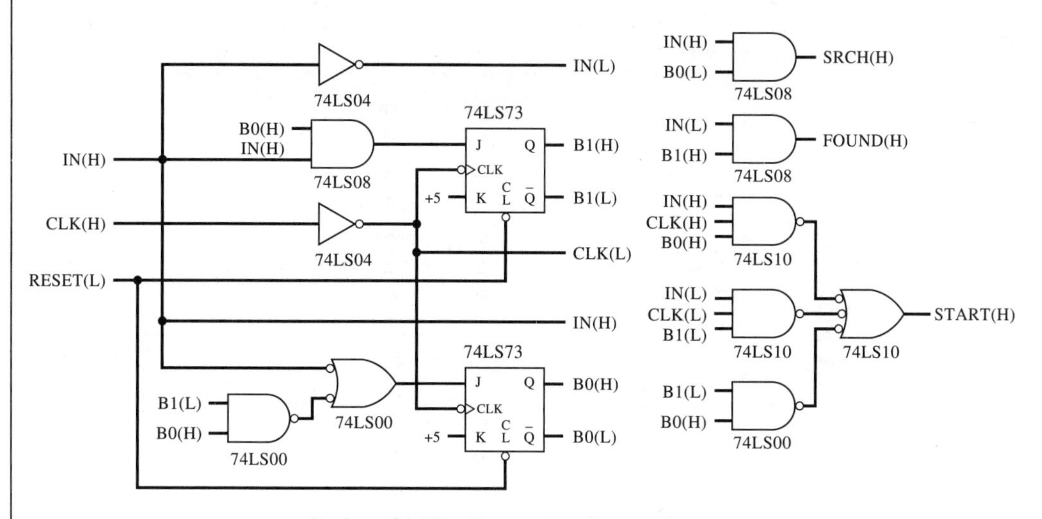

Design with JK memory using discrete elements

As always, we verify the design by simulating the circuits. Following are the simulation outputs for the discrete logic circuits. We haven't simulated the circuit with programmable gate array elements, because it is not possible to get accurate delay for the isolated state machine circuit. The entire gate array would need to be designed and routed to achieve accurate delays.

In both simulations the START signal has problems. In the first, it has short unwanted pulses; in the second, it has short unwanted dropouts. These short anomalies, called *glitches*, are common in the type of output decoder we've designed. We'll learn in Section 9.6 how to avoid glitches by modifying the output decoders.

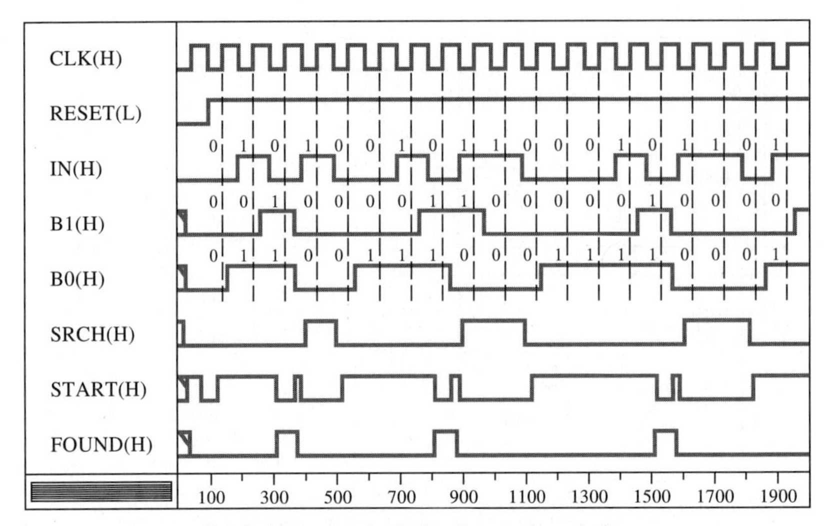

Simulation output for D flip-flop memory design

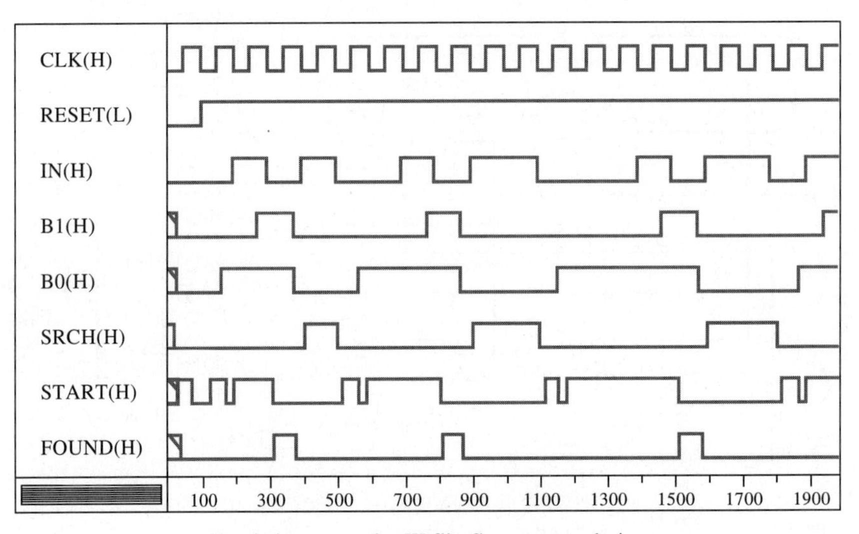

Simulation output for JK flip-flop memory design

Just as in the previous simulation, we've added the values of the state code logic variables and the input logic variables. This aids interpretation of the simulation results. The machine is set to state 00 by the asynchronous reset (RESET), then makes transitions through states 01 and 11 in response to inputs (IN) of 0 and 1, and finally makes transitions back to the state 00. IN 0 and 1 are the first two bits in the desired sequence, but unlike the previous design, the output FOUND is asserted for the last half of the state 11 when IN becomes 0, the last bit in the desired sequence. The input sequence 010 was chosen as the initial simulation input to show that the machine would find this sequence as it should. Other inputs were also chosen to see how the machine responded to their sequences. If we trace through the simulation, it shows us that the machine functions just as we expected except for the glitches in the output. We of course need to look at outputs as well as states to determine that the machine is functioning correctly. In particular, notice that when the machine is in state 00 and IN is not asserted (0), the output START is asserted (with some delay) when the clock is low; furthermore, when the machine is in state 01, START is asserted; and finally, when the machine is in state 11, START is asserted when IN is asserted (1) and the clock is high. As we already noted, FOUND is asserted when the machine is in state 11 and IN is not asserted (0). IN is always 1 for the first half of state 11, so FOUND can be asserted for only one half of a clock cycle. Both START and FOUND behave as expected. If the simulation is analyzed for the behavior of SRCH, it will also be found to behave correctly.

We've explored two examples of state machine design. In the following sections we'll discuss their design steps in more detail. Some of the steps need more explanation, some can be simplified, and some can be eliminated.

As you know, the first step in the design of a state machine is to make a clear verbal description of what the machine is to do, including a listing of all the inputs and the actions (outputs) they cause, and a clear statement of any critical timing constraints on the inputs and outputs. The description of the timing constraints may require timing diagrams. In the state machine design at the beginning of this section, there were one input, IN, and three outputs, SRCH, START, and FOUND. Timing constraints on IN were described in a timing diagram. The effect of IN was that the output SRCH should be asserted until the first bit of the desired sequence was found. The output START was to be asserted when the first bit (0) of the sequence was found and continued if the second bit (1) was found. Finally, the output FOUND was to be asserted when the complete sequence 010 was found. Critical timing constraints were no missed bits and no bit counted in two different sequences.

The second step in the design is to produce a state transition diagram for a machine to perform the functions described. We do not try to assign state codes in this diagram. We should also verify that the state transition diagram is correct by drawing a timing diagram. For the purpose of drawing such a timing diagram or making trial state code assignments, it's useful to name the states A, B, and C or 1, 2, and 3. (The first two steps in the design of a state machine were discussed in detail in Chapter 7.)

9.3 STATE CODE ASSIGNMENT

Once we have a state transition diagram that meets the specifications in the description of the state machine, each state in the state transition diagram should be assigned a code. The number of states required by the machine determines the minimum number of bits in the state code and hence the number of flip-flops in the memory. An n-bit code has 2^n distinct code sequences. A 1-bit code is required for a 2-state state machine; a 2-bit code for a 3- or 4-state state machine; a 3-bit code for a 5-

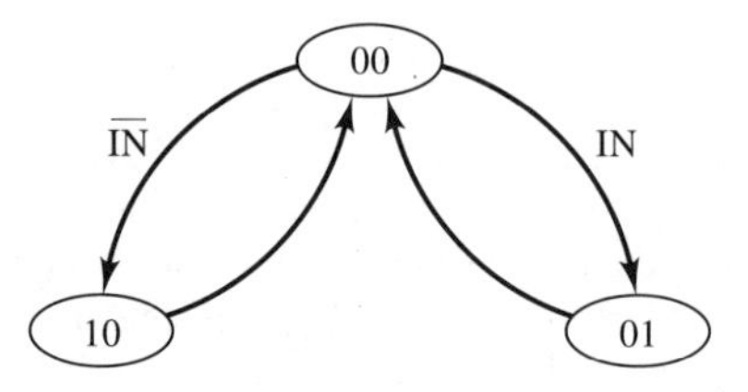

Figure 9.3-1 State transition diagram.

through 8-state state machine; a 4-bit code for a 9- through 16-state state machine; or in general an n-bit code for a $(2^{n-1}+1)$- through 2^n-state state machine.

In addition to determining the size of the code, we need to consider whether there are any reasons for assigning certain codes to certain states. The first and most obvious assignment is of zero to the initial state. This allows an asynchronous reset into this initial state (using the asynchronous CLR of the memory flip-flops) so that the machine can be made to start at the initial state. Such initialization into the proper state requires that initialization circuitry be added to the machine. This circuitry can be as simple as a manual reset switch to force the machine to be set to the initial state, or it can include automatic provisions for reset on startup, using a delay circuit that produces a reset signal after the startup transients have decayed. Single-chip circuits are available that, when first powered up, produce a short pulse after a delay. The circuit that causes the reset into the initial state is often called the *sanity* circuit because it causes the machine to start ''sane'' (in a known state).

There are several other less obvious reasons for assigning state codes in certain ways. States involving transitions that depend on asynchronous inputs need to have their codes assigned in a special fashion to avoid incorrect transitions. State codes need to be assigned in a certain way to avoid short anomalous pulses (glitches) in the output. In addition, state codes can be assigned to simplify the next-state decoder and the output decoder. We'll discuss these state code assignments in the following sections.

9.3.1 Asynchronous Inputs Transitions that depend on asynchronous input must be treated in a special way if they are to be reliable. **Asynchronous inputs** are inputs that are not timed by the clock. They can occur at any time during the clock cycle. For an example of the problem that can result from an asynchronous input, consider a state machine with the state transition diagram in Figure 9.3-1. Notice that the 11 state is unused.

Figure 9.3-2 shows a circuit for implementing this state machine. Because the input, IN, is asynchronous with the clock, it can change very near the clock edge. When it does, the machine can make a transition from the 00 state to the unused 11 state. For example, suppose that the machine is in the 00 state and IN changes from low to high, very near the clock edge. Figure 9.3-3, the timing diagram, shows the effect of this change. Both B1 and B0 are changed to 1 (high), because the change in IN propagates through the D0 decoder and changes the input to the B0 flip-flop

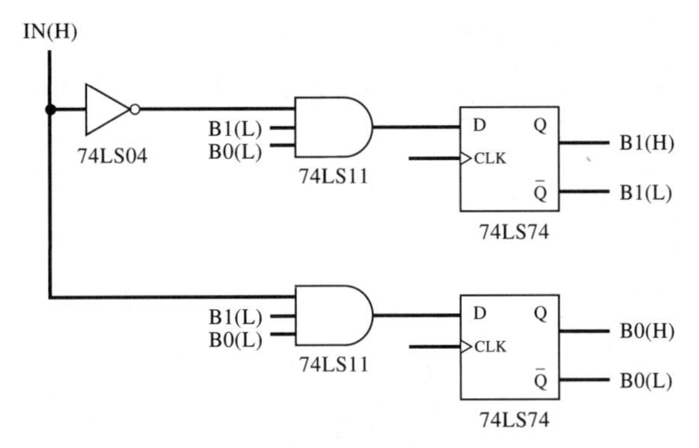

Figure 9.3-2 State machine circuit.

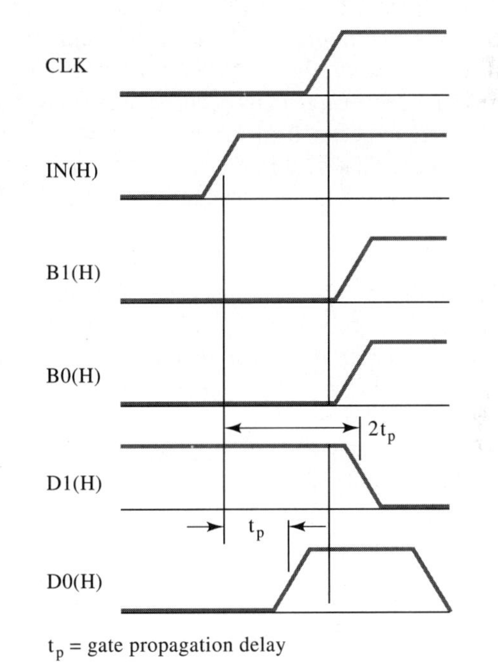

t_p = gate propagation delay

Figure 9.3-3 Timing for an improper state transition.

before the clock edge, but does not propagate through the D1 decoder soon enough to change the input to the B1 flip-flop before the clock edge. The D0 decoder has only one gate delay, and the D1 decoder has two. To simplify the timing diagram, we've assumed that the propagation delays through the *and* gate and the inverter are the same, and the setup time for the flip-flop is small enough that it can be neglected. The problem with the state assignment in this example is that the changes in both B1 and B0 depend on IN.

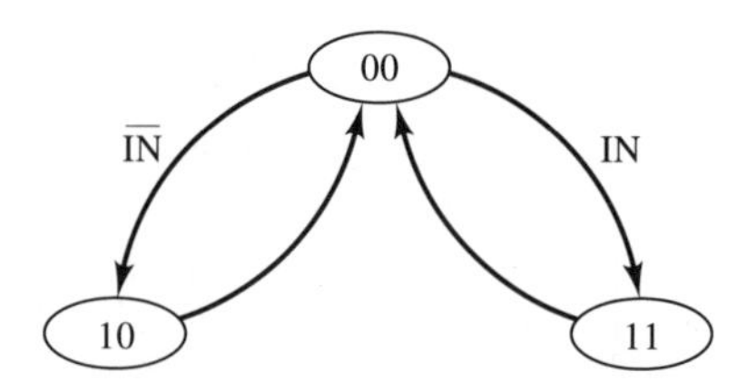

Figure 9.3-4 State transition diagram.

We can change the next-state code assignment and correct this problem. If we change the state code for the 01 state to 11, as shown in Figure 9.3-4, only the change in B0 will depend on IN; B1 will change to 1 (high) for both next states, independent of IN. Figure 9.3-5 shows a circuit for this choice of state code, and Figure 9.3-6 is the timing diagram. The machine starts in the same 00 state, and IN changes from low to high near the clock edge as in the previous circuit. For the timing shown, the change in IN occurs enough before the clock edge for the transition to the 11 state to occur. If the change in IN is slightly later, B1 does not change on this clock edge. That's all right; the machine just branches in the other direction in the state transition diagram because the input is $\overline{\text{IN}}$ and we've missed the change in input for this clock edge. Notice that D1 does not change until after the clock edge (when B1 changes), so there is no possibility that B1 will not change to 1 (high).

The result we've seen in this case can be taken as general, and it leads us to a general rule for choosing the next-state codes for a branch that depends on an asynchronous input: *If branching transitions out of a state depend on a single asynchronous input, the codes for the branch states should be logic-adjacent; that is, they should differ in only one bit.* This rule is illustrated in Figure 9.3-7. A state with one transition back to itself and one transition to another state should be treated as a branch, and the rule should be applied. When the codes for the branch states differ in only one bit, then clearly only one bit depends on the asynchronous input. If we follow the rule, we'll design state machines that branch either one way or the other—not to some unwanted state—even when the input changes near the clock edge.

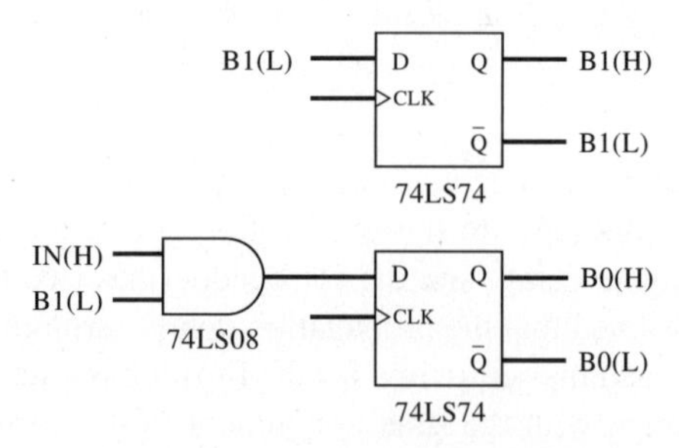

Figure 9.3-5 State machine circuit.

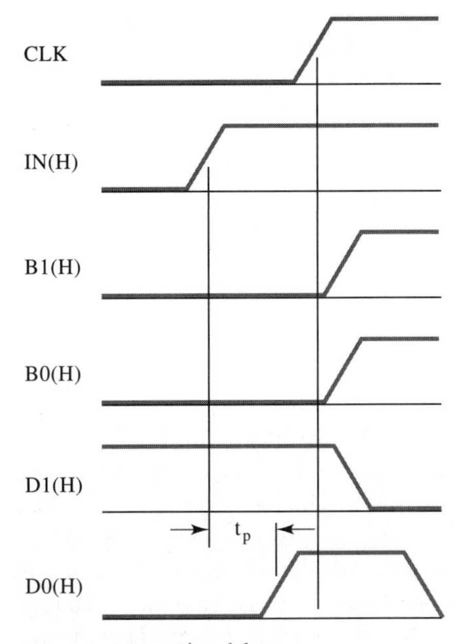

Figure 9.3-6 Timing for normal state transition.

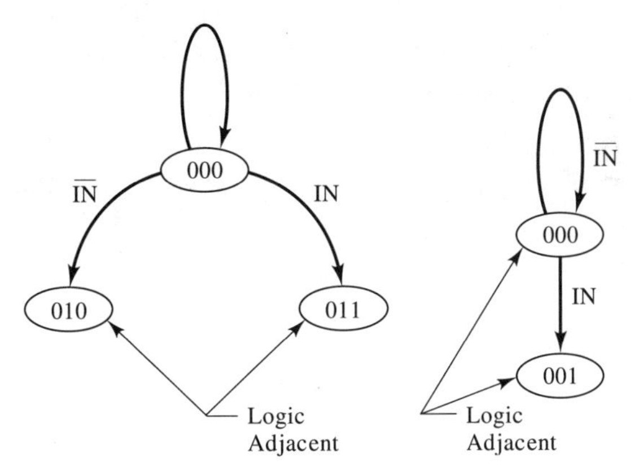

Figure 9.3-7 Examples of next-state assignment.

If a branch depends on two asynchronous inputs, the problems are more severe. Consider the simple *and* gate circuit and timing diagram of Figure 9.3-8. The two asynchronous inputs to the *and* gate can change at almost the same time and result in an output that is a very short, small-amplitude pulse. If this pulse is input to a memory flip-flop just at the clock edge, it can result in a **metastable** condition, in which the flip-flop is neither high nor low but somewhere in between—that is,

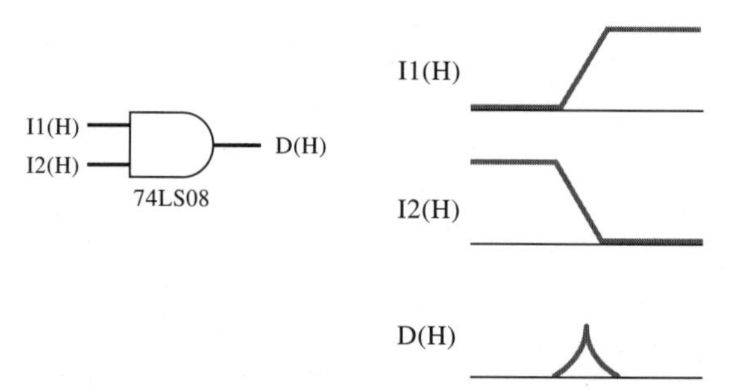

Figure 9.3-8 Effect of coincident input changes.

unstable. A metastable condition lasts for only a short time (less than one clock cycle), but it can clearly cause errors in the state machine.

To avoid metastable conditions, we should avoid branches that depend on two asynchronous inputs. This leads us to a second general rule for branching transitions that depend on asynchronous inputs: *Avoid branches that depend on more than one asynchronous input.* This can be done by adding a state, as shown in Figure 9.3-9.

Besides these two rules, there is another way to handle transitions that depend on asynchronous inputs: design a circuit to synchronize the inputs. In many cases this is the most straightforward thing to do, and in some cases it is the only thing to do. For example, if the asynchronous input is a very short pulse we may miss it entirely, or if it is a very long pulse it may cause multiple transitions, which we may not want. We'll consider synchronizing circuits in Section 9.4, after finishing our discussion of state code assignment.

9.3.2 Output Glitches Output glitches are unwanted, short-duration changes in the output of a state machine that occur when the machine changes state. They come

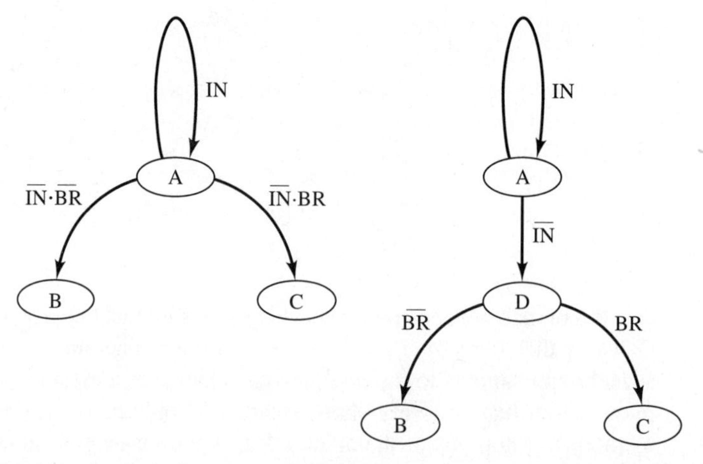

Figure 9.3-9 Correcting a potential asynchronous input problem by adding a state.

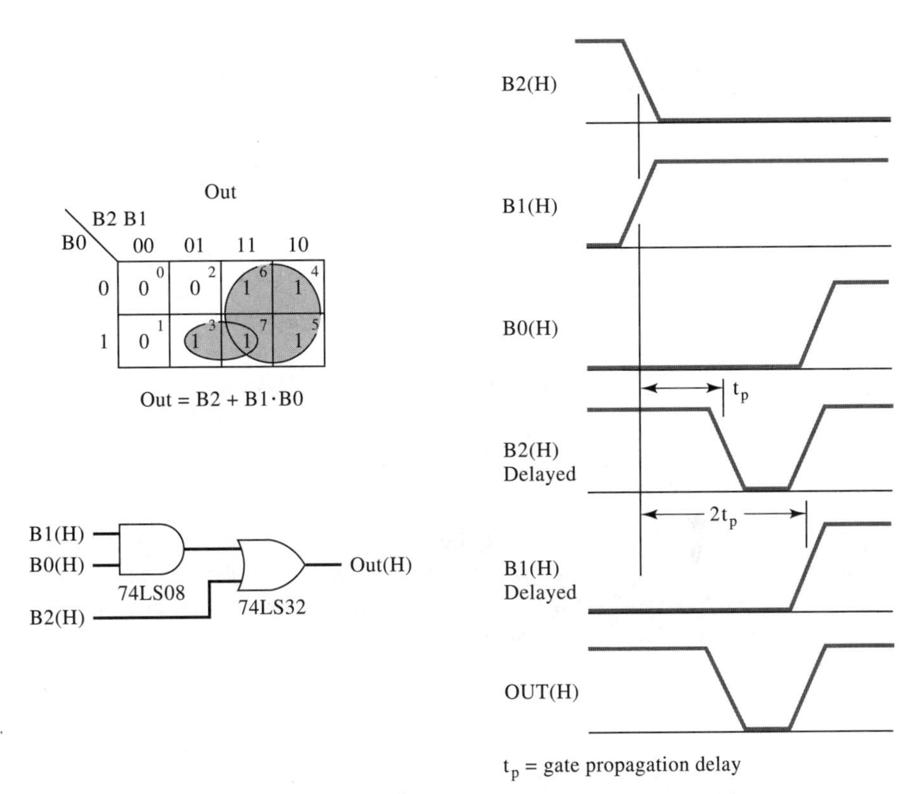

Figure 9.3-10 Output decoder and timing showing an output glitch.

about because the changes in the state variables do not affect the output of the output decoder at the same time. Figure 9.3-10 shows an output decoder, designed for a state machine with a 3-bit memory, that illustrates the problem. The timing diagram assumes that the machine changes from the 101 state to the 011 state. The output is not supposed to change (see the Karnaugh map), but it does. We can see from the circuit diagram that the change in B1 suffers two gate delays before it reaches the output, whereas the change in B2 suffers only one gate delay. This timing difference produces a short pulse during which the output is not asserted (B2 deasserts the output before B1 can assert it). For a state machine that is used as a controller, this false output can be very serious—for example, a machine tool controller might activate the machine too early or too late.

The output glitch problem is similar to the asynchronous input problem. Output glitches occur because two of the state variables change on the same state transition. One way to correct the problem is to allow only one state variable to change in any state transition. For some simple machines this kind of unit distance, or Grey, coding may be possible. For a large machine a unit distance coding is impossible, and other methods of removing glitches from the output must be used; they will be discussed in Section 9.6.

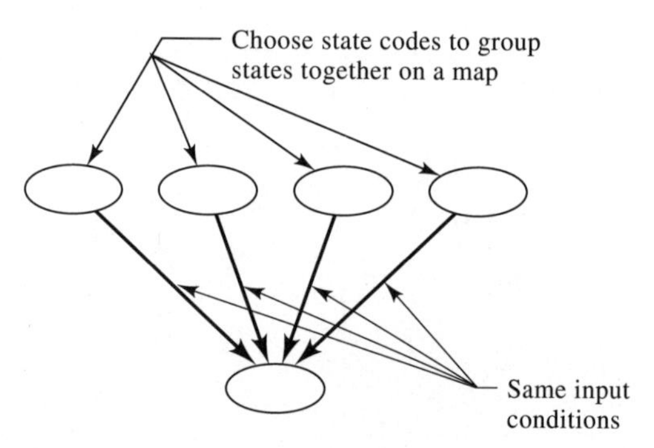

Figure 9.3-11 Choosing state codes for states making a transition to the same next single state.

9.3.3 Simplify the Next-State Decoder One way of arriving at the next-state code is to choose codes that will simplify the next-state decoder design. The challenge is to find a systematic way of doing this. Some complicated methods have been proposed, but we'll present just two simple rules that usually result in significant simplification.

The first rule is illustrated in Figure 9.3-11. *If a number of states all have transitions to a single state for the same input condition, choose state codes that will group these states together into logic-adjacent cells in a map.* To get the most simplification from this rule, it is important that all the input conditions be the same.

The second rule is illustrated in Figure 9.3-12. *If a single state branches to a number of states, choose the state codes that will group these states together into logic-adjacent cells in a map.* In this case, simplification is greatest if the states are made logic-adjacent by making part of the state code equal to the input condition for the transition involved. In Figure 9.3-12, the first two bits of each of the 3-bit

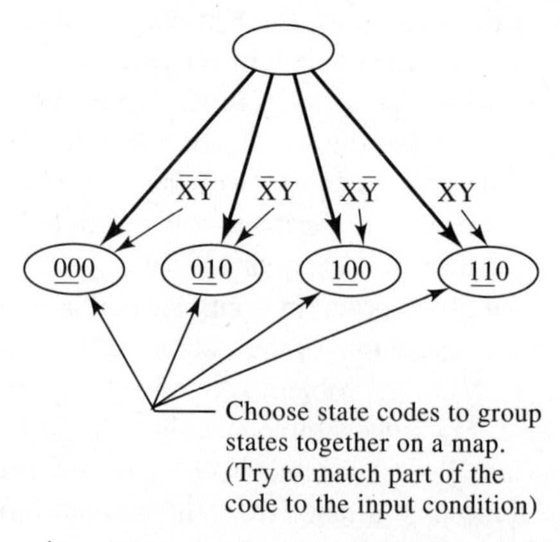

Figure 9.3-12 Choosing state codes for states which branch from the same state.

state codes are made equal to the input condition for the transition. The variables X and Y are inputs. This choice of code to match the input conditions is called **reduced input dependency.**

The two rules we've presented can at times contradict each other. If that is the case, apply the first rule. If a contradiction exists between the conditions that must be met when applying a single rule, just choose between the conditions. We can apply these rules to the designs of Section 9.2, but we will not get much simplification. The codes for that example are already well chosen to satisfy the rules. To really see how the rules work, we'll need to consider a more complicated design, which we'll do after we have developed more tools for simplifying the design procedure.

9.3.4 Simplify the Output Decoder Choosing the state code assignment to simplify the output decoder design is similar to choosing the state code to simplify the next-state decoder design. Some state code assignments will simplify the output decoder, but there is no good, systematic way of finding those assignments. As with the next-state decoder simplification, we'll only offer a rule that usually results in a significant simplification: *Those states which have identical output specifications should have codes that group together into logic-adjacent cells in a map.* Remember that conditional outputs must have the same condition to be identical.

If we examine the simple case we've been considering in this chapter, we'll find that it already has a state code chosen to simplify the output decoder. Look back at Figure 9.2-5. The state codes for the two states with the output START are chosen to be logic-adjacent. If we interchange the state codes for the 10 and 11 states and redesign the output decoder, three gates will be required instead of one.

9.4 SYNCHRONIZING ASYNCHRONOUS INPUTS

Perhaps the simplest way to handle asynchronous inputs to a synchronous machine is to synchronize them with the system clock. We'll examine two circuits (Figs. 9.4-1 and 9.4-3) that synchronize an input signal so that it changes on the negative

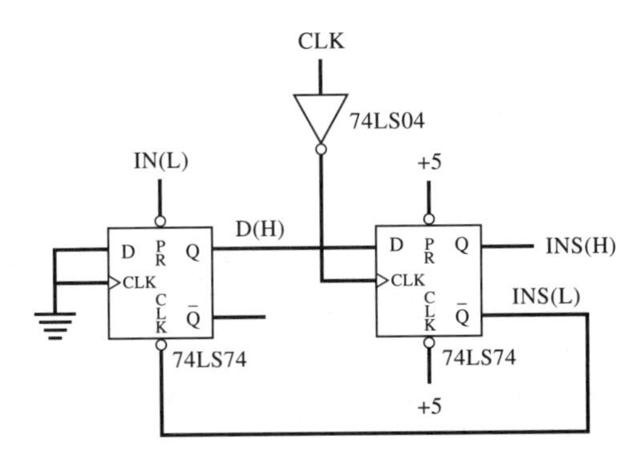

Figure 9.4-1 Input conditioning circuit which uses the SET-RESET function of a D flip-flop to catch and synchronize a short input pulse.

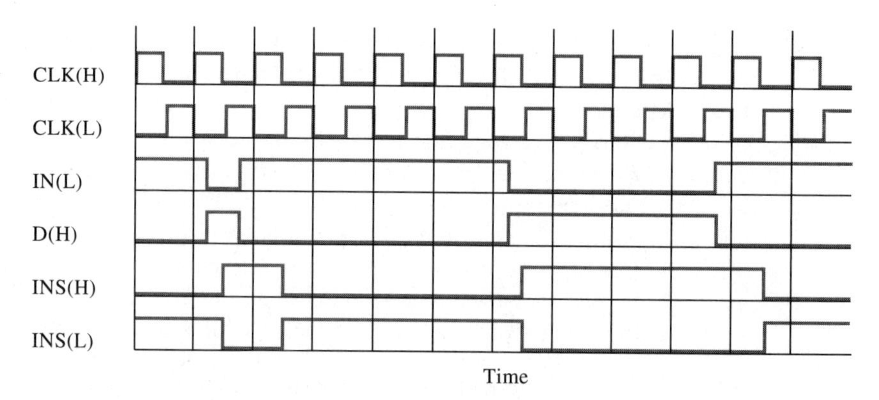

Figure 9.4-2 Timing for the input signal conditioning circuit of Figure 9.4-1.

clock edge. We'll also discuss the modifications needed to synchronize with the positive clock edge. These circuits not only synchronize an input signal, they provide additional input signal conditioning. Both circuits will *catch* an input that is too short to reliably cause a transition in a synchronous machine. The two circuits differ in how long they stretch the synchronized signal. Both circuits have the disadvantage of causing a delay of up to one clock cycle in the response to the input signal.

First consider the circuit of Figure 9.4-1. Figure 9.4-2 is the timing diagram. The first D flip-flop is used as a SET-RESET cell. When the input signal (IN), which is asserted low, goes low, the first flip-flop is set and D(H), the input to the second flip-flop, goes high. The second D flip-flop functions as a normal D flip-flop. When its input is high, it is SET on the next positive edge of its clock, CLK(L). The synchronized input signal (INS) is thus asserted on the first negative system clock edge after IN is asserted. When INS is asserted, INS(L) will send a clear (CL) to the first flip-flop, but this signal will not do its job until IN(L) goes high; this is a peculiarity of the D flip-flop. Because the clear does not function until IN(L) goes high, the synchronized signal is stretched until the first negative system clock edge after IN(L) goes high. As soon as INS is deasserted, INS(L) goes high and the clear is removed from the first flip-flop so that it can again respond to the input signal, IN(L). The circuit synchronizes both the positive- and negative-going edges of the input signal.

The second circuit (Fig. 9.4-3) functions in much the same way as the first, except the timing of the end (negative-going edge) of the synchronized signal is not set by the timing of the unsynchronized signal, but can be adjusted as needed. Figure 9.4-4 is a timing diagram for this circuit. The first flip-flop is wired (D is high) so that each positive-going edge of IN(H), which is connected to the clock input, will cause the flip-flop to be SET.

The principal difference between the circuits is in the CL for the first flip-flop

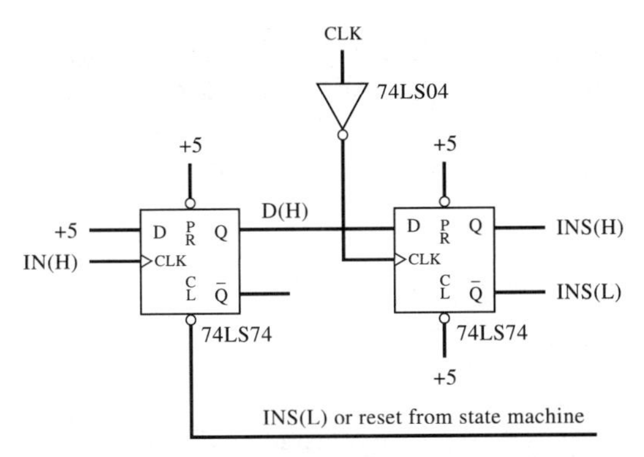

Figure 9.4-3 Input conditioning circuit that uses the clock of a D flip-flop to catch and synchronize the positive going edge of the input signal.

of the second circuit. Because preset (PR) for this flip-flop is not asserted, it can be cleared anytime after it's set, using CL. The time during which the synchronized signal is high can actually be shorter than the time during which the unsynchronized signal is high. If the CL is connected to INS(L), the synchronized signal will be high for one clock cycle, independent of the length of time the unsynchronized signal is high. If the synchronized signal must be stretched, the first flip-flop can be cleared using a signal from the state machine with the proper timing.

The circuits we've described will produce a synchronized signal that changes one-half clock cycle before the positive-going clock edge. In very fast state machines, the time needed for the input to propagate through the next-state decoder, and be available to change the state, may be longer than this half clock cycle. In that case we should omit the inverter in the clock circuit and synchronize with the positive

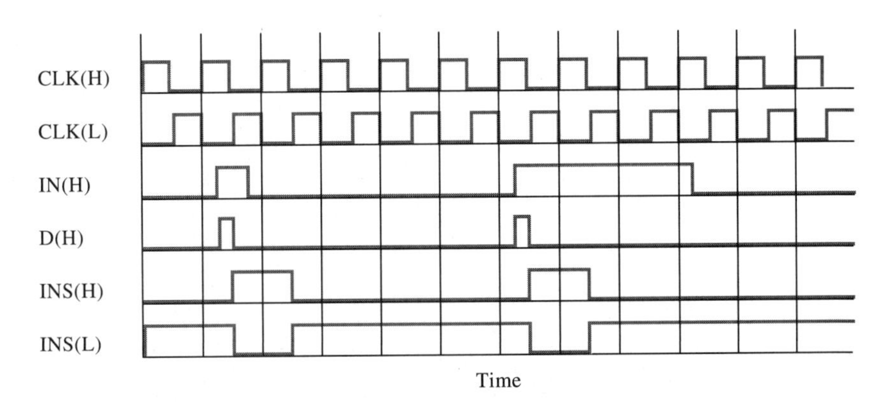

Figure 9.4-4 Timing for the input signal conditioning circuit of Figure 9.4-3.

clock edge, which will produce a synchronized signal that may be delayed almost a full clock cycle. The following section discusses next-state decoder timing considerations in more detail.

9.5 NEXT-STATE DECODER DESIGN

Next-state decoder design consists of four steps. (1) We make a present state–next state table, using the state transition diagram. (2) We add the next-state code to the present state–next state table, using the excitation table for the memory in the design (D or JK). (3) We use maps (Karnaugh or VEM) to simplify the logic expressions for the next-state code. (4) We turn the logic expressions for the next-state code into circuits. Not all of these steps are necessary. It's possible (and for large systems almost essential) to bypass the present state–next state table (and the next-state code table) and go directly to the maps. When we do this, we always produce the maps for D memory first and then transform them, if necessary, to maps for JK memory.

We've been using the present state–next state table to form decoder maps because we need the next-state code that we've made part of the table. For D memory, the next-state code can be taken directly from the state transition diagram because it's the same as the next state; a next-state table isn't needed. To see how to form the maps for D memory directly from the state transition diagram, consider the machine of Section 9.4. Its state transition diagram is repeated in Figure 9.5-1.

The easiest decoder maps to form directly from the state transition diagram are VEMs. In our VEMs, we should make the bits of the state code the map variables,

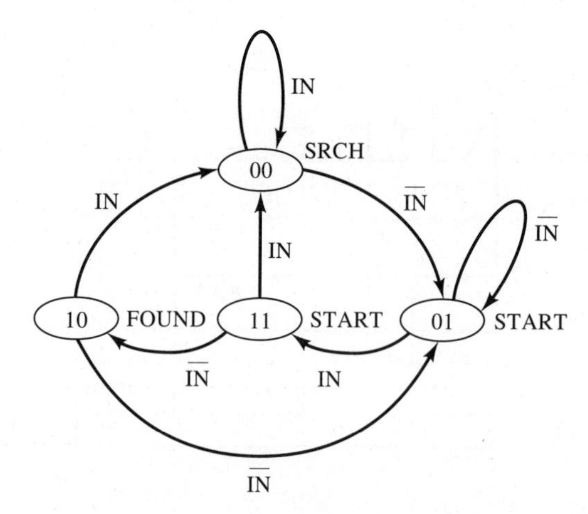

Figure 9.5-1 State transition diagram for example machine.

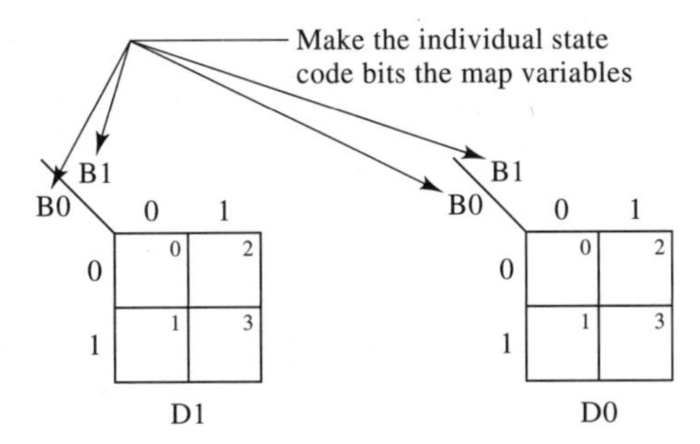

Figure 9.5-2 Bare VEMs for 2-bit state code.

and the inputs to the state machine the entered variables. Figure 9.5-2 shows the bare VEMs for a machine with a 2-bit state code, B1 B0. There are two maps because there are two decoders, the D1 decoder and the D0 decoder. Each state in the state transition diagram corresponds to a cell in each of the VEMs. For example, the 00 state corresponds to the upper left-hand cells (cells 0) of the VEMs.

Figure 9.5-3 shows how to load this cell in each decoder map. We need to investigate each transition out of the 00 state, examining first the high order bit and then the low order bit in each next-state code. The procedure for the high order bit is shown on the left, where we find that the high order bit of each of the next-state codes is 0. In this case, we enter a 0 in the high order (D1) decoder map.

The same procedure for the low order bit is illustrated on the right of Figure 9.5-3, where we find that the low order bit of the code depends on which next-state transition is made. In this case, we enter $\overline{\text{IN}}$—the input condition that causes transitions to the states with a low order bit of 1. If the next-state code had been 1 for all the next states, we would have entered 1.

To complete the maps, we repeat this procedure for each state of the state transition diagram. As further illustration, if we examine the transitions out of the 01 state, we find that we should load cell 1 of the D1 decoder map with IN and cell 1 of the D0 decoder map with 1. We may end up with some of the cells not loaded even though we've considered all the states in the state transition diagram. This means that we have some unused states, and we normally load the cells corresponding to those unused states with don't cares (x).

The design we just discussed had only one input, so there were at most two next states for each state. This meant that only two next-state codes needed to be examined to determine how to load each cell in the decoder maps. With more than one input, there may be more than two next-state codes to examine. If the next-state codes have all 1s or all 0s in the bit being examined, then a 1 or a 0 is entered in the decoder map. If the next-state codes have both 1s and 0s in the bit being examined, then the input expressions entered in the VEMs may get complicated. *All* the input conditions that cause a transition resulting in a 1 in the next-state codes must be entered in the map.

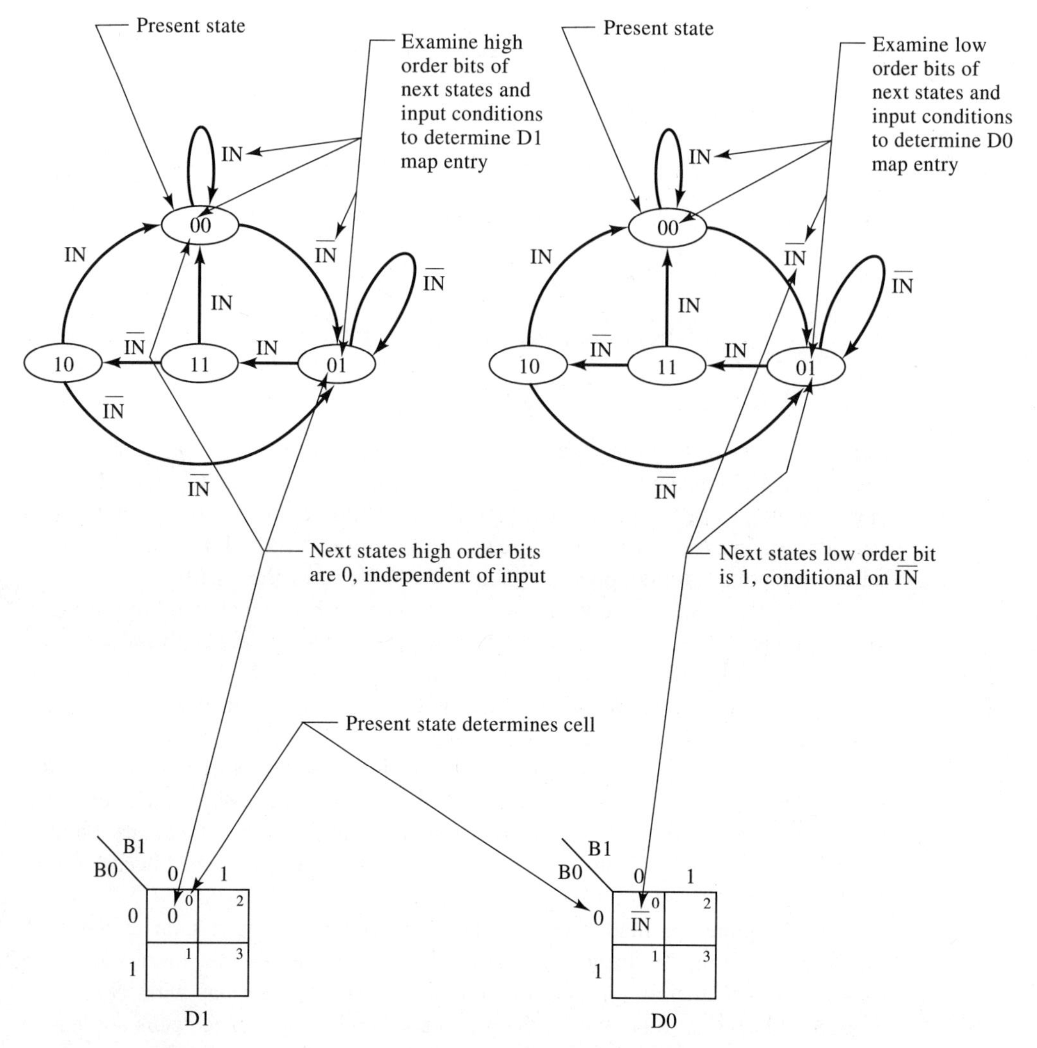

Figure 9.5-3 Illustration showing the loading of the VEM cell corresponding to the 00 state. (The outputs are omitted from the state transition diagram to simplify the illustration.)

If a state machine is small enough, it isn't necessary to use VEMs for the next-state decoder maps. Figure 9.5-4 illustrates how to load a normal Karnaugh map directly from the state transition diagram. Notice that the cell to be loaded depends on both the present state and the input condition. It's generally easier to load VEMs than to load Karnaugh maps directly from the state transition diagram.

Now that you know how to load the next-state decoder maps for D memory directly from the state transition diagram, let's see how to load the next-state decoder

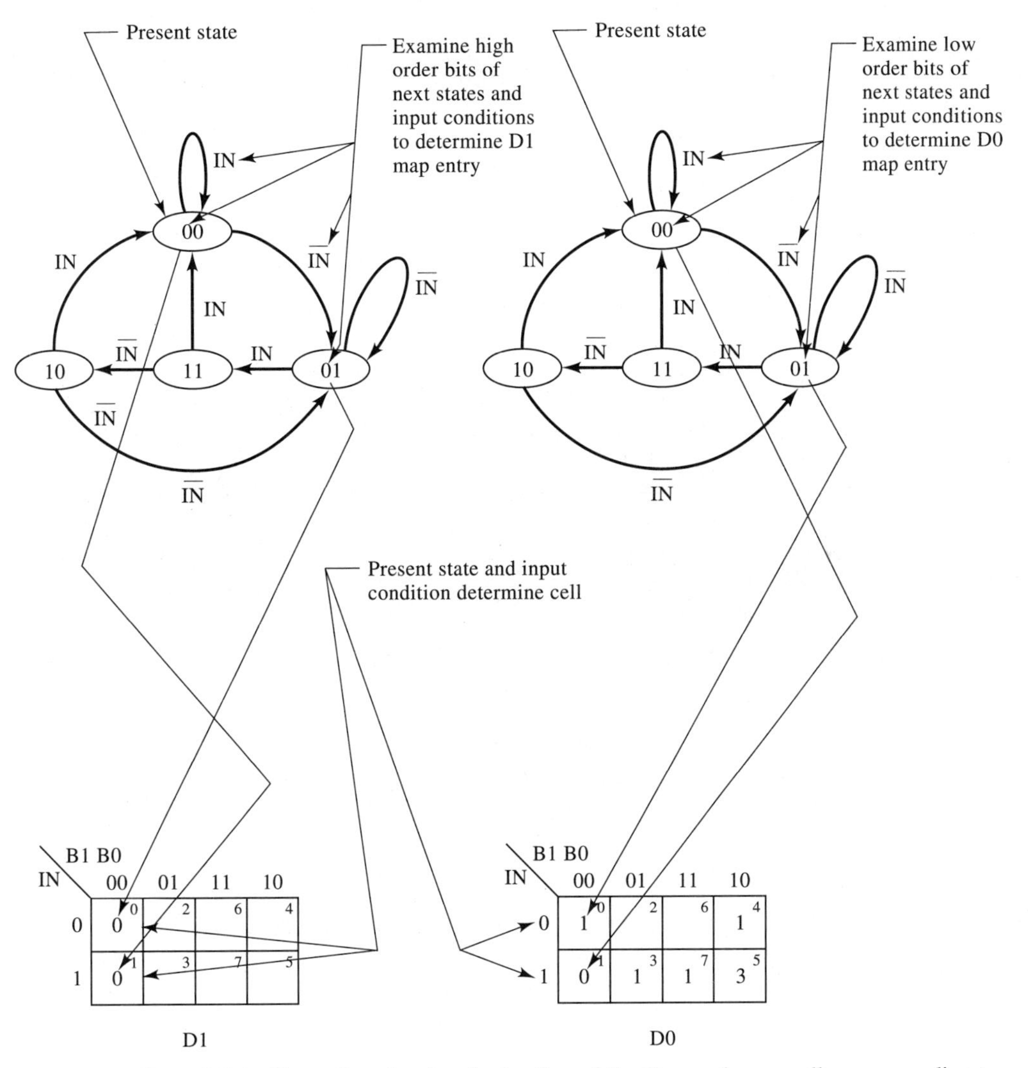

Figure 9.5-4 Illustration showing the loading of the Karnaugh map cells corresponding to the 00 state. (The outputs are omitted from the state transition diagram to simplify the illustration.)

maps for JK memory. As already stated, these maps are loaded from the D maps rather than from the state transition diagram. From each D map we'll form two maps: a J map and a K map. First we need to consider the next-state logic expressions for both D and JK flip-flops. For the D flip-flop,

$$Q_{n+1} = D$$

and for the JK flip-flop,

$$Q_{n+1} = J\overline{Q}_n + \overline{K}Q_n = D$$

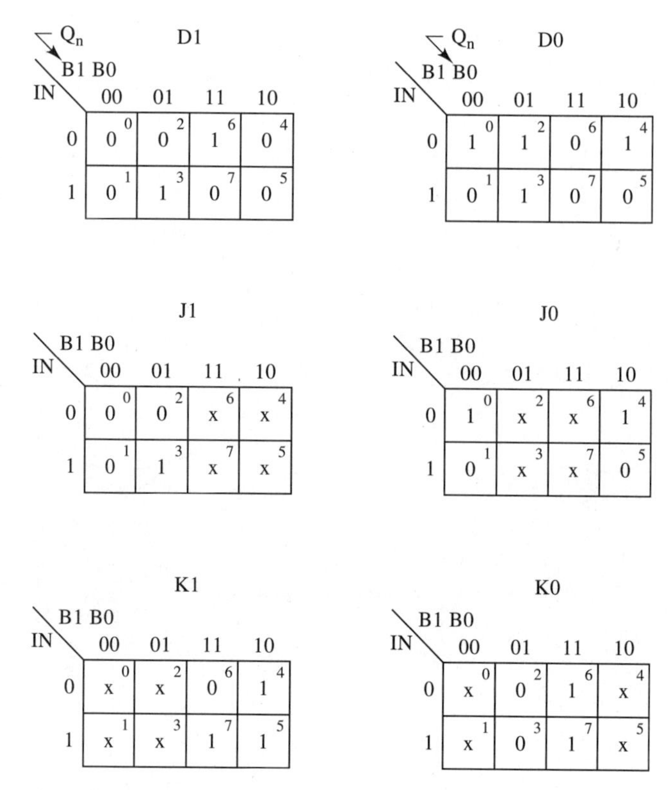

Figure 9.5-5 Forming J and K maps from D maps.

where Q_{n+1} is equal to D from earlier. In solving this equation, we can find J and K in terms of D. Clearly J and K depend on Q_n, the present state of the memory. If $Q_n = 0$, then $J = D$ and it does not matter what K is. If $Q_n = 1$, then $\overline{K} = D$ or $K = \overline{D}$ and it does not matter what J is. These results can be seen more clearly in a table.

Q_n	J	K
0	D	x
1	x	$\overline{D}$

The table tells us exactly how to form the J and the K decoder maps from the D decoder map, but we need to realize that Q_n in the table is the bit in the present-state code that corresponds to the memory whose decoder we're forming. Thus, the J0 and K0 decoder maps are formed from the D0 map, and B0 is Q_n for these maps; the J1 and K1 maps are formed from the D1 map, and B1 is Q_n.

Let's try this process with the Karnaugh maps for D memory we discussed earlier. These maps, completely loaded, appear at the top of Figure 9.5-5. The figure also

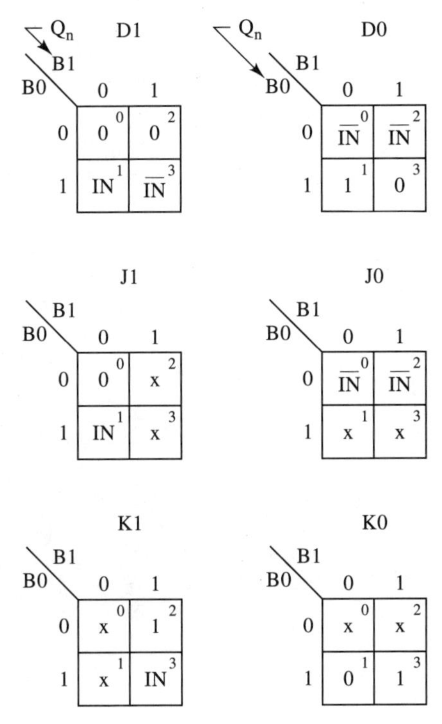

Figure 9.5-6 Forming J and K maps from D maps for VEMs.

shows how we form the J and K decoder maps. In the left-hand maps B1 is Q_n, and in the right-hand maps B0 is Q_n. For the J1 map (on the left), the cells in which B1 = 0 are filled directly from the D1 map, and the cells in which B1 = 1 are filled with don't cares (x). For the J0 map (on the right), the cells in which B0 = 0 are filled directly from the D0 map, and the cells in which B0 = 1 are filled with x. For the K1 map (on the left), the cells in which B1 = 0 are filled with x, and the cells in which B1 = 1 are filled with the complements of the values from the D1 map. For the K0 map (on the right), the cells in which B0 = 0 are filled with x, and the cells in which B0 = 1 are filled with the complements of the values from the D0 map.

Figure 9.5-6 illustrates the same transformation from D decoder maps to J and K decoder maps, but for VEMs. Notice that the entered variable is loaded into the J maps for $Q_n = 0$, and the complement of the entered variable is loaded into the K maps for $Q_n = 1$. It's important to understand this transformation for VEMs, because we'll use VEMs in all of our designs.

Now that we have most of the tools we need to efficiently design the next-state decoders of a state machine, let's consider a more complicated example.

Example 9.5-1

Design a state machine to repetitively sample an input stream, I, in slices of four samples and determine whether the result of the four samples is the sequence 1101. Each time the desired sequence is found, the output signal OUT should be asserted. Changes in the input, I, are synchronous with the negative-going edge of the clock. Use D flip-flop memory. Do not design the output decoder. Make provisions for a reset.

The following figure is a tight state transition diagram for a machine that will meet the specifications of the description; this is the primitive state transition diagram. No state assignments have been made in the figure.

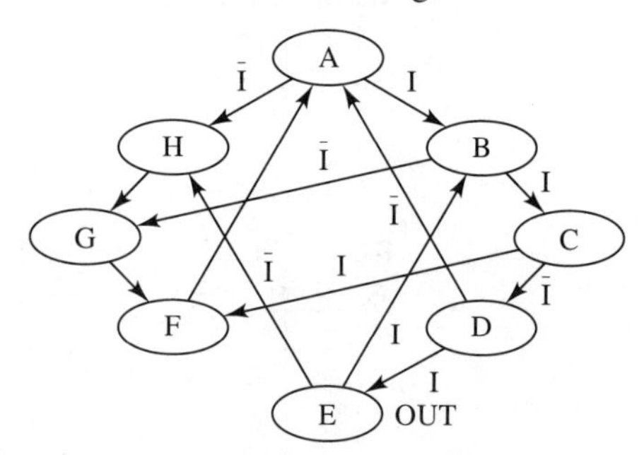

Primitive State Transition Diagram

We'll look at two different designs for the state machine, one with random state code assignments and one with state code assignments that follow the rules for simplifying the next-state decoder. The following figure shows the state transition diagram for a random state code assignment and the resulting VEMs and logic expressions for a D memory design. Notice that we've gone directly from the state transition diagram to the maps, and we've used VEMs, with the bits in the state code used as the map variables and the input used as the entered variable. This speeds the design process.

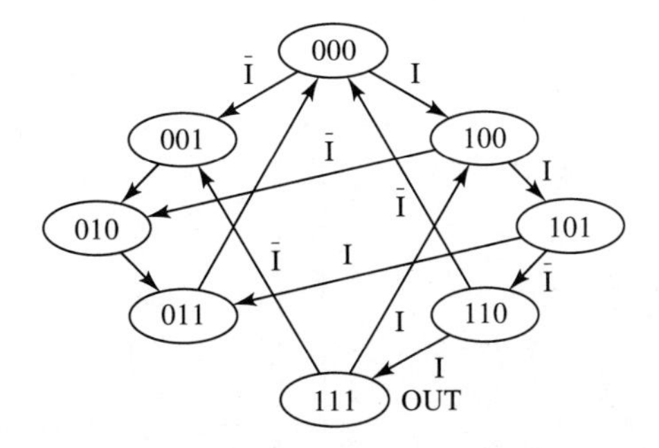

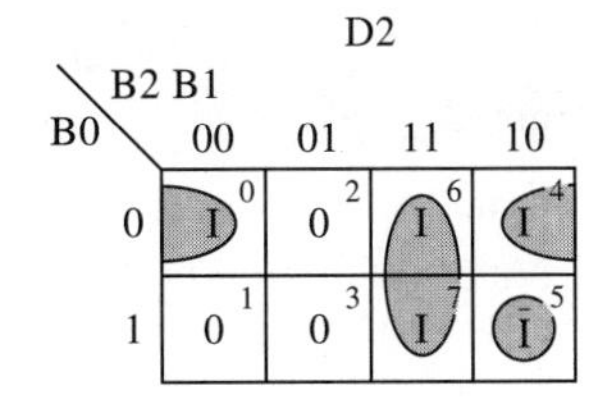

$$D2 = \overline{B1}\cdot\overline{B0}\cdot I + B2\cdot B1\cdot I + B2\cdot\overline{B1}\cdot B0\cdot\overline{I}$$

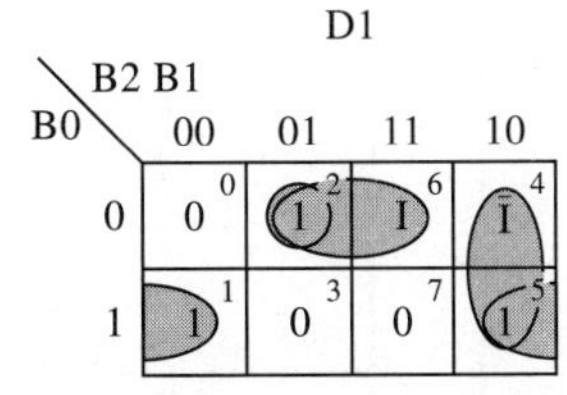

$$D1 = B1\cdot\overline{B0}\cdot I + B2\cdot\overline{B1}\cdot\overline{I} + \overline{B2}\cdot B1\cdot\overline{B0} + \overline{B1}\cdot B0$$

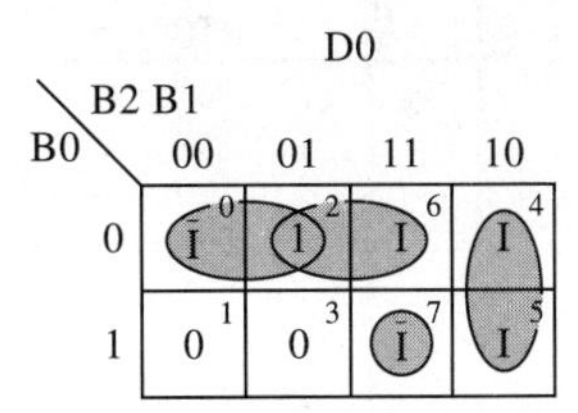

$$D0 = \overline{B2}\cdot\overline{B0}\cdot\overline{I} + B1\cdot\overline{B0}\cdot I + B2\cdot\overline{B1}\cdot I + B2\cdot B1\cdot B0\cdot\overline{I}$$

Following is the circuit resulting from this design. Notice that we needed these gates:

Four 4-input *nand*
Nine 3-input *nand*
One 2-input *nand*

We used *nand* gates because they're available in the variety of configurations needed.

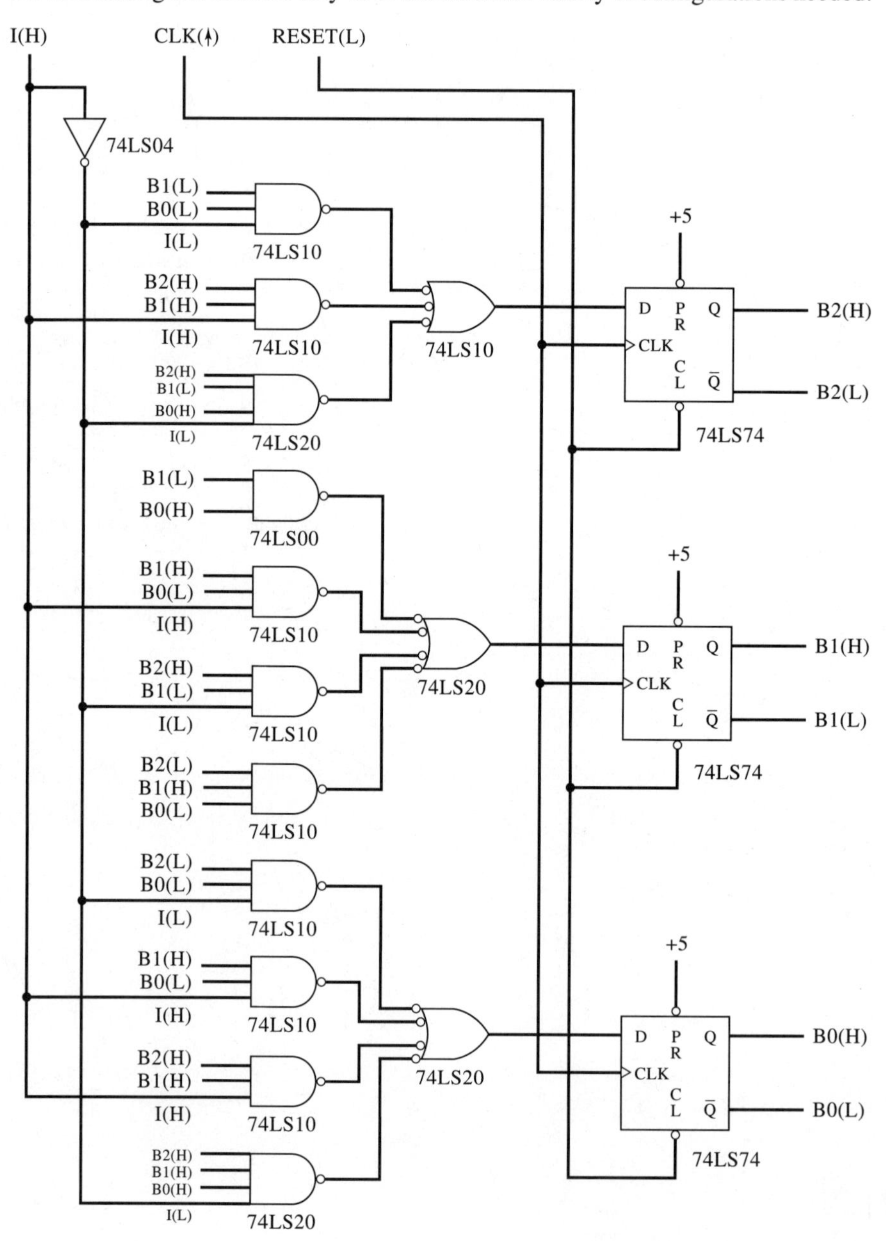

To simplify the decoder design, we return to the primitive state transition diagram without the state code assignment and assign codes, using the first and second rules of Section 9.3.3. We need to examine each state in the diagram, to determine the previous states so that we can use rule 1 and to determine the next states so that we can use rule 2. This can be done in a table, as follows.

Previous State	State	Next State
		I, $\bar{I}$
D, F	A	B, H
A, E	B	C, G
B	C	F, D
C	D	E, A
D	E	B, H
C, G	F	A
B, H	G	F
A, E	H	G

The transitions from A and E to B and from A and E to H have common input conditions. The other transitions, D and F to A, C and G to F, and B and H to G, have input conditions that are partly the same. For example, the transition from D to A occurs for $\bar{I}$, whereas the transition from F to A is independent of I (occurs for I and $\bar{I}$). The next-state branching can only be binary (two-way) because there is only one input.

In the table, the next state listed first is the next state for input I, and the next state listed second is the next state for input $\bar{I}$. We need to know this so we can make part of the state code match the input condition. Each of the states listed first should have a 1 in one of the bits in its code, and each of the states listed second should have a 0 in the same bit. One way to attempt to satisfy all the logic-adjacent groupings required for the previous state and next state is to make tentative assignments of the states in a map, using the letters we've assigned as state names, and see how they group.

The groupings required in this present example are easy to satisfy. The previous-state groupings happen to be the same as the next-state groupings, and only groupings of two occur. The following map shows a possible state assignment.

B0 \ B2 B1	00	01	11	10
0	A (0)	H (2)	G (6)	D (4)
1	E (1)	B (3)	C (7)	F (5)

In this assignment, we've placed the pairs of states to be grouped together in columns in the map so that the state codes in each pair are logic-adjacent. To make part of the state code match the input condition, we always place the state of the pair associated with the branching in which I is asserted in the lower row, where B0 is 1, and the state associated with the transition in which I is not asserted in the upper row, where B0 is 0.

Following are the state transition diagram, the next-state decoder VEMs for these state code assignments, and the circuit. Notice the considerable simplification of the next-state decoder logic for this design.

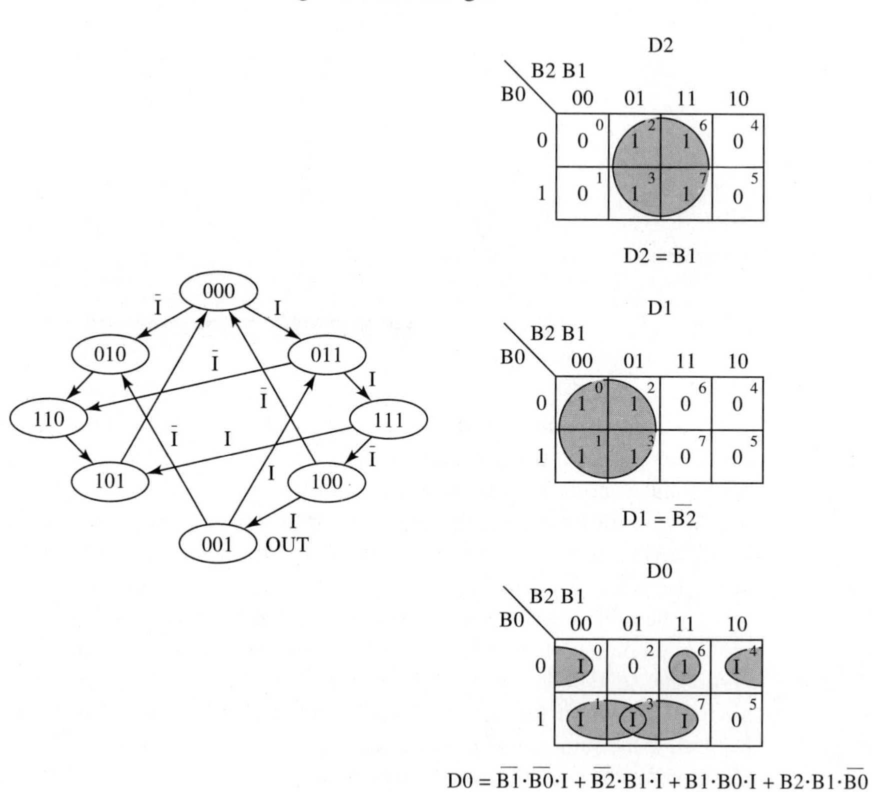

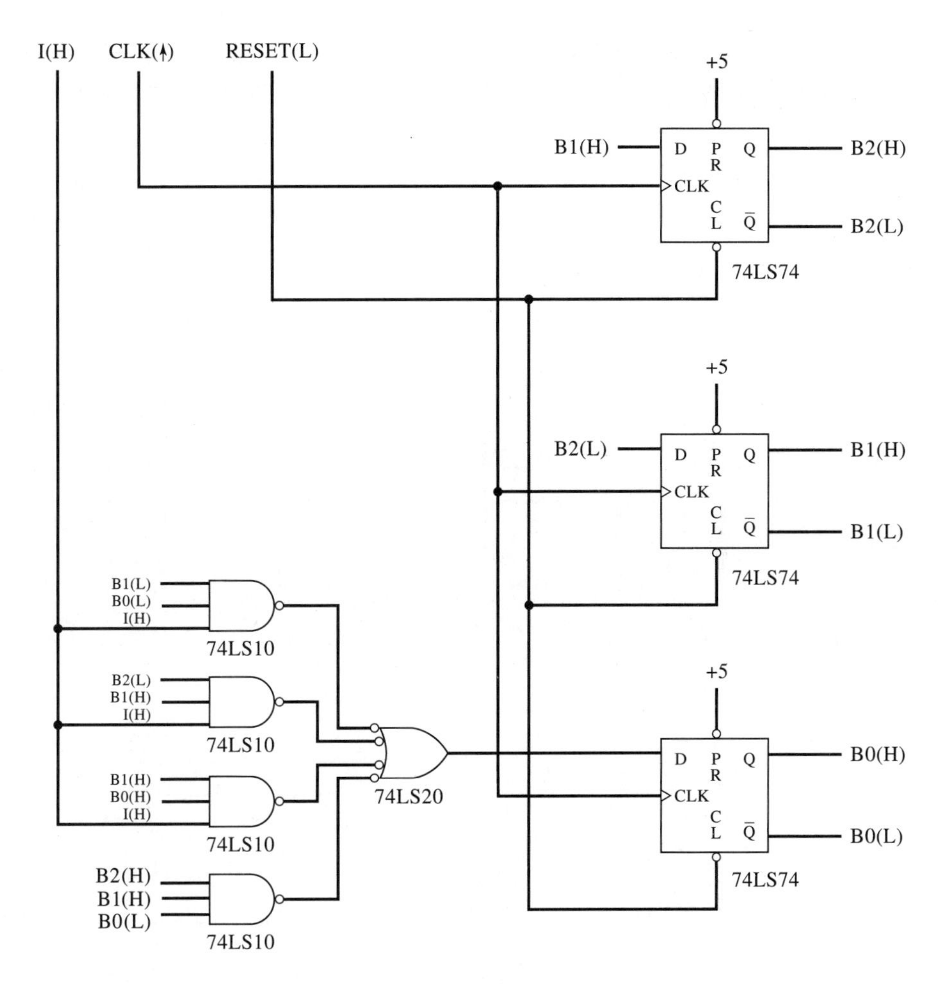

Here the decoder uses the following gates:

Four 3-input *nand*
One 4-input *nand*

Many other possible state code assignments satisfy the first and second rules for grouping states. These other groupings may produce some decoders that are simpler than the one we've produced, and some that are more complicated. In most cases it's not worth the engineering time to find the absolutely best design. It's usually advisable to make a cut at simplification and use it. One exception is the case of a large production run, where it would be worth a large investment of engineering time to get the simplest and least expensive design.

The final step in any design, before building a prototype, is a simulation. Following are the results of a simulation of the circuit with simplified decoder logic. The simulation shows both the next-state code, D2 D1 D0, and the next state, B2 B1 B0. Just as we did in Example 9.7-1 we have added the logic values of the input and the state bits to the simulation to make it easier to follow. We have also marked off the 4-bit input slices at the top of the simulation. The slices start immediately after the reset that puts the state machine into state 000. The first 4-bit input slice, 1001, causes the machine to make transitions through states 011, 110, and 101 and back to 000. This is not the desired sequence, but the machine behaves just as it should. The next 4-bit input slice, 1101, causes the machine to make transitions through states 011, 111, 100, and 001. This is the desired sequence, and if the output decoder were in place the output would be asserted in state 001. If we look at the other 4-bit sequences, we will see that the machine functions just as expected.

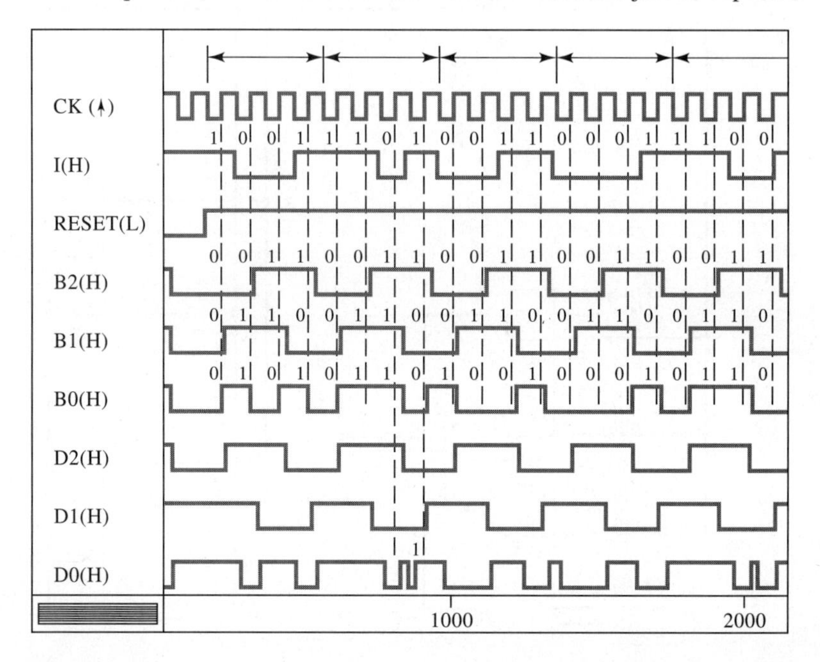

Now we observe the timing behavior of the next-state code. Notice that the next-state code, in particular D0, may change several times between one positive-going clock edge and the next (see where the dashed lines in the figure are extended down to D0). This happens because of the differences in the delays for the various inputs to the next-state decoder. Such differences are no problem as long as the next-state code reaches the correct value before the next clock edge so that the state code bits make the correct transitions. In this design the clock frequency was chosen to be low enough so that this would happen. If we raise the clock to a high enough frequency, the next-state code will still be changing when the clock edge occurs, and the machine may be clocked into an improper state.

As we noted in the example, it takes time for the output of the next-state decoder to stabilize after a change in state or input. This delay limits the maximum frequency of operation of a state machine, because the next-state code must be stable before the clock edge that clocks the next state occurs. In fact, because there is normally a setup time for the memory flip-flop, the next-state code must be stable one setup time before the clock edge occurs. The time at which the state code becomes stable is determined either by the time it takes for a change in the state code to propagate through the memory flip-flops and back through the gates of the next-state decoder, or by the time it takes for a change in an input to propagate through the next-state decoder. In the example, the next-state decoder has at most two gate delays for state code changes and two gate delays for input changes. If we assume a maximum gate delay (t_p) of 15 ns and a maximum propagation delay through the memory flip-flops of 40 ns, the maximum time it takes for a change in the state code to propagate through the flip-flops and next-state decoder is

$$t_d = 15 + 15 + 40 = 70 \text{ ns}$$

Now, because the next-state code must be stable one setup time before the clock edge for the next state (20 ns), we must add this minimum setup time to the delay to find the maximum total time from clock edge to clock edge:

$$T_{CLK} = 70 + 20 = 90 \text{ ns}$$

This means that the maximum clock frequency, based on the time it takes for a change in state to propagate back to the memory inputs, cannot exceed $1/9 \times 10^{-8} = 11.1$ MHz.

For the example design, the timing for an input change turns out to be more critical. This is because the input does not change until the middle of the clock cycle (negative clock edge). To determine the timing constraint due to an input change, we need to add a delay of one-half the clock period to the delay through the decoder gates and the setup time of the flip-flops:

$$T_{CLK} = \frac{T_{CLK}}{2} + 15 + 15 + 20$$

So, solving for T_{CLK},

$$T_{CLK} = 100 \text{ ns}$$

This means that the maximum clock frequency, based on the time it takes for a change in input to propagate to the memory inputs, cannot exceed $1/1 \times 10^{-7} =$ 10 MHz. Because the timing path for a change in an input is more critical than that for a change in state, it is said to be the **critical path**, or the slowest path. One way to analyze the timing constraints of a circuit is to determine the critical path timing. This normally involves finding the timing for a number of paths in order to identify the critical path. The critical path timing shows us that the maximum frequency at which this system will operate correctly is 10 MHz. If we operate at this frequency, there is no margin for error in the timing. In an actual design we would normally limit the frequency to 9 MHz to give us a margin of 10%, or about 10 ns.

Before the development of good simulation packages for computers, critical path analysis was virtually the only way to analyze circuit timing. Now simulation is a much more attractive approach. The simulation of the preceding design shows that the next-state code is stable before the next clock edge, and we can even see

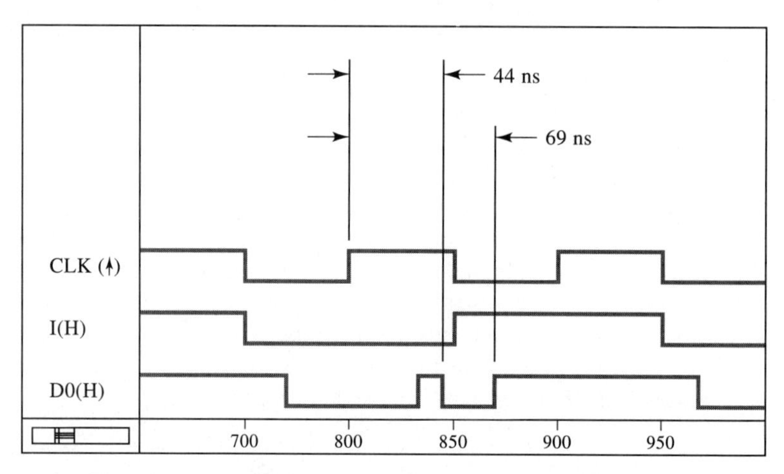

Figure 9.5-7 Time for the next-state code bit D0 to become stable. The smaller time is for a change in the state code. The larger is for a change in the input.

when it becomes stable in Figure 9.5-7. This figure, an enlargement of part of the timing diagram, shows the time at which the next-state code bit D0 becomes stable due to a change in state (44 ns) and the time at which it becomes stable due to a change in input (69 ns). The important time is the interval from the clock edge, as shown. Remember that the next-state code must become stable one setup time (20 ns) before the next clock edge occurs. The next-state code actually becomes stable 31 ns before the clock edge occurs—a margin of 11 ns. The times shown are the maximum delays found in the simulation. For some changes of state and inputs, the next-state code stabilizes more quickly.

The delays in the simulation are smaller than the delays we calculated in the critical path analysis. This is because the simulation uses *typical* propagation delays for the gates and flip-flops. It is also possible to set the simulation to use *maximum* or *minimum* delays. Figure 9.5-8 shows the timing for the same period of time as in

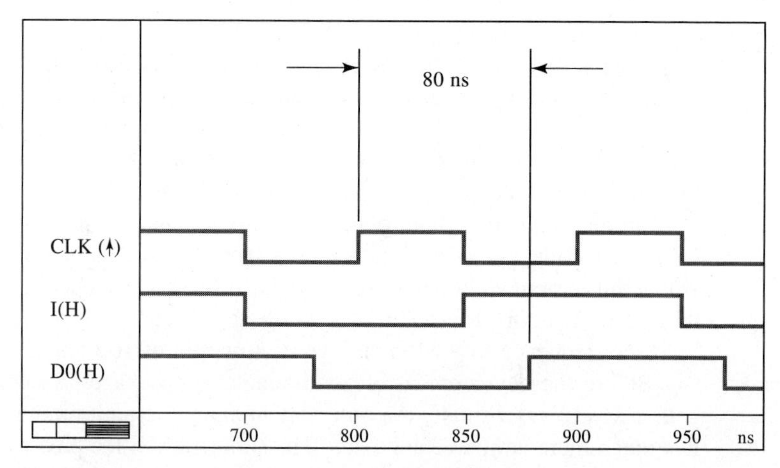

Figure 9.5-8 Time for the next state code bit D0 to become stable when maximum progagation delays for the gate and flip-flops are used.

Figure 9.5-7 when maximum propagation delays are used for the gates and flip-flops. The delay from the clock edge to where D0 becomes stable and correct is 80 ns. The setup time (time to the next clock edge) is thus 20 ns, which is just the minimum setup time for the flip-flop. This agrees with the critical path timing. Although the minimum delays are not critical in this design, they are in some designs.

9.6 OUTPUT DECODER DESIGN

Output decoder design consists of three steps. (1) We make an output table using the state transition diagram, which may simply be added to the present state–next state table. (2) We use maps (Karnaugh or VEM) to simplify the output logic expressions. (3) We turn the output logic expressions into circuits. As with the next-state decoder design, not all of these steps are necessary. It is possible (and, for large systems, almost essential) to bypass the output table and go directly to the maps. This simplification is much easier to implement for the output decoder than it was for the next-state decoder.

To see this simplified output decoder design procedure, we'll continue the design of Section 9.2. We've repeated the state transition diagram in Figure 9.6-1. Because there are three outputs, we'll need three maps—one for each output. This is a Moore machine (the outputs don't depend on the inputs but only on the state), so we'll make Karnaugh maps with the state code bits used as the map variables. For a Moore machine, as long as the state code contains fewer than four bits, Karnaugh maps can be used to simplify the output expressions.

Figure 9.6-2 consists of the bare maps labeled with the outputs. Each cell in each output map corresponds to a state in the state transition diagram. Figure 9.6-3 shows how to load the maps. We consider each state and each output map in turn, and load a 1 into the cell corresponding to that state if the output is asserted in that state and a 0 if it's not asserted. For instance, the output FOUND is not asserted in the 00 state, so 0 is loaded into cell 0 (00) of the FOUND output map. The output

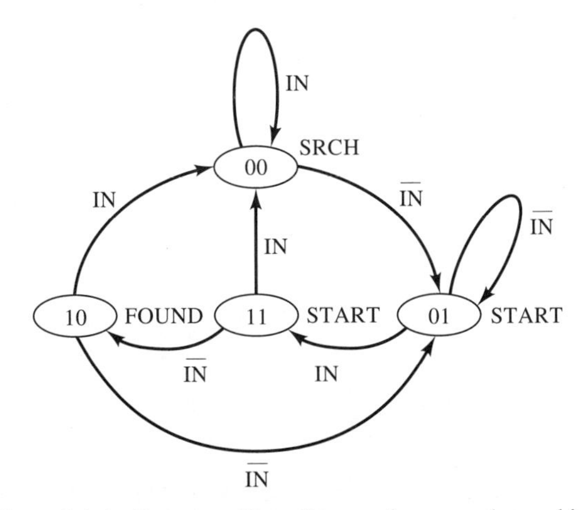

Figure 9.6-1 State transition diagram for example machine.

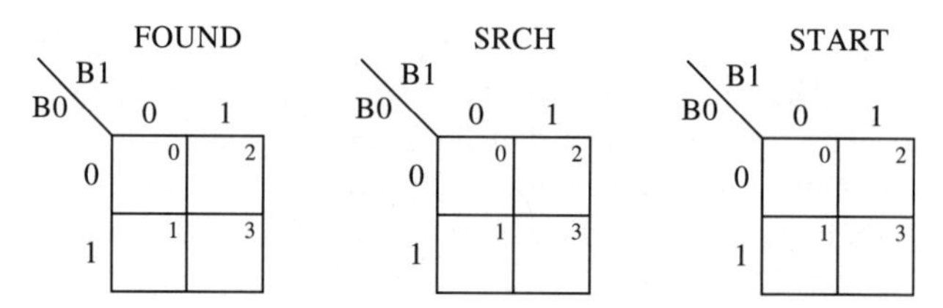

Figure 9.6-2 Bare output maps.

SRCH is asserted in the 00 state, so 1 is loaded into cell 0 of the SRCH output map. The output START is not asserted in the 00 state, so 0 is loaded into cell 0 of the START output map. The output FOUND is not asserted in the 01 state, so 0 is loaded into cell 1 of the FOUND output map. The output SRCH is not asserted in the 01 state, so 0 is loaded into cell 1 of the SRCH output map. The output START is asserted in the 01 state, so 1 is loaded into cell 1 of the START output map. By considering each state and each map in turn, we can load all the required 0s and 1s

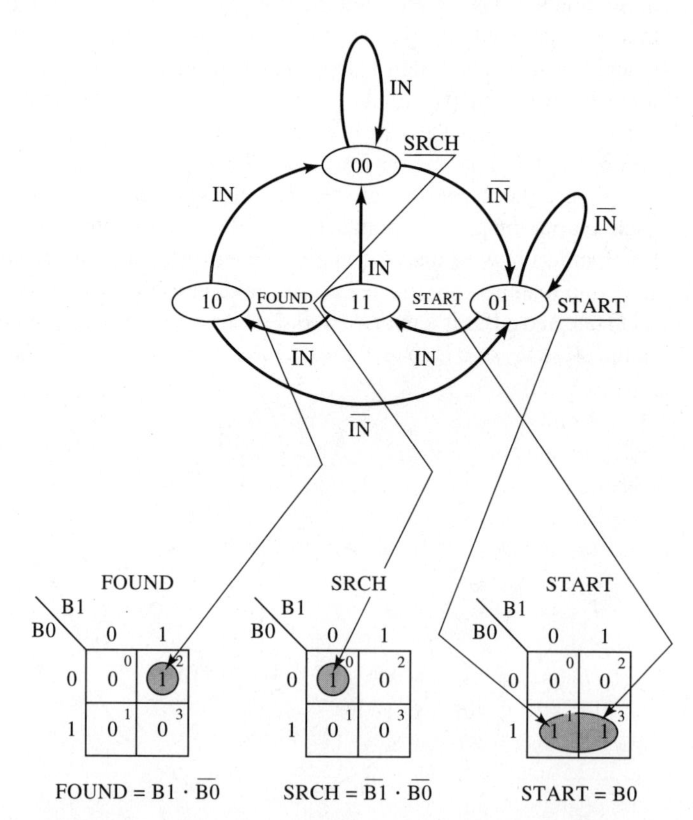

Figure 9.6-3 Illustration showing how to load the output maps.

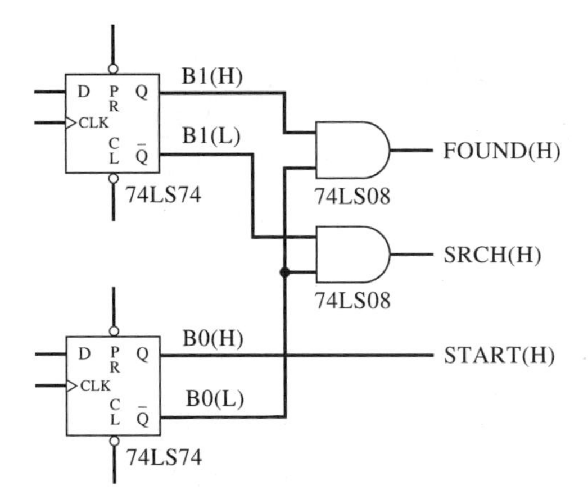

Figure 9.6-4 Output decoder for example machine.

into the output maps. When we've considered all the states, those cells remaining empty represent unused states, and we fill them with x.

To finish the design of the output decoder, we turn each of the simplified logic expressions for the outputs (shown below the maps in Fig. 9.6-3) into circuits. Notice that there is very little simplification of the output logic expressions. This is typical when the output is asserted in only one state. Figure 9.6-4 shows the output decoder circuit.

Before moving on, let's explore two more aspects of output decoder design. First, consider the problem of choosing the state code to simplify the output decoder. The only simplification that can be achieved by choosing the state code assignments has been made. The states with the same outputs (00 and 01) are logic-adjacent. If we were to exchange the 10 and 11 state code assignments, the states with the same outputs would no longer be logic-adjacent, and a more complicated output decoder with two additional gates would be needed.

Second, consider the problem of glitches in the output. The only way we can assure ourselves that there will be no glitches in the output is to have all the state transitions logic-adjacent—differing by only one bit. We've tried to choose the state code assignments so that this condition will be satisfied, but for the state machine under discussion it's not possible—in fact, it's seldom possible for any state machine. We have only two transitions that do not differ by one bit: the 11-to-00 transition and the 10-to-01 transition. We'll look at them and see what glitches we might have. Notice that in order to do this, we must analyze the actual decoder circuit. We can see from the circuit that there will be no glitches in START, because it depends only on B0. If B0 changes, START changes, but B0 will never do more than make a simple change when any state transition is made. This isn't true for the other outputs. We must analyze the circuit in some detail to see what happens to them when the state changes.

First consider the 11-to-00 transition. Figure 9.6-5 shows how the inputs to the

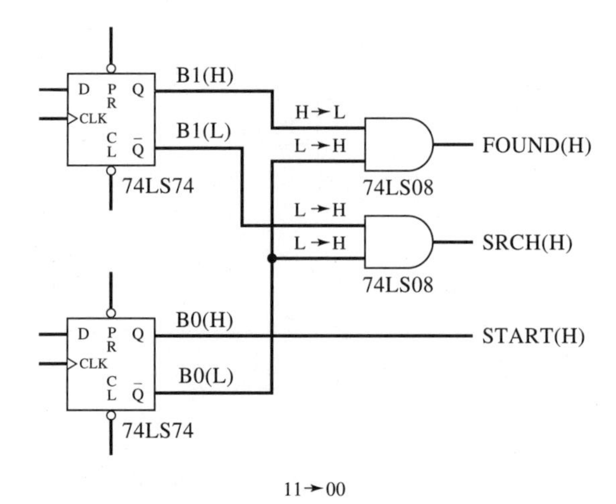

Figure 9.6-5 Changes at the inputs to the output decoder when the state changes from 11 to 01.

SRCH and FOUND decoders change when this transition is made. The notation H → L means the input starts high and changes to low. For this transition, FOUND should start low and remain low. But suppose the low-to-high change on the second gate input occurs before the high changes on the first input. Both of the inputs to the gate will be high, and the output will be high. The output will remain high until the first input changes to low. So we can see that FOUND can start low, go high, and then return to low. This is a glitch. The output FOUND is erroneously asserted on a transition between the 11 and 00 states.

We should *always* assume that because a glitch *can* occur, it *will* occur. In this case, we should consider the circuit that uses the output FOUND to determine whether the glitch will produce undesired results. If it does, then the output of our state machine will require a deglitching circuit, which we'll discuss. *Never ignore a possible glitch and hope that the timing will prevent it.* If you do, you'll have designed a circuit that is potentially unreliable and unreproducible. It may fail when it's heated or cooled, and even though a given circuit may function properly, a different but identical circuit with different components may not.

Observing the input changes on the output decoder for SRCH, we can see that no glitch is possible. Both inputs to the *and* gate change from high to low, and it does not matter which changes first—the output will change to high after both inputs change.

Figure 9.6-6 shows how the inputs to the SRCH and FOUND decoders change when the 10-to-01 transition is made. Analyzing the effect of these changes, we find that a glitch can occur in SRCH. Again we need to determine whether the glitch can cause undesirable results in the circuit that uses the output SRCH.

The glitches we expect in this circuit can actually be seen in a simulation. Figure 9.6-7 shows the results of a simulation where the input is chosen so that the problem

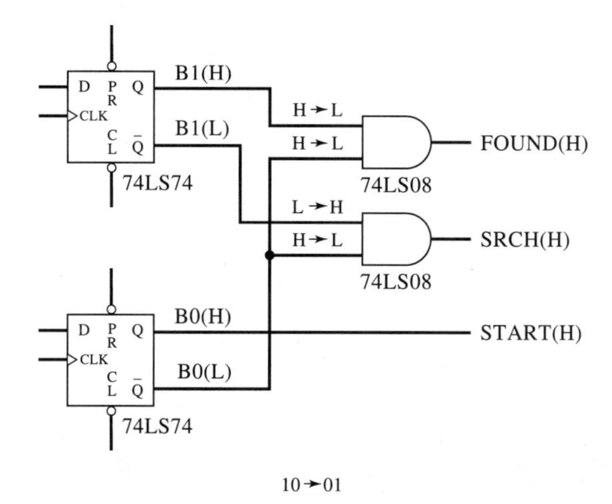

Figure 9.6-6 Changes at the inputs to the output decoder when the state changes from 10 to 01.

transitions occur. Notice the glitches that occur in FOUND on the 11-to-00 transition and in SRCH on the 10-to-01 transition.

We've taken the time to analyze potential glitches in this circuit in order to demonstrate what is possible. Usually, if glitches were a potential problem, we would simply design an output deglitching circuit.

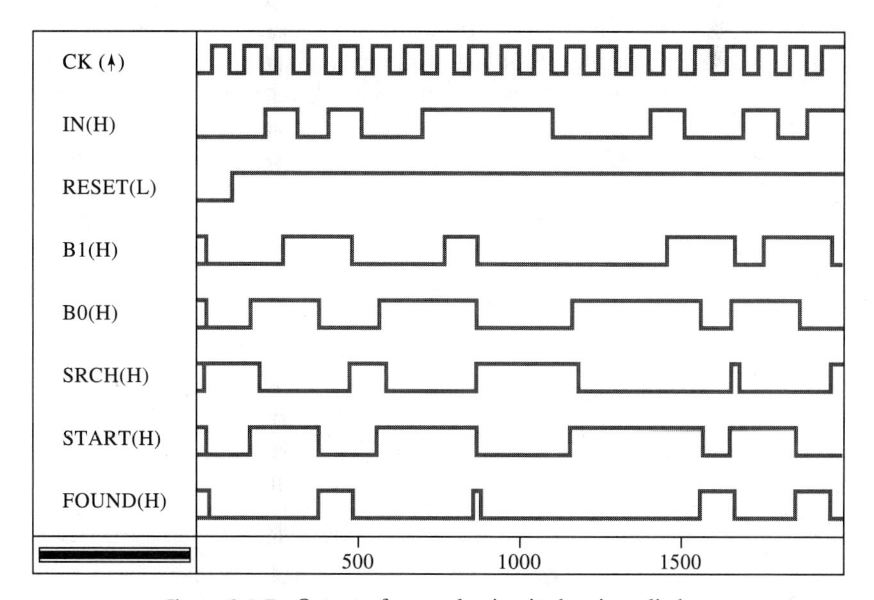

Figure 9.6-7 Output of example circuit showing glitches.

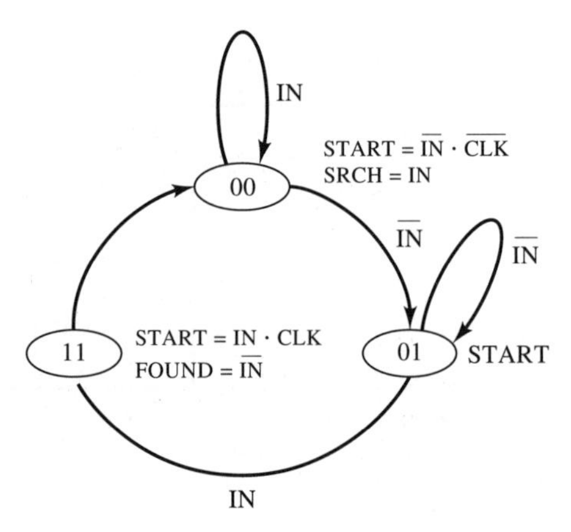

Figure 9.6-8 State transition diagram for the machine of Example 7.2-1.

At the beginning of this section, we saw how to load an output decoder map for a Moore machine directly from the state transition diagram. Now we will achieve the same simplified load for a Mealy machine. Figure 9.6-8 is the state transition diagram for the machine of Example 9.2-1, which we'll use again here. Because there are three outputs, we'll need three maps, one for each. In a Mealy machine the outputs depend on the inputs as well as on the state, so we'll find it convenient to use VEMs to simplify the output logic expressions. In these maps we use the state code bits as the map variables and the inputs as the entered variables. Figure 9.6-9 is the bare maps. Each cell in each map corresponds to a state in the state transition diagram. Figure 9.6-10 shows how to load the maps. We consider each state and each output map in turn, and load the *condition* into the cell corresponding to the state in that output map if the output is conditional, a 1 into the cell if the output is unconditional, and a 0 into the cell if the output is not asserted in that state. For instance, one of the outputs asserted in the 00 state is SRCH, and it is conditional on IN, so IN is loaded into cell 0 (00) of the SRCH output map. Another output asserted in this state is START, and it is conditional on $\overline{IN} \cdot \overline{CLK}$, so $\overline{IN} \cdot \overline{CLK}$ is loaded into cell 0 of the START output map. The third output, FOUND, is not asserted in this state, so a 0 is loaded into cell

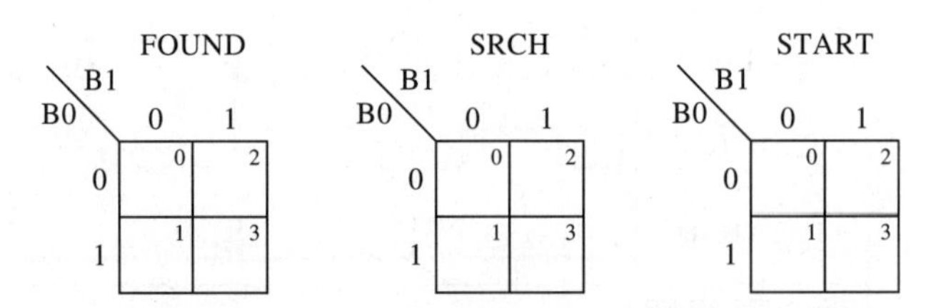

Figure 9.6-9 Bare output decoder maps.

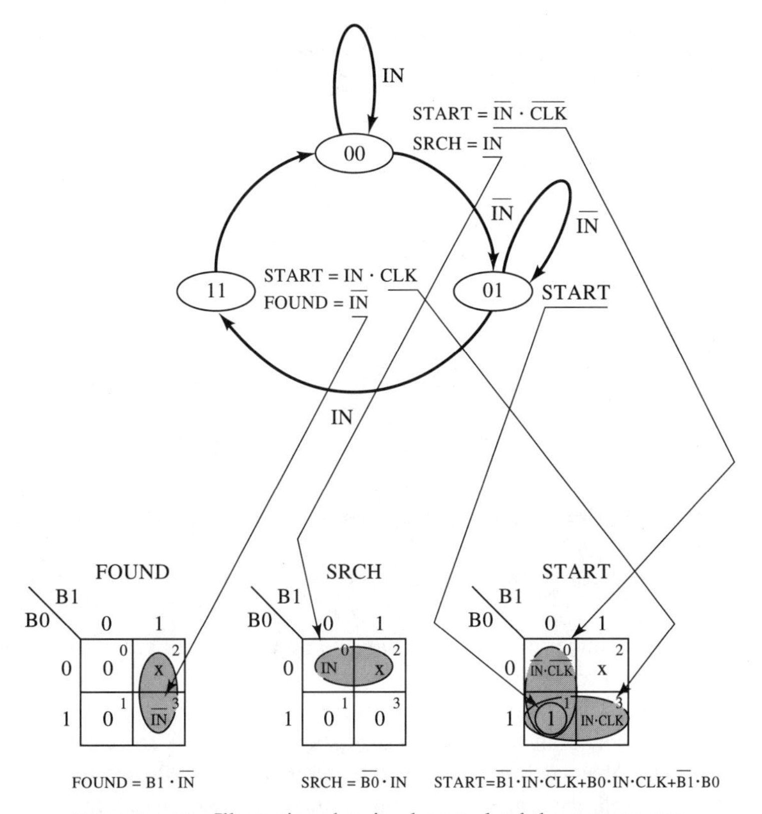

Figure 9.6-10 Illustration showing how to load the output maps.

0 of the FOUND output map. The only output asserted in the 01 state is START, and it is an unconditional output, so 1 is loaded into cell 1 (01) of the START output map. Cell 1 in the other output maps is loaded with 0. By considering all the states and all the maps, we can load all the required 0s, entered variables (conditions), and 1s in the output maps. When all the states and all the maps have been examined, and corresponding 0s, conditions, and 1s loaded into the maps, the remaining cells represent unused states and are filled with x.

To finish the design of the output decoder, we turn each of the simplified logic expressions for the outputs (shown below the VEMs in Fig. 9.6-10) into circuits. Notice that there is very little simplification of the output logic expressions. This is typical when an output is asserted in only one state. Figure 9.6-11 shows the output decoder circuit.

The easiest output maps to generate directly from the state transition diagram for a Mealy machine are VEMs with the bits of the state code used as the map variables and the conditions on the output used as entered variables. If the com-

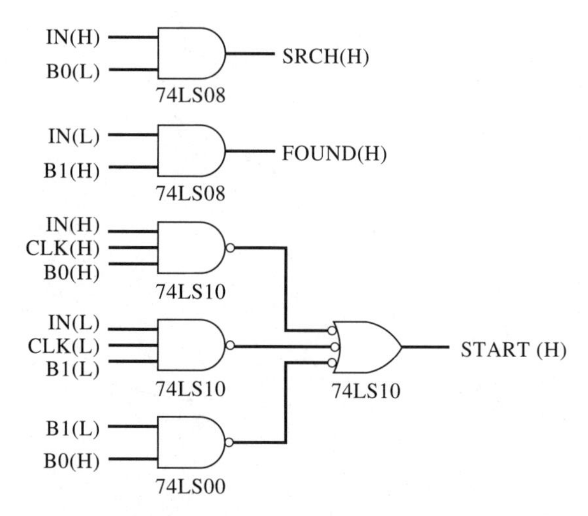

Figure 9.6-11 Output decoder circuit for the machine of Example 7.2-1.

bined number of state code bits and inputs is small enough (less than four), it's also possible to form Karnaugh maps for simplification of the outputs. Figure 9.6-12 shows how this is done. There are still three output maps—one for each output. However, both the state code bits and the inputs are used as map variables. This means that a cell is determined by both the state code and the input. We still examine each cell and each output map in turn and determine which 0s and 1s to load into the map.

For instance, the 00 state has an output, SRCH, that is conditional on IN but independent of CLK, so cells 2 (B1 = 0, B0 = 0, IN = 1, CLK = 0) and 3 (B1 = 0, B0 = 0, IN = 1, CLK = 1) of the SRCH output map are loaded with 1, and cells 0 (B1 = 0, B0 = 0, IN = 0, CLK = 0) and 1 (B1 = 0, B0 = 0, IN = 0, CLK = 1) are loaded with 0. This state also has an output, START, that is conditional on $\overline{\text{IN}} \cdot \overline{\text{CLK}}$, so cell 0 (B1 = 0, B0 = 0, IN = 0, CLK = 0) of the SRCH output map is loaded with 1, and cells 1 (B1 = 0, B0 = 0, IN = 0, CLK = 1), 2 (B1 = 0, B0 = 0, IN = 1, CLK = 0), and 3 (B1 = 0, B0 = 0, IN = 1, CLK = 1) are loaded with 0. If an output is asserted unconditionally, as is START in state 01, then all the cells in the map with these state bits are loaded with 1 (cells 4, 5, 6, and 7). Similarly, if an output is not asserted in a state, as is FOUND in state 00, then all the cells in the map with these state bits are loaded with 0 (cells 0, 1, 2, and 3). When all the states and all the maps have been examined and corresponding 0s and 1s loaded into the map, the remaining cells represent unused states and are filled with x.

The output decoder for this state machine does not allow any simplification based on a proper choice of the state codes. It does have a transition that is not logic-adjacent—the 11-to-00 transition. An analysis would show that this transition could cause a glitch in START. We'll leave the analysis to you.

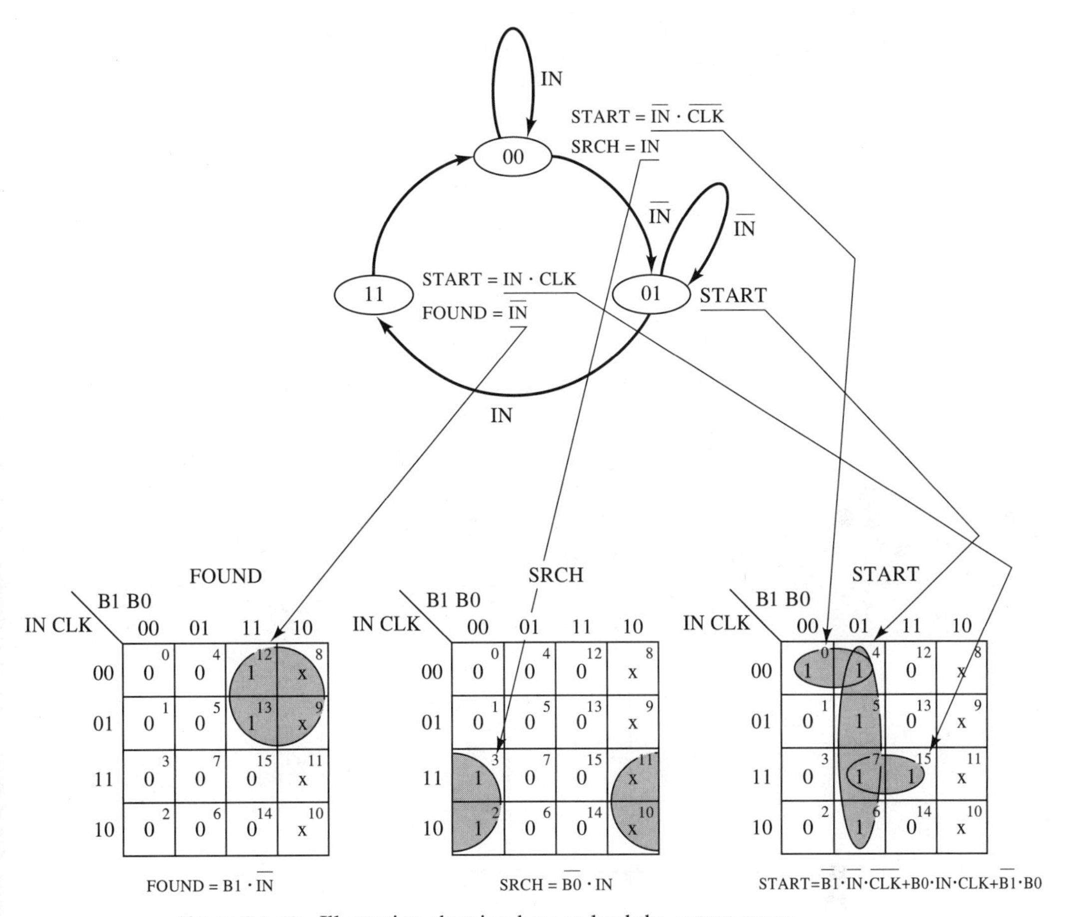

Figure 9.6-12 Illustration showing how to load the output maps.

9.7 DEGLITCHING THE OUTPUT

It should be clear from Section 9.6 that any state machine output decoder more complicated than a simple wire can produce glitches in the output of the machine. If the glitches are unacceptable, they can be removed by adding a D flip-flop to each output, waiting until the output is stable, and clocking the output into the flip-flop. This is called a **register output**. As we've learned, glitches in a state machine output can occur near each transition, just after the clock edge that caused the transition. Figure 9.7-1 shows how glitches would occur in a state machine output (OUT1) if the machine were positive-edge clocked. Because the glitches occur near the positive-going clock edge, the output flip-flops can be clocked on the negative clock edge after the glitches are over and the output is stable.

Figure 9.7-2 shows a deglitching circuit for a state machine with four outputs,

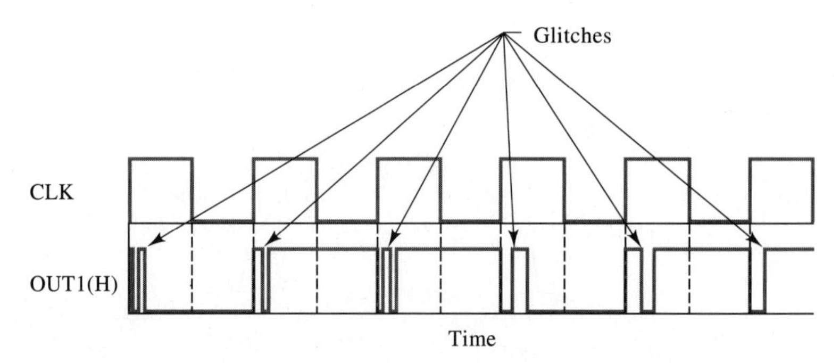

Figure 9.7-1 Position of output glitches near the clock edge.

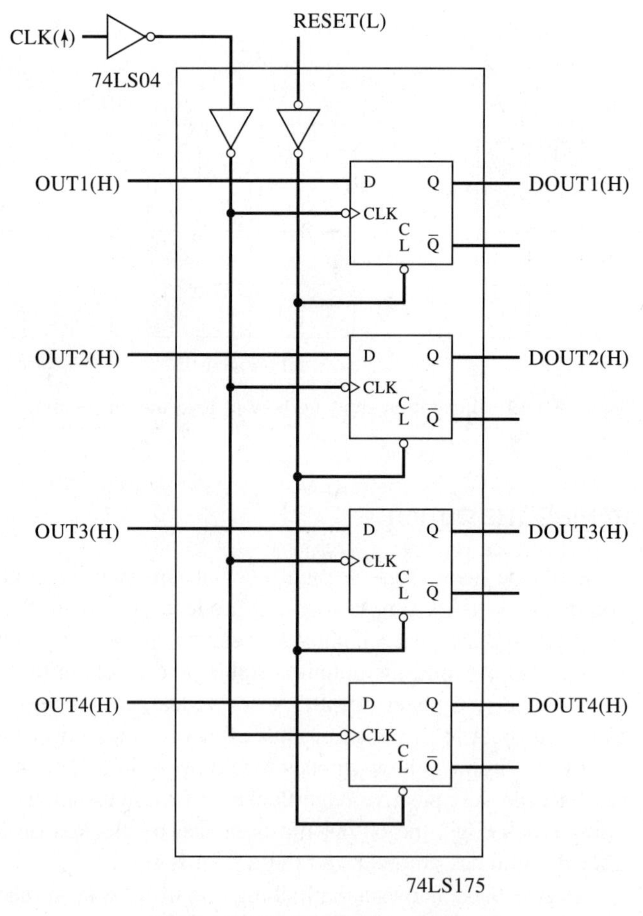

Figure 9.7-2 Deglitching circuit.

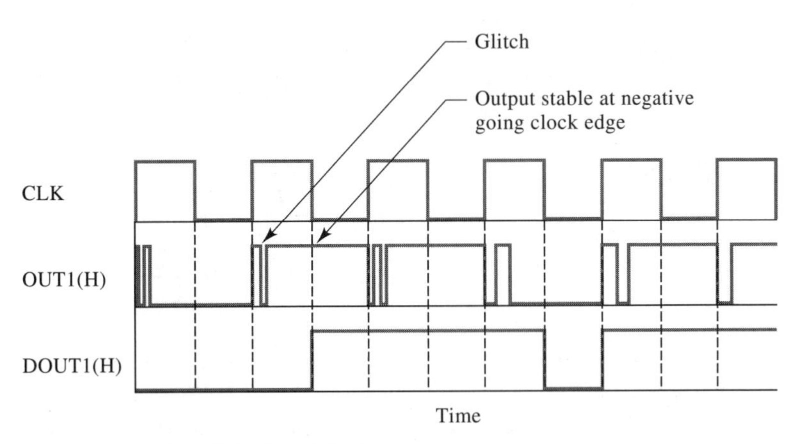

Figure 9.7-3 Deglitched output.

which we've called OUT1, OUT2, OUT3, and OUT4. This circuit uses a register of four D flip-flops with a common clock and reset. The D input for each flip-flop in the register is connected to an output of the state machine. The clock is inverted so that the clocking for the flip-flops will occur on the negative-going clock edge. The outputs of the flip-flops are the deglitched state machine outputs, which we've called DOUT1, DOUT2, DOUT3, and DOUT4. The reset is part of the sanity circuit; it allows us to start the machine with all the outputs low (not asserted).

To see how the deglitching circuit works, consider the timing diagram of Figure 9.7-3. It shows the same output as Figure 9.7-1 (OUT1), which has glitches at each of the positive clock edges. Now look at the deglitched output DOUT1, which is clocked on the negative clock edge. In each case where there were glitches in OUT1, DOUT1 was not clocked until after the glitches were over, so DOUT1 is free of glitches and reflects the correct output. One effect of this registered output is to delay the output signal by one-half clock cycle for the state transitions.

In the deglitching circuit just described, the output data were clocked into the output registered on the clock edge opposite the edge that clocks the state transitions. This was done because it's easier to see how the deglitching circuit works when the clock for the state change is separated from the clock for the output register. It's more common to clock the output register and the states with the same clock edge. The clock edge for the output register will clock the outputs before the glitches, as shown in Figure 9.7-4, because the glitches are generated by signals that are clocked by this same edge but suffer a propagation delay through the state machine memory and output decoder. As seen from the figure, when we clock the outputs on the same edge as the state transitions, there is a complete clock cycle in which the glitches can disappear.

Although this kind of clocking allows more time for the output signal to stabilize after the transition, it causes a full clock cycle of delay in the deglitched outputs. The delay is easily avoided, however. All we need to do is to decode the state just before the one for which we wish to have the output asserted, and feed this decoder

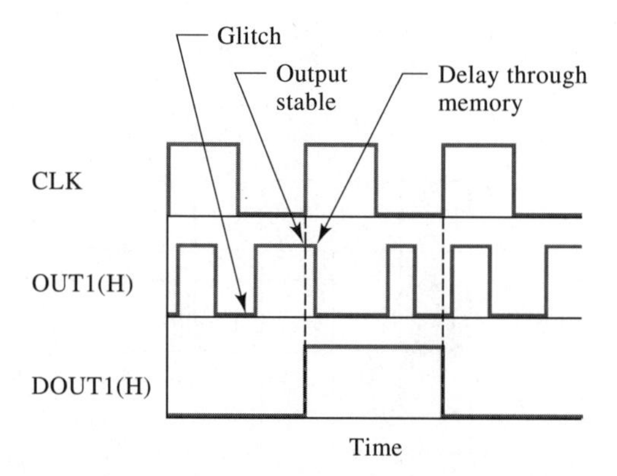

Figure 9.7-4 Deglitching the output with the positive clock edge.

before the one for which we wish to have the output asserted, and feed this decoder output to the output register. The decoded output will be stable at the beginning of the state at which we want the output asserted, and the clock edge will latch the output into the output register. Even though the output register is normally clocked on the same edge as the states, we should not discard the possibility of clocking the output register on the opposite edge and producing a deglitched output delayed by one-half clock cycle from the state transitions.

Let's use our sample state machine to show how to produce deglitched outputs that are not delayed. Figure 9.7-5 is the state transition diagram for this machine. As we noted earlier, when we design the output decoders, we must decode the state or states previous to the state in which we want an output. Figure 9.7-6 shows the

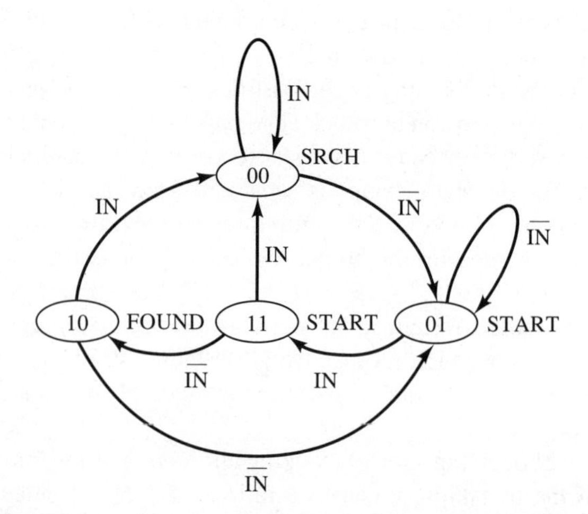

Figure 9.7-5 State transition diagram for example machine.

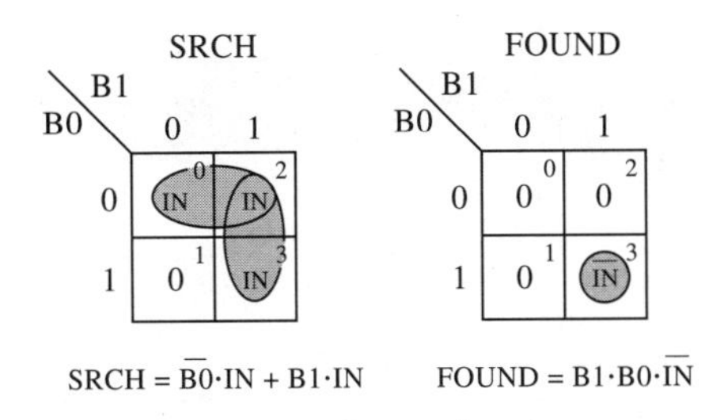

Figure 9.7-6 Output decoder maps.

Only SRCH and FOUND need register outputs; START has no decoder logic and will have no glitches. To load the SRCH map, all the states that have transitions into the 00 state (the state with the output SRCH) must be considered. These are the 00, 10, and 11 states. For each of the cells that represent these states in the SRCH map, the input condition is entered that causes the transition to the state with output SRCH. Thus, IN is entered in cell 0, IN is entered in cell 2, and IN is entered in cell 3. The FOUND map is loaded in the same way. The 10 state (the state with the output FOUND) has only one transition from another state; it is from the 11 state and occurs on the input condition $\overline{\text{IN}}$, so $\overline{\text{IN}}$ is loaded into cell 3 of the FOUND map. The simplified output decoder logic appears below each map in Figure 9.7-6, and the output circuit resulting from this logic is shown in Figure 9.7-7. This circuit has no output glitches, as we can see from the simulation results in Figure 9.7-8.

To deglitch the outputs of the state machine, we've added a register (D flip-flops) to the outputs. As already mentioned, it's common to call these outputs, register outputs. They render the classification of state machines as Mealy or Moore

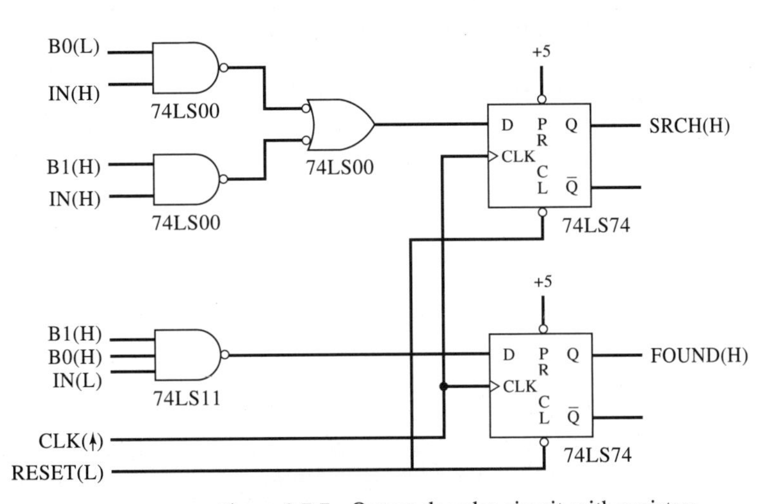

Figure 9.7-7 Output decoder circuit with registers.

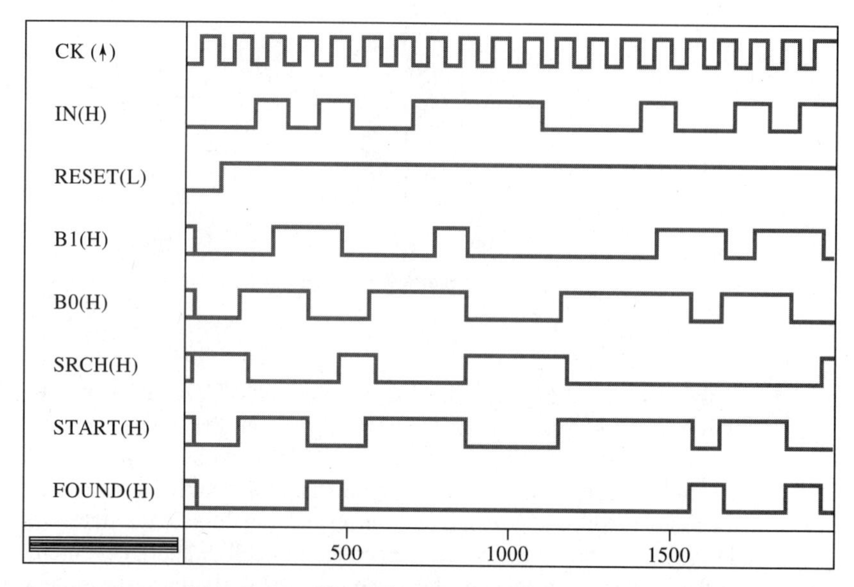

Figure 9.7-8 Simulation results for register output.

machines obsolete. A more meaningful classification would distinguish between machines with combinational outputs and machines with registered outputs. The following further classifications may also prove useful.

1. Machines with combinational outputs (without registers) in which the output does not depend on the inputs, only on the state (Moore machines).
2. Machines with combinational outputs (without registers) in which the output depends on the inputs to the machine as well as on the state (Mealy machines). In these machines, the output can change during the state, but machines in which this occurs should be avoided.
3. Machines with register outputs in which the output depends only on the state and occurs on the same clock cycle as the state. For this type of output the output decoder must decode the previous state. This means that the decoding can involve inputs. (These are sometimes called **Moore machines with register outputs**.)
4. Machines with register outputs in which the output depends on the state and occurs one-half clock cycle after the beginning of the state. The output decoder decodes the present state, and the register is clocked on the half cycle of the clock.
5. Machines with register outputs in which the output depends on the state and occurs one clock cycle after the beginning of the state. The output decoder decodes the present state, and the register is clocked normally on the clock. (These are sometimes called **Mealy machines with register outputs**.)
6. Machines with register outputs in which the output is turned on in one state and

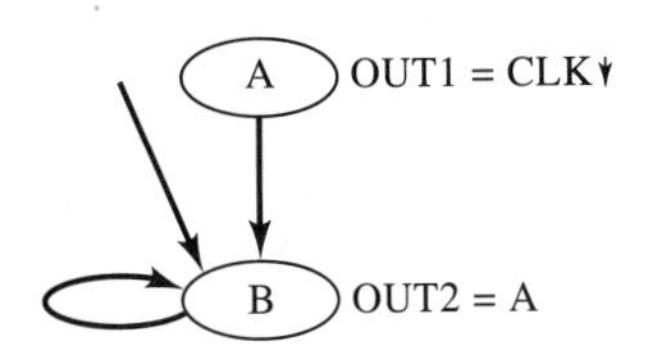

Figure 9.7-9 Partial state transition diagram.

turned off several states later. This type of output will be discussed in Chapter 10, Section 10.4.

Actually, register outputs can depend on inputs as well as states; just remember that, in order to affect the output, the input condition must be asserted before the register is clocked.

An actual state machine may have several different kinds of outputs. For some of the outputs the presence of glitches may not be critical. These outputs can be designed as combinational outputs to simplify the circuit (no flip-flop at the output). Other outputs may be sensitive to glitches; they might be outputs that are counted or outputs that are used to clock data. These outputs must be register outputs. A conservative approach would be to design as a register output *any* output that drives circuitry even remotely suspected of being sensitive to glitches.

We need to supplement our state transition diagram notation to allow us to specify the timing for some of these different kinds of outputs. Figure 9.7-9 shows this new notation in a partial state transition diagram. In the upper state (A), the output OUT1 occurs on the negative clock edge (indicated by the condition CLK↓). In the lower state (B), the output OUT2 occurs for only one clock cycle when the state is entered from the state A. We will not indicate whether outputs are combinational or register on the state transition diagram although the two types of timing indicated above can only be achieved in a register output.

Example 9.7-1

Design a state machine to sample two input signals, S and T, twice (sequentially) and determine each time at which both S and T are 11, each time at which S is 11 but T is not, and each time at which T is 11 and S is not. Every time a sequence 11 is found, the sampling should start over without missing a bit. If neither S nor T is 11, the sampling should start over as soon as the failure is evident. When both S and T are 11, the output FNDB should be asserted for one clock cycle. When only S is 11, the output FND1 should be asserted for one clock cycle. When only T is 11, the output FND2 should be asserted for one clock cycle. The inputs S and T are asserted high, and they change on the negative clock edge. The outputs FNDB, FND1, and FND2 should be asserted low. Use D flip-flop memory and register the outputs to prevent output glitches, but do not delay them.

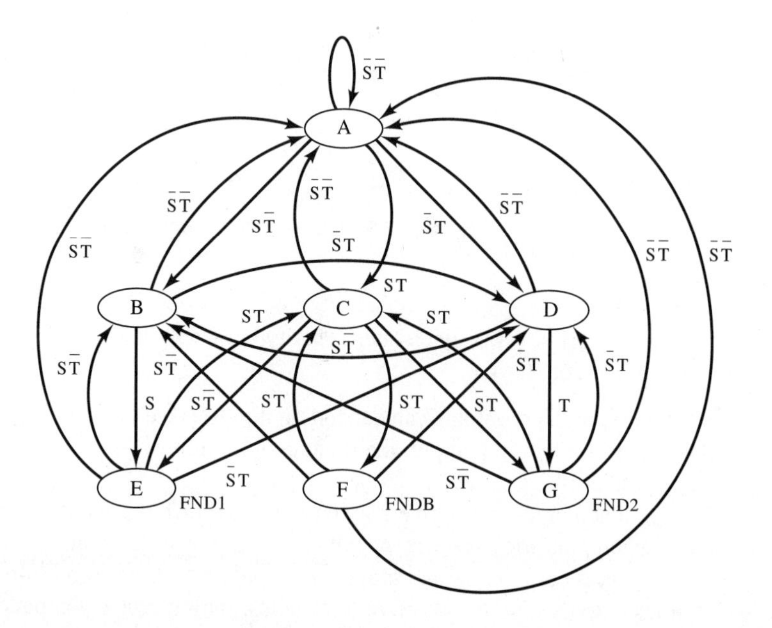

The description of the state machine was just given. The above figure is a primitive state transition diagram for the machine. To form this diagram, we first draw the center branch, where S and T are both 11. Second, we draw the left branch, which starts with $S\overline{T}$ (S = 1 and T = 0). Notice that in the second transition in this branch it does not matter what T is. Third, we draw the right branch, which starts with $\overline{S}T$ (S = 0 and T = 1). Finally, we consider the signal conditions other than the ones we've used to draw the three branches, and add the transitions required for them, one state at a time. With two input signals we have four possible transitions out of each state—one for each of the combinations of input signals, ST, $S\overline{T}$, $\overline{S}T$, and $\overline{S}\,\overline{T}$—and we must show transitions for all of these possibilities.

Now that we have the primitive state transition diagram, we need to assign state codes. We'll assign them to simplify the next-state decoder, using the rules from Section 9.3.3. We'll try to satisfy the first rule by grouping together on a map those states that make a transition into a single state on the same input condition. Then we'll try to satisfy the second rule by grouping together on a map those states that have the same previous state. In this case, we'll also try to make part of the state code match the input condition for the transition. To help us with the rules, we can make a table showing all the previous states and all the next states for each present state of the machine. We also need the input conditions for next states, so we'll show them above the next states. The table follows.

Previous States	Present State	Next States
A, B, C, D, E, F, G	A	00 10 11 01 A, B, C, D
A, D, E, F, G	B	00 01 A, D, E
A, E, F, G	C	00 10 11 01 A, E, F, G
A, B, E, F, G	D	00 10 A, B, G
B, C	E	00 10 11 01 A, B, C, D
C	F	00 10 11 01 A, B, C, D
C, D	G	00 10 11 01 A, B, C, D

We use the table of previous and next states to arrange the states, using the letter designations, in a three-variable Karnaugh map. This gives an opportunity to try to group the desired previous and next states. Aside from A, B, C, D, E, F, and G, which are grouped together no matter what state code assignment is made, there are three large sets of previous states that we should try to group—A, B, E, F, and G; A, D, E, F, and G; and A, E, F, and G. Notice that the set A, E, F, and G is also a set of next states (for the state C). If we put A, G, F, and E across the top of the map (see the following figure), we'll have met the requirement of grouping these large sets of previous states. (No matter where B and D are placed, they'll group with A, G, F, and E across the top row of the map.) This choice not only groups the large sets of previous states together, it also groups one of the sets of next states and even matches part of the state code (the first two bits) for the set with the input conditions for the transitions. The input conditions were matched by choosing the order of the states across the map. (Check the input conditions for the set of next states to C in the table.) We can see that it's not possible to group A, B, C, and D together because of the already fixed position of A. We can, however, group previous states B and C, previous states C and D, and next states B, C, and D. For next states B, C, and D, we can also match part of their state codes (the first two bits) with the input conditions for their transitions, as follows. We haven't managed to group A, D, and E or A, B, and G.

B0 \ B2 B1	00	01	11	10
0	A (0)	G (2)	F (6)	E (4)
1	x (1)	D (3)	C (7)	B (5)

Now that we have a state code assignment, we can redraw the state transition diagram with the state codes:

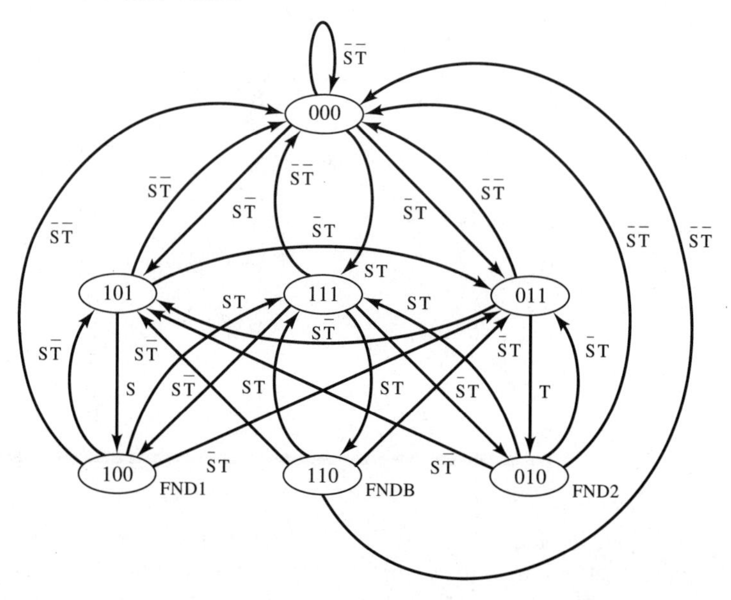

The next step in the design is to make VEMs for the next-state code decoders. We do this directly from the state transition diagram for D flip-flop memory. Remember, the easiest maps to make directly from the state transition diagram are VEMs with the bits of the state code used as the map variables and the input conditions used as the entered variables. The VEMs follow, with the simplified logic expression for each next-state decoder logic circuit below the map.

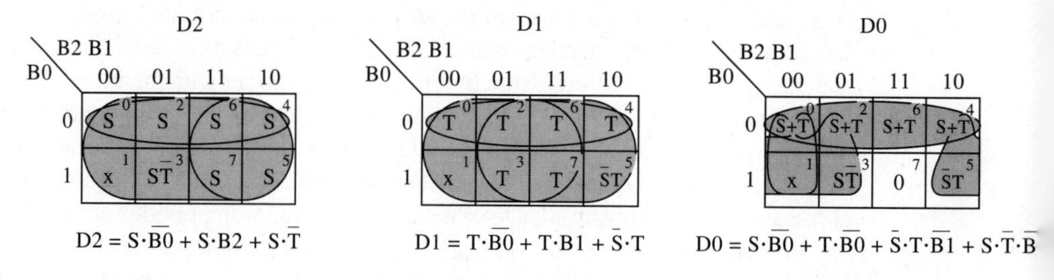

Because the outputs are to be register outputs, each output map must be formed from the states previous to the state with the output asserted. Each map must also indicate the condition that causes this transition. For example, the states before the 100 state (the state with output FND1 asserted) are the 101 and 111 states. The state 101 makes a transition to the 100 state for the condition S, and the state 111 makes a transition to the 100 state for the condition $S\overline{T}$. The output decoder map for FND1 will thus be a VEM with S in the 101 cell and $S\overline{T}$ in the 111 cell, as follows on the left. The other cells in the map are loaded with 0 except for the 001 cell, which is loaded with x because 001 is an unused state. The other output decoder maps are formed in the same way as the map for FND1—from the states previous to the state where the output is to be asserted.

When we simplified the output decoder, we did not use the unused states. Unwanted outputs can occur when an error causes the machine to enter an unused state. In practice, the effect of the unused state needs to be analyzed for the specific problem. The output decoder logic appears below the following output decoder maps.

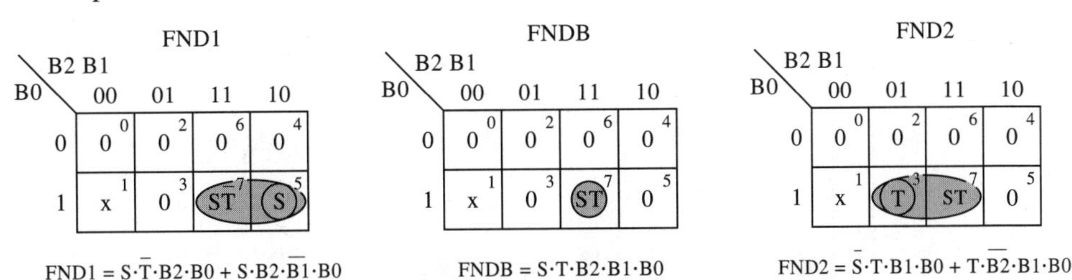

The circuit diagram for the complete state machine follows. The circuit has provisions for resetting the state machine to the 000 state and resetting the output decoder flip-flops. This reset requires a low signal on the RESET(L) input to the machine. Because the circuit has registered outputs, the outputs will be glitch-free. The output flip-flops are clocked with the same positive clock edge as the state machine, but the state decoders decode one state before the state with the output so that the output will be synchronous with the state—that is, it will not be delayed. The output register consists of three D flip-flops. Because the outputs are asserted low, the output flip-flops are initialized high.

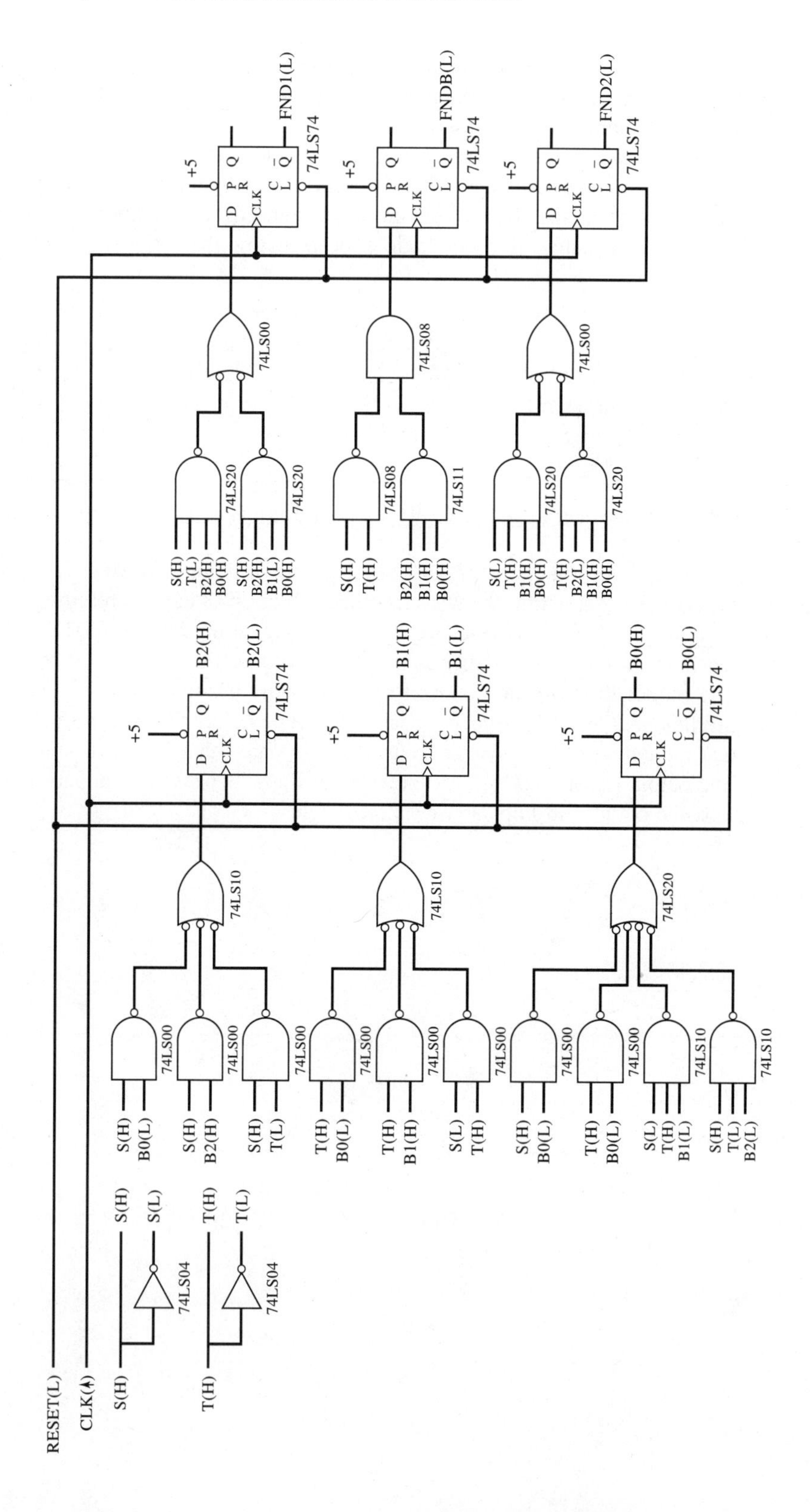
RESET(L)
CLK(↑)
S(H)
S(L)
T(H)
T(L)
74LS04
74LS00
74LS10
74LS20
74LS08
74LS11
74LS74
+5
D
CLK
PR
CL
Q
Q̄
B2(H)
B2(L)
B1(H)
B1(L)
B0(H)
B0(L)
FND1(L)
FNDB(L)
FND2(L)

As the last step in our design, we simulate the circuit to determine whether or not the design is correct. The results from this simulation follow. Notice that we've shown the state code as well as the output; this allows us to make sure that the machine is sequencing through the proper states. If we're concerned with timing, we should also display the next-state code so that we can see what margin we have for the setup times of the memory flip-flops. A good simulator will warn us if the setup time is violated, but we can tell how close to the minimum it is only by actually looking at a simulated wave form.

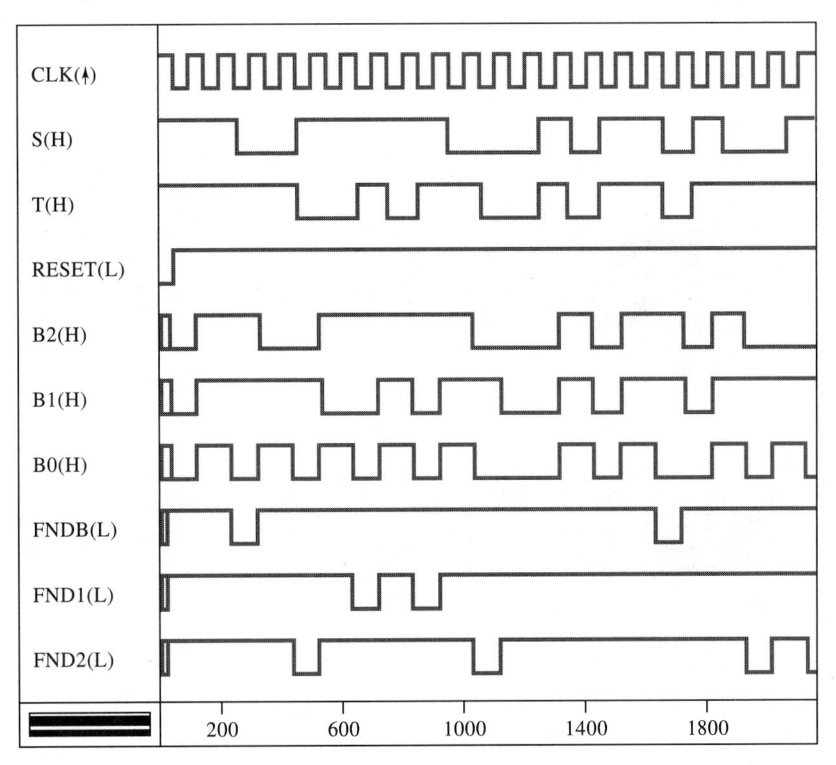

Example 9.7-2

Repeat Example 9.7-1 for JK flip-flop memory.

We can transform the D memory decoder map directly to JK memory decoder maps using the following table.

Q_n	J	K
0	D	x
1	x	$\overline{D}$

We'll get two maps, a J map and a K map, for each D map. The D maps from Example 9.7-1 are repeated as follows, and the J and K maps that are formed from those D maps appear directly below them.

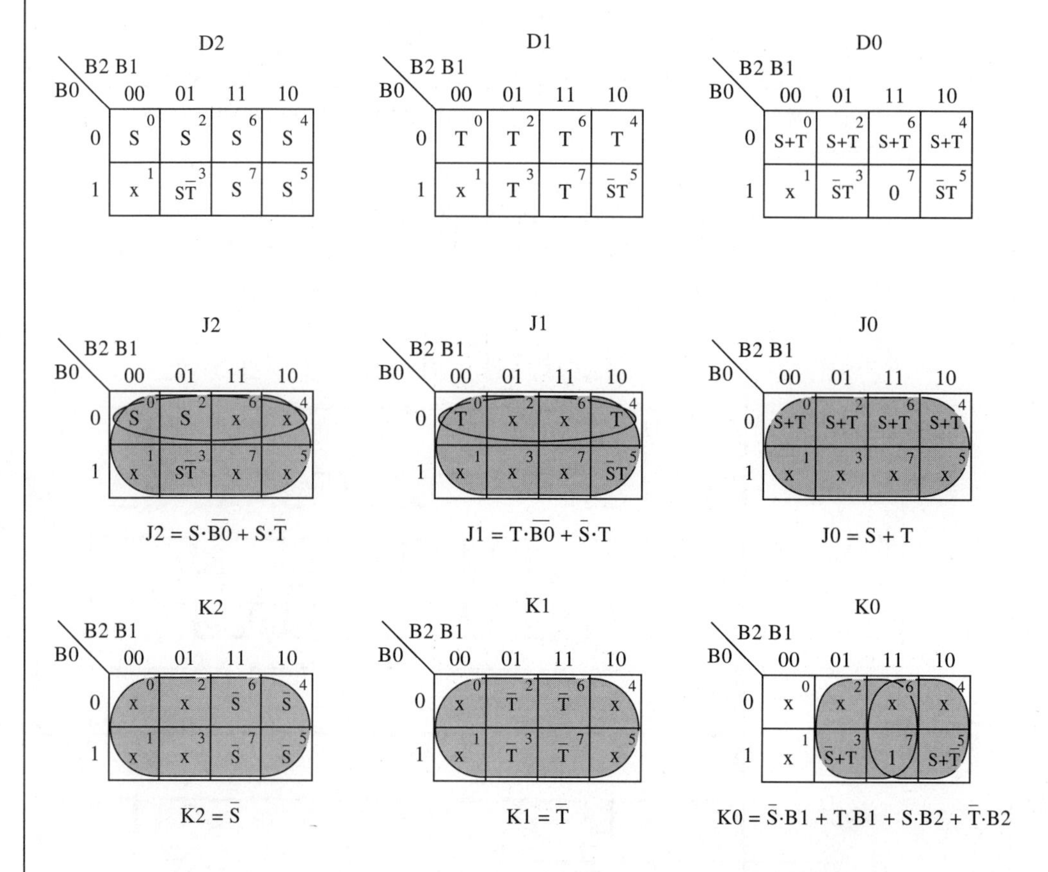

Notice that the present state for the memory variable (Q_n) is B2 in the first map, B1 in the second map, and B0 in the third map. The logic expressions for the decoder logic circuits appear below their respective maps.

The output design is the same as for Example 9.7-1. Actually, we could make the output register from JK flip-flops, and transform the output decoders to JK decoders in the same way we did for next-state decoders. This may result in a simplification of the output decoder. Try it. Following is a complete circuit for the state machine, using D flip-flops in the output register.

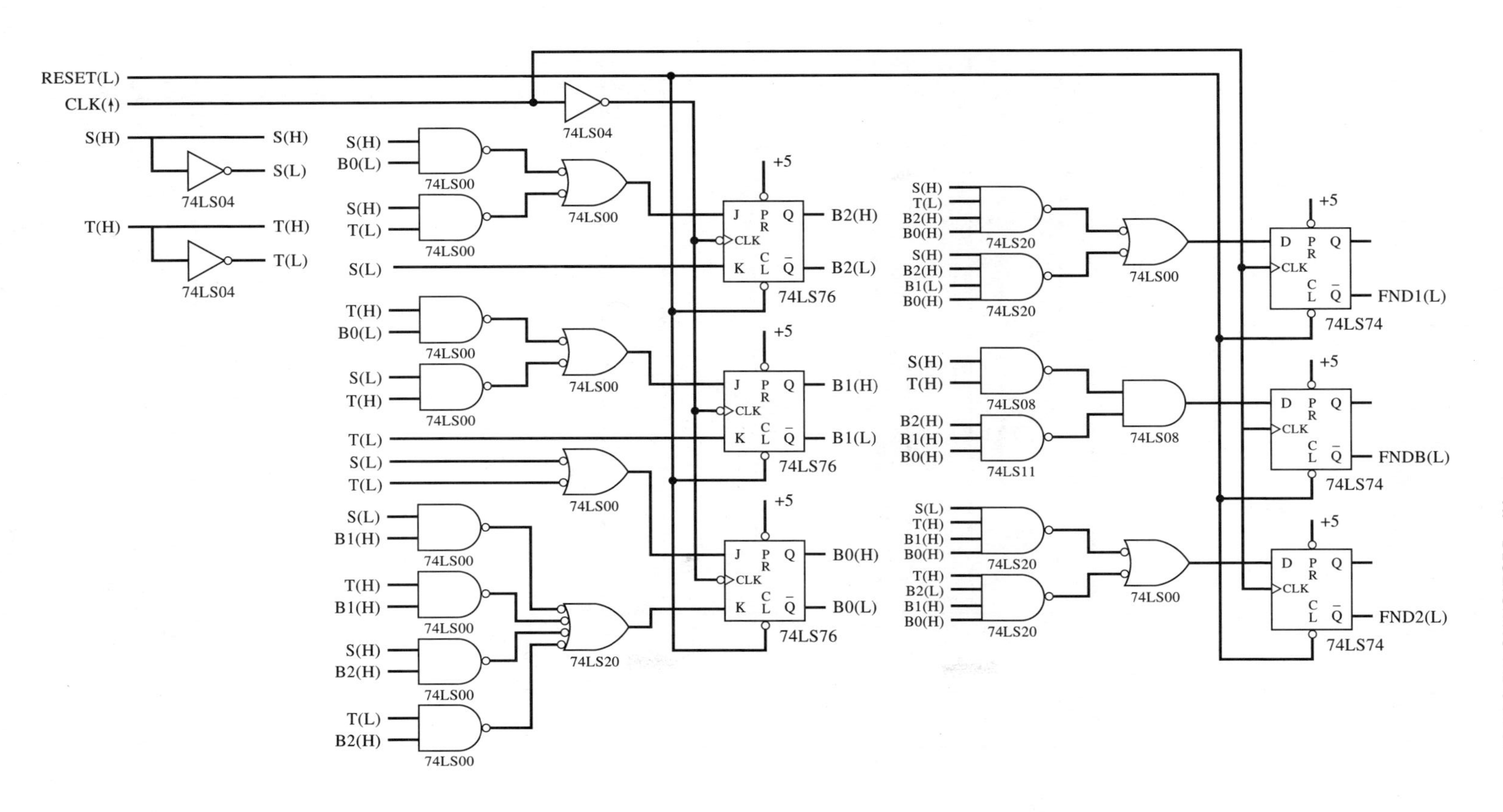
RESET(L)
CLK(↑)
S(H)
S(L)
T(H)
T(L)
74LS04
74LS00
74LS20
74LS08
74LS11
74LS76
74LS74
B0(H)
B0(L)
B1(H)
B1(L)
B2(H)
B2(L)
+5
J
K
D
CLK
P R
C L
Q
Q̄
FND1(L)
FNDB(L)
FND2(L)

As the last step in our design, we simulate the circuit to determine whether or not the design is correct. The results of this simulation follow. Notice that we've shown the state code as well as the output; this allows us to make sure that the machine is sequencing through the proper states. If we're concerned with timing, we should also display the next-state code so that we can see what margins we have for the setup times of the memory flip-flops. Again, a good simulator will warn us if the setup time is violated, but we can tell how close to the minimum it is only by actually looking at a simulated wave form.

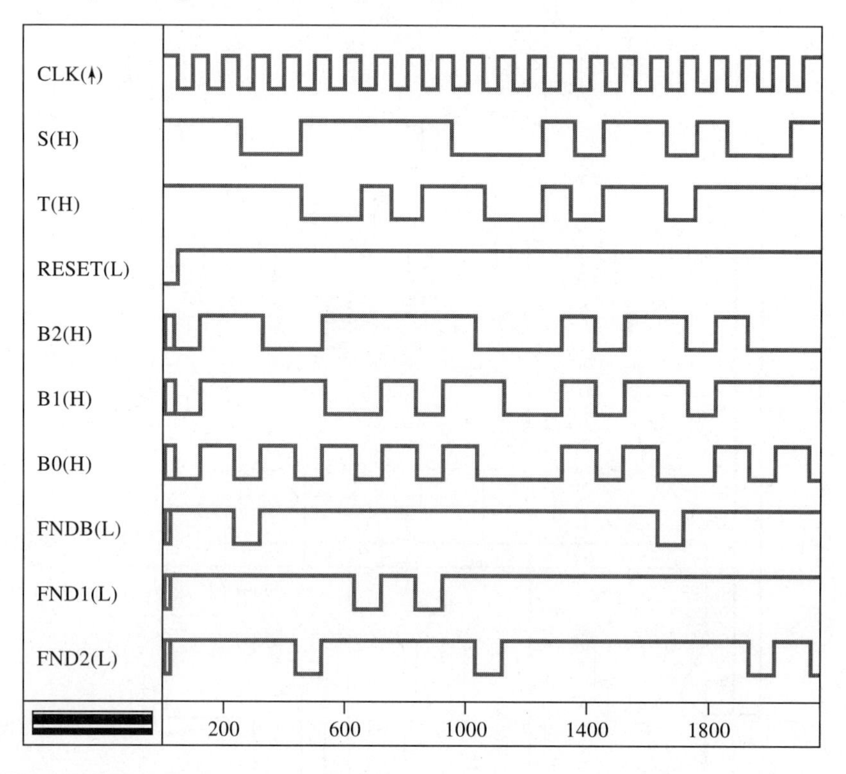

Example 9.7-3

State machines are commonly used as controllers for larger digital systems. We'll discuss this use of state machines in Chapter 12. These controllers are difficult to describe apart from the digital system they control so in this example of such a state machine, we'll start the design with a state transition diagram rather than a word description. We want to look at this kind of example because the timing requirement on the outputs produced by controllers can be more stringent than we've seen in the state machines we've designed so far. Design a state machine for the state transition diagram shown below. Use discrete logic elements. Make all the outputs glitch free.

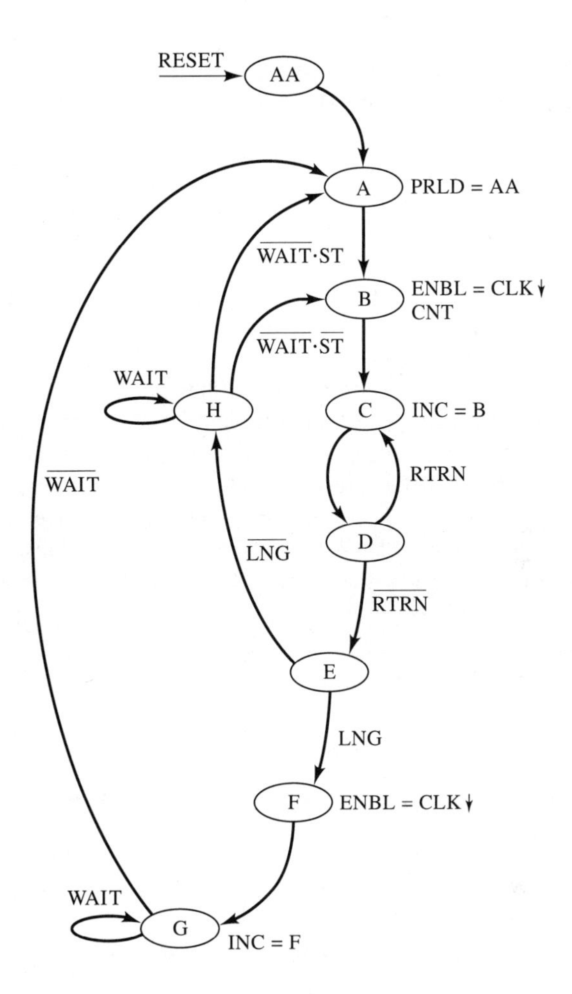

The inputs are: RESET, WAIT, RTRN (return), LNG (long), ST (start). RESET is an asynchronous reset that starts the machine. The special timing on the outputs needs some explanation. In state A, the output notation

$$\text{PRLD (preload)} = \text{AA}$$

means that PRLD is conditional on the previous state being AA. This output is not asserted when A is entered from G or H. Just for interest we've given the signal names in parentheses after the mnemonic. Similar kinds of outputs occur in states C and F.

$$\text{INC (increment)} = \text{B} \quad \text{(state C)}$$
$$\text{INC} = \text{F} \quad \text{(state G)}$$

Notice that in state G, INC is asserted for only one clock cycle when G is entered from F. If we wanted INC to be asserted for the entire time the state machine is in state G, we would not use the condition E, but simply place INC by state G. The output notation in states B and F

$$\text{ENBL (enable)} = \text{CLK}\downarrow \quad \text{(state B)}$$
$$\text{ENBL} = \text{CLK}\downarrow \quad \text{(state F)}$$

shows that the output is asserted on the negative going clock edge and lasts for one clock cycle.

Because we're starting with the state transition diagram, the next step in our design is to assign state codes. We'll assign states to simplify the next state decoder logic. We use the table of previous states and next states (shown below) to do this. Remember that we want to choose the state codes to group those states that have transitions into a single state (previous states) together on a map and to similarly group those states that have transition out of a single state (next states). In the latter case, we try to make part of the state code match the input conditions that cause the transitions. We've shown the input conditions that we're trying to match above the next states in the table. We've grouped all the previous states together, but we have not grouped the next states H and F, and we have not matched the input conditions very well.

Previous States	Present State	Next State
	AA	A
AA, G, H	A	B
A, H	B	C
B, D	C	D
C	D	0 1 E, C
D	E	0 1 H, F
E	F	G
F, G	G	0 1 A, G
E, G	H	00 01 1 A, B, H

CD \ AB	00	01	11	10
00	AA (0)	H (4)	A (12)	G (8)
01	x (1)	x (5)	x (13)	F (9)
11	x (3)	G (7)	x (15)	x (11)
10	x (2)	B (6)	C (14)	E (10)

The state transition diagram with state codes is shown below. We keep the letter name of the state and add the state code because the outputs are defined in terms of letter names. The design of the next state decoder follows the usual pattern. We make VEMs for the next state decoder logic directly from the state transition diagram and then simplify the next state code. Both the VEMs and the simplified logic expressions for the next state decoder are shown following the state transition diagram.

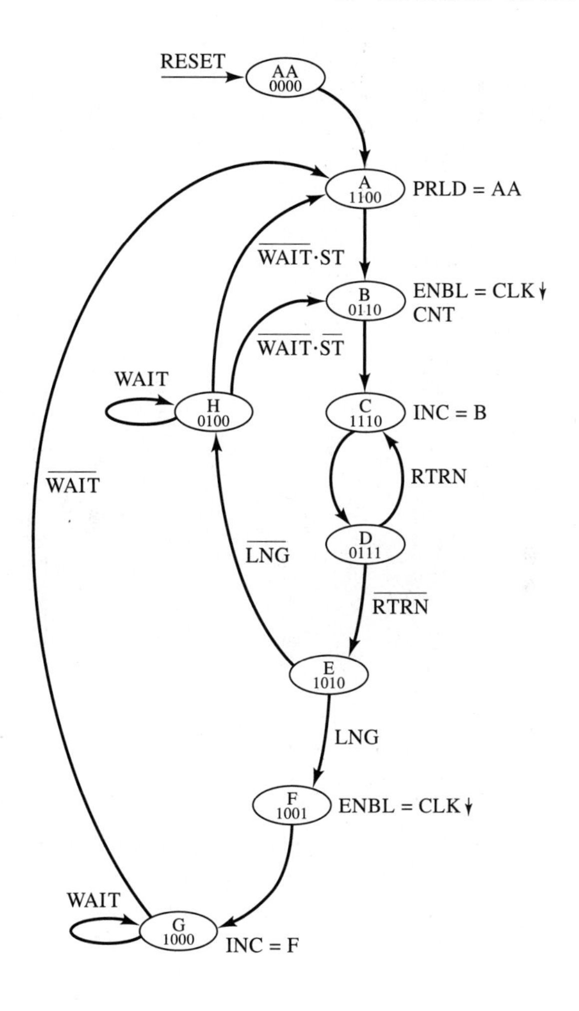

RESET
AA 0000
A 1100
PRLD = AA
$\overline{\text{WAIT}}\cdot$ST
B 0110
ENBL = CLK↓
CNT
$\overline{\text{WAIT}}\cdot\overline{\text{ST}}$
WAIT
H 0100
C 1110
INC = B
RTRN
$\overline{\text{WAIT}}$
D 0111
$\overline{\text{LNG}}$
$\overline{\text{RTRN}}$
E 1010
LNG
F 1001
ENBL = CLK↓
WAIT
G 1000
INC = F

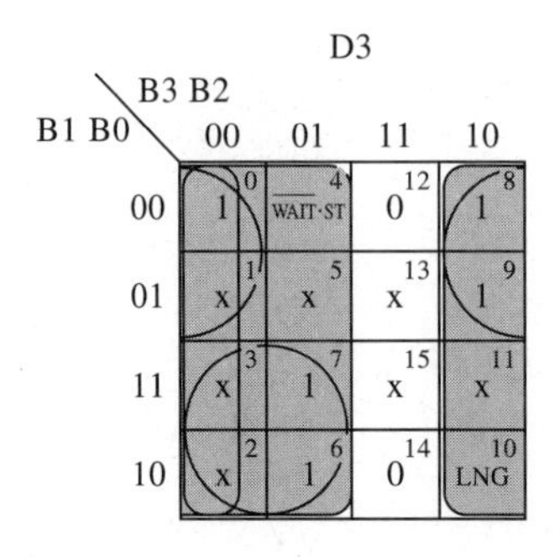

$D3 = \overline{WAIT}\cdot ST\cdot\overline{B3}+LNG\cdot\overline{B2}+\overline{B2}\cdot\overline{B1}+\overline{B3}\cdot B1$

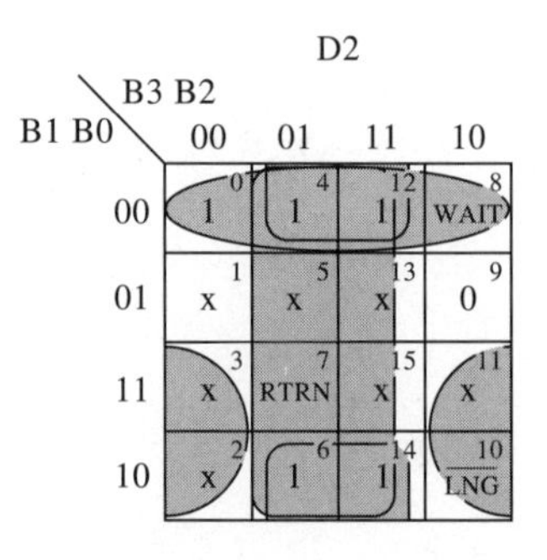

$D2 = \overline{WAIT}\cdot\overline{B1}\cdot\overline{B0}+\overline{LNG}\cdot\overline{B2}\cdot B1+RTRN\cdot B2+B3\cdot\overline{B1}+B2\cdot\overline{B0}$

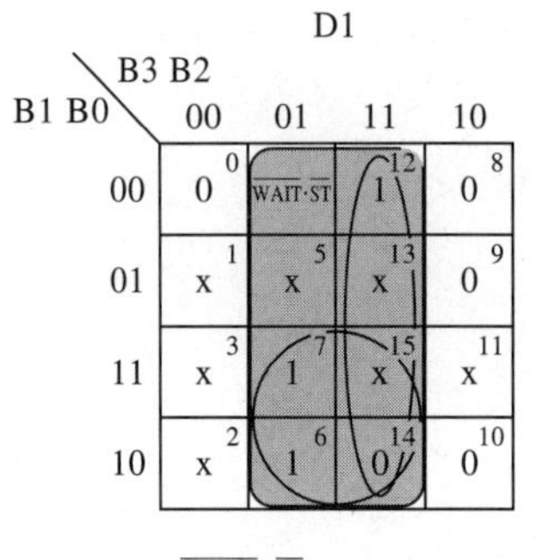

$D1 = \overline{WAIT}\cdot\overline{ST}\cdot B2+B3\cdot B2+B2\cdot B1$

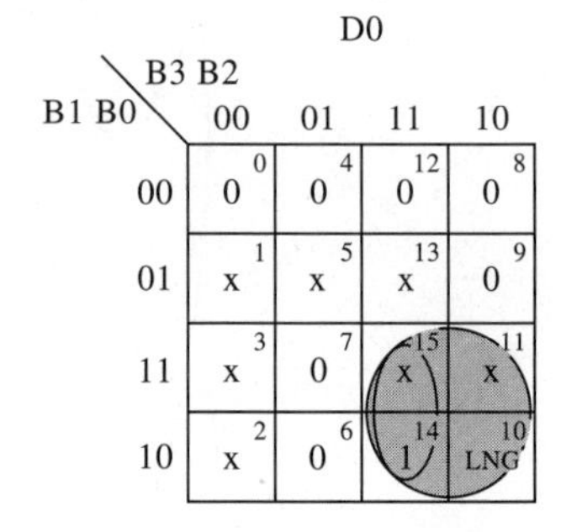

$D0 = LNG\cdot B3\cdot B1+B3+B2\cdot B1$

Now we need to consider how to handle the output conditions. PRLD is asserted during state A but only when it is entered from AA. This means that we must use a register (D flip-flop) and decode the previous state (AA). The map for the decoder logic is in the upper left corner of the output maps shown below.

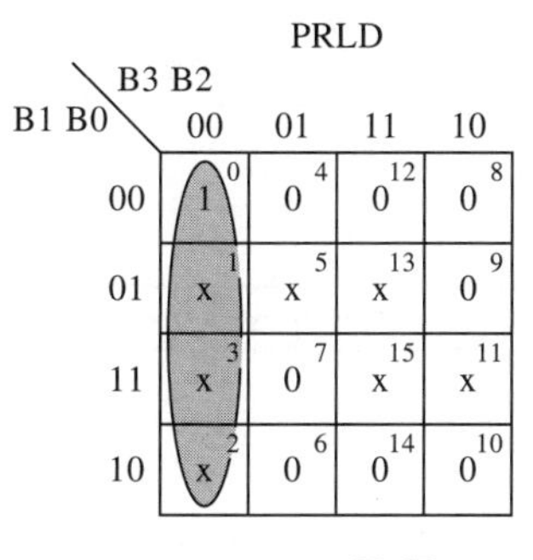

$PRLD = \overline{B3} \cdot \overline{B2}$

INC ENBL

B1 B0 \ B3 B2	00	01	11	10
00	0 (0)	0 (4)	0 (12)	0 (8)
01	X (1)	X (5)	X (13)	1 (9)
11	X (3)	0 (7)	X (15)	X (11)
10	X (2)	1 (6)	0 (14)	0 (10)

$INC = ENBL = \overline{B3} \cdot B1 \cdot \overline{B0} + \overline{B1} \cdot B0$

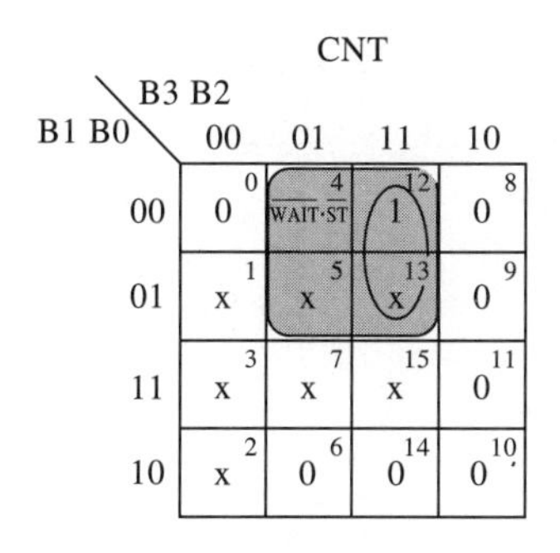

$CNT = \overline{WAIT} \cdot \overline{ST} \cdot B2 \cdot \overline{B1} + B3 \cdot B2 \cdot \overline{B1}$

INC is asserted in state C but only when it is entered from B, and in state G but only when it is entered from F. Again we must use a register and decode the previous states. The map for the decoder logic is in the upper right corner of the output maps. ENBL is to be asserted in states B and F on the negative going clock edge. Here we use a register output, but we decode the present state and clock the output on the negative clock edge (invert the clock). It just happens that these are the same states we decoded for INC so we can use the same decoder. The final output CNT is a normal registered output that is to be asserted in state B. For a registered output we must decode the previous state (A and H) including any input conditions. The map for the CNT output decoder is at the bottom center of the output decoder maps. The circuit for the state machine is shown on page 458. The simulation results follow the circuit. The machine functions just as we've designed it to function. We do find one problem when we simulate. The simulation software warns us that there is insufficient setup time for the register flip-flop on the ENBL output. Our clock period is 100 ns and this is the flip-flop that is clocked on the negative going clock edge so there is only 50 ns after the state changes for the input to this register to reach a correct value. The propagation time through the state memory flip-flops and the output decoder plus the set-up time on the output decoder is simply longer than 50 ns. If we wanted to run this state machine as fast as the simulation we would have to change to a different (faster, perhaps F) logic family of components to build the circuit. We should do this and simulate again as a check before we build.

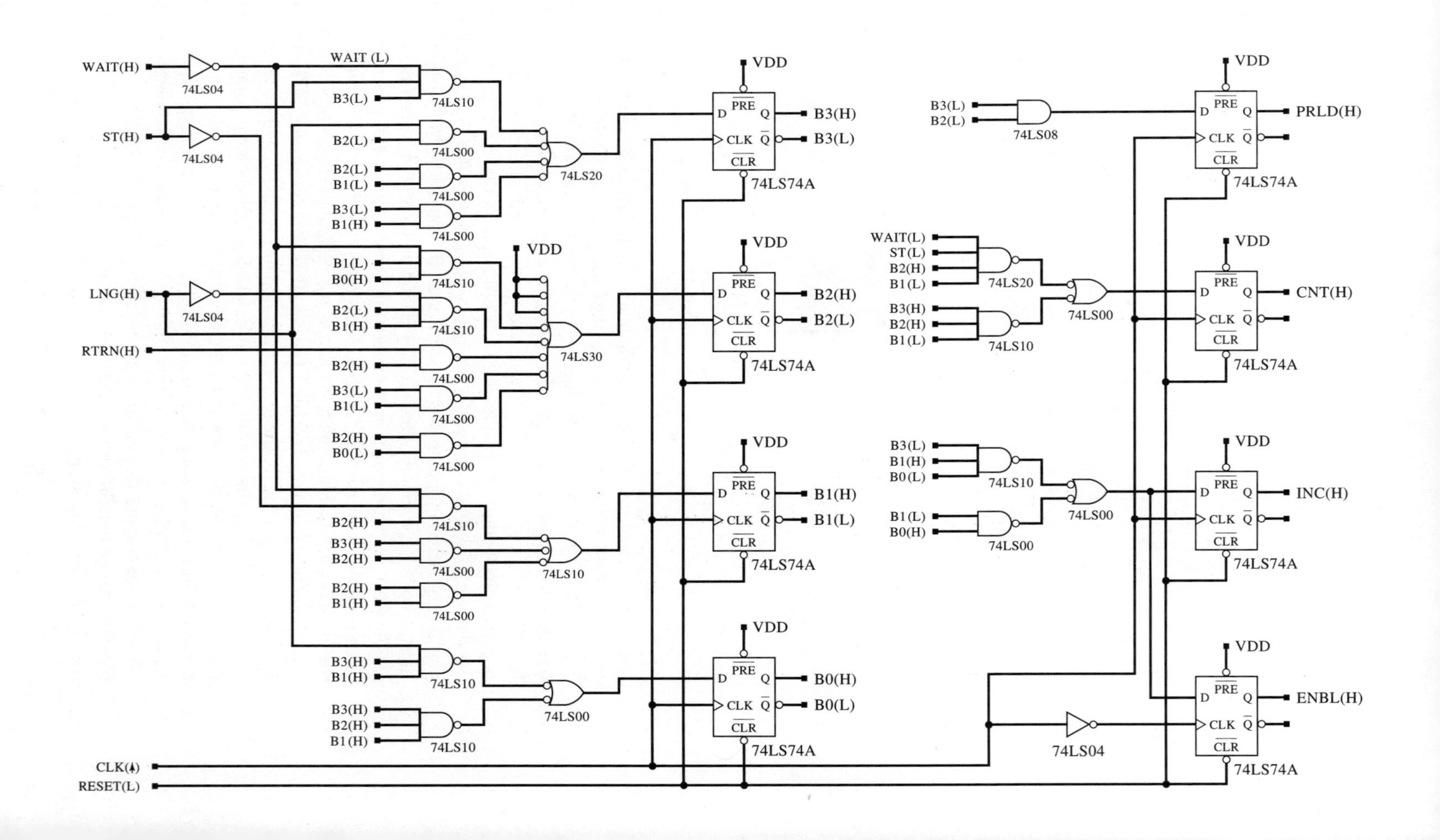

WAIT(H)
ST(H)
LNG(H)
RTRN(H)
CLK(↑)
RESET(L)
WAIT (L)
B3(L)
B2(L)
B2(L) B1(L)
B3(L) B1(H)
B1(L) B0(H)
B2(L) B1(H)
B2(H)
B3(L) B1(L)
B2(H) B0(L)
B2(H)
B3(H) B2(H)
B2(H) B1(H)
B3(H) B1(H)
B3(H) B2(H) B1(H)
74LS04
74LS10
74LS00
74LS20
74LS30
74LS08
74LS74A
VDD
D
PRE
Q
CLK
CLR
B3(H) B3(L)
B2(H) B2(L)
B1(H) B1(L)
B0(H) B0(L)
B3(L) B2(L)
WAIT(L) ST(L) B2(H) B1(L)
B3(H) B2(H) B1(L)
B3(L) B1(H) B0(L)
B1(L) B0(H)
PRLD(H)
CNT(H)
INC(H)
ENBL(H)

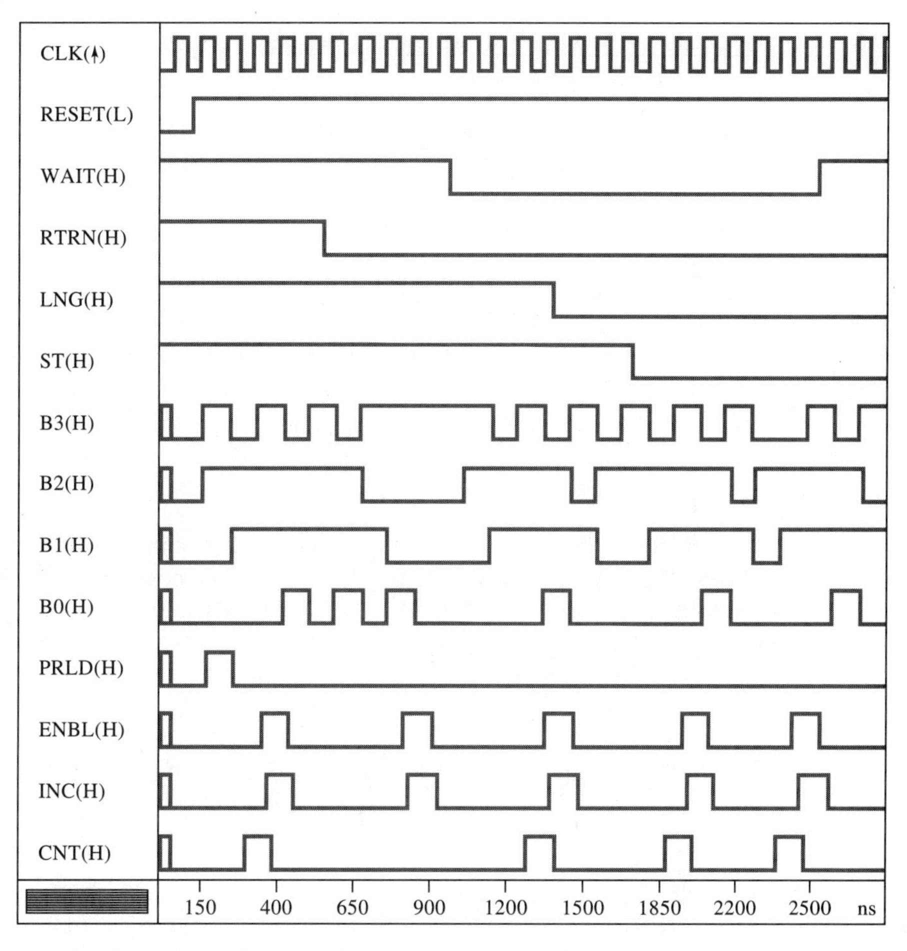

To determine whether this state machine is functioning properly from the simulation results actually requires some work. We must consider each positive going clock edge and read off the state just before this edge by determining the values of B3, B2, B1, and B0. Then we must compare the sequence of state with the state transition diagram. If we do this we see that the machine passes through states AA(0000), A(1100), B(0110), C(1110) and then it bounces between C and D(0111) because RTRN is asserted. When RTRN is no longer asserted it then continues through states E(1010), F(1001) (because LNG is asserted), and G(1000). It stays in G because WAIT is asserted and finally when WAIT is no longer asserted it returns to A. The simulation input is also designed to check whether the machine correctly traverses the two smaller loops. Check that it does.

Finally, we must check the outputs. For instance, we need to see that PRLD occurs only on the first time the machine passes through the state A. It does. We must determine that INC, ENBL, and CNT are asserted in the proper state and remain asserted for only one clock cycle.

9.8 OTHER FORMS OF STATE TRANSITION DIAGRAMS

The form of the state transition diagram we're using is not the only form of diagram we may encounter. The most common form of diagram is somewhat more terse than the form we've been using. It actually has two versions, depending on whether the machine is a Moore machine or a Mealy machine.

Let's first consider the Mealy machine version. The example state transition diagram we first used in Chapter 7 to explain our form of diagram is repeated in Figure 9.8-1 on the left, along with the same diagram in the terse form on the right. Each transition in the new diagram is marked with a notation of the form

$$\frac{\text{Conditions}}{\text{Outputs}}$$

The conditions are the input conditions that cause the transition. The outputs are the outputs that are asserted in the next state if the conditions are met. In order to interpret the conditions and outputs, we need ordered lists of both inputs and outputs. Sometimes these lists are actually given on the diagram as we've given them here. In any case, we must know the inputs and outputs and also their order of use in the diagram. For instance, in the self-transition at the top with the notation

$$\frac{\begin{array}{c}100,101,110,\\111,000,001\end{array}}{00}$$

the transition occurs if the inputs RESET, START, and GO are 100, 101, 110, 111, 000, or 001 from above the line. The outputs READY and DONE, that occur in the next state, are 00 from below the line. The combination of the input is the same as RESET + $\overline{\text{START}}$ on the left diagram. We can list all the conditions for the transition together because the outputs are the same. In the transition between states A and B with the notation

$$\frac{010}{10}, \frac{011}{00}$$

we have an output in state B that is conditional on an input. Here we split the input conditions 010 and 011 and place each with the proper outputs 10 and 00, respectively.

This type of notation can be concise if the number of inputs and outputs is small, but it can be very unwieldy if the number of inputs and outputs, particularly inputs, is large. For instance, the transition from state C to A has eight different input conditions and the machine has only three inputs. The Mealy machine diagram will

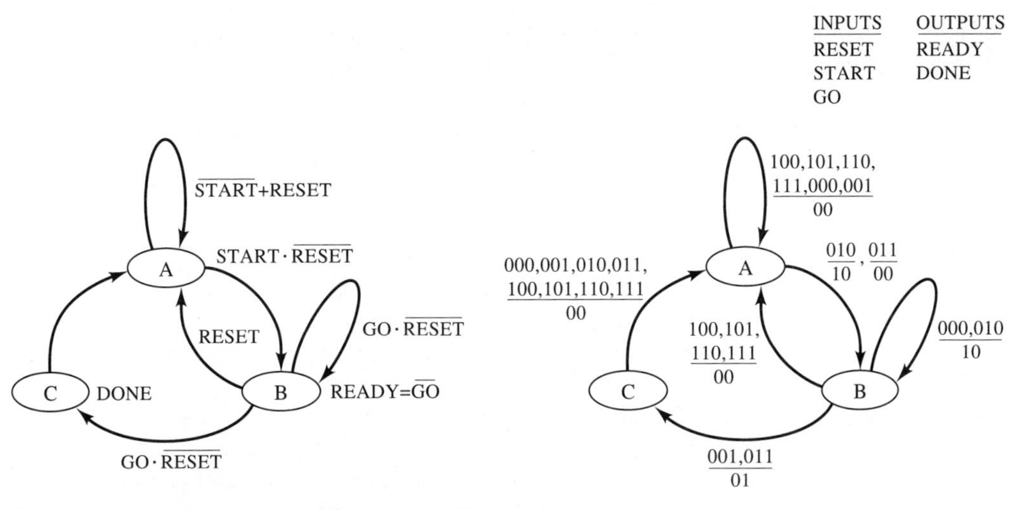

Figure 9.8-1 Example state transition diagram.

handle most output conditions, including Moore machine outputs, but it will not handle outputs on the negative going clock edge or delayed output without modifications.

To show the Moore machine variation of the new notation, we'll modify the state diagram so it represents a Moore machine by making the output READY occur in state B, unconditionally. Figure 9.8-2 shows the state transition diagram for this modified state machine on the left in our usual notation and on the right using the new notation. In the new notation, only the conditions for transition are listed on the transitions; the outputs are given within the state below the state name.

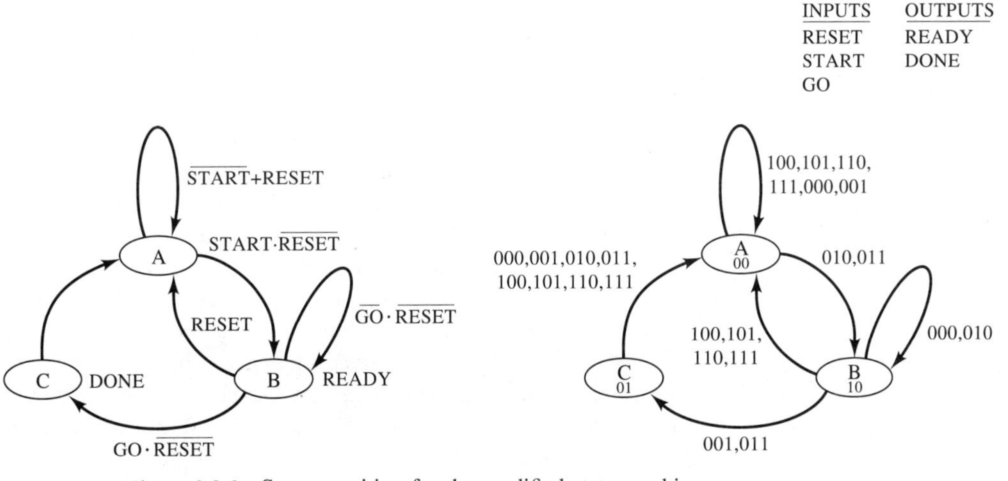

Figure 9.8-2 State transition for the modified state machine.

There is another notation for documenting state machines that is quite different from the diagrams we've seen. It uses what is called flow control language. The notation is very similar to a computer flow chart. It results in a very detailed documentation of the state machine, but it can be cumbersome and slow down the design process. We'll discuss this language in Appendix A.

KEY TERMS

Asynchronous Inputs Inputs which are not clocked.
Critical Path Slowest timing path for a change in input.
Many-State Synchronous State Machine State machine with more than two states.
Mealy Machine Machine with combinational outputs in which the outputs depend on the inputs to the machine, as well as on the state.
Mealy Machine with Register Outputs Machine with register outputs in which the output depends on the state and input and occurs one clock cycle after the beginning of the state.
Metastable Conditions Unstable condition in which the flip-flop is neither high nor low.
Moore Machine Machine with combination outputs in which the output does not depend on the inputs, only the state.
Moore Machine with Register Outputs Machine with register outputs in which the output depends only on the state and occurs on the same clock cycle as the state.
Primitive State Transition Diagram State transition diagram without state code assignments.
Reduced Input Dependency Choice of state code to match input conditions.
Register Output Adding a D flip-flop to each output, waiting until output is stable, and clocking the output into the flip-flop.
Sanity Circuit A circuit causing a machine to start in a known condition.

EXERCISES

1. Design a state machine, using D memory and discrete logic elements, to find each of the following sequences in a serial input signal, SIG(H). Each time the sequence is found, the output FND should be asserted. When a complete sequence is found, the machine should start looking for the sequence again without missing a bit. When the sequence is not found, the machine should immediately start looking again. That is, if the sequence 001 is to be found and an improper sequence such as 1 or 01 is found, the machine should start looking for a new correct sequence on the next bit. Similarly, if a sequence such as 001 is to be found and a sequence of more than two 0s is found, the machine should consider that the first 1 after the 0s completes the desired sequence. Changes in SIG are synchronous with the negative-going edge of the clock, as shown in the following timing diagram. Make the designs Moore machine designs. Make arbitrary state code assignments.

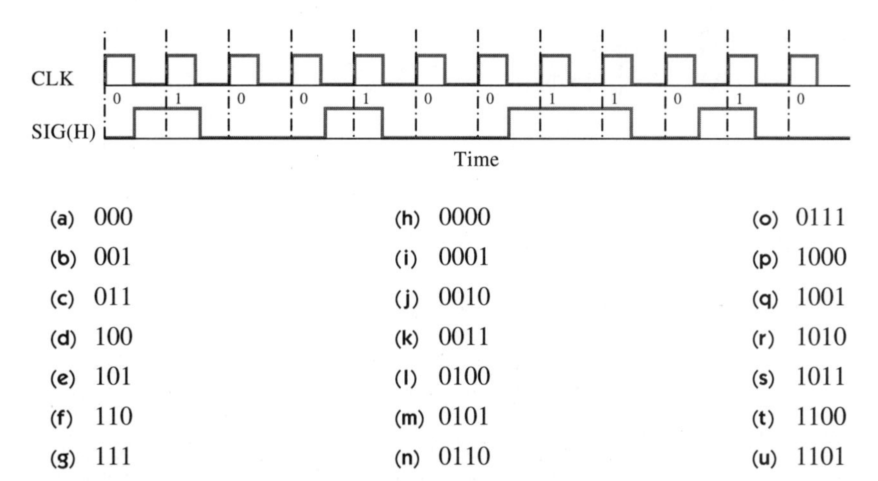

(a) 000
(b) 001
(c) 011
(d) 100
(e) 101
(f) 110
(g) 111
(h) 0000
(i) 0001
(j) 0010
(k) 0011
(l) 0100
(m) 0101
(n) 0110
(o) 0111
(p) 1000
(q) 1001
(r) 1010
(s) 1011
(t) 1100
(u) 1101

2. Make a Mealy machine design with D memory and discrete logic elements to find one of the sequences of exercise 3. In some cases the output will need to be dependent on the clock in order for the Mealy machine to function properly. You should be able to make a Mealy machine with fewer states than a Moore machine. Use arbitrary state code assignments.

3. Repeat exercise 1, but use JK memory and discrete logic elements.

4. Design a circuit to implement a state machine with one of the following state transition diagrams. Use D memory. Make arbitrary state code assignments.

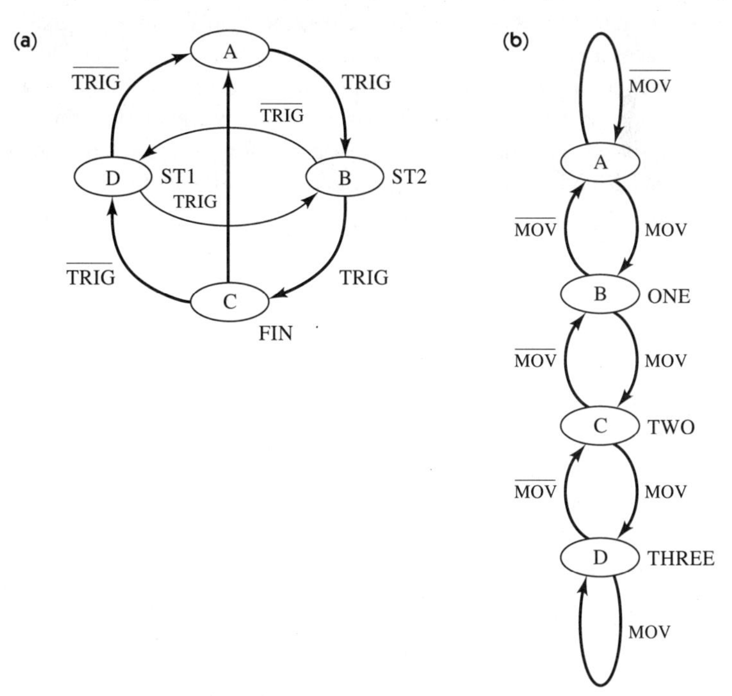

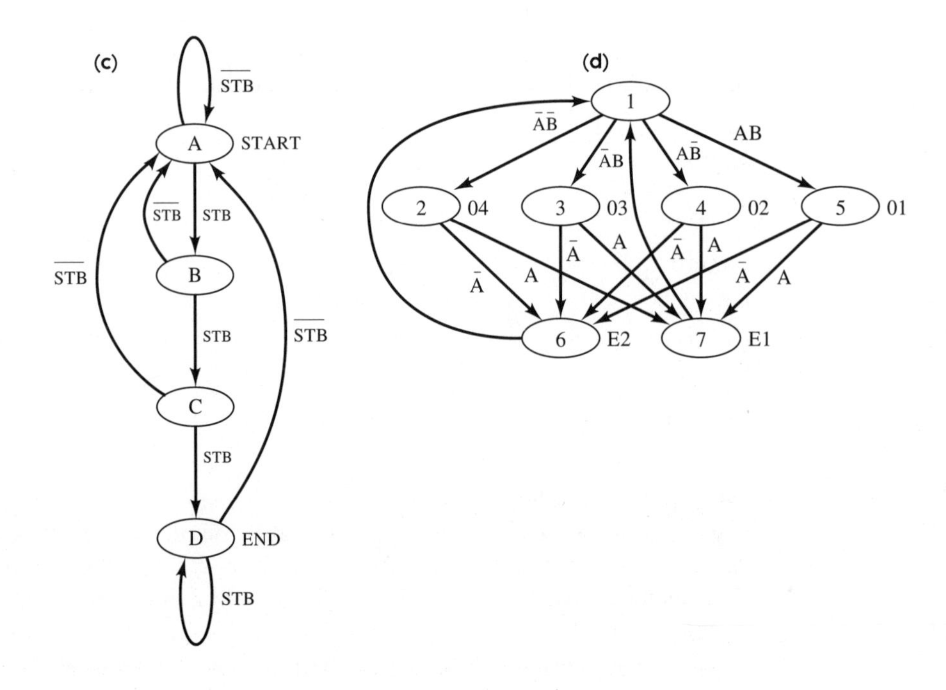

5. Repeat exercise 4 using JK memory.

6. Design a state machine that will sample an incoming signal, SIG(H), in 3-bit slices and determine whether the three bits are as specified below. Each time the bits are as specified, the machine should assert the output CNT for a full clock cycle. The machine should start sampling when an incoming signal SYNC(L) is asserted. The machine should stop sampling when it has completed a 3-bit slice and SYNC is no longer asserted. Both SIG and SYNC are synchronized to change on the negative-going clock edge. Use D memory and discrete logic elements. Choose the state code assignment to simplify the next-state decoder.

(a) 000
(b) 001
(c) 010
(d) 100
(e) 101
(f) 110
(g) 111

7. Repeat exercise 6, but find two different sequences, as follows. Each time the first sequence is found, assert output CNT1. Each time the second is found, assert output CNT2. Use D memory and discrete logic elements. Choose the state code assignment to simplify the next-state decoder.

(a) 000 001
(b) 000 010
(c) 000 011
(h) 001 010
(i) 001 011
(j) 001 100
(o) 010 100
(p) 010 101
(q) 010 110

(d) 000 100	(k) 001 101	(r) 010 111
(e) 000 101	(l) 001 110	(s) 011 100
(f) 000 110	(m) 001 111	(t) 011 101
(g) 000 111	(n) 010 011	(u) 011 110

8. Repeat exercise 6 using JK memory and discrete logic elements.

9. Repeat exercise 7 using JK memory and discrete logic elements.

10. Repeat exercise 6 using D memory and the gate array modules discussed in the text.

11. Repeat exercise 7 using D memory and the gate array modules discussed in the text.

12. Repeat exercise 6 with D memory and register outputs that are not delayed. Use discrete logic elements.

13. Repeat exercise 7 with D memory and register outputs that are not delayed. Use discrete logic elements.

14. Design a state machine to find the first positive-going edge of an incoming signal, SIG(H), that occurs after the machine is *armed*. The machine is to have three inputs: FDEDGE(H), a signal that arms the machine and causes it to start looking for an edge; SIG, as specified earlier; and RESET(L), a signal that causes the machine to get ready to find another edge. The machine should assert an output signal, FND(H), when the edge is found. FND should remain asserted until the input RESET is asserted. Be sure to consider both the case where SIG is high when FDEDGE is asserted and the case where SIG is low then FDEDGE is asserted. Use D memory.

15. Design a state machine to determine whether an input signal, SIG(L), is asserted one, two, or three clock cycles after the input START(L) is asserted. Assert output ONE, TWO, or THR depending on which condition is found. Repeat each time START is asserted. Assume that changes in SIG and START are synchronous with the negative-going edge of the clock. Use a design with D memory and discrete logic elements. Make the outputs glitch-free with no delay.

16. Repeat exercise 15 using the gate array modules discussed in the text.

17. Design a state machine to control processes that require the following sets of inputs with the timings shown. In each case, choose a suitable clock frequency, but make it as low as possible. The output should be glitch-free.

    (a) (Start the sequence when RESET is asserted.)
    START—for 10 ms
    (Wait for 5 ms.)
    STOP—for 1 ms
    (Wait for 5 ms.)

INC—until GO is asserted
(Repeat when RESET is asserted.)

(b) (Start the sequence when START is asserted.)
FWD—for 6 ms
(Wait for 1 ms.)
RVS—until STP is asserted
(Wait for 2 ms.)
INC—until NEW is asserted
(Repeat sequence until START is deasserted.)

(c) (Start the sequence when NCT is asserted.)
DOWN—until BTM is asserted
(Wait for 5 ms.)
STOP—for 10 ms
(Wait for 5 ms.)
UP—for 5 ms
(Repeat if NCT is still asserted, or wait until NCT is asserted and repeat.)

(d) (Start the sequence when RESET is asserted.)
START—for 6 ms
(Starting at thc same time as START) LEVEL—for 3 ms
(Wait for 3 ms after the end of START.)
STOP—for 3 ms
(Wait for 3 ms.)
STOP—for 1 ms
(Wait for 3 ms.)
(Repeat the sequence endlessly.)

(e) (Start the sequence when GO is asserted.)
START—until STOP is asserted
(Starting at the same time as START) NEWCT—for 2 ms
FNL—for 1 ms after STOP is asserted
(Wait for 4 ms.)
NEW—for 4 ms
(Repeat until GO is deasserted.)

18. For each of the designs of exercise 6, determine the maximum clock frequency. Assume for every gate that it has a delay of 10 ns, the delay through the flip-flops is 40 ns, and the minimum setup time for the flip-flop is 15 ns.

19. For each of the designs of exercise 7, determine whether the machine will operate at a clock rate of 10 MHz and at a clock rate of 5 MHz. Assume for every gate that it has a delay of 10 ns, the delay through the flip-flops is 40 ns, and the minimum setup time for the flip-flop is 15 ns.

10 State Machine Design and Analysis—Additional Topics

10.1 INTRODUCTION

We're going to look now at a number of different topics related to state machines. We've been designing state machines using both discrete logic and field programmable gate arrays (FPGAs). FPGAs are *flexible-architecture* programmable logic devices. Now we'll explore the design of state machines using *fixed-architecture* programmable logic devices (PLDs). These are the PROMs, PALs, and PLAs we introduced in Chapter 6 for designing combinational logic. Their architecture is exactly what we need for state machines, but because of the software aids used, we cannot conveniently include their design in a discussion of state machine designs that use discrete logic elements.

Second, we'll examine a special kind of state machine that is particularly useful in FPGAs, called a **one–flip-flop–per–state state machine**. Third, we'll address the formal analysis of state machines and how to proceed from a circuit for a state machine (or, in the case of a PLD, the fuse map) to a present state–next state and output table or the equivalent state transition diagram. Finally, we'll discuss the design of asynchronous state machines. Although normally these machines are not needed, on occasion they are sometimes indispensable. We'll consider some of the problems involved in their design and present a design method which is similar to the design method we've used for synchronous machines.

10.2 DESIGNING STATE MACHINES WITH FIXED-ARCHITECTURE PROGRAMMABLE LOGIC

In Chapter 6, Section 6.1, we discussed the use of programmable logic devices to evaluate combinational logic expressions. In Chapters 8 and 9, we found that we

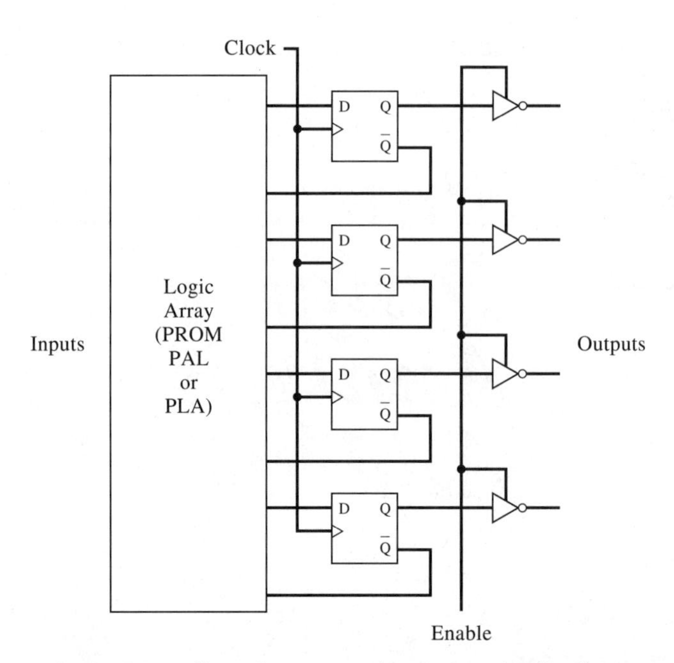

Figure 10.2-1 General programmable device with flip-flops.

can design a state machine using a set of combinational logic circuits (next state decoder) and a set of flip-flops (memory). A state machine has one or two combinational logic circuits associated with each flip-flop and one flip-flop for each binary bit in the state code. To make the programmable logic devices of Chapter 6 useful for state machine design, we need only add a flip-flop to each of the programmable combinational logic arrays. Figure 10.2-1 is a general programmable logic device with four flip-flops. The logic array can be configured in a PROM, PAL, or PLA architecture. The flip-flops are most often D flip-flops, as shown. A programmable logic device with memory elements is called a *register* device (register PROM, register PAL, or register PLA). Notice in Figure 10.2-1 how the output of the memory element is fed back into the logic array to allow for the next-state decoding.

As examples of the several different types of programmable logic devices, let's examine a few simple register PALs to see how they are used in state machine design. Figure 10.2-2 is a small register PAL. It consists of eight programmable logic circuit sections. The top two sections and bottom two sections are simple combinational logic circuits. As in all programmable devices with PAL architecture, the logic array for each section consists of several programmable *and* gates followed by an *or* gate that is not programmable. Figure 10.2-3 is a single section of this kind, isolated for clarity. Each of the four middle sections of Figure 10.2-2 has a D flip-flop in the output; Figure 10.2-4 is one of these sections. Notice that the feedback is from the $\overline{Q}$ output of the flip-flop. Actually, both Q and $\overline{Q}$ are available for feedback because of the inverter in the feedback path. Presently we'll design a state machine using this PAL.

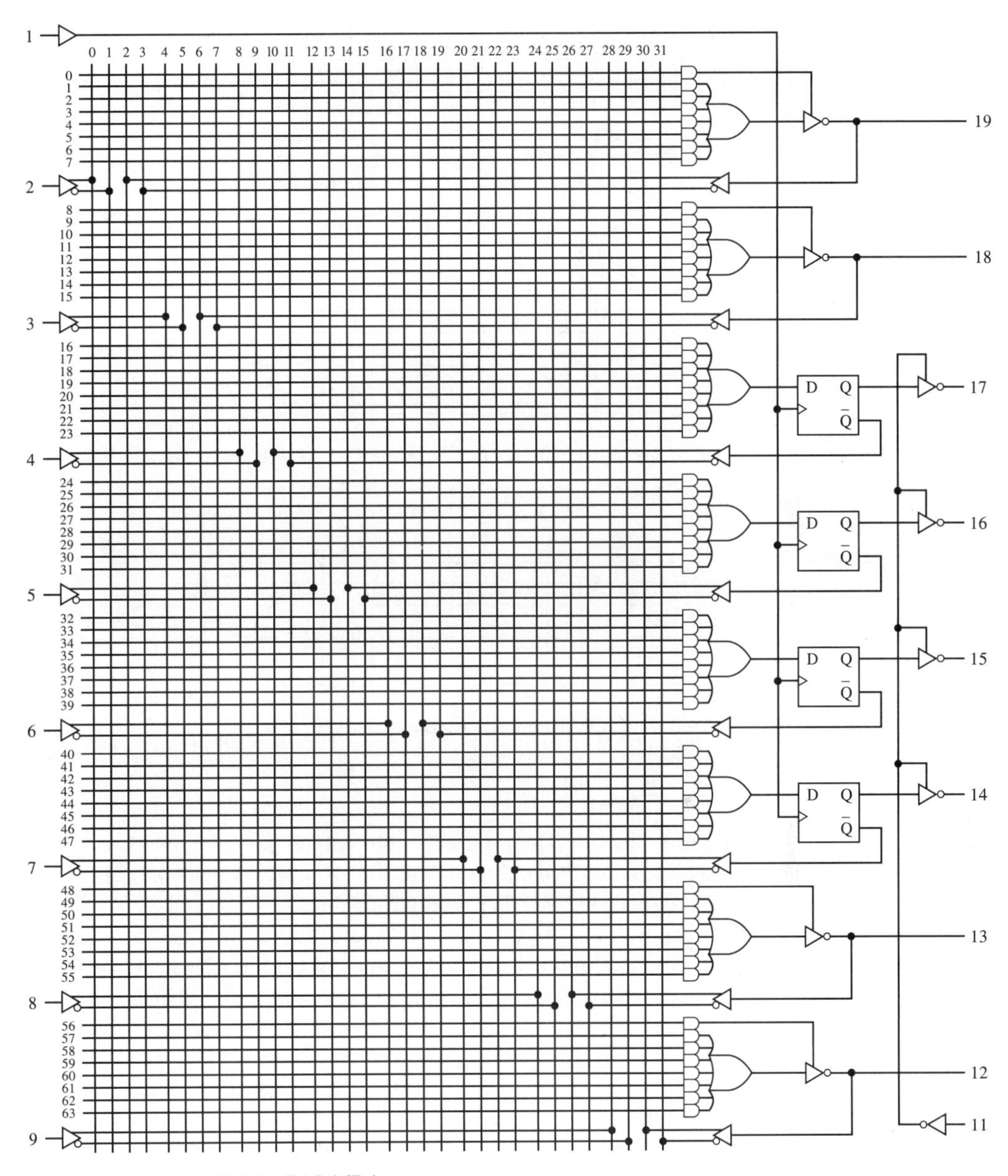

Figure 10.2-2 PAL16R4.

The PAL we're considering is simpler than most. It has a simple fixed output circuit that consists of either an inverting output buffer with a programmable enable for the combinational circuits or a flip-flop and inverting output buffer with an enable for the circuits with memory. Most PALs (and PLAs) have programmable outputs (Fig. 10.2-5) that can be configured so that they can be used as simple combinational

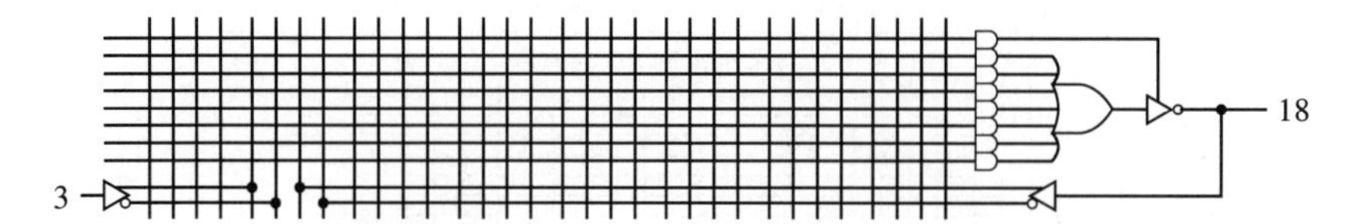

Figure 10.2-3 One of the combinational logic circuits in the PAL of Figure 10.2-2.

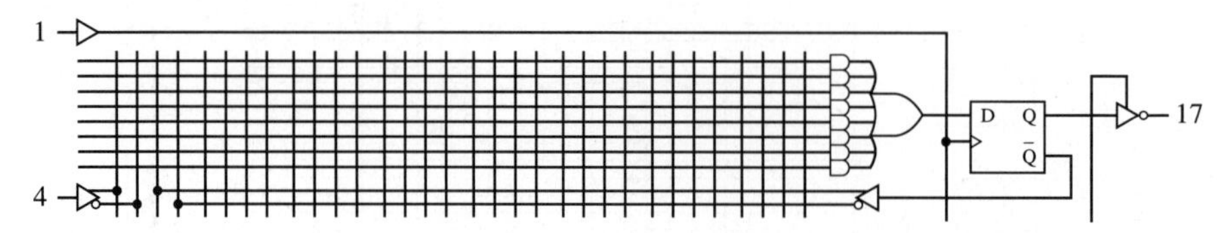

Figure 10.2-4 One of the register logic circuits in the PAL of Figure 10.2-2.

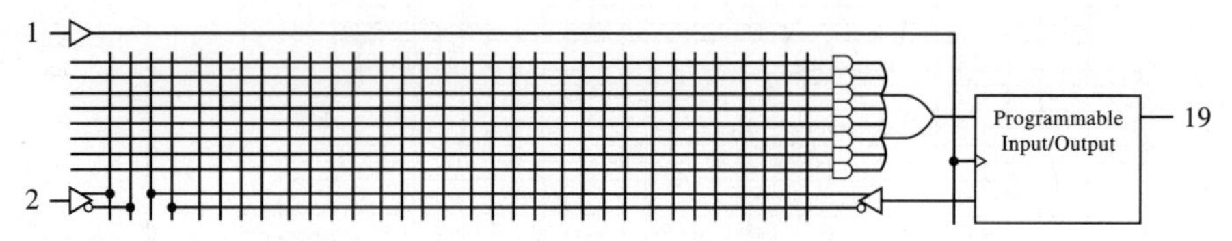

Figure 10.2-5 A section of a PAL with a programmable output.

circuits or as register circuits. The programmable output usually allows the output assertion level to be programmed either high or low. Register outputs may also have presets and clears associated with their output flip-flops.

The timing specifications for a register PAL are similar to those for a D flip-flop with a few additions. Figure 10.2-6 shows the typical timing waveforms for a PAL with both combinational and register outputs, as would be found in a data book. In the data book, such a figure is accompanied by a table of values for the times shown on the waveforms. Usually, a maximum or minimum value and a typical value are given. The values, of course, depend on the particular device; they are normally in the low 10s of nanoseconds.

Let's examine the meanings of the timing specifications. At the upper left of Figure 10.2-6 we show the time from an input or feedback change to the corresponding output change (propagation delay t_{PD}) for a combinational section of the PAL. By *feedback* we mean that $\overline{Q}$ from one of the register outputs is fed back to one of the combinational sections. The hatching (crisscross) pattern on part of the output waveform means that the value during that period is unknown.

At the middle right we show the input change to clock setup time (t_{su}) and the input change to clock hold time (t_h) for one of the registered sections of the PAL.

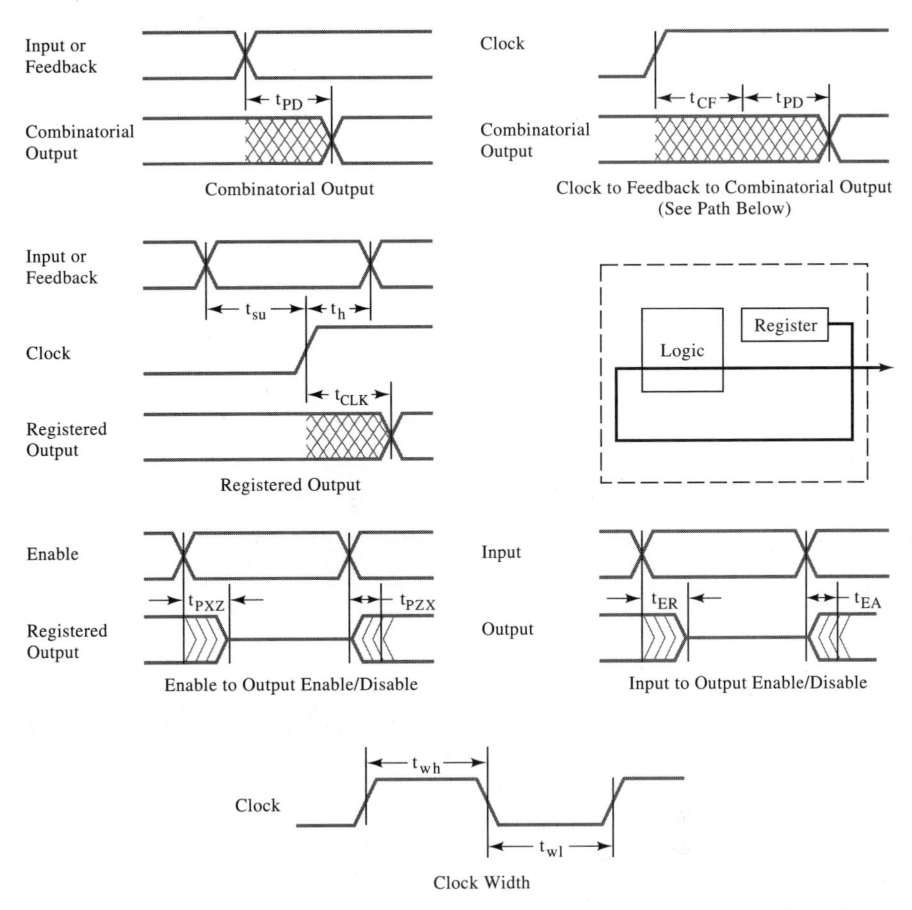

Figure 10.2-6 Timing waveforms for a typical PAL with combinational and registered outputs.

These are just like the setup and hold times for a D flip-flop. If input values change between the setup and hold times, the output value is unpredictable. We also show the time from the rising clock edge to valid output or feedback (t_{CLK}).

At the bottom left we show the times from enable deasserted to output tristated (t_{PXZ}) and from enable asserted to output enabled (t_{PZX}). Here the enable is the pin that enables and disables the register outputs. Recall that the tristate condition is a high impedance condition of the output. The fishbone pattern shows the region where the condition of the output is undetermined.

At the upper right we show the time between the rising clock edge and valid combinatorial (combinational) output where the signal path is shown, just below the

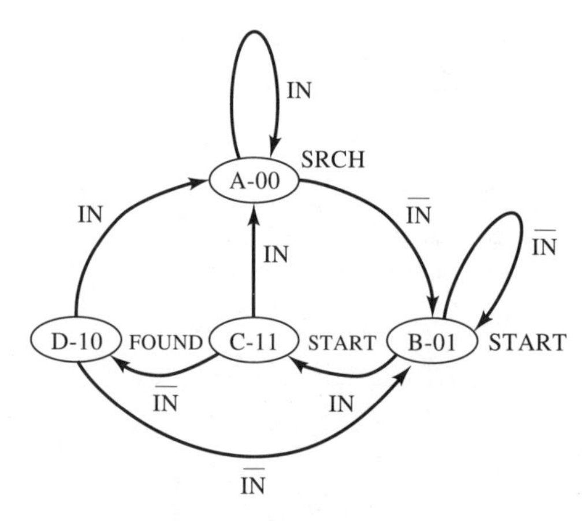

Figure 10.2-7 State machine used to illustrate design.

waveform. This time period consists of two parts, the time from the rising clock edge to valid feedback values (t_{CF}) and the time from valid feedback values to valid combinatorial output values (t_{PD}).

At the bottom right we show the time between an input change and output tristate or output disable for one of the combinatorial output sections of the PAL. These enables are the result of a programmable *and* of the inputs to the PAL.

At the bottom center we show the clock timing (t_{wh} and t_{wl}). Sometimes a PAL has synchronous or asynchronous presets and/or resets, which have time delays associated with them that can be found by consulting data books.

Computer software is available to do the actual programming of any programmable logic device. Usually, all that is necessary is to present the software with a set of equations describing the behavior of the state machine. The software then produces a map showing the connections that should be removed in the programmable device. The map is contained in a standard file, called a JEDEC file that is formatted in a special way; it's used in a device programmer to actually program the device.

The first steps in a design using a programmable logic device are the same as those we've been using: describe the problem concisely, and produce a state transition diagram. After the state transition diagram is produced, the only steps remaining are to assign the state codes and to describe the state machine in a software source program. Some software can even assign state codes for you.

To illustrate the design procedure, we'll use the PAL of Figure 10.2-2; software called PALASM; and the problem we introduced in Chapter 7, Section 7.6, and have been using throughout the text. The design procedure with PALASM requires that each state have both a name and a state code. The software can assign the state code, but this is not recommended. Figure 10.2-7 is the state transition diagram with each state assigned both a name and a code. Table 10.2-1 is the PALASM program that describes this machine. The program is divided into the DECLARATION, STATE, CONDITIONS, and SIMULATION sections. Comments (anything preceded by a ;)

are used to make the program more readable. Examples of comments are the title at the top and the lines of dashes used to delineate the sections.

The declaration section contains information to document the design and the device identification, pin assignments, and string substitutions. The first six lines are called the DESIGN HEADER. Each line starts with a keyword (TITLE, PATTERN, and so on). The intended use of the line should be evident from the keyword, although a line can consist of any text or can be blank. The next line after the header, starting with the keyword CHIP, contains two entries. The first is a name the user can assign to the CHIP. The second is the manufacturer's designation for the device to be used in the design (for instance, PAL16R4). These two items are separated by at least one space.

After the line describing the chip is a listing of pin numbers and the signal names that are assigned to those pins. Something must be listed for each pin. The polarity of the signal is specified by adding a / in front of the name if the assertion level is low. The signals ENABLE, SRCH, START, B1, B0, and FOUND are all asserted low. As we can see from Figure 10.2-2, these assertion levels match the outputs of the device. Actually, choosing assertion levels that match is not necessary, but it makes it easier to understand the logic produced. The pins that are not used are labeled NC (no connection). The power pin is labeled VCC and the ground GND. To determine which pin is the power pin and which is the ground pin, it may be necessary to consult a data book for the device. In this program we've assigned the input signal to pin 2 and the output signals to pins 12, 13, and 18. These outputs must be assigned to circuits with simple combinational outputs. The state code bits must be assigned to register outputs (pins 14 and 15). Each pin can also be assigned a STORAGE type and have a comment added that describes the signal on that pin. The STORAGE types are none (blank), COMBINATORIAL, REGISTERED, and LATCHED. The comments are none (blank), INPUT, OUTPUT, IO, CLOCK, and ENABLE.

If we use the PALASM editor to produce the source program, the declaration section is just a matter of filling in blanks on the screen, as shown in Table 10.2-2. All the keywords are supplied by the editor, and many of the blanks are filled by selecting from a list of appropriate entries. Lists are available for DEVICE (all the PALs that can be programmed by the software), P/N (PIN or NODE), storage, and comment. The selection lists appear in response to the F2 key.

The remainder of the source code is entered by using a text editor. (Normally, PALASM comes with a Wordstar-like editor.) This remaining code includes the final lines of the declaration section, which can contain string substitutions that replace complex statements with simple names. String substitutions are a convenience in writing the subsequent program but are not necessary. The syntax is

```
STRING name    'string'
```

The string is delimited by single quotes. A sample of a string substitution statement is

```
STRING IINRS    ' INPUT1 * INPUT2 + RST '
```

Table 10.2-1 PALASM Program for Sample State Machine

```
;PALASM Design Description

;------------------------------------ Declaration Segment ------------------------------------
TITLE      Sequence detector 010
PATTERN    None
REVISION   None
AUTHOR     A W Shaw
COMPANY    Utah State University
DATE       05/25/92

CHIP    PAL_REG   PAL16R4

;--------------------------------------- PIN Declarations ---------------------------------------
PIN   1       CLK                                  ; CLOCK
PIN   2       IN                                   ; INPUT
PIN   3       NC                                   ;
PIN   4       NC                                   ;
PIN   5       NC                                   ;
PIN   7       NC                                   ;
PIN   8       NC                                   ;
PIN   9       NC                                   ;
PIN   10      GND                                  ;
PIN   11      /ENBL                                ; ENABLE
PIN   12      /FOUND            COMBINATORIAL; OUTPUT
PIN   13      /SRCH             COMBINATORIAL; OUTPUT
PIN   14      /B1                 REGISTERED ; OUTPUT
PIN   15      /B0                 REGISTERED ; OUTPUT
PIN   16      NC                                   ;
PIN   17      NC                                   ;
PIN   18      /START            COMBINATORIAL; OUTPUT
PIN   19      NC                                   ;
PIN   20      VCC                                  ;

;--------------------------------------- State Segment ---------------------------------------
STATE

MOORE_MACHINE

A = /B1 * /B0
B = /B1 *  B0
C =  B1 *  B0
D =  B1 * /B0

A := NOT_IN_C → B
     + IN_C → A
B := NOT_IN_C → B
     + IN_C → C
```

```
C := NOT_IN_C → D
      + IN_C → A
D := NOT_IN_C → B
      + IN_C → A

A.OUTF = SRCH*/START*/FOUND
B.OUTF = /SRCH*START*/FOUND
C.OUTF = /SRCH*START*/FOUND
D.OUTF = /SRCH*/START*FOUND

CONDITIONS

NOT_IN_C = /IN
IN_C     =  IN

;------------------------------------- Simulation Segment -------------------------------------
SIMULATION

SETF /CLK IN ENBL
TRACE_ON CLK IN /B1 /B0 /SRCH /START /FOUND
CLOCKF CLK
CLOCKF CLK
SETF /IN
CLOCKF CLK
SETF IN
CLOCKF CLK
SETF /IN
CLOCKF CLK
CLOCKF CLK
CLOCKF CLK
SETF IN
CLOCKF CLK
CLOCKF CLK
SETF /IN
CLOCKF CLK
SETF IN
CLOCKF CLK
SETF /IN
CLOCKF CLK
SETF IN
TRACE_OFF

;----------------------------------------------------------------------------------------------------
```

Table 10.2-2 PALASM New Design Editor Screen

PDS Declaration Segment

Title	
Pattern	
Revision	
Author	
Company	
Date	05/28/92

CHIP ChipName = ______ Device = ______

P/N	Number	Name	Paired with PIN	Storage	;comment
					1
					2
					3
					4
					5
					6
					7
					8
					9

Enter PIN/NODE Data. [Press ⟨ESC⟩ = abort, F1 = help, F2 = choices, F10 = save & exit]

This substitution allows us to use IINRS instead of the longer INPUT1 * INPUT2 + RST anywhere in the subsequent program. There are no string statements in the sample program we're considering.

The state section of the program starts with the keyword STATE (see the sample program). The keyword STATE is followed by either MOORE_MACHINE or MEALY_MACHINE to specify the type of machine to be designed. The machine type declaration is followed by global default declarations, which specify default values for the signals produced by the machine when they cannot be determined from the design specifications. Syntax and meanings for default declarations are given in Table 10.2-3.

Global default declarations are followed by state code assignments. The syntax is

state_name = high_order_bit * . . . * low_order_bit

Table 10.2-3 Global Defaults

Default Syntax	Meaning
OUTPUT_HOLD signal list	List of signals that hold their previous values when their values cannot be determined from the state machine specifications.
DEFAULT_OUTPUT signal list	List of signals that default to the value specified when the output is not specified in a particular state. The default value is specified by a / or % in front of the signal name. The absence of a symbol in front of the signal name means the signal is asserted; / means the signal is not asserted; and % means *don't care.*
DEFAULT_BRANCH state name	Specifies the next state when the next state cannot be determined from state specifications.
DEFAULT_BRANCH HOLD_STATE	Holds the machine in the same state when the next state cannot be determined from the state specifications.
DEFAULT_BRANCH NEXT_STATE	Moves the machine to the following state equation when the next state cannot be determined from the state specifications. Notice that there is no following state for the last state.

where the bit is specified by giving the name of the signal that represents this bit with or without a / to specify whether the signal is not asserted (0) or asserted (1). Each state must have a name, and each state should be given a state code assignment, although the software will supply a state code assignment if none is given (register pins assigned NC labels will be used as state code bits). In the sample program, state A is assigned the code /B1*/B0—that is, B1 not asserted (0) and B0 not asserted (0), as specified in the state transition diagram.

The state assignments are followed by the state equations, which describe the transitions from one state to the next. The syntax is

state_name_1 := condition_1 → state_name_2 + condition_2 → state_name_3

This equation says that the state with state_name_1 makes a transition to the state with state_name_2 when condition_1 is true, or it makes a transition to the state with state_name_3 when condition_2 is true. We can add as many *or* statements as we need to describe all the transitions that are made from a state. When the transition is unconditional (always made on the next clock), the condition is VCC, which is always true. The last transition can be a default condition that specifies the state to go to if no other conditions are met. The default state is specified using no condition (+ → state name). The symbol := signifies a register output. State outputs must be register outputs. The conditions are logic conditions involving the inputs to the machine. They are defined in the CONDITIONS section. The first state equation shows that state A makes a transition to state B on the condition NOT_IN_C or a transition to state A on the condition IN_C. To see whether these conditions match the state transition diagram, we'll need to look at the CONDITIONS section of the program and determine the conditions NOT_IN_C (= $\overline{IN}$) and IN_C (= IN).

The state equations are followed by the output equations. Output equations can be specified as either combinational outputs or register outputs (for deglitching); furthermore, they can depend on input conditions if the machine is a Mealy machine.

The syntax for the Moore machine is

state_name.OUTF = output_1 * output_2
(combinational output)

state_name.OUTF := output_1 * output_2
(register output)

The syntax for the Mealy machine is

state_name.OUTF = condition_1 → output_1 * output_2 + Condition_2 → output_3 * output_4
(combinational output)

state_name.OUTF := condition_1 → output_1 * output_2 + condition_2 → output_3 * output_4
(register output)

State_name is the name of the state that is to produce the output. Output_1, output_2, and so on are the names of the output signals to be produced. They must be signal names assigned to output pins. Condition_1 and condition_2 are the names of input logic conditions, which are specified in the CONDITIONS section of the program. Combinational outputs occur as soon as the state is entered or as soon as the state is entered and the condition is met. Register outputs should be used if an output with no glitches is needed. Register outputs for a Moore machine occur as soon as the state is entered; register outputs for a Mealy machine occur one clocked cycle after the state is entered. The values of all the outputs must be specified in each state unless it does not matter. An unspecified output may be either asserted or not asserted. The global default output can be used to specify the output for register outputs which are otherwise not specified.

The outputs here must be combinational because they are physically combinational (no register) outputs of the device. The PAL does not have programmable outputs. The first output statement specifies that in state A the output SRCH is asserted and the outputs START and FOUND are not asserted. No condition is given, because the machine is a Moore machine.

The conditions section of the program starts with the keyword CONDITIONS.

It defines all the combinations of input conditions that were used in the STATE section. The syntax for a condition is

condition = logic expression involving input signals

Condition is a name for the particular condition. The logic expression involves *ands* (*), *ors* (+), and complements (/) of the input signals. Samples of logic expressions are given in the program in Table 10.2-1. Conditions cannot have the same names as inputs. The first conditions statement defines NOT_IN_C to be equal to *not* IN. Naming the conditions with mnemonics that describe them, as was done here, is often a helpful memory aid.

The final section of the program is the simulation section. It starts with the keyword SIMULATION and describes a sequence of input signals that are to be applied to the machine. The syntax for all the simulation commands is

command   signal list

The first command is SETF, which causes the signals listed to be asserted or not asserted (/ means not asserted). The second command, TRACE_ON, specifies the signals that are to be traced and can be viewed separately. The assertion level of the signal is specified by using a / to mean asserted low. (Notice that this use of / is different from that in the other simulation commands.) The TRACE_ON command has a companion command, TRACE_OFF, that turns off the signals to be traced. Tracing allows select signals to be viewed. All the signals are recorded and can be viewed, but without any selection unless the preceding commands are used.

The CLOCKF command causes one clock cycle to be simulated. The signal list used with the CLOCKF command should be the name of the clock signal (usually there is only one). A CLOCKF causes a positive-going transition followed by a negative-going transition.

The simulation allows loops and conditional branching, which we will not cover. If this sample program is processed using PALASM, a number of outputs are produced, including a JEDEC output file for programming the actual programmable device, a fuse map (Table 10.2-4), and a timing diagram (Fig. 10.2-8). Notice that the time scale in the timing diagram is not truly proportional to time, but it does show the relative positions of the transitions. State changes are delayed from the clock edge, but the distance is not proportional to the actual time delay.

If we want to check the logic circuits that have been programmed into the programmable device, we can mark the programmed connections on the circuit for the programmable device, as shown in Figure 10.2-9. The programmed connections are taken from the fuse map; they are marked with an X on the circuit. An X in one of the programmable *and* gates means that all the inputs to the gate are connected and that this particular gate is *not asserted.* As a sample of the programmed logic, consider the next-state decoder logic for B0 (D0):

$$\mathrm{D0} = \overline{\mathrm{IN}} \cdot \overline{\mathrm{B0}} + \overline{\mathrm{B1}} \cdot \mathrm{B0}$$

This is the same decoder logic we found in Chapter 7, Section 7.2. Finding the logic expression is a little tricky because B1(L) and B0(L) are the signals that are fed back to the next-state decoder.

Table 10.2-4 Fuse Map

PALASM4 PAL Assembler—Market Version 1.2 (5-31-91)
(c)—Copyright Advanced Micro Devices Inc., 1991

TITLE :Sequence detector 010
PATTERN:None
REVISION:None

AUTHOR :A W Shaw
COMPANY:Utah State University
DATE :05/25/92

PAL16R4
PAL_REG

			11	1111	1111	2222	2222	2233
	0123	4567	8901	2345	6789	0123	4567	8901
0	XXXX	XXXX	XXXX	XXXX	XXXX	XXXX	XXXX	XXXX
1	XXXX	XXXX	XXXX	XXXX	XXXX	XXXX	XXXX	XXXX
2	XXXX	XXXX	XXXX	XXXX	XXXX	XXXX	XXXX	XXXX
3	XXXX	XXXX	XXXX	XXXX	XXXX	XXXX	XXXX	XXXX
4	XXXX	XXXX	XXXX	XXXX	XXXX	XXXX	XXXX	XXXX
5	XXXX	XXXX	XXXX	XXXX	XXXX	XXXX	XXXX	XXXX
6	XXXX	XXXX	XXXX	XXXX	XXXX	XXXX	XXXX	XXXX
7	XXXX	XXXX	XXXX	XXXX	XXXX	XXXX	XXXX	XXXX
8	----	----	----	----	----	----	----	----
9	----	----	----	----	---X	----	----	----
10	XXXX	XXXX	XXXX	XXXX	XXXX	XXXX	XXXX	XXXX
11	XXXX	XXXX	XXXX	XXXX	XXXX	XXXX	XXXX	XXXX
12	XXXX	XXXX	XXXX	XXXX	XXXX	XXXX	XXXX	XXXX
13	XXXX	XXXX	XXXX	XXXX	XXXX	XXXX	XXXX	XXXX
14	XXXX	XXXX	XXXX	XXXX	XXXX	XXXX	XXXX	XXXX
15	XXXX	XXXX	XXXX	XXXX	XXXX	XXXX	XXXX	XXXX
16	XXXX	XXXX	XXXX	XXXX	XXXX	XXXX	XXXX	XXXX
17	XXXX	XXXX	XXXX	XXXX	XXXX	XXXX	XXXX	XXXX
18	XXXX	XXXX	XXXX	XXXX	XXXX	XXXX	XXXX	XXXX
19	XXXX	XXXX	XXXX	XXXX	XXXX	XXXX	XXXX	XXXX
20	XXXX	XXXX	XXXX	XXXX	XXXX	XXXX	XXXX	XXXX
21	XXXX	XXXX	XXXX	XXXX	XXXX	XXXX	XXXX	XXXX
22	XXXX	XXXX	XXXX	XXXX	XXXX	XXXX	XXXX	XXXX
23	XXXX	XXXX	XXXX	XXXX	XXXX	XXXX	XXXX	XXXX
24	XXXX	XXXX	XXXX	XXXX	XXXX	XXXX	XXXX	XXXX
25	XXXX	XXXX	XXXX	XXXX	XXXX	XXXX	XXXX	XXXX
26	XXXX	XXXX	XXXX	XXXX	XXXX	XXXX	XXXX	XXXX
27	XXXX	XXXX	XXXX	XXXX	XXXX	XXXX	XXXX	XXXX
28	XXXX	XXXX	XXXX	XXXX	XXXX	XXXX	XXXX	XXXX
29	XXXX	XXXX	XXXX	XXXX	XXXX	XXXX	XXXX	XXXX
30	XXXX	XXXX	XXXX	XXXX	XXXX	XXXX	XXXX	XXXX
31	XXXX	XXXX	XXXX	XXXX	XXXX	XXXX	XXXX	XXXX

```
32  -X--  ----  ----  ----  --X-  ----  ----  ----
33  ----  ----  ----  ----  ---X  --X-  ----  ----
34  XXXX  XXXX  XXXX  XXXX  XXXX  XXXX  XXXX  XXXX
35  XXXX  XXXX  XXXX  XXXX  XXXX  XXXX  XXXX  XXXX
36  XXXX  XXXX  XXXX  XXXX  XXXX  XXXX  XXXX  XXXX
37  XXXX  XXXX  XXXX  XXXX  XXXX  XXXX  XXXX  XXXX
38  XXXX  XXXX  XXXX  XXXX  XXXX  XXXX  XXXX  XXXX
39  XXXX  XXXX  XXXX  XXXX  XXXX  XXXX  XXXX  XXXX

40  X---  ----  ----  ----  ---X  --X-  ----  ----
41  -X--  ----  ----  ----  ---X  ---X  ----  ----
42  XXXX  XXXX  XXXX  XXXX  XXXX  XXXX  XXXX  XXXX
43  XXXX  XXXX  XXXX  XXXX  XXXX  XXXX  XXXX  XXXX
44  XXXX  XXXX  XXXX  XXXX  XXXX  XXXX  XXXX  XXXX
45  XXXX  XXXX  XXXX  XXXX  XXXX  XXXX  XXXX  XXXX
46  XXXX  XXXX  XXXX  XXXX  XXXX  XXXX  XXXX  XXXX
47  XXXX  XXXX  XXXX  XXXX  XXXX  XXXX  XXXX  XXXX

48  ----  ----  ----  ----  ----  ----  ----  ----
49  ----  ----  ----  ----  --X-  --X-  ----  ----
50  XXXX  XXXX  XXXX  XXXX  XXXX  XXXX  XXXX  XXXX
51  XXXX  XXXX  XXXX  XXXX  XXXX  XXXX  XXXX  XXXX
52  XXXX  XXXX  XXXX  XXXX  XXXX  XXXX  XXXX  XXXX
53  XXXX  XXXX  XXXX  XXXX  XXXX  XXXX  XXXX  XXXX
54  XXXX  XXXX  XXXX  XXXX  XXXX  XXXX  XXXX  XXXX
55  XXXX  XXXX  XXXX  XXXX  XXXX  XXXX  XXXX  XXXX

56  ----  ----  ----  ----  ----  ----  ----  ----
57  ----  ----  ----  ----  --X-  ---X  ----  ----
58  XXXX  XXXX  XXXX  XXXX  XXXX  XXXX  XXXX  XXXX
59  XXXX  XXXX  XXXX  XXXX  XXXX  XXXX  XXXX  XXXX
60  XXXX  XXXX  XXXX  XXXX  XXXX  XXXX  XXXX  XXXX
61  XXXX  XXXX  XXXX  XXXX  XXXX  XXXX  XXXX  XXXX
62  XXXX  XXXX  XXXX  XXXX  XXXX  XXXX  XXXX  XXXX
63  XXXX  XXXX  XXXX  XXXX  XXXX  XXXX  XXXX  XXXX

SUMMARY
      TOTAL FUSES BLOWN = 305
```

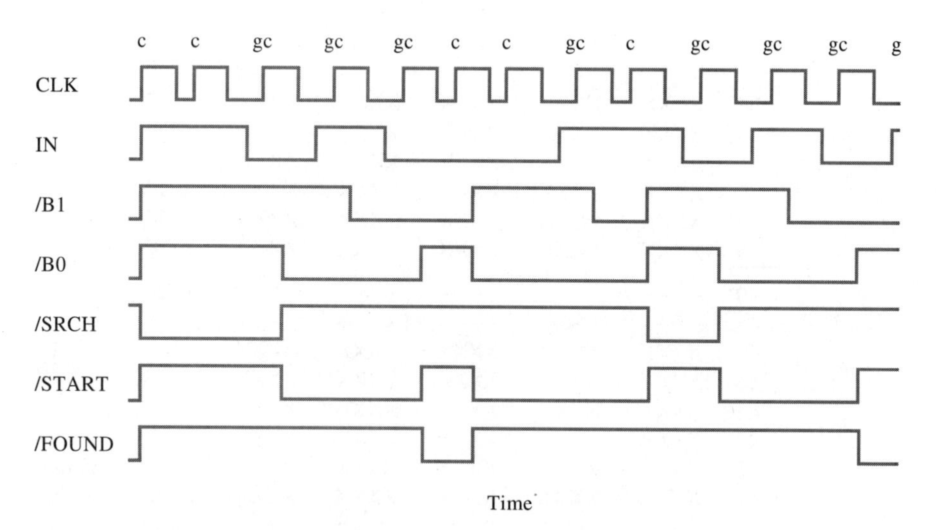

Figure 10.2-8 Timing diagram for selected signals.

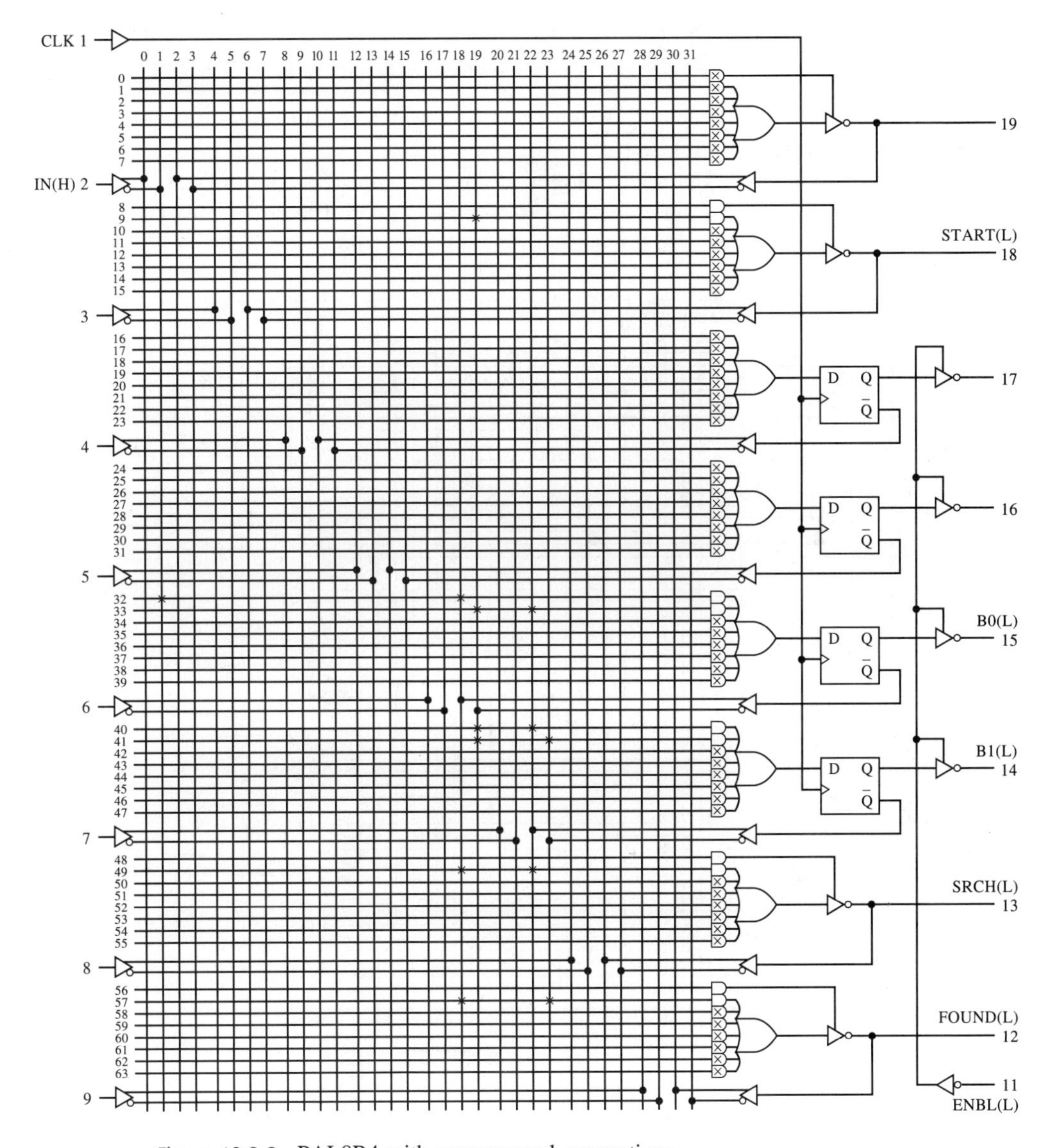

Figure 10.2-9 PAL8R4 with programmed connections.

Example 10.2-1

Repeat the state machine design described in this section of the text, but add an output deglitching register.

To add an output register to this design, we'll need three flip-flops. This means we'll need a PAL with at least five flip-flops — two for the state code and three for the output register. We'll use the following PAL16R8.

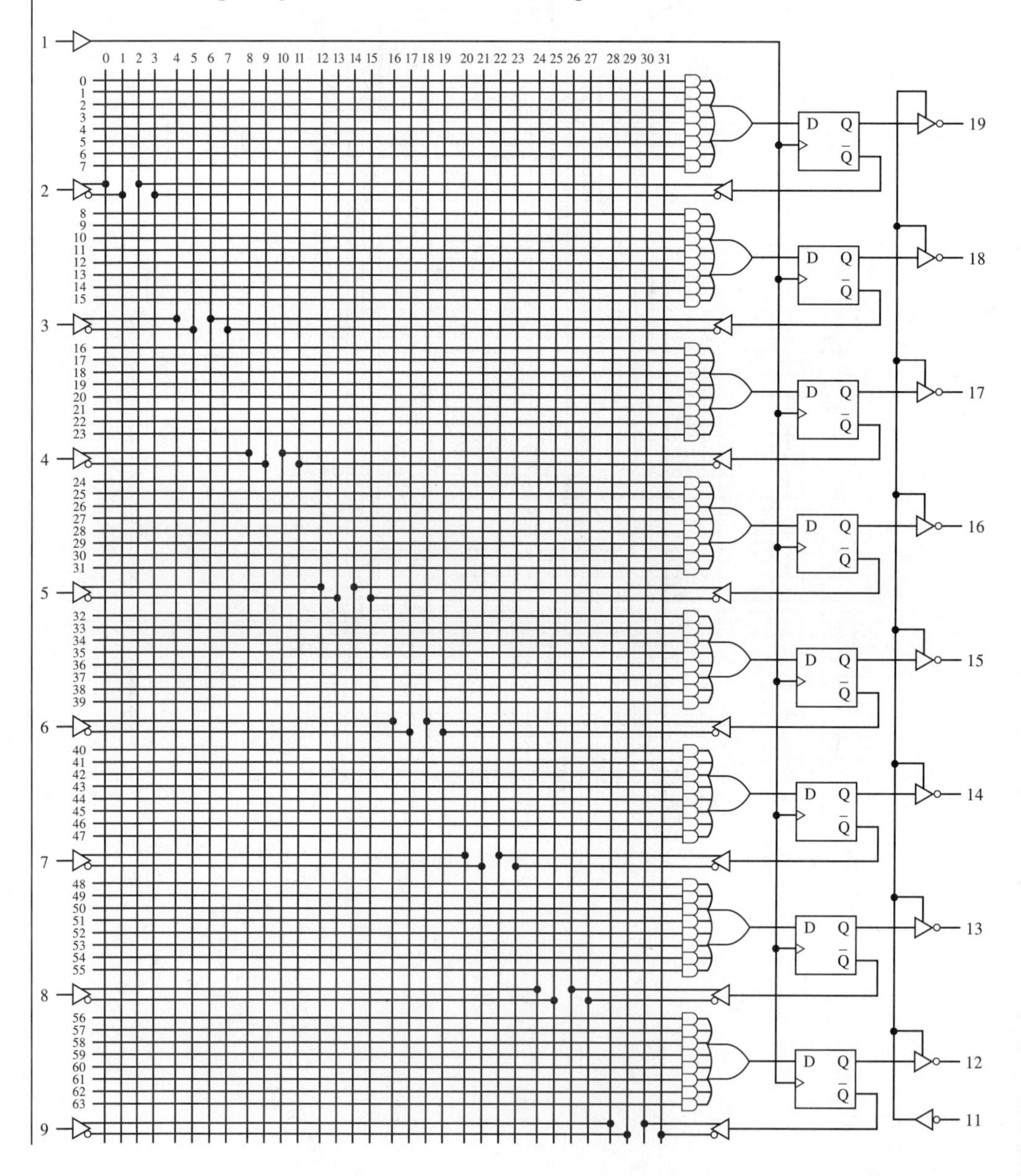

The programming of this device is almost the same as that of the device described at the beginning of this section. One difference is that here the ouputs are specified as register outputs rather than combinational outputs. Notice that the = in the output section is replaced with a :=. A subtlety in this machine is the difference in the behavior of the output, depending on whether the machine is a Moore machine or Mealy machine. If it is specified as a Moore machine, the logic is programmed to make the outputs occur during the states in which they are specified as occurring; if the machine is specified as a Mealy machine, the logic is programmed so that the output is delayed one clock cycle.

Following is the program for this state machine design, along with the fuse map and the simulation output for selected signals.

Program

```
;PALASM Design Description

;---------------------------------------- Declaration Segment ----------------------------------------
TITLE       Sequence detector 010
PATTERN     None
REVISION    None
AUTHOR      A W Shaw
COMPANY     Utah State University
DATE        06/30/92

CHIP     PAL_REG   PAL16R8

;------------------------------------------ PIN Declarations ------------------------------------------
PIN    1          CLK                              ; CLOCK
PIN    2          IN                               ; INPUT
PIN    3          NC                               ;
PIN    4          NC                               ;
PIN    5          NC                               ;
PIN    6          NC                               ;
PIN    7          NC                               ;
PIN    8          NC                               ;
PIN    9          NC                               ;
PIN    10         GND                              ;
PIN    11         /ENBL                            ; ENABLE
PIN    12         /FOUND               REGISTERED ; OUTPUT
PIN    13         /SRCH                REGISTERED ; OUTPUT
PIN    14         /B1                  REGISTERED ; OUTPUT
PIN    15         /B0                  REGISTERED ; OUTPUT
PIN    16         NC                               ;
PIN    17         NC                               ;
PIN    18         /START               REGISTERED ; OUTPUT
PIN    19         NC                               ;
PIN    20         VCC                              ;
```

```
;------------------------------------ State Equation Segment ------------------------------------
STATE

MEALY_MACHINE

DEFAULT_OUTPUT /SRCH /START /FOUND

A = /B1 * /B0
B = /B1 *  B0
C =  B1 *  B0
D =  B1 * /B0

A := NOT_IN_C → B
      + IN_C → A
B := NOT_IN_C → B
      + IN_C → C
C := NOT_IN_C → D
      + IN_C → A
D := NOT_IN_C → B
      + IN_C → A

A.OUTF := VCC → SRCH*/START*/FOUND
B.OUTF := VCC → START*/SRCH*/FOUND
C.OUTF := VCC → START*/SRCH*/FOUND
D.OUTF := VCC → FOUND*/START*/SRCH

CONDITIONS

NOT_IN_C = /IN
IN_C     =  IN

;-------------------------------------- Simulation Segment ----------------------------------------
SIMULATION

SETF /CLK IN ENBL
TRACE_ON CLK IN /B1 /B0 /SRCH /START /FOUND
CLOCKF CLK
CLOCKF CLK
SETF /IN
CLOCKF CLK
SETF IN
CLOCKF CLK
SETF /IN
CLOCKF CLK
CLOCKF CLK
CLOCKF CLK
SETF IN
CLOCKF CLK
CLOCKF CLK
SETF /IN
```

```
CLOCKF CLK
SETF IN
CLOCKF CLK
SETF /IN
CLOCKF CLK
SETF IN
TRACE_OFF
```

;---

FUSE MAP

PALASM4 PAL Assembler—Market Version 1.2 (5-31-91)
(c)—Copyright Advanced Micro Devices Inc., 1991

```
TITLE     Sequence detector 010     AUTHOR  :A W Shaw
PATTERN   None                      COMPANY:Utah State University
REVISION  None                      DATE    :06/30/92

PAL16R8
PAL_REG
```

	0123	4567	11 8901	1111 2345	1111 6789	2222 0123	2222 4567	2233 8901
0	XXXX	XXXX	XXXX	XXXX	XXXX	XXXX	XXXX	XXXX
1	XXXX	XXXX	XXXX	XXXX	XXXX	XXXX	XXXX	XXXX
2	XXXX	XXXX	XXXX	XXXX	XXXX	XXXX	XXXX	XXXX
3	XXXX	XXXX	XXXX	XXXX	XXXX	XXXX	XXXX	XXXX
4	XXXX	XXXX	XXXX	XXXX	XXXX	XXXX	XXXX	XXXX
5	XXXX	XXXX	XXXX	XXXX	XXXX	XXXX	XXXX	XXXX
6	XXXX	XXXX	XXXX	XXXX	XXXX	XXXX	XXXX	XXXX
7	XXXX	XXXX	XXXX	XXXX	XXXX	XXXX	XXXX	XXXX
8	----	----	----	----	---X	----	----	----
9	XXXX	XXXX	XXXX	XXXX	XXXX	XXXX	XXXX	XXXX
10	XXXX	XXXX	XXXX	XXXX	XXXX	XXXX	XXXX	XXXX
11	XXXX	XXXX	XXXX	XXXX	XXXX	XXXX	XXXX	XXXX
12	XXXX	XXXX	XXXX	XXXX	XXXX	XXXX	XXXX	XXXX
13	XXXX	XXXX	XXXX	XXXX	XXXX	XXXX	XXXX	XXXX
14	XXXX	XXXX	XXXX	XXXX	XXXX	XXXX	XXXX	XXXX
15	XXXX	XXXX	XXXX	XXXX	XXXX	XXXX	XXXX	XXXX
16	XXXX	XXXX	XXXX	XXXX	XXXX	XXXX	XXXX	XXXX
17	XXXX	XXXX	XXXX	XXXX	XXXX	XXXX	XXXX	XXXX
18	XXXX	XXXX	XXXX	XXXX	XXXX	XXXX	XXXX	XXXX
19	XXXX	XXXX	XXXX	XXXX	XXXX	XXXX	XXXX	XXXX
20	XXXX	XXXX	XXXX	XXXX	XXXX	XXXX	XXXX	XXXX
21	XXXX	XXXX	XXXX	XXXX	XXXX	XXXX	XXXX	XXXX
22	XXXX	XXXX	XXXX	XXXX	XXXX	XXXX	XXXX	XXXX
23	XXXX	XXXX	XXXX	XXXX	XXXX	XXXX	XXXX	XXXX

```
24   XXXX XXXX XXXX XXXX XXXX XXXX XXXX XXXX
25   XXXX XXXX XXXX XXXX XXXX XXXX XXXX XXXX
26   XXXX XXXX XXXX XXXX XXXX XXXX XXXX XXXX
27   XXXX XXXX XXXX XXXX XXXX XXXX XXXX XXXX
28   XXXX XXXX XXXX XXXX XXXX XXXX XXXX XXXX
29   XXXX XXXX XXXX XXXX XXXX XXXX XXXX XXXX
30   XXXX XXXX XXXX XXXX XXXX XXXX XXXX XXXX
31   XXXX XXXX XXXX XXXX XXXX XXXX XXXX XXXX

32   -X-- ---- ---- ---- --X- ---- ---- ----
33   ---- ---- ---- ---- ---X --X- ---- ----
34   XXXX XXXX XXXX XXXX XXXX XXXX XXXX XXXX
35   XXXX XXXX XXXX XXXX XXXX XXXX XXXX XXXX
36   XXXX XXXX XXXX XXXX XXXX XXXX XXXX XXXX
37   XXXX XXXX XXXX XXXX XXXX XXXX XXXX XXXX
38   XXXX XXXX XXXX XXXX XXXX XXXX XXXX XXXX
39   XXXX XXXX XXXX XXXX XXXX XXXX XXXX XXXX

40   X--- ---- ---- ---- ---X --X- ---- ----
41   -X-- ---- ---- ---- ---X ---X ---- ----
42   XXXX XXXX XXXX XXXX XXXX XXXX XXXX XXXX
43   XXXX XXXX XXXX XXXX XXXX XXXX XXXX XXXX
44   XXXX XXXX XXXX XXXX XXXX XXXX XXXX XXXX
45   XXXX XXXX XXXX XXXX XXXX XXXX XXXX XXXX
46   XXXX XXXX XXXX XXXX XXXX XXXX XXXX XXXX
47   XXXX XXXX XXXX XXXX XXXX XXXX XXXX XXXX

48   ---- ---- ---- ---- --X- --X- ---- ----
49   XXXX XXXX XXXX XXXX XXXX XXXX XXXX XXXX
50   XXXX XXXX XXXX XXXX XXXX XXXX XXXX XXXX
51   XXXX XXXX XXXX XXXX XXXX XXXX XXXX XXXX
52   XXXX XXXX XXXX XXXX XXXX XXXX XXXX XXXX
53   XXXX XXXX XXXX XXXX XXXX XXXX XXXX XXXX
54   XXXX XXXX XXXX XXXX XXXX XXXX XXXX XXXX
55   XXXX XXXX XXXX XXXX XXXX XXXX XXXX XXXX

56   ---- ---- ---- ---- --X- ---X ---- ----
57   XXXX XXXX XXXX XXXX XXXX XXXX XXXX XXXX
58   XXXX XXXX XXXX XXXX XXXX XXXX XXXX XXXX
59   XXXX XXXX XXXX XXXX XXXX XXXX XXXX XXXX
60   XXXX XXXX XXXX XXXX XXXX XXXX XXXX XXXX
61   XXXX XXXX XXXX XXXX XXXX XXXX XXXX XXXX
62   XXXX XXXX XXXX XXXX XXXX XXXX XXXX XXXX
63   XXXX XXXX XXXX XXXX XXXX XXXX XXXX XXXX

SUMMARY
     TOTAL FUSES BLOWN = 209
```

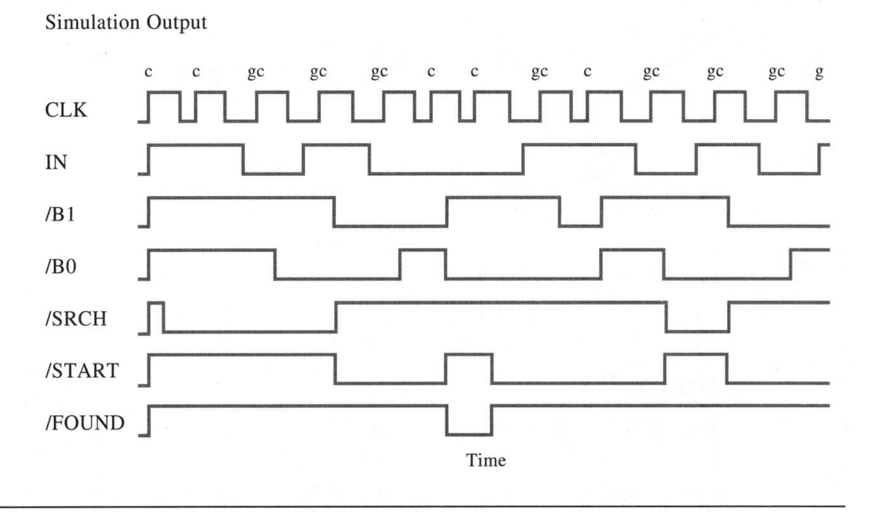

Example 10.2-2

Neither the PAL16R4 nor the PAL16R8 is equipped to reset the flip-flops, so some provision must be made to start a state machine using these devices in a known state. Redesign the machine of this section to include a provision for setting the initial state.

The goal is to add an input reset (RST) that will cause the machine to make a transition to state A (00) from any other state. This will reset the machine to state A on the first clock cycle following an asserted RST. Notice that the reset is synchronous. There is no way to make an asynchronous reset with a device that does not have a hardware provision for a reset.

We choose to modify the program for the PAL16R4 to make a provision for reset. We do not need to redraw the state transition diagram, but only modify the description of the transitions in the programming of the device to make sure that when RST is asserted, there is a transition to state A. This, of course, necessitates adding the condition that RST not be asserted to all the other transitions. The condition

NOT_IN_C = /IN

thus becomes

N_IN_N_R = /IN*/RST

and the condition

IN_C = IN

becomes

IN_R = IN + RST

or

$$IN\_N\_R = IN*/RST$$

depending on whether the transition is to state A or a state other than A. The condition

$$R = RST$$

must be added to allow transitions to state A that do not already exist. The program for the PAL16R4 with provisions for reset follows, along with the fuse map and simulation output. Notice in the fuse map that the logic to handle the reset is more complicated than the logic in the previous design without reset, and in the simulation that the initial reset starts the simulation in state A.

Program

```
;PALASM Design Description

;--------------------------------------- Declaration Segment ---------------------------------------
TITLE       Sequence detector 010
PATTERN     None
REVISION    None
AUTHOR      A W Shaw
COMPANY     Utah State University
DATE        05/25/92

CHIP      PAL_REG   PAL16R4

;------------------------------------------ PIN Declarations ------------------------------------------
PIN   1     CLK                                ; CLOCK
PIN   2     IN                                 ; INPUT
PIN   3     RST                                ;
PIN   4     NC                                 ;
PIN   5     NC                                 ;
PIN   7     NC                                 ;
PIN   8     NC                                 ;
PIN   9     NC                                 ;
PIN   10    GND                                ;
PIN   11    /ENBL                              ; ENABLE
PIN   12    /FOUND          COMBINATORIAL; OUTPUT
PIN   13    /SRCH           COMBINATORIAL; OUTPUT
PIN   14    /B1             REGISTERED   ; OUTPUT
PIN   15    /B0             REGISTERED   ; OUTPUT
PIN   16    NC                                 ;
PIN   17    NC                                 ;
PIN   18    /START          COMBINATORIAL; OUTPUT
PIN   19    NC                                 ;
PIN   20    VCC                                ;
```

```
;------------------------------------------ State Segment ------------------------------------------
STATE

MOORE_MACHINE

A = /B1 * /B0
B = /B1 *  B0
C =  B1 *  B0
D =  B1 * /B0

A := N_IN_N_R → B
     + IN_R → A
B := N_IN_N_R → B
     + IN_N_R → C
     + R → A
C := N_IN_N_R → D
     + IN_R → A
D := N_IN_N_R → B
     + IN_R → A

A.OUTF = SRCH*/START*/FOUND
B.OUTF = /SRCH*START*/FOUND
C.OUTF = /SRCH*START*/FOUND
D.OUTF = /SRCH*/START*FOUND

CONDITIONS

N_IN_N_R = /IN * /RST
IN_R     =  IN + RST
IN_N_R   = IN * /RST
R        = RST
;--------------------------------------- Simulation Segment ---------------------------------------
SIMULATION

SETF /CLK IN ENBL RST
TRACE_ON CLK IN  RST /B1 /B0 /SRCH /START /FOUND
CLOCKF CLK
SETF /RST
CLOCKF CLK
SETF /IN
CLOCKF CLK
SETF IN
CLOCKF CLK
SETF /IN
CLOCKF CLK
CLOCKF CLK
CLOCKF CLK
SETF IN
CLOCKF CLK
```

```
CLOCKF CLK
SETF /IN
CLOCKF CLK
SETF IN
CLOCKF CLK
SETF /IN
TRACE_OFF
```

FUSE MAP

```
;-----------------------------------------------------------------------------------------------------------

PALASM4   PAL Assembler—Market Version 1.2 (5-31-91)
  (c)—Copyright Advanced Micro Devices Inc., 1991

TITLE      Sequence detector 010          AUTHOR  :A W Shaw
PATTERN    None                           COMPANY:Utah State University
REVISION   None                           DATE    :05/25/92

PAL16R4
PAL_REG

                            11    1111    1111    2222    2222    2233
        0123    4567      8901    2345    6789    0123    4567    8901

0       XXXX    XXXX      XXXX    XXXX    XXXX    XXXX    XXXX    XXXX
1       XXXX    XXXX      XXXX    XXXX    XXXX    XXXX    XXXX    XXXX
2       XXXX    XXXX      XXXX    XXXX    XXXX    XXXX    XXXX    XXXX
3       XXXX    XXXX      XXXX    XXXX    XXXX    XXXX    XXXX    XXXX
4       XXXX    XXXX      XXXX    XXXX    XXXX    XXXX    XXXX    XXXX
5       XXXX    XXXX      XXXX    XXXX    XXXX    XXXX    XXXX    XXXX
6       XXXX    XXXX      XXXX    XXXX    XXXX    XXXX    XXXX    XXXX
7       XXXX    XXXX      XXXX    XXXX    XXXX    XXXX    XXXX    XXXX

8       ----    ----      ----    ----    ----    ----    ----    ----
9       ----    ----      ----    ----    ---X    ----    ----    ----
10      XXXX    XXXX      XXXX    XXXX    XXXX    XXXX    XXXX    XXXX
11      XXXX    XXXX      XXXX    XXXX    XXXX    XXXX    XXXX    XXXX
12      XXXX    XXXX      XXXX    XXXX    XXXX    XXXX    XXXX    XXXX
13      XXXX    XXXX      XXXX    XXXX    XXXX    XXXX    XXXX    XXXX
14      XXXX    XXXX      XXXX    XXXX    XXXX    XXXX    XXXX    XXXX
15      XXXX    XXXX      XXXX    XXXX    XXXX    XXXX    XXXX    XXXX

16      XXXX    XXXX      XXXX    XXXX    XXXX    XXXX    XXXX    XXXX
17      XXXX    XXXX      XXXX    XXXX    XXXX    XXXX    XXXX    XXXX
18      XXXX    XXXX      XXXX    XXXX    XXXX    XXXX    XXXX    XXXX
19      XXXX    XXXX      XXXX    XXXX    XXXX    XXXX    XXXX    XXXX
20      XXXX    XXXX      XXXX    XXXX    XXXX    XXXX    XXXX    XXXX
21      XXXX    XXXX      XXXX    XXXX    XXXX    XXXX    XXXX    XXXX
22      XXXX    XXXX      XXXX    XXXX    XXXX    XXXX    XXXX    XXXX
23      XXXX    XXXX      XXXX    XXXX    XXXX    XXXX    XXXX    XXXX
```

```
24   XXXX XXXX XXXX XXXX XXXX XXXX XXXX XXXX
25   XXXX XXXX XXXX XXXX XXXX XXXX XXXX XXXX
26   XXXX XXXX XXXX XXXX XXXX XXXX XXXX XXXX
27   XXXX XXXX XXXX XXXX XXXX XXXX XXXX XXXX
28   XXXX XXXX XXXX XXXX XXXX XXXX XXXX XXXX
29   XXXX XXXX XXXX XXXX XXXX XXXX XXXX XXXX
30   XXXX XXXX XXXX XXXX XXXX XXXX XXXX XXXX
31   XXXX XXXX XXXX XXXX XXXX XXXX XXXX XXXX

32   -X-- -X-- ---- ---- --X- ---- ---- ----
33   ---- -X-- ---- ---- ---X --X- ---- ----
34   XXXX XXXX XXXX XXXX XXXX XXXX XXXX XXXX
35   XXXX XXXX XXXX XXXX XXXX XXXX XXXX XXXX
36   XXXX XXXX XXXX XXXX XXXX XXXX XXXX XXXX
37   XXXX XXXX XXXX XXXX XXXX XXXX XXXX XXXX
38   XXXX XXXX XXXX XXXX XXXX XXXX XXXX XXXX
39   XXXX XXXX XXXX XXXX XXXX XXXX XXXX XXXX

40   X--- -X-- ---- ---- ---X --X- ---- ----
41   -X-- -X-- ---- ---- ---X ---X ---- ----
42   XXXX XXXX XXXX XXXX XXXX XXXX XXXX XXXX
43   XXXX XXXX XXXX XXXX XXXX XXXX XXXX XXXX
44   XXXX XXXX XXXX XXXX XXXX XXXX XXXX XXXX
45   XXXX XXXX XXXX XXXX XXXX XXXX XXXX XXXX
46   XXXX XXXX XXXX XXXX XXXX XXXX XXXX XXXX
47   XXXX XXXX XXXX XXXX XXXX XXXX XXXX XXXX

48   ---- ---- ---- ---- ---- ---- ---- ----
49   ---- ---- ---- ---- --X- --X- ---- ----
50   XXXX XXXX XXXX XXXX XXXX XXXX XXXX XXXX
51   XXXX XXXX XXXX XXXX XXXX XXXX XXXX XXXX
52   XXXX XXXX XXXX XXXX XXXX XXXX XXXX XXXX
53   XXXX XXXX XXXX XXXX XXXX XXXX XXXX XXXX
54   XXXX XXXX XXXX XXXX XXXX XXXX XXXX XXXX
55   XXXX XXXX XXXX XXXX XXXX XXXX XXXX XXXX

56   ---- ---- ---- ---- ---- ---- ---- ----
57   ---- ---- ---- ---- --X- ---X ---- ----
58   XXXX XXXX XXXX XXXX XXXX XXXX XXXX XXXX
59   XXXX XXXX XXXX XXXX XXXX XXXX XXXX XXXX
60   XXXX XXXX XXXX XXXX XXXX XXXX XXXX XXXX
61   XXXX XXXX XXXX XXXX XXXX XXXX XXXX XXXX
62   XXXX XXXX XXXX XXXX XXXX XXXX XXXX XXXX
63   XXXX XXXX XXXX XXXX XXXX XXXX XXXX XXXX

SUMMARY
     TOTAL FUSES BLOWN = 301
```

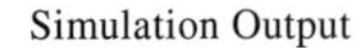

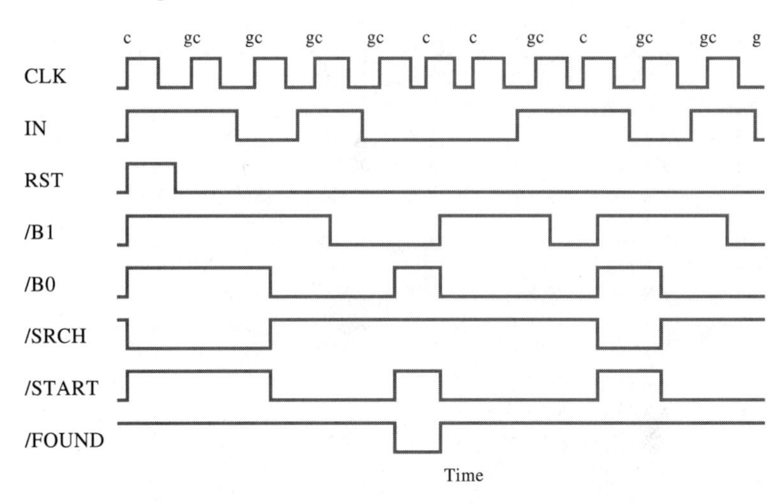

We have illustrated one way of adding synchronous reset to the PAL16R4 design. There is a second and often preferable way: change to a PAL with an asynchronous reset. The PAL22V10, as seen in the circuit diagram on page 495, is a 24-pin PAL with 10 logic sections; each section has from 8 to 16 *and* gates (they differ in size) and each has an independent programmable enable. The PAL22V10 has 12 inputs (one of which must be used for the clock when the clock is needed), 10 programmable input/outputs (programmable as inputs or as combinatorial or register outputs with the assertion levels programmable as high or low), and provisions for an asynchronous reset and a synchronous preset of all registered outputs. With this device, we can program the outputs to be register or combinatorial by simply declaring them one or the other, and make the assertion level high or low by simply declaring it as one or the other. We'll make all the outputs (SRCH, START, and FOUND) register and asserted high in this design. The source program, fuse map, and simulation output for a PAL22V10-based design follow.

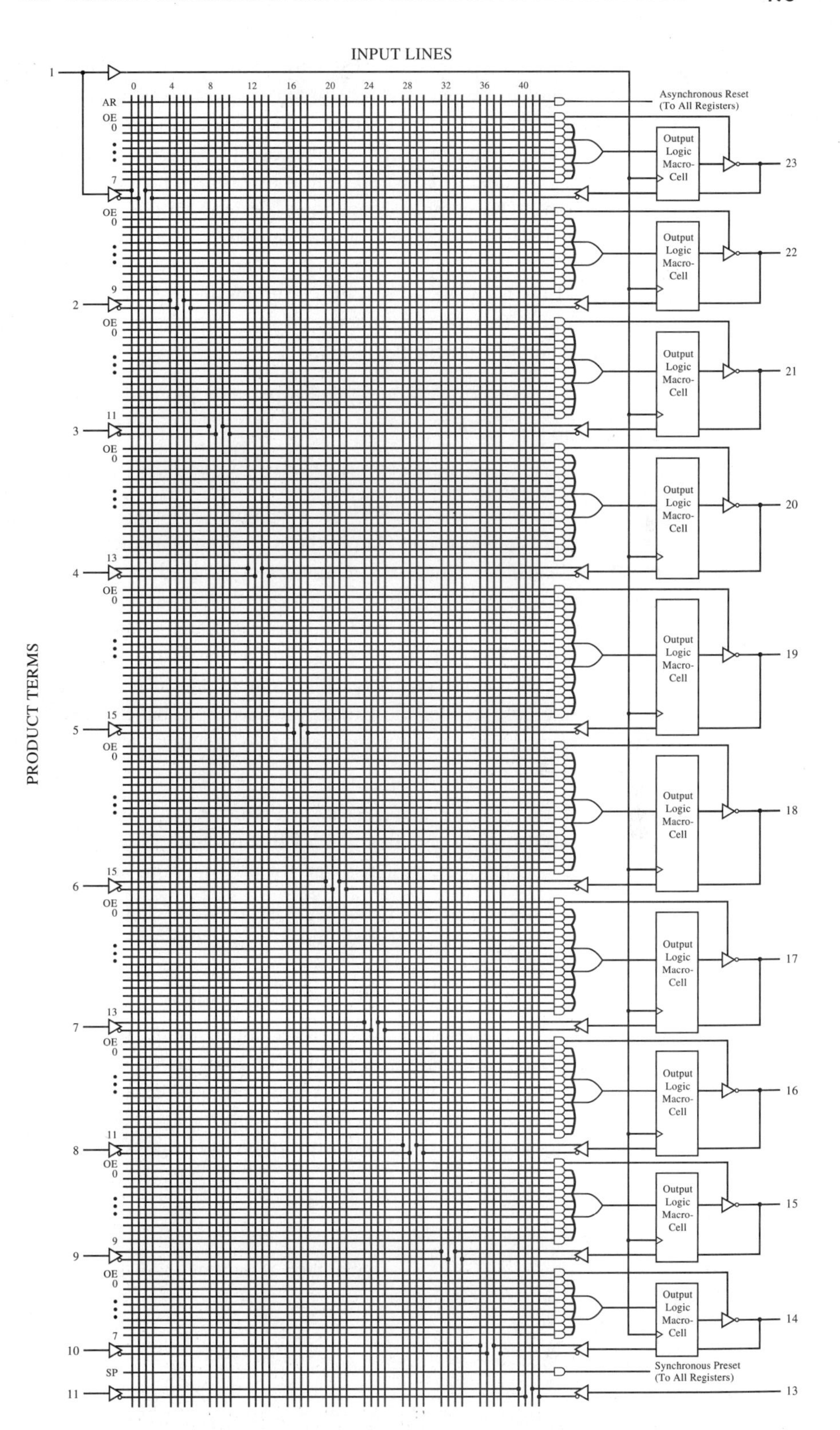
INPUT LINES
PRODUCT TERMS
Asynchronous Reset
(To All Registers)
Synchronous Preset
(To All Registers)
AR
SP
OE
Output
Logic
Macro-
Cell
0
4
8
12
16
20
24
28
32
36
40
1
2
3
4
5
6
7
8
9
10
11
13
14
15
16
17
18
19
20
21
22
23

The PAL22V10 has a programmable enable (which we learned to program in Chapter 6) for each of its outputs and a common programmable asynchronous reset that is programmed much like the enables. (The programmable preset, also a feature, will not be used in this design.) The enables and the reset are programmable only as a simple *and* of inputs to the PAL. They are programmed in the equations section of the program. The combinational logic designs with PALs in Chapter 6 had equations sections, but we have not needed such sections in the state machines designed so far in this chapter. We must add an equations section, beginning with the keyword EQUATIONS, to properly program the PAL22V10 (see the following source program). This section contains logic expressions for the enables and the reset. The syntax for the enables is

output.TRST = logic expression (simple *and*)

and an example is

SRCH.TRST = VCC

SRCH is an output signal, .TRST is a required keyword, and VCC is a logic expression that is always true. One of these program statements is needed for each output to be enabled. When the output is not enabled, it is tristated (essentially disconnected).

The syntax for the reset is much the same, but, unlike the output, the reset does not have a name. We give it one by adding a line to the pin declaration section. Because the reset is not a pin, it is called a node, and if we consult the PALASM manuals or the help file for the PAL22V10, we find that the reset is node 1. The syntax for naming node 1 in the pin declaration section of the program is

NODE 1 GLOBAL

It is common to name the signal associated with node 1, GLOBAL, because it is a global reset signal. Actually, it can be given any name. Now that we know how to name the reset signal, here is the syntax for programming it:

reset name.RSTF = logic expression (simple and)

An example is

GLOBAL.RSTF = RST

where GLOBAL is the name of the reset signal (node 1), .RSTF is a required keyword, and RST is a logic expression. The logic expression is particularly simple — it is merely a PAL input RST. The equations section can also contain logic expressions for programming combinational and register outputs that are best programmed using logic equations rather than the more usual state machine equations.

Program

```
;PALASM Design Description

;------------------------------------ Declaration Segment ------------------------------------
TITLE       Sequence detector 010
PATTERN     None
REVISION    None
```

```
AUTHOR    A W Shaw
COMPANY   Utah State University
DATE      07/02/92

CHIP      PAL_REG   PAL22V10

;----------------------------------------- PIN Declarations -----------------------------------------
PIN    1        CLK                             ; CLOCK
PIN    2        IN                              ; INPUT
PIN    3        RST                             ; INPUT
PIN    4        ENBL                            ;
PIN    5        NC                              ;
PIN    7        NC                              ;
PIN    8        NC                              ;
PIN    9        NC                              ;
PIN    10       NC                              ;
PIN    11       NC                              ;
PIN    12       GND                             ;
PIN    13       NC                              ;
PIN    14       NC                              ;
PIN    15       FOUND             REGISTERED    ; OUTPUT
PIN    16       SRCH              REGISTERED    ; OUTPUT
PIN    17       /B1               REGISTERED    ; OUTPUT
PIN    18       /B0               REGISTERED    ; OUTPUT
PIN    19       NC                              ;
PIN    20       NC                              ;
PIN    21       START             REGISTERED    ; OUTPUT
PIN    22       NC                              ;
PIN    23       NC                              ;
PIN    24       VCC                             ;
NODE 1          GLOBAL                          ;

;---------------------------------- State Equations Segment ----------------------------------
STATE

MOORE_MACHINE

A = /B1 * /B0
B = /B1 *  B0
C =  B1 *  B0
D =  B1 * /B0

A := N_IN_N_R → B
     + IN_R → A
B := N_IN_N_R → B
     + IN_N_R → C
     + R → A
C := N_IN_N_R → D
     + IN_R → A
D := N_IN_N_R → B
     + IN_R → A
```

```
A.OUTF = SRCH*/START*/FOUND
B.OUTF = /SRCH*START*/FOUND
C.OUTF = /SRCH*START*/FOUND
D.OUTF = /SRCH*/START*FOUND

CONDITIONS

N_IN_N_R = /IN * /RST
IN_R     = IN + RST
IN_N_R   = IN * /RST
R        = RST

;---------------------------------- Boolean Equations Segment ----------------------------------
EQUATIONS

SRCH.TRST = ENBL
START.TRST = ENBL
FOUND.TRST = ENBL
B1.TRST = VCC
B0.TRST = VCC
GLOBAL.RSTF = RST

;--------------------------------------- Simulation Segment ----------------------------------------
SIMULATION

SETF /CLK IN ENBL RST
TRACE_ON CLK IN RST ENBL /B1 /B0 SRCH START FOUND
CLOCKF CLK
SETF /RST
CLOCKF CLK
SETF /IN
CLOCKF CLK
SETF IN
CLOCKF CLK
SETF /IN
CLOCKF CLK
CLOCKF CLK
CLOCKF CLK
SETF IN
CLOCKF CLK
CLOCKF CLK
SETF /IN
CLOCKF CLK
SETF IN
CLOCKF CLK
SETF /IN
TRACE_OFF

;-------------------------------------------------------------------------------------------------------------
```

FUSE MAP

```
PALASM4  PAL Assembler—Market Version 1.2 (5-31-91)
  (c)—Copyright Advanced Micro Devices Inc., 1991

TITLE      Sequence detector 010        AUTHOR  :A W Shaw
PATTERN    None                         COMPANY:Utah State University
REVISION   None                         DATE    :07/02/92

PAL22V10
PAL_REG

                11 1111 1111 2222 2222 2233 3333 3333 4444
     0123 4567 8901 2345 6789 0123 4567 8901 2345 6789 0123

0    ---- ---- X--- ---- ---- ---- ---- ----  --- ---- ----

1    XXXX XXXX XXXX XXXX XXXX XXXX XXXX XXXX XXXX XXXX XXXX
2    XXXX XXXX XXXX XXXX XXXX XXXX XXXX XXXX XXXX XXXX XXXX
3    XXXX XXXX XXXX XXXX XXXX XXXX XXXX XXXX XXXX XXXX XXXX
4    XXXX XXXX XXXX XXXX XXXX XXXX XXXX XXXX XXXX XXXX XXXX
5    XXXX XXXX XXXX XXXX XXXX XXXX XXXX XXXX XXXX XXXX XXXX
6    XXXX XXXX XXXX XXXX XXXX XXXX XXXX XXXX XXXX XXXX XXXX
7    XXXX XXXX XXXX XXXX XXXX XXXX XXXX XXXX XXXX XXXX XXXX
8    XXXX XXXX XXXX XXXX XXXX XXXX XXXX XXXX XXXX XXXX XXXX
9    XXXX XXXX XXXX XXXX XXXX XXXX XXXX XXXX XXXX XXXX XXXX

10   XXXX XXXX XXXX XXXX XXXX XXXX XXXX XXXX XXXX XXXX XXXX
11   XXXX XXXX XXXX XXXX XXXX XXXX XXXX XXXX XXXX XXXX XXXX
12   XXXX XXXX XXXX XXXX XXXX XXXX XXXX XXXX XXXX XXXX XXXX
13   XXXX XXXX XXXX XXXX XXXX XXXX XXXX XXXX XXXX XXXX XXXX
14   XXXX XXXX XXXX XXXX XXXX XXXX XXXX XXXX XXXX XXXX XXXX
15   XXXX XXXX XXXX XXXX XXXX XXXX XXXX XXXX XXXX XXXX XXXX
16   XXXX XXXX XXXX XXXX XXXX XXXX XXXX XXXX XXXX XXXX XXXX
17   XXXX XXXX XXXX XXXX XXXX XXXX XXXX XXXX XXXX XXXX XXXX
18   XXXX XXXX XXXX XXXX XXXX XXXX XXXX XXXX XXXX XXXX XXXX
19   XXXX XXXX XXXX XXXX XXXX XXXX XXXX XXXX XXXX XXXX XXXX
20   XXXX XXXX XXXX XXXX XXXX XXXX XXXX XXXX XXXX XXXX XXXX

21   ---- ---- ---- X--- ---- ---- ---- ---- ---- ---- ----
22   ---- ---- ---- ---- ---- ---X ---- ---- ---- ---- ----
23   XXXX XXXX XXXX XXXX XXXX XXXX XXXX XXXX XXXX XXXX XXXX
24   XXXX XXXX XXXX XXXX XXXX XXXX XXXX XXXX XXXX XXXX XXXX
25   XXXX XXXX XXXX XXXX XXXX XXXX XXXX XXXX XXXX XXXX XXXX
26   XXXX XXXX XXXX XXXX XXXX XXXX XXXX XXXX XXXX XXXX XXXX
27   XXXX XXXX XXXX XXXX XXXX XXXX XXXX XXXX XXXX XXXX XXXX
28   XXXX XXXX XXXX XXXX XXXX XXXX XXXX XXXX XXXX XXXX XXXX
29   XXXX XXXX XXXX XXXX XXXX XXXX XXXX XXXX XXXX XXXX XXXX
30   XXXX XXXX XXXX XXXX XXXX XXXX XXXX XXXX XXXX XXXX XXXX
31   XXXX XXXX XXXX XXXX XXXX XXXX XXXX XXXX XXXX XXXX XXXX
32   XXXX XXXX XXXX XXXX XXXX XXXX XXXX XXXX XXXX XXXX XXXX
33   XXXX XXXX XXXX XXXX XXXX XXXX XXXX XXXX XXXX XXXX XXXX
```

```
34   XXXX XXXX XXXX XXXX XXXX XXXX XXXX XXXX XXXX XXXX XXXX
35   XXXX XXXX XXXX XXXX XXXX XXXX XXXX XXXX XXXX XXXX XXXX
36   XXXX XXXX XXXX XXXX XXXX XXXX XXXX XXXX XXXX XXXX XXXX
37   XXXX XXXX XXXX XXXX XXXX XXXX XXXX XXXX XXXX XXXX XXXX
38   XXXX XXXX XXXX XXXX XXXX XXXX XXXX XXXX XXXX XXXX XXXX
39   XXXX XXXX XXXX XXXX XXXX XXXX XXXX XXXX XXXX XXXX XXXX
40   XXXX XXXX XXXX XXXX XXXX XXXX XXXX XXXX XXXX XXXX XXXX
41   XXXX XXXX XXXX XXXX XXXX XXXX XXXX XXXX XXXX XXXX XXXX
42   XXXX XXXX XXXX XXXX XXXX XXXX XXXX XXXX XXXX XXXX XXXX
43   XXXX XXXX XXXX XXXX XXXX XXXX XXXX XXXX XXXX XXXX XXXX
44   XXXX XXXX XXXX XXXX XXXX XXXX XXXX XXXX XXXX XXXX XXXX
45   XXXX XXXX XXXX XXXX XXXX XXXX XXXX XXXX XXXX XXXX XXXX
46   XXXX XXXX XXXX XXXX XXXX XXXX XXXX XXXX XXXX XXXX XXXX
47   XXXX XXXX XXXX XXXX XXXX XXXX XXXX XXXX XXXX XXXX XXXX
48   XXXX XXXX XXXX XXXX XXXX XXXX XXXX XXXX XXXX XXXX XXXX

49   XXXX XXXX XXXX XXXX XXXX XXXX XXXX XXXX XXXX XXXX XXXX
50   XXXX XXXX XXXX XXXX XXXX XXXX XXXX XXXX XXXX XXXX XXXX
51   XXXX XXXX XXXX XXXX XXXX XXXX XXXX XXXX XXXX XXXX XXXX
52   XXXX XXXX XXXX XXXX XXXX XXXX XXXX XXXX XXXX XXXX XXXX
53   XXXX XXXX XXXX XXXX XXXX XXXX XXXX XXXX XXXX XXXX XXXX
54   XXXX XXXX XXXX XXXX XXXX XXXX XXXX XXXX XXXX XXXX XXXX
55   XXXX XXXX XXXX XXXX XXXX XXXX XXXX XXXX XXXX XXXX XXXX
56   XXXX XXXX XXXX XXXX XXXX XXXX XXXX XXXX XXXX XXXX XXXX
57   XXXX XXXX XXXX XXXX XXXX XXXX XXXX XXXX XXXX XXXX XXXX
58   XXXX XXXX XXXX XXXX XXXX XXXX XXXX XXXX XXXX XXXX XXXX
59   XXXX XXXX XXXX XXXX XXXX XXXX XXXX XXXX XXXX XXXX XXXX
60   XXXX XXXX XXXX XXXX XXXX XXXX XXXX XXXX XXXX XXXX XXXX
61   XXXX XXXX XXXX XXXX XXXX XXXX XXXX XXXX XXXX XXXX XXXX
62   XXXX XXXX XXXX XXXX XXXX XXXX XXXX XXXX XXXX XXXX XXXX
63   XXXX XXXX XXXX XXXX XXXX XXXX XXXX XXXX XXXX XXXX XXXX
64   XXXX XXXX XXXX XXXX XXXX XXXX XXXX XXXX XXXX XXXX XXXX
65   XXXX XXXX XXXX XXXX XXXX XXXX XXXX XXXX XXXX XXXX XXXX

66   ---- ---- ---- ---- ---- ---- ---- ---- ---- ---- ----
67   ---- -X-- -X-- ---- ---- --X- ---- ---- ---- ---- ----
68   ---- ---- -X-- ---- ---- ---X --X- ---- ---- ---- ----
69   XXXX XXXX XXXX XXXX XXXX XXXX XXXX XXXX XXXX XXXX XXXX
70   XXXX XXXX XXXX XXXX XXXX XXXX XXXX XXXX XXXX XXXX XXXX
71   XXXX XXXX XXXX XXXX XXXX XXXX XXXX XXXX XXXX XXXX XXXX
72   XXXX XXXX XXXX XXXX XXXX XXXX XXXX XXXX XXXX XXXX XXXX
73   XXXX XXXX XXXX XXXX XXXX XXXX XXXX XXXX XXXX XXXX XXXX
74   XXXX XXXX XXXX XXXX XXXX XXXX XXXX XXXX XXXX XXXX XXXX
75   XXXX XXXX XXXX XXXX XXXX XXXX XXXX XXXX XXXX XXXX XXXX
76   XXXX XXXX XXXX XXXX XXXX XXXX XXXX XXXX XXXX XXXX XXXX
77   XXXX XXXX XXXX XXXX XXXX XXXX XXXX XXXX XXXX XXXX XXXX
78   XXXX XXXX XXXX XXXX XXXX XXXX XXXX XXXX XXXX XXXX XXXX
79   XXXX XXXX XXXX XXXX XXXX XXXX XXXX XXXX XXXX XXXX XXXX
80   XXXX XXXX XXXX XXXX XXXX XXXX XXXX XXXX XXXX XXXX XXXX
81   XXXX XXXX XXXX XXXX XXXX XXXX XXXX XXXX XXXX XXXX XXXX
82   XXXX XXXX XXXX XXXX XXXX XXXX XXXX XXXX XXXX XXXX XXXX
```

```
83   --------------------------------------------
84   ----X----X-------------X--X-----------------
85   -----X---X-------------X---X----------------
86   XXXX XXXX XXXX XXXX XXXX XXXX XXXX XXXX XXXX XXXX XXXX
87   XXXX XXXX XXXX XXXX XXXX XXXX XXXX XXXX XXXX XXXX XXXX
88   XXXX XXXX XXXX XXXX XXXX XXXX XXXX XXXX XXXX XXXX XXXX
89   XXXX XXXX XXXX XXXX XXXX XXXX XXXX XXXX XXXX XXXX XXXX
90   XXXX XXXX XXXX XXXX XXXX XXXX XXXX XXXX XXXX XXXX XXXX
91   XXXX XXXX XXXX XXXX XXXX XXXX XXXX XXXX XXXX XXXX XXXX
92   XXXX XXXX XXXX XXXX XXXX XXXX XXXX XXXX XXXX XXXX XXXX
93   XXXX XXXX XXXX XXXX XXXX XXXX XXXX XXXX XXXX XXXX XXXX
94   XXXX XXXX XXXX XXXX XXXX XXXX XXXX XXXX XXXX XXXX XXXX
95   XXXX XXXX XXXX XXXX XXXX XXXX XXXX XXXX XXXX XXXX XXXX
96   XXXX XXXX XXXX XXXX XXXX XXXX XXXX XXXX XXXX XXXX XXXX
97   XXXX XXXX XXXX XXXX XXXX XXXX XXXX XXXX XXXX XXXX XXXX

98   ------------X-------------------------------
99   ----------------------X---X-----------------
100  XXXX XXXX XXXX XXXX XXXX XXXX XXXX XXXX XXXX XXXX XXXX
101  XXXX XXXX XXXX XXXX XXXX XXXX XXXX XXXX XXXX XXXX XXXX
102  XXXX XXXX XXXX XXXX XXXX XXXX XXXX XXXX XXXX XXXX XXXX
103  XXXX XXXX XXXX XXXX XXXX XXXX XXXX XXXX XXXX XXXX XXXX
104  XXXX XXXX XXXX XXXX XXXX XXXX XXXX XXXX XXXX XXXX XXXX
105  XXXX XXXX XXXX XXXX XXXX XXXX XXXX XXXX XXXX XXXX XXXX
106  XXXX XXXX XXXX XXXX XXXX XXXX XXXX XXXX XXXX XXXX XXXX
107  XXXX XXXX XXXX XXXX XXXX XXXX XXXX XXXX XXXX XXXX XXXX
108  XXXX XXXX XXXX XXXX XXXX XXXX XXXX XXXX XXXX XXXX XXXX
109  XXXX XXXX XXXX XXXX XXXX XXXX XXXX XXXX XXXX XXXX XXXX
110  XXXX XXXX XXXX XXXX XXXX XXXX XXXX XXXX XXXX XXXX XXXX

111  ------------X-------------------------------
112  ----------------------X----X----------------
113  XXXX XXXX XXXX XXXX XXXX XXXX XXXX XXXX XXXX XXXX XXXX
114  XXXX XXXX XXXX XXXX XXXX XXXX XXXX XXXX XXXX XXXX XXXX
115  XXXX XXXX XXXX XXXX XXXX XXXX XXXX XXXX XXXX XXXX XXXX
116  XXXX XXXX XXXX XXXX XXXX XXXX XXXX XXXX XXXX XXXX XXXX
117  XXXX XXXX XXXX XXXX XXXX XXXX XXXX XXXX XXXX XXXX XXXX
118  XXXX XXXX XXXX XXXX XXXX XXXX XXXX XXXX XXXX XXXX XXXX
119  XXXX XXXX XXXX XXXX XXXX XXXX XXXX XXXX XXXX XXXX XXXX
120  XXXX XXXX XXXX XXXX XXXX XXXX XXXX XXXX XXXX XXXX XXXX
121  XXXX XXXX XXXX XXXX XXXX XXXX XXXX XXXX XXXX XXXX XXXX

122  XXXX XXXX XXXX XXXX XXXX XXXX XXXX XXXX XXXX XXXX XXXX
123  XXXX XXXX XXXX XXXX XXXX XXXX XXXX XXXX XXXX XXXX XXXX
124  XXXX XXXX XXXX XXXX XXXX XXXX XXXX XXXX XXXX XXXX XXXX
125  XXXX XXXX XXXX XXXX XXXX XXXX XXXX XXXX XXXX XXXX XXXX
126  XXXX XXXX XXXX XXXX XXXX XXXX XXXX XXXX XXXX XXXX XXXX
127  XXXX XXXX XXXX XXXX XXXX XXXX XXXX XXXX XXXX XXXX XXXX
128  XXXX XXXX XXXX XXXX XXXX XXXX XXXX XXXX XXXX XXXX XXXX
129  XXXX XXXX XXXX XXXX XXXX XXXX XXXX XXXX XXXX XXXX XXXX
130  XXXX XXXX XXXX XXXX XXXX XXXX XXXX XXXX XXXX XXXX XXXX
```

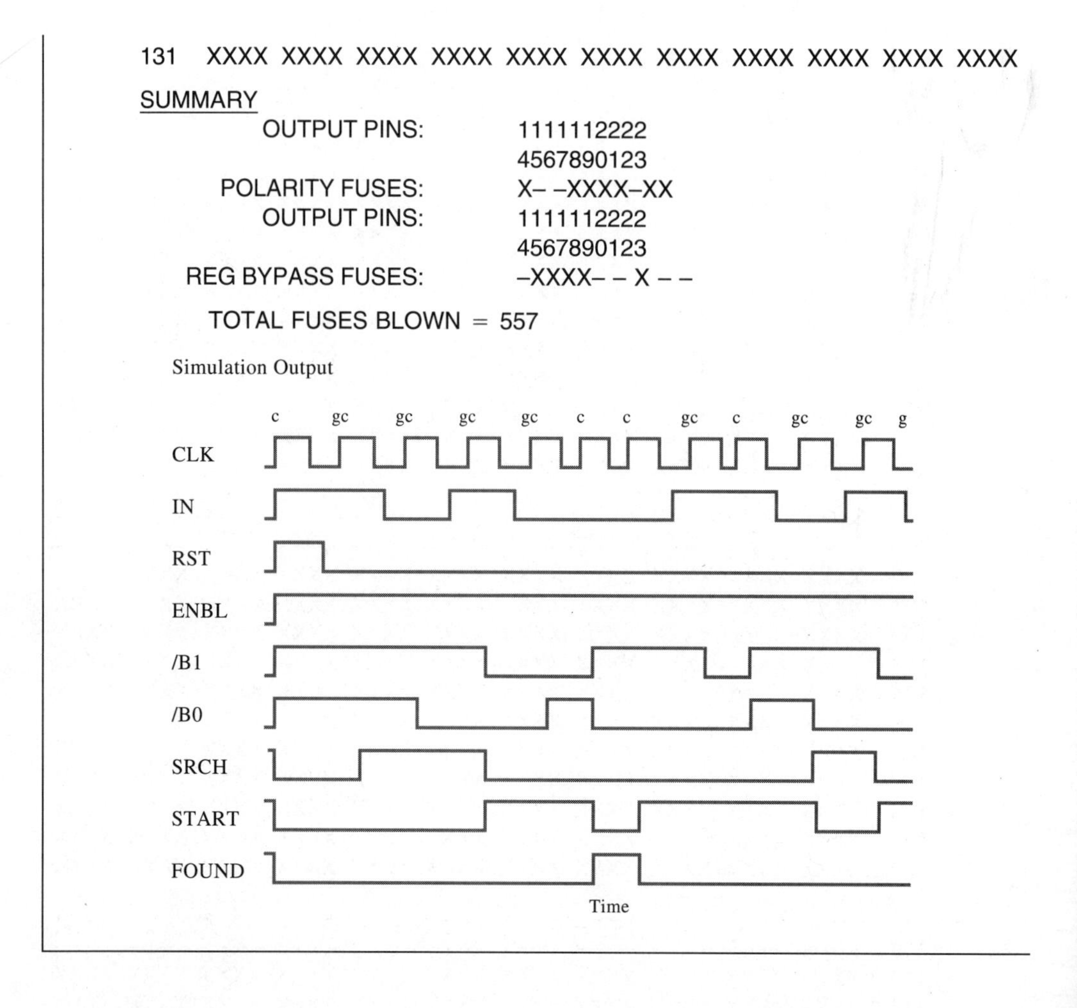

```
131   XXXX XXXX XXXX XXXX XXXX XXXX XXXX XXXX XXXX XXXX XXXX
SUMMARY
           OUTPUT PINS:          1111112222
                                 4567890123
       POLARITY FUSES:           X– –XXXX–XX
           OUTPUT PINS:          1111112222
                                 4567890123
    REG BYPASS FUSES:            –XXXX– – X – –

      TOTAL FUSES BLOWN = 557
```

Simulation Output

10.3 ANALYZING SYNCHRONOUS STATE MACHINES

We've been designing state machines and analyzing them to the extent needed to confirm our designs. In this section we'll formalize our analysis technique.

Analysis of a state machine is the reverse of design, and the steps are almost the reverse of the design steps. When we analyze a state machine, we start with the circuit (or the JEDEC file, in the case of a PAL or PLA design). Suppose we have a PAL design and start with the JEDEC file. First we obtain (from a data book) a circuit diagram for the PAL and mark the programmed connections on the diagram as we did in Figure 10.2-9. At this point we have a circuit for the PAL design, and analysis proceeds as though we had started with a circuit. (Actually, there is an alternative to marking the PAL circuit. Software such as PALASM can disassemble

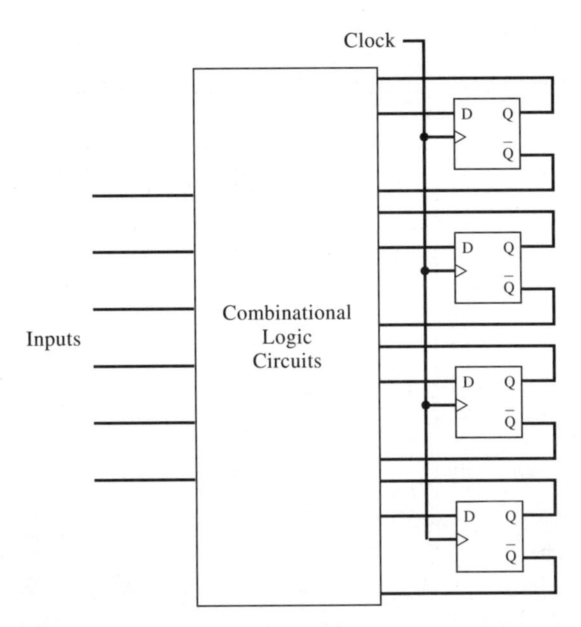

Figure 10.3-1 State machine architecture.

the JEDEC file and produce the next-state decoder logic expressions directly so that we can skip the first part of the analysis that follows.)

Suppose we start with a circuit that is not a PAL or PLA. The state machine may be part of a larger circuit, and the first steps of the analysis should be to isolate that state machine and identify its parts. (In a PAL design, although we need to identify the parts of the state machine, we do not need to isolate it from some larger circuit.) The parts of a state machine are the memory, the next-state code decoders, and the output decoders. The most conspicuous part of the machine is the memory. A circuit can be analyzed as a state machine if it contains memory (flip-flops) and if the output of the memory feeds back through logic circuits into the inputs of the memory, as shown in Figure 10.3-1. Here the memory elements are D flip-flops; they could equally well be JK flip-flops. A PAL or a PLA has exactly the architecture of Figure 10.3-1.

Once we've isolated the state machine and located its memory, we should label the memory bits by labeling the outputs of the memory flip-flops—B0 (H), B0 (L), B1 (H), and so on. Actually, we may have difficulty determining which flip-flops are memory elements if the machine has some register outputs. If we suspect that some of the flip-flops are output register flip-flops, we should examine the feedback. Flip-flops with no feedback can be taken as part of the output register. Identification of the output flip-flops is not necessary for a successful analysis, but it simplifies the analysis. If output flip-flops are taken to be memory elements, the analysis can still be carried out.

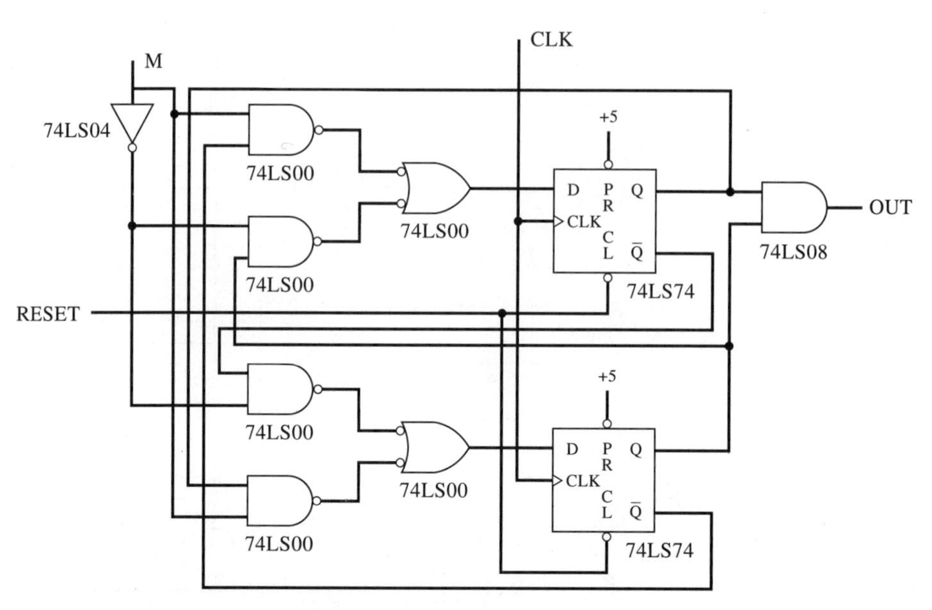

Figure 10.3-2 Example circuit for analysis.

Now that the memory is labeled we can locate the next-state decoder logic circuits—the combinational logic circuits that produce the inputs to the memory flip-flops. One logic circuit is associated with each D memory element (the D input); two logic circuits are associated with each JK memory element (the J and K inputs).

When the decoder logic circuits have been identified, we can find the inputs to the state machine. They are the inputs to the next-state decoder logic that are produced externally to the state machine. We need to label all of them.

The final part that must be identified is the output decoder. For a machine with no output register, the decoder consists of combinational logic circuits that produce outputs used externally to the state machine. For a machine with a register output, the output decoder logic circuit consists of combinational logic circuits that produce the input to the output register flip-flops.

We can see the reverse design process, or analysis, by considering the state machine with the circuit in Figure 10.3-2. This is a machine with D memory elements. None of the D flip-flops is part of any register output, because the output of each D flip-flop is fed back through a logic circuit to the flip-flop's input. The machine has, in addition to the clock, one input, labeled M; an asynchronous reset, labeled RESET; and one output, labeled OUT.

The first step in the analysis is to assign the state code bits to the memory by labeling the outputs of the memory flip-flops as in Figure 10.3-3. The labeling of the upper flip-flop as the high order memory element is arbitrary; the memory elements can be labeled in any order. We've also assigned somewhat arbitrary assertion levels to the state machine's inputs and a matching assertion level to the output.

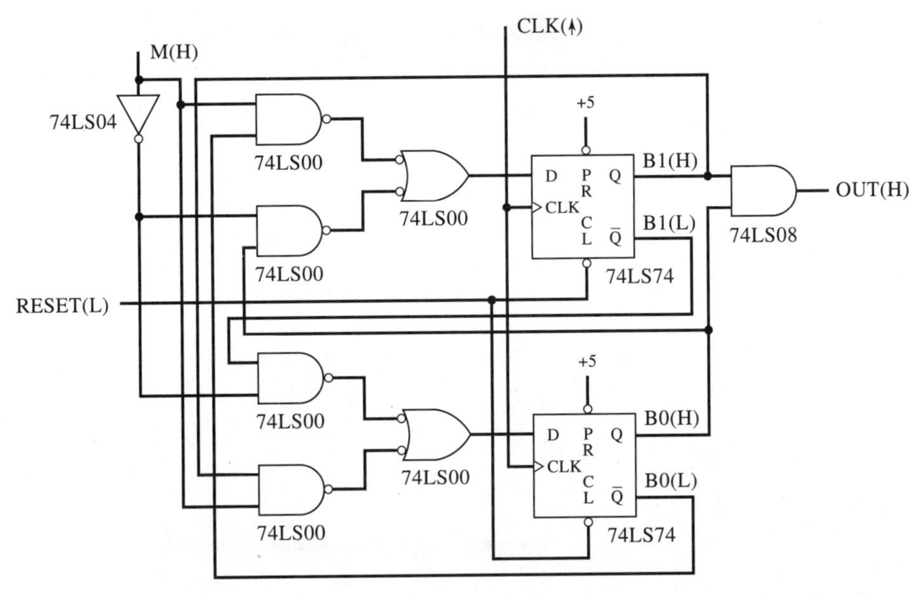

Figure 10.3-3 Circuit with the memory elements labeled and assertion levels assigned to the inputs and output.

All of the foregoing identification must be done preliminary to the actual analysis. When we say that the design process is reversed, we actually mean that we'll make a present state–next state table and reverse the process of filling it. Table 10.3-1 is the present state–next state table with only the present state and input filled in. Normally, to fill this table in a reverse order, we would find the next state

Table 10.3-1 Present State–Next State Table

Present State			Next State	
B1	B0	M	B1	B0
0	0	0		
0	0	1		
0	1	0		
0	1	1		
1	0	0		
1	0	1		
1	1	0		
1	1	1		

Table 10.3-2 Complete Present State–Next State and Output Table

Present State			Next State		
B1	B0	M	B1	B0	OUT
0	0	0	0	1	0
0	0	1	1	0	0
0	1	0	1	1	0
0	1	1	0	0	0
1	0	0	0	0	0
1	0	1	1	1	0
1	1	0	1	0	1
1	1	1	0	1	1

code and then the next state. For D memory the next state is the same as the next state code, so we find the next state directly. The next state code is the output of the decoder logic. To find it, we need to know the next state decoder logic expressions. These are the logic expressions that are evaluated by the next state decoder circuits, so they can be found by analyzing these circuits. Analysis of the upper decoder logic circuit of our sample state machine gives us a next state decoder logic expression:

$$D1 = M \cdot \overline{B0} + \overline{M} \cdot B0$$

Analyzing the lower decoder logic circuit gives us

$$D0 = M \cdot B1 + \overline{M} \cdot \overline{B1}$$

Because the next state B1 is the same as D1 and the next state B0 is the same as D0, we can fill in the next state in the present state–next state table by using these logic expressions. Table 10.3-2 is the present state–next state table with the next state filled in. The easiest way to finish filling in the table is to first consider the logic expression for B1 (D1), noticing that it is 1 for M = 1 and B0 = 0 (first term) and for M = 0 and B0 = 1 (second term). Therefore, we place 1 in all the rows of the next state B1 column for which M = 1 and B = 0 and for which M = 0 and B0 = 1, and 0 in all the other rows. We repeat the same process for the next state decoder B0 column.

Table 10.3-2 also contains an output table, which we fill by using the output logic expressions found from the output decoder logic circuit. In this circuit there is one output, OUT, and the output decoder shows that

$$OUT = B1 \cdot B0$$

This expression is 1 only when B1 = 1 and B0 = 1—the 11 state—as reflected in the table. Once we have a present state–next state and output table, we can make a state transition diagram if it's needed. Here we'll assume the table is all the result we need.

In this analysis we did not need the present state code, because it was the same as the next state. For JK memory elements we do need the next state code. Keeping this in mind, we can summarize our analysis procedure as follows:

1. **(a)** If the analysis is to be of a PAL and thus is to start from a JEDEC file, obtain a circuit diagram for the PAL and mark the programmed connections on the diagram. Using this circuit diagram, identify the parts of the state machine as follows in (b).
   (b) If the analysis is to start from a circuit, isolate the part of the circuit that is the state machine and identify its parts. A state machine is most easily identified by finding the potential memory elements (flip-flops) and determining whether they have feedback associated with them. Flip-flops without feedback can be taken as part of the output register. After identifying the memory elements, identify the next state decoder logic circuits, output decoder logic circuits, state machine outputs, and state machine inputs.
2. Label each memory element with the label of the bit in the state code that the memory element is to store. Label all inputs and outputs. Memory elements can be labeled in any order.
3. Make a present state–next state and output table that is empty except for the present state code and the inputs. All possible present states and inputs should be listed in the table. If the memory elements are JK flip-flops, include a place in the table for the next state code.
4. Find expressions for the next state decoder logic from the next state decoder circuits.
5. Fill in the next state code using the next state decoder logic expressions.
6. Find the next state from the next state code and present state, using a present state–next state table for the memory element of the state machine. This step can be omitted with D memory elements because the next state code is the same as the next state.
7. Find expressions for the output decoder logic from the output decoder logic circuits.
8. Fill in the output table using the output logic expressions.
9. If a state transition diagram is needed, make it from the present state–next state and output table.

Now let's analyze the same circuit as in Figure 10.3-3 but with an output register. Figure 10.3-4 is a circuit with D memory elements, but we need to decide whether any of the potential memory elements is part of an output register. We do this by considering the feedback from each flip-flop. Obviously, the bottom flip-flop has no feedback into the inputs of any of the flip-flops. In fact, its only output is an external output, OUT. This bottom flip-flop is part of an output register—the only part. Once the output register is identified, the rest of the circuit is identical to the circuit we just analyzed, and the analysis follows identically.

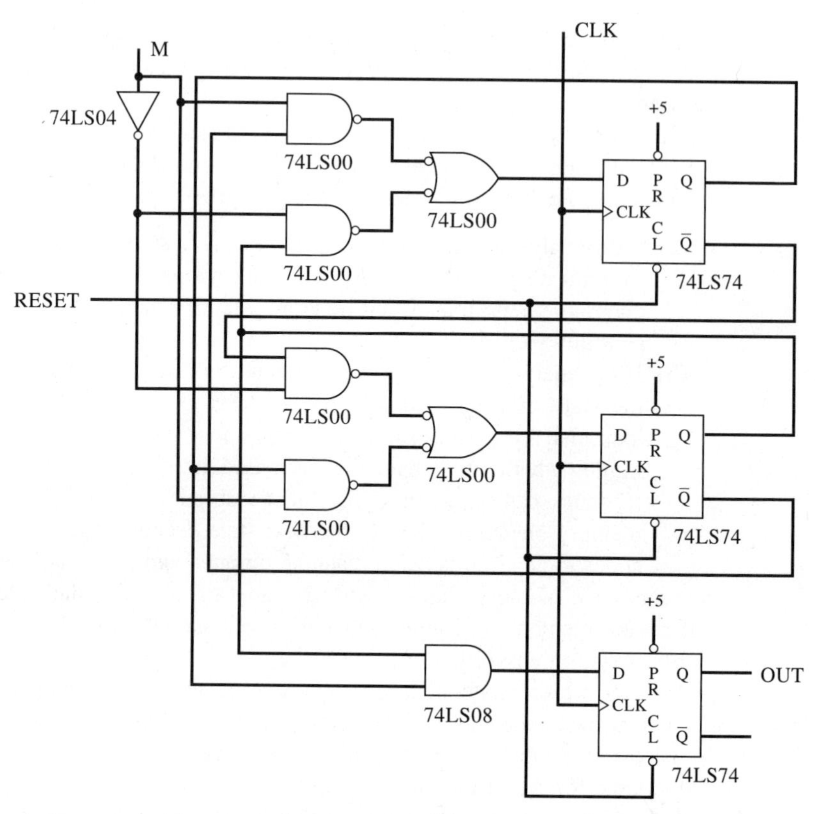

Figure 10.3-4 State machine circuit for a state machine with an output register having one flip-flop.

Example 10.3-1

Analyze a state machine with the circuit shown.

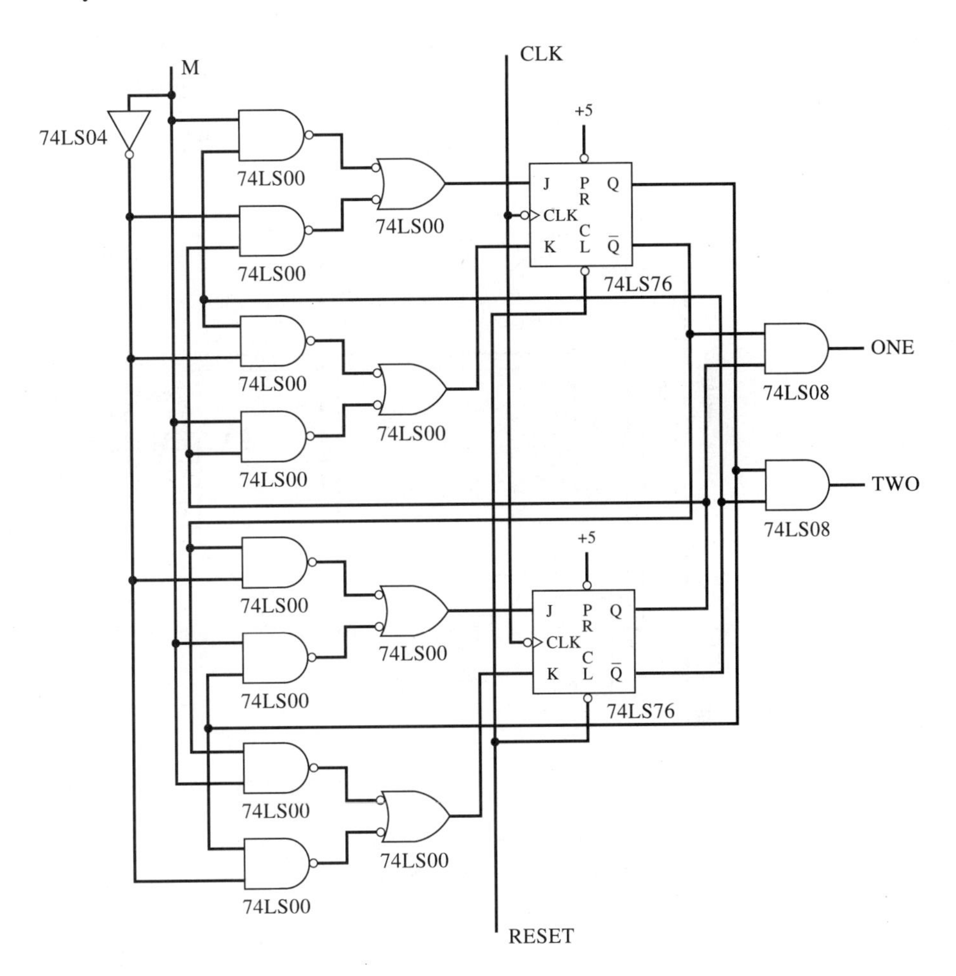

We can see that this circuit has JK memory elements and no output register. The next state decoder logic circuits are the combinational logic circuits that feed the J and K inputs of each JK flip-flop. The only external input is M. The output decoder circuits are the two *and* gates on the right, with external outputs ONE and TWO.

Having identified the parts of the state machine, we need to label the memory elements and fix the assertion levels of the inputs and outputs. We've done that in the following circuit, in which we've made the input and output assertion levels compatible with the circuit elements to which they connect.

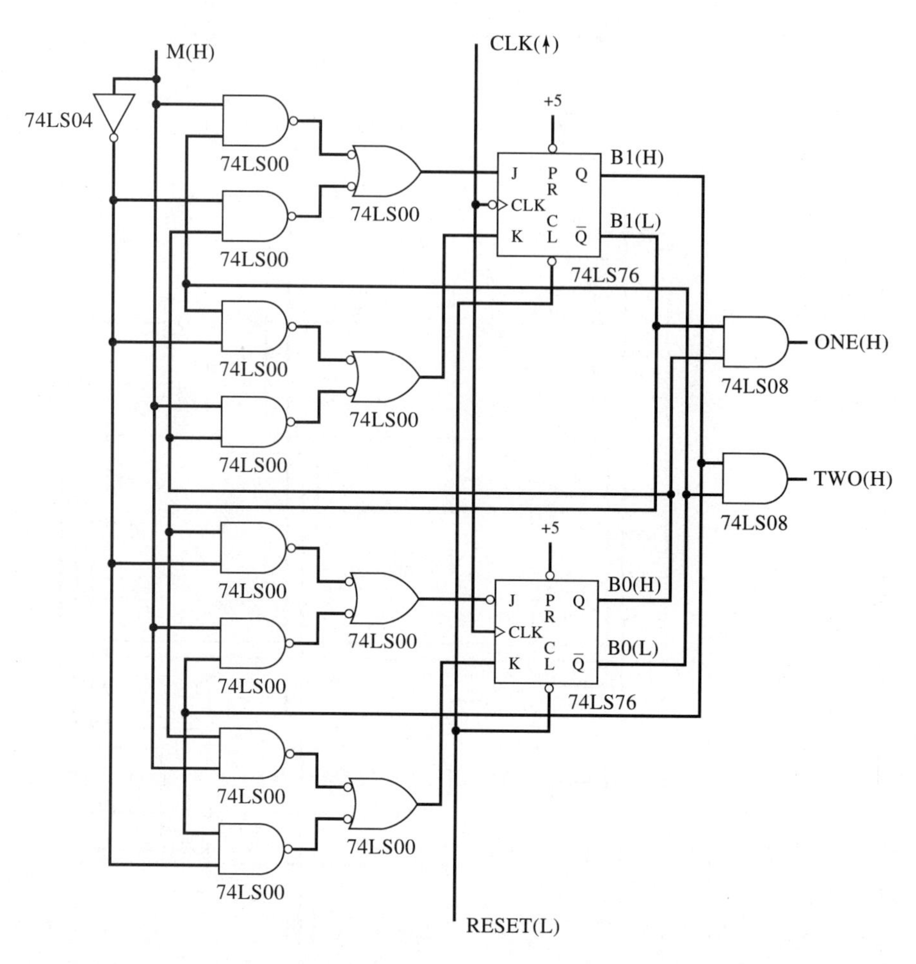

The next step in our analysis is to make a present state–next state and output table, which follows, with all possible combinations of present states and inputs filled in. The table has spaces for the next state code because this state machine uses JK memory.

Present State			Next State		Next State Code				Output	
B1	B0	M	B1	B0	J1	K1	J0	K0	ONE	TWO
0	0	0								
0	0	1								
0	1	0								
0	1	1								
1	0	0								
1	0	1								
1	1	0								
1	1	1								

Now we need the next state code logic expressions, which we can find from the next state code decoder circuits. These circuits show us that

$$J1 = M \cdot \overline{B0} + \overline{M} \cdot B0$$

$$K1 = \overline{M} \cdot \overline{B0} + M \cdot B0$$

$$J0 = \overline{M} \cdot \overline{B1} + M \cdot B1$$

$$K0 = M \cdot \overline{B1} + \overline{M} \cdot B1$$

Using these logic expressions, we can fill in the next state code in the present state–next state table as follows.

Present State			Next State		Next State Code				Output	
B1	B0	M	B1	B0	J1	K1	J0	K0	ONE	TWO
0	0	0			0	1	1	0		
0	0	1			1	0	0	1		
0	1	0			1	0	1	0		
0	1	1			0	1	0	1		
1	0	0			0	1	0	1		
1	0	1			1	0	1	0		
1	1	0			1	0	0	1		
1	1	1			0	1	1	0		

Next, we use the following present state–next state table for JK memory to fill in the next state.

Q_n J K	Q_{n+1}
0 0 x	0
0 1 x	1
1 x 0	1
1 x 1	0

The table is arranged in a form that is particularly convenient for finding the next state. It is important that the use of B1 and B0 in this table be consistent. To find the next state B1, we use the present state B1 for Q_n and find Q_{n+1}, which is the next state B1. To find the next state B0, we use the present state B0 for Q_n and find Q_{n+1}, which is the next state B0.

Present State			Next State		Next State Code				Output	
B1	B0	M	B1	B0	J1	K1	J0	K0	ONE	TWO
0	0	0	0	1	0	1	1	0		
0	0	1	1	0	1	0	0	1		
0	1	0	1	1	1	0	1	0		
0	1	1	0	0	0	1	0	1		
1	0	0	0	0	0	1	0	1		
1	0	1	1	1	1	0	1	0		
1	1	0	1	0	1	0	0	1		
1	1	1	0	1	0	1	1	0		

It remains for us to find the output logic expressions and fill in the output. The output logic expressions are found from the output decoders to be

$$\text{ONE} = \overline{\text{B1}} \cdot \text{B0}$$

$$\text{TWO} = \text{B1} \cdot \overline{\text{B0}}$$

Using these expressions to find ONE and TWO for the output table, we have the following complete present state–next state and output table.

Present State			Next State		Next State Code				Output	
B1	B0	M	B1	B0	J1	K1	J0	K0	ONE	TWO
0	0	0	0	1	0	1	1	0	0	0
0	0	1	1	0	1	0	0	1	0	0
0	1	0	1	1	1	0	1	0	1	0
0	1	1	0	0	0	1	0	1	1	0
1	0	0	0	0	0	1	0	1	0	1
1	0	1	1	1	1	0	1	0	0	1
1	1	0	1	0	1	0	0	1	0	0
1	1	1	0	1	0	1	1	0	0	0

Example 10.3-2
Analyze a PAL-based state machine design that uses a PAL16R8 with the following JEDEC file.

```
PALASM4   PAL Assembler—Market Version 1.2 (5-31-91)
  (c)—Copyright Advanced Micro Devices Inc., 1991


TITLE       Analysis Example          AUTHOR  :A W Shaw
PATTERN                               COMPANY:Utah State University
REVISION                              DATE     :07/05/92

PAL16R8
_SEQ5*
QP20*
QF2048*
G0*F0*
L0000  00000000000000000000000000000000*
L0032  00000000000000000000000000000000*
L0064  00000000000000000000000000000000*
L0096  00000000000000000000000000000000*
L0128  00000000000000000000000000000000*
L0160  00000000000000000000000000000000*
L0192  00000000000000000000000000000000*
L0224  00000000000000000000000000000000*
L0256  00000000000000000000000000000000*
L0288  00000000000000000000000000000000*
L0320  00000000000000000000000000000000*
L0352  00000000000000000000000000000000*
L0384  00000000000000000000000000000000*
L0416  00000000000000000000000000000000*
L0448  00000000000000000000000000000000*
L0480  00000000000000000000000000000000*
L0512  00000000000000000000000000000000*
L0544  00000000000000000000000000000000*
L0576  00000000000000000000000000000000*
L0608  00000000000000000000000000000000*
L0640  00000000000000000000000000000000*
L0672  00000000000000000000000000000000*
L0704  00000000000000000000000000000000*
L0736  00000000000000000000000000000000*
L0768  00000000000000000000000000000000*
L0800  00000000000000000000000000000000*
L0832  00000000000000000000000000000000*
L0864  00000000000000000000000000000000*
L0896  00000000000000000000000000000000*
L0928  00000000000000000000000000000000*
L0960  00000000000000000000000000000000*
L0992  00000000000000000000000000000000*
```

```
L1024  11111111111111111111111111111110*
L1056  00000000000000000000000000000000*
L1088  00000000000000000000000000000000*
L1120  00000000000000000000000000000000*
L1152  00000000000000000000000000000000*
L1184  00000000000000000000000000000000*
L1216  00000000000000000000000000000000*
L1248  00000000000000000000000000000000*
L1280  11111111111111111111111111101101*
L1312  00000000000000000000000000000000*
L1344  00000000000000000000000000000000*
L1376  00000000000000000000000000000000*
L1408  00000000000000000000000000000000*
L1440  00000000000000000000000000000000*
L1472  00000000000000000000000000000000*
L1504  00000000000000000000000000000000*
L1536  01111011111111111111111111111101*
L1568  00000000000000000000000000000000*
L1600  00000000000000000000000000000000*
L1632  00000000000000000000000000000000*
L1664  00000000000000000000000000000000*
L1696  00000000000000000000000000000000*
L1728  00000000000000000000000000000000*
L1760  00000000000000000000000000000000*
L1792  01111011111111111111111111101111*
L1824  10111011111111111111111111011110*
L1856  11111011111111111111111111101101*
L1888  00000000000000000000000000000000*
L1920  00000000000000000000000000000000*
L1952  00000000000000000000000000000000*
L1984  00000000000000000000000000000000*
L2016  00000000000000000000000000000000*
C1588*
E8A2
```

The JEDEC file is arranged in a matrix much like the fuse files we saw in Section 10.2. In the JEDEC file, 0 represents a connection and 1 represents no connection. If we mark the programmed connection represented by the JEDEC file on a PAL16R8 schematic, we obtain the following circuit.

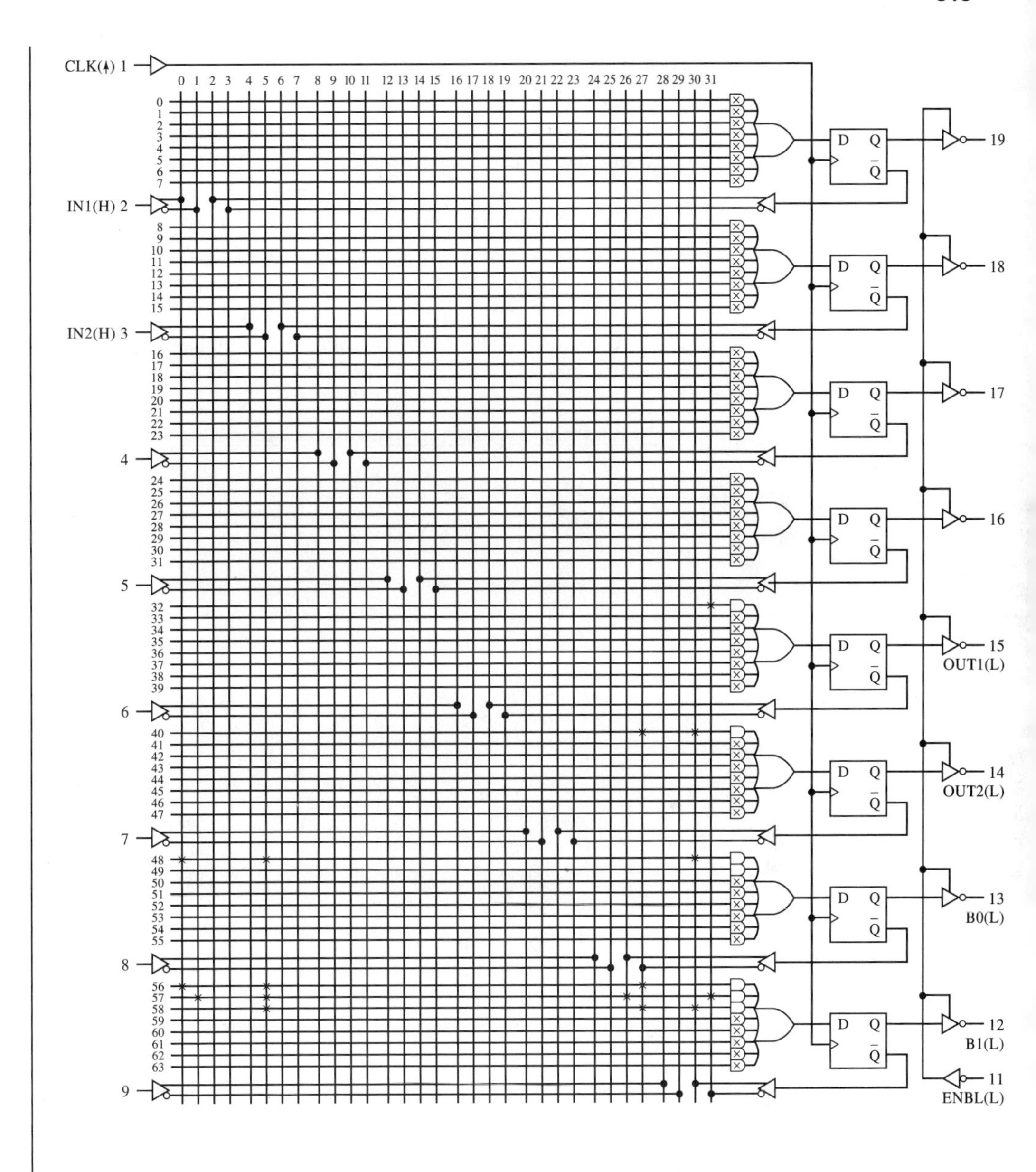

Only the four bottom logic sections are used in this design. We can see that, of these four, the upper two have no feedback from the flip-flop outputs back into any of the logic circuits that drive flip-flops. These flip-flop outputs can be taken as register outputs. We've labeled the outputs of these circuits OUT1 and OUT2 and made their assertion levels low to match the output drivers. The bottom two logic sections do have feedback into the logic circuits that drive flip-flops. The flip-flops of these sections are the state code memory. We've labeled the outputs of these sections (the state code bits) B0 and B1 and made their assertion levels low to match the output drivers. The logic circuits that drive the D inputs of the memory flip-flops are the next state decoder.

Now, if we look for external inputs, we find that pin 1 is the clock. We've labeled this pin CLK and shown it as positive-edge triggered to match the circuit. Pins 2 and 3 are inputs to the next state decoder, labeled IN1 and IN2. We've assumed that they are asserted high, but we could equally well assume that they are asserted low. Finally, pin 11 is an output enable that is asserted low. We have labeled it ENBL. The other pins of the PAL are not used.

Because this machine has D memory flip-flops, we don't need a provision for the next state code in our present state–next state and output table—the next state code is the next state. The table follows, with the next state and output blank.

Present State				Next State			
B1	B0	IN1	IN2	B1	B0	OUT1	OUT2
0	0	0	0				
0	0	0	1				
0	0	1	0				
0	0	1	1				
0	1	0	0				
0	1	0	1				
0	1	1	0				
0	1	1	1				
1	0	0	0				
1	0	0	1				
1	0	1	0				
1	0	1	1				
1	1	0	0				
1	1	0	1				
1	1	1	0				
1	1	1	1				

To fill this table, we need the logic expressions for the next state code, which we can find from the next state decoder circuit (we can also find these expressions by disassembling the JEDEC file if software is available):

$$D1 = IN1 \cdot \overline{IN2} \cdot B0 + \overline{IN1} \cdot \overline{IN2} \cdot B1 \cdot \overline{B0} + \overline{IN2} \cdot \overline{B1} \cdot B0$$

$$D0 = IN1 \cdot \overline{IN2} \cdot \overline{B1}$$

Using these logic expressions, we can complete the next state code in the table as follows. Remember that it is easier to fill this table in columns by first finding the rows that are 1 from the terms in the logic expressions and then filling the remainder of the rows with 0s. When you write the logic expression, be careful to notice that the feedback is from $\overline{Q}$, not Q.

Present State				Next State			
B1	B0	IN1	IN2	B1	B0	OUT1	OUT2
0	0	0	0	0	0	0	0
0	0	0	1	0	0	0	0
0	0	1	0	0	1	0	0
0	0	1	1	0	0	0	0
0	1	0	0	1	0	0	1
0	1	0	1	0	0	0	1
0	1	1	0	1	1	0	1
0	1	1	1	0	0	0	1
1	0	0	0	1	0	1	0
1	0	0	1	0	0	1	0
1	0	1	0	0	0	1	0
1	0	1	1	0	0	1	0
1	1	0	0	0	0	1	0
1	1	0	1	0	0	1	0
1	1	1	0	1	0	1	0
1	1	1	1	0	0	1	0

We've finished the table by finding logic expressions for the output decoders for OUT1 and OUT2 and using these expressions to fill in the output columns of the table. The logic expressions for the inputs to the output register flip-flops are

$$\text{DOUT1} = \text{B1}$$

$$\text{DOUT2} = \overline{\text{B1}} \cdot \text{B0}$$

Again it is easier to first fill in the 1s in a column, using the terms of the logic expression, and then fill the remainder of the column with 0s. Because the outputs are register outputs, they occur one state after the state shown in the table.

The technique just given for analyzing a state machine does not make use of simulation. Good simulation software is available that can simplify the analysis of logic circuits. The analysis of a state machine using a simulator still requires that the parts of the state machine be identified and labeled, but after that all that need be done is to enter the circuit into the simulator with the signals labeled, create test inputs, then simulate the circuit and examine the output to determine what the machine does. Simulation can also be used to analyze PAL designs. Some simulation software will accept the JEDEC file as an input and do a complete simulation of a PAL design.

10.4 UNCONVENTIONAL STATE MACHINES

We've been designing machines that have the minimum number of memory elements by making the state code as small as possible (making it contain the least number of bits). Machines with state codes chosen to minimize the memory elements tend

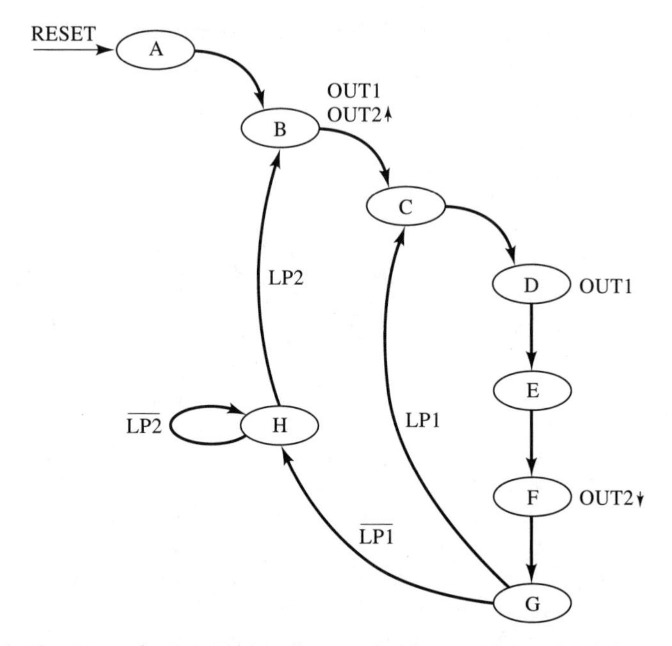

Figure 10.4-1 State transition diagram for the machine to be designed.

to have complicated next state and output decoders, and because of the delay through the decoders, they can have timing limitations at high clock frequencies. Sometimes the state code can be made longer and the next state and output decoders simplified in such a way as to streamline the overall design of the machine. One thing we can do, for instance, is to make one of the bits in the state code match one of the outputs so that an output decoder is not needed. This kind of simplification is generally a matter of judgment and trial.

One particularly useful choice is to have only one asserted bit in each state code. The simplest code of this type is 100000 . . . , 010000 . . . , 001000, and so on for the various states. Notice that this state code uses one memory flip-flop for each state. For moderately large machines with a number of state transitions that do not depend on inputs (other than the clock), a machine with this kind of state code can be an attractive alternative to a conventional machine. These *one–flip-flop–per–state* state machines could be designed using conventional techniques, but there is a much simpler approach.

Suppose we wish to design a state machine with the state transition diagram shown in Figure 10.4-1. This machine has three inputs. The first, RESET, is used to start the machine in state A. It will be treated in a special way. The other two inputs, LP1 and LP2, are conventional inputs. The output OUT1 is asserted in two isolated states. This type of output is particularly easy to implement in a one-bit-per-state state machine. The other output, OUT2, is asserted in state B and stays asserted until state F. We've used special symbols in Figure 10.4-1 to show this. The up arrow indicates the assertion on, and the down arrow indicates the assertion off. Notice that when the state machine follows the inner loop, OUT2 is not asserted.

One–flip-flop–per–state state machines are most easily designed using D flip-flops. The machine in Figure 10.4-1 will require eight of these flip-flops because it has eight states. We'll give each flip-flop the same designation as the state. Thus, the first flip-flop will be designated A to correspond to the first state, A, and when the output of this flip-flop is asserted, the machine will be in state A. Notice that only one flip-flop is asserted at a time. When the output of A is asserted, the outputs of all the other flip-flops should be not asserted. We'll assume that the uncomplemented outputs of the flip-flops corresponding to the states are all asserted high. We could equally well assume that they are asserted low or even that the assertion levels are a mixture of highs and lows.

The machine is started by an asynchronous input, RESET. This input should put the machine in state A; that is, the flip-flop that represents state A should be set to a high output level by the input RESET. One way to do this is to use the first flip-flop as a SET-RESET cell. To make a D flip-flop into a SET-RESET cell, we disable the clock and D inputs by connecting them to ground (or +5) and use the PR and CL to *set* and *reset* the resulting cell. We can thus use RESET, which we'll assume to be asserted low, to preset the A flip-flop and start the machine in state A. Notice how RESET is connected to PR on the A flip-flop in Figure 10.4-2, which shows the complete circuit for the state machine.

At the same time as we preset the first flip-flop, we need to clear all the other flip-flops. We do this by connecting RESET to CL on all the flip-flops except the first, as in the figure. Clearing all the flip-flops can present a practical problem. The fanout may be too large for a single output driver. Here the fanout is only 8, so we don't have a problem. If the fanout were too large, we would need to divide the CL inputs between two or more different circuits, each with a separate driver.

We do not want to clear the first flip-flop (the A flip-flop) with RESET. We must clear it, however, when the state machine makes a transition into state B. To clear it we must use its CL, because its clock and D are disabled. The complemented output of the B flip-flop furnishes a signal that clears the first flip-flop as soon as the machine enters state B. This signal is connected back to the CL on the A flip-flop, as shown in Figure 10.4-2.

Now let's address the next state decoder design for the remaining states. The next-state decoder logic can be found directly from the state transition diagram. The machine makes a transition to state B when it's in state A or when it's in state H and LP1 is asserted. Thus, if we call the next state code for the state B DB,

$$DB = A + H \cdot LP2$$

When we continue this process through the remainder of the states, we find

$$C = B + G \cdot LP1$$

$$DD = C$$

$$DE = C$$

$$DF = D$$

$$DG = F$$

$$DH = G \cdot \overline{LP1} + H \cdot \overline{LP2}$$

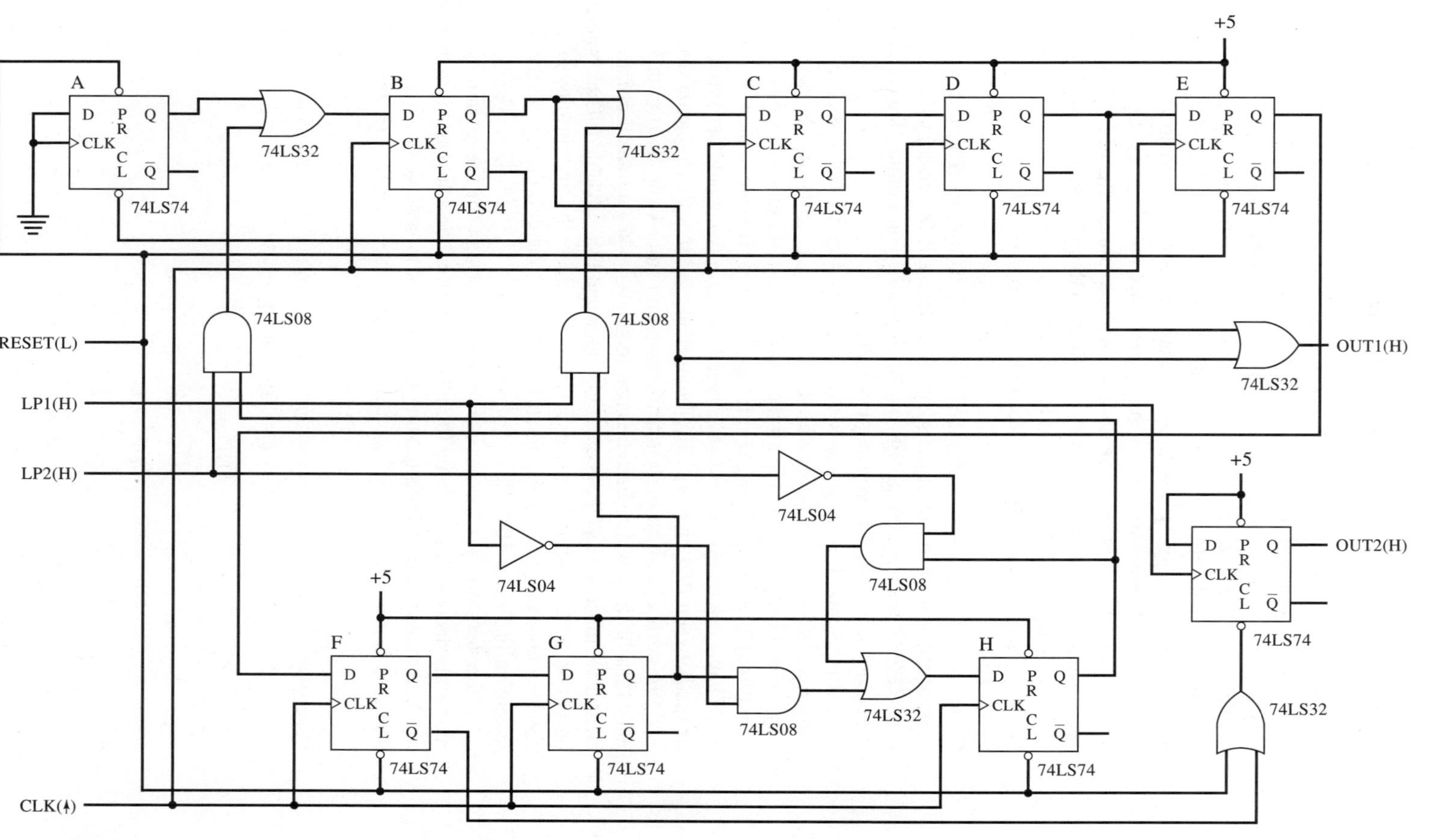

Figure 10.4-2 Circuit for one flip-flop state machine example.

Notice the simplicity of the next state decoders. They can be implemented using simple wires in some cases or, at most, simple *and-or* logic, as shown in the figure. If the expression for the next state code is complicated, we can make a Karnaugh map or a VEM to simplify the logic.

The output decoders are equally simple. The first output, OUT1, which we'll assume is to be asserted high, is the *or* of the outputs from states B and D:

$$\text{OUT1} = \text{B} + \text{D}$$

The second output, OUT2, which we'll also assume is asserted high, must be asserted at the beginning of state B and deasserted at the beginning of state F. We can do this with a D flip-flop. If we connect the D input of the flip-flop high, then a positive-going clock edge will cause the flip-flop's output to go high. Thus, we can use the output of the B state flip-flop to clock the output flip-flop to a high when the state machine enters state B. We can use the complemented output from state F to clear the output flip-flop and drive its output low when the state machine enters state F. We also need a provision for clearing the output flip-flop when RESET is asserted. To see the output circuits, refer to Figure 10.4-2.

Notice that the output of a one–flip-flop–per–state machine can often be formed from a simple *or* of the state outputs. Further, these outputs will be glitch-free as long as the states are disjoint (*not timed to occur one after another*). If an output is to be formed from states that are not disjoint, then the states previous to those that require outputs can be *or*ed and registered just as for conventional machines. We can also use a flip-flop that is first turned on and then turned off, as we did for OUT2.

Now we'll test the design by simulating it. Figure 10.4-3 shows the results from a simulation of the circuit. Representative inputs, resultant output, and all the states are provided in the figure. Notice that the inner loop of the state transition diagram is traversed when LP1 is asserted, that the machine remains in state H when neither LP1 nor LP2 is asserted, and that the outer loop of the state transition diagram is traversed when LP2 is asserted. OUT1 occurs only once for each inner loop traversed, but twice for each outer loop traversed; OUT2 occurs only when the outer loop is traversed.

The preceding discussion introduced a new type of register output circuit: one that turns the output on (asserts the output) in one state and turns it off (deasserts it) in another. There are actually three ways of doing this. We used one of them in our discussion. We turned the output on by using a state output to clock the output flip-flop with the D input to the flip-flop tied high. We turned the output off using the clear (CL) of the output flip-flop. We can equally well use the preset (PR) of the output flip-flop to turn the output on and use the clock with the D input tied low to turn it off, or use the preset to turn it on and the clear to turn it off. The three alternatives are shown in Figure 10.4-4: CLK is the flip-flop clock, ON is the state that turns the output [OUT (H)] on, and OFF is the state that turns the output off. In the first circuit, ON is the clock for the output circuit. In the second, OFF is the clock for the output circuit. These three circuits have subtle differences that can be important. To see the differences, consider the timing diagram in Figure 10.4-5. It shows the timing for the first circuit, which uses the rising edge of the input signal

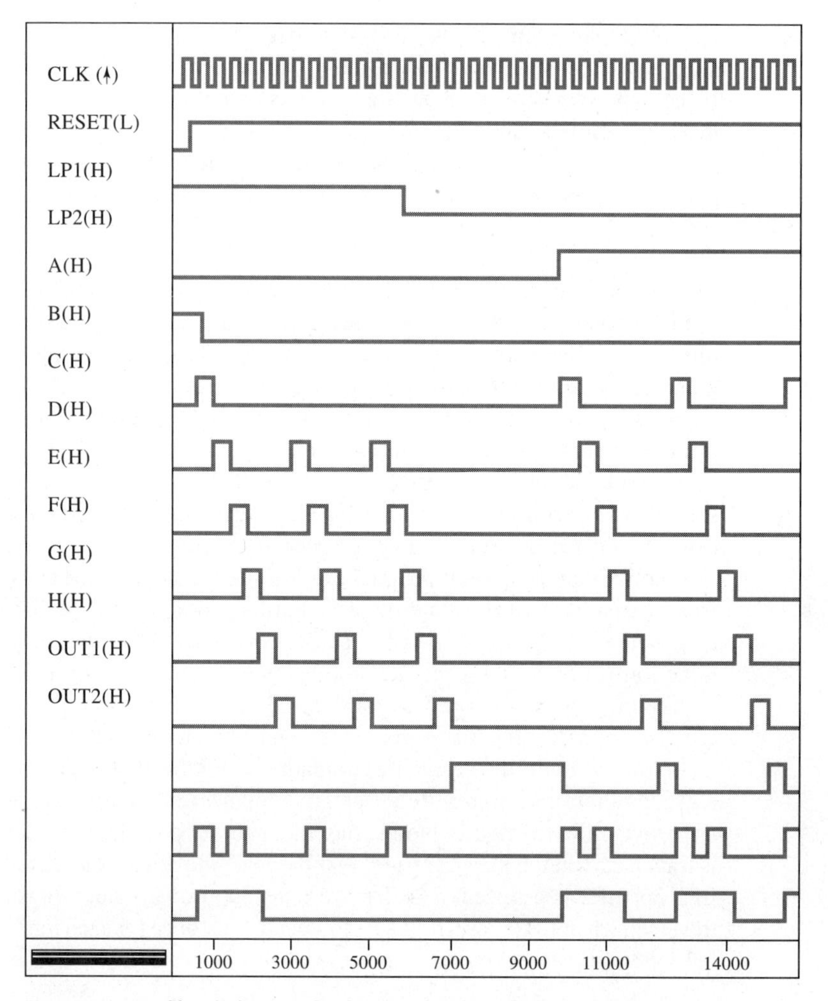

Figure 10.4-3 Simulation results for the circuit of Figure 10.4-2.

ON to clock the output on and uses the input signal OFF(L) to turn the output off via the clear. Both ON and OFF(L) have a duration of one clock cycle of the state machine. A problem can occur when it is necessary to turn the output on and it has been off for only one state. The clear may still be asserted, and ON will not cause the output to be turned on.

We can alleviate the problem found in the first output circuit by using the second circuit, where the output is turned on using the signal ON(L) via the preset, and turned off using the positive-going clock edge with D tied low. The timing for this circuit is shown in Figure 10.4-6. The circuit can have an output that turns on after being off for only one clock cycle, but it cannot have an output that is on for only one state.

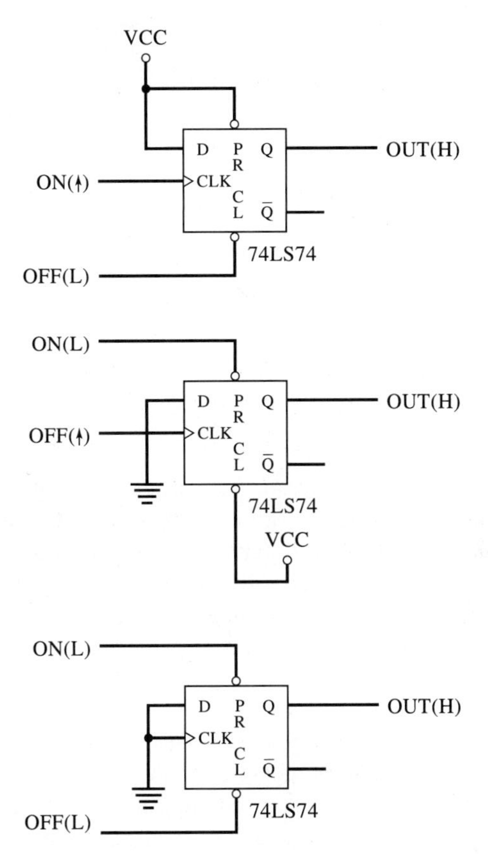

Figure 10.4-4 Output circuit that turns the output on and off.

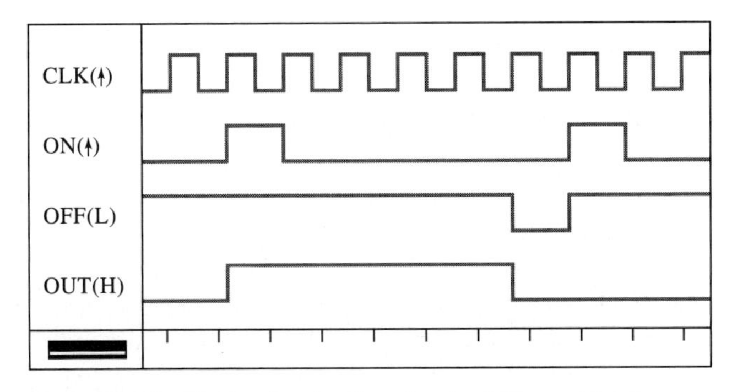

Figure 10.4-5 Timing for the first circuit of Figure 10.4-4.

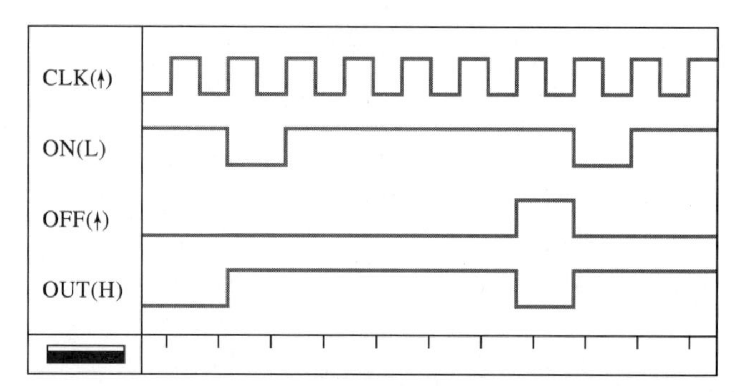

Figure 10.4-6 Timing for the second circuit of Figure 10.4-4.

The final output circuit that uses the flip-flop as an asynchronous SET-RESET cell may have a conflict between the preset (SET) and the clear (RESET). But the conflict is of only short duration and will be resolved when either the preset or the clear is no longer asserted. The three different circuits have slightly different timings that may be important, depending on the application. These timing differences should be investigated if timing is a problem with one or another of the circuits.

We need to consider whether this type of register output circuit offers any advantage in conventional state machine design. It can be used in conventional machines, but not nearly as conveniently as in one–flip-flop–per–state machines. The turn on–turn off output circuits already described work because their inputs are the states of a one–flip-flop state machine and thus are glitch-free. For this type of circuit to work in a conventional machine, it is necessary to use three flip-flops. As we can see in Figure 10.4-7, two flip-flops are needed to deglitch the decoded ON and OFF signals that turn the output on and off, and a third to turn the output on and off. As noted, such added complexity limits the usefulness of this type of output circuit for conventional state machines. The circuit will be useful only if the number of states between the turn on and turn off is large enough to complicate conventional output decoding.

One–flip-flop–per–state state machine can result in simplification of certain types of state machines. They can also be attractive during debug, because it's immediately evident which state they are in. Two serious errors can occur in this kind of state machine, however. One occurs when more than one of the state flip-flops has an asserted output. This type of error can be corrected by periodically restarting the machine using a reset that clears all the states except one, as in the sample design. We can also design the machine to clear all the other state flip-flops as it passes through one of the states.

The other error occurs when none of the flip-flops has an asserted output. This type of error is commonly associated with an asynchronous input that occurs near a transition time. Suppose that, in the case under discussion, the machine is in state G and LP1 is not asserted early enough to cause a transition to C but is asserted early enough to prevent a transition to G. In this case, the machine will stop. This

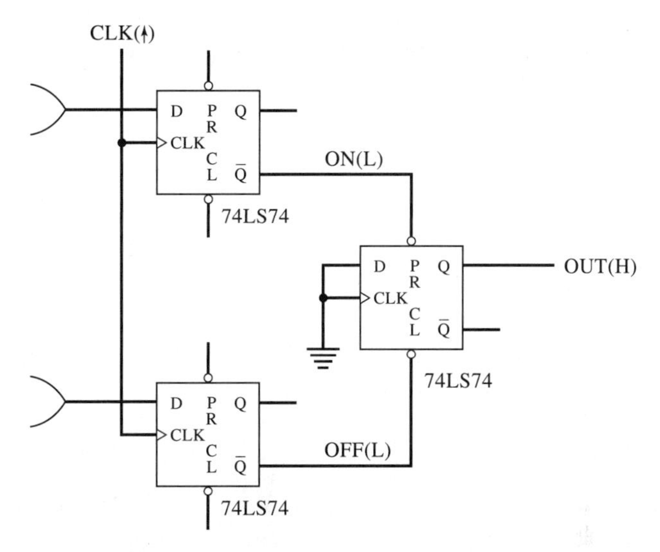

Figure 10.4-7 Circuit needed to turn on and turn off an output in a conventional state machine.

error is best corrected by changing the timing of the input so that it does not occur near the transition.

Before leaving one–flip-flop–per–state state machines, let's look at a final example. It is the same as Example 9.6-2 in Chapter 9—the design of a state machine that we might use to control a digital system such as we will discuss in Chapter 12. These controllers are difficult to describe apart from the digital systems they control, so we will start the design with a state transition diagram rather than a word description.

Example 10.4-1

Design a one–flip-flop–per–state state machine from the following state transition diagram. Use the types of logic elements found in the FPGA used as an example throughout this text. Make all the outputs glitch-free. The inputs are RESET, WAIT, RTRN (return), LNG (long), and ST (start). RESET is an asynchronous reset that starts the machine.

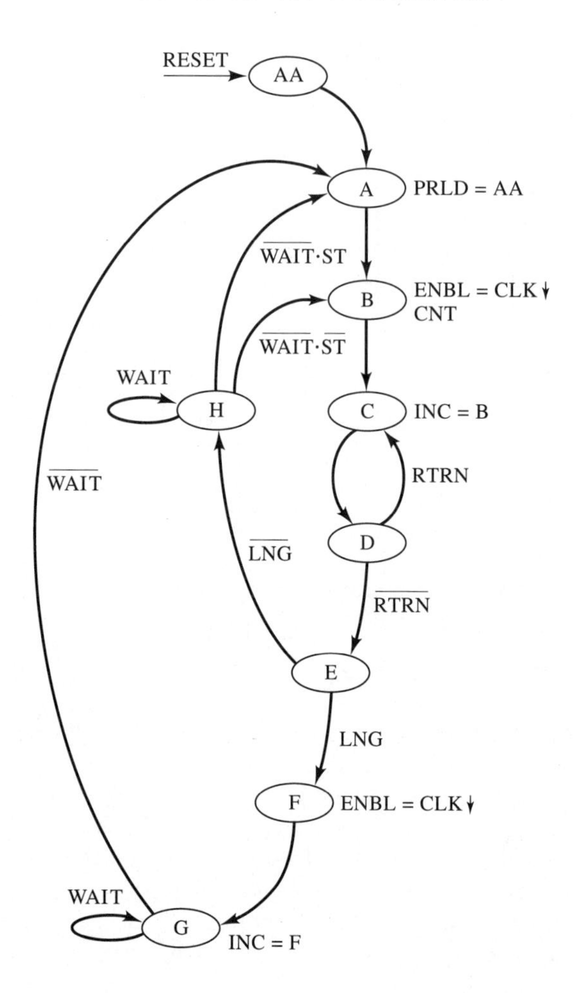

Notice the special timing on most of the outputs. In state A, the output notation

$$\text{PRLD (preload)} = \text{AA}$$

means that PRLD is conditional on the previous state being AA. This output is not asserted when A is entered from G or H. Just for interest, the signal names are given in parentheses after the mnemonic. Similar kinds of outputs occur in states C and F.

$$\text{INC (increment)} = \text{B (state C)}$$

$$\text{INC} = \text{E (state G)}$$

Notice that in state G, INC is asserted for only one clock cycle when G is entered from F. If we wanted INC to be asserted for the entire time the state machine is in state G, we would not use the condition E, but simply place INC by state G. The output in states B and F

$$\text{ENBL (enable)} = \text{CLK}\downarrow \text{(state B)}$$

$$\text{ENBL} = \text{CLK}\downarrow\ (\text{state F})$$

shows that the output is asserted on the negative-going clock edge and lasts for one clock cycle.

The design of the state machine is much the same as the design we have discussed. The first state is special and uses a D flip-flop with preset (PR) and clear (CL) as an asynchronous SET-RESET cell. The next-state decoder logic for the remainder of the states can be found directly from the state transition diagram.

$$\text{DA} = \text{AA} + \overline{\text{WAIT}} \cdot \text{G} + \overline{\text{WAIT}} \cdot \text{ST} \cdot \text{H}$$

$$\text{DB} = \text{A} + \overline{\text{WAIT}} \cdot \overline{\text{ST}} \cdot \text{H}$$

$$\text{DC} = \text{B} + \text{RTRN} \cdot \text{D}$$

$$\text{DD} = \text{C}$$

$$\text{DE} = \overline{\text{RTRN}} \cdot \text{D}$$

$$\text{DF} = \text{LNG} \cdot \text{E}$$

$$\text{DG} = \text{F} + \text{WAIT} \cdot \text{G}$$

$$\text{DH} = \overline{\text{LNG}} \cdot \text{E} + \text{WAIT} \cdot \text{H}$$

Following is the circuit. Inputs are on the left; outputs are on the right. For clarity, it is useful to organize a one–flip-flop–per–state circuit to reflect the signal flow, as has been done in this diagram. The state memory flip-flops are labeled with the states they represent. They start in the upper left corner of the diagram with state AA. In normal operation, the state machines generally sequence through the states represented by the top row of flip-flops and then through the states represented by the bottom row of flip-flops. At state E the bottom row of flip-flops splits into two rows to reflect the branch in the state transition diagram. The circuit for the first state (AA) is a SET-RESET cell, just as in the previous design. The input RESET that starts the machine in state AA also clears all the other flip-flops. In this circuit there are too many of these inputs for a single driver, so we've split the clears into two circuits and used a driver for each of these circuits. The other next state decoder circuits follow directly from the logic equations.

Many of these circuits could be simplified by using a flip-flop with a multiplexed input and absorbing an *and* or *or* gate into the flip-flop as described in Chapter 8, Section 8.8. We have not simplified, because it's easier to see the next state decoder without this simplification.

The following simulated output shows how the state machine sequences through the states, depending on the inputs. This is a simulation without delays (a functional simulation). To get accurate delays in an FPGA design, it's necessary to do the complete design, including routing. The machine starts in state AA on RESET. It bounces back and forth between states C and D as long as RTRN is asserted, and then goes around through state G if LNG is asserted. It stays in state G as long as WAIT is asserted, and then goes back to state A. It circles back through states E, H, and A if LNG is not asserted and ST is asserted and through states E, H, and B if

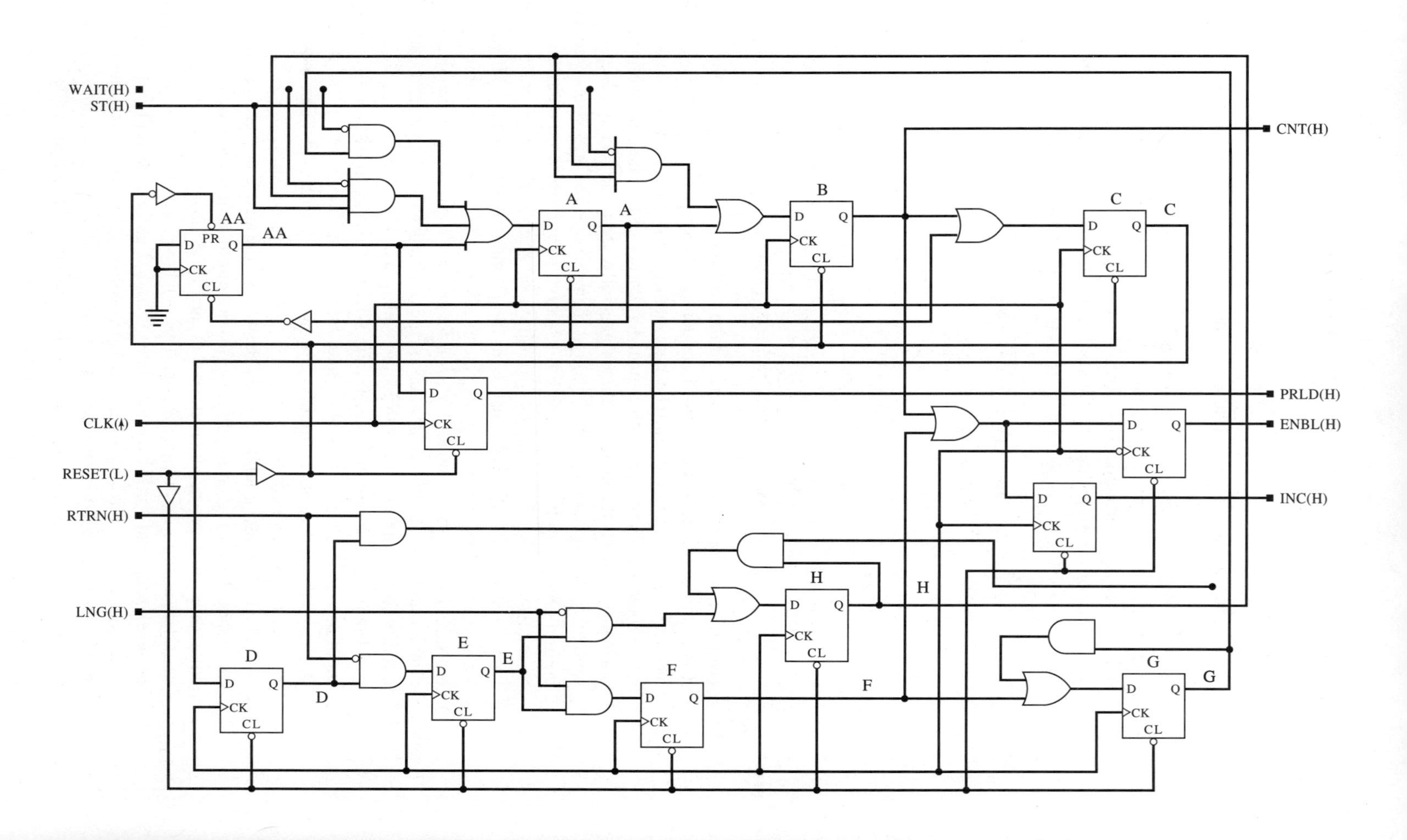

WAIT(H)
ST(H)
CLK(↓)
RESET(L)
RTRN(H)
LNG(H)
CNT(H)
PRLD(H)
ENBL(H)
INC(H)
AA
A
B
C
D
E
F
G
H

LNG is not asserted and ST is not asserted. Finally, it stays in state H if WAIT is asserted.

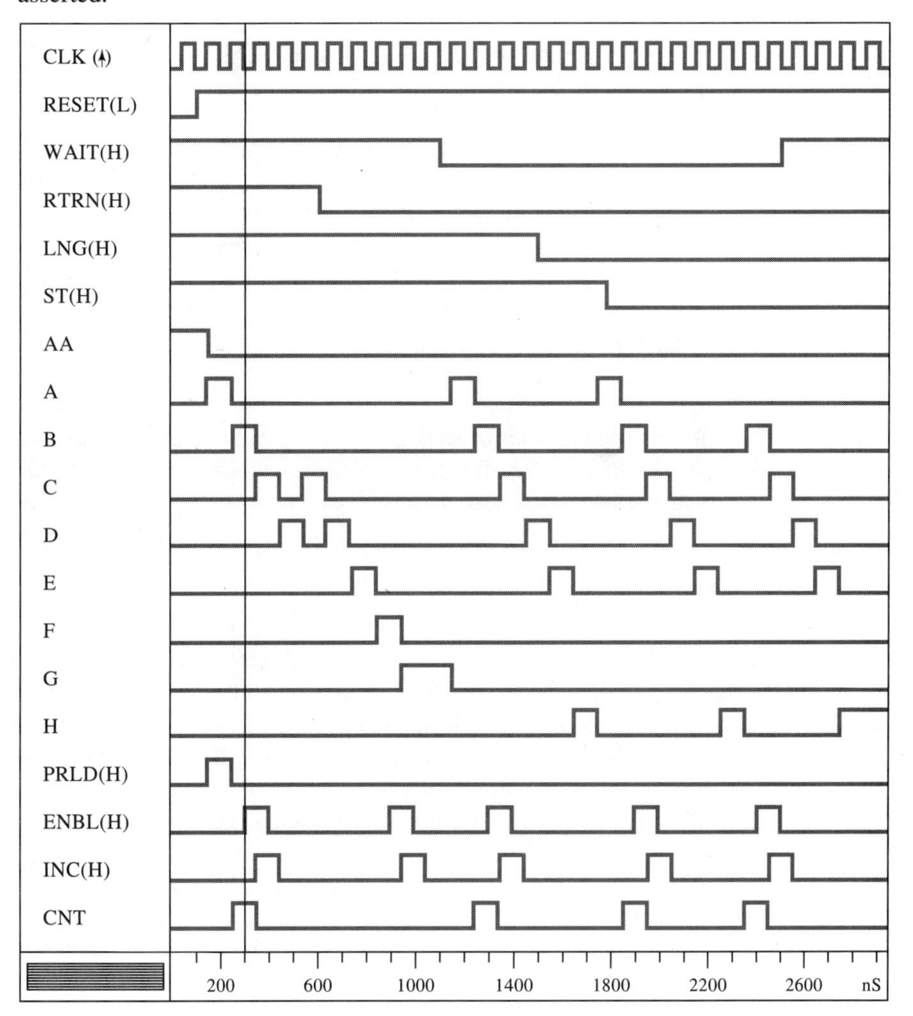

The output circuits require some explanation. CNT is straightforward. It is just equal to state B and is automatically deglitched in a one–flip-flop design. The remainder of the outputs require both register outputs and decoders. To produce PRLD in state A only when it is entered from state AA, we must decode this previous state (AA) and use a flip-flop that delays the output to the next state. Decoding AA is simple — it is just AA. In the circuit diagram, look back along the output line for PRLD and you can see the PRLD output circuit. The output INC occurs in state C when it is entered from state B and in state G when it is entered from state F. Again we must decode the previous states and use a register output. The previous states are easy to decode; they are simply the states, and they must be *or*ed. To see the INC output circuit, look back along the output line INC in the circuit diagram. The output ENBL occurs on the negative clock edge in states B and F. Here we do not need to decode the previous state — we just *or* B and F. Because this is already done for the INC decoder, we can use that same decoder. (This is a coincidence; normally we would need another decoder gate.) Now, to make the output occur at the negative clock edge, we use a flip-flop that is negative-edge triggered; such flip-flops are readily available in FPGAs.

The simulation shows that the outputs function as we have designed them. In particular, PRLD is asserted in state A only when it is entered from state AA; INC is asserted in state C only when it is entered from state B, and in state G only when it is entered from state F. Notice in particular that INC is asserted for only one clock cycle even though the state machine remains in state G for more than one clock cycle. Finally, ENBL is asserted in the proper states on the negative clock edge. The beginning of the first ENBL is marked with a vertical line to make this easy to see.

The preceding example shows how we design outputs that have many different kinds of timing in a one–flip-flop–per–state state machine and how this is generally easier than with a conventional machine.

10.5 ASYNCHRONOUS STATE MACHINE DESIGN

Most problems of design requiring a state machine can be handled using synchronous machines, which are easier to design and debug than asynchronous machines. There are situations, however, in which an asynchronous machine either greatly simplifies the design or is the only way to accomplish it. We will discuss some simple guidelines that will enable us to design asynchronous machines in much the same way as synchronous machines. Asynchronous machines are timing-critical; to successfully design them, one should have accurate simulation software or else expect to spend time tuning the hardware.

The basic architecture of an asynchronous state machine is the same as that of a synchronous machine (Fig. 10.5-1). The difference is that in the asynchronous design there is no clock to sample the next state code; if the next state code changes, the next state changes almost immediately.

This feature has strong implications for the ways in which the state code can change. Suppose we have a state machine with state codes 00, 01, 11, and 10 and we want it to make a transition from state 00 to state 11. We have to change both bits of the state code. In an asynchronous machine, that is almost impossible. Either the first bit will change to 1 before the second bit can change, putting the machine in state 10, or the second bit will change to 1 before the first, putting the machine in state 01. Once it is in state 10 or 01, the input to the next-state decoder has been changed, and we may no longer have the proper logic to send the machine to state 11. Normally, in an asynchronous state machine, *we can change only one bit of the state code at a time.* This does not mean that it is impossible to make a transition from state 00 to state 11, but it does mean that we must go by way of state 01 or 10.

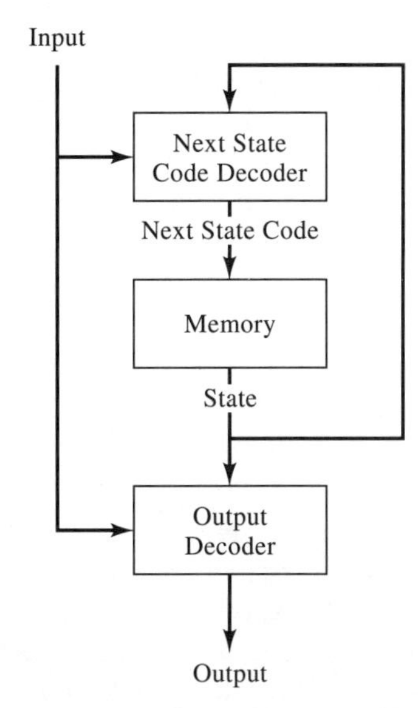

Figure 10.5-1 General state machine.

This is a serious limitation on design; we must produce a design in which only one bit changes in any transition.

The first state machines we designed in Chapter 8 were asynchronous. These 2-state state machines always met the condition that only one bit of the state code change in any transition, because there was only one bit to change. We met the 1-bit condition then, so we did not need to worry about it; but now we do.

Let's see what the asynchronous designs of Chapter 8 tell us about general asynchronous designs. First, we found that for some sets of input conditions it is possible to have an unstable machine—a machine that will change back and forth between states. This is because the machine can move through either state without stopping, which it will do if the input conditions that cause a transition *out of* that state contain the input conditions that cause a transition *into* that state. We call this an *unstable state.* Figure 10.5-2(a) shows what this kind of state looks like on a state transition diagram. In our unstable 2-state machines of Chapter 8, both states were unstable; the machines continuously moved through first one state and then the other.

We'll find unstable states to be useful in our designs of general asynchronous state machines, but a machine that contains *only* unstable states would not be very useful. Fortunately, it is not difficult to see how to make a state stable. We must make the input conditions for staying in the state the same as the input conditions for entering, and we must make the input conditions for exiting the state different. Figure 10.5-2(b) shows how this kind of state looks in a state transition diagram. The machine will stay in this state as long as input conditions COND1 are met, and will make a transition out of the state as soon as input conditions COND2 are met.

To design an asynchronous state machine, we need three of the ideas developed

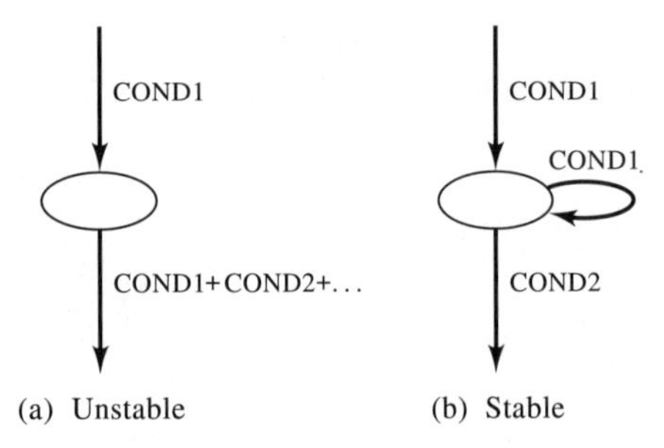

Figure 10.5-2 Conditions for unstable and stable asynchronous states.

in the previous paragraphs. First, we need to know that only one bit of the state code can change in a state transition. Second, we need to know how to make an unstable state. Third, we need to know how to make a stable state. With these three ideas we can use the next state decoder design process developed for synchronous state machines to design next state decoders for asynchronous state machines.

Suppose we want to design a positive-edge triggered D flip-flop using the state transition diagram suggested in Chapter 8, which is repeated in Figure 10.5-3. This diagram is designed so that only one bit changes on any transition and so that all the states are stable. The machine actually has two states for Q = 1 and two states for Q = 0. Q is the low order bit in the state code and is shown at the bottom of the diagram. For each value of Q, we can think of one set of states as the armed states and one as the holding states. The armed and holding states are labeled on the right in the diagram. The idea is to put the state machine into one of the armed states when the clock is not asserted and then allow it to make the proper transition, based on D, when the clock is asserted.

We'll design our asynchronous state machine using SET-RESET memory cells.

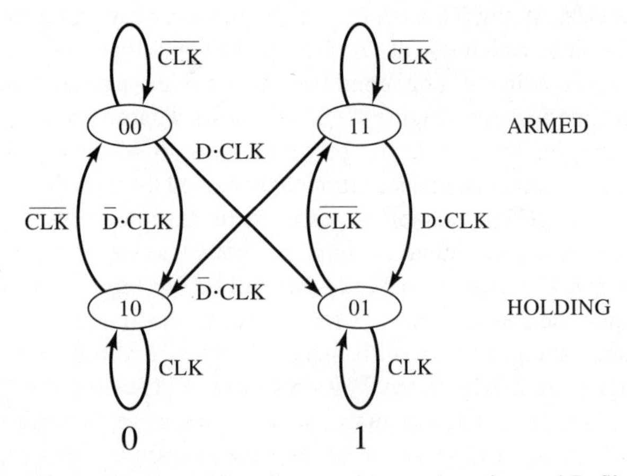

Figure 10.5-3 State transition diagram for an edge triggered D flip-flop.

Table 10.5-1 Present State–Next State and Next State Code for D Flip-Flop Design

Present State				Next State		Next State Code			
B1	B0	CLK	D	B1	B0	S1	R1	S0	R0
0	0	0	0	0	0	0	x	0	x
0	0	0	1	0	0	0	x	0	x
0	0	1	0	1	0	1	0	0	x
0	0	1	1	0	1	0	x	1	0
0	1	0	0	1	1	1	0	x	0
0	1	0	1	1	1	1	0	x	0
0	1	1	0	0	1	0	x	x	0
0	1	1	1	0	1	0	x	x	0
1	0	0	0	0	0	0	1	0	x
1	0	0	1	0	0	0	1	0	x
1	0	1	0	1	0	x	0	0	x
1	0	1	1	1	0	x	0	0	x
1	1	0	0	1	1	x	0	x	0
1	1	0	1	1	1	x	0	x	0
1	1	1	0	1	0	x	0	0	1
1	1	1	1	0	1	0	1	x	0

Q_n	Q_{n+1}	SET	RESET
0	0	0	x
0	1	1	0
1	0	0	1
1	1	x	0

This means that we do not have a procedure for going directly from the state transition diagram to the maps, as we do with D and JK memory; instead we'll make a present state–next state table (Table 10.5-1) and find the next state code from that. Table 10.5-1 also includes the excitation table for the SET-RESET memory cell that we used as an aid in determining the next state code. We've shortened SET and RESET to S and R and, because there are two memory cells, labeled them 0 and 1.

From this point on, the design of an asynchronous machine follows the same steps as the design of a synchronous machine. The next step in the design is to make Karnaugh maps from the table for the next state code and get simplified logic expressions for the next state decoder logic. Figure 10.5-4 shows the four next state decoder maps and their logic expressions.

A subtle point can be raised here: how should we treat the don't cares in the map? If we choose to use the don't cares in our groupings, we would force them to be 1. This could mean that we would be forcing either a SET or a RESET when we did not need it, and maybe on the next state change we would actually have to simultaneously change the assertion levels of SET and RESET on one of the memory cells, rather than simply assert either SET or RESET. Changing both SET and RESET at the same time could result in a timing problem; they could both be asserted at the same time.

Nevertheless, we've chosen to use the don't cares to achieve as much simplifica-

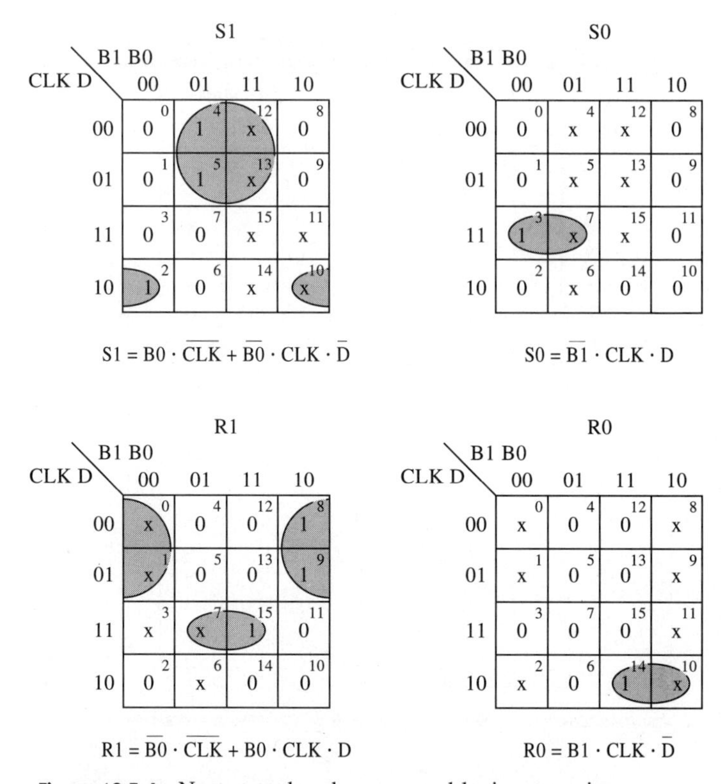

Figure 10.5-4 Next state decoder maps and logic expressions.

tion as possible. We're assuming that we'll find any timing problems in the simulation or in the prototype. A more conservative design would provide for checking whether or not SET and RESET were both changing for any of the transitions. In any case, this problem of SET and RESET changing at the same time is unavoidable in some unstable states.

Figure 10.5-5 is the circuit of the D flip-flop, using discrete logic. Notice the SET-RESET memory cells. There is no output decoder—Q is B0 (H) and $\overline{Q}$ is B0 (L). Simulation of an asynchronous design is very important because timing can be critical. Figure 10.5-6 shows the simulation results for the flip-flop. If we remember that B0 (H) is Q in this timing diagram, the circuit functions just as we would expect a positive-edge triggered D flip-flop to function; that is Q [B (H)] is set equal to D just after each positive-going clock edge.

The simulation input we have used is not very a very critical test of the circuit. We need to look very carefully at all conditions where D is changing near the positive-going clock edge. If any condition shows that the machine does not function as it should, we can then analyze the condition to see if we can improve it. We might need to change our method of choosing the don't cares in the next state decoder simplification; we might need to change the relative delay of D with respect to CLK

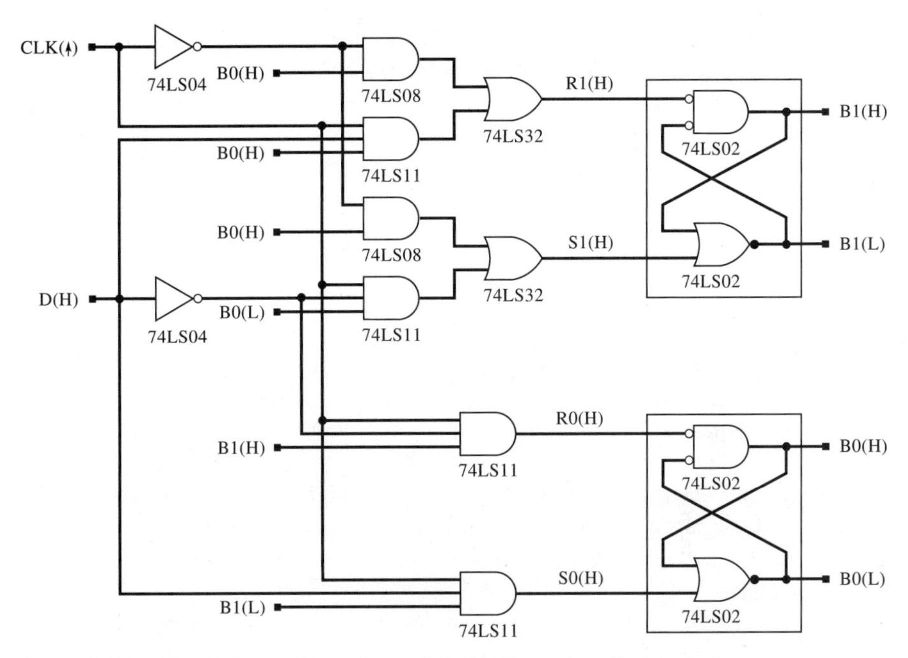

Figure 10.5-5 Circuit for positive triggered D flip-flop using discrete logic.

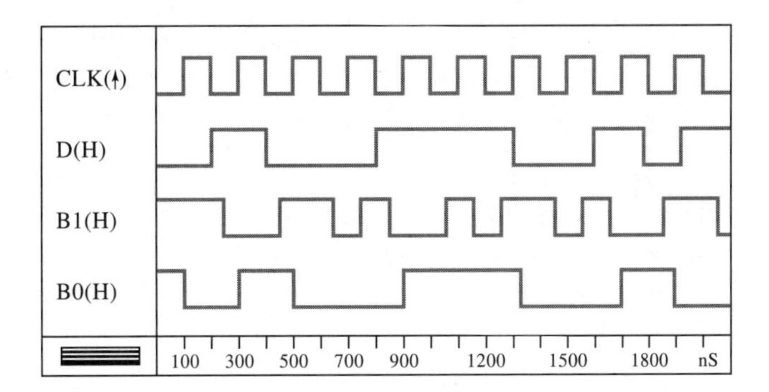

Figure 10.5-6 Simulation for edge triggered D flip-flop.

through the next state decoder, by using different kinds of gates. In any case, there will be a region where D is too close to CLK, and the machine will not function in this region. (The commercial D flip-flop has a setup time.)

Actually, we would have no reason to design a D flip-flop other than to demonstrate the technique used to design an asynchronous state machine, and if we did design and build a D flip-flop, we would use not discrete logic but rather some kind of custom logic.

The preceding case shows that synchronous and asynchronous state machines require the same design steps. Their difference lies in the restrictions on the state transitions and the use of asynchronous SET-RESET memory cells. We must choose the state code so that only one of its bits changes in a transition, and we must be careful to make the states in which we wish to stop stable, and the states through which we wish to move unstable.

We'll consider a final, more complicated example to clarify these ideas.

Example 10.5-1

Design an asynchronous state machine to find the pattern in the inputs I1, I2, and I3 in the following figure; when this pattern is found, assert an output SF (sequence found). All the input signals must start low. First, I1 must go high; second, I2 must go high; and, finally, I3 must go high. I1 must remain high until I2 and I3 go high, and I2 must remain high until I3 goes high. When the pattern is not found, a new pattern must start with the three input signals all low.

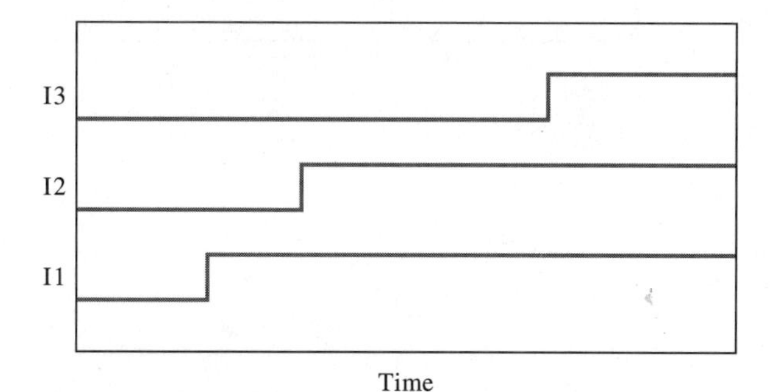

The difficult part of the asynchronous design is the state transition diagram. We must make a diagram in which only one state code bit changes in a given transition. We must also make sure the state the machine makes a transition into when all the inputs are low is stable; the state the machine makes a transition into when I1 goes high is stable; and so on.

Following is the complete state transition diagram when we assume I1, I2, and I3 are all asserted high. We start the diagram by considering the normal sequence we want to find. This normal sequence gives rise to the right side of the state transition diagram. The machine stays in state 000 as long as one or more of the inputs are high. As soon as all the inputs are low, the machine makes a transition to state 001, where it stays as long as all the inputs are low. We've made 001 a stable state. Now, if I1 goes high and I2 and I3 are both still low, the machine makes a transition to state 011, where it stays as long as I1 is high and I2 and I3 are low. We've made 011 a stable state. We complete the right side of the diagram by considering that I2 goes high and then that I3 goes high. This gives rise to the stable states 111 and 110. Notice that we have chosen the state codes so that only one bit changes for a given transition.

The output SF is asserted in state 110 because at that state the desired sequence has been found. Now we need to consider what happens if we fail to find the sequence. In state 001, if either I2 or I3 goes high, then we have a failure, and we need a transition back to state 000; this is a proper transition because only one bit of the state code changes. In state 011, if I1 goes low or I3 goes high, then we have a failure, and we need a transition back to state 000; this is an improper transition. To make proper transitions, we add an *unstable* state with state code 010 between states 011 and 000. The transition out of state 010 shows no condition because it should always occur. We move directly through state 010 with only enough time in the state to allow for the propagation delay through the next-state decoders. We finish the diagram by considering a failure to find the desired sequence in state 111 or a change out of state 110. With these transitions we again need unstable states to get proper transitions back to state 000.

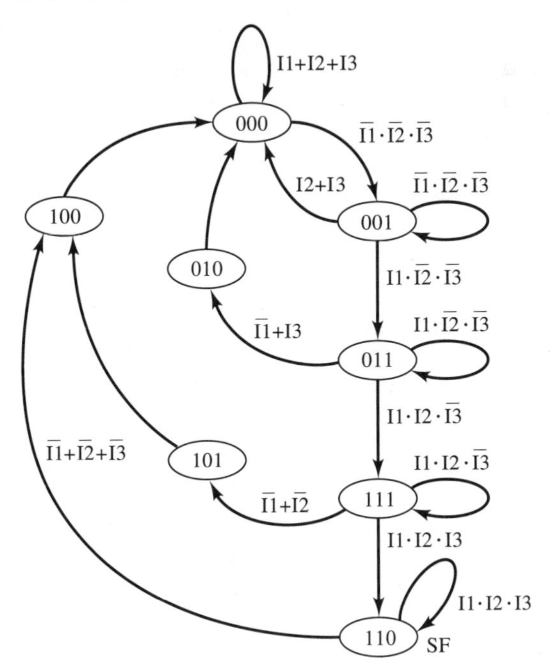

The diagram shown is the result of a lot of juggling of state codes and unstable states. It looks easy only after it's finished. We were able to make a correct state diagram using only a 3-bit state code. If we had encountered trouble attempting to make all the transitions proper and attempting to make the state machine function as it should, we would have tried a 4-bit or 5-bit code.

Next we need to make a present state–next state table and from it find the next state code. We use the table because the memory elements are SET-RESET cells that make it difficult to go directly to maps. For this machine there are three inputs and three state codes. If we don't simplify the table, it will be very large. We'll make a variable entered table as follows.

Present State			Next State			Next State Code	
B2	B1	B0	B2	B1	B0	S2	R2
0	0	0	0	0	$\overline{I1} \cdot \overline{I2} \cdot \overline{I3}$	0	x
0	0	1	0	$I1 \cdot \overline{I2} \cdot \overline{I3}$	$\overline{I2} \cdot \overline{I3}$	0	x
0	1	0	0	0	0	0	x
0	1	1	$I1 \cdot I2 \cdot \overline{I3}$	1	$I1 \cdot \overline{I3}$	$I1 \cdot I2 \cdot \overline{I3}$	$x\overline{I1} + x\overline{I2} + xI3$
1	0	0	0	0	0	0	1
1	0	1	1	0	0	x	0
1	1	0	1	$I1 \cdot I2 \cdot I3$	0	x	0
1	1	1	1	$I1 \cdot I2$	$\overline{I1} + \overline{I2} + \overline{I3}$	x	0

Present State			Next State Code			
B2	B1	B0	S1	R1	S0	R0
0	0	0	0	0	$\overline{I1} \cdot \overline{I2} \cdot \overline{I3}$	$xI1 + xI2 + xI3$
0	0	1	$I1 \cdot \overline{I2} \cdot \overline{I3}$	$x\overline{I1} + xI2 + xI3$	$x\overline{I2} \cdot \overline{I3}$	$I2 + I3$
0	1	0	0	1	0	x
0	1	1	x	0	$xI1 \cdot \overline{I3}$	$\overline{I1} + I3$
1	0	0	0	x	0	x
1	0	1	0	x	0	1
1	1	0	$xI1 \cdot I2 \cdot I3$	$\overline{I1} + \overline{I2} + \overline{I3}$	0	x
1	1	1	$xI1 \cdot I2$	$\overline{I1} + \overline{I2}$	$x\overline{I1} + x\overline{I2} + x\overline{I3}$	$I1 \cdot I2 \cdot I3$

Q_n	Q_{n+1}	SET	RESET
0	0	0	x
0	1	1	0
1	0	0	1
1	1	x	0

This table is similar to a variable entered map. The state variable code bits are shown explicitly, and the input variables are entered variables. For instance, in the fourth row, the entered condition $I1 \cdot I2 \cdot \overline{I3}$ in the B2 column of the next state means that B2 is 1 if the condition $I1 \cdot I2 \cdot \overline{I3}$ is true.

The next state code is found from the excitation table for the SET-RESET memory cell, which follows the next state code table. It is a little difficult to find this code because of the don't cares (x) in the excitation table and the entered variables. Keep in mind that the excitation table tells us that if we start with a state code bit that is 0 and we don't want to change it, we can reset or not as we wish, but we cannot set. Similarly, if we start with a bit that is 1 and we don't want to change it, we can set or not as we wish, but we cannot reset. On the other hand, if we want to change a bit, we must either SET and *not* RESET to change from 0 to 1, or RESET and *not* SET to change from 1 to 0.

SET and RESET have been shortened to S and R in the table, and the appropriate state bit number has been appended (for example, S2). The table is no problem as long as we do not have entered variables. For instance, in the first row the present state B2 is 0, and it does not change in the next state; therefore, S1 is 0, but R2 is don't care (x) because we can RESET or not as we wish. Now let's look at an instance with an entered condition. In the fourth row of the table, the present state B2 is 0, but it changes to 1 in the next state if the condition $I1 \cdot I2 \cdot \overline{I3}$ is true. This means that S2 must be 1 if $I1 \cdot I2 \cdot \overline{I3}$ is true ($S2 = I1 \cdot I2 \cdot \overline{I3}$), and R2 must be 0. That's not all; we must also handle the fact that B2 does not change if $I1 \cdot I2 \cdot \overline{I3}$ is not true. Because B2 does not change, R2 is don't care for $\overline{I1 \cdot I2 \cdot \overline{I3}}$ ($R2 = x\overline{I1} + x\overline{I2} + xI3$). The remainder of the entered conditions in the table are handled in the same way.

Once we have the table, we make a VEM for the next state logic and get simplified logic expressions. We again use don't cares without checking to see whether they cause simultaneous transitions in any of the SETs and RESETs for the memory cells. The simplified maps follow, along with the logic expressions.

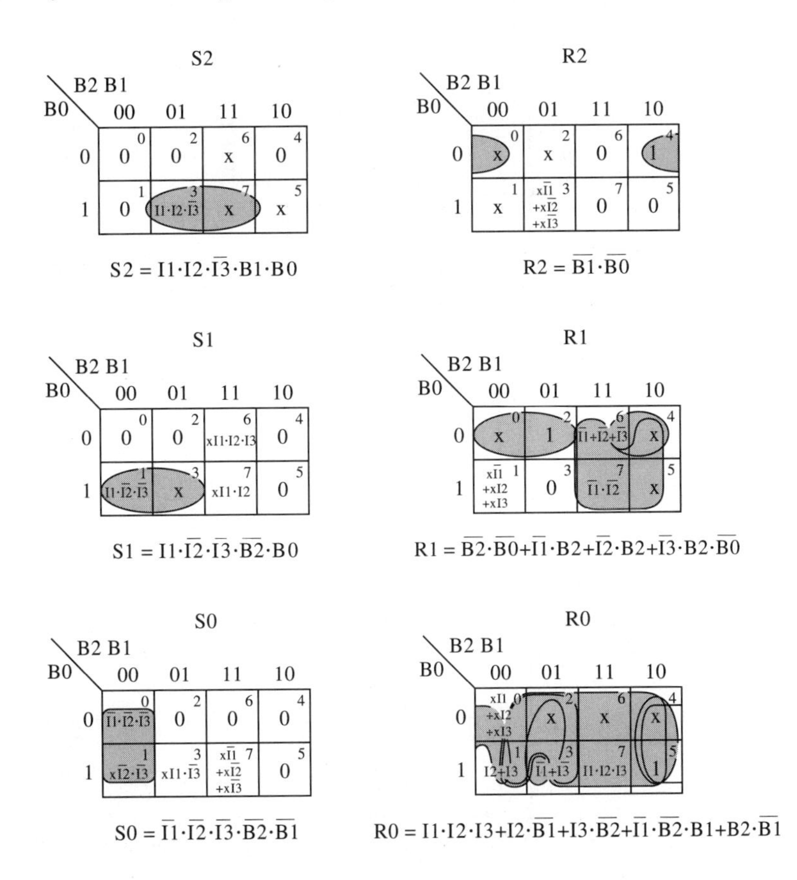

The output occurs in only one state (011), so the decoder needs to decode that state:

$$SF = B2 \cdot B1 \cdot \overline{B0}$$

The circuit that results from these logic expressions follows. It is rather complicated. Presently we'll see how to simplify it using a PAL for the next state and output decoder. The simulation for a test input follows the circuit.

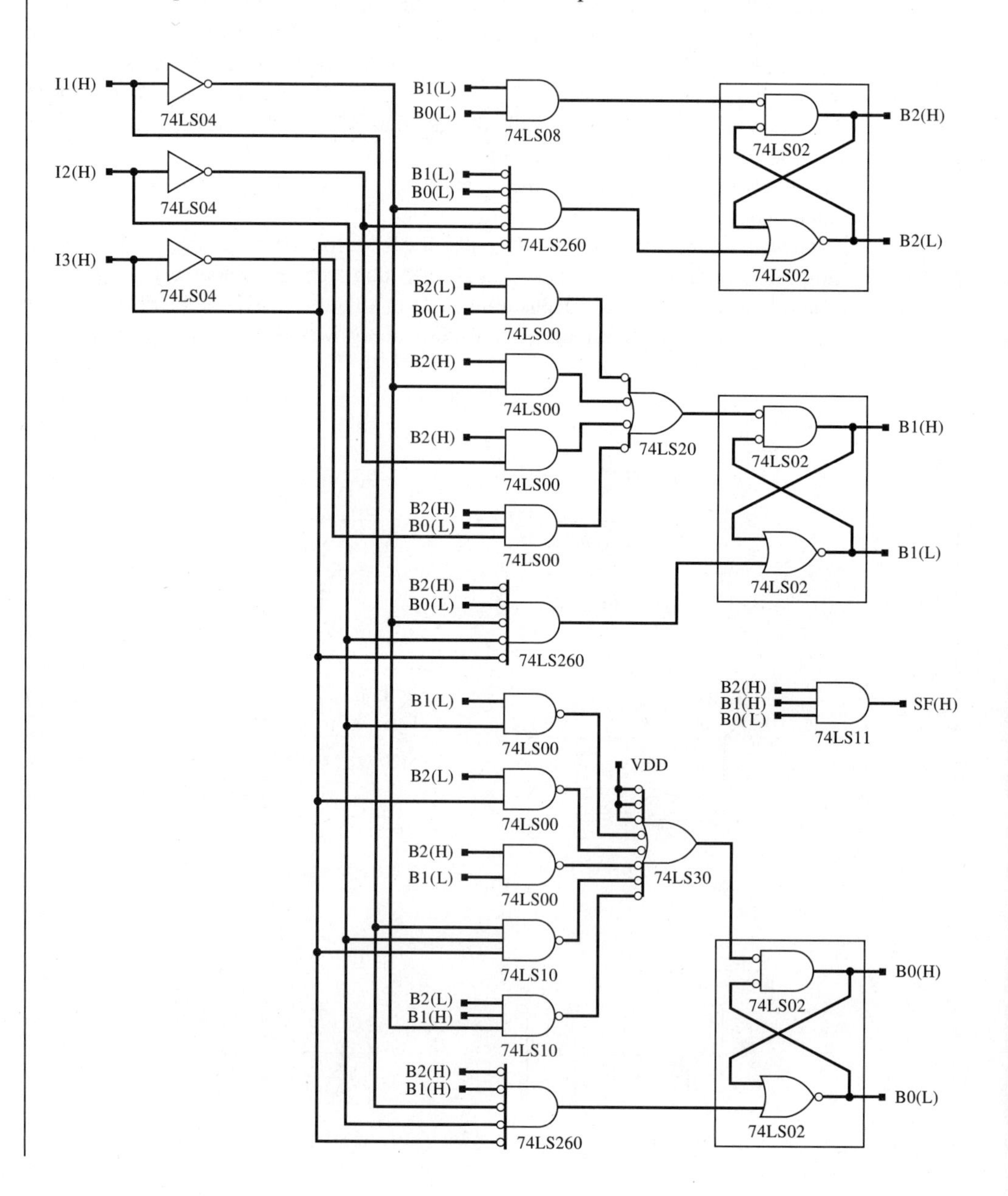

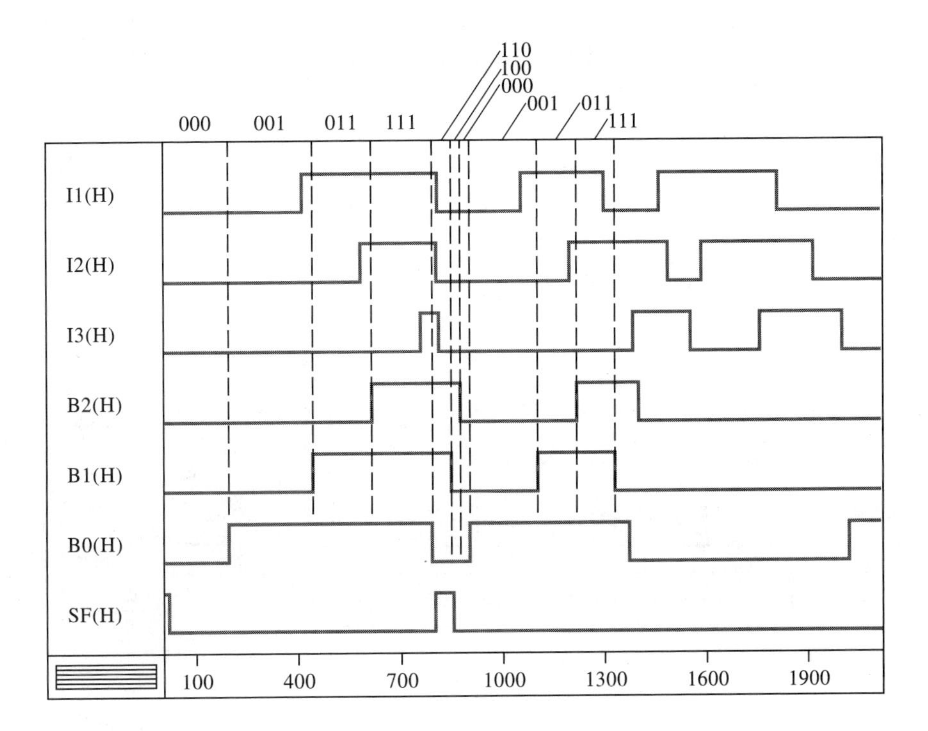

The circuit we've designed has no provision for RESET; we've made the simulator force the state to be 000 for the initial 200 ns of the simulation and then release it (the states appear at the top of the simulation results). Notice that the state changes to 001 as soon as it is released, because all the inputs are low. The state changes to 011 when I1 goes high, to 111 when I2 goes high, and to 011 when I3 goes high. In state 011 the output SF is asserted.

The machine does not stay in state 011 long, because I1 almost immediately goes low, which causes a transition to state 100. But 100 is unstable; the machine remains in this state only long enough for the next state to be decoded, then it makes a transition to 000. So far we've shown that the machine finds the desired transitions. Now we need to see whether or not it acts properly when it does not find them. We have not performed enough simulation to verify all the possible transitions. We have shown how it starts to find another sequence and then finds where the sequence is incorrect and returns to the 000 state.

The complicated decoders of this design are natural candidates for a PAL. If we use a PAL (such as the 22V10), the circuit becomes very simple, as the following figure shows, and all the complication goes into the PAL.

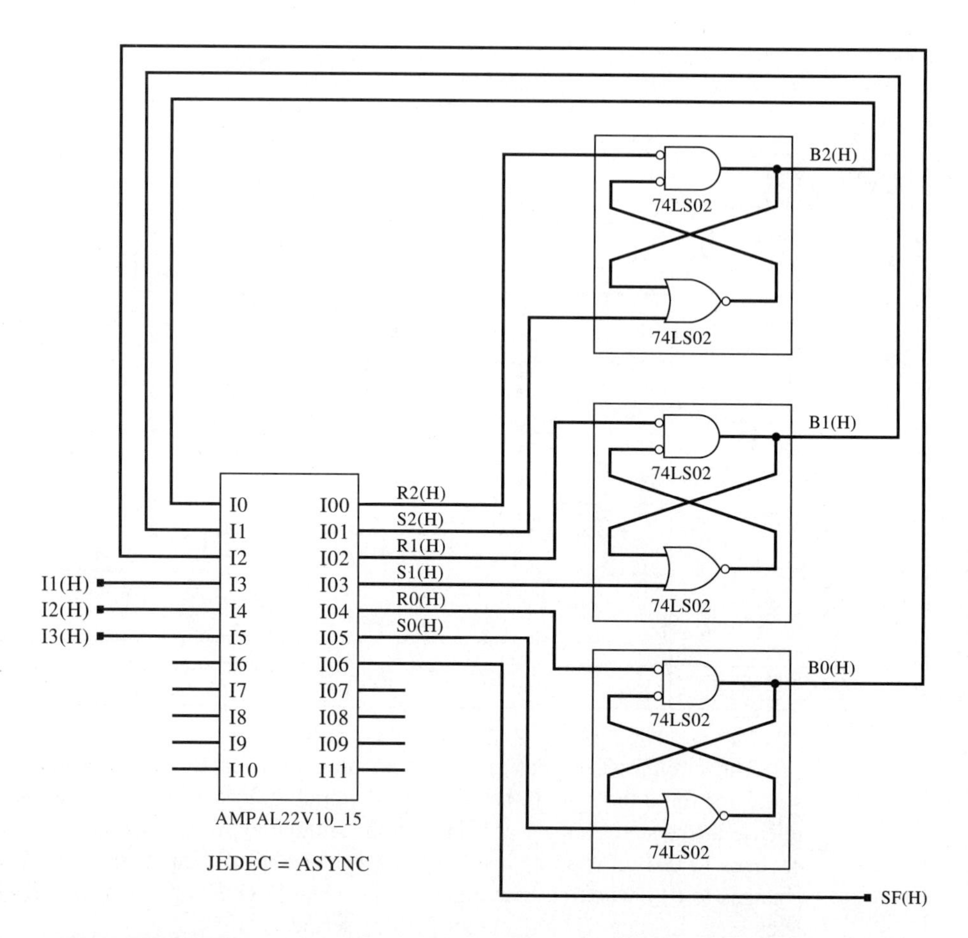

The AMPAL22V10_15 denotes a 15_ns version of the 22V10. The PAL program using PALASM follows the circuit. Simulation involves both discrete logic and a PAL, but good simulation software can simulate a PAL embedded in a circuit from the JEDEC file for the PAL. The simulation result is not shown, because it is just the same as the simulation for the discrete logic circuit (except for subtle timing differences).

Program

```
;PALASM Design Description

;---------------------------------------- Declaration Segment ----------------------------------------
TITLE      Asynchronous design
PATTERN
REVISION
AUTHOR     A W Shaw
COMPANY    Utah State University
DATE       07/11/92

CHIP   _async   PAL22V10
```

```
;------------------------------------------ PIN Declarations ------------------------------------------
PIN 1     B2                        ; INPUT
PIN 2     B1                        ; INPUT
PIN 3     B0                        ; INPUT
PIN 4     I1                        ; INPUT
PIN 5     I2                        ; INPUT
PIN 6     I3                        ; INPUT
PIN 7     NC                        ;
PIN 8     NC                        ;
PIN 9     NC                        ;
PIN 10    NC                        ;
PIN 11    NC                        ;
PIN 12    GND                       ;
PIN 13    NC                        ;
PIN 14    NC                        ;
PIN 15    NC                        ;
PIN 16    NC                        ;
PIN 17    SF     COMBINATORIAL ; OUTPUT
PIN 18    S0     COMBINATORIAL ; OUTPUT
PIN 19    R0     COMBINATORIAL ; OUTPUT
PIN 20    S1     COMBINATORIAL ; OUTPUT
PIN 21    R1     COMBINATORIAL ; OUTPUT
PIN 22    S2     COMBINATORIAL ; OUTPUT
PIN 23    R2     COMBINATORIAL ; OUTPUT
PIN 24    VCC                       ;
;---------------------------------- Boolean Equation Segment ----------------------------------
EQUATIONS

S2 = I1*I2*/I3*B1*B0
R2 = /B1*/B0
S1 = I1*/I2*/I3*/B2*B0
R1 = /B2*/B0 + /I1*B2 + /I2*B2 + /I3*B2*/B0
S0 = /I1*/I2*/I3*/B2*/B1
R0 = I1*I2*I3 + I2*/B1 + I3*/B2 + /I1*/B2*B1 + B2*/B1
SF = B2*B1*/B0
;-------------------------------------- Simulation Segment --------------------------------------
SIMULATION

;-----------------------------------------------------------------------------------------------------
```

The purpose of this short discussion of asynchronous state machines is to serve as a starting point for design. The design of any working machine will require a design on paper, using the approach outlined; extensive simulation, if accurate simulation software is available; and finally the building and testing of the circuit. The design procedure outlined does not consider all the subtleties of timing that are involved in the actual circuit, and even the best simulation software may not find all the timing problems that exist. All the timing issues can be resolved only by building a prototype of the design and testing it. In the next chapter we return to our first concern—synchronous circuits.

KEY TERMS

Flexible Architecture Programmable device architecture which allows both the type of elements and their interconnections to be programmed.
Fixed Architecture Programmable device architecture in which the types of element and their pattern of interconnection is fixed.
One Flip-Flop per State State machines in which each state is coded in a separate D flip-flop.
Register Device A device with flip-flops at each output.
Program For fixed architecture programmable devices, the code which describes the desired function to the device.
Declaration Segment The part of a programmable device program which documents the design and specifies the signal names for the device pins.
State Segment The part of a programmable device program which describes the state machine to be implemented in the device.
Simulation Segment The part of a programmable device program which describes the simulation to be preformed on the design which is created in the device.
Combinatorial A programmable device output which has no storage element.
Register A programmable device output which has a D flip-flop storage element.
Latched A programmable device output which has a D latch storage element.
PALASM A particular software for programming programmable logic devices.
Moore Machine A declaration in the program for a programmable logic device which specifies that the output is not dependent on the inputs.
Mealy Machine A declaration in the program for a programmable logic device which specifies that the output is not dependent on the inputs.
Fuse Map The output from a processed programmable logic device software which specifies the interconnection to be made in the device.
JEDEC File The output from programmable logic software which specifies to the programmer or burner how the device is to be programmed.
Memory Elements The elements in a state machine which store the state code bits.
Output Register The flip-flops associated with the output of a programmable logic device.
Decoder Logic Circuit The part of a state machine which determines the next state code from the present state and the inputs.
Output Decoder Circuit The part of a state machine which determines the output from the state and the inputs.
Disjoint States which are not next to each other in time.
Unstable State A state in an asynchronous state machine which is passed through quickly.

EXERCISES

1. Design a state machine, using a PAL16R8 or a PAL22V10 to find each of the following sequences in a serial input signal, SIG(H). The design should consist of a complete source code (including simulation input) in PALASM format (or other format if specified by the instructor). Each time the sequence is found, the output FND should be asserted. When a complete sequence is found, the machine should start looking for the sequence again without missing a bit. When the sequence isn't found, the machine should immediately start looking again. That is, if the sequence 001 is to be found and an improper sequence such as 1 or 01 is found, the machine should start looking for a new correct sequence on the next bit. Similarly, if a sequence such as 001 is to be found and a sequence of more than two 0s is found, the machine should consider that the first 1 after the

0s completes the desired sequence. Changes in SIG are synchronous with the negative-going edge of the clock, as shown in the following timing diagram.

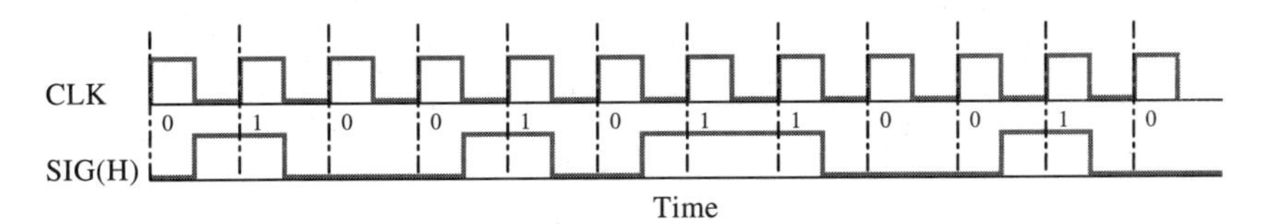

Make the designs Moore machine designs with register outputs. Make arbitrary state code assignments. Process the design, and simulate it if you have software available.

(a) 000
(b) 001
(c) 011
(d) 100
(e) 101
(f) 110
(g) 111
(h) 0000
(i) 0001
(j) 0010
(k) 0011
(l) 0100
(m) 0101
(n) 0110
(o) 0111
(p) 1000
(q) 1001
(r) 1010
(s) 1011
(t) 1100
(u) 1101

2. Make a Mealy machine design, using a PAL16R4 or PAL22V10 to find one of the sequences of exercise 2. The design should consist of a source code (including simulation input) in PALASM format (or other format if specified by the instructor). In some cases, the output will need to depend on the clock in order for the Mealy machine to function properly. You should be able to make a Mealy machine with fewer states than a Moore machine. Use arbitrary state code assignments. You will need to use combinational outputs. (*Hint:* It may be necessary to use the clock on some of the conditional outputs.) Process the design, and simulate it if you have software available.

3. Design a circuit, using a PAL22V10 to implement a state machine with one of the following state transition diagrams. The design should consist of a source code (including simulation input) in PALASM format (or other format if specified by the instructor). Make arbitrary state code assignments. Make the outputs register. Process the design, and simulate it if you have software available.

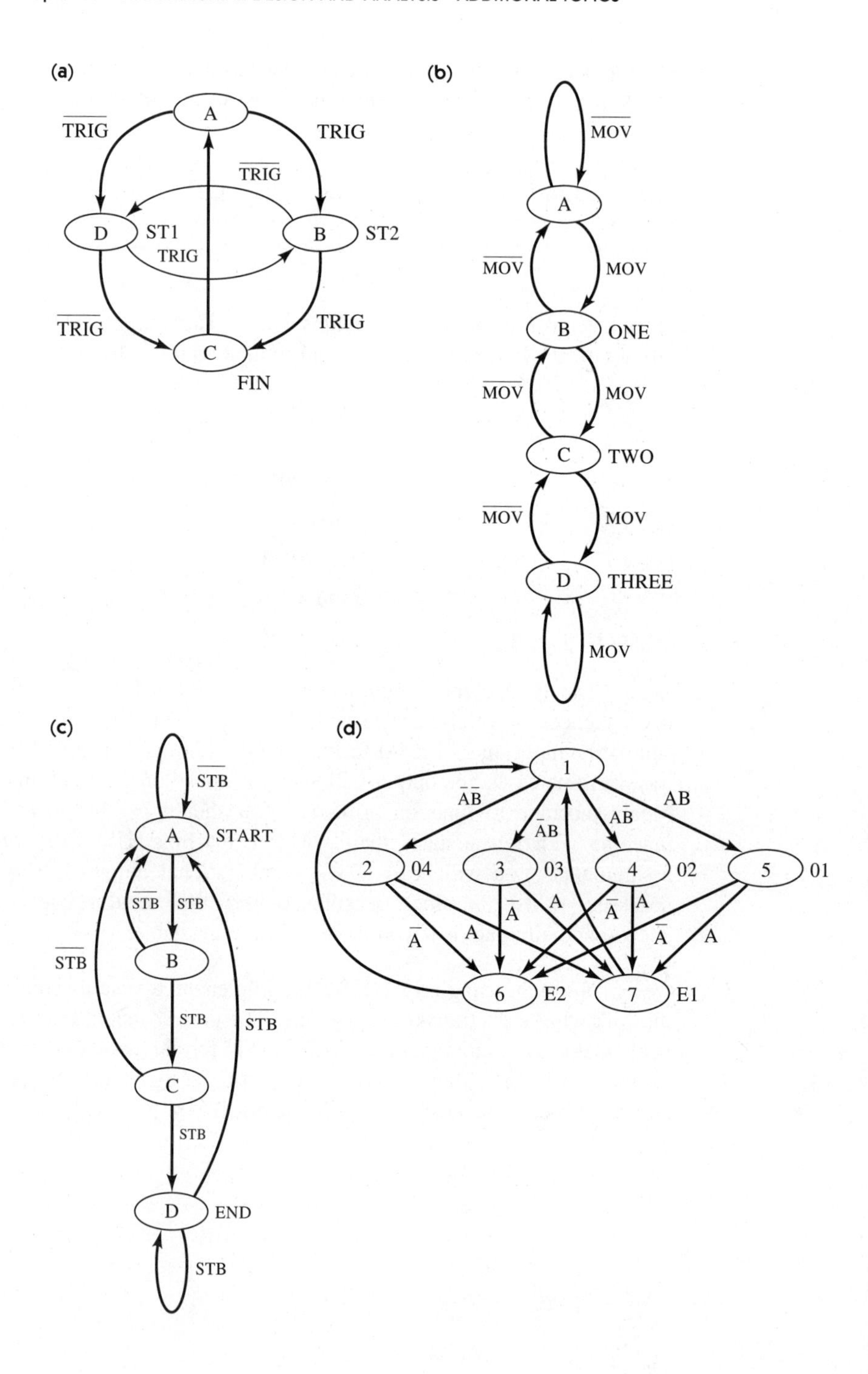

(a)
A
$\overline{\text{TRIG}}$
TRIG
$\overline{\text{TRIG}}$
D
ST1
TRIG
B
ST2
$\overline{\text{TRIG}}$
TRIG
C
FIN
(b)
$\overline{\text{MOV}}$
A
$\overline{\text{MOV}}$
MOV
B
ONE
$\overline{\text{MOV}}$
MOV
C
TWO
$\overline{\text{MOV}}$
MOV
D
THREE
MOV
(c)
$\overline{\text{STB}}$
A
START
$\overline{\text{STB}}$
STB
$\overline{\text{STB}}$
B
STB
$\overline{\text{STB}}$
C
STB
D
END
STB
(d)
1
$\bar{A}\bar{B}$
AB
$A\bar{B}$
$\bar{A}B$
2
04
3
03
4
02
5
01
$\bar{A}$
A
$\bar{A}$
A
$\bar{A}$
A
$\bar{A}$
A
6
E2
7
E1

4. Design a state machine that will sample an incoming signal, SIG(H), in 3-bit slices and determine whether the three bits are as specified in each of the following sequences. The design should consist of a source code (including simulation input) in PALASM format (or other format if specified by the instructor). Each time the bits are as specified, the machine should assert the output CNT for a full clock cycle. The machine should start sampling when an incoming signal, SYNC(L), is asserted. The machine should stop sampling when it has both completed a 3-bit slice and SYNC is no longer asserted. Both SIG and SYNC are synchronized to change on the negative-going clock edge. If you have trouble fitting the design onto a single PAL, choose the state code assignment to simplify the next state decoder. Process the design, and simulate it if you have software available.

   (a) 000
   (b) 001
   (c) 010
   (d) 011
   (e) 100
   (f) 101
   (g) 110
   (h) 111

5. Repeat exercise 4, but find the two sequences shown. Each time the first sequence is found, assert output CNT1. Each time the second is found, assert output CNT2. If you have trouble fitting the design onto a single PAL, choose the state code assignment to simplify the next state decoder. Process the design, and simulate it if you have software available.

   (a) 000 001
   (b) 000 010
   (c) 000 011
   (d) 000 100
   (e) 000 101
   (f) 000 110
   (g) 000 111
   (h) 001 010
   (i) 001 011
   (j) 001 100
   (k) 001 101
   (l) 001 110
   (m) 001 111
   (n) 010 011
   (o) 010 100
   (p) 010 101
   (q) 010 110
   (r) 010 111
   (s) 011 100
   (t) 011 101
   (u) 011 110

6. Design a state machine to find the first positive-going edge of an incoming signal SIG(H) that occurs after the machine is *armed.* The machine is to have three inputs: FDEDGE(H), a signal that arms the machine, that is, causes it to start looking for an edge; SIG, as specified above; and RESET(L), a signal that causes the machine to get ready to find another edge. The machine should assert the output signal FND(H) when the edge is found. FND should remain asserted until the input RESET is asserted. Be sure to consider both the case where SIG is high when FDEDGE is asserted and the case where SIG is low when FDEDGE is asserted. Use a PAL22V10. The design should consist of a source code (including simulation input) in PALASM format (or other format if specified by the instructor). Process the design, and simulate it if you have software available.

7. Design a state machine to determine whether an input signal, SIG(L), is asserted one, two, or three clock cycles after the input START(L) is asserted. Assert

output ONE, TWO, or THR depending on which condition is found. Repeat each time START is asserted. Assume that changes in SIG and START are synchronous with the negative-going edge of the clock. Use a one–flip-flop–per–state design with D memory and discrete logic elements. Make the outputs glitch-free with no delay. Use a PAL22V10. The design should consist of a source code (including simulation input) in PALASM format (or other format if specified by the instructor). Process the design, and simulate it if you have software available.

8. Design a state machine to control processes that require each of the following sets of outputs with the timings shown. In each case, choose a suitable clock frequency but make it as low as possible. The output should be glitch-free. Use a PAL22V10. The design should consist of a source code (including simulation input) in PALASM format (or other format if specified by the instructor). Process the design, and simulate it if you have software available.

   (a) (Start the sequence when RESET is asserted.)
   START—for 10 ms
   (Wait for 5 ms.)
   STOP—for 1 ms
   (Wait for 5 ms.)
   INC—until GO is asserted
   (Repeat when RESET is asserted.)

   (b) (Start the sequence when START is asserted.)
   FWD—for 6 ms
   (Wait for 1 ms.)
   RVS—until STP is asserted
   (Wait for 2 ms.)
   INC—until NEW is asserted
   (Repeat sequence until START is deasserted.)

   (c) (Start the sequence when NCT is asserted.)
   DOWN—until BTM is asserted
   (Wait for 5 ms.)
   STOP—for 10 ms
   (Wait for 5 ms.)
   UP—for 5 ms
   (Repeat if NCT is still asserted, or wait until NCT is asserted and repeat.)

   (d) (Start the sequence when RESET is asserted.)
   START—for 6 ms
   Starting at the same time as START, LEVEL—for 3 ms
   (Wait for 3 ms after the end of START.)
   STOP—for 3 ms
   (Wait for 3 ms.)
   STOP—for 1 ms

(Wait for 3 ms.)
(Repeat the sequence endlessly.)

(e) (Start the sequence when GO is asserted.)
START—until STOP is asserted
Starting at the same time as START, NEWCT—for 2 ms
FNL—for 1 ms after STOP is asserted
(Wait for 4 ms.)
NEW—for 4 ms
(Repeat until GO is deasserted.)

9. Analyze the following state machines. That is, make a present state–next state and output table for each.

(a)

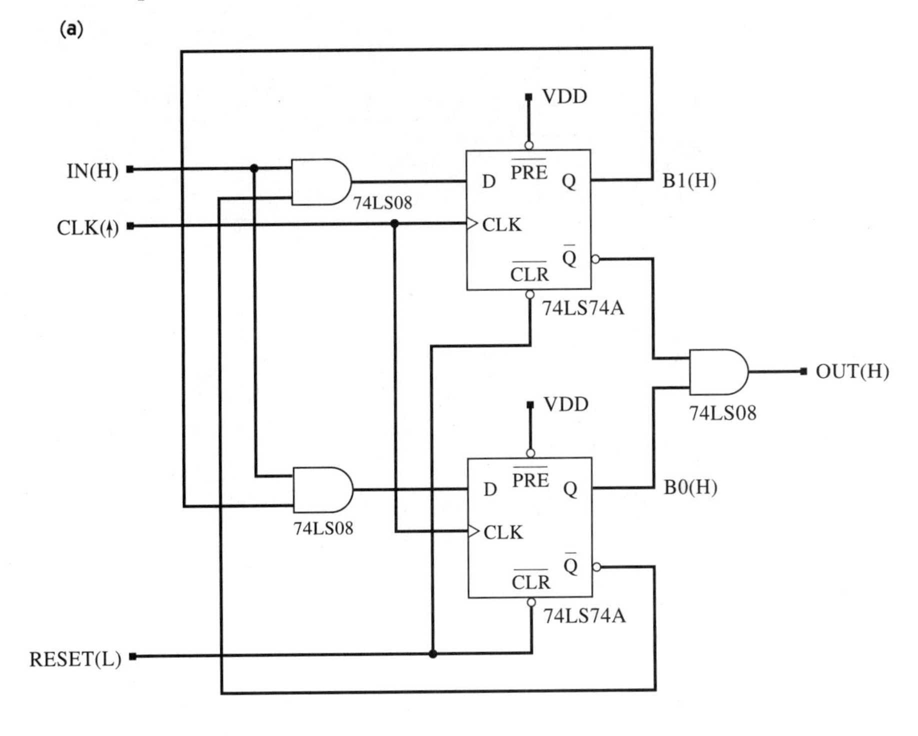

(b)

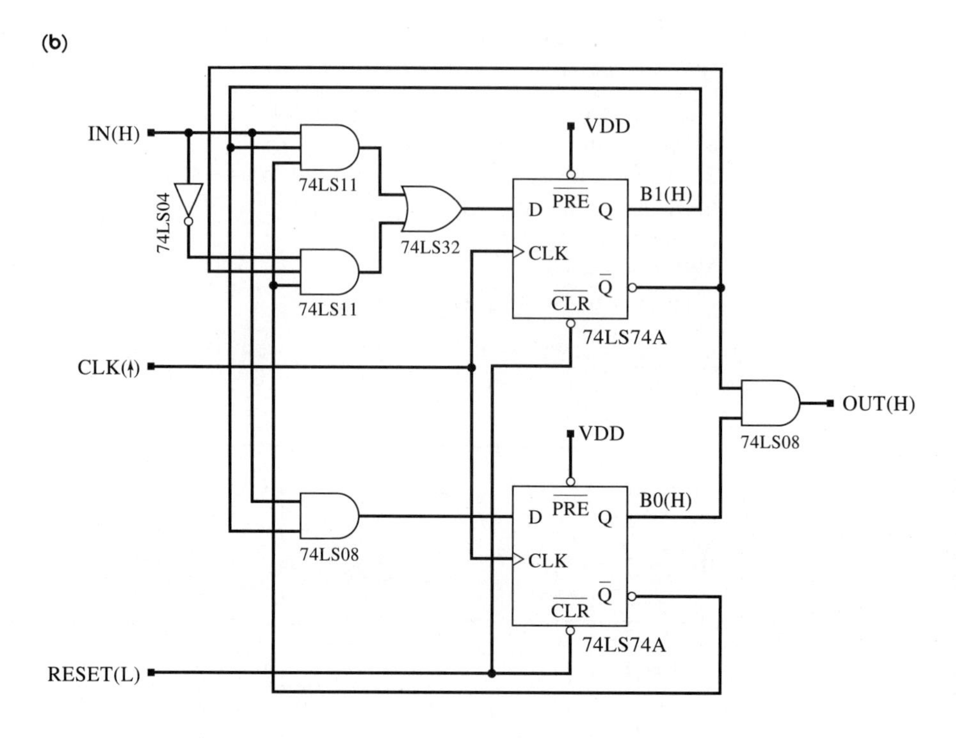
IN(H)
74LS04
74LS11
74LS11
74LS32
VDD
D
PRE
Q
B1(H)
CLK
Q
CLR
74LS74A
CLK(↑)
OUT(H)
74LS08
VDD
B0(H)
D
PRE
Q
74LS08
CLK
Q
CLR
74LS74A
RESET(L)

(c)

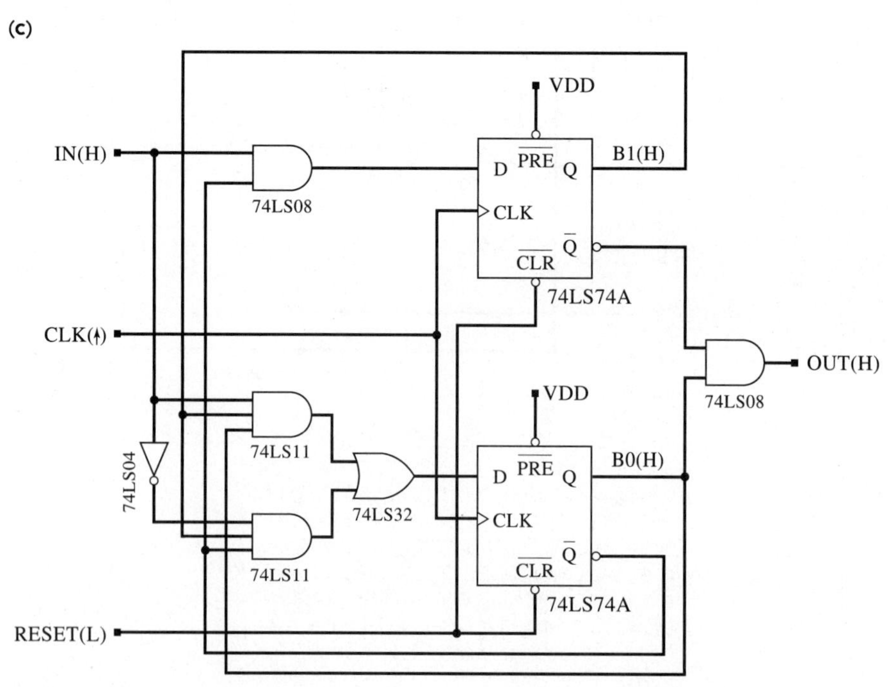
VDD
IN(H)
D
PRE
Q
B1(H)
74LS08
CLK
Q
CLR
74LS74A
CLK(↑)
OUT(H)
74LS08
VDD
74LS11
74LS04
B0(H)
D
PRE
Q
74LS32
CLK
74LS11
Q
CLR
74LS74A
RESET(L)

(d)

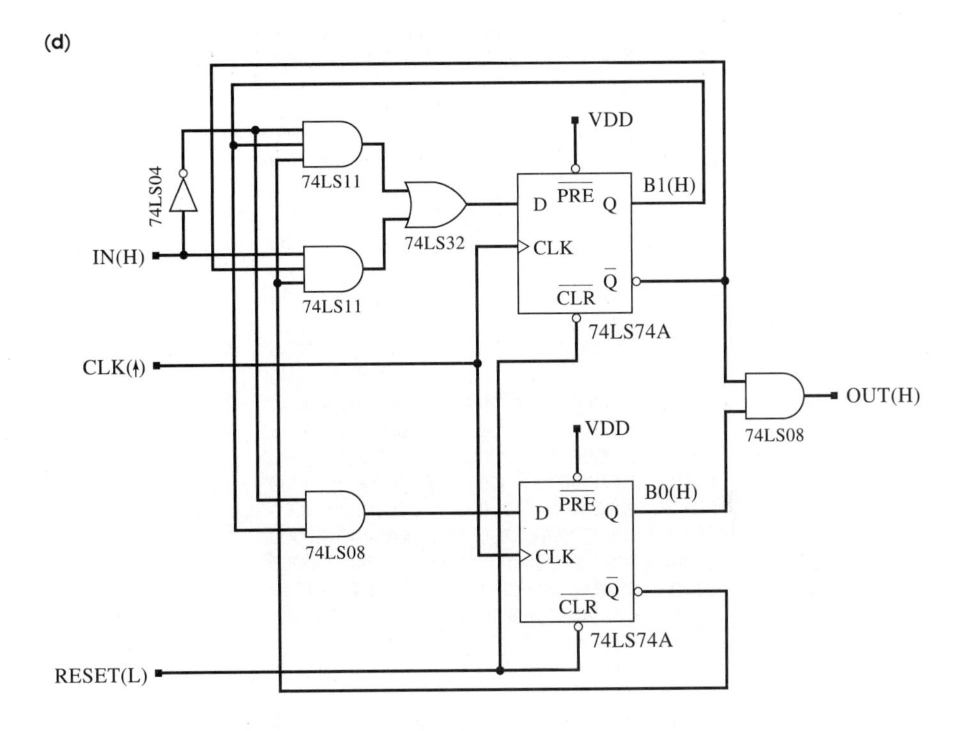

10. Analyze the following state machines, which were designed using a PAL16R4 from the JEDEC files given.

(a)

```
PALASM4  PAL Assembler—Market Version 1.2 (5-31-91)
  (c)—Copyright Advanced Micro Devices Inc., 1991

TITLE      :                          AUTHOR  :A W Shaw
PATTERN :                             COMPANY:Utah State University
REVISION:                             DATE      :07/14/92

866
PAL16R4
QP20*
QF2048*
G0*F0*
L0000  00000000000000000000000000000000*
L0032  00000000000000000000000000000000*
L0064  00000000000000000000000000000000*
L0096  00000000000000000000000000000000*
L0128  00000000000000000000000000000000*
L0160  00000000000000000000000000000000*
L0192  00000000000000000000000000000000*
L0224  00000000000000000000000000000000*
L0256  00000000000000000000000000000000*
L0288  00000000000000000000000000000000*
L0320  00000000000000000000000000000000*
```

```
L0352 00000000000000000000000000000000*
L0384 00000000000000000000000000000000*
L0416 00000000000000000000000000000000*
L0448 00000000000000000000000000000000*
L0480 00000000000000000000000000000000*
L0512 00000000000000000000000000000000*
L0544 00000000000000000000000000000000*
L0576 00000000000000000000000000000000*
L0608 00000000000000000000000000000000*
L0640 00000000000000000000000000000000*
L0672 00000000000000000000000000000000*
L0704 00000000000000000000000000000000*
L0736 00000000000000000000000000000000*
L0768 00000000000000000000000000000000*
L0800 00000000000000000000000000000000*
L0832 00000000000000000000000000000000*
L0864 00000000000000000000000000000000*
L0896 00000000000000000000000000000000*
L0928 00000000000000000000000000000000*
L0960 00000000000000000000000000000000*
L0992 00000000000000000000000000000000*
L1024 01111111111111111111111011111111*
L1056 00000000000000000000000000000000*
L1088 00000000000000000000000000000000*
L1120 00000000000000000000000000000000*
L1152 00000000000000000000000000000000*
L1184 00000000000000000000000000000000*
L1216 00000000000000000000000000000000*
L1248 00000000000000000000000000000000*
L1280 01111111111111111101111111111111*
L1312 00000000000000000000000000000000*
L1344 00000000000000000000000000000000*
L1376 00000000000000000000000000000000*
L1408 00000000000000000000000000000000*
L1440 00000000000000000000000000000000*
L1472 00000000000000000000000000000000*
L1504 00000000000000000000000000000000*
L1536 00000000000000000000000000000000*
L1568 00000000000000000000000000000000*
L1600 00000000000000000000000000000000*
L1632 00000000000000000000000000000000*
L1664 00000000000000000000000000000000*
L1696 00000000000000000000000000000000*
L1728 00000000000000000000000000000000*
L1760 00000000000000000000000000000000*
L1792 11111111111111111111111111111111*
L1824 11111111111111111110110111111111*
L1856 00000000000000000000000000000000*
L1888 00000000000000000000000000000000*
L1920 00000000000000000000000000000000*
L1952 00000000000000000000000000000000*
L1984 00000000000000000000000000000000*
L2016 00000000000000000000000000000000*
C0F22*
^CE8CA
```

(b)

```
PALASM4     PAL Assembler—Market Version 1.2 (5-31-91)
  (c)—Copyright Advanced Micro Devices Inc., 1991

TITLE      :                    AUTHOR   :A W Shaw
PATTERN :                       COMPANY:Utah State University
REVISION:                       DATE     :07/14/92

869
PAL16R4
QP20*
QF2048*
G0*F0*
L0000   00000000000000000000000000000000*
L0032   00000000000000000000000000000000*
L0064   00000000000000000000000000000000*
L0096   00000000000000000000000000000000*
L0128   00000000000000000000000000000000*
L0160   00000000000000000000000000000000*
L0192   00000000000000000000000000000000*
L0224   00000000000000000000000000000000*
L0256   00000000000000000000000000000000*
L0288   00000000000000000000000000000000*
L0320   00000000000000000000000000000000*
L0352   00000000000000000000000000000000*
L0384   00000000000000000000000000000000*
L0416   00000000000000000000000000000000*
L0448   00000000000000000000000000000000*
L0480   00000000000000000000000000000000*
L0512   00000000000000000000000000000000*
L0544   00000000000000000000000000000000*
L0576   00000000000000000000000000000000*
L0608   00000000000000000000000000000000*
L0640   00000000000000000000000000000000*
L0672   00000000000000000000000000000000*
L0704   00000000000000000000000000000000*
L0736   00000000000000000000000000000000*
L0768   00000000000000000000000000000000*
L0800   00000000000000000000000000000000*
L0832   00000000000000000000000000000000*
L0864   00000000000000000000000000000000*
L0896   00000000000000000000000000000000*
L0928   00000000000000000000000000000000*
L0960   00000000000000000000000000000000*
L0992   00000000000000000000000000000000*
L1024   10111111111111111101111011111111*
L1056   01111111111111111101111011111111*
L1088   00000000000000000000000000000000*
L1120   00000000000000000000000000000000*
L1152   00000000000000000000000000000000*
L1184   00000000000000000000000000000000*
L1216   00000000000000000000000000000000*
L1248   00000000000000000000000000000000*
L1280   10111111111111111101111111111111*
L1312   00000000000000000000000000000000*
```

```
L1344 00000000000000000000000000000000*
L1376 00000000000000000000000000000000*
L1408 00000000000000000000000000000000*
L1440 00000000000000000000000000000000*
L1472 00000000000000000000000000000000*
L1504 00000000000000000000000000000000*
L1536 00000000000000000000000000000000*
L1568 00000000000000000000000000000000*
L1600 00000000000000000000000000000000*
L1632 00000000000000000000000000000000*
L1664 00000000000000000000000000000000*
L1696 00000000000000000000000000000000*
L1728 00000000000000000000000000000000*
L1760 00000000000000000000000000000000*
L1792 11111111111111111111111111111111*
L1824 11111111111111111110110111111111*
L1856 00000000000000000000000000000000*
L1888 00000000000000000000000000000000*
L1920 00000000000000000000000000000000*
L1952 00000000000000000000000000000000*
L1984 00000000000000000000000000000000*
L2016 00000000000000000000000000000000*
C128F*
^CE8EB
```

(c)

```
PALASM4     PAL Assembler—Market Version 1.2 (5-31-91)
  (c)—Copyright Advanced Micro Devices Inc., 1991

TITLE    :                    AUTHOR  :A W Shaw
PATTERN  :                    COMPANY :Utah State University
REVISION :                    DATE    :07/14/92

871
PAL16R4
QP20*
QF2048*
G0*F0*
L0000 00000000000000000000000000000000*
L0032 00000000000000000000000000000000*
L0064 00000000000000000000000000000000*
L0096 00000000000000000000000000000000*
L0128 00000000000000000000000000000000*
L0160 00000000000000000000000000000000*
L0192 00000000000000000000000000000000*
L0224 00000000000000000000000000000000*
L0256 00000000000000000000000000000000*
L0288 00000000000000000000000000000000*
L0320 00000000000000000000000000000000*
L0352 00000000000000000000000000000000*
L0384 00000000000000000000000000000000*
L0416 00000000000000000000000000000000*
L0448 00000000000000000000000000000000*
L0480 00000000000000000000000000000000*
L0512 00000000000000000000000000000000*
L0544 00000000000000000000000000000000*
L0576 00000000000000000000000000000000*
```

```
L0608  00000000000000000000000000000000*
L0640  00000000000000000000000000000000*
L0672  00000000000000000000000000000000*
L0704  00000000000000000000000000000000*
L0736  00000000000000000000000000000000*
L0768  00000000000000000000000000000000*
L0800  00000000000000000000000000000000*
L0832  00000000000000000000000000000000*
L0864  00000000000000000000000000000000*
L0896  00000000000000000000000000000000*
L0928  00000000000000000000000000000000*
L0960  00000000000000000000000000000000*
L0992  00000000000000000000000000000000*
L1024  01111111111111111101111011111111*
L1056  10111111111111111101111011111111*
L1088  00000000000000000000000000000000*
L1120  00000000000000000000000000000000*
L1152  00000000000000000000000000000000*
L1184  00000000000000000000000000000000*
L1216  00000000000000000000000000000000*
L1248  00000000000000000000000000000000*
L1280  01111111111111111101111111111111*
L1312  00000000000000000000000000000000*
L1344  00000000000000000000000000000000*
L1376  00000000000000000000000000000000*
L1408  00000000000000000000000000000000*
L1440  00000000000000000000000000000000*
L1472  00000000000000000000000000000000*
L1504  00000000000000000000000000000000*
L1536  00000000000000000000000000000000*
L1568  00000000000000000000000000000000*
L1600  00000000000000000000000000000000*
L1632  00000000000000000000000000000000*
L1664  00000000000000000000000000000000*
L1696  00000000000000000000000000000000*
L1728  00000000000000000000000000000000*
L1760  00000000000000000000000000000000*
L1792  11111111111111111111111111111111*
L1824  11111111111111111110110111111111*
L1856  00000000000000000000000000000000*
L1888  00000000000000000000000000000000*
L1920  00000000000000000000000000000000*
L1952  00000000000000000000000000000000*
L1984  00000000000000000000000000000000*
L2016  00000000000000000000000000000000*
C1290*
^CE8F6
```

(d)

```
PALASM4     PAL Assembler—Market Version 1.2 (5-31-91)
  (c)—Copyright Advanced Micro Devices Inc., 1991

TITLE      :Seq                  AUTHOR   :A W Shaw
PATTERN :                        COMPANY:Utah State University
REVISION:                        DATE     :07/14/92

874
PAL16R4
QP20*
```

```
QF2048*
G0*F0*
L0000 00000000000000000000000000000000*
L0032 00000000000000000000000000000000*
L0064 00000000000000000000000000000000*
L0096 00000000000000000000000000000000*
L0128 00000000000000000000000000000000*
L0160 00000000000000000000000000000000*
L0192 00000000000000000000000000000000*
L0224 00000000000000000000000000000000*
L0256 00000000000000000000000000000000*
L0288 00000000000000000000000000000000*
L0320 00000000000000000000000000000000*
L0352 00000000000000000000000000000000*
L0384 00000000000000000000000000000000*
L0416 00000000000000000000000000000000*
L0448 00000000000000000000000000000000*
L0480 00000000000000000000000000000000*
L0512 00000000000000000000000000000000*
L0544 00000000000000000000000000000000*
L0576 00000000000000000000000000000000*
L0608 00000000000000000000000000000000*
L0640 00000000000000000000000000000000*
L0672 00000000000000000000000000000000*
L0704 00000000000000000000000000000000*
L0736 00000000000000000000000000000000*
L0768 00000000000000000000000000000000*
L0800 00000000000000000000000000000000*
L0832 00000000000000000000000000000000*
L0864 00000000000000000000000000000000*
L0896 00000000000000000000000000000000*
L0928 00000000000000000000000000000000*
L0960 00000000000000000000000000000000*
L0992 00000000000000000000000000000000*
L1024 01111111111111111111110111111111*
L1056 00000000000000000000000000000000*
L1088 00000000000000000000000000000000*
L1120 00000000000000000000000000000000*
L1152 00000000000000000000000000000000*
L1184 00000000000000000000000000000000*
L1216 00000000000000000000000000000000*
L1248 00000000000000000000000000000000*
L1280 10111111111111111101110111111111*
L1312 01111111111111111101110111111111*
L1344 00000000000000000000000000000000*
L1376 00000000000000000000000000000000*
L1408 00000000000000000000000000000000*
L1440 00000000000000000000000000000000*
L1472 00000000000000000000000000000000*
L1504 00000000000000000000000000000000*
L1536 00000000000000000000000000000000*
L1568 00000000000000000000000000000000*
L1600 00000000000000000000000000000000*
L1632 00000000000000000000000000000000*
```

```
L1664 00000000000000000000000000000000*
L1696 00000000000000000000000000000000*
L1728 00000000000000000000000000000000*
L1760 00000000000000000000000000000000*
L1792 11111111111111111111111111111111*
L1824 11111111111111111110110111111111*
L1856 00000000000000000000000000000000*
L1888 00000000000000000000000000000000*
L1920 00000000000000000000000000000000*
L1952 00000000000000000000000000000000*
L1984 00000000000000000000000000000000*
L2016 00000000000000000000000000000000*
C1258*
^CE8DB
```

11. Repeat exercise 8 but design one–flip-flop–per–state machines. Remember that this type of machine must have a provision for periodically resetting all the flip-flops but one. Use discrete logic. Make the outputs register outputs.

12. Repeat exercise 10 for one–flip-flop–per–state machines using the gate array modules discussed in the text.

13. Design a positive-edge triggered JK flip-flop, using the state transition diagram of Figure 8.6-2.

14. Design a positive-edge triggered T flip-flop, using an asynchronous 4-state state machine.

Examples of Sequential Logic Circuits

11.1 INTRODUCTION

Chapter 5 described useful combinational logic circuits that are available as discrete logic devices. This chapter will do the same for sequential logic circuits, which include **counters**, **registers**, and **memory**. We'll also show how to design very large state machines using counters and read-only memory.

The sequential circuits we're describing are also available as predesigned subcircuits in FPGA designs. We'll provide some examples of these circuits during our descriptions.

11.2 COUNTERS

Counters are state machines that clock through the same set of states endlessly. Figure 11.2-1 shows the state transition diagram for (a) the simplest possible counter, with only a clock input, and (b) a more versatile *multimode* counter that can count either up or down depending on the input M. Both are 3-bit binary counters; they count through the binary numbers 000, 001, 010, and so on.

Counters are classified in three ways: as simple counters or multimode counters; as ripple or synchronous counters; and by the **modulus** of the counter—that is, by the number of states they count through.

The counter with the state transition diagram of Figure 11.2-1(a) is a simple, 3-bit binary counter, and the counter with the state transition diagram of Figure 11.2-1(b) is a multimode, 3-bit binary counter. The fact that they are 3-bit binary counters implies that the counter modulus is $2^3 = 8$. Counters are usually either

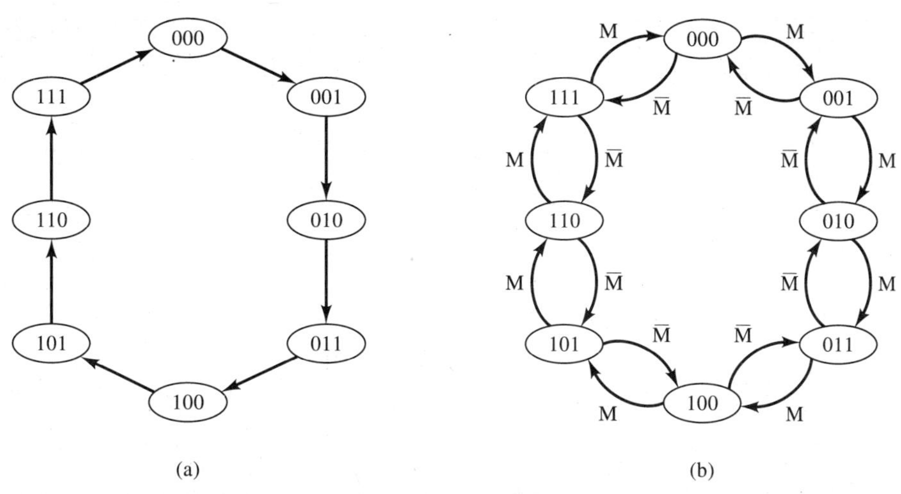

Figure 11.2-1 State transition diagrams for (a) a simple 3-bit binary counter and (b) a 3-bit multimode (up/down) binary counter.

binary counters with a modulus that is some power of 2, or decade counters with a modulus of 10 or 100.

There are two counters with special counting sequences, called the ring counter and the twisted ring counter. A 4-bit ring counter counts through the sequence 0000, 0001, 0010, 0100, 1000. A 3-bit twisted ring counter counts through the sequence 000, 001, 011, 111, 110, 100 or 000, 100, 110, 111, 011, 001. Notice that the output of a twisted ring counter is a unit distance code. The outputs are logic-adjacent. The extension of these counter types to different numbers of bits should be obvious.

Standard counters are available as either ripple (asynchronous) or synchronous counters. Ripple counters are not state machines, although they use sequential logic. Section 11.2.1 will describe some of the ripple counters that are available as discrete devices. Synchronous counters are state machines. We could design synchronous counters of any type we desire, using the state machine design techniques developed in previous chapters. In Section 11.2.2 we'll design a simple synchronous counter and then discuss some of the counters that are available as discrete devices. We'll also give examples of predesigned counters that are available for use in FPGA designs.

11.2.1 Ripple Counters *Ripple counters* consist of cascade JK master-slave flip-flops, as shown in Figure 11.2.1-2 (74LS93). JK master-slave flip-flops differ from edge triggered JK flip-flops in that they are two-stage devices. The first stage, called the master section, is loaded depending on the JK inputs when the clock is high. This type of flip-flop requires that the inputs not change while the clock is high. When the clock makes the high-to-low transition, the state of the master section is transferred to the second stage, the slave section, and is output on Q and $\overline{Q}$.

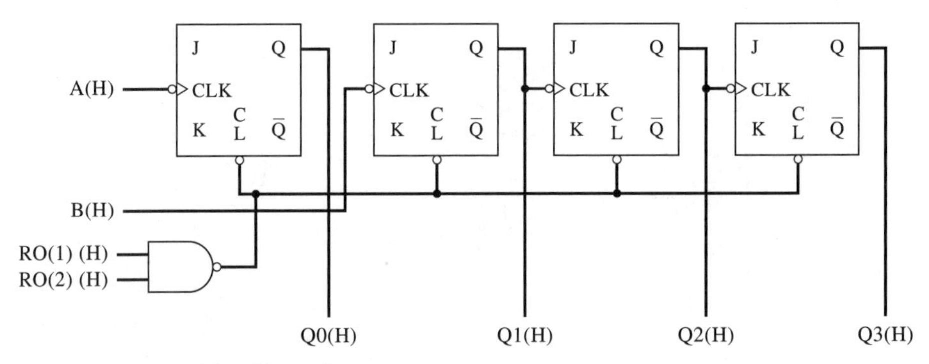

Figure 11.2.1-2 Four-bit ripple counter.

The timing diagram in Figure 11.2.1-3 shows the function of the Q1, Q2, Q3 part of the ripple counter. All the J and K inputs in the counter are tied high (here, an absence of a connection in the circuit diagram means tied high). If both J and K are tied high, these negative clocked JK flip-flops toggle on each high-to-low clock transition. This means that each time the clock (B) makes a high-to-low transition, Q1 toggles, so Q1 toggles half as fast as the clock (see the timing diagram). Now, because Q1 is the clock for Q2, Q2 toggles on each high-to-low transition of Q1 so that Q2 toggles half as fast as Q1. Finally, Q3 toggles on each high-to-low transition of Q2 so that Q3 toggles half as fast as Q2. The counter is called a ripple counter because the count *ripples through* the JK flip-flops.

The counter in Figure 11.2.1-2 is a 4-bit binary ripple counter with clear (R0). It actually consists of two independent counters—a 1-bit counter and a 3-bit counter. (We just analyzed the action of the 3-bit counter.) It's common to describe the 1-bit counter as a divide-by-2 counter, since the output frequency is one-half the clock frequency, and the 3-bit counter as a divide-by-8 counter since the output of the high order bit is one-eighth the clock frequency. If the output of the 1-bit counter is connected to the clock of the 3-bit counter, then the device functions as a 4-bit counter, but it can also be used as two separate counters. Notice that it has two clears (or resets), called R0(1) and R0(2), both of which must be asserted to reset the counter to 0.

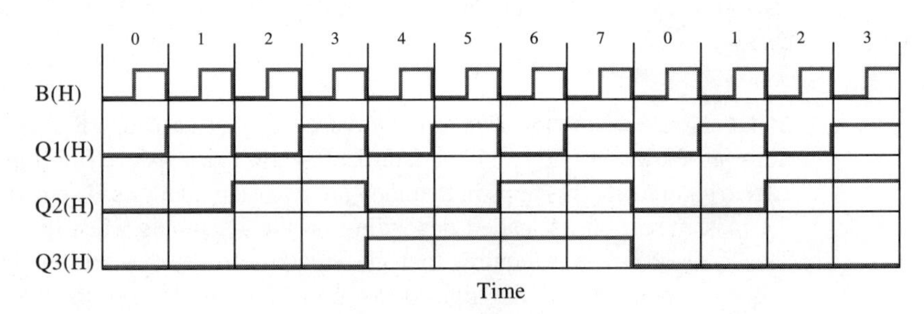

Figure 11.2.1-3 Timing diagram for 3-bit ripple counter.

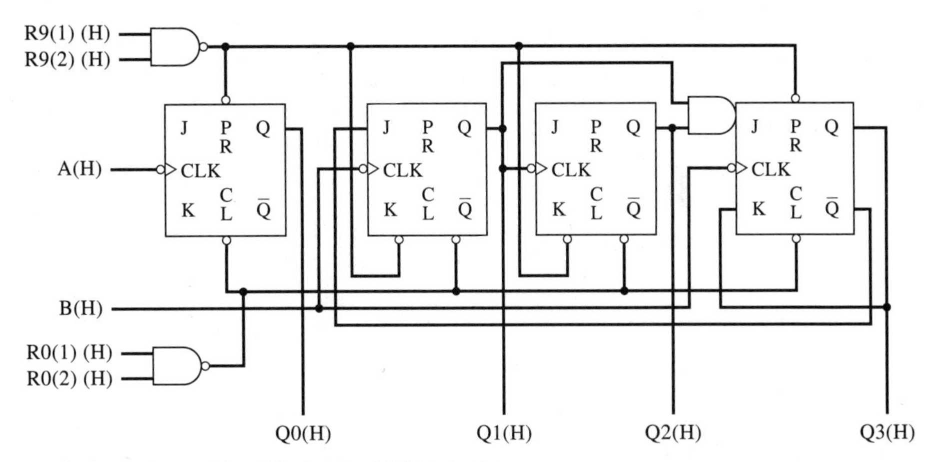

Figure 11.2.1-4 Four-bit decade (BCD) ripple counter.

The 3-bit section of the 4-bit binary ripple counter is a modulus 8 counter. It can count from 0 to 7. To create a decade counter, we need to make the 3-bit section a modulus 5 counter so that it can be cascaded with the 1-bit counter to form a modulus 10 counter. A modulus 5 counter counts from 0 to 4 and then starts over. Figure 11.2.1-4 shows a 4-bit decade (BCD) ripple counter (74LS90) with both clear (R0), that initializes the counter to 0, and preset (R9), which initializes the counter to 9.

Figure 11.2.1-5 is a timing diagram for the Q1, Q2, Q3 part of this counter. It does not have all the JK inputs high. J1 is tied to $\overline{Q3}$, and the Q3 flip-flop has inputs that depend on Q1, Q2, and Q3. To see how this counter works, notice that J1 = $\overline{Q3}$ is high during the 0, 1, 2, and 3 counts, so Q1 simply toggles on each high-to-low clock transition. Because Q1 is the clock for Q2, Q2 functions as a simple ripple counter and toggles on each high-to-low transition of Q1. During the 0, 1, 2, and 3 counts, J3 is low, and because the Q3 flip-flop is in the 0 state, Q3 doesn't change. When the count reaches 3, Q1 and Q2 are both high, so J3 is high and Q3 changes state (Q3 goes high) on the next positive-to-negative clock edge. (Notice that Q3 is

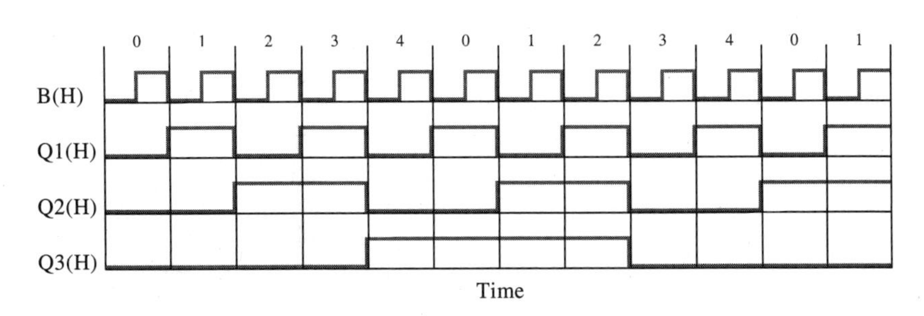

Figure 11.2.1-5 Timing diagram for 3-bit modulus 5 counter.

clocked by the clock, not Q2.) Because both Q1 and Q2 toggle to low on this same clock transition, the count is 4, as it should be. When the count is 4, J1 = $\overline{\text{Q3}}$ is low, and because Q1 is in the 0 state, it doesn't change. Because Q1 doesn't change, neither does Q2. Finally, K3 = Q3 is high and Q3 is in the 1 state, so Q3 toggles low on the next high-to-low clock transition. Because both Q1 and Q2 stay low and Q3 changes to low on the next high-to-low clock transition after the 4 count, the next count after 4 is 0, as desired.

A ripple counter isn't exactly a synchronous machine, but Q1 and Q3 function as though they're part of a synchronous state machine, and their input decoders could actually be designed using the ideas we've developed for such machines.

To make the 4-bit ripple counter function as a decade counter (counting from 0 to 9), the Q0 output must be connected to the clock (B) of the last three stages. The two sections can also be used as two independent counters except for the common clear and preset. Notice that there are two clear inputs, M0(1) and M0(2), and both must be asserted to clear the counter. There are also two preset inputs, M9(1) and M9(2), and both must be asserted to preset the counter (to 9).

Ripple counters have a serious defect: there is a time delay between the outputs due to the delay through the flip-flops. This means that any decoded output will have serious glitches and should not be used for functions where glitches are a problem, such as clocking and counting. In addition to the 4-bit binary (divide-by-16) and decade (divide-by-10) ripple counters already described, divide-by-12 counters (74LS92) are also standard.

11.2.2 Synchronous Counters *Synchronous counters* are normal synchronous state machines. They use edge triggered D or JK memory, and they can be designed and analyzed using the techniques we've developed for state machines. Standard synchronous counters have a number of convenient features in addition to counting. They have provisions for clearing, loading, and in some cases up/down counting. The load feature allows a number to be initially loaded into the counter. The clear feature allows the counter to be set to 0. The up/down feature allows the counter to count in either direction. These features make the counter circuits rather complicated.

Let's first see how we would design a synchronous 4-bit (modulus 16) counter. We'll use flip-flops with asynchronous resets so that we can add an asynchronous reset to the counter. A synchronous counter counts its clock. The counter is *cascadable* if it can be turned off and on (enabled) and can produce an output when it reaches its terminal count (1111). We'll design the counter to count when an input called ENBL (enable) is asserted, and stop counting when ENBL is not asserted. We'll further design it to produce an output, TC, that is asserted when a count of 1111 is reached. We can then cascade two 4-bit counters by connecting the TC of the lower order counter to the ENBL of the high order counter to make an 8-bit counter. The TC output will be asserted when the low order counter reaches a count of 1111, and it will enable the high order counter to count once each time the low order counter reaches 1111 and starts over at 0000.

Figure 11.2.2-1 is the state transition diagram for a 4-bit counter with ENBL and TC. Figure 11.2.2-2 shows the VEMs for the next-state decoder and the resulting

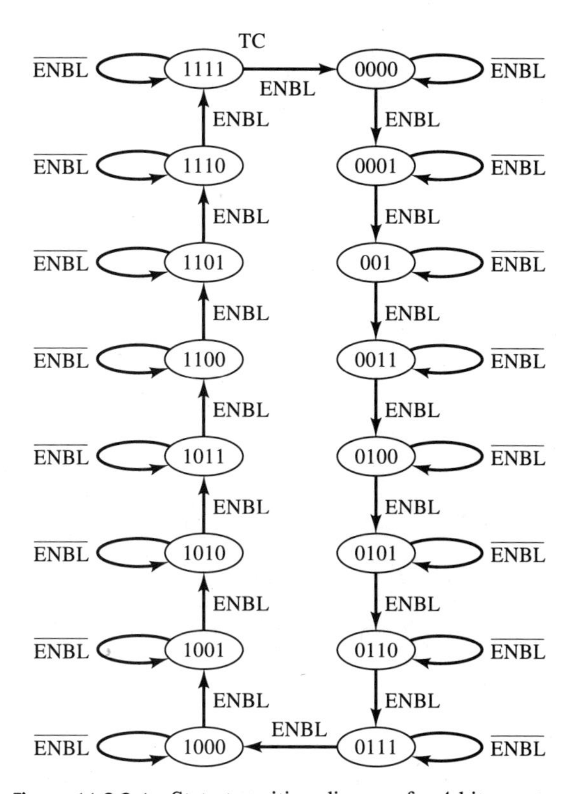

Figure 11.2.2-1 State transition diagram for 4-bit counter.

simplified next-state decoder logic expressions. We've labeled the state code bits Q3, Q2, Q1, and Q0 because they are the output count and it is customary to label the bits of a counter output using Q. The output count obviously doesn't need a decoder. The output TC does need a decoder, but the logic expression is easy because TC is asserted only in the 1111 state.

$$\text{TC} = \text{Q3} \cdot \text{Q2} \cdot \text{Q1} \cdot \text{Q0}$$

Figure 11.2.2-3 is the circuit for the 4-bit counter, and Figure 11.2.2-4 shows the results of a simulation. The first thing we do in the simulation is to reset to 0000. Notice that the counter then counts the clock as long as ENBL is asserted. The output TC is asserted in the 1111 state, but it has a number of glitches in other states, a condition we might expect for a combinational output. Fortunately these glitches do not occur at the positive-going clock edge, so they're not important. The purpose of TC is to enable a cascaded counter to count once when the terminal count is reached, and the cascaded counter only changes state at the positive-going clock edge.

We've chosen to design a 4-bit (modulus 16) counter with enable and terminal count. We could design a counter with any modulus and any controls, such as an enable or even no control. We could also design a counter in a PAL, and sometimes for large counters this is a good idea.

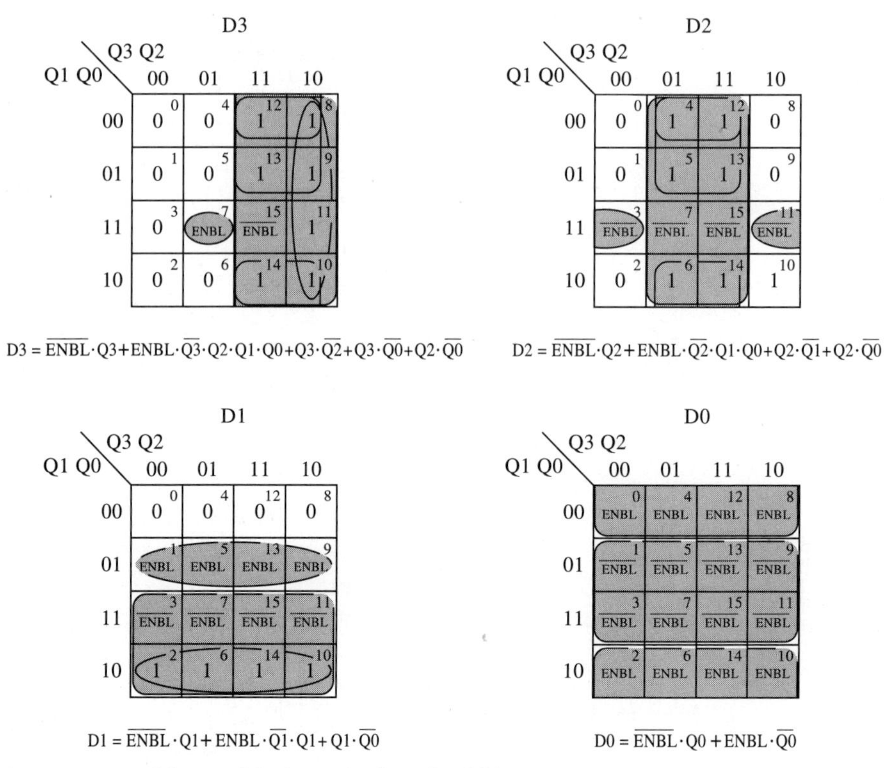

Figure 11.2.2-2 Maps and logic expressions for 4-bit counter.

The counter we have designed functions about the same as some of the counters available as discrete devices, but it requires a large parts count and lacks the features of these discrete devices. One of the simplest synchronous counters is a 4-bit binary counter with load and clear (74LS161). It is similar to the counter we've designed, but with additional inputs to give it more functions. The circuit for this counter is shown in Figure 11.2.2-5. The counter has nine inputs—CLK, ENP, ENT, LOAD, CLR, D3, D2, D1, and D0. In the count mode, the counter counts the low-to-high transitions of the clock, CLK. The two enables, ENP and ENT, must be asserted in order for the counter to count. The LOAD input is asserted to load a binary number into the counter. LOAD is a synchronous function. When it is asserted, the input on D3, D2, D1, and D0 is loaded into the counter on the positive-going clock edge. LOAD has to be deasserted in order for the counter to count. The clear input, CLR, sets the counter to zero. The counter has five outputs, Q3, Q2, Q1, Q0, and RCO. Four of these outputs, Q3, Q2, Q1, and Q0, indicate the count. RCO is a carry output (like TC in the counter we designed) that is asserted when the count rolls over from 15 to 0. The rollover output is useful if counters are to be cascaded.

We can see the function of the counter most easily in a timing diagram. Figure 11.2.2-6 shows a timing diagram for this counter—showing first a load and then several counts, including a rollover and an inhibit.

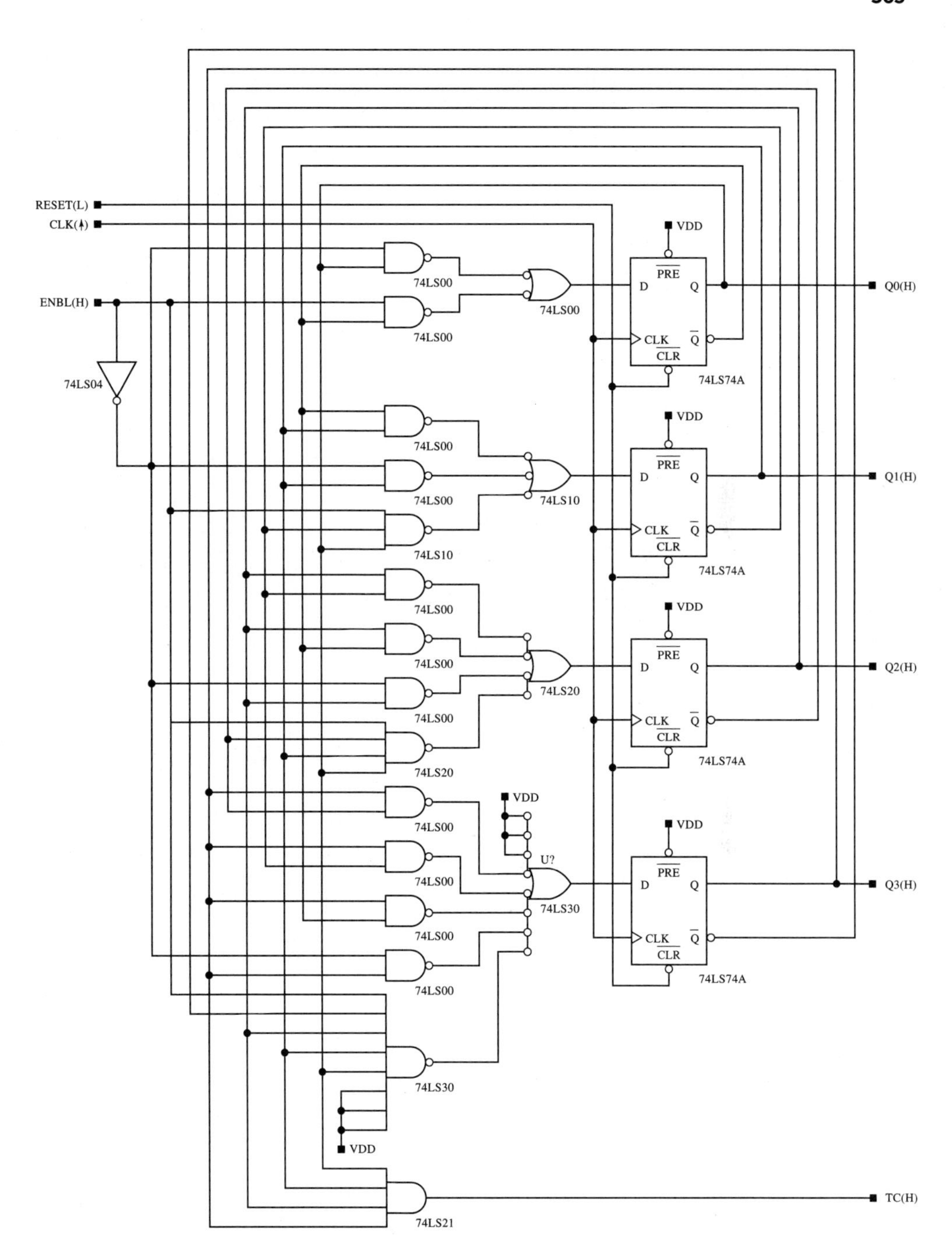

Figure 11.2.2-3 Circuit for 4-bit counter.

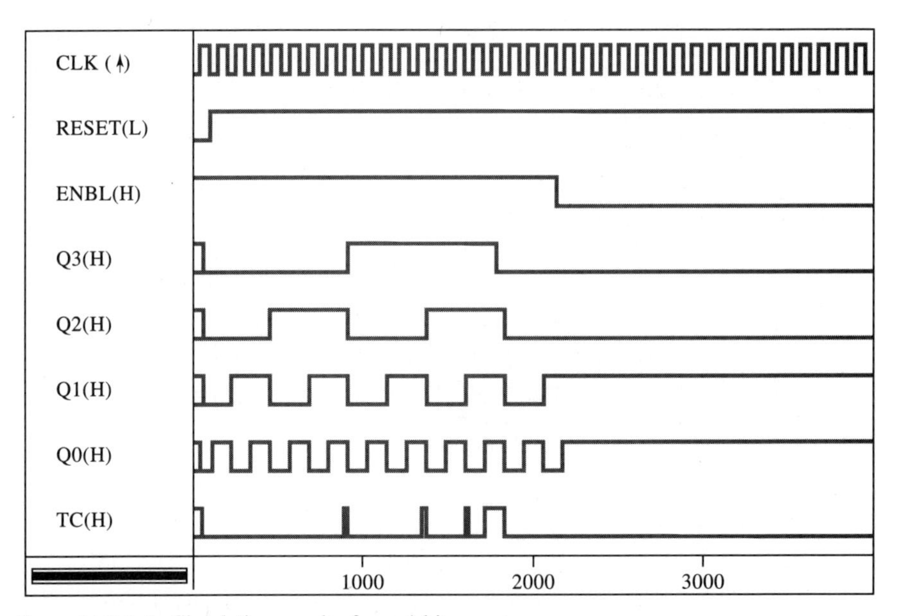

Figure 11.2.2-4 Simulation results for a 4-bit counter.

We can analyze this counter most easily by considering the functions one at a time. First, notice that CLR is a simple asynchronous clear that acts directly on the asynchronous CLR inputs of the D flip-flops. Next, notice that LOAD forces a low on each of the left two *and* gates of each of the next-state decoders. The outputs of these gates are thus low, and the output of the *xnor* gate is high. This means that whatever signal is placed on the D input (D3, D2, D1, or D0) will be fed directly through the *nand* gate and the inverter to the D input of the flip-flop. Thus, the D flip-flop will be loaded with this signal on the next high-to-low clock transition.

Now consider the counting function. For the counter to function in the counting mode, both the enables (ENP and ENT) must be asserted, LOAD must not be asserted, and CLR must not be asserted. When both ENP and ENT are asserted and LOAD is not asserted (all high), the inputs from the enable-load circuit (across the top of the diagram) to each of the two left *and* gates of each next-state decoder will be high. This means we can ignore these inputs to the first two gates when we analyze the count function. Also, when LOAD is not asserted (high), a low is fed to the *nand* gate in each next-state decoder. This low at the input of the *nand* gate produces a high at the output so that the output of the *xnor* gate is fed directly through the inverter to the D input of the flip-flop. This means that the inverter can be combined with the *xnor* gate to make an *xor* gate. The next-state decoder for each of the D flip-flops in the counting mode thus reduces to two *and* gates and an *xor* gate.

The circuit for the counting mode (with the simplification just described) is shown in Figure 11.2.2-7. In the simplified circuit, all the inputs to the next-state decoder *and* gates that are high in the counting mode have been left off. The gate with no inputs has an output that is always high. The gate with one input has an

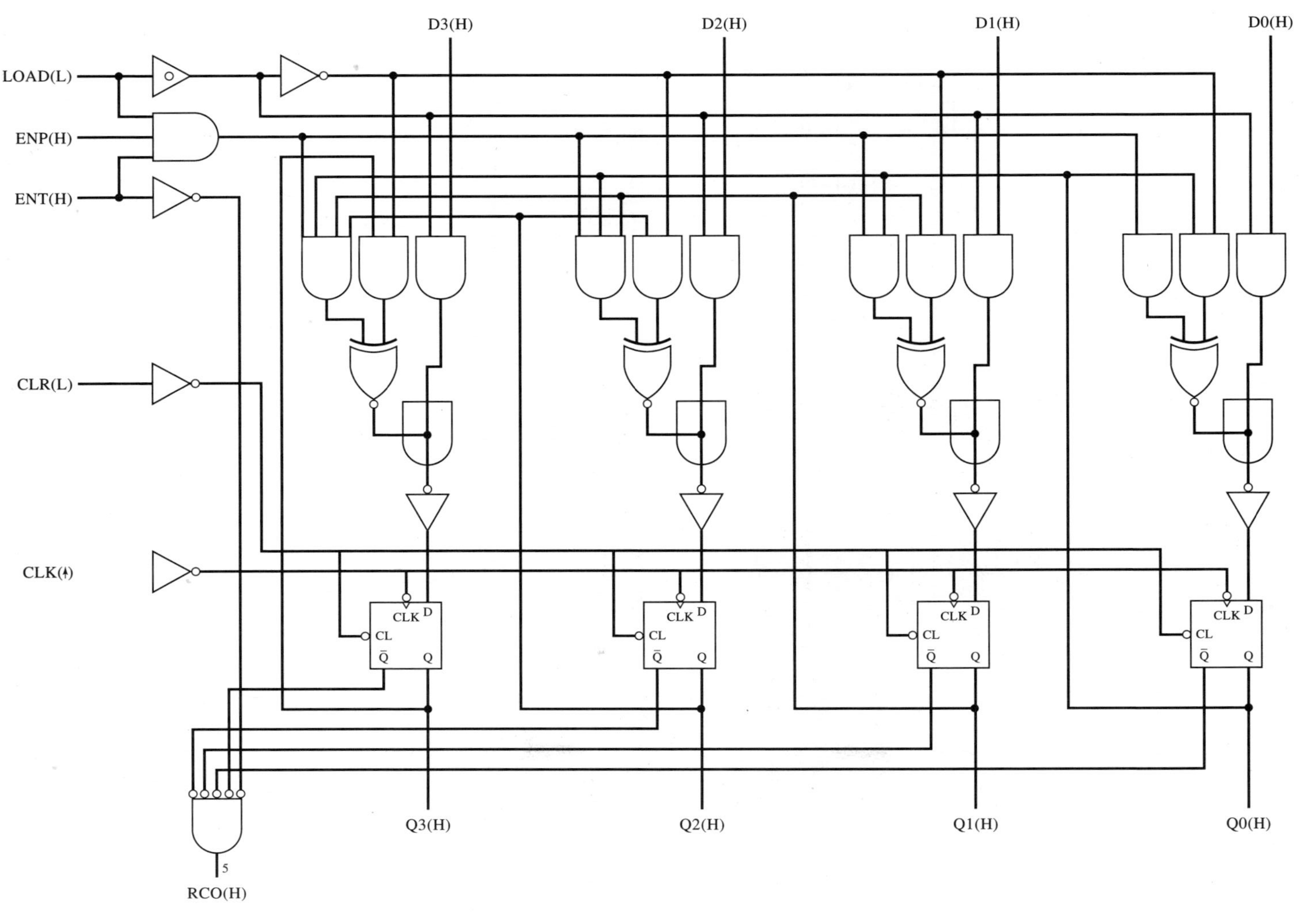

Figure 11.2.2-5 Four-bit synchronous counter.

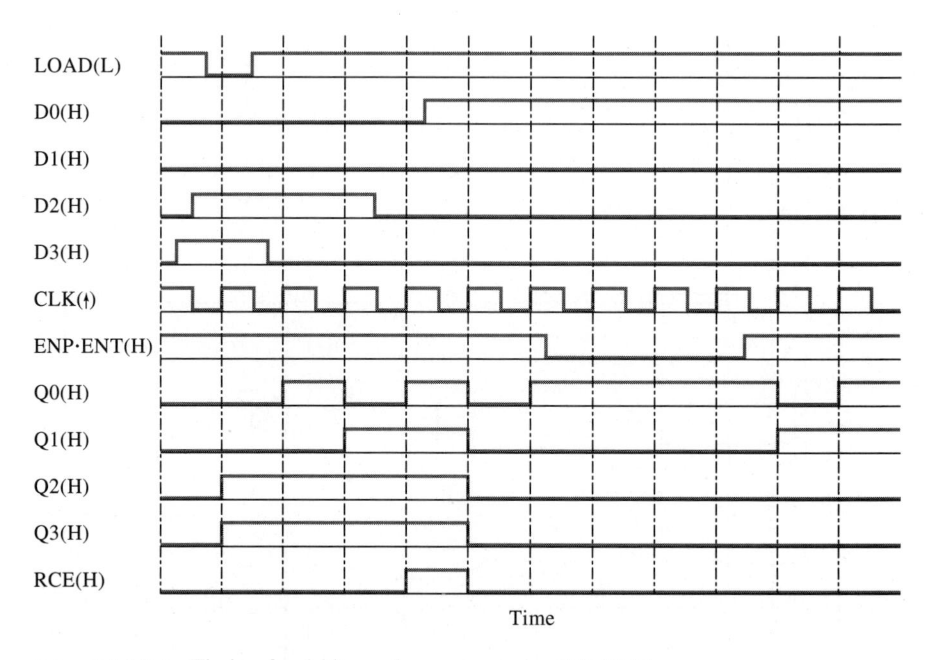

Figure 11.2.2-6 Timing for 4-bit synchronous counter (74LS161).

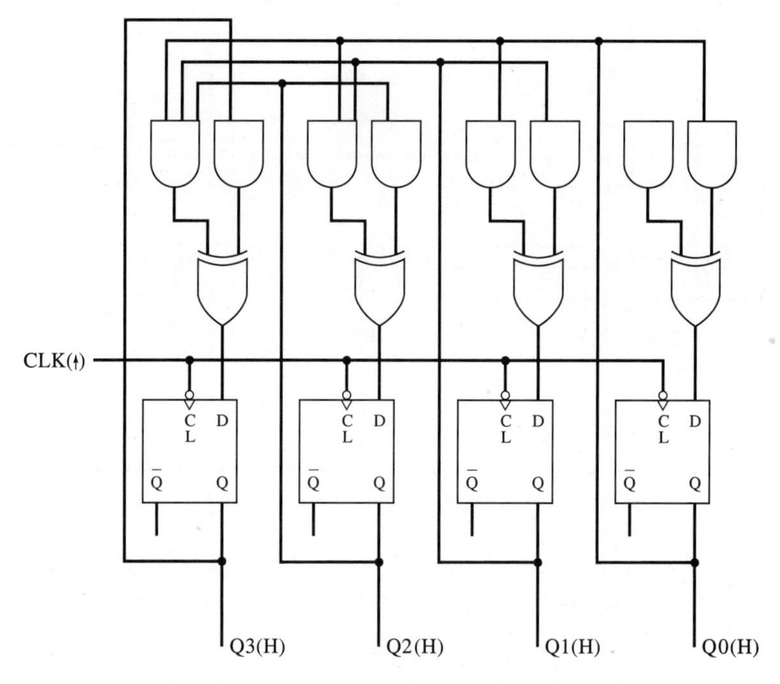

Figure 11.2.2-7 Four-bit synchronous counter simplified to the counting circuit.

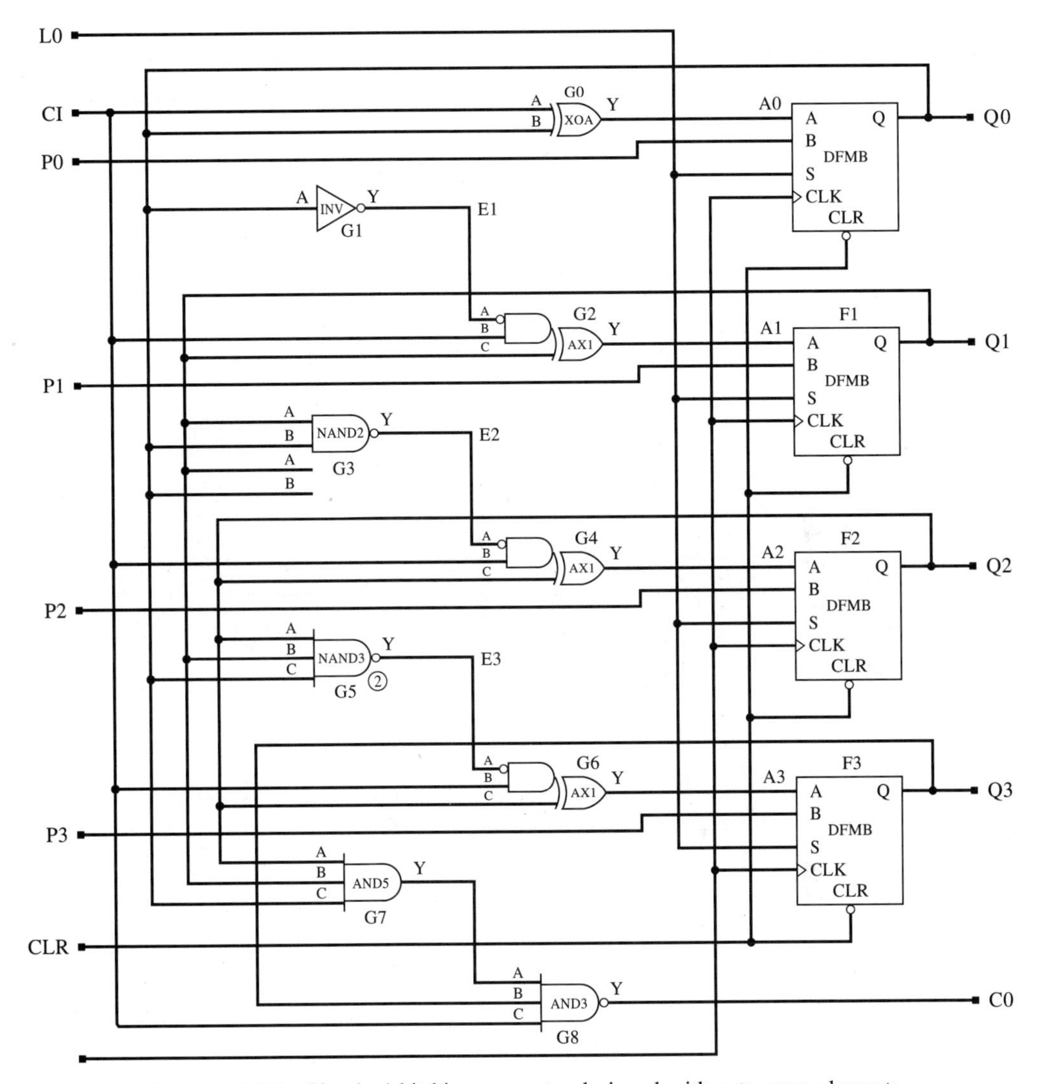

Figure 11.2.2-8 Simple 4-bit binary counter designed with gate array elements.

output that always follows the input. We can analyze this simplified circuit using standard state machine analysis methods. The only complication is the *xor* gates. Remember that the logic expression for an *xor* function is $F = A\overline{B} + \overline{A}B$. The completion of the analysis is left as an exercise for the reader.

Synchronous counters can easily be designed using gate array modules. In fact, gate array design software usually has predesigned counters of various types. These predesigned circuits are called soft macros. Figure 11.2.2-8 shows a simple 4-bit binary counter that uses gate array modules. The circuit of this counter differs in some details from that of the counter just described, but it performs the same function.

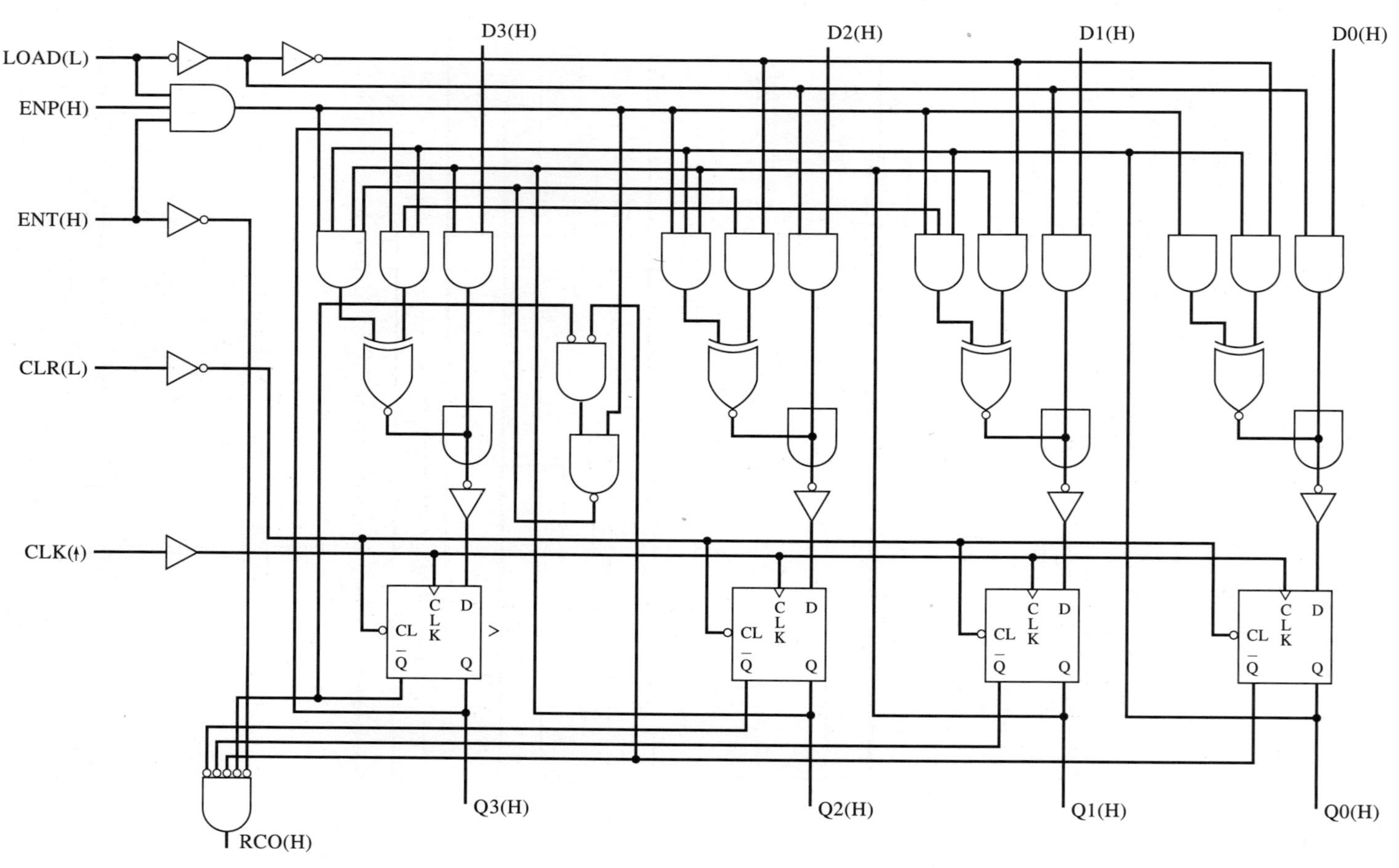

Figure 11.2.2-9 Synchronous decade counter.

Notice the use of flip-flops with multiplexed inputs (described in Chapter 6, Section 6.8). Here we use the multiplexers to select the parallel inputs when we want to do a parallel load.

The synchronous binary counter can easily be modified to make a decade (BCD) counter. Figure 11.2.2-9 is the circuit for a 4-bit synchronous decade counter with load and clear (74LS160). This circuit is almost identical to the circuit for the binary counter just analyzed. The only difference is the two gates between the Q3 and Q2 next-state decoders. They decode the 9 (1001) count and force the next count to 0.

The idea of the decade counter can be used to make a counter with any modulus from a standard counter. All we need to do is decode the terminal count we desire with gates external to the counter and use the decoded count to clear the counter so that it will start over at 0. This procedure works better if the counter has a synchronous clear, because the terminal count decoder may have glitches. Fortunately, both decade and binary counters with synchronous clears are available (74LS162 and 74LS163). If we use the binary counter with synchronous clear (74LS163) to make a counter with modulus 8, the terminal count should be 7, and the terminal count decoder logic is

$$TC = CLR = \overline{Q3} \cdot Q2 \cdot Q1 \cdot Q0$$

Figure 11.2.2-10 shows the circuit for this counter and the terminal count decoder.

Simple synchronous counters count in only one direction. Up/down counters that count in either direction are also available. Most up/down counters have an input that determines the direction of the count. Figure 11.2.2-11 shows the circuit for a synchronous 4-bit binary up/down counter (74LS191) with asynchronous load

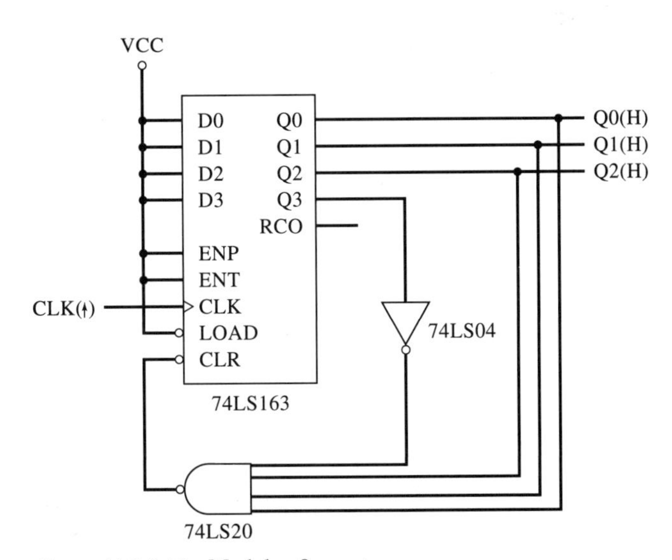

Figure 11.2.2-10 Modulus 8 counter.

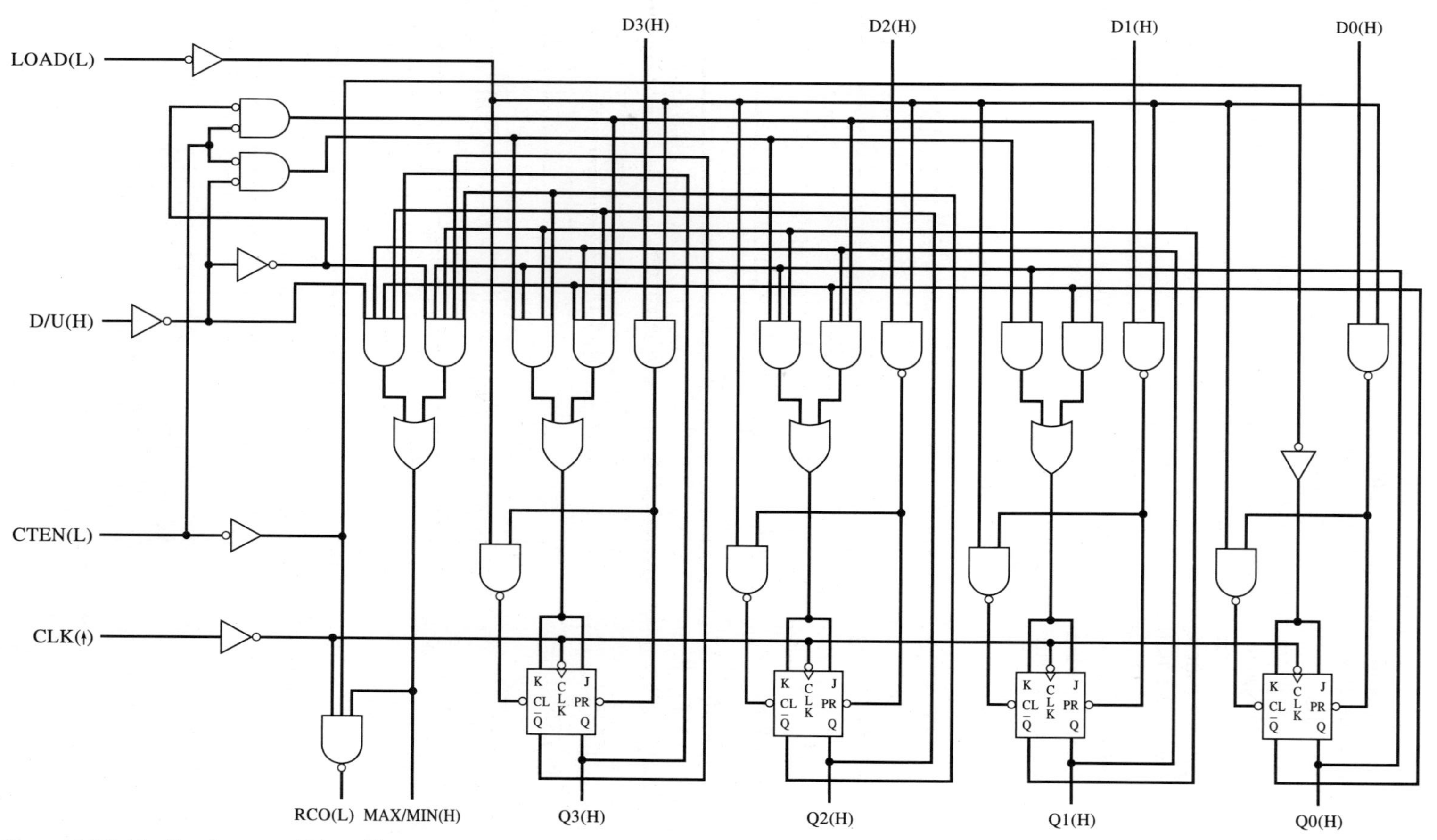

Figure 11.2.2-11 Synchronous 4-bit up/down counter.

and no clear. The 74LS190 is a decade version; the 74LS169 has a synchronous load. The implementation of the count function of the 74LS191 has an interesting difference from the 4-bit synchronous counter of Figure 11.2.2-5. The memory flip-flops are configured as T flip-flops (JK flip-flops with J and K tied together). T flip-flops, which toggle on the clock when T is asserted, make the next-state decoder simple because they match the function needed in a counter. The timing (Figure 11.2.2-12) is much the same as the timing of Figure 11.2.2-6, with an addition—the counter counts *down* when the D/$\overline{\text{U}}$(H) is asserted and *up* when it isn't asserted. Other minor differences are a single enable, CTEN(L); a MAX/MIN(H) output that is asserted as the counter reaches its maximum count when counting up, and as it reaches zero when counting down; and a rollover count output RCO that is asserted low (rather than high) and for only half a clock cycle. RCO, rather than MAX/MIN, should be used in cascaded applications.

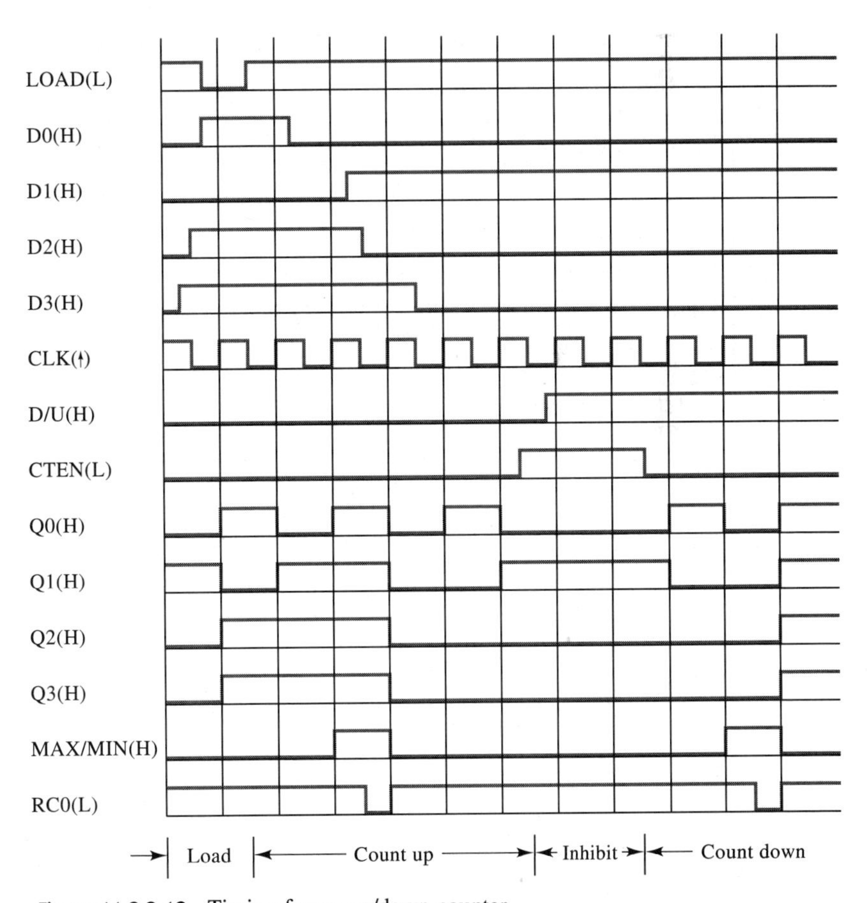

Figure 11.2.2-12 Timing for an up/down counter.

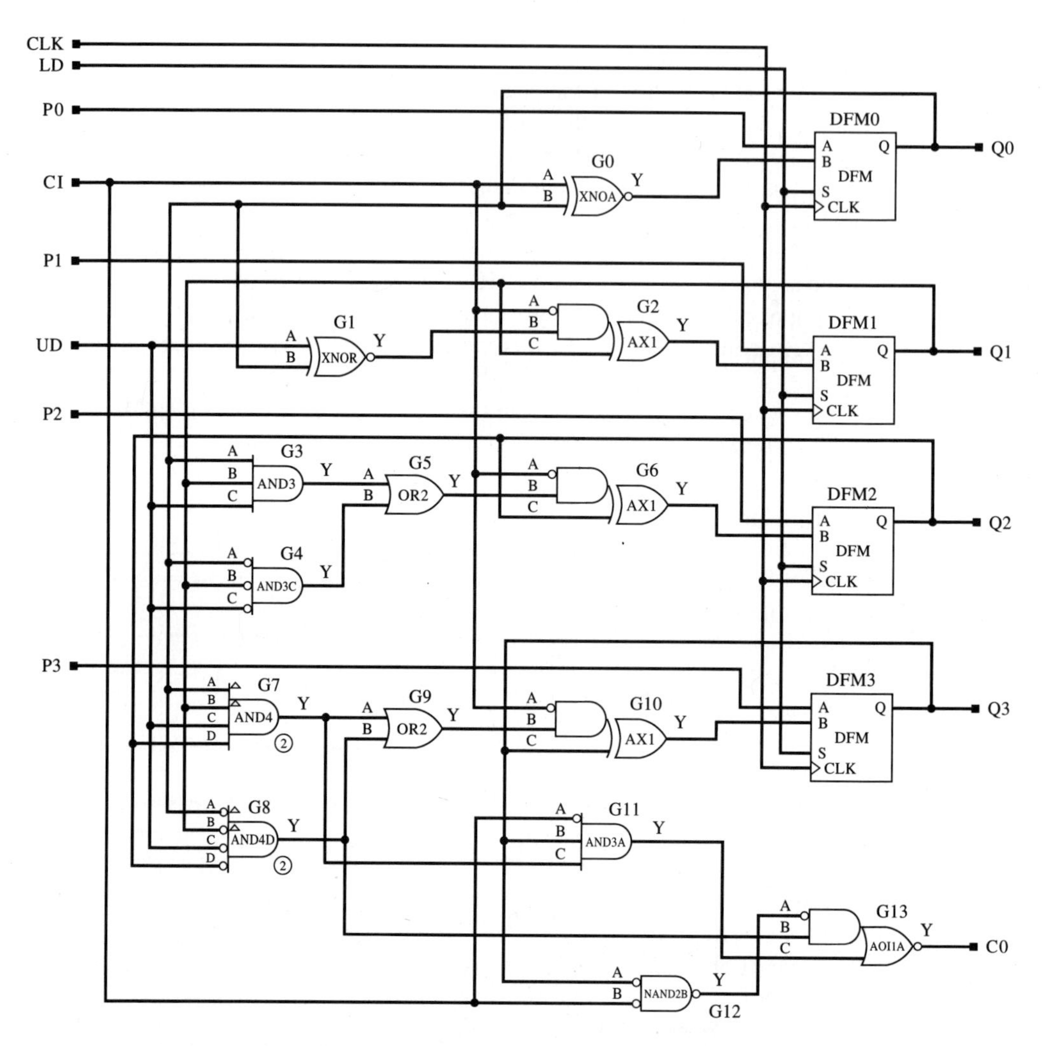

Figure 11.2.2-13 Synchronous up/down counter designed using gate array modules.

Up/down counters can also be designed using gate array modules. Figure 11.2.2-13 shows the circuit of such a soft macro. Although this counter functions much the same as that in Figure 11.2.2-11, its design details are quite different. It uses D flip-flops, rather than T-configured JK flip-flops, which make the next-state decoders completely different. Notice that flip-flops with multiplexed inputs are used. This makes it very easy to do a parallel load: all that is necessary is to select the parallel inputs with the select, S.

11.3 REGISTERS

In Chapter 8, Section 8.6, we discussed *simple registers*—collections of D flip-flops clocked from a common clock. Sometimes they have output drivers that allow the output to be tristated (turned off). They're essentially storage devices. This kind of register is called a parallel in–parallel out (PIPO) register because all the inputs can be fed to it simultaneously (each to a different pin) and the outputs can be read simultaneously.

Registers, called *shift registers*, that will store data presented serially on a single input can also be made. They have a variety of forms. To see how they function, let's look at a very simple 8-bit serial in–serial out (SISO) register (74LS91). Figure 11.3-1 is its circuit diagram. Like most shift registers, this one consists of a number of edge triggered SR flip-flops connected in cascade. The state of each flip-flop is transferred to the next flip-flop to the right each time a low-to-high clock transition occurs. This register is said to ''do a shift right'' on a low-to-high clock transition. We can see this shift to the right in the timing diagram (Fig. 11.3-2). A single pulse is input on A(H) and transferred into the first flip-flop (Q0) on the low-to-high clock transition. This pulse then shifts right on each successive low-to-high clock transition. The timing shows the output of each flip-flop, even though only the last one (Q7) is actually available as an output to the device. Both Q7(H) and Q7(L) are outputs. A string of eight bits (signal levels) can be shifted into this register from the left, using the clock, and then shifted out on the right (again using the clock).

The 8-bit shift register just described is a state machine with $2^8 = 256$ states—one for each possible combination of bits in the register. If we were to design this state machine using our usual procedure, which starts with a description and a state transition diagram, we would find that the state transition diagram was just too complicated to draw reasonably. (We'll draw the 16-state state transition diagram for a 4-bit shift register in an exercise.) Fortunately, there is a description of the shift register that does not use a state transition diagram and is simple to formulate; this is the following logic description.

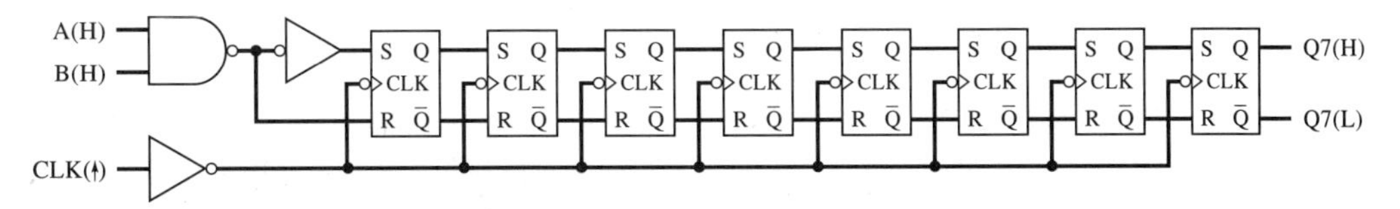

Figure 11.3-1 SISO register.

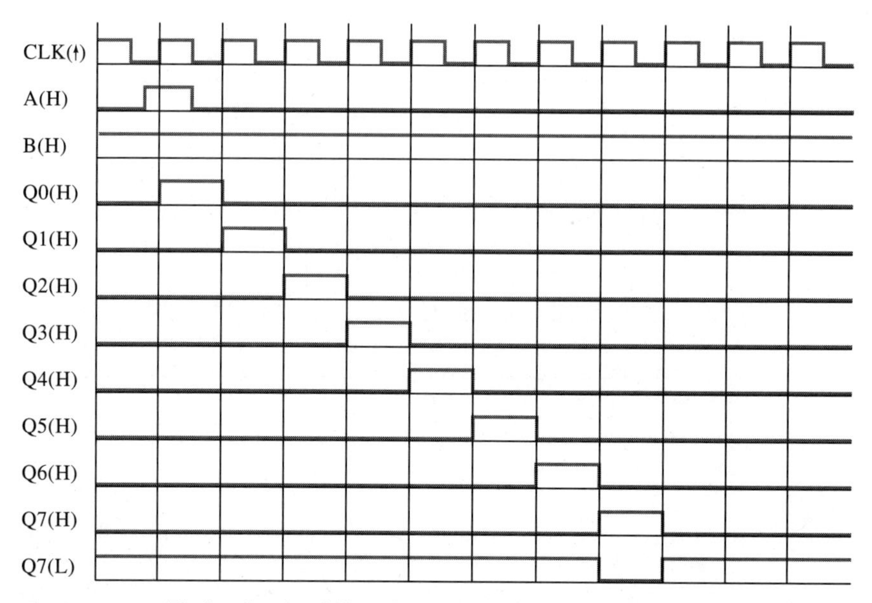

Figure 11.3-2 Timing for the shift register of Figure 11.3-1 (74LS91).

$$
\begin{aligned}
Q0(n+1) &= A \\
Q1(n+1) &= Q0(n) \\
Q2(n+1) &= Q1(n) \\
&\ \vdots \\
Q6(n+1) &= Q5(n) \\
Q7(n+1) &= Q6(n)
\end{aligned}
$$

where the Qi(n) are the present-state code bits and the Qi(n + 1) are the next-state code bits. These equations tell us that the next state is the present state shifted right one bit, with the next Q0 taken from the serial input A. If we need to design a shift register, we can start with these equations. In fact, they are the next-state decoder logic equations if the memory elements are D flip-flops. They are also the equations to use if we want a PAL design.

The register we've considered is rather limited. Other, more flexible shift registers have provisions for a parallel load and a parallel output. Figure 11.3-3 shows a 5-bit shift register (74LS96) with serial in–parallel/serial out data transfer. The register also has provisions for a kind of parallel loading of the flip-flops. Data are loaded serially from the SER(H) input and clocked through the register on the positive-going edge of CLK, as shown in the function table (Table 11.3-1) and the timing diagram (Fig. 11.3-4). The CLR(L) input allows an asynchronous clear to 0

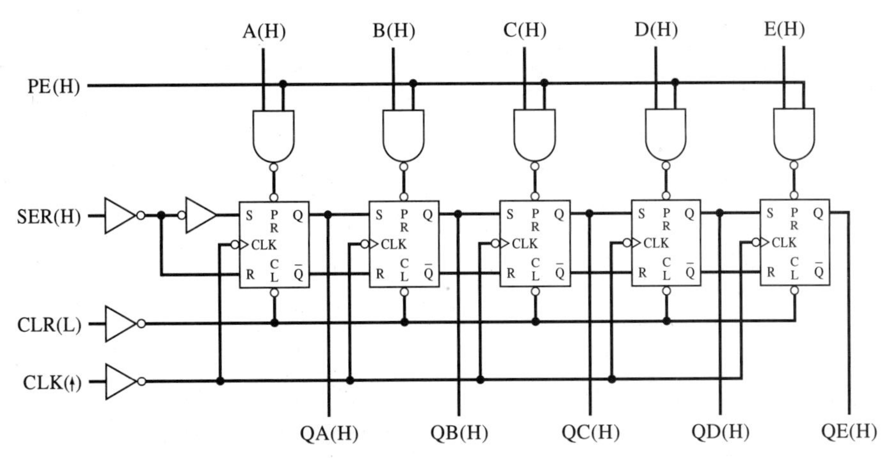

Figure 11.3-3 Five-bit SIPO shift register.

Table 11.3-1 Function Table for 5-Bit Shift Register

Inputs										Outputs				
		Preset												
Clear	Preset Enable	A	B	C	D	E	Clock	Serial		QA	QB	QC	QD	QE
L	L	x	x	x	x	x	x	x		x	x	x	x	x
L	x	L	L	L	L	L	x	x		L	L	L	L	L
H	H	H	H	H	H	H	x	x		H	H	H	H	H
H	H	L	L	L	L	L	L	x		QA0	QB0	QC0	QD0	QE0
H	H	H	L	H	L	H	L	x		H	QB0	H	QD0	H
H	L	x	x	x	x	x	L	x		QA0	QB0	QC0	QD0	QE0
H	L	x	x	x	x	x	↑	H		H	QAn	QBn	QCn	QDn
H	L	x	x	x	x	x	↑	L		L	QAn	QBn	QCn	QDn

H = high level (steady state); L = low level (steady state).
x = irrelevant (any input i, including transition).
↑ = transition from low to high.
QA0, QB0, and so on = the level of QA, QB, and so on, respectively, before the indicated steady-state input conditions were established.
QAn, QBn, and so on = the level of QA, QB, and so on, respectively, before the most recent ↑ transition of the clock.

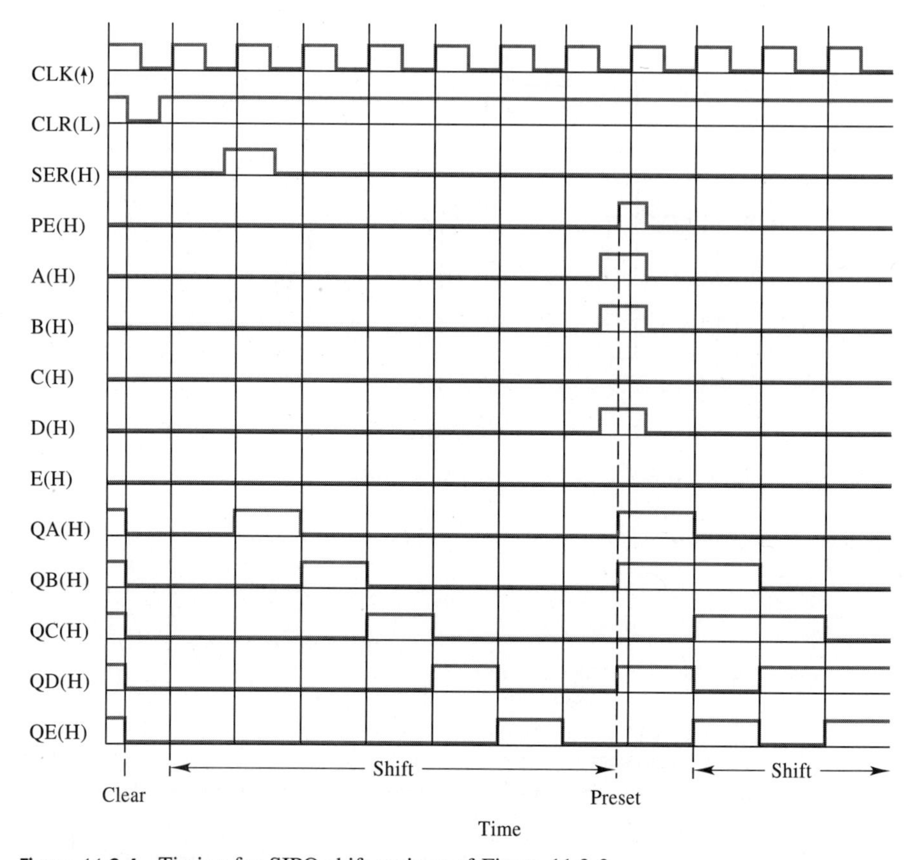

Figure 11.3-4 Timing for SIPO shift register of Figure 11.3-3.

of all the flip-flops in the register. Notice the initial clear in the timing diagram. The PE input, in conjunction with the A through E inputs, allows each of the flip-flops to be selectively set high. Notice the load in the function table and the timing diagram. This selective load feature can only load a high; it cannot change a high to a low. All of the outputs from the flip-flops are outputs of the device, so parallel output of the contents of the register is available. Serial output is also available at the final flip-flop output (QE).

Figure 11.3-5 shows the circuit for a 4-bit register with true parallel load (74LS95); notice that this register requires more next-state decoder logic circuitry. This particular shift register has a mode control (MODE) that determines whether data are to be loaded serially from IN or in parallel from D0 through D3. It also has two clocks—one for shifting right (CLK1) and one for loading parallel data (CLK2). The device is clocked on the negative-going clock edge. The action of the mode control and clocks can be seen in the function table (Table 11.3-2) and the timing diagram (Fig. 11.3-6). Notice the anomalous behavior that occurs with some mode

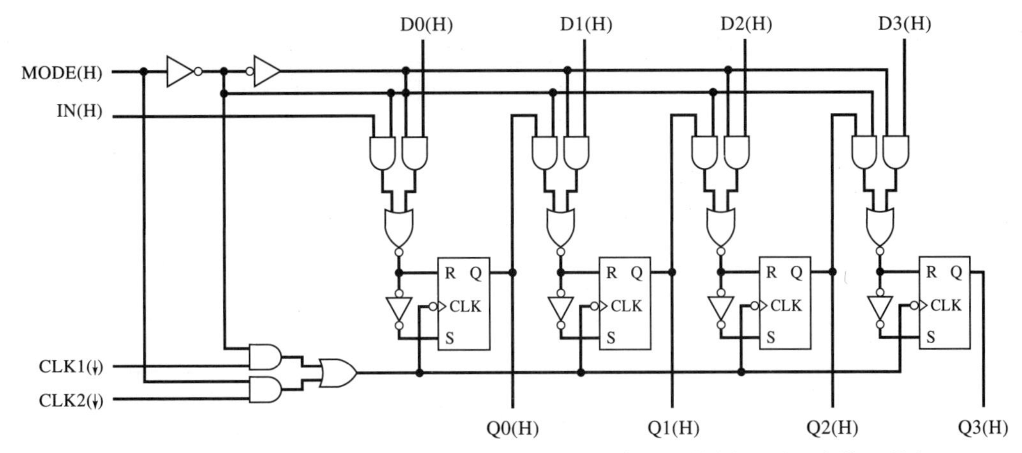

Figure 11.3-5 Four-bit shift register which allows serial/parallel in and serial/parallel out.

Table 11.3-2 Function Table for 4-Bit PIPO Shift Register

	Inputs					Outputs			
Operating Mode	MODE	CLK1	CLK2	IN	Dn	Q0	Q1	Q2	Q3
Parallel load	H	x	↓	x	l	L	L	L	L
	H	x	↓	x	h	H	H	H	H
Shift right	L	↓	x	l	x	L	q0	q1	q2
	L	↓	x	h	x	H	q0	q1	q2
Mode change	↑	L	x	x	x	No change			
	↑	H	x	x	x	Undetermined			
	↓	x	L	x	x	No change			
	↓	x	H	x	x	Undetermined			

H = high-voltage-level steady state.
h = high voltage level one setup time prior to the high-to-low clock transition.
L = low-voltage-level steady state.
l = low voltage level one setup time prior to the high-to-low clock transition.
q = Lowercase letters indicate the state of the referenced output one setup time prior to the high-to-low clock transition.
x = don't care.
↓ = high-to-low transition of clock or mode.
↑ = low-to-high transition of clock or mode.

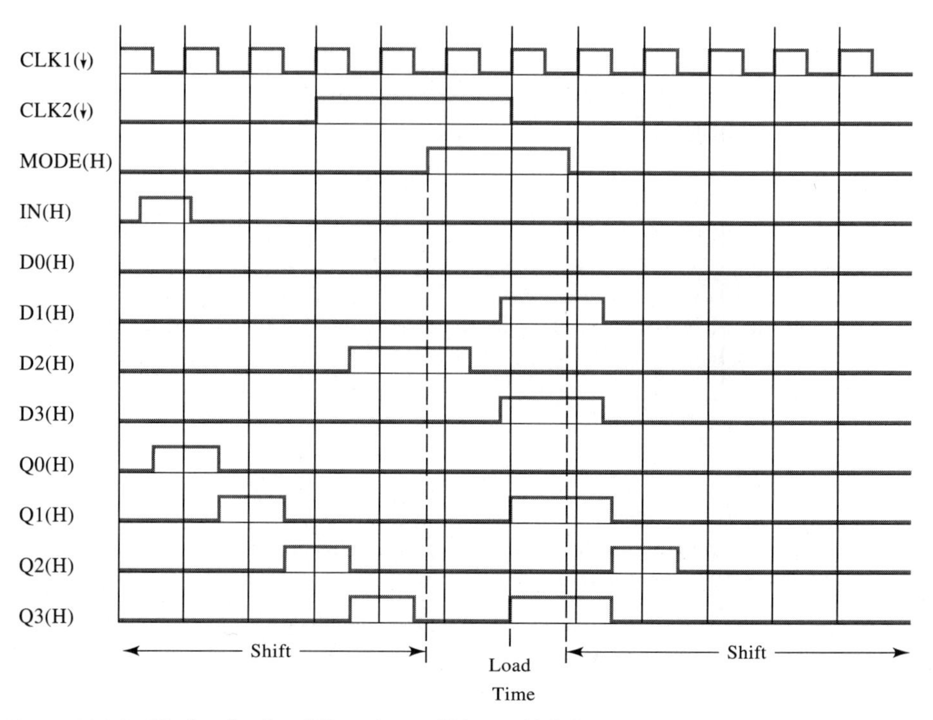

Figure 11.3-6 Timing for the shift register of Figure 11.3-5.

changes, shown at the end of the truth table. It occurs because the clock is *and*ed with MODE, and when MODE changes it can generate an unwanted clock edge. *And*ing a clock with some other signal often creates problems. The diagram shows how a single bit is shifted through the register by CLK1 when MODE is not asserted, and then how a parallel load is made using CLK2 when MODE is asserted.

The shift registers just described are samples of the shift-right shift registers that are commercially available. All but the first of these are short—four bits. Four-bit registers were chosen because they are easier to describe. Eight-bit registers with much the same functions are also available, as are registers that are bidirectional (can shift either right or left). Registers that have serial/parallel in and parallel out, and in addition are bidirectional, are called *universal shift registers.* Before examining a typical 4-bit universal shift register, let's design a simple 4-bit bidirectional register with discrete logic elements. A more realistic design might be an 8- or 10-bit shift register in a PAL, but a discrete logic design will be more instructive. This will give us an example of a design in which we can describe a state machine as a set of logic equations more readily than by means of a state transition diagram.

We'll design the shift register with an input signal called MODE to control the direction of shift. When MODE(H) is asserted, the register will shift left, and when MODE is not asserted, it will shift right. The register will also have an asynchronous preset (PE) and an asynchronous clear (CLR), which will add little complexity because D flip-flops will be used for memory, and they already have presets and clears. We'll need two serial inputs—one input for shifting right, which we'll call SR_IN (shift-right input), and one input for shifting left, which we'll call SL_IN (shift-left input).

We need to describe the shift register concisely, as was done earlier in this section for a shift register that shifts right. We used logic expressions to describe how the state changes, and we can extend the same idea to a bidirectional shift register. Let the four state bits be Q0, Q1, Q2, and Q3. For the shift-right mode ($\overline{\text{MODE}}$), each time the positive-going clock edge occurs, the value in Q0 is shifted to Q1 [Q1(n + 1) = Q0(n)], the value in Q1 is shifted to Q2 [Q2(n + 1) = Q1(n)], the value in Q2 is shifted to Q3 [Q3(n + 1) = Q3(n)], and Q0 is loaded from the shift-*right* serial input; but for the shift-left mode (MODE), the value in Q3 is shifted to Q2 [Q2(n + 1) = Q3(n)], the value in Q2 is shifted to Q1 [Q1(n + 1) = Q2(n)], the value in Q1 is shifted to Q0 [Q0(n + 1) = Q1(n)], and Q3 is loaded from the shift-*left* serial input. For the shift register we're designing, either a shift right or a shift left must occur, and that shift is conditional on MODE. The logic expression for each of the Qi(n + 1) in the next-state state code is the *or* of the two foregoing expressions for that Qi(n + 1) conditioned on MODE:

$$Q0(n + 1) = SR\_SER \cdot \overline{MODE} + Q1(n) \cdot MODE$$

$$Q1(n + 1) = Q0(n) \cdot \overline{MODE} + Q2(n) \cdot MODE$$

$$Q2(n + 1) = Q1(n) \cdot \overline{MODE} + Q3(n) \cdot MODE$$

$$Q3(n + 1) = Q2(n) \cdot \overline{MODE} + SL\_SER \cdot MODE$$

These conditions describe how the state bits change. For D memory, they also describe the next-state decoder logic, because Di = Qi. So the next-state decoder logic is

$$D0 = SR\_SER \cdot \overline{MODE} + Q1 \cdot MODE$$

$$D1 = Q0 \cdot \overline{MODE} + Q2 \cdot MODE$$

$$D2 = Q1 \cdot \overline{MODE} + Q3 \cdot MODE$$

$$D3 = Q2 \cdot \overline{MODE} + SL\_SER \cdot MODE$$

The n + 1 and n arguments are not needed to distinguish between the present state (n) and next state (n + 1) in these expressions, so they're dropped.

Once we have the next-state decoder logic expression the design process is the same as in any of our previous designs. Figure 11.3-7 is the circuit for the shift register we've designed, and Figure 11.3-8 shows the results of a simulation of this circuit. In the simulation the register is first cleared, then a single bit is loaded on the SR_IN input and shifted right until it passes through the four bits of the register. MODE is not asserted (low) during this shift right. A single bit is then loaded on the SL SER input and shifted left through the four bits of the register. MODE is asserted (high) during this shift right. Finally, a single bit is loaded on the SR_SER

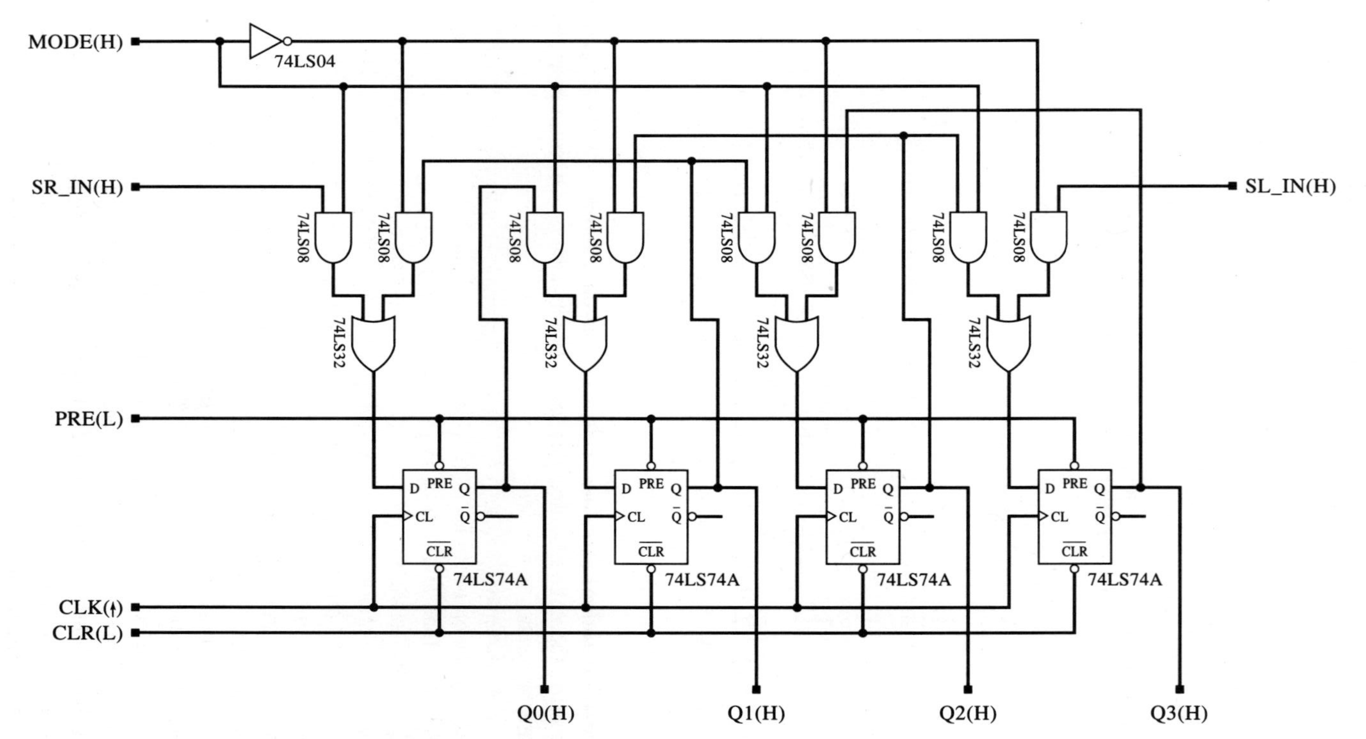

Figure 11.3-7 Four-bit bidirection shift register design.

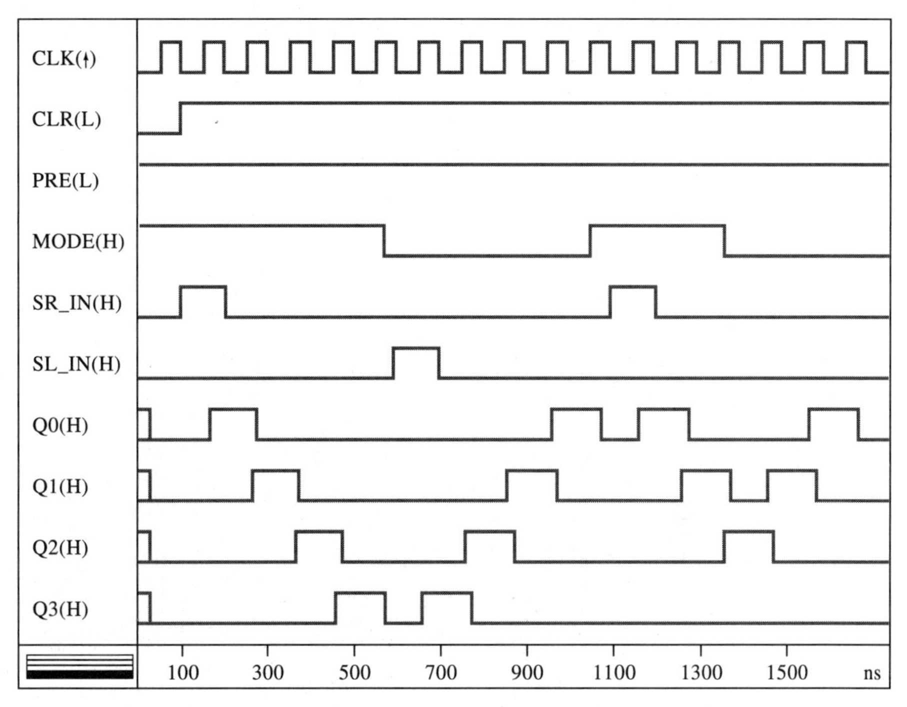

Figure 11.3-8 Simulation results for the 4-bit bidirectional shift register of Figure 11.2.2-8.

input and shifted three bits right, and it is then shifted left until it goes off the end of the register. During this part of the simulation, MODE is first not asserted for the shift right and then asserted for the shift left.

The ultimate shift register for versatility is the universal register. It is much the same as the bidirectional register just designed, but it has additional modes, including a parallel load and an inhibit mode. The circuit for a 4-bit universal shift register (74LS194) is shown in Figure 11.3-9. Serial output is available at QD for the shift-right mode and at QA for the shift-left mode. The mode of the shift register is controlled by S1 and S0, as shown in Table 11.3-3. Notice that the shift register can hold, shift right, shift left, and parallel-load. Separate inputs are available for serial input for the shift-right function, SR_SER(H), and the shift-left function, SL_SER(H). The master reset, MR(L), clears all the flip-flops. The function of the shift register can be most readily seen in the timing diagram (Fig. 11.3-10). The diagram shows a master reset, a parallel load, a shift right for six clock cycles, a shift left for five clock cycles, a hold (inhibit) for four clock cycles, and a master reset.

Shift registers of various types can be designed using gate array modules. Figure 11.3-11 shows a 4-bit shift register with parallel load that is available as a soft module in one type of gate array. Notice that this shift register uses D flip-flops, because SR flip-flops are not available in this gate array. The multiplexed inputs to the D flip-flops make it convenient to parallel-load the register. All that is necessary is to select the parallel inputs. This register behaves in the same way as the 4-bit

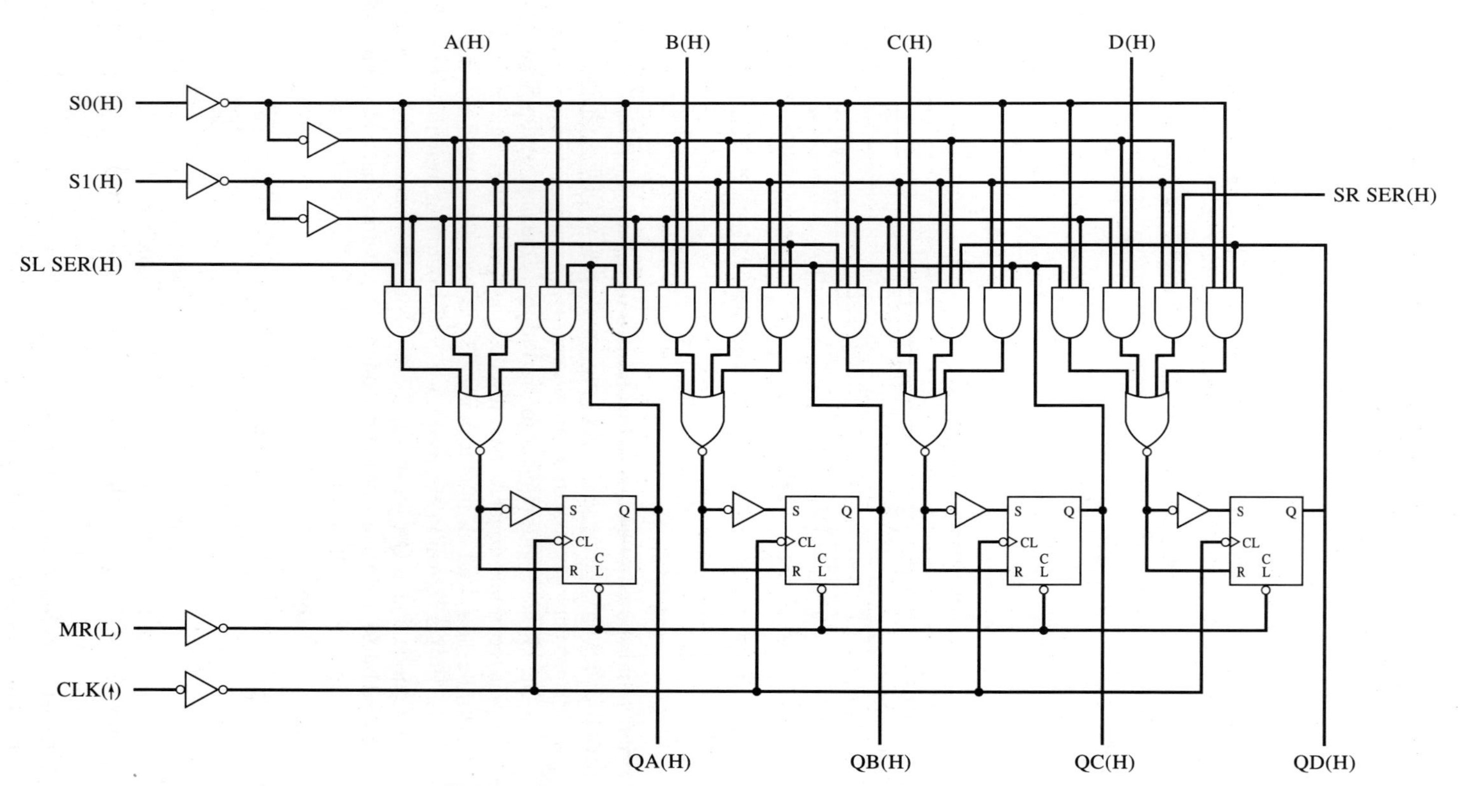

Figure 11.3-9 Four-bit universal shift register (74LS194).

Table 11.3-3 Modes

S1	S0	Mode
L	L	Hold
L	H	Shift right
H	L	Shift left
H	H	Parallel load

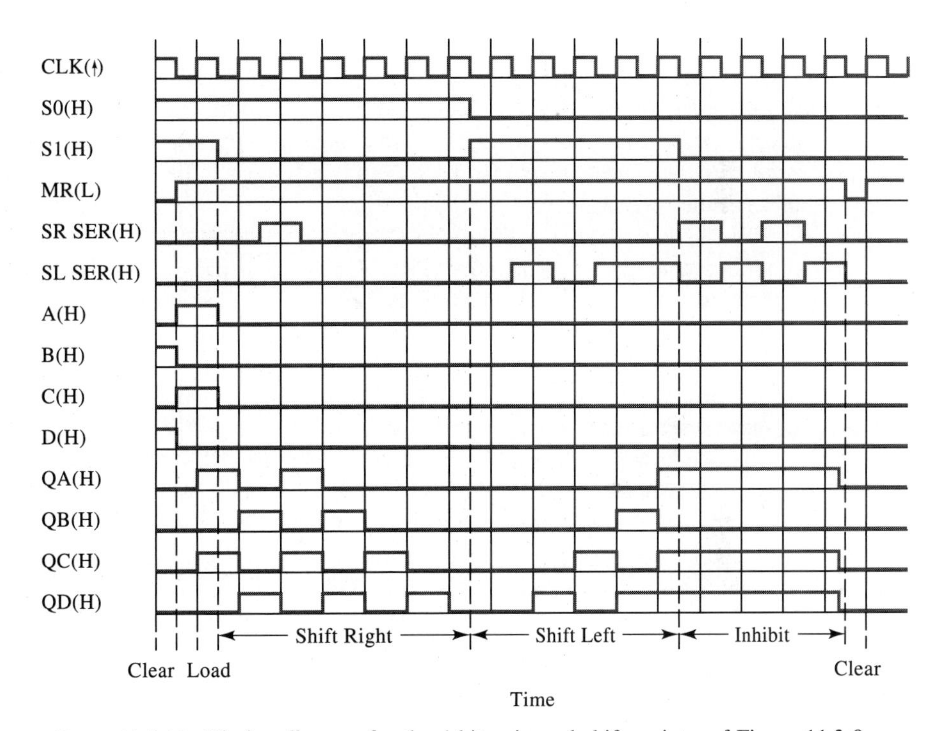

Figure 11.3-10 Timing diagram for the 4-bit universal shift register of Figure 11.3-9.

register (74LS95) already described, except that it uses the same clock for both serial and parallel loads, and SHLD selects the type of load. SHLD asserted (high) selects a parallel load, and SHLD not asserted (low) selects a shift right.

11.4 MEMORY

Registers, especially PIPO registers, are useful for storing information; but if a large amount of information needs to be stored, the registers must be organized for easy

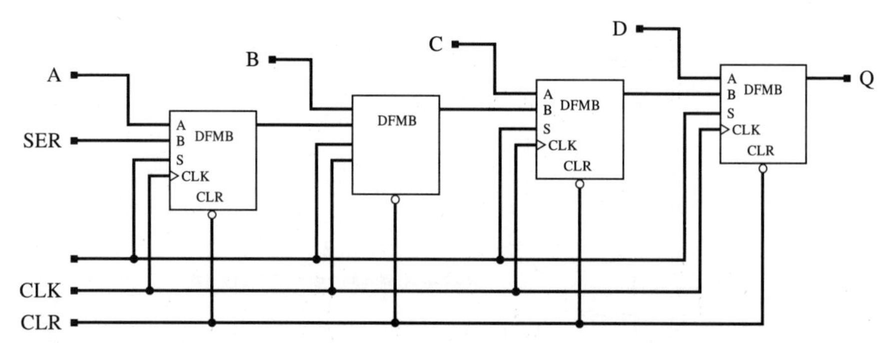

Figure 11.3-11 Four-bit shift register with parallel load using gate array modules.

access. Organized arrays of registers are called memory. More specifically, they're *random access memory (RAM)*, because any register can be accessed independent of any other register.

We first considered memory in Chapter 6, Section 6.3, which showed how programmable read-only memory (PROM) is used as programmable logic. In that section we were interested in memory only as a collection of storage elements (registers), because we needed to know how to program the memory to act as logic. In this section we consider the organization of memory to store large quantities of data. Many digital systems are used for data processing, and the ability to store data in these systems can be important.

To see how general read/write memory functions, let's design a small memory using logic elements. *Read/write* (*R/W*) *memory* is memory that we can load (write) data into and subsequently recover (read) the data. Suppose we want to be able to store eight *words* of eight bits each (a tiny memory). We can store the eight words in eight 8-bit registers, such as the 74LS374 described in Chapter 8, Section 8.6. The circuit and symbol for these registers are repeated in Figure 11.4-1. The circuit consists of eight D flip-flops with a common clock. It also has output buffers that can be tristated using the output control ($\overline{OC}$). We'll find the tristate output quite useful in designing our memory. To transfer data to and from the memory registers, we need eight data lines, which we'll call D0, D1, D2, D3, D4, D5, D6, and D7 (the data bus). We'll connect these lines to both the inputs (Ds) and the outputs (Qs) of all the registers, and use the CLK and $\overline{OC}$ to steer data to and from the proper register.

The data are normally steered by addressing the register by number, starting at 0. The addresses for our memory would thus be 000 (0) through 111 (7) and would require three address lines, A0, A1, and A2 (the address bus). To decode the address, we'll use a 74LS138 decoder, described in Section 5.4. Figure 11.4-2 shows the memory we've designed with eight registers and a decoder. It has an address bus A[2:0]; a data bus D[7:0] for both input and output; and three control lines, a read (RD) and a write (WR), which indicate whether we want to read from the memory or write to it, and a chip select, CS, that selects or deselects the entire memory. (Buses

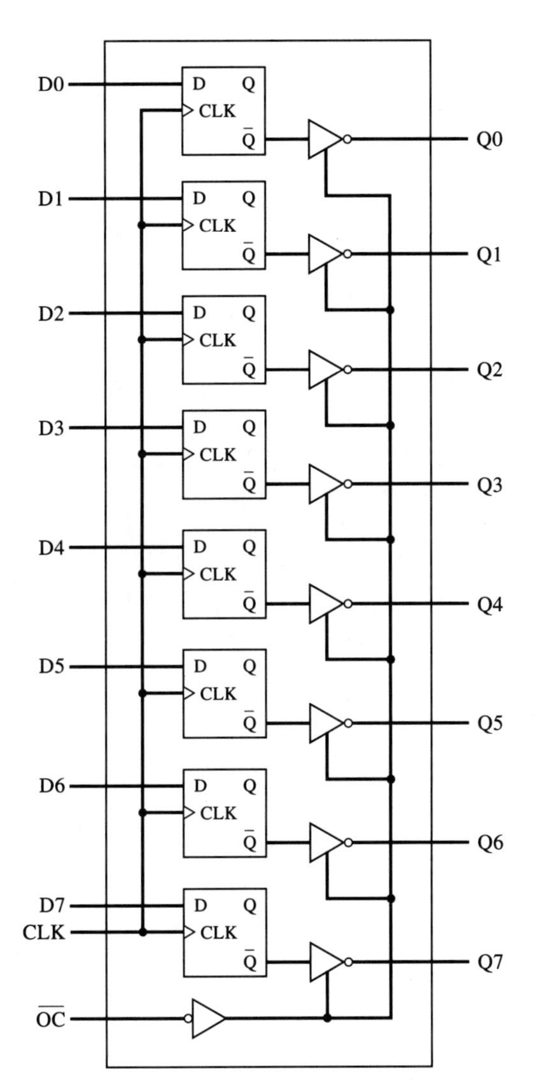

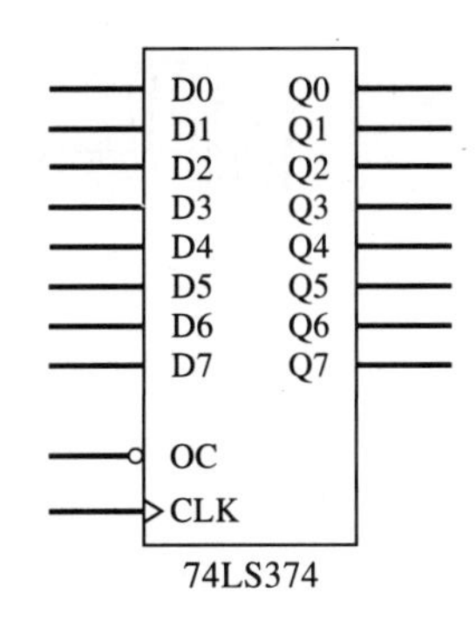

Figure 11.4-1 Octal D flip-flop (register) with tristate output.

are heavy lines in the circuit diagram). We have labeled each of the memory registers at the top of the register in the circuit diagram. Sometimes we need a memory that has separate input and output. We can easily modify the memory we've designed for separate inputs DI[7:0] and outputs DO[7:0], as shown in Figure 11.4-3.

To see how the memory functions, we need the function table for the decoder (Table 11.4-1). Suppose the address is 001 (LLH) and CS is asserted (low); we find for these inputs that the Y1 output of the decoder (74LS138) is asserted (low). This pulls one of the inputs low on the gates that control the MEM1 output control ($\overline{\text{OC}}$) and the MEM1 clock. If, in addition, RD is asserted (low), both inputs to the

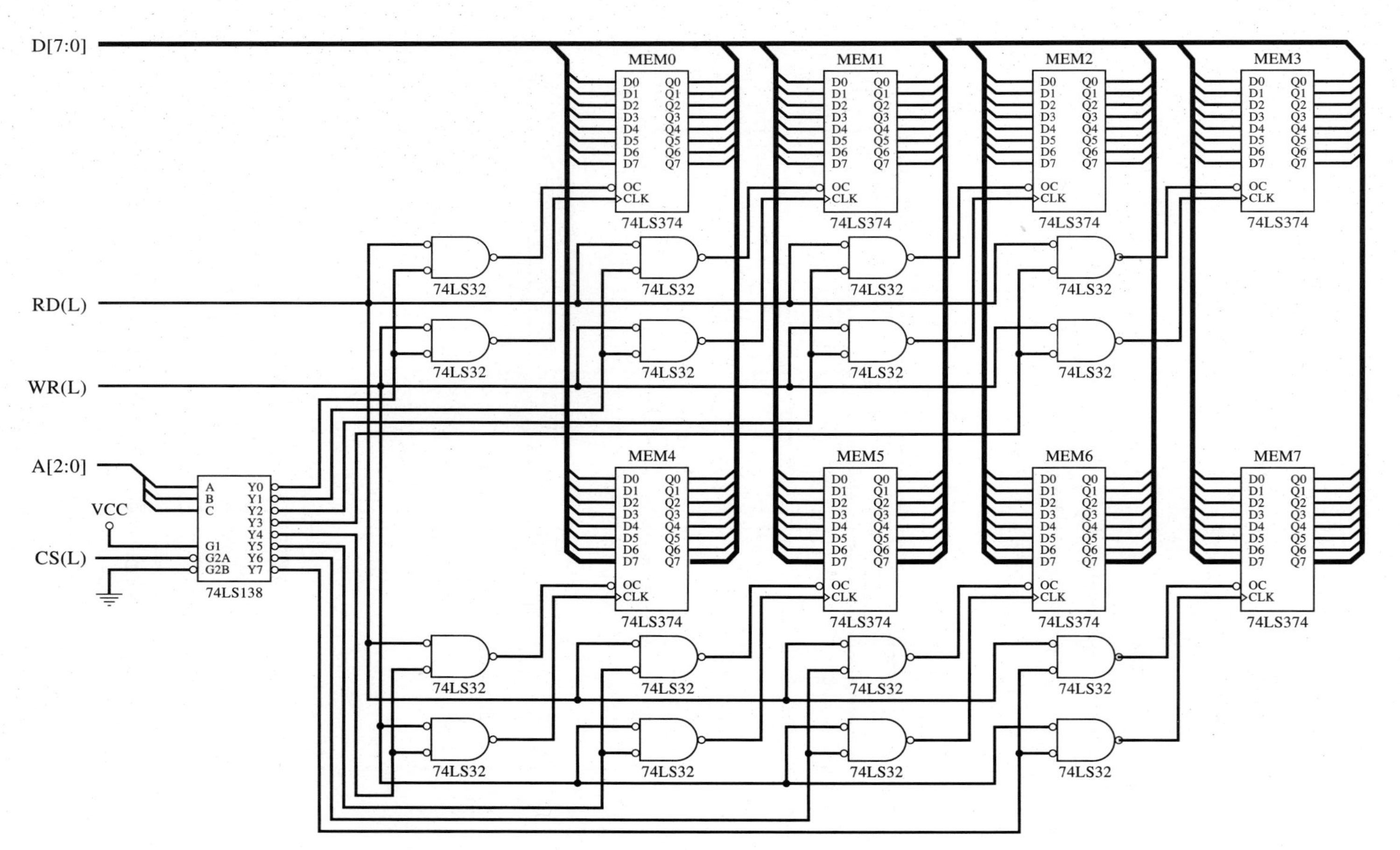

Figure 11.4-2 Memory design.

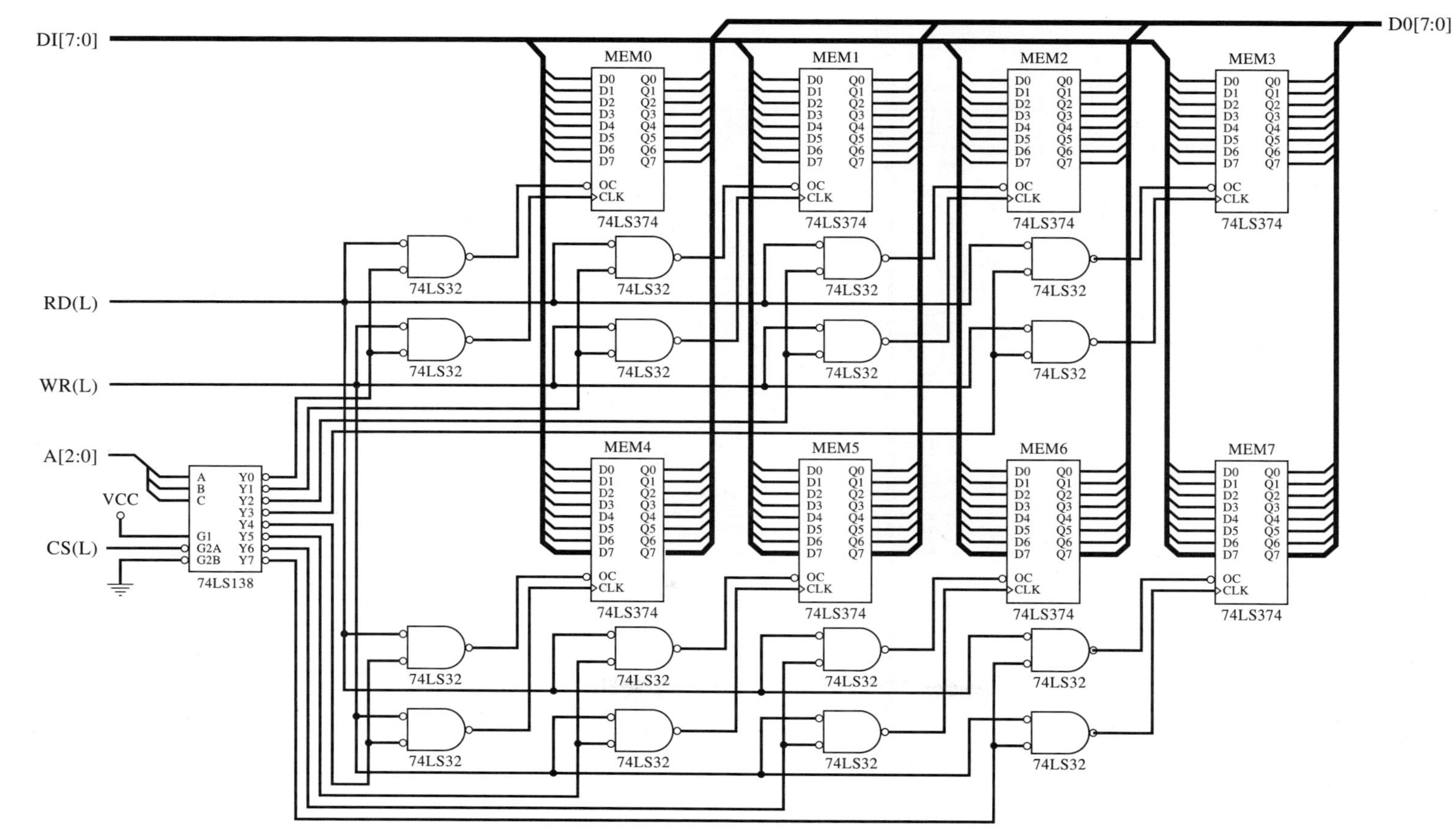

Figure 11.4-3 Memory design with separate I/O (dual port memory).

Table 11.4-1 Function Table for a 3-Line to 8-Line Decoder/Demultiplexer

Inputs						Outputs							
Enable			Select										
G1	G2A	G2B	C	B	A	Y0	Y1	Y2	Y3	Y4	Y5	Y6	Y7
L	x	x	x	x	x	H	H	H	H	H	H	H	H
x	H	x	x	x	x	H	H	H	H	H	H	H	H
x	x	H	x	x	x	H	H	H	H	H	H	H	H
H	L	L	L	L	L	L	H	H	H	H	H	H	H
H	L	L	L	L	H	H	L	H	H	H	H	H	H
H	L	L	L	H	L	H	H	L	H	H	H	H	H
H	L	L	L	H	H	H	H	H	L	H	H	H	H
H	L	L	H	L	L	H	H	H	H	L	H	H	H
H	L	L	H	L	H	H	H	H	H	H	L	H	H
H	L	L	H	H	L	H	H	H	H	H	H	L	H
H	L	L	H	H	H	H	H	H	H	H	H	H	L

gate that controls the MEM1 output control ($\overline{OC}$) will be low, and the output of the gate will be low. $\overline{OC}$ for MEM1 will thus be asserted (low), and the output of MEM1 will be turned on to the data bus D[7:0]. Alternatively, if WR is asserted (low), both inputs to the gate that control the MEM1 clock will be low and the output will be low. The clock for MEM1 will thus be low, and when it goes high, the data on the data bus D[7:0] will be clocked into MEM1. WR must be taken low and then high in order to produce a positive-going clock edge to clock the register. The decoded address (A[2:0]) controls which register is active. RD and WR control whether the data are output onto the data bus from the register or clocked into the register from the data bus. Because of the way this memory is designed, we can also control a write using CS. If we set RD low and then pulse CS low, data will be clocked into the register on the low-to-high transition of CS.

A read from the memory is executed by selecting the memory (asserting CS), inputting the memory address on the address bus A[2:0], and pulling RD low. Data are output to the data bus D[7:0] when the change in the address has propagated through the decoder and RD is low. The data stay valid until either the address changes or RD is taken high.

A write to the memory is executed by selecting the memory (asserting CS), inputting the memory address on the address bus A[2:0], inputting the data to be stored on the data bus D[7:0], and taking WR low and then high. Data are stored on the low-to-high transition of WR. The data must be stable for a sufficient period of time before this transition to meet the setup time on the register, and the address must be stable for a sufficient period of time before this transition for the decoder address to become stable at the register. (Later in this section we'll look more closely at the read and write cycles and the timing constraints involved for an actual memory.)

The memory we've designed is a small static read/write RAM. Here we're concerned only with RAM as distinguished from sequential memory such as disk or tape. Random access memory is classified as either read/write memory (R/W) or read-only memory (ROM). It is more common but less descriptive to call R/W memory random access memory (RAM) and to call read-only random access memory simply read-only memory (ROM).

R/W memory is further classified as either static or dynamic. In the first part of this section, we designed a *static RAM* using D flip-flops as the basic memory cell. The individual memory cells in all static R/W RAM are essentially simplified D flip-flops. Static RAM will store a value as long as it has power applied to it, and it can be made to function very fast.

Dynamic RAM uses small capacitors as storage elements. Because the charge on a capacitor slowly leaks off, dynamic RAM has to be refreshed periodically. This is a significant disadvantage, but dynamic RAM has the advantages of small physical size, low cost, and low power consumption and is therefore used extensively in computers. Because of the need for refreshing and the large capacity of dynamic RAM, the address is split into two parts (called the row and column address), which are separately input to the memory on the same pins but at different times (time multiplexed). The complications of time multiplexing and refreshing make it necessary to control dynamic RAM with a special controller (called, appropriately, a dynamic RAM controller), so we will not present the details of the use of dynamic RAM here.

ROM is memory that permanently or semipermanently stores data. *Masked* ROM has the stored values fixed in it at the time of manufacture. *Fused* ROM (programmable read-only memory—PROM) can be programmed once (is one-time programmable) by blowing fuses in the memory. Erasable programmable read-only memory (EPROM) is many-times programmable. It is normally erased by exposure to ultraviolet light through a window in the top of the package, and programmed in a special PROM programmer. (Electronically erasable ROM is also available but is less common and much more expensive.) Figure 11.4-4 summarizes the classification of memory.

We could design a tiny ROM memory in much the same way as we've designed the R/W memory. In place of the registers we would need only buffers. The values would be permanently stored by tying the inputs to the buffers high or low, as needed. The $\overline{\text{OC}}$ for the buffers would control the read, just as in the W/R memory. There would, or course, be no write. A single memory location with the stored value 00110010 would take the form of Figure 11.4-5; this would be similar to masked ROM. Programmable ROM would be functionally the same but more complicated, to allow for programming.

Memory is specified by size as well as type. The basic size of a memory is the number of bits it can store. A 1-Meg memory can store 1,048,576 bits. As we've

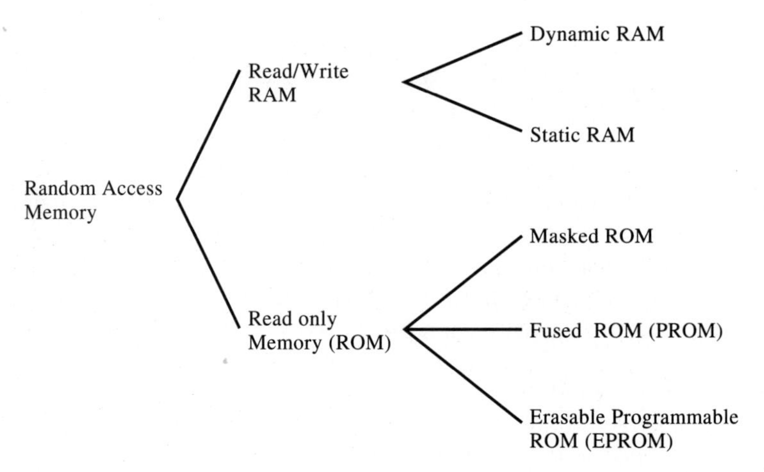

Figure 11.4-4 Memory classification.

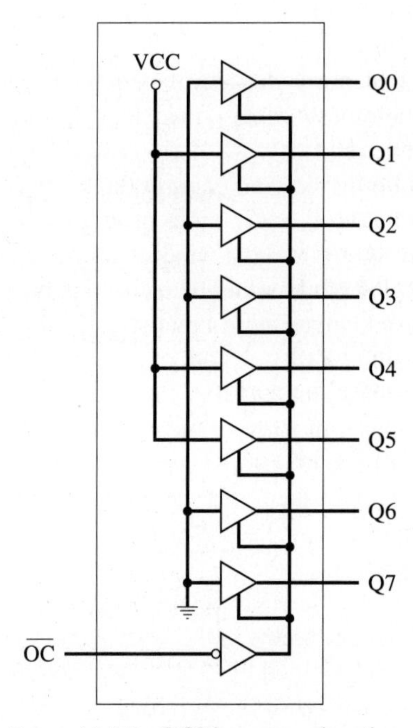

Figure 11.4-5 ROM memory location.

seen, memory is usually organized in terms of the number of storage locations (words) and the size of data in each location (size of the words). Based on this kind of organization, it is common to specify memory as

$$(\text{number of words}) \times (\text{size of word})$$

A small static RAM or ROM might be 512×8; that is, it contains 512 words, and each word is eight bits long. A large dynamic RAM might be 1 Meg $\times$ 1; that is, it contains 1,048,576 words, and each word is one bit long. The number of words can always be expressed as a power of 2 and is related to the number of address lines by

$$\text{number of words} = 2^{\text{number of address lines}}$$

For example, a memory that can store 1024 words has ten address lines. Word lengths for ROM and static R/W RAM are generally 4, 8, or 16 bits. Word length for dynamic R/W RAM is generally 1 bit.

In order to use memory in our digital designs, we'll need to examine the read and write cycles for actual memory—especially the timing—more carefully. Figure 11.4-6 shows the timing for both the read and write cycles of a typical static RAM. The typical read cycle for a ROM is similar to this read cycle. For actual memory, a table of values would be associated with this diagram. Table 11.4-2 is a sample table for the 7C170. It contains three sets of timing values because this memory is manufactured in three different timing versions. The 7C170-25 is a 25-ns version of the memory; the 7C170-35 is a 35-ns version; and the 7C170-45 is a 45-ns version. Only the critical value—maximum (max) or minimum (min)—is given for any particular timing. This particular RAM has a single data bus; an address bus; a chip select ($\overline{\text{CS}}$) that selects the entire chip; an output enable ($\overline{\text{OE}}$) that turns the output on and off, similar to $\overline{\text{RD}}$ in the memory designed in the first part of this section; and a write ($\overline{\text{WE}}$). (The term WR is also used for "write" and appears in the memory designed in the first part of this section.)

These diagrams appear complicated because they contain a lot of information. But focusing on the information a little at a time can clarify it. First, there are two diagrams for the read cycle and two for the write cycle. At the top of each diagram, t_{RC} and t_{WC} are the respective cycle times for the read and write. These are the times from the beginning of one read or write cycle until the beginning of the next—the basic speed of the memory. A 50-ns memory should have read and write cycles of 50 ns. (The read and write cycle times are usually the same for any given memory.)

To read from the memory, we

1. Place the proper address on the address bus
2. Assert $\overline{\text{CS}}$
3. Assert $\overline{\text{OE}}$

The first read is a simplified cycle in which both $\overline{\text{CS}}$ and $\overline{\text{OE}}$ are assumed to be asserted. It shows the delay, t_{AA}, from the time the address is placed on the address bus until data are available on the data bus ("address to data valid"). This is the single most important timing constraint of the memory. The second read is also a

Switching Waveforms

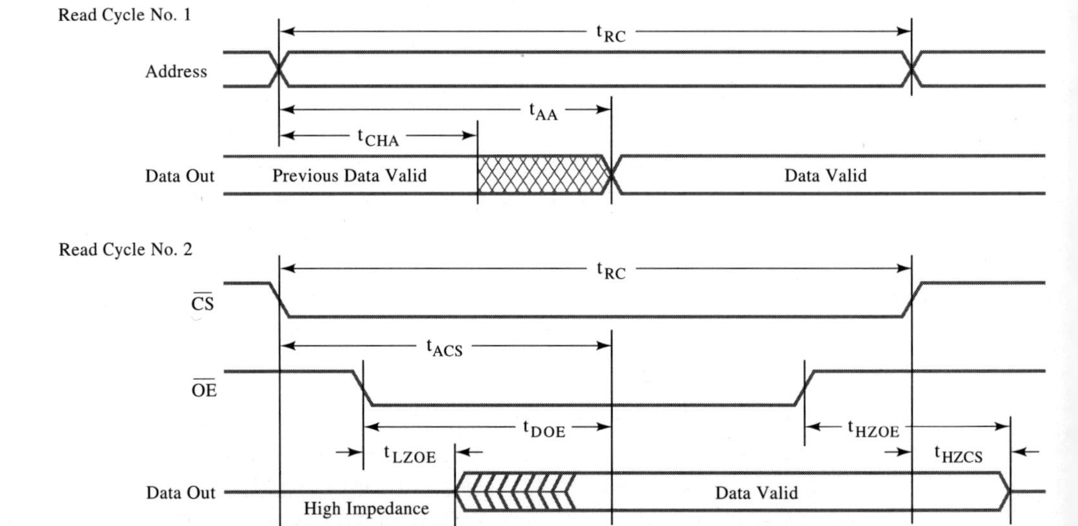

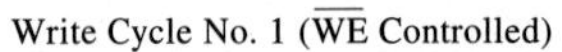

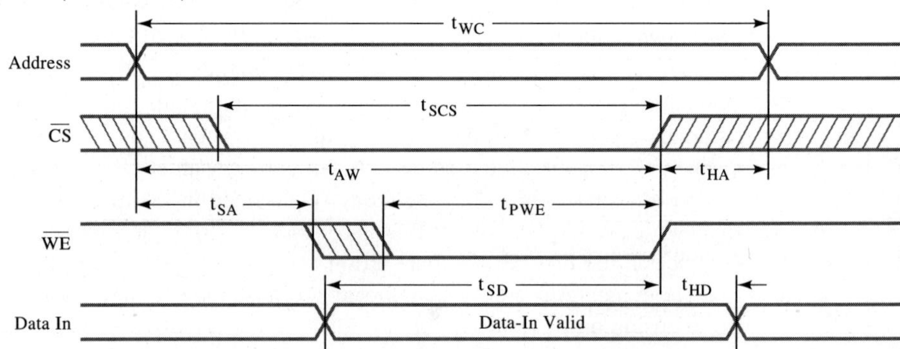

Write Cycle No. 2 ($\overline{CS}$ Controlled)

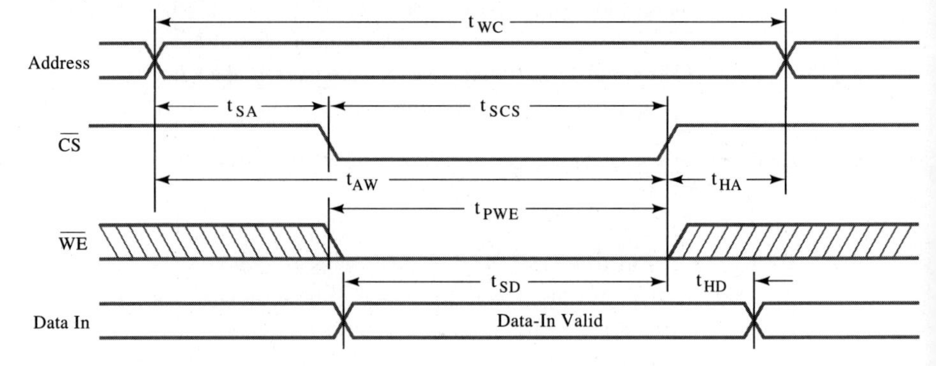

Figure 11.4-6 Typical memory timing.

Table 11.4-2 Values for Sample Memory Timing—Switching Characteristics over the Operating Range

Parameter	Description	7C170A-25		7C170A-35		7C170A-45		Unit
		Min.	Max.	Min.	Max.	Min.	Max.	
Read Cycle								
t_{RC}	Read cycle time	25		35		45		ns
t_{AA}	Address to data valid		25		35		45	ns
t_{OHA}	Output hold from address change	3		3		3		ns
t_{ACS}	$\overline{CS}$ low to data valid		15		25		30	ns
t_{DOE}	$\overline{OE}$ low to data valid		15		15		20	ns
t_{LZOE}	$\overline{OE}$ low to low Z	0		0		0		ns
t_{HZOE}	$\overline{OE}$ high to high Z		15		15		15	ns
t_{LZCS}	$\overline{CS}$ low to low Z	3		5		5		ns
t_{HZCS}	$\overline{CS}$ high to high Z		15		20		25	ns
Write Cycle								
t_{WC}	Write cycle time	25		35		45		ns
t_{SCE}	$\overline{CS}$ low to write end	25		35		40		ns
t_{AW}	Address setup to write end	20		30		35		ns
t_{HA}	Address hold from write end	0		0		0		ns
t_{SA}	Address setup to write start	0		0		0		ns
t_{PWE}	$\overline{WE}$ pulse width	20		30		35		ns
t_{SD}	Data setup to write end	10		15		15		ns
t_{HD}	Data hold to write end	0		0		0		ns

simplified read cycle in which the timing constraints on the address are assumed to have been met. It shows us the delay from the time $\overline{CS}$ is asserted until data are available on the data bus t_{ACS} ("$\overline{CS}$ low to data valid") or from the time $\overline{OE}$ is asserted until data are available on the data bus t_{DOE} ("$\overline{OE}$ low to data valid").

The remaining times in the second read cycle tell us how soon the high impedance condition of the data bus turns off after $\overline{CS}$ is asserted (t_{LZCS}) or after $\overline{OE}$ is asserted (t_{LZOE}) and how soon the high impedance condition turns on after $\overline{CS}$ is deasserted (t_{HZCS}) or after $\overline{OE}$ is deasserted (t_{HZOE}). These timing parameters are important because they determine when the data bus is available for other data transfer—which only is when all outputs but the one in use are tristated (high impedance).

To write to the memory, we

1. Place the address on the address bus
2. Place the proper data on the data bus
3. Assert $\overline{CS}$
4. Assert $\overline{WE}$

We saw in our simple memory design (Fig. 11.4-3) that either $\overline{CS}$ or $\overline{WE}$ (or WR) could control the memory write cycle, which is also true for this typical RAM. The first write cycle shows the $\overline{WE}$-controlled cycle; the second shows the $\overline{CS}$-controlled cycle. The key to understanding the write cycle is to realize that the data are clocked into the memory by the positive-going edge at the end of the $\overline{WE}$ pulse or $\overline{CS}$ pulse. The first of these edges to occur clocks the data. In Figure 11.4-6 look at the first write cycle, which is $\overline{WE}$ controlled. The write occurs where $\overline{WE}$ goes from low to high. Now look at the data (Data In). The memory acts like a D flip-flop—data must be valid a setup time, t_{SD}, before the write clock edge and a hold time, t_{HD}, after the write clock edge.

Four other important timing constraints apply in writing to memory. The address must be valid t_{AW} before the write clock edge and t_{SA} before the start of the write pulse. It must stay valid until t_{HA} after the write clock edge. These constraints allow the address decoding to become stable (no glitches) before write occurs. If these constraints are not met, a glitch may clock the data too soon or the address may not have time to be decoded correctly. Finally, the write pulse must be long enough, t_{PWE}. This write cycle assumes that the $\overline{CS}$ pulse starts before the $\overline{WE}$ pulse and ends after it. In addition, the $\overline{CS}$ pulse must be at least t_{SCS} long.

The second write cycle is the same as the first except that $\overline{CS}$ controls the cycle and the timing constraints are on $\overline{CS}$. In this cycle the $\overline{WE}$ pulse must start before $\overline{CS}$ and end after it. Table 11.4-2 gives short descriptions of all the timing parameters. In a data book, a number of notes are usually associated with the timing parameters; it is important to consult these notes. Remember that this is a sample of how the timing associated with a memory device is given in a data book. The format of the timing data varies among manufacturers, but the information contained is generally as in our sample.

11.5 DESIGNING VERY LARGE STATE MACHINES

In principle we could design large state machines by taking the approach we have used throughout this book. In practice, a state machine with, say, more than 64 states would be incredibly complicated. There is another approach that allows us to design certain very large state machines that are used only for timing and do not have branches. The idea is simple: we use a large counter, and the count output is the state. To decode the states for outputs, we use ROM.

Suppose we want to design a timer that will repeatedly time a 31,000-ns cycle. Starting at 500 ns, we want to assert the output PROC1 for 100 ns every 500 ns until we reach 20,000 ns. We also want to assert the output PROC2 at 500 ns, 21,000 ns, and 28,000 ns for 200 ns; the output PROC3 at 800 ns and 15,000 ns for 500 ns; and the output END at 31,000 ns for 100 ns. The timer should have a provision for a reset (RST) to start the timer at 0. The outputs should be glitch-free and asserted high; RST should be asserted low.

From the specification we can see that we need a timing resolution of 100 ns, so we'll choose a clock frequency of

$$f_{CLK} = 1/100 \times 10^{-9} = 1 \times 10^{7} = 10 \text{ MHz}$$

and in order to time to 31,000 ns, we'll need 310 states. That requires a counter with 9 bits, because $2^9 = 512$ but $2^8 = 256$, which is less than 310. By cascading three 4-bit counters (74LS163) we will have a 12-bit counter, which is more than adequate. We'll need to decode the terminal count (310) and use it to reset the counter to 0; we've chosen counters with a synchronous reset to make this easier.

The output and terminal count decoding will be done in a 2k × 8 ROM (CY7C291). This ROM is larger than necessary (we need 310 × 5), but smaller ROMs, especially EPROMs, are not readily available. The 2k × 8 ROM is available in both 35-ns and 50-ns versions. The 50-ns version will be adequate with a 100-ns timing resolution. We'll use a quad D flip-flop (74LS175) to deglitch the outputs. Figure 11.5-1 shows the circuit for the timer. We *or* the terminal count and RST to add an external reset. Either the terminal count or RST will start the counter at 0. Notice that RST is a synchronous reset. When it is low and a positive-going clock edge occurs, the counter is reset. RST must be long enough to catch a clock edge. If RST is held low, the counter resets to 0 at each positive-going clock edge and never counts. This circuit will function correctly only if the PROM is programmed (loaded) correctly.

Because the machine has a register output for all the outputs except the terminal count, the count (state) just before the output should occur must be decoded. To properly assert PROC1 on output O0 of the memory, we have to load a 1 into the O0 bit of the memory at addresses 4, 9, 14, and so on, up to 199. To properly assert PROC2 on output O1 of the memory, we have to load a 1 into the O1 bit of the memory at addresses 4, 5, 209, 210, 279, and 280. We load two consecutive addresses so that the output will have a 200-ns duration. To properly assert PROC3 on output O2 of the memory, we have to load a 1 into the O2 bit of memory at addresses 7, 8, 9, 10, 11, 149, 150, 151, 152, and 153. We load five consecutive addresses so that the output will have a 500-ns duration. To properly assert END on output O3 of the memory, we have to load a 1 into the O3 bit of memory at address 309. The terminal count on output O4 of the memory is not registered, so we don't decode the previous state. It's asserted low, however, so we have to load a 1 into the O4 bit of memory at all the addresses except 310, into which we load a 0.

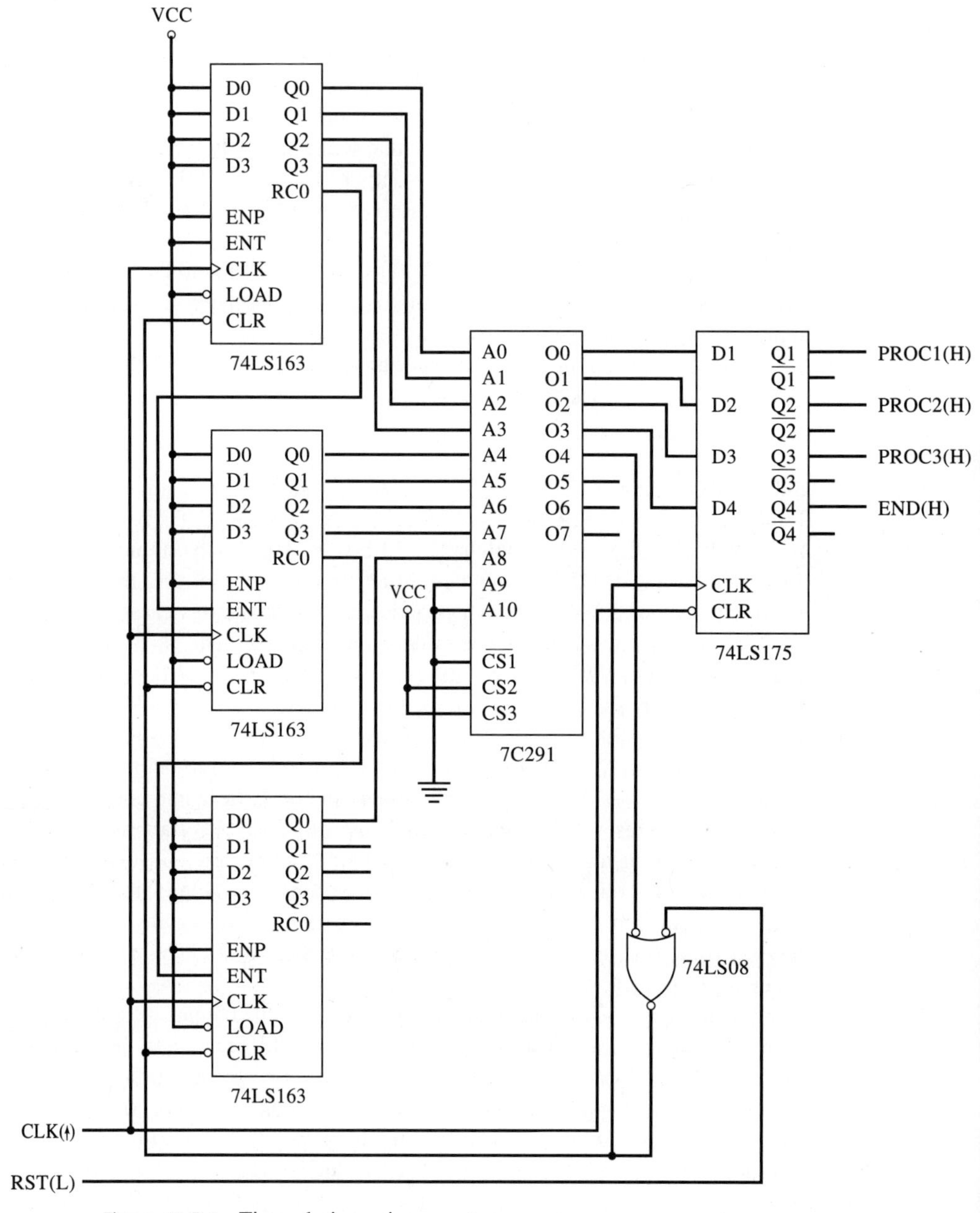

Figure 11.5-1 Timer design using counters.

This programming description must be translated into a table that shows the 8-bit word to be loaded into each memory location, starting at location 0. The first several entries in the table are 00010000, 00010000, 00010000, 00010000, 00010011, 00010010, and so on. Because we address only the first 512 locations in the memory (A9 and A10 are always 0), and we do not use any address above 310, it does not matter what we load into address 311 and above. Tabulating the memory by hand can be tedious even in this simple machine; normally we would write a computer program to do this tabulation.

We can increase the flexibility of the simple timer in two ways. First, we can make the output depend on inputs. If we connect the inputs to address lines of the memory, then we can decode these inputs in just the same way as we decode count bits. Second, we can make the terminal count depend on one or more inputs. To do this, we again connect the input or inputs to address lines of the memory and make different terminal counts conditional on the appropriate inputs.

Example 11.5-1

Modify the timer just designed so that it will count to a terminal count of 31,000 ns (as it does now) if the input LONG is asserted and to only 25,000 ns if LONG is not asserted. Also, make output PROC2 conditional on the input PRE. Assume LONG and PRE are asserted high.

This is an easy modification. The circuit is essentially the same. We only remove A9 and A10 from ground and connect LONG to A9 and PRE to A10. The programming changes to the PROM are more extensive. PROC1, PROC3, and END should be asserted independent of PRE and LOAD. This means they should be asserted for all combinations of A10 A9—00, 01, 10, and 11. In the previous design, they were asserted only for 00 since A9 and A10 were grounded. Now they must be asserted when A10 A9 is 01, 10, or 11, also. The positional value of A9 is 512, and the positional value of A10 is 1024, so 01 corresponds to 512, 10 corresponds to 1024, and 11 corresponds to 1536. We must load 1s not only into the addresses we previously loaded but also into addresses offset by 512, 1024, and 1536 from those addresses. For instance, for PROC1, we not only have to load a 1 into bit O0 of the ROM every fifth address, starting at 4 and ending at 199, but we also have to load a 1 at every fifth address, starting at $4 + 512 = 516$ and ending at $199 + 512 = 711$, and starting at $4 + 1024 = 1028$ and ending at $199 + 1024 = 1223$, and starting at $4 + 1536 = 1540$ and ending at $199 + 1536 = 1735$. We must follow the same kind of procedure for PROC3 and END.

PROC2 is conditional on PRE and independent of LOAD, so PROC2 must be asserted when A10 A9 is 10 or 11. As we saw, 10 and 11 correspond to offsets of 1024 and 1536, so we need to load 1s in the O1 bit of memory at addresses $4 + 1024 = 1028$, $5 + 1024 = 1029$, $209 + 1024 = 1233$, $210 + 1024 = 1234$, $279 + 1024 = 1303$, $280 + 1024 = 1304$, $4 + 1536 = 1540$, $5 + 1536 = 1541$, $209 + 1536 = 1745$, $210 + 1536 = 1746$, $279 + 1536 = 1815$, and

280 + 1536 = 1816. Notice that we don't load 1s in the addresses with no offsets (as we did in the previous design), because they correspond to PRE not asserted.

The terminal count must be 310 when LONG is asserted independent of PRE—that is, when A10 A9 is 01 or 11. These correspond to offsets of 512 and 1536. Now we have to remember that the terminal count is asserted low, so we must load *zeros* in bit O4 of the ROM at addresses 310 + 512 = 822 and 310 + 1536 = 1846. The terminal count must be 250 when LONG is not asserted independent of PRE—that is, when A10 A9 is 00 or 10, which correspond to offsets of 0 and 1024. Since the terminal count is asserted low, we must load *zeros* in bit O4 of the ROM at addresses 250 + 0 = 250 and 250 + 1024 = 1274. We must load 1s in bit O4 at all the other addresses.

The table for loading the ROM would be messy to tabulate by hand. A computer program is definitely the proper way to make the table.

KEY TERMS

Counters Sequential logic circuits that progress through the binary or decimal number in response to a clock.

Dynamic Memory Memory that stores information in small capacitors and has the disadvantage of needing a refresh but the advantage of small size, low cost, and low power consumption.

Memory A device containing an array of addressable storage locations.

Modulus The maximum count of a counter.

Predesigned Subcircuits FPGA circuits which have been designed and placed in a library.

Random Access Memory (RAM) Memory in which any storage location can be addressed independently of any other.

Read Only Memory (ROM) Memory in which the values in the storage locations are permanently stored and can only be read.

Registers An array of flip-flops, with a common clock, used to store digital information.

Ripple Counter A counter consisting of flip-flops in which the first flip-flop is a clock for the second, the second a clock for the third, and so on. Because of the delay through each flip-flop the output count is badly skewed.

Shift Registers A register in which each clock edge causes the state of every flip-flop to be passed to the next flip-flop on the right (left) and the first flip-flop to be loaded for an external source.

Simple Registers A parallel in–parallel out (PIPO) register in which the data are loaded simultaneously to all the flip-flops and then output simultaneously.

Static Memory Memory that stores information in simplified D flip-flops. Fast and easy to use, but larger and more costly than dynamic RAM.

Synchronous Counter A counter in which all the bits change simultaneously with the clock.

Universal Shift Register A bidirectional shift register with parallel in and parallel out.

EXERCISES

1. (a) Show how to connect two 4-bit ripple counters (74LS93) to make an 8-bit ripple counter. Consult a data book if you need to.
   (b) For the 74LS93, the maximum propagation delay from CLK A to Q0 is 18 ns, and that from CLK A to Q3 is 70 ns. For the 8-bit ripple counter you have designed in (a), what is the maximum frequency at which this counter can be operated before a change in the high order bit lags one clock cycle behind a change in the low order bit? This is the highest frequency at which the output count has meaning; beyond this frequency, the high order bit would be incorrect.
2. Show how to connect two 4-bit synchronous binary counters (74LS161) to make an 8-bit synchronous counter. Consult a data book if you need to.
3. Show how to connect three 4-bit synchronous binary counters (74LS161) to make a 12-bit synchronous counter. Consult a data book if you need to.
4. Design counters, using one or more 74LS163s and gates as needed, that will count to the following terminal counts and then start over at 0 and continue counting:

(a) 7	(j) 18	(s) 287
(b) 8	(k) 35	(t) 310
(c) 9	(l) 90	(u) 325
(d) 10	(m) 100	(v) 385
(e) 11	(n) 123	(w) 402
(f) 12	(o) 189	(x) 500
(g) 13	(p) 210	(y) 510
(h) 14	(q) 240	(z) 550
(i) 17	(r) 254	(aa) 600

5. Show how to connect two 4-bit synchronous up/down binary counters (74LS191) to make an 8-bit synchronous counter. Consult a data book if you need to.
6. Show how to connect three 4-bit synchronous binary counters (74LS191) to make a 12-bit synchronous counter. Consult a data book if you need to.
7. Design a 4-bit binary counter using a PAL16R8. The design should consist of the PALASM program needed to program the PAL. Assemble the program and simulate the design if you have access to the necessary software.
8. Make a state transition diagram for a 4-bit shift-right shift register.

9. (a) Show how to connect two 5-bit shift registers (74LS96) to make a 10-bit shift register. Consult a data book if you need to.

   (b) The minimum input data setup time to clock and the maximum output data propagation delay from clock for the 74LS96 are 30 ns and 40 ns, respectively. What is the highest frequency at which the setup time in the second shift register will be met?

10. (a) Show how to connect three 5-bit shift registers (74LS96) to make a 15-bit shift register. Consult a data book if you need to.

    (b) The minimum input data setup time to clock and the maximum output data propagation delay from clock for the 74LS96 are 30 ns and 40 ns, respectively. What is the highest frequency at which the setup time at the third shift register will be met?

11. (a) Show how to connect two 4-bit shift registers (74LS95) to make an 8-bit shift register. Consult a data book if you need to.

    (b) The minimum input data setup time to clock and the maximum output data propagation delay from clock for the 74LS95 are 20 ns and 32 ns, respectively. What is the highest frequency at which the setup time at the second shift register will be met?

12. (a) Show how to connect three 4-bit shift registers (74LS95) to make a 12-bit shift register. Consult a data book if you need to.

    (b) The minimum input data setup time to clock and the maximum output data propagation delay from clock for the 74LS95 are 20 ns and 32 ns, respectively. What is the highest frequency at which the setup time at the third shift register will be met?

13. (a) Show how to connect two 4-bit universal shift registers (74LS194) to make an 8-bit universal shift register. Consult a data book if you need to.

    (b) The minimum input data setup time to clock and the maximum output data propagation delay from clock for the 74LS194 are 20 ns and 26 ns, respectively. What is the highest frequency at which the setup time at the second shift register will be met?

14. (a) Show how to connect three 4-bit universal shift registers (74LS194) to make a 12-bit shift register. Consult a data book if you need to.

    (b) The minimum input data setup time to clock and the maximum output data propagation delay from clock for the 74LS194 are 20 ns and 26 ns, respectively. What is the highest frequency at which the setup time at the third shift register will be met?

15. Design a 4-bit shift-right shift register with parallel out, using a PAL16R8. The design should consist of the PALASM program needed to program the PAL. Assemble the program and simulate the design if you have access to the necessary software.

16. Design a 6-bit shift-right shift register with parallel out, using a PAL16R8. The design should consist of the PALASM program needed to program the PAL. Assemble the program and simulate the design if you have access to the necessary software.

17. Design an 8-bit shift-right shift register with parallel out, using a PAL16R8. The design should consist of the PALASM program needed to program the PAL. Assemble the program and simulate the design if you have access to the necessary software.

18. Design a 10-bit shift-right shift register using a PAL22V10. The design should consist of the PALASM program needed to program the PAL. Assemble the program and simulate the design if you have access to the necessary software.

19. Design a 4-bit bidirectional shift register with parallel out, using a PAL16R8. Make it shift right when MODE(H) is asserted and left when MODE is not asserted. The design should consist of the PALASM program needed to program the PAL. Assemble the program and simulate the design if you have access to the necessary software.

20. Design a 6-bit bidirectional shift register with parallel out, using a PAL16R8. Make it shift right when MODE(H) is asserted and left when MODE is not asserted. The design should consist of the PALASM program needed to program the PAL. Assemble the program and simulate the design if you have access to the necessary software.

21. Design an 8-bit bidirectional shift register with parallel out, using a PAL16R8. Make it shift right when MODE(H) is asserted and left when MODE is not asserted. The design should consist of the PALASM program needed to program the PAL. Assemble the program and simulate the design if you have access to the necessary software.

22. Design a 10-bit bidirectional shift register, using a PAL22V10. Make it shift right when MODE(H) is asserted and left when MODE is not asserted. The design should consist of the PALASM program needed to program the PAL. Assemble the program and simulate the design if you have access to the necessary software.

23. Design timers, using counters (74LS161) and 2k $\times$ 8 ROM (7C291), that do the following. In addition to counters and ROM, use any other circuit elements you need. Draw each circuit and specify how the ROM must be programmed. You need not make a complete memory table for the ROM. Make the outputs glitch-free.

    (a) A terminal count of 23,000 ns; output OUT1(H) asserted every 800 ns for 200-ns duration, starting at 800 ns; output OUT2(L) asserted at 10,000 ns for 600 ns.

    (b) A terminal count of 15,000 ns; output STOR(H) asserted every 200 ns for a

100-ns duration, starting at 4000 ns and ending at 12,000 ns; output END(H) asserted for a 100-ns duration at 15,000 ns.

(c) A terminal count of 12,000 ns; output STR(H) asserted every 300 ns for a 100-ns duration, starting at 900 ns and ending at 8100 ns; output FIN(H) asserted for a 100-ns duration at 10,000 ns.

(d) A terminal count of 41,000 ns; output HLP(H) asserted every 500 ns for a 400-ns duration, starting at 4000 ns and ending at 35,000 ns; output EDHLP(H) asserted for a 100-ns duration at 35,000 ns.

(e) A terminal count of 35,000 ns if input HLD(H) is asserted or a terminal count of 18,000 if HLD is not asserted; output PRE(L) at 10,000, 20,000, and 33,000 ns for a 100-ns duration; output POST(H) at 500, 8000, 12,000, and 17,000 ns for a duration of 400 ns—except at 17,000 ns the duration should only be 2000 ns on the short timing cycle.

(f) A terminal count of 12,000 ns if input LONG(L) is asserted or a terminal count of 6000 if LONG is not asserted; output STOR1(H) at 1000, 2000, and 3300 ns for a 500-ns duration; output STOR2(H) at 500, 8000, 10,000, and 12,000 ns for a duration of 100 ns.

(g) A terminal count of 43,000 ns if input BRIEF(L) is asserted or a terminal count of 23,000 if BRIEF is not asserted; output TERM1(H) every 600 ns for a duration of 200 ns, starting at 12,000 ns and ending at 38,000 ns; output TERM2(H) at 500, 8000, 10,000, and 12,000 ns for a duration of 100 ns.

(h) A terminal count of 18,300 ns if input LONG(L) is asserted or a terminal count of 7200 if LONG is not asserted; output STOR1(H) at 1300, 2500, and 17,000 ns for a 500-ns duration; output STOR2(H) at 500, 8000, 10,000, and 12,000 ns for a duration of 100 ns.

12 Digital Systems Design

12.1 INTRODUCTION

In this chapter we're going to study the design of **systems** that perform a sequence of operations just like a state machine but are more complicated to design. Our approach to these systems will be to **partition** them into more tractable pieces called **subsystems**. By "more tractable," we mean that the subsystems are simple enough to be designed from components we're familiar with. For some large or complicated systems the partitioning will have a number of layers. That is, we'll first partition the system into subsystems, then partition those subsystems into smaller subsystems, until we can readily see how to design the subsystems of the final, bottom-most layer. This kind of continual subdivision is called a **top down** or **hierarchical** design structure.

The variety of final subsystems that comprise the entire system can be almost endless. We will, however, limit the final subsystems in our designs to the following—combinational logic circuits that we'll have to design; combinational logic circuits, such as those we discussed in Chapter 5, that are standard and need not be designed; state machines that we'll have to design; and sequential logic circuits, such as those we discussed in Chapter 11, that are standard and need not be designed. In all cases, there will be at least one subsystem that controls the other subsystems, which we will call the **system controller**. In a large system there can be several controllers with one of them—also called the *system controller*—controlling all others. The controllers (or controller, if there's only one) will normally be state machines that we have to design. In order to accomplish the coordinated task of the system, the system controller will have to communicate with the other subsystems and some of the subsystems may have to communicate with each other. Often this communication will require a **handshake** between two subsystems.

We'll approach our design study by presenting several examples of digital system designs. During these examples we'll consider the problems of partitioning, communications, and controller design. Even the simplest digital system isn't simple, but we'll start with a fairly uncomplicated example and progress to something more complicated—a simple computer.

12.2 A SIMPLE DIGITAL SYSTEM

Suppose we must design a digital system to change parallel input data to serial output data. The parallel data are presented to us as 4-bit parallel words on a 4-line input bus A[3:0](H). The serial data SDAT(H) are output on a single output line. The serial data should be output at 1 Megabit/sec. Each slice of four serial data bits (output word) is to be followed by at least three zero bits before the next 4-bit word is started. The most significant bit should be output first.

The data input is under the control of two **handshake** lines—an input RDY(L) and an output NEXT(H) as shown in Figure 12.2-1. We use NEXT(H) to request new data. When NEXT is asserted, it causes valid input data to be presented on the input bus. The hatched part of A3 through A0 in the figure indicates where the data on these lines are not valid. Data become valid in response to NEXT, and RDY is asserted to indicate that the data are valid. When NEXT is deasserted, the data become invalid and RDY is deasserted until NEXT is asserted again. Notice how we show the cause-effect relation of the signals in the figure. A circle on a signal level (or an edge) indicates it's a cause. An arrow from the circle to another signal level indicates it's an effect. Thus, NEXT going high causes valid data. If a combination of signal levels is the cause, these signals are connected by lines and a dot is placed on each of the signals. Thus, having all four input signals valid causes RDY to go low. We'll find it useful to indicate a cause-effect relation in this same way on many of our future timing diagrams.

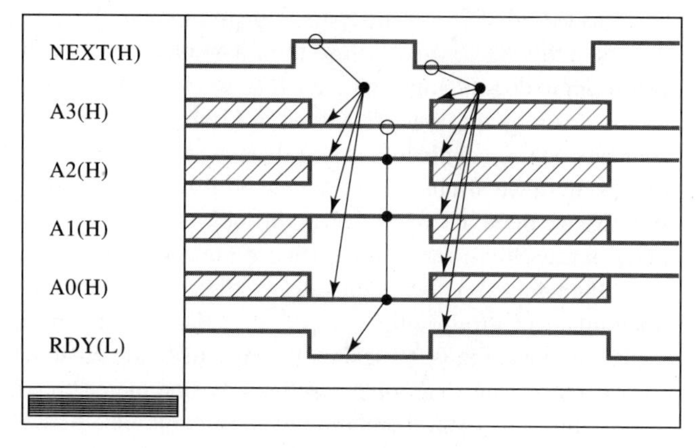

Figure 12.2-1 Handshake for input data.

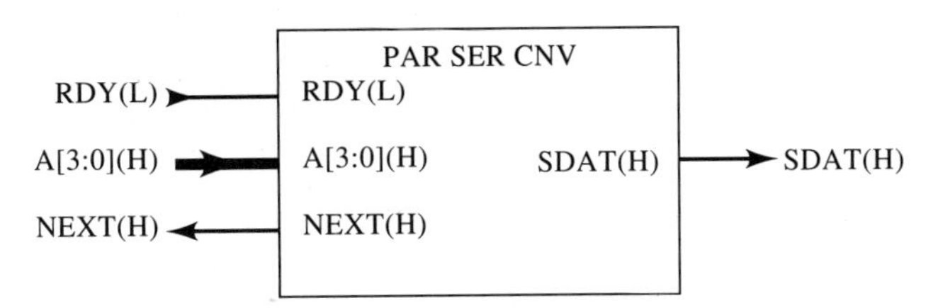

Figure 12.2-2 The overall system with inputs and outputs.

The first thing we need to do in this or any design is to make a clear word description of the problem. Actually, we'll include at least one block diagram in our description, and it may be necessary to include timing diagrams as well. The description should include a list of inputs to the digital system, a list of outputs, and a description of which inputs cause which outputs along with any timing constraints. In the preceding specification, there are five inputs—the input data bus A[3:0] (four lines) and the handshake RDY—and two outputs—the serial data stream SDAT and the handshake NEXT. The relation of the input data to the handshake was shown in Figure 12.2-1. The output words are 4-bit serial words with a 1 MHz clock and at least three zero bits in between each word. We'll find it helpful to represent the system at this point in the design as a single block (one big block) with inputs and outputs as shown in Figure 12.2-2. Notice that we use arrows to show the direction of signal flow on the block diagram.

The system we are designing would be complicated if a straight state machine design approach were taken. If it's partitioned into subsystems, however, it becomes a simple system. One of the subsystems in any digital system is always the system controller that directs the other subsystems. It's important that we start the design by first considering these other subsystems. To find the other subsystems, we examine the functions that the digital system must perform and make a partition for each function (a **functional partition**). In the system we're discussing now, it's easy to see that one of the subsystems should be a 4-bit shift register with parallel input (such as the 74LS95 described in Section 11.3). Also, because we need to count the four data bits and the three zero bits, we need a counter. We might use the system controller to do this counting, but this would add six or seven states to the controller—four states to count the four serial data bits out and two or three to count the three zero bits out. We'll choose instead to make a very simple system controller and add a subsystem—a counter such as the 74LS161 counter described in Section 11.2.2. Finally, we need a 1 MHz clock with a square wave output. We make the clock 1 MHz because this is the output data rate and we can use the clock to clock the output data. Figure 12.2-3 shows our first attempt at partitioning the system. In all of our diagrams, we'll try to put inputs on the top and left, put outputs on the bottom and right, and make the signals generally pass from the upper left to the lower right. This format makes our diagrams easier to interpret.

In this first partition, we do not attempt to label all signal paths interconnecting the subsystems. We do label those that we know such as the inputs and outputs. We

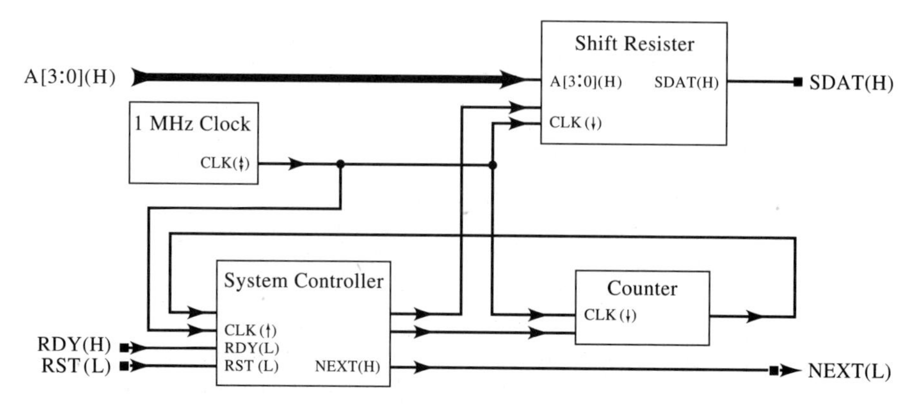

Figure 12.2-3 Preliminary functional partition.

also attempt to show the interconnections between the subsystems as much as possible. As we did with our first block diagram in this section (one big block), we show the directions of signals here with arrows. The input bus A[3:0] is connected to the shift register because this is where the actual parallel-to-serial conversion is made. The output is taken from the shift register for the same reason. The input handshake RDY(L), on the other hand, is connected to the system controller, as is the RST(L) line. One or more control lines (unlabeled at this point in the design and shown as a single line) are connected from the system controller to both the shift register and the counter. A control line connects the counter to the system controller to signal when the counter has counted the desired number of output bits. The output handshake NEXT(H) is taken from the system controller because it will be generated there. Because this system is slow, we can avoid confusion in timing if we let the system controller change state on low-to-high clock transitions and let the shift register and counter perform their functions on the high-to-low clock transitions. We show these choices in the diagram by using an up arrow to represent a low-to-high transition and a down arrow to represent a high-to-low transition. The clock subsystem output has an arrow at each end because both the low-to-high and high-to-low transitions are used in other subsystems. Actually, it's not important to the clock design which edges are active, so this is an embellishment that is not really needed.

As the next step in the design, we refine the block diagram. It's important to recognize that before we can design the system controller we have to design all the other subsystems (at least partly). In fact, we'll be led to the functions the system controller must perform by determining exactly what inputs the subsystems need for their required functions and what outputs they produce.

By using the devices suggested earlier for the subsystems, it's easy to determine the exact inputs needed and the exact outputs produced by each of the subsystems. The task is made easier still by inserting the device symbols for the shift register and counter in the block diagram as we've done in Figure 12.2-4. Now we can look at all the inputs to the shift register and determine if they are generated by the system

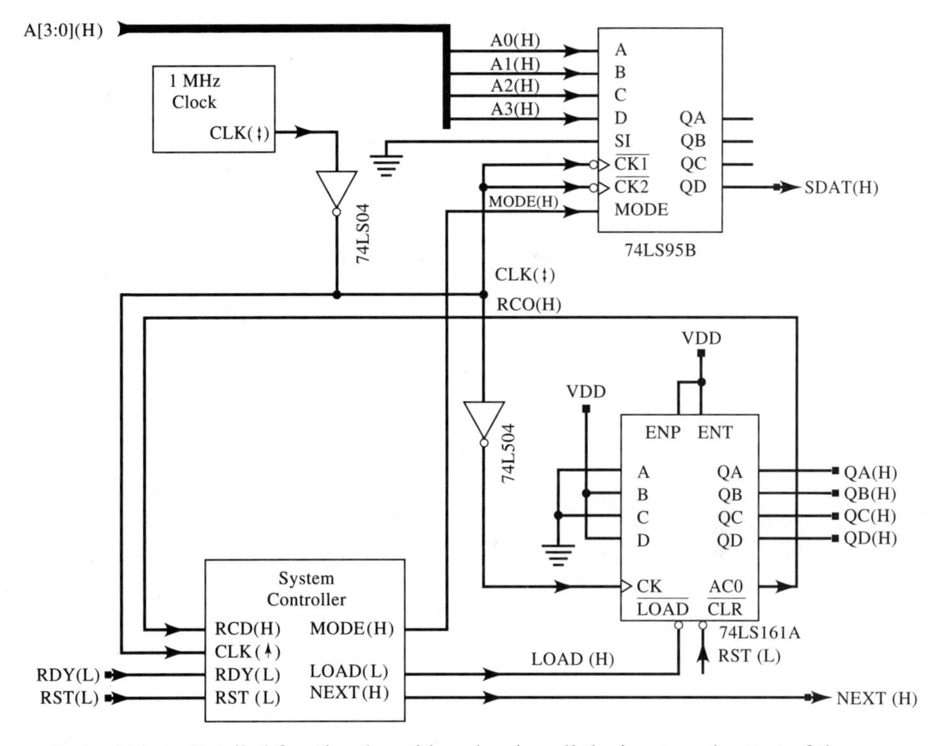

Figure 12.2-4 Detailed functional partition showing all the inputs and output of the subsystems.

controller, by one of the other subsystems, or by inputs to the system from the outside.

The 74LS95 shift register needs inputs to each of the parallel inputs A through D. These inputs come directly from the input bus A[3:0]. On the diagram, notice how we break the bus out into the single separate input of the shift register. To control the loading of the shift register from the parallel input or the shifting of data through the shift register, we need a MODE control line from the system controller. We need a clock on both clock inputs ($\overline{\text{CKL1}}$ and $\overline{\text{CLK2}}$)—one to load parallel data and one to shift the register (MODE tells us which function to perform). This shift register is clocked on the negative going clock edge, just as we require. We connect the serial input (SI) to ground so that we can shift in zeros between the serial output data words. The parallel outputs QA through QC are not used, and QD is the serial output SDAT(H). This completes the input and outputs to the shift register. Now let's look at the counter.

The counter has a clear input $\overline{\text{CLR}}$ connected to RST(L). We need the counter to count the number of clock cycles needed to shift the serial data word and three zeros out of the shift register. We'll look at the details of the exact count later, but first we'll look at the two ways to count. One way is to reset the counter to zero and count to the desired value, then produce an output to the controller. This requires a count decoder (a gate). A better way is to use the ripple carry-out (RCO) output of the counter to indicate that we have reached the desired value. RCO is output when the counter reaches a count of 15. If we load the counter with the difference between our desired value and 15, RCO will be asserted when we reach the desired value. This means we need to load the difference into the counter with a LOAD control from the system controller as shown in the figure. The actual number to be loaded is hard-wired on A through D. We've hard-wired a 10—D and B high, and C and A low—so we count five clock cycles from LOAD asserted to RCO asserted. ENT and ENP control whether the counter counts or not (holds the previous count). Both ENT and ENP must be asserted for the counter to count. We might suspect that we need to control the count with a control line from the system counter, but this is not the case. We tie both ENT and ENP high so that the counter counts continuously. We can do this because the counter is reset by loading the initial count each time we do a parallel load. We do not use the outputs QA through QD, although we've labeled them in our simulation. RCO from the counter is connected to the system control so that the system controller will know when the output cycle of four serial bits and three zeros is completed. The clock into the counter must be inverted because the counter counts on the positive going clock edge and we want it to count on the negative going edge. This completes the input and output connections to the counter.

The final subsystem, excluding the system controller, is the clock. A 1 MHz clock is available as a single device in a package much like a TTL device. The principal difference is that the clock has a metal cover. The clock has no inputs except power and ground and has one output—the 1 MHz signal. Clock circuits usually have very poor (low fan-out) output drivers, yet they often drive all the other subsystems in a system. In our circuit, the clock drives five inputs—two in the shift register, one in the counter, and two in the system controller. (We'll have to design the system controller to verify its two clock inputs, and we haven't done this yet.) A fan-out of five is probably not too many, but we'll add an inverter as a driver on the clock just for safety. There will be an inverter chip in the design with unused inverters available that we can use.

By determining inputs and outputs for all the subsystems except the system controller, we've already designed the interconnections of all subsystems *including* those of the system controller. The inputs and outputs of the system controller, however, are only part of the information we need to design the controller itself. We also need to know the *timing relationships* between the system controller inputs and outputs. We find these relationships by constructing a timing diagram that shows the timing constraints between the system controller inputs and outputs, and we find these constraints by examining the input requirements of each of the subsystems.

We can see the timing relationships between the system controller inputs and outputs in Figure 12.2-5. This diagram shows the clock at the top, and the input handshake and date next. At first, RDY is not asserted (high) and parallel input data A[3:0] is not valid. The X on A[3:0] indicates that it is unknown. Nothing should happen until RDY is asserted and the input data are valid, so we assert RDY and

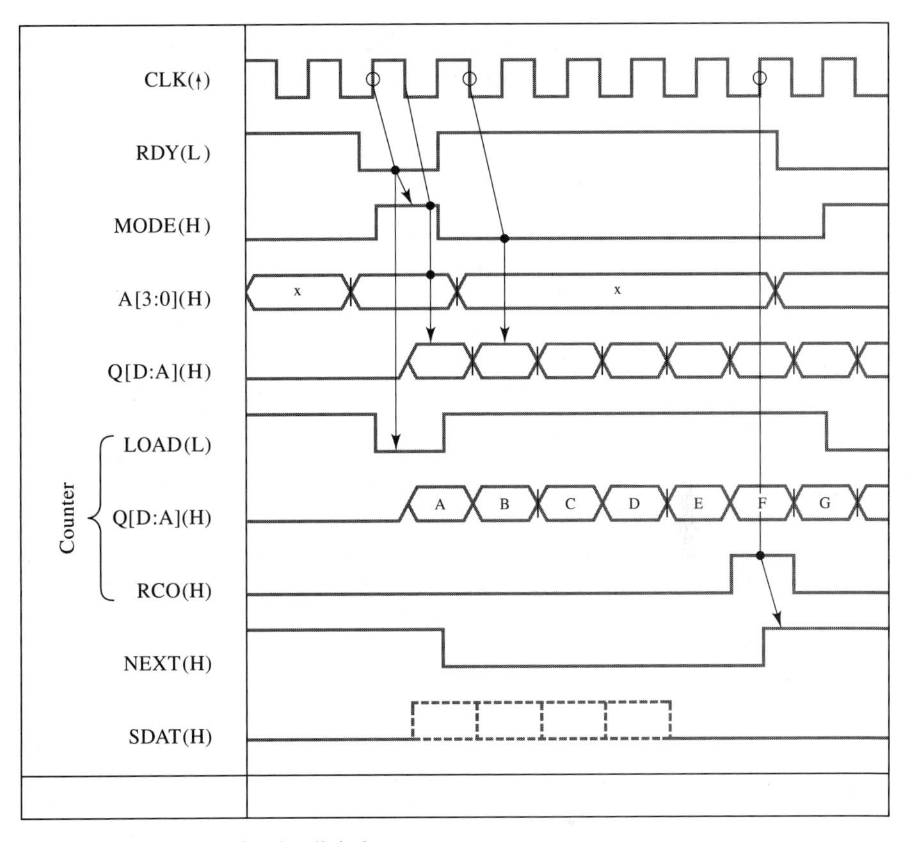

Figure 12.2-5 Timing for the digital system.

make the data valid. That is, on the diagram, we take RDY low and show that the parallel input data change. It doesn't matter in this diagram what the parallel input data are, simply when they change. When RDY is asserted and the input data are valid, we want to load this parallel input data into the shift register. To make the shift register load the data, we must first assert MODE in response to RDY and then let $\overline{\text{CLK2}}$ go through one high-to-low transition. MODE is an output that the system controller must produce. Notice how we show on the timing diagram that MODE is set high in response to the positive clock edge and RDY—with a line from the clock edge to RDY (low), which are the causes, and an arrow to MODE (high), which is the effect. Now we see that if MODE is asserted for one clock cycle (see the timing diagram), the clock will make the high-to-low clock transition we need to load the parallel data. Notice how we show that this negative going clock edge, MODE (high), and the parallel input levels cause the shift register to be loaded.

To make the shift register shift the data out on the QD output, we must deassert MODE and clock the CLK1 input three times (high to low). The load of the shift register already causes an output of the high order bit. To output the three zero bits, we would expect that we need to clock three more times. Because the serial input to the shift register is tied low, these three clock cycles will shift three zero bits from the serial input through the shift register to the output. Actually, we only need to count two clock cycles because the system controller timing allows an additional shift after the counter reaches its final value. We probably would not see this on our initial construction of the timing diagram until we finish the timing cycle. We would then have to adjust the count to make our system function correctly. In the fantasy world of a textbook, the designer is perfect and can always foresee these subtleties (primarily because he has already finished the design).

To count five clock cycles (three data and two zeros), we load the 74LS161 counter with 10 and allow it to count the clock (high-to-low transitions) until it reaches 15. RCO is asserted by the counter when its terminal count (15) is reached, and we use this RCO to indicate that a count of five has been reached. To load the counter when we start processing the serial word output, we must assert LOAD at the same time MODE is asserted. We load 10 (1010) into the counter by tying the counter parallel inputs D and B to VDD (+5) volts and C and D to ground. In the timing diagram, we've shown the details of the counter timing so that we can be sure that there is no end effect that might cause us to gain or lose a count as we start or stop the counter. The value of the count Q[D:A] (for the counter, not the shift register) is shown in hexadecimal in the diagram. Notice how we have shown that the count starts at A (10) on the positive-to-negative clock edge when LOAD is asserted and changes on each subsequent negative clock edge until it reaches F (15), at which time we assert RCO. Because the counter counts continuously, RCO is only asserted for one clock cycle. This is adequate. We have one additional function to complete the processing of a word. We have to assert NEXT and ask for a new word.

Notice that in order to assert NEXT we must first deassert it. This can be done any time after we loaded the parallel input data into the shift register and before the processing of the word is completed. We have deasserted it at the same time that MODE and LOAD are deasserted. We actually need to continue the timing for the start of processing for the next word to make sure we have three zeros between each serial word. When NEXT is asserted, the input handshake RDY can be asserted and the input data can become available with only a small delay if the system inputting data is fast. When RDY is asserted, we start processing a new word. MODE and LOAD are asserted on the next negative-to-positive clock edge, and the new parallel data are loaded on the following positive-to-negative clock edge. If we show these changes on the diagram, we can see that there are three zeros between the end of the first serial data word and the loading of the next word. The bottom signal in the timing diagram is the serial output. This gives us the same information as Q[D:A] but spread serially in time. It's shown dashed because we don't know if it's high or low; its value depends on the parallel input. Actually, the value is not important, but the transition times are.

We've arrived at a satisfactory partitioning of the system and satisfactory timing on our first attempt in this system because any trial and error involved has already been done by the designer. Because most designers are not perfect, we would

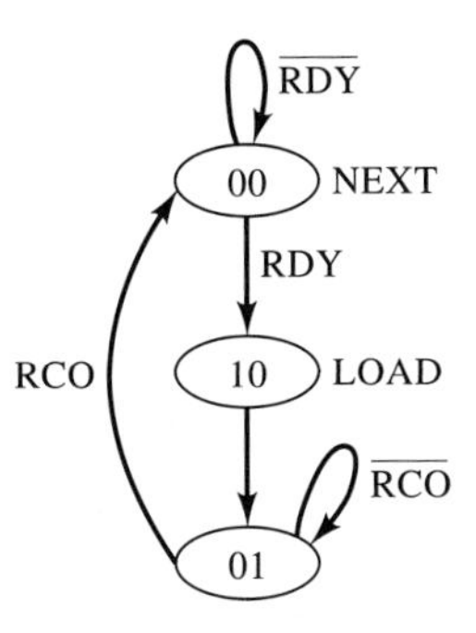

Figure 12.2-6 State transition diagram for system controller.

normally have to make a trial partition, then check this partition by looking at the details of the interconnections of the subsystems and the associated timing constraints. We would then resolve any problems by changing the partitioning, interconnections, and/or timing. This cycle may well have to be repeated several times before a satisfactory design is achieved.

Now that we have the interconnections of the subsystems, we can see what the system controller must do. From the interconnection and timing diagrams, we see that the system controller must assert outputs MODE and LOAD for one clock cycle in response to input RDY. It must then wait for RCO, and in response to RCO, it must assert NEXT. The state transition diagram for a system controller that will do this is shown in Figure 12.2-6. Notice that because this state transition diagram is a simple one, we haven't tried to simplify the next state decoder design. We have, however, chosen the state code so that both NEXT and MODE will be glitch-free—neither of them has an output decoder. This is important because glitches in NEXT can cause new parallel input data to be loaded at the wrong time and glitches in MODE can produce false clocks in this particular shift register because of its design.

The next state decoder maps for D memory elements and the output decoder maps for the system controller are shown in Figure 12.2-7. The simplified next-state and output decoder logic is given below the map. There is no map for LOAD because it's the same as MODE (except that it is asserted low). Notice that there is no decoder

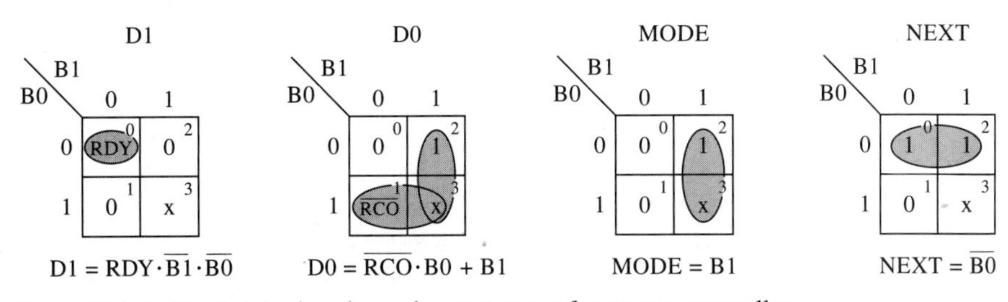

Figure 12.2-7 Next state decoder and output maps for system controller.

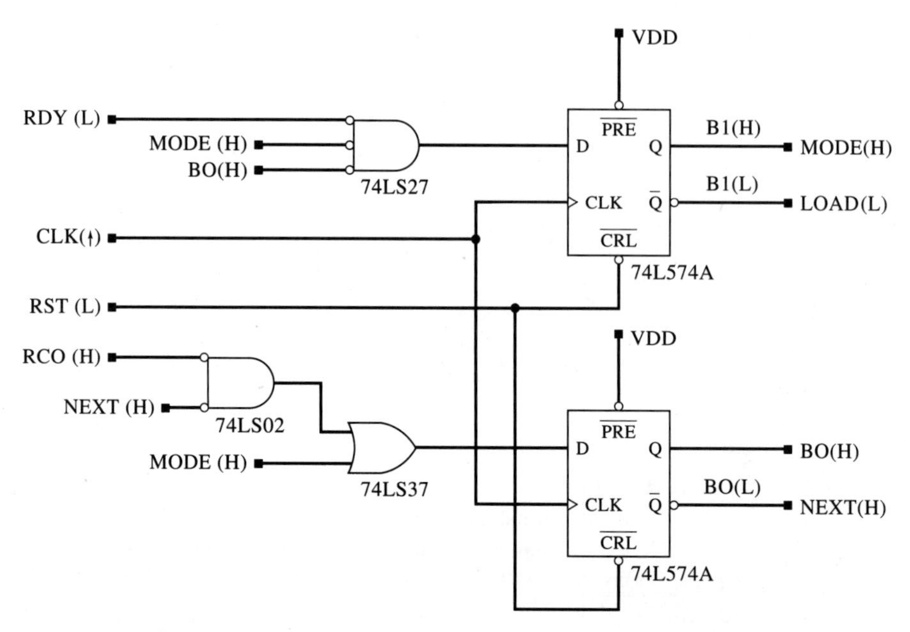

Figure 12.2-8 System controller.

needed for NEXT and MODE. Figure 12.2-8 shows the circuit for the system controller. This circuit along with the circuit of Figure 12.2-4 completes the design. In the documentation we're using, Figure 12.2-4 is not only the second-level partition of the system, it's also a second-level circuit diagram. This is a real advantage. As we complete the partitioning of the system, we've also completed the circuit diagrams. There is no need to combine the two circuits; the **hierarchical** circuits are entirely satisfactory. In fact, a hierarchical set of circuit diagrams gives us a clearer description of the system than a combined or **flat** circuit diagram.

Now let's simulate the system we've designed and see if it performs properly. Figure 12.2-9 shows the results of our simulation. In this simulation, the first parallel input is 0101. We could have used any input for the simulation test. Notice that the serial output is 0101, as it should be. There are five serial zeros following the 0101 before the next input. The reason for five zeros and not three is that the parallel input wasn't available quickly enough. This is because of the simulation input we chose. RCO actually occurred after three zeros. The second parallel word is 0001, and again the serial output is the same, 0001. We've included the state code bits B1 and B0 as outputs in the simulation so we can make sure that the system controller is traversing the proper states. Notice that it starts in the 00 state, remains there until RDY is asserted, then goes through the 10 state in one clock cycle and stays in the 01 state until RCO is asserted.

A modern logic circuit designer would not design the system we've just finished in discrete logic. She or he would most likely design it in a single PAL as in the following example.

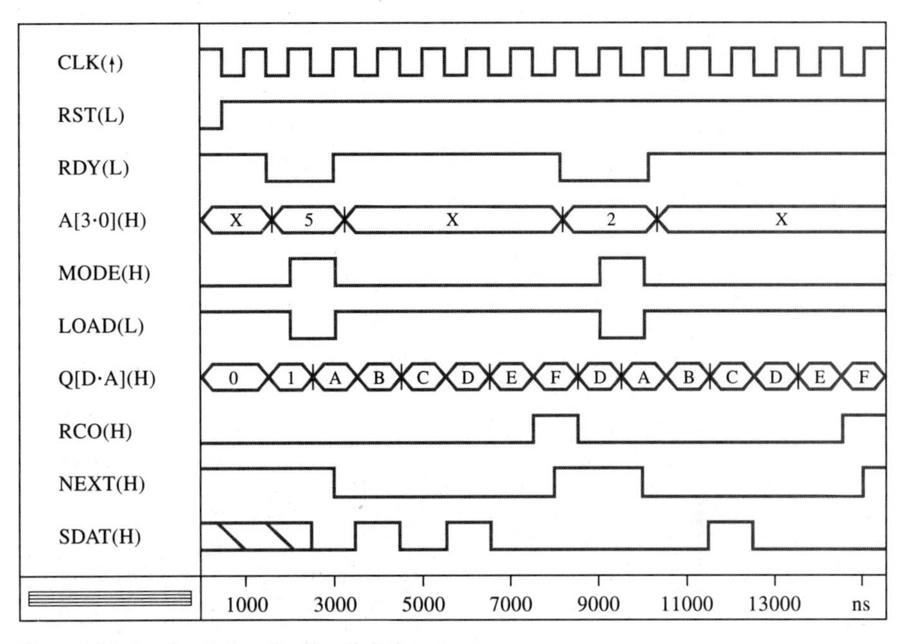

Figure 12.2-9 Simulation for the digital system.

Example 12.2-1

Redesign the digital system described in this section in one or more PLDs.

The design we just completed required 2 flip-flops in the controller, 4 in the shift register, and 4 in the counter. It should just fit in one 22V10 because the 22V10 has 10 flip-flops (see Example 10.2-2). Here we're assuming that we can design a 4-bit shift register and 4-bit counter in the PAL, and we should be able to do this. Actually, we could encounter difficulty fitting the entire system in the PAL if we have to deglitch the output NEXT with a flip-flop, because this adds a flip-flop to our previous design. (The previous design did not require a registered NEXT because it was already glitch-free.)

Let's try putting the entire system in the single PAL anyway by dispensing with the counter and adding the count function to the system controller. By doing this, the first-level (functional) block diagram for the system (the single big block) is unchanged, but the second-level diagram becomes as follows.

We need a new timing diagram for this design because both the controller and the shift register are different. First, the controller does not interface with a counter but rather does the count function internally. Second, the PAL has only one clock, so the transitions of the shift register must occur on the low-to-high clock transition. The timing we need is shown below.

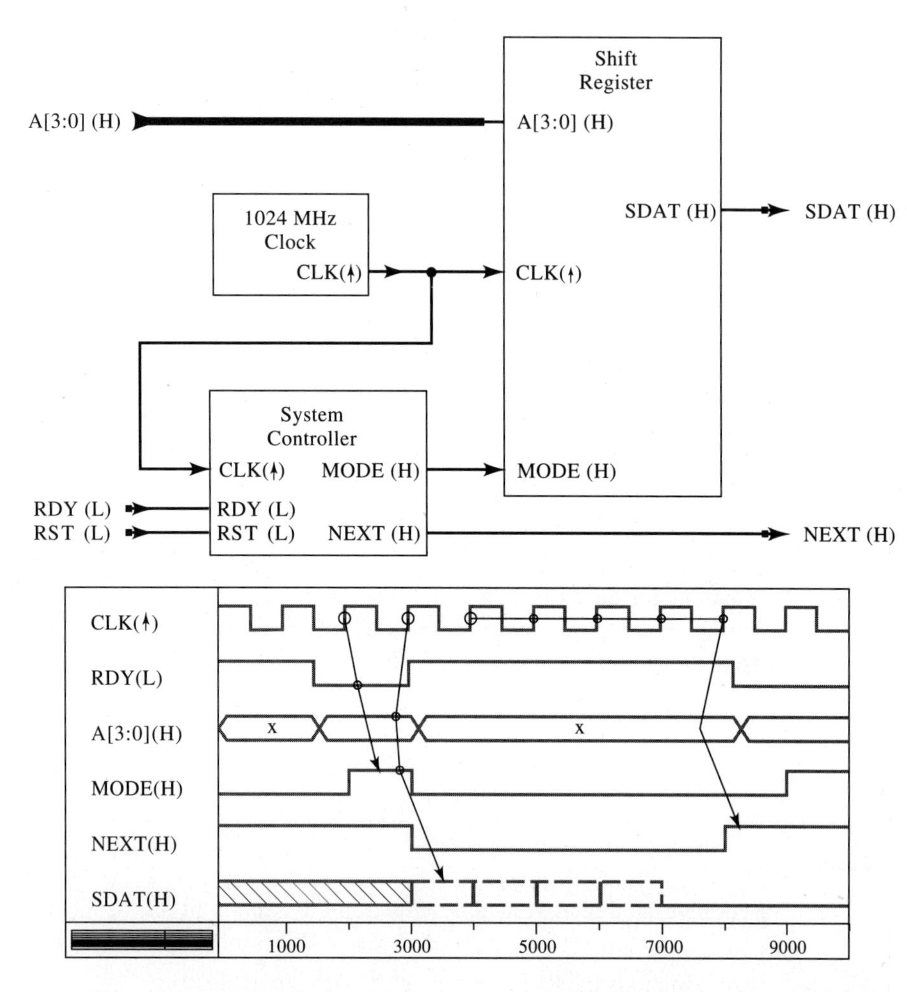

Notice the cause-effect relationships in this diagram. RDY and a clock edge cause MODE to be asserted. MODE, valid data, and a clock edge cause the parallel word to be loaded into the shift register and the first bit to be output. Finally, five clock cycles after the shift register is loaded, NEXT is asserted to request the next word. Seven clock cycles are needed to output four bits of data and three bits of zeros, but the new data are not loaded for at least two clock cycles after NEXT is asserted.

The block diagram showing the inputs and outputs of the system controller and the timing diagram showing their timing allows us to design the system controller. The controller has only one input RDY besides the clock, and two outputs NEXT and MODE. It does require five timing states to get the time delay needed before asserting NEXT. The state transition diagram for the system controller is shown below. In this diagram, we have given each state both a name and a state code because we are going to program it into a PAL. The state transition diagram is all we need to do the programming. This programming can be seen in the PALASM source program below. Remember that the programming of a state machine has state

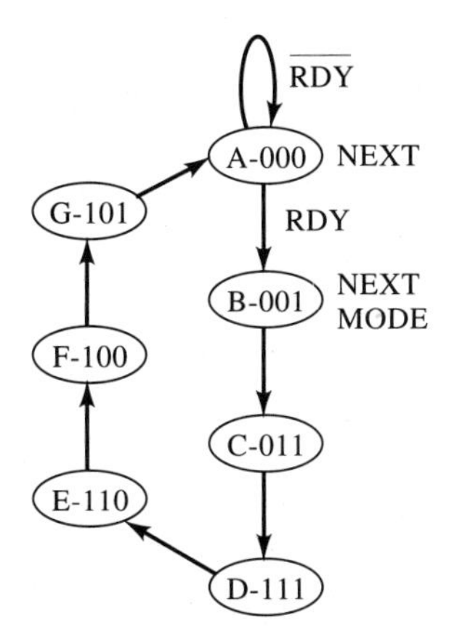

assignments, state transitions, outputs, and conditions. Notice that the only output from the controller is NEXT; MODE is internal to the PAL and is treated differently from an external output.

Besides programming the state equations in the PAL, we must also program the equations for the shift register. We'll call the bits of the shift register Q0, Q1, Q2, and Q3 and use our results from Chapter 11 (Section 11.3) to program the shift register. The logic expressions from that section that describe a shift register are

$$Q0(n + 1) = A$$

$$Q1(n + 1) = Q0(n)$$

$$Q2(n + 1) = Q1(n)$$

$$Q3(n + 1) = Q2(n)$$

where we have shortened the shift register to 4 bits. These equations need to be changed for the present context. First, we can drop the n and n + 1 arguments because they simply tell us how the Qs change each time the register is clocked. The PAL logic equations have exactly this form. The expression on the right tells us what the variable on the left will become at the next clock edge. These expressions are for a simple shift register with no parallel load. We need to add a parallel load that depends on MODE. If MODE is not asserted, the equations should be as above. If mode is asserted, the variables should be loaded from A0, A1, A2, and A3. With this addition and changing to PALASM notation, in which := denotes a register output,

$$Q0 := A0 \cdot MODE$$

$$Q1 := A1 \cdot MODE + Q0 \cdot \overline{MODE}$$

$$Q2 := A2 \cdot MODE + Q1 \cdot \overline{MODE}$$

$$Q3 := A3 \cdot MODE + Q2 \cdot \overline{MODE}$$

The serial input for this design is always 0, so it drops out of the first expression. Finally, we need to see how to handle the internal variable MODE. If we consult the state transition diagram, we see that MODE is asserted in state 001, so

$$MODE = \overline{B2} \cdot \overline{B1} \cdot B0$$

which means that everywhere MODE occurs in the equations for the shift register it can be replaced by $\overline{B2} \cdot \overline{B1} \cdot B0$. In the source program we've given the shift register its own section of code and labeled it the shift register equations.

The 22V10 has an asynchronous reset that we use to reset all the flip-flops to zero via the GLOBAL.RST equation in the source program; it has a separate synchronous preset for each of the shift registers that we are not using; and it has tristate outputs that are programmed to always be enabled via the ∗.TRST equations in the source program. All the outputs are enabled so they can be measured for debugging.

```
;PALASM Design Description

;---------------------------------------- Declaration Segment ----------------------------------------

TITLE      SERIAL TO PARALLEL CONVERTER
PATTERN
REVISION   A
AUTHOR     ALAN W SHAW
COMPANY    UTAH STATE UNIVERSITY
DATE       08/28/92

CHIP     STPCNV   PAL22V10

;------------------------------------------ PIN Declarations ------------------------------------------

PIN    1      CLK                          ; CLOCK
PIN    2      /RST                         ; INPUT
PIN    3      A0                           ; INPUT
PIN    4      A1                           ; INPUT
PIN    5      A2                           ; INPUT
PIN    6      A3                           ; INPUT
PIN    7      /RDY                         ; INPUT
PIN    8      NC                           ;
PIN    9      NC                           ;
PIN    10     NC                           ;
PIN    11     NC                           ;
PIN    12     GND                          ;
PIN    13     NC                           ;
PIN    14     Q0               REGISTERED  ; OUTPUT
PIN    15     Q1               REGISTERED  ; OUTPUT
PIN    16     Q2               REGISTERED  ; OUTPUT
PIN    17     Q3               REGISTERED  ; OUTPUT
PIN    18     B0               REGISTERED  ; OUTPUT
PIN    19     B1               REGISTERED  ; OUTPUT
PIN    20     B2               REGISTERED  ; OUTPUT
PIN    21     NEXT             REGISTERED  ; OUTPUT
PIN    22     NC                           ;
PIN    23     NC                           ;
PIN    24     VCC                          ;
NODE 1        GLOBAL                       ;
```

```
;------------------------------------- State Equation Segment -------------------------------------
STATE

MOORE_MACHINE

A = /B2*/B1*/B0
B = /B2*/B1* B0
C = /B2* B1* B0
D =  B2* B1* B0
E =  B2* B1*/B0
F =  B2*/B1*/B0
G =  B2*/B1* B0

A := NOT_RDY → A
      + _RDY → B

B := VCC → C

C := VCC → D

D := VCC → E

E := VCC → F

F := VCC → G

G := VCC → A

A.OUTF := NEXT
B.OUTF := NEXT
C.OUTF := /NEXT
D.OUTF := /NEXT
E.OUTF := /NEXT
F.OUTF := /NEXT
G.OUTF := /NEXT

CONDITIONS

_RDY = RDY
NOT_RDY = /RDY

;------------------------------------ Boolean Equation Segment ------------------------------------
EQUATIONS

B0.TRST = VCC
B1.TRST = VCC
B2.TRST = VCC
Q0.TRST = VCC
Q1.TRST = VCC
```

```
Q2.TRST = VCC
Q3.TRST = VCC
GLOBAL.RSTF = RST

;------------------------------------- Shift Register Equations -------------------------------------

Q0 := A0*/B2*/B1*B0
Q1 := A1*/B2*/B1*B0 + Q0*/(/B2*/B1*B0)
Q2 := A2*/B2*/B1*B0 + Q1*/(/B2*/B1*B0)
Q3 := A3*/B2*/B1*B0 + Q2*/(/B2*/B1*B0)

;--------------------------------------- Simulation Segment ----------------------------------------
SIMULATION

TRACE_ON CLK /RST A0 A1 A2 A3 NEXT /RDY Q0 Q1 Q2 Q3 B0 B1 B2
SETF RST /RDY
CLOCKF CLK
SETF /RST
CLOCKF CLK
SETF RDY A3 /A2 A1 /A0
CLOCKF CLK
CLOCKF CLK
CLOCKF CLK
SETF /RDY
CLOCKF CLK
CLOCKF CLK
CLOCKF CLK
CLOCKF CLK
CLOCKF CLK
SETF RDY /A3 /A2 /A1 A0
CLOCKF CLK
CLOCKF CLK
CLOCKF CLK
SETF /RDY
CLOCKF CLK
CLOCKF CLK
CLOCKF CLK
CLOCKF CLK
CLOCKF CLK

;----------------------------------------------------------------------------------------------------
```

The PALASM simulation results for this design are given below. Following the reset (RST), the first parallel input is 1010 and the output (Q3) is indeed a serial 1010 followed by 000. The second parallel input is 0001, and it gives a serial output 0001 followed by 000. The simulation inputs RDY and A[3:0] must be furnished by the designer, and they are not entirely arbitrary. In this simulation, we input the first parallel input and asserted RDY; then we simulated for several clock cycles to determine when NEXT was asserted. Once NEXT is asserted we know where to input new parallel data and assert RDY. We also need to realize that RDY is

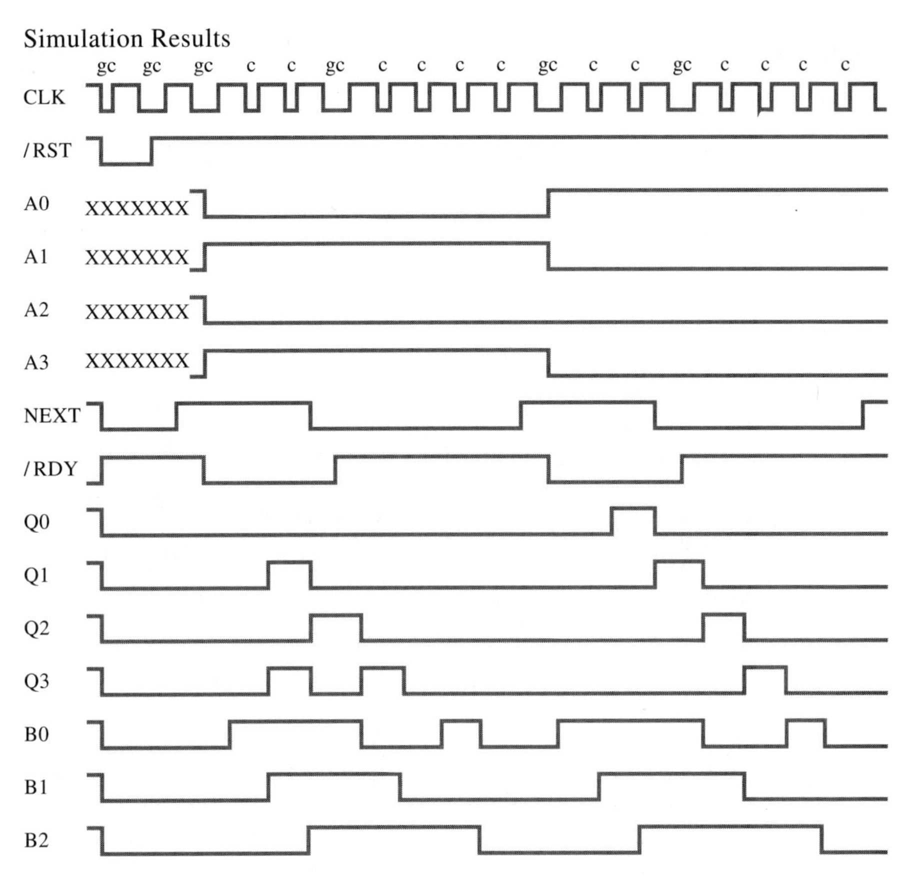

deasserted in response to NEXT deasserted. PALASM has no good way to show the inputs as unknown as does the other simulator we've used, so we leave the inputs the same even though RDY is not asserted.

The system we've designed consists of the single PAL22V10 with a clock connected to the clock input. The pin configuration of the PAL22V10 follows.

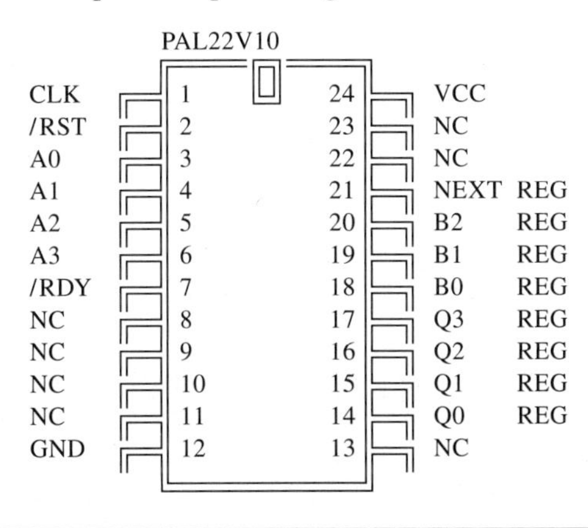

12.3 DIGITAL SYSTEMS DESIGN

In the last section we designed a simple digital system (controller) using both discrete logic and a PAL. Before we look at further system examples, let's look at the structure of the design we created and summarize the steps we used. First of all, we've made a top down or hierarchical design. The general structure of a hierarchical design is shown in Figure 12.3-1. At the top of the hierarchy, we have the entire system (one big block)—*the zeroth-level functional partition.* Below that we partition the system into a number of subsystems—this is the first-level functional partition or just **first-level functional**. The number of subsystems in this or any partition depends on the design, but should be limited to a maximum of seven, with no more than five desirable. Now, one or more of the first-level subsystems may have to be partitioned again into second-level subsystems. The partitioning should continue until the subsystems at the last or bottom level of the partition are simple enough to be designed using known components.

The system we just designed had only a first-level functional partition with four subsystems. Three of the subsystems were discrete devices. The fourth, the system controller, was a state machine that we designed using standard state machine design techniques. (The resulting design, a schematic circuit diagram, could be considered a second-level partition.) The software we used to design the system was a schematic capture package that supported hierarchical schematics. In this particular software, subsystems are represented by symbols that can be custom produced by the user. For example, the system controller in the detailed first-level partition of Figure 12.2-4 is represented by a symbol. This symbol is the bottom subsystem in the partitioning. It can be designed as a state machine and has a circuit schematic associated with it. In a more complicated design, this sym-

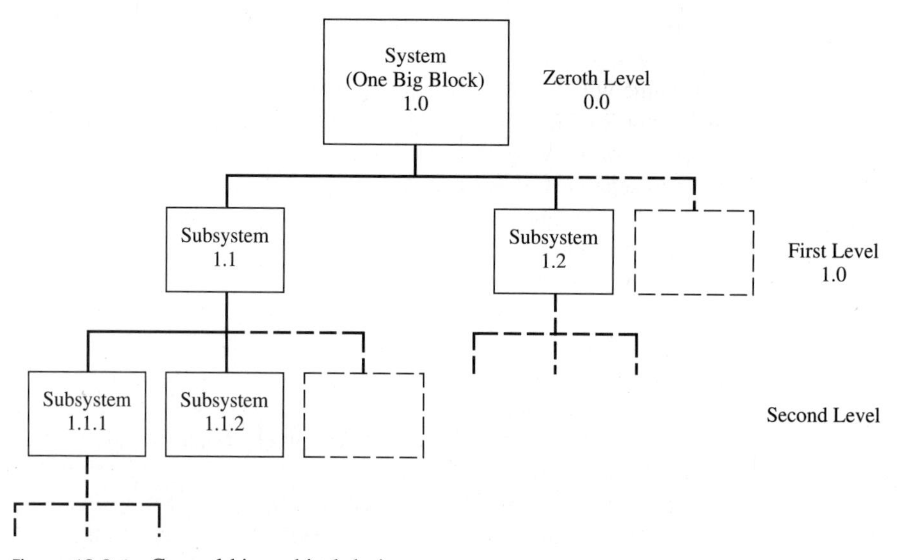

Figure 12.3-1 General hierarchical design.

bol might not be the bottom subsystem in the functional partition and could have a function partition associated with it—composed of further subsystems (symbols).

The software used in the design was one of many available schematic packages. For top down design, one requirement of the software must be that it can support hierarchical schematics. The details of how the hierarchical scheme is supported are unimportant, although some schematic capture programs will be more readily adapted to drawing functional partitions than others.

With this brief discussion of the structure of top down or hierarchical design, let's list the steps we used in our design and will use in future designs.

1. Make a careful description of the desired digital system. Be sure to identify all the inputs to the system and all the outputs the system must produce. The description of the system may require timing diagrams to show the critical timing constraints on the inputs and outputs. In this first step, you should think of the system as one big block and depict it as a single-block, block diagram—the zeroth-level functional.
2. If the system is too complicated to design as a single state machine, partition it into subsystems. To see how the partitioning can be done, list the subfunctions that the system must perform in order to accomplish its overall function. When breaking down the overall function, try to choose subfunctions that can be constructed from standard devices. Remember that one of the subsystems will always be a controller to control the other subsystems. Make a simple block diagram showing the subsystems—a preliminary functional partition.
3. Fill in the details of the functional partition. Do this by determining the inputs and outputs of all the subsystems (other than the controller). Look at each of the subsystems to see what inputs it needs to perform the subfunction required of it in the system. Also determine what outputs it must produce and feed either to the controller, to other subsystems, or to the world external to the system. If a subsystem is a standard device, look at the specifications for the device and determine what inputs it needs and what outputs it produces. To find the inputs and outputs of the subsystems, it may be necessary to design them. If it's not evident how to design them with standard devices or as state machines, then treat them as separate designs and go through these steps of the design process for that subsystem. Continue down the design hierarchy until you reach a subsystem that you can design completely from known devices or as a state machine. Then *go back up the hierarchy* filling in the details of the upper levels. The final result should be a set of detailed functional partition diagrams at all levels of the hierarchy. These diagrams should show all the inputs and outputs of all the subsystems, including the (system) controller if one is required. The functional partition at any level can be a combination block diagram and schematic circuit diagram because some of the subsystems may be identified as standard devices. At the bottom level, the partition will be a schematic circuit diagram.
4. For those partitions that require a controller, make a timing diagram showing the timing of the inputs and outputs of each of the subsystems in the partition. Concentrate on those signals that must be produced by the controller and those

signals that will be used by the controller as inputs. The first-level partition will always require a controller which we've called the *system controller.*

5. Determine the function that the controller must perform by considering the detailed partitioning and the timing diagram. Draw a state transition diagram for the system controller.
6. Design the controller using standard state machine design techniques. Be sure to consider the problems we described in the chapter on state machine design. Pay particular attention to asynchronous inputs and to output glitches. Don't neglect the possibility that a one–flip-flop–per–state state machine might be a good choice for a controller.
7. Put all the pieces of the design together in a set of hierarchical circuit diagrams.

The second, third, fourth, and fifth steps may need to be repeated several times to get a design that will work. The hierarchical design process should be a process of subdividing the system until the subsystems can be constructed using standard devices or can be designed as state machines. The detailed functional partition diagram at any level of the hierarchy is simply a matter of adding details to the preliminary diagram. The fifth step is both a check that the partitioning we've made will perform the function we want for the overall system and a timing specification for the controller. If the partitioning does not perform the function we wish, we must modify it.

For a large system that has more than a first-level partition, the process of design goes down the hierarchy to the bottom subsystem in the partitioning, and as the design of this bottom-level subsystem is completed it will be necessary to change the details of the partition diagrams above this lowest level. In fact, this is how we determine the details for the upper levels of the hierarchy. If we keep these changes current at all the levels of the design, we'll finish with a complete and accurate set of combination functional-schematics for the design. This set of hierarchical drawings (normally stored in a computer and sometimes never printed) allows us to produce one or more printed circuit boards.

Now let's look at a more complicated digital system design.

Example 12.3-1

Design a digital system to measure the difference between the time that a square wave SIG (H) is low and high as shown below. Notice that the absolute difference is required (t_2 may be greater than t_1). The frequency of the square wave can vary

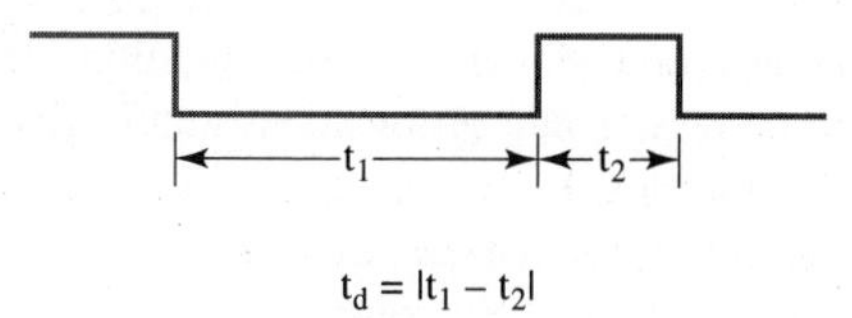

$t_d = |t_1 - t_2|$

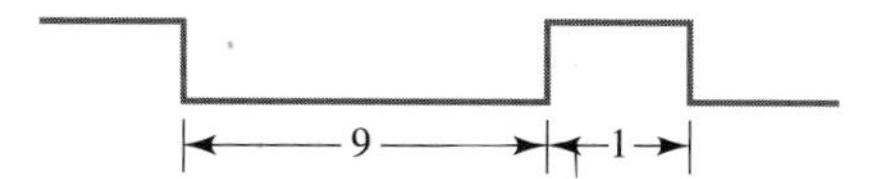

from 1 Hz to 100 Hz. A ratio of times as large as 9 to 1 as shown above should be detectable. Measurements should start on the first high-to-low transition after an input START(L) is asserted. The output should be presented as a binary number on a parallel output bus with high representing 1 for the individual bits. The system should have four modes which are controlled by two inputs X(H) and Y(H) as follows:

X	Y	Function
0	0	Measure one difference, latch the difference, and stop.
0	1	Time only the high portion of the wave. Latch the time and stop.
1	0	Time only the low portion of the wave. Latch the time and stop.
1	1	Continuously measure the time difference. Latch the difference and hold it until the next measurement is completed.

Don't worry about errors that might be caused by mode changes during a measurement.

The first step in the design is to make a detailed description of the system. This was done in the statement of the problem. We complete this description with a zeroth-level functional block diagram (one big block) showing the inputs and output for the system. The inputs are the clock (CLK), the signal (SIG), a systems reset (RESET), the start signal (START), and the two mode signals (X and Y). Assertion levels are indicated on the diagram. The clock is actually part of the system, but for now we'll consider it as an external input and defer its design until later. The only output is the binary difference between the time that the signal is low and high (DIFF). DIFF will be a bus signal, but we have not determined how many bits it will have.

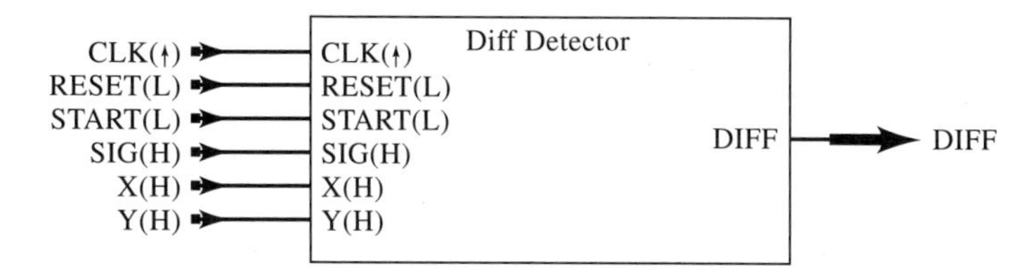

This system is too complicated to be designed as a simple state machine. For the timing involved, the machine would require a very large number of states. Because it cannot be designed as a simple state machine, the second step in the design is to partition the system. To partition the system, we need to consider how we're going to accomplish the functions of the system. The first function is to find the first high-to-low transition after START is asserted. This is where we start our measurements. We'll do this with a separate state machine which we'll call an *edge finder*.

The timing function of the system suggests that we might count a clock. Suppose we count a clock during the time that SIG is low. This will give us a count that is proportional to the period of time that SIG is low. We could also count the clock for the period of time that SIG is high and subtract the two numbers. If we do this we'll need a means of subtraction in our system. Using an up/down counter there is a simpler way to subtract. We can count up during the period of time that SIG is low and count down during the period of time that SIG is high, as shown in sketch (a). This works as long as the period of time SIG is low is longer than the period of

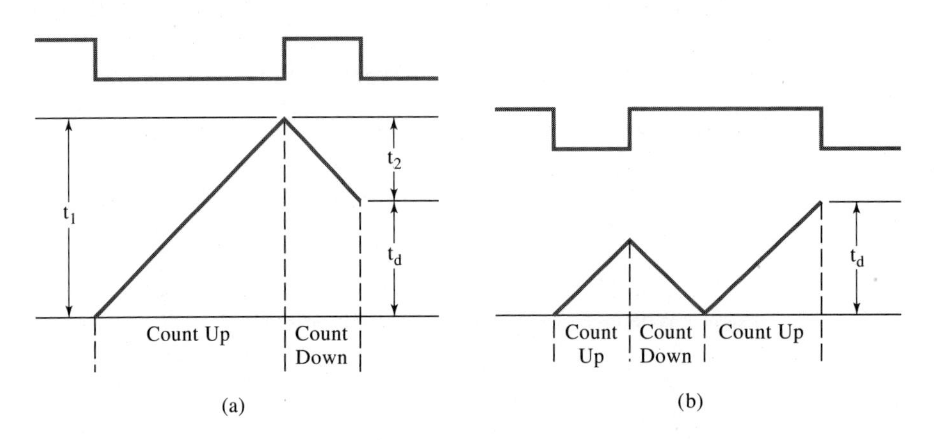

time SIG is high. Fortunately, we can handle the reverse situation, when the period of time that SIG is low is shorter than the period of time it's high, with a simple modification of the up/down counting scheme. If when SIG is high, we count down until we reach a zero count, and then count up again, we'll find the absolute difference as shown in sketch (b). This suggests that we should use one or more up/down counters to accomplish our difference measurement. To hold the difference count we obtain, we'll also need one or more registers. A preliminary first-level functional partition for the system is shown on page 627. It consists of an edge finder, a counter, a register, and a system controller.

We address the third step in the design by considering how to design the subsystems that make up the first-level functions. We need to complete their design to the point where we can determine the detailed inputs and outputs of each of these subsystems. In this system, we can actually design these first-level subsystems from counters and registers or, in the case of the edge finder, as a simple state machine. To design the counter and the register subsystems, we need to determine the size of the counters and registers we need for the system. This will allow us to choose the actual counter and register devices. To discriminate a time difference of 9 to 1 in the high-to-low portions of SIG, we need 10 counts, because we need to have 9 counts while SIG is low and 1 while it's high or vice versa. The smallest period of the signal we are to measure is 1/100 sec (at 100 Hz). To count 10 counts in 1/100 sec requires a counting frequency of

$$f = (10 \text{ counts})/(1/100 \text{ sec}) = 1000 \text{ counts/sec}$$

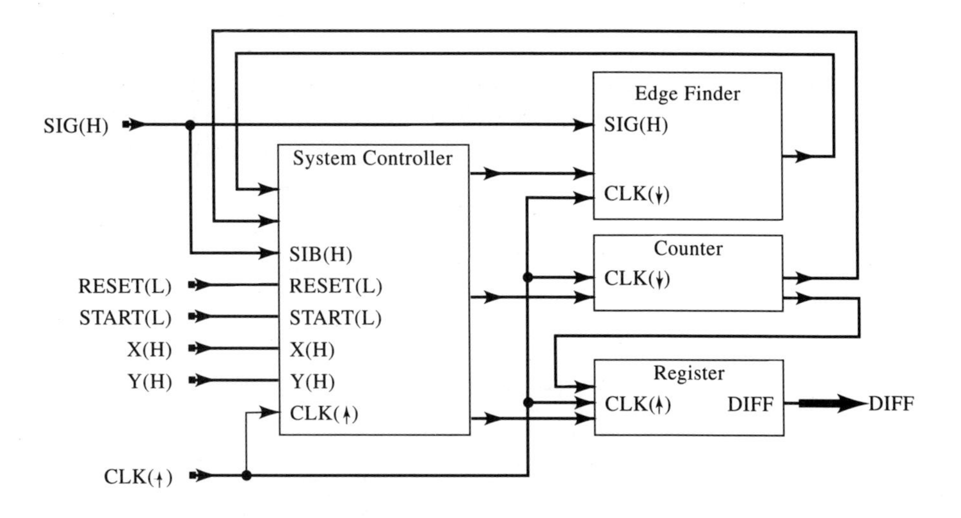

Thus, because we're going to count a clock, the clock frequency must be at least 1000 Hz. When we use this 1000 Hz clock and the period of SIG is long, we'll count to more than 10. In fact, at 1 Hz the period of SIG is 1 sec and we will count to as high as 900 during the long portion of SIG. To count to 900 requires $2^n = 900$ or $n = 10$ bits. (2^{10} actually equals 1024, but 2^9 is only 512, which is too small to count to 900.) We therefore require at least 10 bits in our counters and registers. Counters are made in a 4-bit size (74LS191). We need three of these. If we use three counters, we'll have 12 bits, and we can count as high as $2^{12} - 1 = 4047$, which is a more than adequate resolution. Our detailed schematic for the counter is shown below. Registers are made in an 8-bit size (74LS374). We'll need two of these. Our detailed schematic for the register is also shown below.

These schematics can be considered the second-level partition for the system. They are the bottom of the partition. In our scheme of design, the bottom level of the partition will always be a detailed schematic. These detailed schematics identify all of the inputs and outputs for the counter and register subsystems and allow us to go back and complete the details in the first-level functional.

The counter subsystem is comprised of 74LS191 up-down counters, which we discussed in Section 11.2.1. Three of these counters are cascaded by connecting the RCO from one counter to the G of the next and using a common clock. (G denotes gate—it *gates* the counter off and on.) As we can see, the counter subsystem requires the following inputs—CLK(↓); CNT(L), which enables the counting function of the counter; $D/\overline{U}$(H), which causes the counter to count down when it is asserted and up when it is not asserted; and LOAD(L), an asynchronous parallel load for the counter. Inputs internal to the counter subsystem allow us to load the initial count of the counter. We suspect that we'll lose a count at the start of our measurement, so we have hard-wired the parallel input to be 1 rather than the 0 that would be our unconsidered choice. We have inverted the clock so that the counter counts on the negative going clock edge. This allows us to separate some of the timing functions of the system. The counter subsystem produces two outputs: MX/MN(H), which

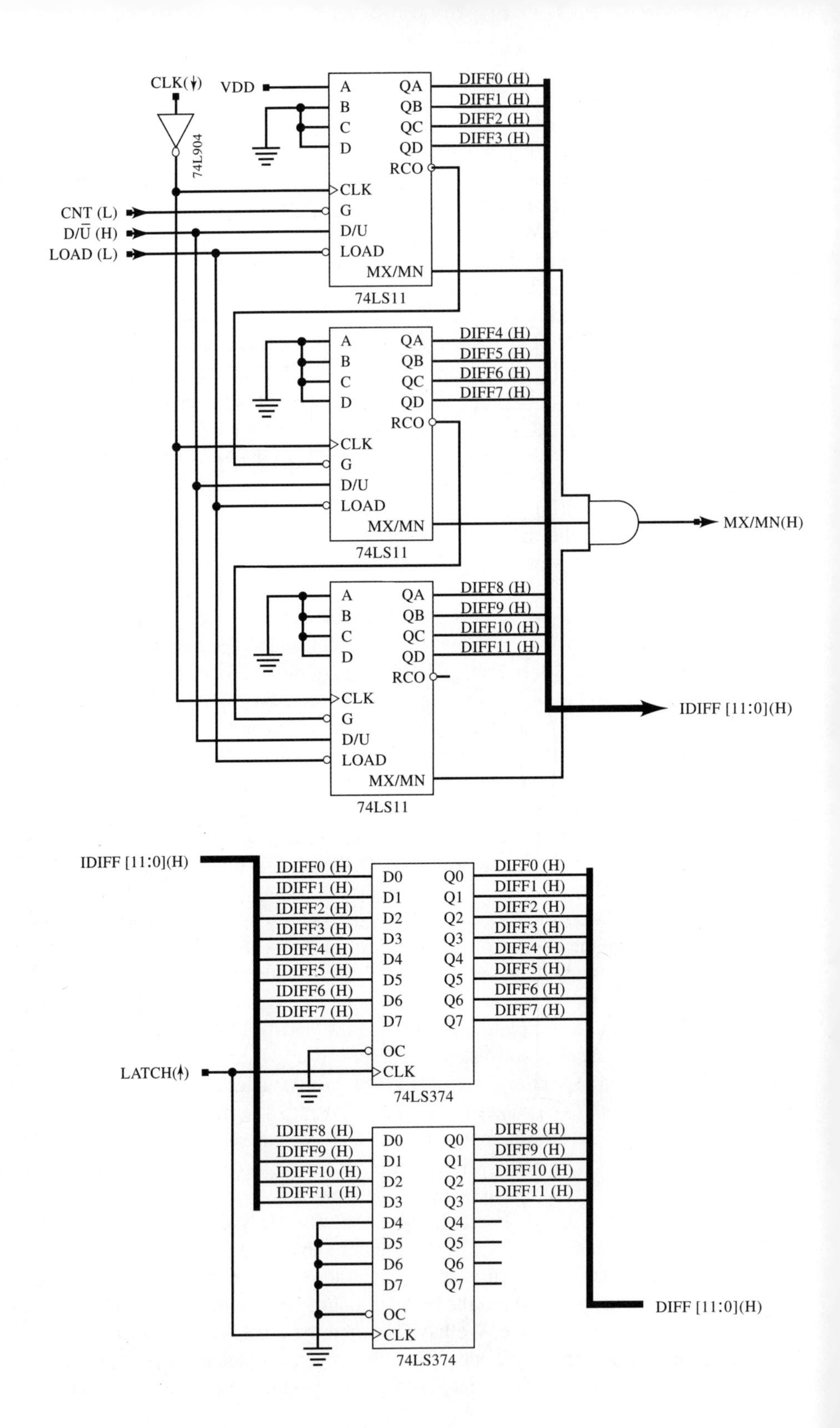
CLK(↓)
VDD
74L904
CNT (L)
D/Ū (H)
LOAD (L)
A B C D
QA QB QC QD
RCO
CLK
G
D/U
LOAD
MX/MN
74LS11
DIFF0 (H)
DIFF1 (H)
DIFF2 (H)
DIFF3 (H)
DIFF4 (H)
DIFF5 (H)
DIFF6 (H)
DIFF7 (H)
DIFF8 (H)
DIFF9 (H)
DIFF10 (H)
DIFF11 (H)
MX/MN(H)
IDIFF [11:0](H)
IDIFF0 (H)
IDIFF1 (H)
IDIFF2 (H)
IDIFF3 (H)
IDIFF4 (H)
IDIFF5 (H)
IDIFF6 (H)
IDIFF7 (H)
IDIFF8 (H)
IDIFF9 (H)
IDIFF10 (H)
IDIFF11 (H)
D0 D1 D2 D3 D4 D5 D6 D7
Q0 Q1 Q2 Q3 Q4 Q5 Q6 Q7
OC
CLK
LATCH(↑)
74LS374
DIFF [11:0](H)

tells us when the counter reaches zero, and IDIFF[11:0](H), the binary difference between the low and high signal times. Notice that we have *and*ed the MX/MNs from each of the three counters together because we want all the counters to have reached zero. We've named the binary difference IDIFF to distinguish it from the final registered output DIFF.

The register subsystem is comprised of 74LS374 PIPO register devices, which we discussed in Section 8.6. These registers are positive-edge clocked and have an output enable OC(L) that we always enable. We also follow the accepted practice of tying all the unused inputs low. (We could also tie them high.) The inputs to this subsystem are IDIFF[11:0](H), the 12-bit binary difference that is to be registered, and LOAD, which clocks IDIFF into the register on a positive going clock edge. Notice that LOAD is indeed an edge triggered clock—only the positive going edge is important. The output from the register subsystem is the 12-bit binary difference DIFF[11:0](H). Before we use the result of the second-level design to complete the details of the first-level design, we need to design the edge finder, at least to the extent that we can specify its input and outputs.

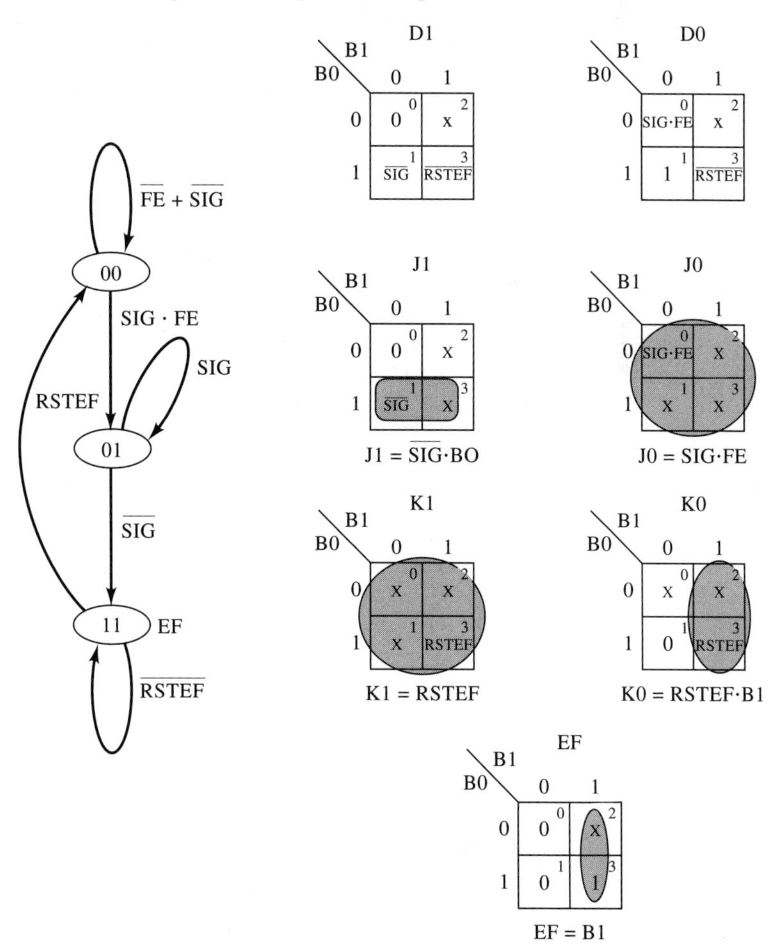

We'll design the edge finder completely. The edge finder finds the first negative going edge of SIG after it's activated. We need an input signal to activate it, which we'll call FE(H)—(Find Edge). As soon as it finds an edge, it should assert an output that we'll call EF(H)—(Edge Found). Finally, it needs an input that acknowledges that EF has been asserted and then resets the edge finder to get ready to find another edge. We'll call this input RSTEF(H)—(ReSeT Edge Finder). Because the edge finder is a state machine, we also need a sanity reset. For this we'll use the system reset RESET(L). The state transition diagram and next-state decoder maps for our edge finder are on the previous page. We've chosen to make the edge finder clock on the negative going clock edge, like the counter, to separate the timing functions. To this end, and because it results in a simpler next-state decoder, we've used JK flip-flops as the memory in the state machine. The detailed schematic and a simulation for the edge finder subsystem are below. Notice that EF(H) is B1(H) in the simulation diagram. The inputs for the edge finder subsystem are SIG, FE, RSTEF, and RESET. The output is EF.

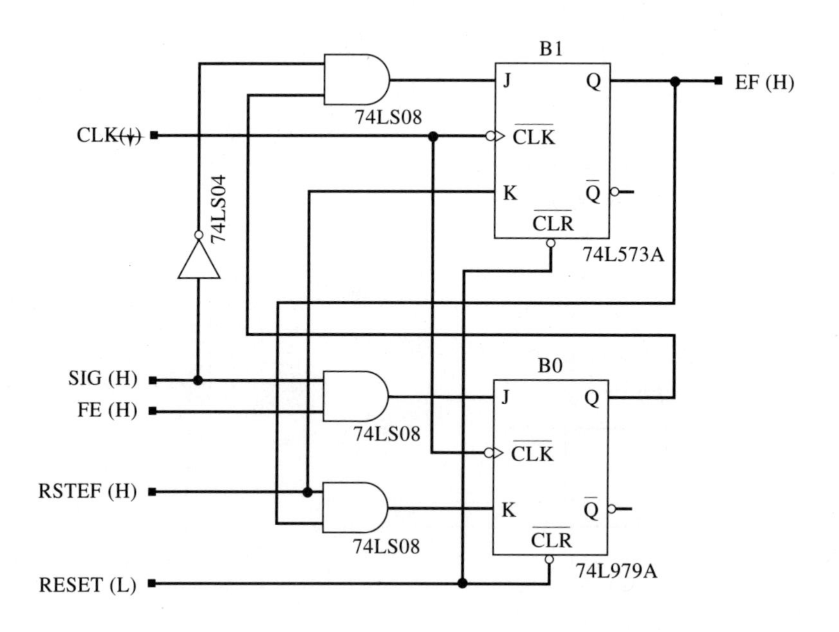

The detailed first-level function partition showing all the inputs and outputs of the subsystems is given below. We've assumed that the state controller is positive-edge triggered, and we've made the edge finder and counter negative-edge triggered to avoid having the control signals to the edge finder and counter changing at the same time as that of the edge finder and counter clock. The register does not use the clock, so we remove the clock connection we showed in the preliminary functional partition. Notice how the design of the subsystems other than the system controller leads to the specification of the system controller inputs and outputs.

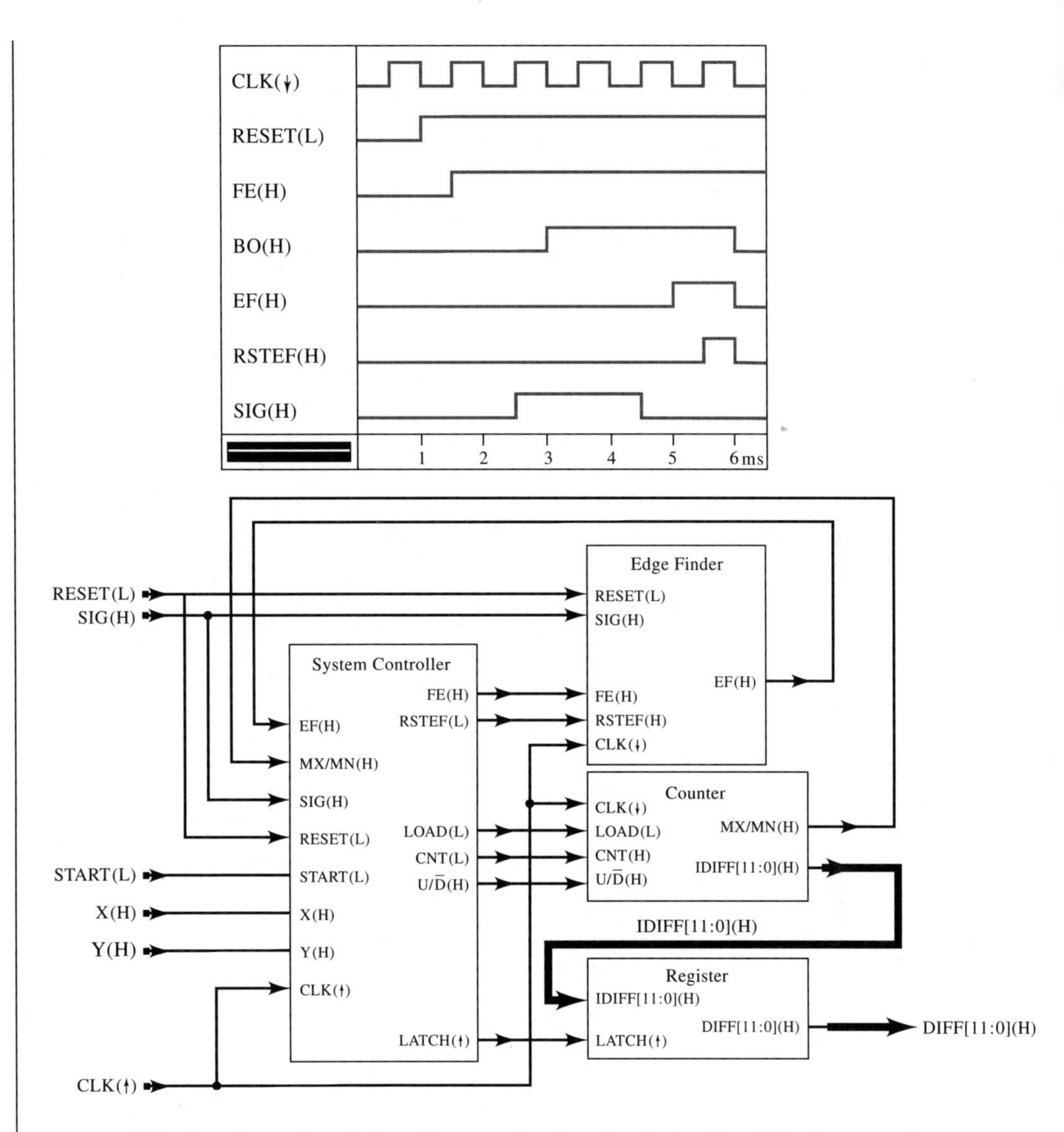

The fourth step is a timing diagram showing the timing for all the inputs and outputs of the subsystems. We'll do the timing for operation in the mode that makes repeated measurements. This will give us all the capability we need to handle the other modes. If we consider the timing for a measurement when the low period is longer than the high period, the timing must be as shown below. A measurement is initiated when START is asserted (goes low). START causes FE and LOAD to be asserted (FE goes high and LOAD goes low) as shown. When FE is asserted in response to START, the edge finder (which we design to be negative-edge triggered) searches for the first negative going edge of SIG and asserts its EF output when this edge is found. EF is asserted on the negative going *clock* edge. As soon as the system controller sees that EF is asserted, it asserts RSTEF. This causes FE to be deasserted and puts the edge finder in a state in which it can start looking for another edge.

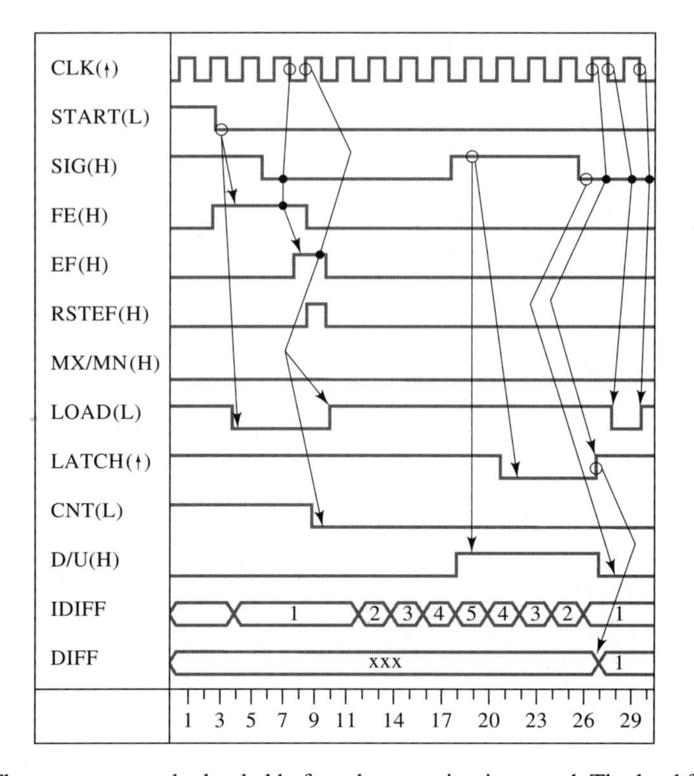

The counter must be loaded before the counting is started. The load function for the counter is asynchronous and occurs when LOAD is asserted. For the first measurement we assert LOAD in response to START. As soon as the first negative going edge of SIG is found (EF is asserted by the edge finder), the counter must start counting (up). This means that LOAD must be deasserted, and the counters must be enabled by asserting CNT. We assume that $D/\overline{U}$ isn't initially asserted, and it must remain so in order for the counter to count up. Notice that we miss the first clock count because of the edge finding function, so we actually must initialize to one rather than zero (tie the low order load bit high as shown in the counter subsystem schematic).

The counter continues to count up until SIG goes high, then it reverses and counts down. $D/\overline{U}$ is asserted when SIG goes high. When SIG again goes low, the count IDIFF is a measure of the time difference. This is the desired output. To hold this output while another difference is measured, we parallel load it into the register. A parallel load into the register is initiated by a positive going edge on its clock LATCH. When SIG goes high, LATCH is taken low; it remains low until the first positive going clock edge after SIG goes low. It's immaterial when LATCH is taken low; only the positive going edge, which clocks the IDIFF into the register, is important.

At this point, we're ready to start over and make a new difference measurement. In the continuous measurement mode we don't need to find a new negative going edge of SIG. We've already found it. To start counting again, we initialize the counter by asserting LOAD for one clock cycle and repeating the counting sequence in the same way we've just described. To make sure that the IDIFF is latched before the counter is initialized, we delay LOAD by half of a clock cycle from the first positive going clock edge after SIG goes low. We can do this by making LOAD a register output from the system controller and using an inverted clock on the output flip-flop. Actually, both LATCH and LOAD must be glitch-free—LATCH because it functions as a clock for the register and any glitch will cause the register to be loaded, LOAD because it is an asynchronous parallel load and any time it goes low it will cause the counter to be loaded with 1.

Just as a check on the timing we've added the difference IDIFF and DIFF at the bottom of the diagram. IDIFF, the output of the counter, is loaded with 1 as soon as LOAD is asserted (goes low). It stays at 1 until LOAD is deasserted. It then increments (counts up) by 1 on each negative going clock edge. At the end of the timing cycle for a difference when SIG goes low, the positive going edge of LATCH clocks IDIFF into the register and its output is DIFF. The counts shown on the diagram are for a representative SIG. SIG necessarily has a very short period, so the diagram is of manageable size. This does not change the timing at the transitions, which is the important timing.

The timing sequence is slightly different when the time of the low period is less than the time of the high period. The timing diagram below shows how the counter first counts up, then counts down, and finally counts up again in response to the MX/MN output from the counter. MX/MN signals that the counter has reached a zero count when counting down. It causes D/$\overline{\text{U}}$ to be asserted so that the counter will count up during the end of the high part of SIG.

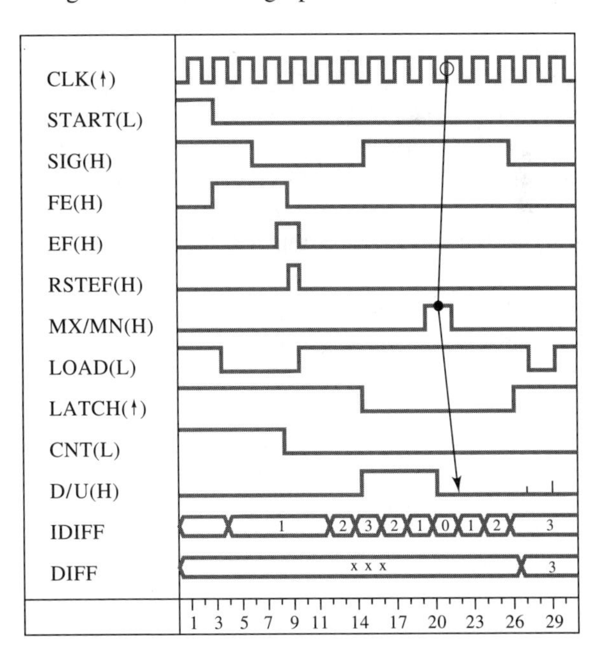

At this point we'll go to the fifth step in the design and start the design of the system controller by making a state transition diagram for it. This will be the system controller for the continuous measurement mode only. When we have the system controller for this mode, we'll modify it for the other modes.

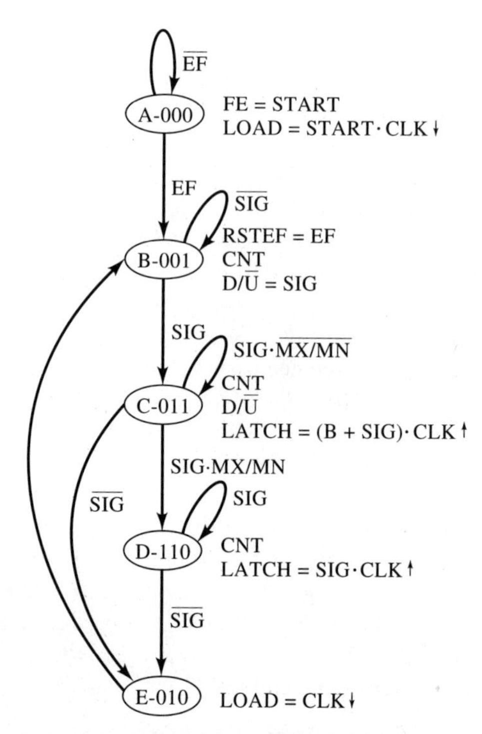

The state transition diagram for the system controller is constructed using the timing diagram for the signals needed by the subsystems. The system controller must produce these signals and it must produce them at the proper time. Our state transition diagram is shown above. In state A, we start the edge finder looking for a negative going edge as soon as START is asserted by asserting FE. We also assert LOAD, which initializes the counter. Notice that START asserted is not an explicit condition for leaving the first state, but rather an implicit condition. We stay in the first state as long as EF is not asserted, but EF will not be asserted until after START is asserted.

As soon as the edge is found (EF is asserted), we go to state B. We stay in state B as long as SIG is not asserted (low). In state B, we reset the edge finder (RSTEF) and start the counter (CNT). The counter counts up because $D/\overline{U}$ is not asserted. We'll get a more accurate count if we assert $D/\overline{U}$ as soon as SIG is asserted (high) and start counting down, rather than waiting until the state changes.

We go to state C when SIG is asserted. In state C, we continue to count, but we assert $D/\overline{U}$ so that we'll count down. We also assert LATCH. We count down until either SIG is not asserted (goes low) or until we reach zero count (MX/MN asserted). If SIG goes low (not asserted) before we reach zero, we've finished a measurement. We then deassert LATCH to clock the difference we've measured into the output registers and go to state E.

In state E, we assert LOAD for one clock cycle delayed by a half of a clock cycle to initialize the counter for the next count sequence. This finishes a measurement sequence, and because we're in the continuous mode, we then go to state B to start a new measurement. We don't need to find the edge because it's already found.

In state C, if we reach zero before SIG goes low (not asserted), we go to state D. In state D, we continue to count, but we count up again ($D/\overline{U}$ is not asserted). We also continue to assert LATCH. In this state, when SIG goes low, we've again finished a measurement and we deassert LATCH to clock the difference we've measured into the output registers. We can now go to state E and initialize the counter and start a new measurement.

Now how do we modify the state transition diagram so that we can implement the other three modes? All the functions we need have been implemented; we just need to inhibit some of the counting for these other modes and make a provision for stopping the machine after one measurement cycle. We show this modification in the state transition diagram below. We've added a transition so that the measurement will start over in state A if the mode is 00 ($\overline{X}\,\overline{Y}$), 01 ($\overline{X}Y$), or 10 ($X\overline{Y}$), but will continue in state B if the mode is 11 (XY). Notice that $\overline{X}\,\overline{Y} + \overline{X}Y + X\overline{Y} = \overline{X} + \overline{Y}$.

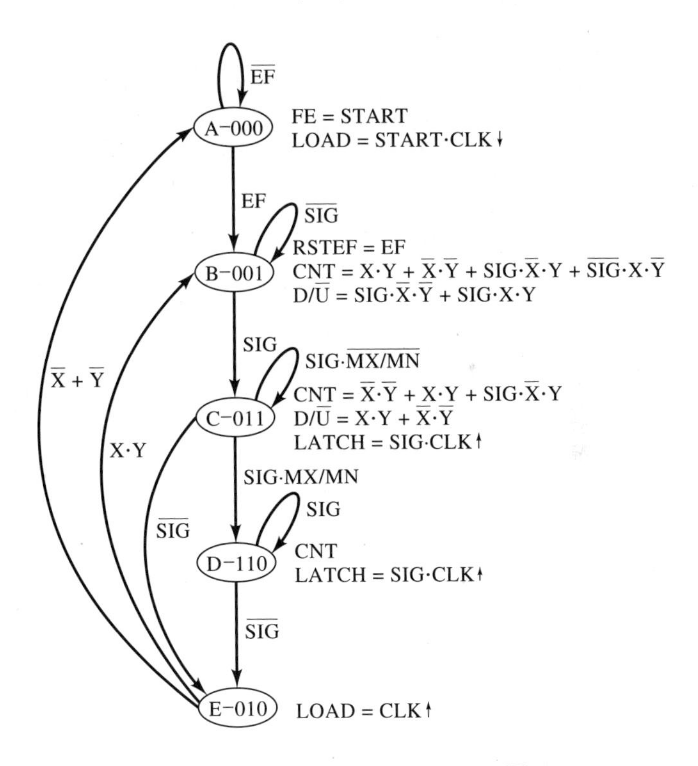

If we wish to measure the period that SIG is low ($\overline{X}Y$), we only count during state B. If we wish to measure the period that SIG is high ($X\overline{Y}$), we only count in state C and we change the direction (count up) by not asserting $D/\overline{U}$. To finish the system controller design, we need to make next-state decoder and output decoder maps and find the decoder logic. The maps and logic are shown below. All of the outputs except LOAD and LATCH are combinational outputs. LOAD is a register output but it uses the negative going clock edge, so the decoder decodes the present not the previous state. LATCH is a normal register output, so the decoder must decode the previous state. To cause LATCH to be asserted at the beginning of state

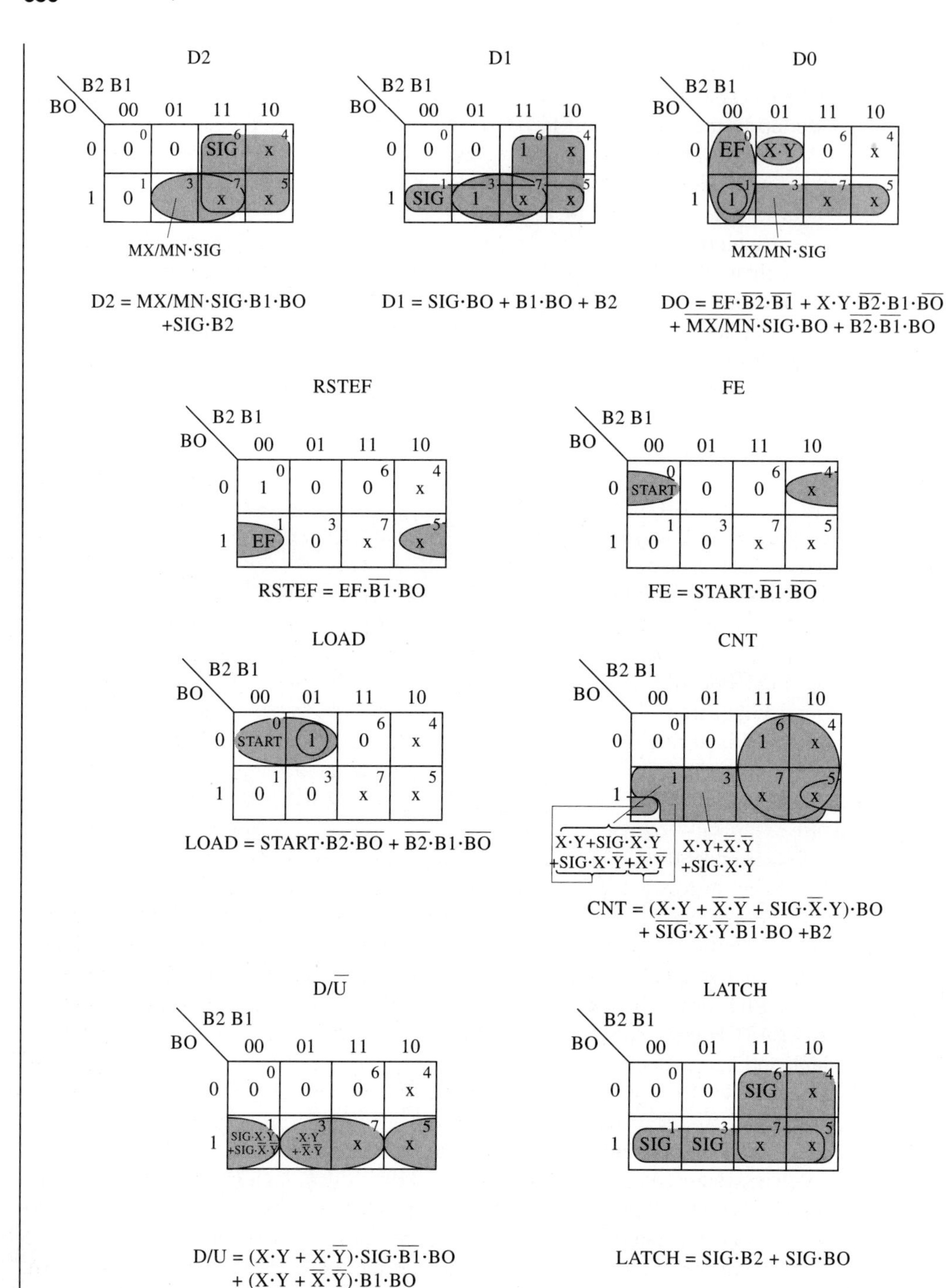

D2
B2 B1
BO
00 01 11 10
SIG
MX/MN·SIG
$D2 = MX/MN \cdot SIG \cdot B1 \cdot BO + SIG \cdot B2$
D1
SIG
$D1 = SIG \cdot BO + B1 \cdot BO + B2$
D0
EF
X·Y
$\overline{MX/MN} \cdot SIG$
$DO = EF \cdot \overline{B2} \cdot \overline{B1} + X \cdot Y \cdot \overline{B2} \cdot B1 \cdot \overline{BO} + \overline{MX/MN} \cdot SIG \cdot BO + \overline{B2} \cdot \overline{B1} \cdot BO$
RSTEF
EF
$RSTEF = EF \cdot \overline{B1} \cdot BO$
FE
START
$FE = START \cdot \overline{B1} \cdot \overline{BO}$
LOAD
START
$LOAD = START \cdot \overline{B2} \cdot \overline{BO} + \overline{B2} \cdot B1 \cdot \overline{BO}$
CNT
$X \cdot Y + SIG \cdot \overline{X} \cdot Y + SIG \cdot X \cdot \overline{Y} + \overline{X} \cdot \overline{Y}$
$X \cdot Y + \overline{X} \cdot \overline{Y} + SIG \cdot X \cdot Y$
$CNT = (X \cdot Y + \overline{X} \cdot \overline{Y} + SIG \cdot \overline{X} \cdot Y) \cdot BO + \overline{SIG} \cdot X \cdot \overline{Y} \cdot \overline{B1} \cdot BO + B2$
$D/\overline{U}$
$SIG \cdot X \cdot \overline{Y} + SIG \cdot \overline{X} \cdot Y$
$X \cdot Y + \overline{X} \cdot \overline{Y}$
$D/U = (X \cdot Y + X \cdot \overline{Y}) \cdot SIG \cdot \overline{B1} \cdot BO + (X \cdot Y + \overline{X} \cdot \overline{Y}) \cdot B1 \cdot BO$
LATCH
SIG
$LATCH = SIG \cdot B2 + SIG \cdot BO$

C, we must decode state B with the condition that SIG is asserted. To make sure that we do not assert LATCH beyond state C, we include the condition that SIG is asserted in the decoder logic; then if SIG is deasserted before we leave the state, LATCH will not be clocked on the clock edge that caused the transition out of the state. We do a similar thing in state D.

The circuit diagram for the system controller is shown on the next page. The state machine is on the left and bottom of the diagram. The next state decoders are on the right. LOAD and LATCH have register outputs. We've reduced the complexity of the decoders by noting that the inputs X and Y only occur in the combinations XY, $X\overline{Y}$, $\overline{X}Y$, and XY + $\overline{X}\overline{Y}$. These are the combinations we can decode in a 2-line–to–4-line decoder (74LS139). The decoder for decoding the combinations of X and Y that we need is in the upper left of the diagram along with the rail.

Simulation results for the system controller in the 11 mode follow the controller. In this diagram, remember that CLK, RESET, START, X, Y, SIG, EF, and MX/MN are inputs we have to furnish the simulation. The inputs through SIG are not too difficult. CLK is a square wave with period 1 ms (freq 1024 Hz). RESET must be asserted for a short period at the beinning of the simulation to reset the state to 000, and START must be asserted shortly after that to start the differencing cycle. X and Y are both asserted in the 11 mode. We've chosen to simulate for two cycles of SIG. First, a cycle with the low duration portion longer than the high duration portion, and second, a cycle with the low duration shorter than the high duration. The signal EF (Edge Found) is a signal returned by the edge finder in response to the signal FE (Find Edge) from the controller. The easiest way to determine the proper time to assert EF is to run the simulation and see where FE is asserted. Then look back at the simulation of the edge finder and see where EF should be asserted. MX/MN is a signal generated by the counter that is asserted when the counter counts down to zero. This will occur in the high portion of the cycle with the low portion shorter than the high portion. The exact timing of MX/MN does not matter for the simulation. The effect of MX/MN is what is important.

We can see from the simulation that when the input FE is asserted, a timing cycle is initiated by causing LOAD to be deasserted and count to be asserted. $D/\overline{U}$ is not asserted. Counting continues and nothing changes until SIG goes high. When SIG goes high, counting continues, but $D/\overline{U}$ is asserted to cause down counting and LATCH goes low. The counting cycle ends when SIG goes low. LATCH goes high and half a clock cycle later LOAD is asserted for one clock cycle. This starts a new cycle. This cycle is the same as the first except that $D/\overline{U}$ is deasserted when MX/MN is asserted. We should simulate the timing for other modes of the system. These simulations were done and found to be correct but are not included in the text.

The system controller completes the system design. The complete set of schematics for the system consists of the first-level functional and the schematics for the edge finder, the counter, the register, and the system controller.

We've done four simulations of the entire system—one for each mode—but we only present the results for two of these. The simulations of the entire system are easier to implement than the simulation of the system controller because none of the internal signals need be specified, only the external signals CLK, RESET, START,

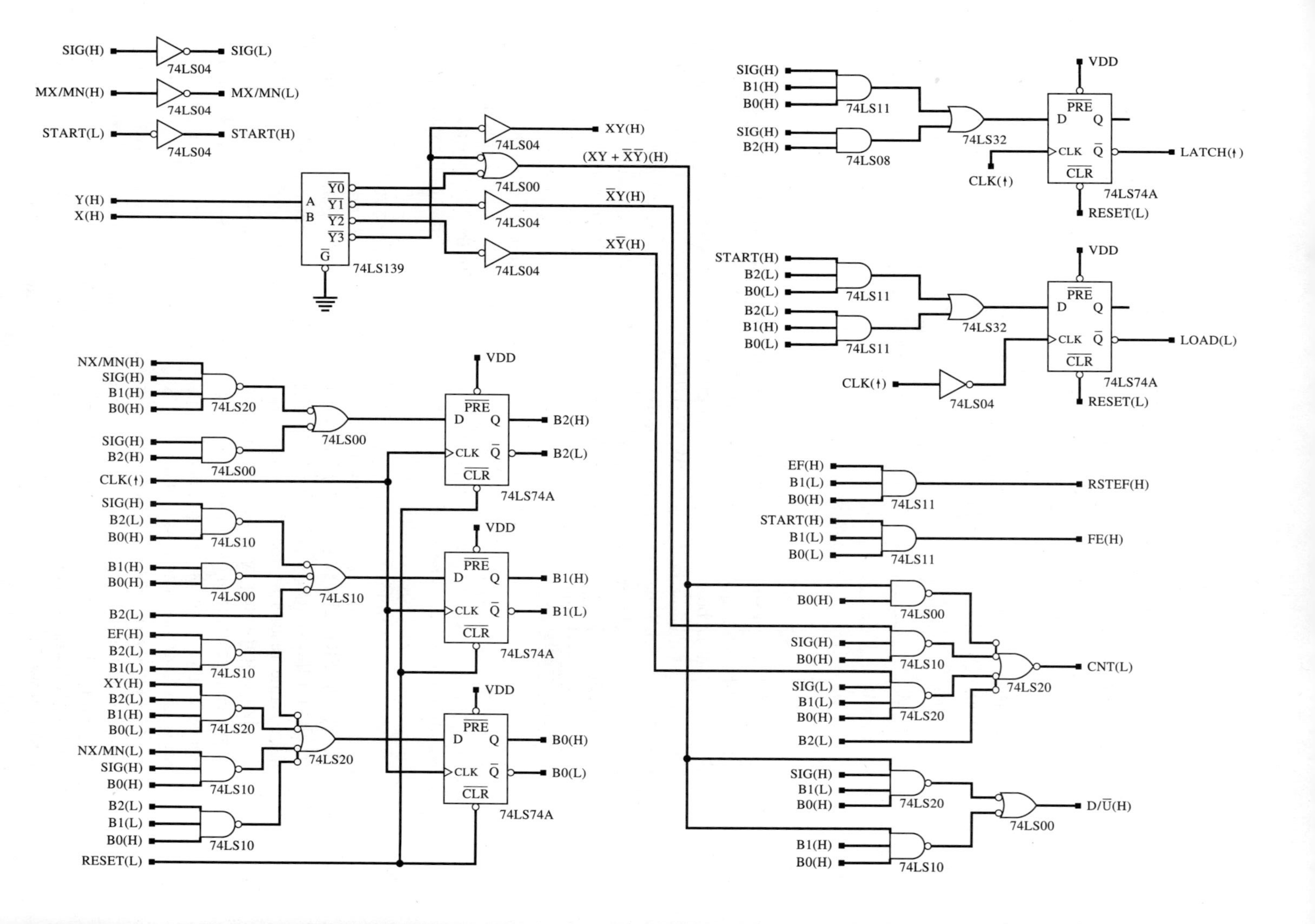
SIG(H)
74LS04
SIG(L)
MX/MN(H)
74LS04
MX/MN(L)
START(L)
74LS04
START(H)
Y(H)
X(H)
A
B
Y0
Y1
Y2
Y3
G
74LS139
74LS04
XY(H)
74LS00
(XY + X̄Ȳ)(H)
74LS04
X̄Y(H)
74LS04
XȲ(H)
SIG(H)
B1(H)
B0(H)
74LS11
SIG(H)
B2(H)
74LS08
74LS32
CLK(↑)
VDD
PRE
D
Q
CLK
Q̄
CLR
74LS74A
RESET(L)
LATCH(↑)
START(H)
B2(L)
B0(L)
74LS11
B2(L)
B1(H)
B0(L)
74LS11
74LS32
CLK(↑)
74LS04
VDD
74LS74A
RESET(L)
LOAD(L)
NX/MN(H)
SIG(H)
B1(H)
B0(H)
74LS20
SIG(H)
B2(H)
74LS00
74LS00
CLK(↑)
VDD
74LS74A
B2(H)
B2(L)
SIG(H)
B2(L)
B0(H)
74LS10
B1(H)
B0(H)
74LS00
B2(L)
74LS10
VDD
74LS74A
B1(H)
B1(L)
EF(H)
B2(L)
B1(L)
74LS10
XY(H)
B2(L)
B1(H)
B0(L)
74LS20
NX/MN(L)
SIG(H)
B0(H)
74LS10
B2(L)
B1(L)
B0(H)
74LS10
74LS20
RESET(L)
VDD
74LS74A
B0(H)
B0(L)
EF(H)
B1(L)
B0(H)
74LS11
RSTEF(H)
START(H)
B1(L)
B0(L)
74LS11
FE(H)
B0(H)
74LS00
SIG(H)
B0(H)
74LS10
SIG(L)
B1(L)
B0(H)
74LS20
B2(L)
74LS20
CNT(L)
SIG(H)
B1(L)
B0(H)
74LS20
B1(H)
B0(H)
74LS10
74LS00
D/Ū(H)

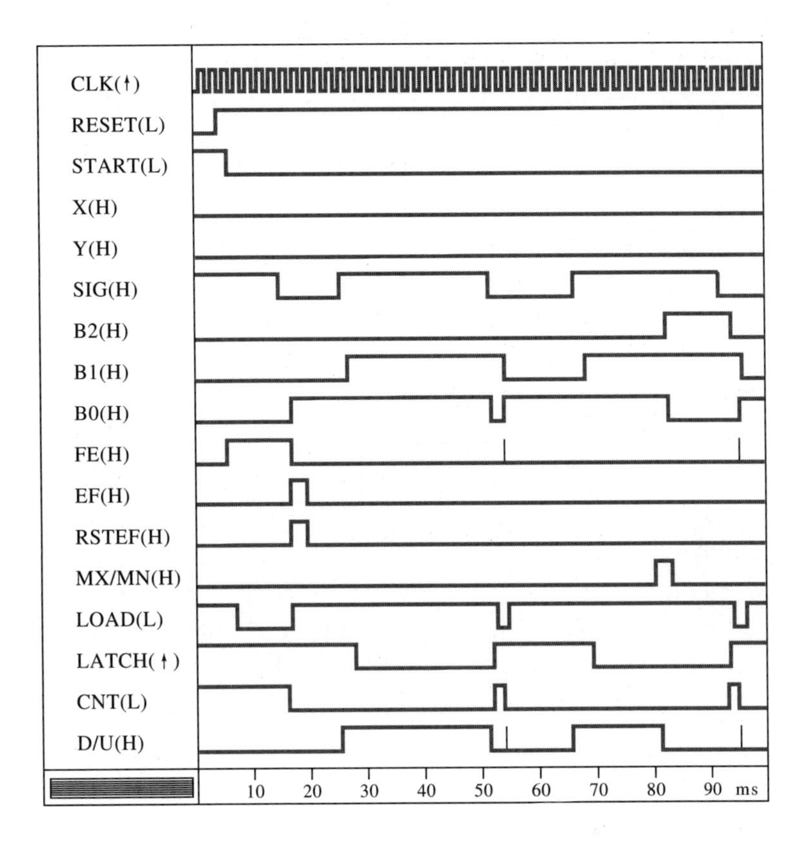

X, Y, and SIG. The results of the simulations follow. First, we show simulation results for the system running in the 11 mode. Second, we show an enlargement of these same results so that the actual counting can be seen. For bus signals the simulation software prints the value in binary on the signal, if there's room between signal changes. The value is in hexadecimal because we chose this option. A decimal option is also available. Notice how the counting first goes up, then down, and back up when MX/MN is asserted. Finally, we show simulation results for the 10 mode, which only counts the high portion of SIG.

The simulation reveals at least one problem. We actually miss two counts at the beginning of the differencing cycle rather than the one count we had anticipated. This only happens on the first differencing cycle; after that we lose one count. The problem with the initial cycle is the time spent in handshaking between the system controller and the edge finder. Perhaps a design with the edge finder incorporated in the system controller would correct this problem.

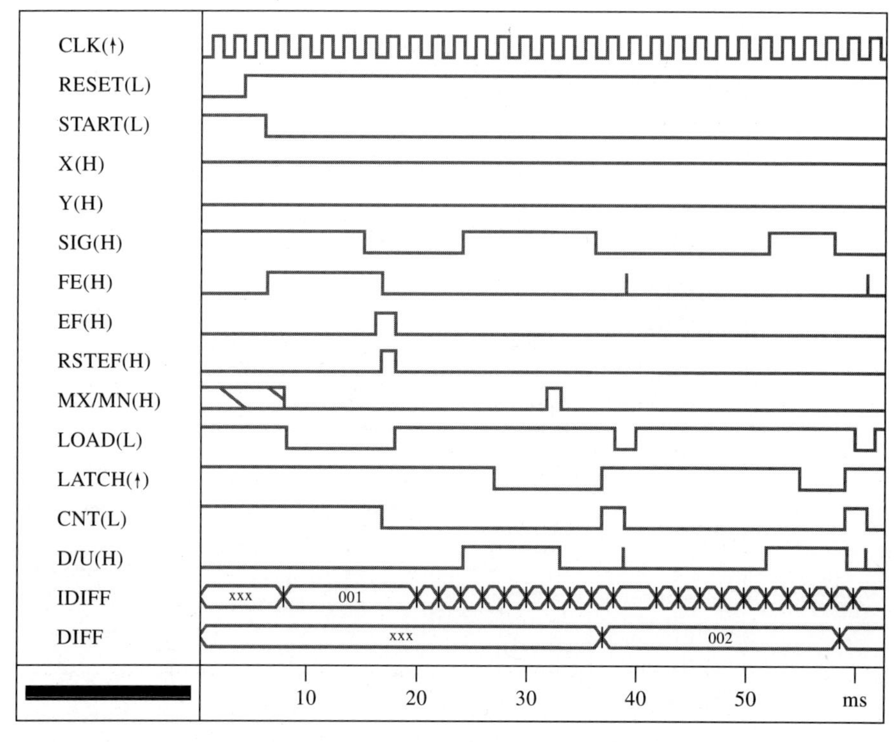
CLK(↑)
RESET(L)
START(L)
X(H)
Y(H)
SIG(H)
FE(H)
EF(H)
RSTEF(H)
MX/MN(H)
LOAD(L)
LATCH(↑)
CNT(L)
D/U(H)
IDIFF
DIFF
xxx
001
002
10
20
30
40
50
ms

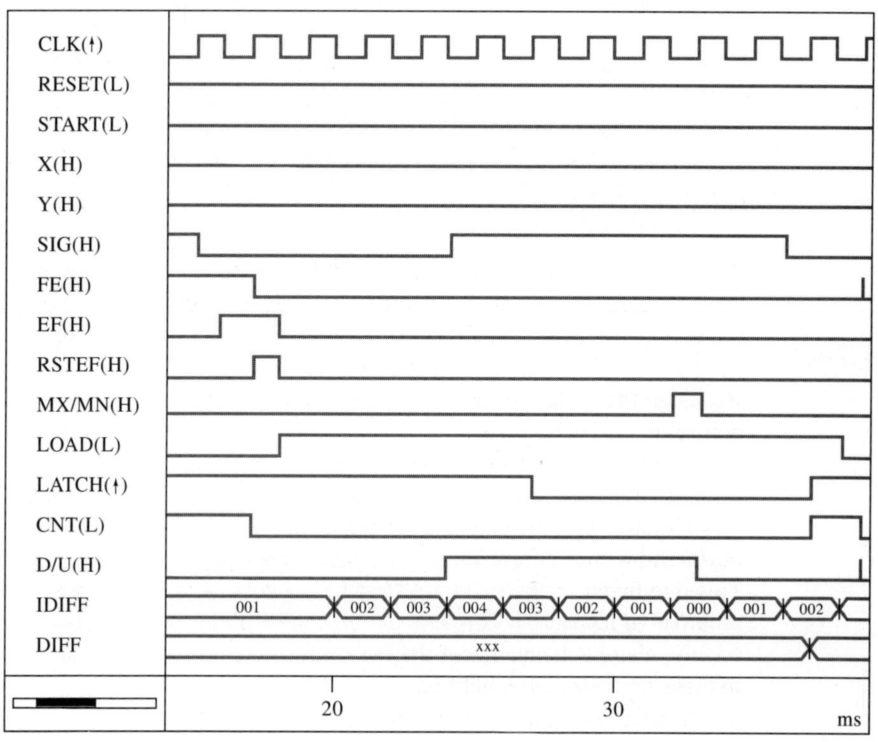
CLK(↑)
RESET(L)
START(L)
X(H)
Y(H)
SIG(H)
FE(H)
EF(H)
RSTEF(H)
MX/MN(H)
LOAD(L)
LATCH(↑)
CNT(L)
D/U(H)
IDIFF
DIFF
001
002
003
004
003
002
001
000
001
002
xxx
20
30
ms

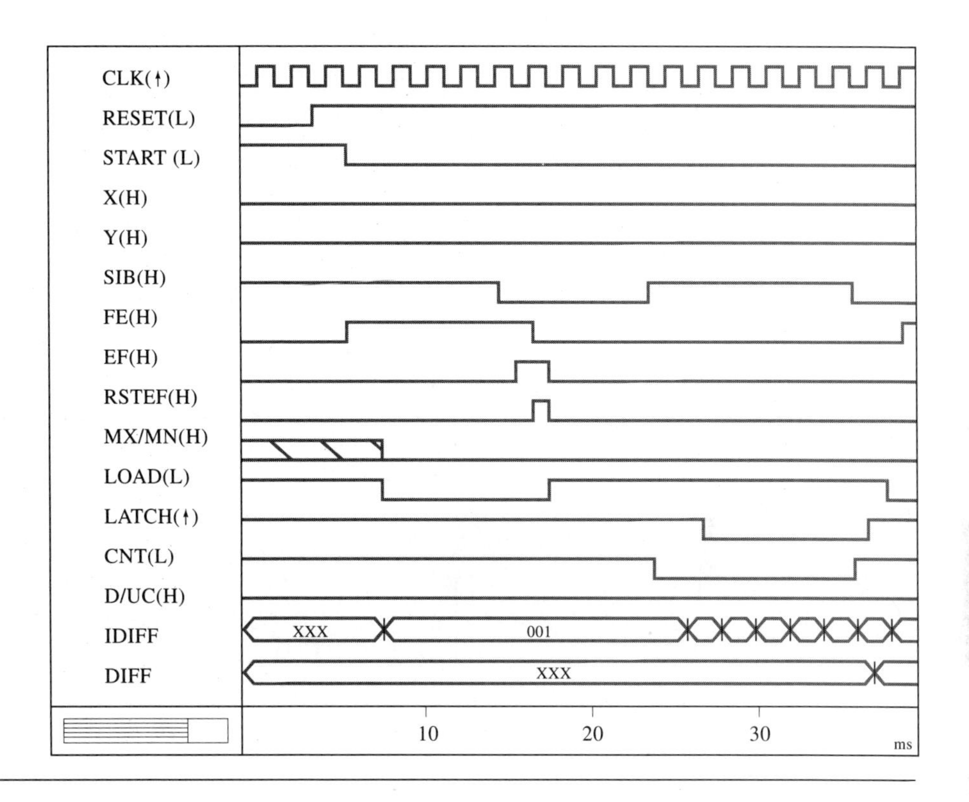

This example suffers from the same general problem as the first example in this chapter. No modern designer would design a system controller of this complexity using discrete elements unless he or she had some constraints that were not indicated in our specifications. The modern designer would use a PAL for the system controller and perhaps the edge finder, or maybe a FPGA for the entire system. We'll look at the PAL design in the next example.

Example 12.3-2

Redesign the system of the previous example using a PAL (22V10) for the system controller and edge finder.

The system controller in the previous example uses three state code bits and six outputs. If we use a PAL to design it, the PAL must have nine logic sections. The 22V10 has 10 logic sections. It would only have room for the edge finder if it's combined with the system controller. The state transition diagram below incorporates the edge finder in the system controller.

We've named the first two states of the diagram AAA and AA, so we did not have to change the names of the remaining states. We've also assigned state codes to these first two states and have reassigned the state code for state A. The state code assignments were made this way in a rough attempt to simplify the next state decoder.

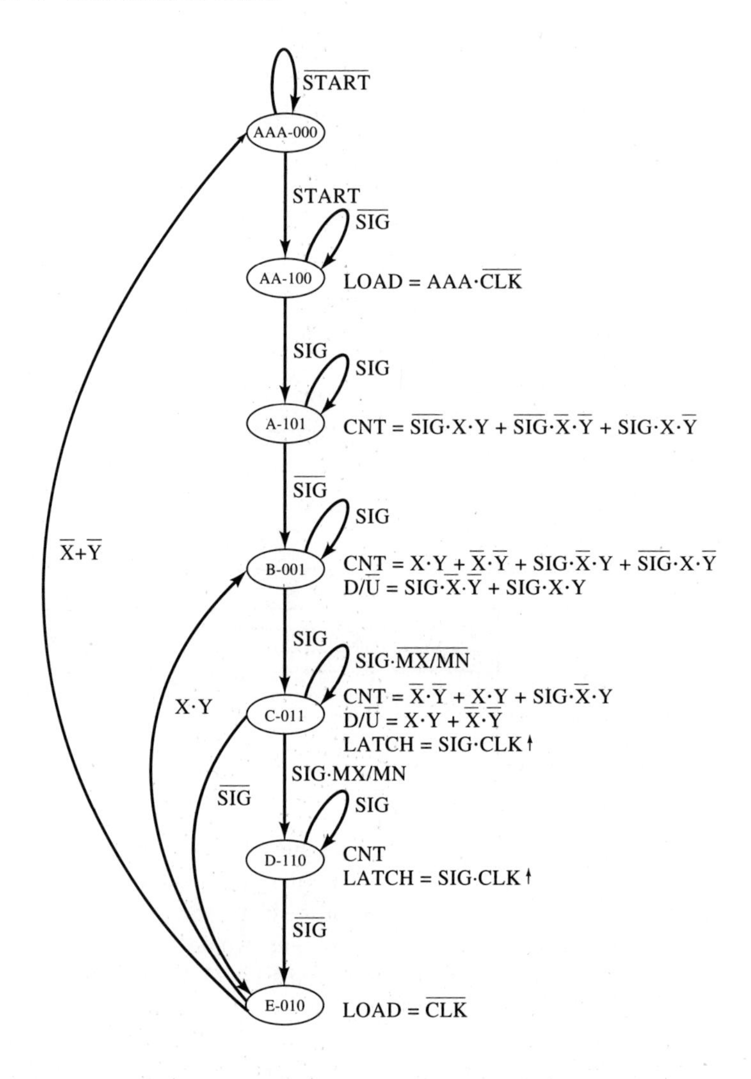

Including the edge finder in the system controller actually results in a cleaner design. It reduces the number of subsystems in the design by one, and it reduces the number of outputs from the system controller while only adding two states (but no state code bits). Since the system controller is to be designed in a PAL, LOAD cannot be clocked by the negative going clock edge without adding a flip-flop external to the PAL. A solution to this problem is shown in the state transition diagram—LOAD is only asserted for the low part of the clock. To do this, a registered output ILOAD is asserted for the entire state where the output is desired, then ILOAD is *and*ed with $\overline{CLK}$. This produces a glitch-free output, which is asserted for the last half of the clock cycle. (Be aware that *and*ing with CLK to get an output that is asserted over the first half of the clock cycle will result in a glitch at the start of the next clock cycle.)

A PALASM source program to produce the systems controller in a 22V10 is given below. The name LATCH was changed to LTCH and the name D/$\overline{U}$ was changed to DU because LATCH is a reserved word in PALASM and / is used by PALASM to denote a signal which is asserted low or a *not*.

```
;PALASM Design Description

;--------------------------------------- Declaration Segment ---------------------------------------

TITLE       PULSE DIFFERENCE MEASUREMENT
PATTERN
REVISION
AUTHOR      AW SHAW
COMPANY     UTAH STATE UNIVERSITY
DATE        09/10/92

CHIP   diffpal   PAL22V10

;------------------------------------------ PIN Declarations ------------------------------------------

PIN    1      CLK                            ; CLOCK
PIN    2      /RESET                         ; INPUT
PIN    3      /START                         ; INPUT
PIN    4      X                              ; INPUT
PIN    5      Y                              ; INPUT
PIN    6      SIG                            ; INPUT
PIN    7      MN                             ; INPUT
PIN    8      NC                             ;
PIN    9      NC                             ;
PIN   10      NC                             ;
PIN   11      NC                             ;
PIN   12      GND                            ;
PIN   13      NC                             ;
PIN   14      NC                             ;
PIN   15      B2              REGISTERED     ; OUTPUT
PIN   16      B1              REGISTERED     ; OUTPUT
PIN   17      B0              REGISTERED     ; OUTPUT
PIN   18      /LTCH           REGISTERED     ; OUTPUT
PIN   19      /CNT            COMBINATORIAL  ; OUTPUT
PIN   20      DU              COMBINATORIAL  ; OUTPUT
PIN   21      ILOAD           REGISTERED     ; OUTPUT
PIN   22      /LOAD           COMBINATORIAL  ; OUTPUT
PIN   23      NC                             ;
PIN   24      VCC                            ;
NODE 1        GLOBAL                         ;

;------------------------------------ State Equations Segment ------------------------------------

STATE

MEALY_MACHINE
```

```
AAA = /B2*/B1*/B0
AA  =  B2*/B1*/B0
A   =  B2*/B1* B0
B   = /B2*/B1* B0
C   = /B2* B1* B0
D   =  B2* B1*/B0
E   = /B2* B1*/B0

AAA := SC → AA
       + NSC → AAA
AA  := HSIG → A
       + LSIG → AA
A   := LSIG → B
       + HSIG → A
B   := HSIG → C
       + LSIG → B
C   := HSIG_MN → D
       + LSIG → E
       + HSIG_NMN → C
D   := LSIG → E
       + HSIG → D
E   := X_AND_Y → B
       + NX_OR_NY → AAA

AAA.OUTF = VCC → /CNT*/DU
AA.OUTF  = VCC → /CNT*/DU
A.OUTF   = CD1 → CNT*/DU
           + → /CNT*/DU
B.OUTF   = CD2 → CNT*DU
         + CD3 → CNT*/DU
           + → /CNT*/DU
C.OUTF   = CD4 → CNT*DU
           + CD5 → CNT*/DU
           + → /CNT*/DU
D.OUTF   = VCC → CNT*/DU
E.OUTF   = VCC → /CNT*/DU

CONDITIONS

SC = START
NSC = /START
HSIG = SIG
LSIG = /SIG
HSIG_MN = SIG*MN
HSIG_NMN = SIG*/MN
CD1 = /SIG*X*Y + /SIG*/X*/Y + /SIG*X*/Y
CD2 = SIG*X*Y + SIG*/X*/Y
CD3 = /SIG*X*Y + /SIG*/X*/Y + SIG*/X*Y + /SIG*X*/Y
CD4 = X*Y + /X*/Y
```

```
CD5 = SIG*/X*Y
X_AND_Y = X*Y
NX_OR_NY = /X + /Y

;---------------------------------- Boolean Equation Segment ----------------------------------

EQUATIONS

ILOAD := /B2*/B1*/B0 + /B2*B1*/B0
LOAD = ILOAD*/CLK
LTCH := /B2*/B1*B0*SIG + /B2*B1*B0*SIG + B2*B1*/B0*SIG
GLOBAL.RSTF = RESET
B2.TRST = VCC
B1.TRST = VCC
B0.TRST = VCC
ILOAD.TRST = VCC
LOAD.TRST = VCC
LTCH.TRST = VCC
CNT.TRST = VCC
DU.TRST = VCC

;-------------------------------------- Simulation Segment ----------------------------------------

SIMULATION

TRACE_ON CLK /RESET /START SIG MN B2 B1 B0 Y /LOAD /CNT DU /LTCH
SETF RESET /START X Y SIG /MN
CLOCKF CLK
SETF /RESET START
CLOCKF CLK
SETF START
CLOCKF CLK
SETF /SIG
CLOCKF CLK
CLOCKF CLK
CLOCKF CLK
SETF SIG
CLOCKF CLK
CLOCKF CLK
CLOCKF CLK
CLOCKF CLK
SETF MN
CLOCKF CLK
SETF /MN
CLOCKF CLK
CLOCKF CLK
CLOCKF CLK
SETF /SIG
CLOCKF CLK
CLOCKF CLK
CLOCKF CLK
TRACE_OFF

;----------------------------------------------------------------------------------------------------
```

The program was written directly from the state transition diagram in state transition language. The design is a Mealy machine with both combinational and register outputs. The combinational outputs are programmed in the usual state machine output form

$$\begin{aligned} \text{state.OUTF} &= \text{cond1} \rightarrow \text{output1} \\ &+ \text{cond2} \rightarrow \text{output2} \\ &\ \ \ldots \end{aligned}$$

Remember that = signifies a combinational output. The conditions can be a little tricky for these Mealy outputs. Notice that in state B there are three conditions and three outputs—first, the condition that both CNT and UD are asserted; second, the condition that CNT is asserted and UD is not; finally, the condition that neither CNT nor UD is asserted. The last output line for state B with the symbol $+\rightarrow$ means that this is the output when none of the other conditions are met (the local default condition).

Because the software will not handle mixed outputs with these output equations, the register outputs cannot be programmed in this part of the program. Thus, the remainder of the outputs (the register outputs) are programmed in the EQUATIONS segment of the program by writing explicit logic expressions for the output decoders. These logic expressions must decode the previous state as with any register outputs. For instance, the first term in the expression for LTCH, /B2*/B1*B0*SIG, decodes state B with the condition that SIG is asserted. The logic expressions for the decoders were produced directly from the state transition diagram without any simplification using maps. It's not necessary to simplify the logic expression, because PALASM does it. Notice how LOAD is produced in two steps. First, the register output ILOAD is produced, then ILOAD is *and*ed with $\overline{\text{CLK}}$. A PALASM simulation of the controller is shown below. The inputs were chosen to simulate through a RESET

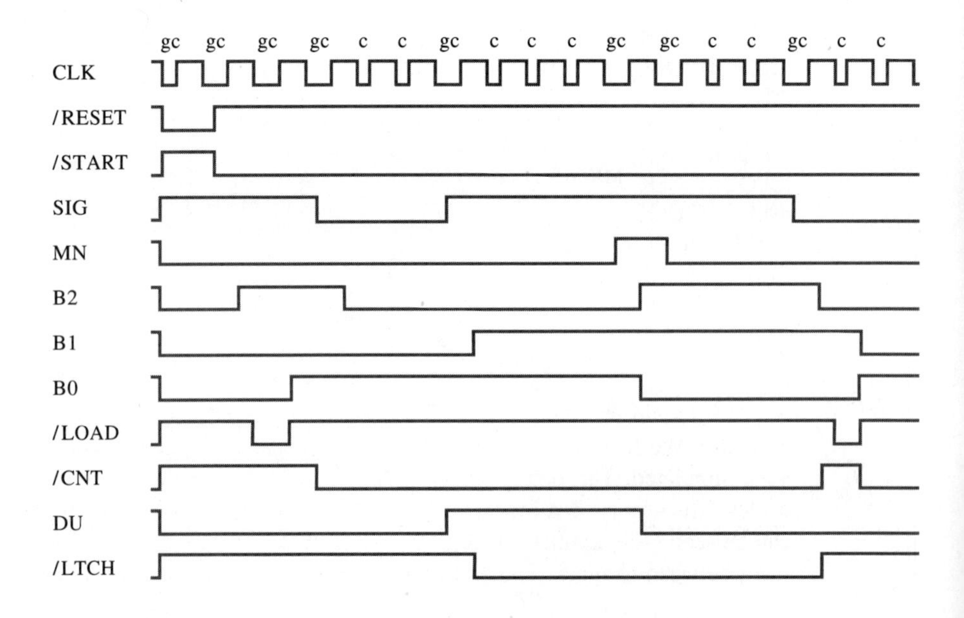

and a START followed by a low SIG for several clock cycles and then a high SIG for several cycles. MX was asserted during the high portion of SIG to check the action of the controller in response to this input.

To complete the design using a PAL as the system controller, we need to change the first-level functional to reflect the removal of the edge finder subsystem, and change the system controller schematic. The counter and the register subsystems are unchanged. The modified first-level functional is shown below.

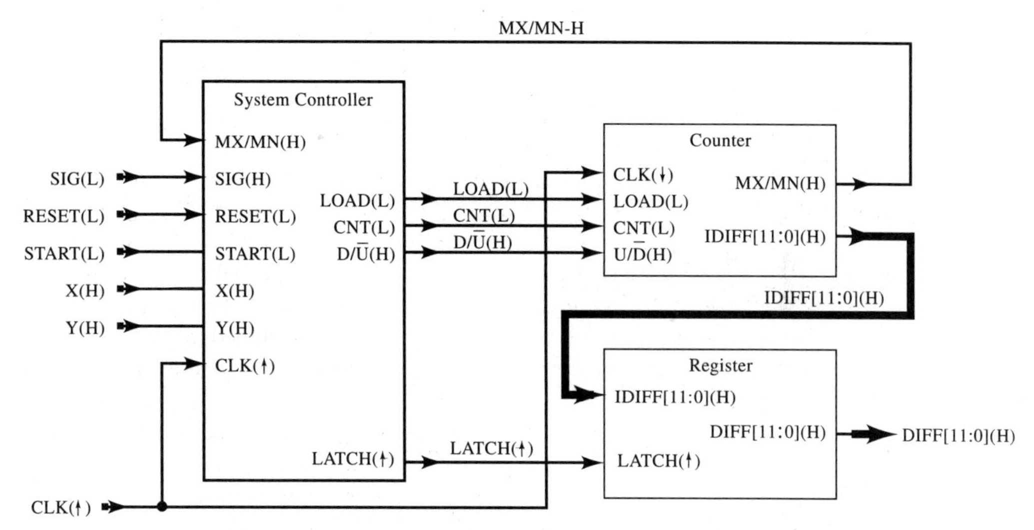

The system controller subsystem, which was a complicated many-element schematic for the previous design, becomes a single PAL as shown below. Furthermore, this PAL included the edge finder subsystem.

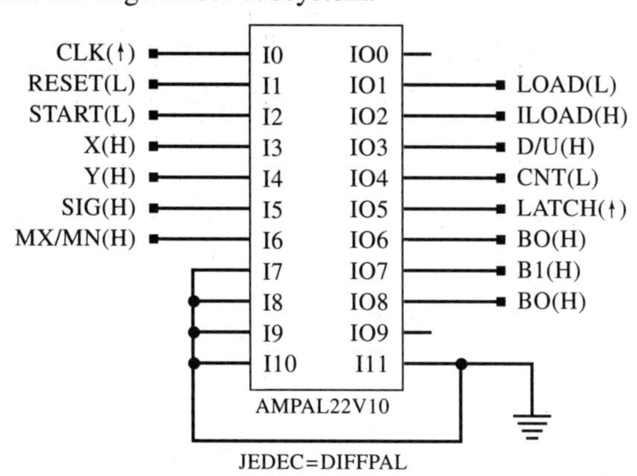

We should simulate the entire redesigned system to determine if it functions correctly. We'll only show simulation results for the 11 mode although all modes were simulated. The first timing diagram shows a simulation for three difference cycles. The second is a blowup of one of the cycles so that the counting sequence can be seen. The simulation package will show bus values if there's room to write them between changes. The values are in hexadecimal. This design is much better

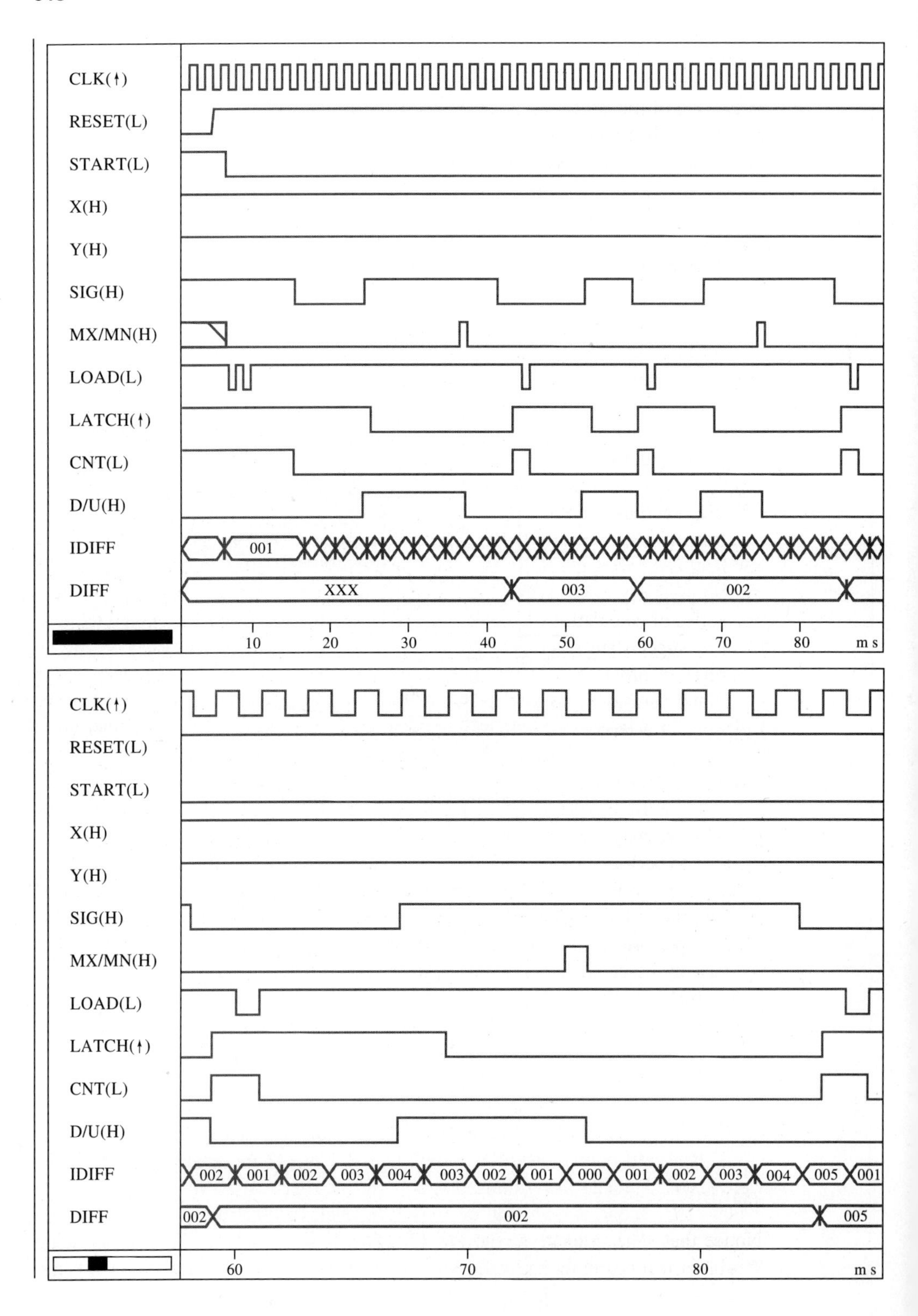
CLK(↑)
RESET(L)
START(L)
X(H)
Y(H)
SIG(H)
MX/MN(H)
LOAD(L)
LATCH(↑)
CNT(L)
D/U(H)
IDIFF
001
DIFF
XXX
003
002
10
20
30
40
50
60
70
80
m s
CLK(↑)
RESET(L)
START(L)
X(H)
Y(H)
SIG(H)
MX/MN(H)
LOAD(L)
LATCH(↑)
CNT(L)
D/U(H)
IDIFF
002 001 002 003 004 003 002 001 000 001 002 003 004 005 001
DIFF
002
002
005
60
70
80
m s

than the previous design as far as timing is concerned at the start of the first difference cycle and at the start of the subsequent cycles. This better timing is the result of getting rid of the edge finder, which required a handshake, and using a LOAD that was only half a clock wide. Using two subsystems requiring a handshake usually costs a timing delay and for this design was not necessary.

12.4 FUNCTIONAL PARTITION

The crucial part of any large system design is the functional partition. It's also the most difficult part of the design. In the last two sections, we saw that partitioning a system involves dividing it into subsystems, which are easier to design. This is an orderly subdivision. We first divide the overall system into several subsystems (usually less than seven). If necessary, we then further divide the subsystems into subsystems, and we continue this subdividing until we arrive at subsystems that are simple enough to design as state machines or as combinations of known logic elements such as counters, registers, adders, and so on.

But how do we subdivide the overall system or one of the subsystems? The key here is the *functional* in functional partition. The systems we're designing do something. They control; they process data; they perform an arithmetic calculation or a series of calculations. For the purpose of partitioning them, we should think of them as super state machines. They step through an orderly sequence of smaller *functions* in order to perform their desired overall function. The process of the functional partition is to determine the smaller functions which make up the larger function. In a large and/or complicated system the smaller functions may still be pretty complicated. In simpler systems, such as the ones in the last two sections, the smaller functions are simple enough to design from available logic devices.

Ultimately, the partitions in a functional partition are some kind of hardware. Initially, we may want to forget that we want hardware and just list the functions we need to perform. This can be done using a diagram like a state transition diagram where each circle (or ellipse) is some function and the arrows between the circles indicate the order in which the functions are performed. For instance, suppose we want to perform a multiplication of binary numbers. To recall how this is done (it was discussed in Section 2.8) consider the multiplication of A = 1101 and B = 1001.

```
   1101
   1001
   ----
   1101
  0000
 0000
1101
-------
1110101
```

Notice that we first examine the low order bit of B and added A to the product if this bit is 1. It is 1 in the example. We then shift A one bit left and examine the next higher order bit of B. If this bit is 1 we add the shifted A to the product. It is 0 in the example so we have shown that we add zero. We continue shifting A and examining the next bit of B and adding the shifted A if the bit of B is 1 until we have exhausted all the bit in B.

If we generalize A and B to be any n-bit binary numbers, we can list the function to be performed in a binary multiplication as:

1. Determine if the low order bit of B is 0 or 1.
2. If it's 1, add A to the product.
3. If it's 0, don't add A.
4. Shift A left 1 bit.
5. Determine if the next order bit of B is 0 or 1.
6. If it's 1, add the shifted A to the product.
7. If it's 0, don't add the shifted A.
8. Increment a bit count.
9. If the count is equal to n, stop.
10. Shift A 1 more bit left.

This is a repetitive pattern and can be put in a loop. It can also be nicely represented by the diagram in Figure 12.4-1. If you like lists use lists; if you like to think more graphically use a diagram.

In any case, get the functions down some way or other. Now what kinds of hardware systems do we need to perform the functions? For the multiplier, we need to look at the low order bit of B, then the next low order bit, and so on. This can be done with a shift register for B. We need to determine if the low order bit is 0 or 1. This is a function we can perform with a branch in our system controller. We need to shift A left 1 bit. This requires a shift register for A. We need to add the shifted A to a partial sum. This requires both an adder to do the addition and a register to store the partial sum. Finally, we need to count bits. This requires a counter. The list of functions leads us to the kinds of partitions we need. We need to remember that the problem of partitioning is a design problem. There is no single answer. There may not even be a best answer, but several answers that are equally good. When faced with a problem requiring a functional partition, try something; any functional partition is better than none.

When doing the partition, we can try to make the relative complexity of the partitions at a given level equal, but it shouldn't be a strong constraint. It's more important that the partitioning follow the functions. In any case, subsequent design may show that some of the partitions are trivial and some require several additional partitions. This constitutes a perfectly good design.

Notice that the interconnection and timing details at any level of a functional partition mature as the lower levels of the partition are developed. First, we make a preliminary partition with interconnections as detailed as possible. Then we complete the lower levels of the partition, which allows us to go back and complete the details of the higher levels. This is where experience helps us. With experience, we'll be able to make the interconnections in the preliminary functional partition more

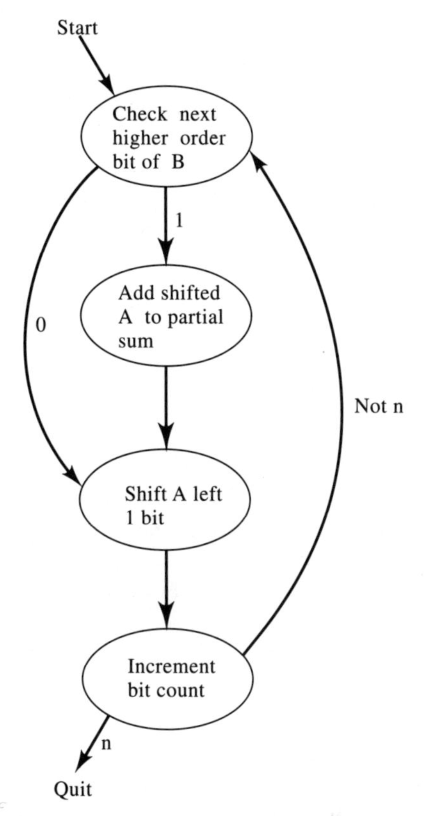

Figure 12.4-1 A diagram showing functions.

detailed because we'll be able to visualize what interconnections are needed in the subsequent partitions of our design.

Now let's look at another aspect of functional partitions. Figure 12.3-1, which shows the structure of the hierarchical partitions we're learning how to design, suggests a way to number the functional drawings produced in the design. The zeroth-level drawing is numbered 0.0, the first-level drawing is numbered 1.0, the first subsystem of the first level 1.1, the second subsystem 1.2, and so on. The first subsystem of the first subsystem is numbered 1.1.1, and the pattern continues. Thus, drawing 1.2.4.1 is the first subsystem, of the fourth subsystem, of the second subsystem, of the first functional.

Functional partitions perform a service in large designs besides producing subsystems of manageable complexity. They allow the design to be divided between several engineers. Each engineer can do the design of one of the subsystems or partitions. In this case, the *communications* between the subsystems must be more formal than when the entire system is designed by one engineer. To avoid misunderstandings, a formal description of the interface between the subsystem (an interface specification) should be made. We consider communications between subsystems in Section 12.5.

12.5 COMMUNICATIONS

By *communications* we mean the interchange of signals between a system and the external world or between two subsystems in a system. Part of communications is the handshake we have already seen. Communications can be simple or complex. One of the considerations in the partitioning of a system should be to choose partitions that will make the communications between the subsystems as simple as possible.

Communications can be broadly classified as *synchronous* and *asynchronous.* **Synchronous** signals are signals that are clocked by a common clock. Communications that are synchronous are the easier to handle. **Asynchronous** signals are signals that are clocked by unrelated clocks or not clocked at all.

Our model for communications is shown in Figure 12.5-1. Generally, communication with a subsystem involves an input, a handshake, and a response. We show a full handshake consisting of a *request*, *reply*, and *acknowledgement*. The **request** asks the subsystem to do something (it is a message to the subsystem to "act"). This something may involve an input (a single signal or a bus signal). The response is doing what is requested. It normally involves an output (a single signal or a bus signal). The **reply** is a reply to the request, which says that the request has been accomplished (it's a message from the subsystem, "I have acted"). Finally, the **acknowledgement** acknowledges the reply (it's a message to the subsystem, "I see that you have acted").

Not all of the elements of the model are necessary for communication. The simplest meaningful communication must involve a request, a response, and a reply. This kind of communication occurs in a simple unclocked subsystem. Unclocked systems are usually combinational logic circuits such as adders. In this kind of system, the request is the input and the response is the output, which occurs after a propagation delay (t_p). The reply is implicit, since the response always occurs after the specified propagation delay. After this fixed time, the requesting system can assume that a reply has occurred.

The simplest kind of synchronous communication is a completely synchronous

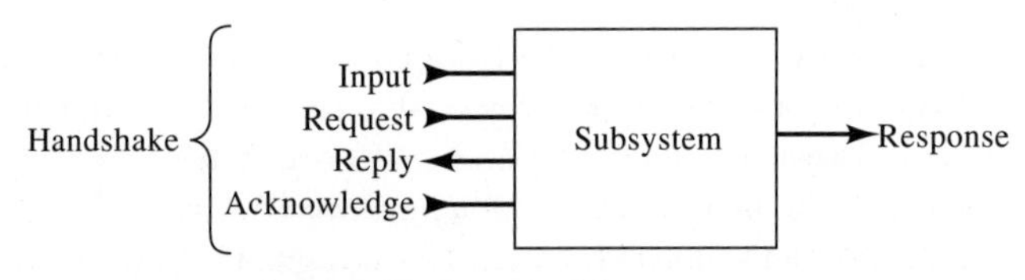

Figure 12.5-1 Communications model.

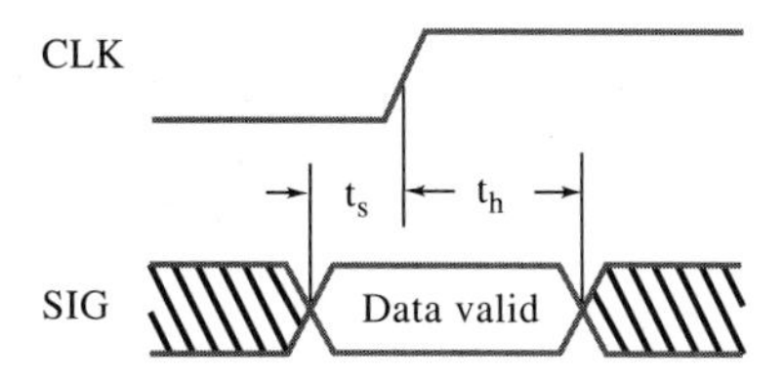

Figure 12.5-2 One-way synchronous communications.

(completely timed) one-way communication. This kind of communication with a subsystem involves an input, a request, a response, and a reply (implicit). An example of this communication is the serial input data stream we've used in many of our sample state machine designs. This communication requires two lines. The data (input) are presented on one line, and a clock is presented on the other. Each positive (or negative) clock edge signifies that data on the data line are valid. What we really mean is that the data are valid for some setup time (t_s) before the clock edge and remain valid for some hold time (t_h) after the clock edge, as shown in Figure 12.5-2. The clock edge is a request for the system to take data. The response is the system taking data, and the reply is implicit because the system is assumed to take data after a time t_h.

Another example of this kind of communication is the data transfer into the parallel output register in the sample design of Section 12.3. In this transfer, the data to be transferred are presented on a bus. Thus, IDIFF is valid at the positive going edge of LATCH. LATCH is the request for the system to take the data. The reply is implicit. After a given time (t_h), the assumption is that the data have been taken. This kind of transfer is synchronous only if the input clock clocks the transfer. If the input clock is used to signal that data are valid and the data are then clocked by another clock unrelated to the input clock, the transfer is really asynchronous.

Completely synchronous communications do not need a full handshake. No explicit reply is needed because the result always happens after a given time, and no acknowledgement is needed for the same reason.

The communication to the edge finder in the design of Section 12.3 is an example of communication that is not entirely synchronous and thus needs at least an explicit reply. The design actually includes a full three-wire handshake with an acknowledgment, just to illustrate this full handshake. Figure 12.5-3 illustrates the cause-effect relationship of a full handshake. The request causes the result. The result causes the reply to be asserted. The reply causes the acknowledgement to be asserted and the request to be deasserted. Finally, the acknowledgement causes the reply to be deasserted.

The full handshake can be either synchronous, partially synchronous, or asynchronous, depending on the relation of the various signals in the handshake to the clock. However, there's no need for a full handshake in a completely synchronous system. The handshake in the design of Section 12.3 is partially synchronous. All the handshake signals are derived from the same clock, so their timing with respect to the clock is fixed; but the number of clock cycles between the various handshake

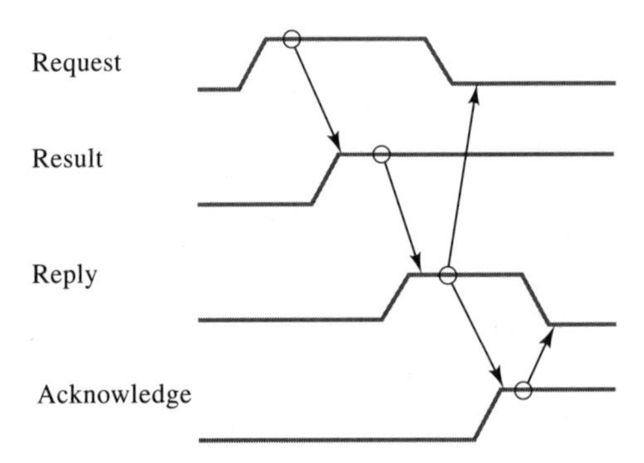

Figure 12.5-3 Full handshake.

signals is not fixed. The request to the edge finder is to find the edge (FE). The response to this request can happen at any time, depending on the negative edge of SIG; therefore, the reply must be explicit. This response EF (edge found) is also the reply handshake. It specifies that the response has occurred. The acknowledgement part of the handshake RSTEF (reset edge finder) tells the edge finder that the EF has been noted and can be deasserted. Generally, a three-wire handshake is an overkill. The acknowledgement can be made by deasserting the request. In this example, FE is asserted to tell the edge finder to find an edge. When the edge finder finds an edge and asserts EF, FE can be deasserted to acknowledge that the edge is found.

Actually, this communication does not need an acknowledgment because it's synchronous. EF is synchronous with the clock of the system controller, which acknowledges EF, and EF is always used half a clock cycle after it's asserted. It only needs to be asserted for a full clock cycle of the edge finder clock. If EF was not used at a given time but depended on some input signal to the system controller, then the acknowledgement would be required to tell the edge finder when to deassert EF.

If two subsystems have unrelated (asynchronous) clocks, a complete handshake is usually necessary. There is no way to tell when the response will occur and no way to determine when to deassert the reply. Subsystems that are asynchronous have a further complication. In the edge finder, all the handshake signals are generated from the same clock. Their timing with respect to the clock is thus fixed, and if the system is properly designed, they will not change too close to a clock edge and violate a setup or hold time constraint. In an asynchronous system, this is no longer true. Handshake signals can occur near a clock edge. Problems due to setup and hold time violations can be solved by using the rules for asynchronous state machine inputs and by using the synchronizing circuit described in Sections 9.3.1 and 9.4.

Before leaving communications, let's summarize the kinds of communications that can occur.

1. Communications in which the response occurs a fixed time after the request. These types of communications only require a request.

2. Communications in which the response can occur at a random time after the request, but the reply is acknowledged at a fixed time after it's received. These types of communications require both a request and a reply.
3. Communications in which the response can occur at a random time after the request and the acknowledgement of the reply can occur at a random time after the reply. These types of communications require a full handshake—request, reply, and acknowledgement.

All three types of communications can be synchronous or asynchronous. Asynchronous handshakes have the further problem that setup and hold time violations can occur.

12.6 A SIMPLE COMPUTER DESIGN

In this last section of the text, we'll design a simple stored program computer. First, we'll have to get enough understanding of computers to describe the computer we're going to design. A computer consists of two basic parts—a memory and a processor. The memory stores a large block of binary numbers and codes. The processor performs arithmetic and logic operations on the binary numbers stored in the memory.

The memory is organized like the memory devices we described in Section 11.4. It consists of a large array of fixed-length binary registers, each of which is independently addressable. Normally, in a small computer, memory is organized to store 8-bit or 16-bit words. The computer we're going to design will be an 8-bit coumputer. The maximum size of the memory is determined by the size of the address. We'll design for an address size of 16 bits, so our processor will be capable of addressing $2^{16} = 65{,}536$ locations in memory.

The processor processes data using a few simple operations that it can do very rapidly. We'll design our computer to be able to execute the following instructions.

Read from memory

Write to memory

Read for input/output (I/O)

Write to I/O

Add

Subtract

In order to conveniently handle data, which are often stored sequentially in memory, we also need to be able to

Increment address

Decrement address

A stored program computer is a computer that stores instructions as well as data in its memory. It then reads the instructions to determine the arithmetic operations it will perform. Normally it starts reading instructions at some memory location and reads sequentially from there on. This does not allow for much flexibility. Often we

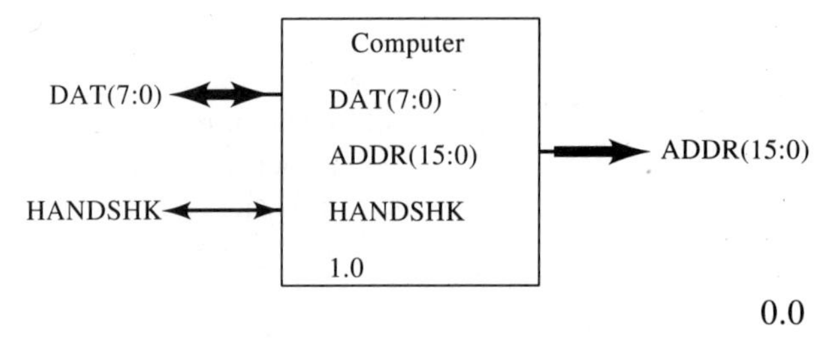

Figure 12.6-1 Computer as large block (zero level functional).

want to execute one set of instructions if a certain condition is true and another if it's not true. We can do this if we have a way to break into the normal address sequencing for reading instructions and go to a specified address. Breaking the normal sequencing and going to a specified address is called **jumping**. We'll make our machine so that it's able to execute four jump instructions.

Jump

Jump on zero

Jump on positive

Jump on carry

Where *zero, positive,* and *carry* refer to the results of the previous arithmetic operation performed by the computer.

This brief description of a computer should get us started. It's our *big block description* of the computer, shown in Figure 12.6-1. To the outside world, a computer is very simple. It consists of the computer with an input/output bus for passing data in and out of the computer, an address bus for directing where the data should go, and handshake lines to control the data transfer. Both instructions and data are input on the bus, and results from the computations performed in the computer are output on the same bus. We might consider building our computer using discrete logic, but it will be more realistic to use PALs as much as possible. Since we would like a fast computer, we'll use the fast (F) logic series where we need discrete logic and fast PALs where we need PALs. Just to give us an idea of speed, we should note that fast logic performs operations such as a 4-bit add in less than 10 ns, and we will use fast PAL22V10s, which perform most operations in 15 ns or less. A computer constructed using discrete logic elements and PALs is pretty unlikely, given present technology. We might make such a computer as a step in the design of a computer using custom logic. In any case, it will give us an interesting look at design.

The data bus in our description of the computer is a bidirectional bus. It passes data in both directions. This is accomplished by making all devices that drive the bus tristate devices. This means they can be turned off—made high impedance so they no longer drive the bus. Bidirectional busses, like all busses (except simple

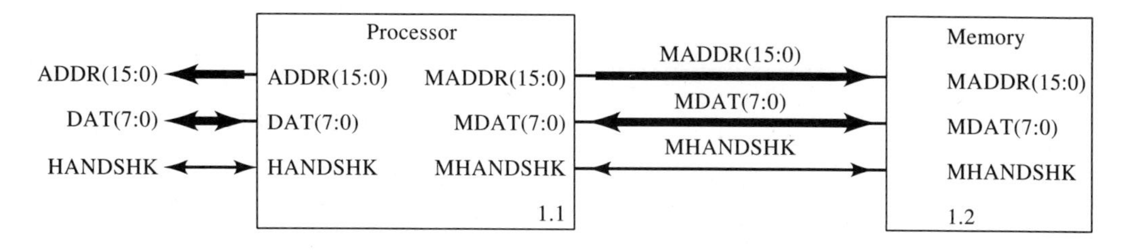

Figure 12.6-2 Preliminary first level functional.

point-to-point busses), must be managed so that only one device drives the bus at any time. The design of even a simple computer makes heavy use of busses.

Our preliminary first-level functional partition of the computer follows the breakdown into processor and memory we've already suggested (Figure 12.6-2). In this diagram, we show bidirection busses and handshakes to the external world as well as to the memory. These are the communications to the outside world we've shown on the zeroth-level functional. Our first-level partitioning does not have a system controller. We're considering the processor as controlling the memory, with the system controller embedded in the processor. Notice that we are referencing each of the functionals with an index number in the lower-right corner of the drawing.

Now we need to add detail to the functional partition. We can do this by completing the design of the memory using memory devices that we discussed in Section 11.4. There are a number of ways to organize the memory. We'll choose to use two 64K $\times$ 4 static R/W memory devices such as the CY7C196 (Cypress). We need two memory devices because each device only stores 4 bits in each memory location and we need 8 bits. The memory we've chosen is high performance memory and would be expensive. Figure 12.6-3 shows the detailed memory design and Figures 12.6-4 and Table 12.6-1 show the timing of the memory. We've included a 2K $\times$ 8 PROM in our memory because we need some way to start the computer. Normally, a startup program is loaded into PROM and the computer is made to start in this program on a reset. The computer only uses this memory once each time it's reset. After startup, the PROM control line is made to select the R/W memory. The ROM timing for the memory device we've chosen (CY7C291) is similar in speed to, but slightly faster than, the R/W memory.

By completing the memory design, we've not only determined the interconnection details of the first-level functional partition, but the timing details as well—at least the timing limits. We see that we need a write ($\overline{\text{WE}}$) clock, read ($\overline{\text{RD}}$), and a ROM handshake to the memory as well as an address bus and a data bus. Now we can go back and add these interconnection details to the preliminary first-level functional partition and get a detailed functional partition. Before we do this, let's consider what to do about the interface to the external world. One way to handle this interface is to notice that it's similar to the interface to the memory subsystem and make the interconnections and timing the same as the interconnections and

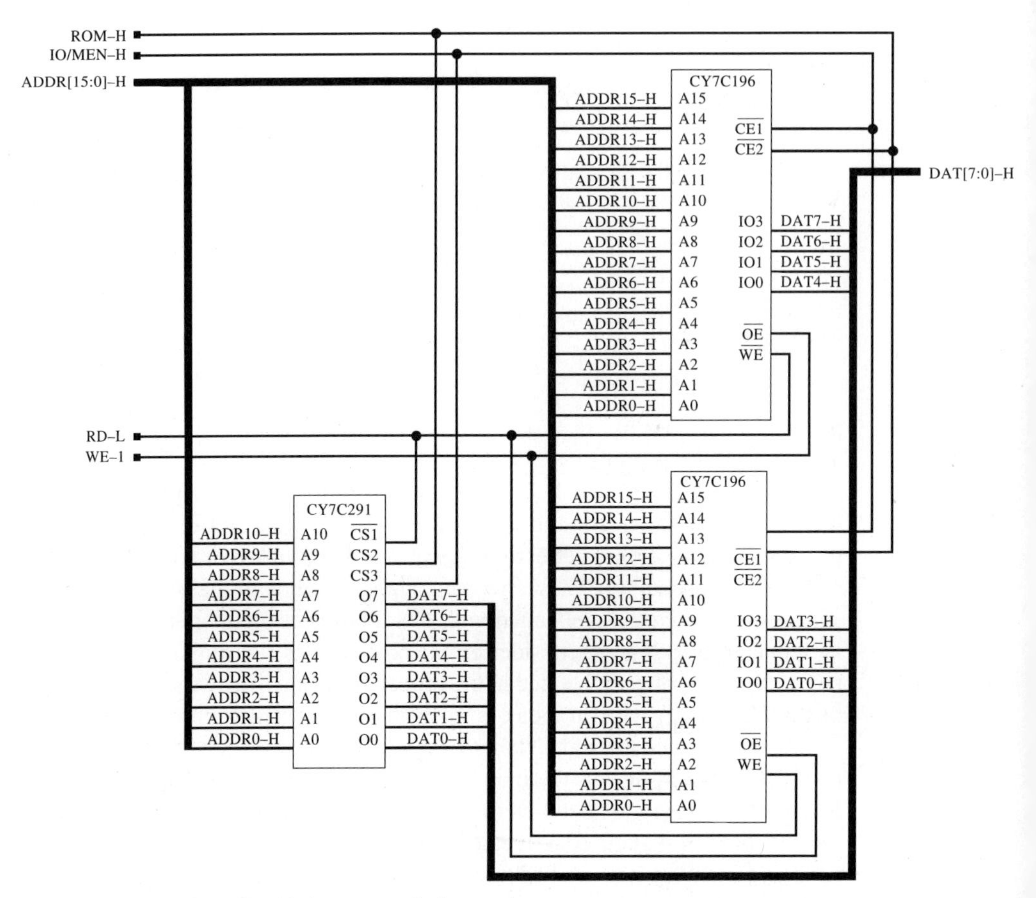

Figure 12.6-3 Detailed memory design.

timing to the memory. If we do this, then we can use the same bus for memory and input/output and just have an additional control line (IO/MEM), which determines whether the data on the bus are data to or from memory or the external world (I/O). We'll do this in our design for simplicity. If we make our computer fast enough to properly utilize the memory we have chosen, the I/O timing will be much too fast, but we're going to do it anyway. With this provision for I/O, the preliminary first-level functional partition of Figure 12.6-2 becomes the detailed first-level functional partition shown in Figure 12.6-5. At this point we can also go back and add detail to our zeroth-level (big block) functional partition (Figure 12.6-6).

The memory we've chosen has three speeds. Which should we choose? We would like a fast computer, so we might choose the 25 ns memory. To properly utilize this memory speed, we would need a clock speed of at least 40 MHz (25 ns period). We could then possibly read or write in one clock cycle. Both the clock speed and reading or writing in one clock cycle are tight constraints. The maximum

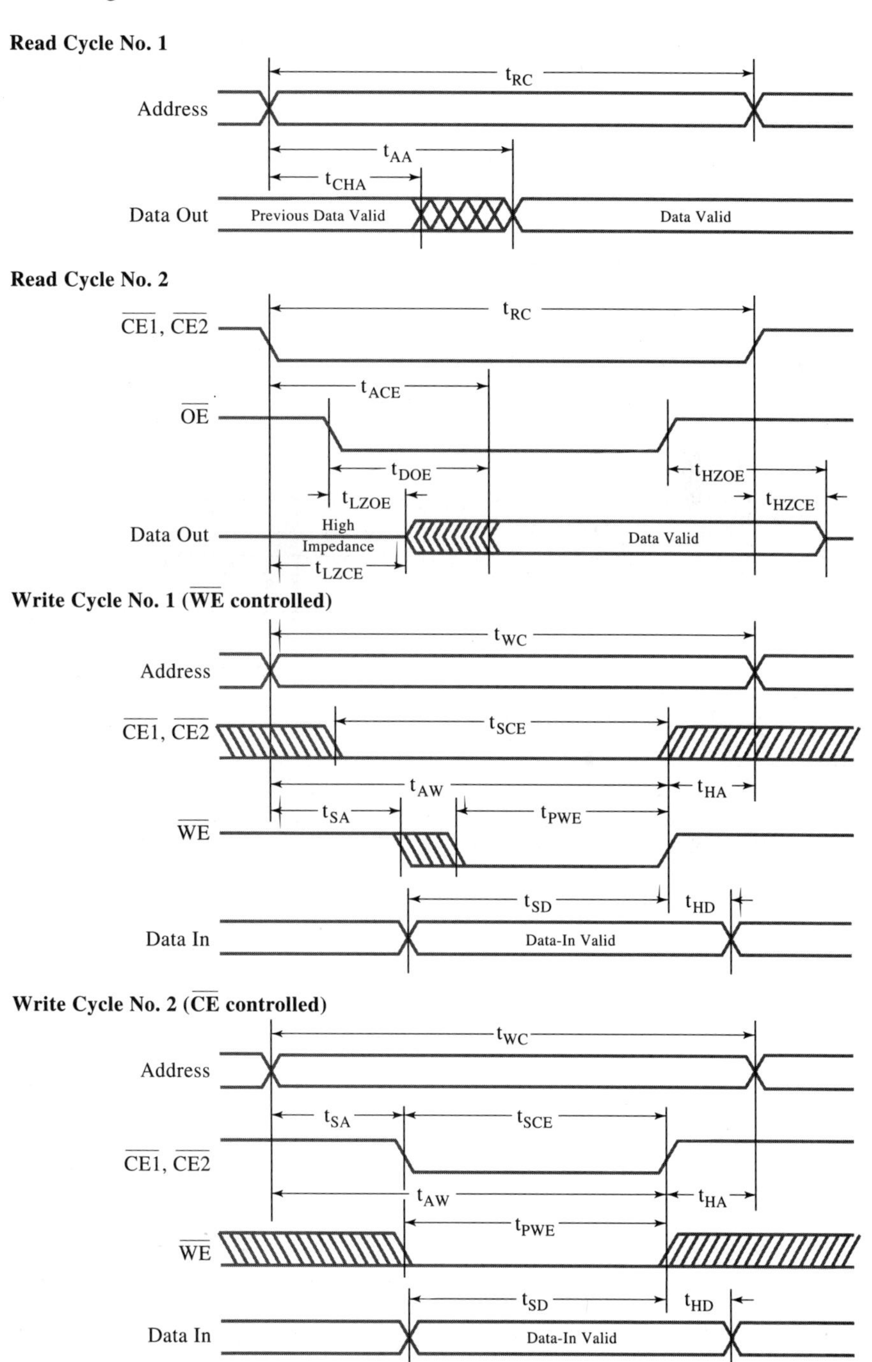

Figure 12.6-4 Memory timing for CY7C196.

Table 12.6-1 Values for CY7C196 Memory Timing

Switching Characteristics over the Operating Range

Parameter	Description	7C196-25		7C196-35		7C196-45		Unit
		Min.	Max.	Min.	Max.	Min.	Max.	
Read Cycle:								
t_{RC}	Read Cycle Time	25		35		45		ns
t_{AA}	Address to Data Valid		25		35		45	ns
t_{OHA}	Output Hold from Address Change	3		3		3		ns
t_{ACE}	$\overline{CE}$ LOW to Data Valid		25		35		45	ns
t_{DOE}	$\overline{OE}$ LOW to Data Valid		15		20		20	ns
t_{LZOE}	$\overline{OE}$ LOW to Low Z	3		3		3		ns
t_{HZOE}	$\overline{OE}$ HIGH to High Z		13		15		20	ns
t_{LZCE}	$\overline{CE}$ LOW to Low Z	0		0		0		ns
t_{HZCE}	$\overline{CE}$ HIGH to High Z		13		15		20	ns
Write Cycle:								
t_{WC}	Write Cycle Time	20		30		40		ns
t_{SCE}	$\overline{CE}$ LOW to Write End	20		30		40		ns
t_{AW}	Address Set-Up to Write End	20		25		35		ns
t_{HA}	Address Hold from Write End	0		0		0		ns
t_{SA}	Address Set-Up to Write Start	0		0		0		ns
t_{PWE}	$\overline{WE}$ Pulse Width	20		25		30		ns
t_{SD}	Data Set-Up to Write End	15		17		20		ns
t_{HD}	Data Hold to Write End	0		0		0		ns

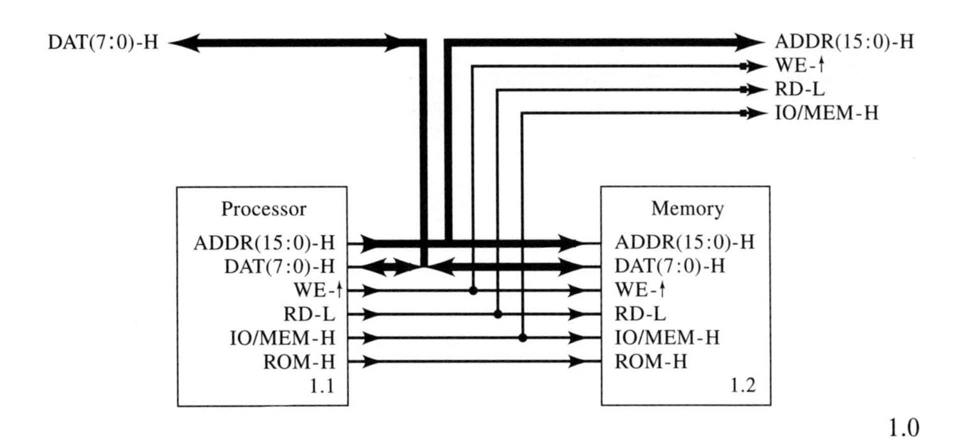

Figure 12.6-5 Detailed first level functional partition.

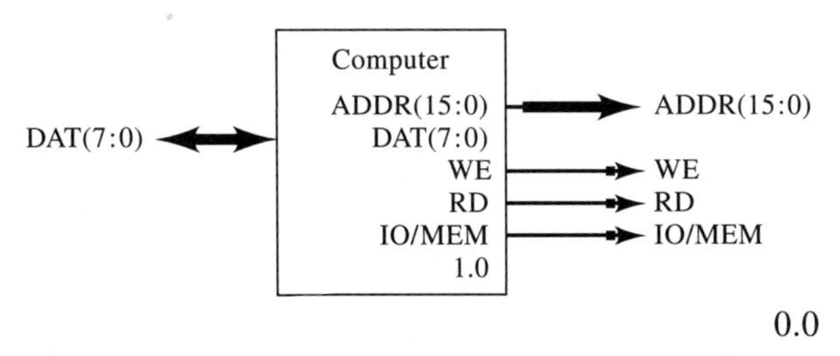

Figure 12.6-6 Detailed zero level functional.

clock speed for a fast PAL22V10 is 40 MHz, and one clock cycle for a read or write is possible but allows only one clock edge (the negative going edge) for timing internal to the read or write cycle [see Figure 12.6-7(a)]. We'll be conservative and choose 45 ns memory and relax the clock speed to 25 MHz. This means that if we take two clock cycles to complete a read or write [see Figure 12.6-7(b)], we have 80 ns, and this does not push the memory speed. Two clock cycles allow more flexibility than one because there are more timing points. We could actually delay the consideration of clock speed until later in our design, but it's well to make a tentative choice and get something to think about as we complete the processor part of the design.

Now let's get on with the design of the processor. The question here is, what does the processor do? In a broad sense it cycles through two steps.

1. Fetch an instruction from memory.
2. Execute the instruction.

This is the fetch and execute cycle, which is basic to all computers. This idea leads

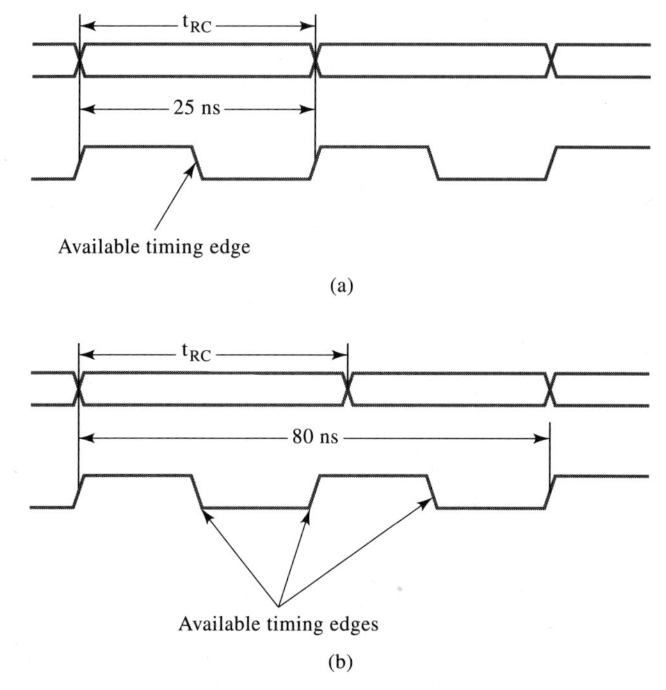

Figure 12.6-7 Memory speed considerations. (a) 40 MHz clock and 25ns memory cycle. (b) 40 MHz clock and 50 ns memory cycle.

us to the two-block functional partition of Figure 12.6-8. The instruction processor fetches the instruction from memory and interprets it. The arithmetic processor does the add and subtract. We'll also make it store temporary data as opposed to instructions. We need a way to store temporary data because we want to make a machine that only gets one set of data from memory for each instruction. In a computer, the register that stores data which is being operated on (temporary storage) is called the **accumulator**. We want our computer to get data from memory and store them in the accumulator or get data from memory and add them to, or subtract them from, the data in the accumulator. We also want our computer to take data from the accumulator and store them into memory.

The instruction processor fetches instructions from memory and interprets them. It passes the interpretation of the instruction on to the arithmetic process when necessary. It then addresses any data that are to be used in the instruction and signals the arithmetic processor to store these data. We'll let the instruction processor handle all memory and I/O addressing and handshakes. This will simplify the communication between the instruction and arithmetic processor. To properly execute jump instructions, which are conditional on the results of an arithmetic operation, the instruction processor must get these results (ZERO, POS, and CRY) as input from the arithmetic processor. We actually would need to complete the design of

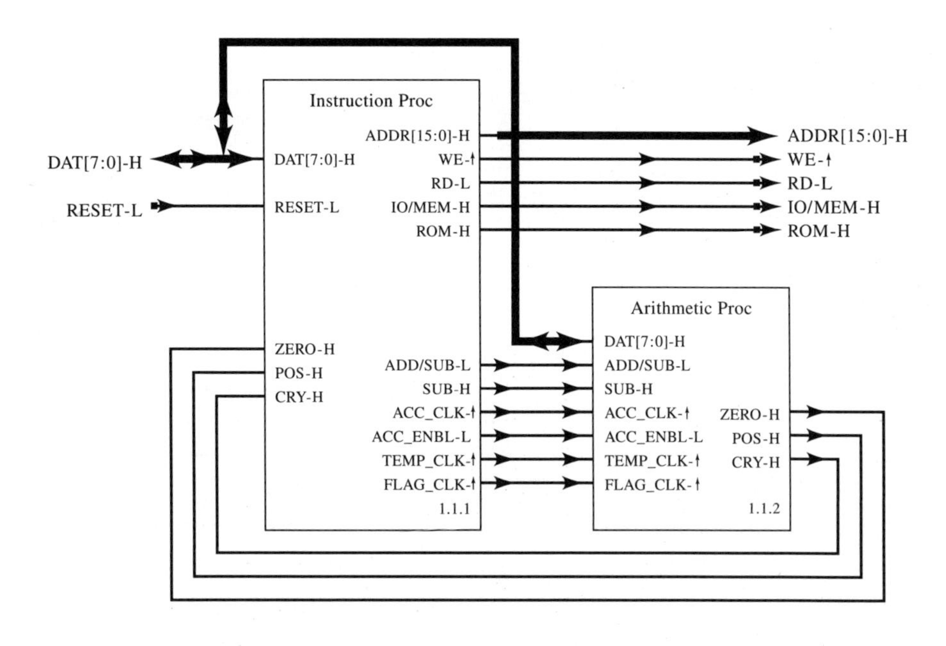

Figure 12.6-8 Functional partition for the processor subsystem.

the arithmetic processor to find all the interface lines between the two processors. We already have them on the figure, but let's look at the arithmetic processor and see how they come about.

Figure 12.6-9 shows a functional partition of the arithmetic processor. To perform the addition and the subtraction, we require two registers—one register, which is usually called the **accumulator** as we saw earlier, for storing one of the numbers to be added or subtracted, and another register, which we'll call the **temporary register**, for storing the other. We also need an adder. We'll do the addition with a simple binary adder. We'll do the subtraction by complementing the number to be subtracted and adding. The complementing will be done in the temporary register by inverting all of the bits. Remember that in order to do a 2s complement, we complement all the bits, but we also have to add 1. We'll do this in the adder by adding 1 into the carry. We'll make adding a two-step process. First, data are moved from memory to the accumulator. Second, data are moved from memory to the temporary register and added or subtracted from the data in the accumulator. The adder subsystem will also be made to determine the three conditions ZERO, POS, and CRY for an arithmetic operation and store these conditions. Notice how the result of an addition or subtraction is fed back into the accumulator so it can be stored.

The next level of partition for each subsystem of the arithmetic processor is the bottom of the hierarchy and is thus a circuit diagram. The three circuit diagrams are

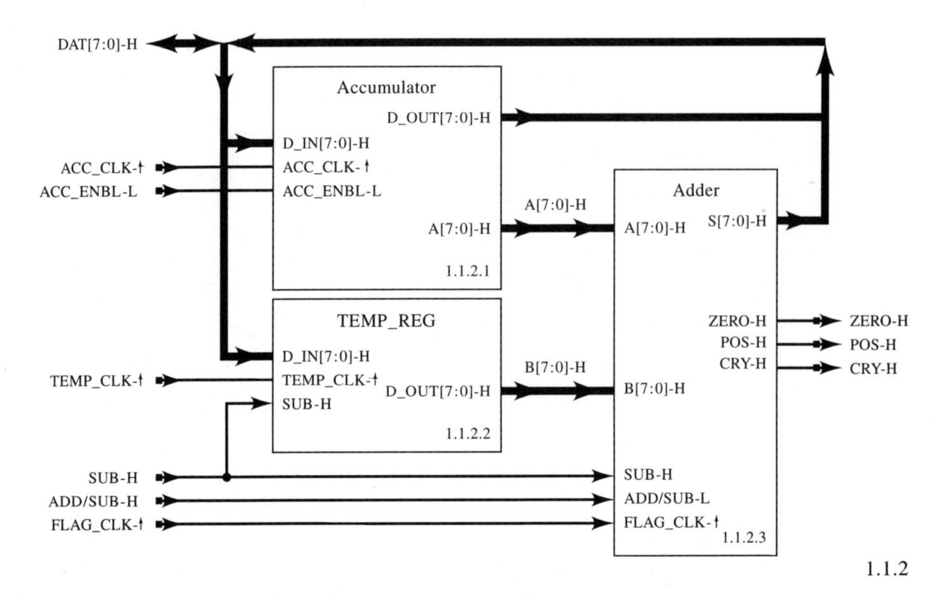

Figure 12.6-9 Functional partition for the arithmetic processor partition subsystem.

shown in Figures 12.6-10 through 12.6-12. The accumulator consists of an 8-bit register (74F574) and an output buffer (74F244). Data are clocked into the register by ACC_CLK and enabled to the data bus by ACC_ENBL. The output to the adder is always enabled.

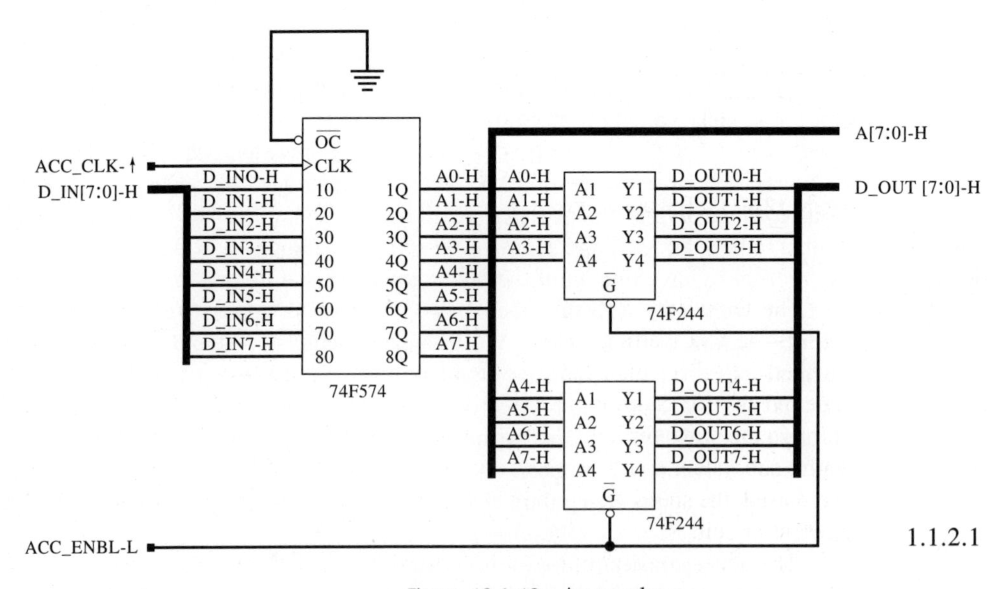

Figure 12.6-10 Accumulator.

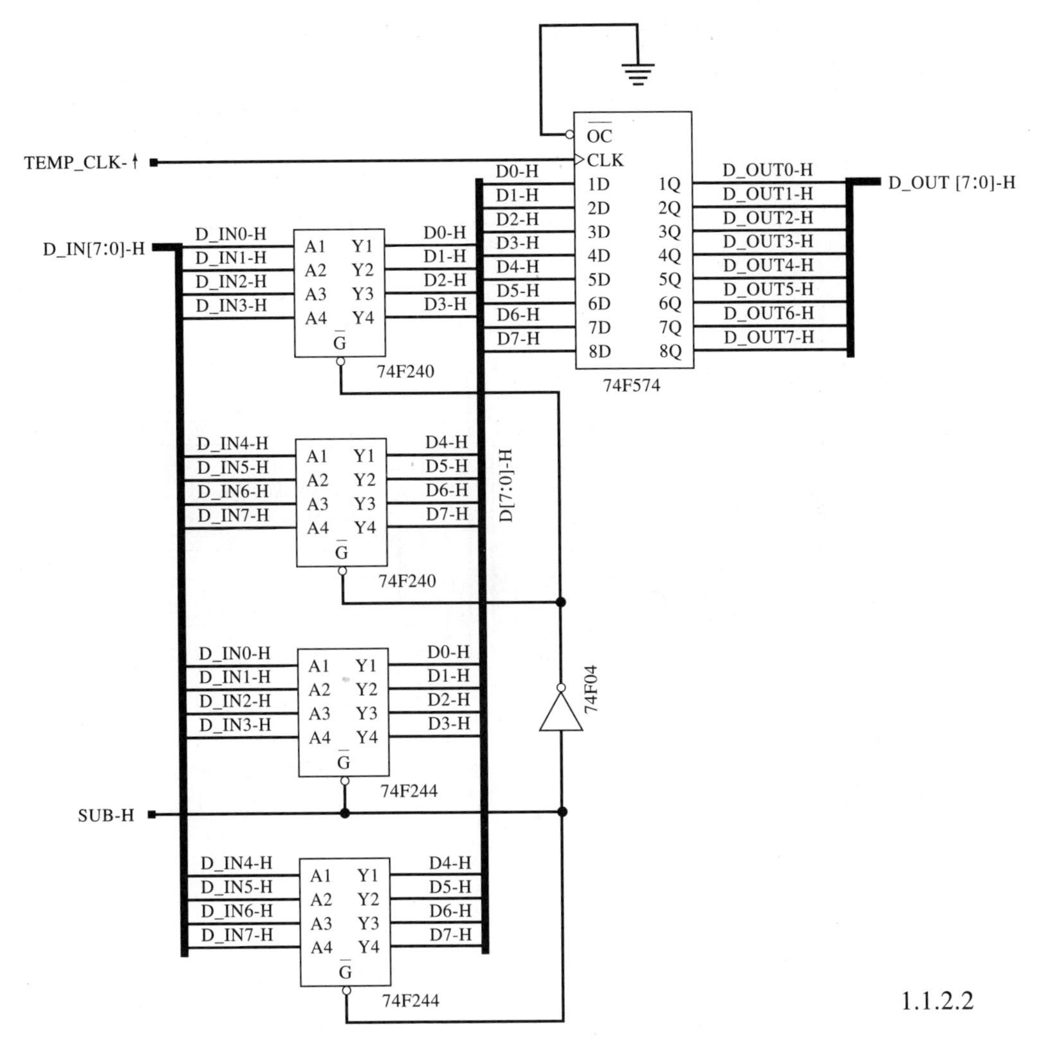

Figure 12.6-11 Temporary register.

The temporary register consists of an 8-bit register (74F574) and two input buffers—a noninverting buffer (74F244) and an inverting buffer (74F240). Data are clocked into the register by TEMP_CLK. Data are inverted for subtraction (SUB asserted) by steering them through the inverting buffer. Otherwise they're steered through the noninverting buffer (SUB not asserted). Just as a matter of interest, this entire subsystem could be programmed into a single PAL22V10. If a 15 ns 22V10 were used, the speed would only be slightly slower than the fast logic devices of the present circuit.

The adder consists of two 4-bit fast adders (74F283) with the low order carry cascaded to form an 8-bit adder, a tristate buffer (74F244) to control the output to

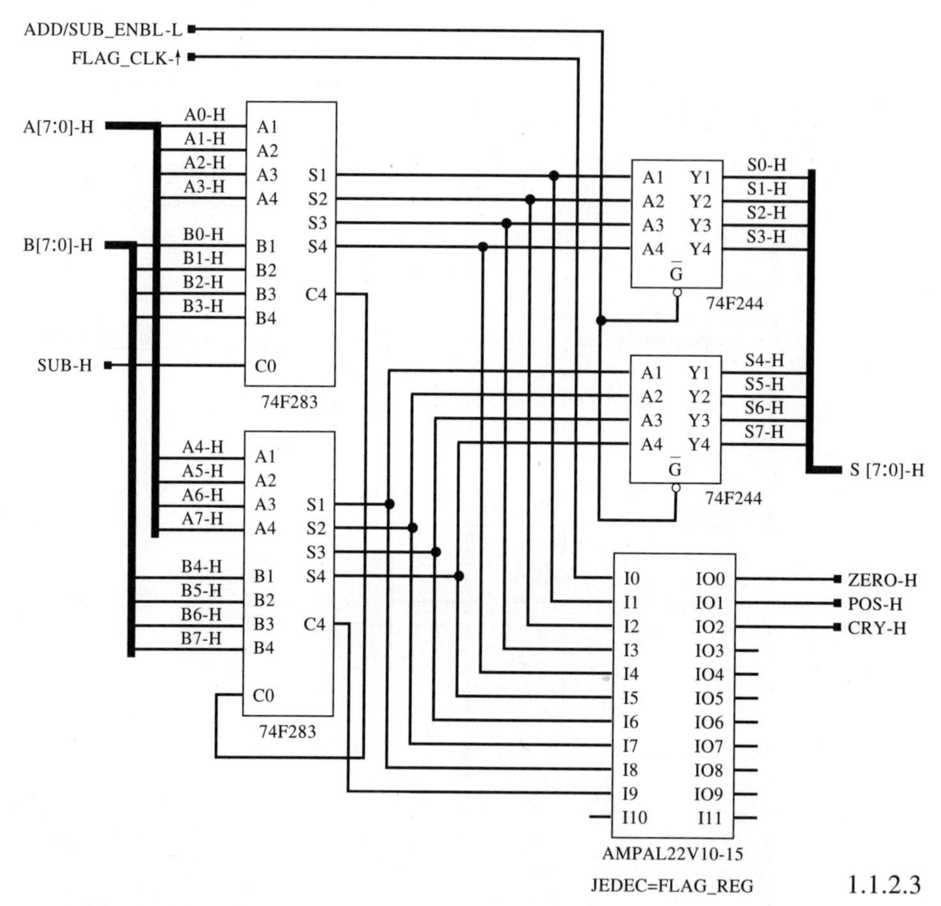

Figure 12.6-12 Adder.

the data bus, and a PAL used to generate and store the ZERO, POS, and CRY flags. A PAL is used because it can not only store the flags but also performs the logic needed to generate the ZERO flag. The ZERO flag is asserted when all the bits of the sum or difference are zero.

$$\text{ZERO} = \overline{S7} \cdot \overline{S6} \cdot \overline{S5} \cdot \overline{S4} \cdot \overline{S3} \cdot \overline{S2} \cdot \overline{S1} \cdot \overline{S0}$$

The POS flag is asserted when the most significant bit of the sum or difference (S7) is zero, and the CRY flag is asserted when the carry out (C4) from the most significant adder is asserted. Data are enabled to the data bus by ADD/SUB_ENBL. Data are clocked into the flag register by FLAG_CLK. The carry needed for a subtraction is provided by SUB. The source program for programming the flag register is given in Table 12.6-2.

This completes the arithmetic processor except for the timing and allows us to see how we completed the interface between the instruction process and the arithmetic processor. The steps of going down the hierarchy and then back up were left out in the interest of keeping the description to a reasonable size.

We still have the most difficult subsystem left to design—the instruction processor. The instruction processor is the most difficult part because it contains the system controller. To design the instruction processor, we ask the question, What does it

Table 12.6-2 Source Program for Flag Register PAL

```
;PALASM Design Description

;------------------------------------ Declaration Segment ------------------------------------
TITLE       FLAG REGISTER
PATTERN
REVISION
AUTHOR      A W SHAW
COMPANY     UTAH STATE UNIVERSITY
DATE        09/26/92

CHIP    FLAG_REG    PAL22V10

;---------------------------------------- PIN Declarations ----------------------------------------
PIN   1      CLK                              ; CLOCK
PIN   2      D0                               ; INPUT
PIN   3      D1                               ; INPUT
PIN   4      D2                               ; INPUT
PIN   5      D3                               ; INPUT
PIN   6      D4                               ; INPUT
PIN   7      D5                               ; INPUT
PIN   8      D6                               ; INPUT
PIN   9      D7                               ; INPUT
PIN   10     C4                               ; INPUT
PIN   11     NC                               ;
PIN   12     GND                              ;
PIN   13     NC                               ;
PIN   14     NC                               ;
PIN   15     NC                               ;
PIN   16     NC                               ;
PIN   17     NC                               ;
PIN   18     NC                               ;
PIN   19     NC                               ;
PIN   20     NC                               ;
PIN   21     CRY                   REGISTERED ; OUTPUT
PIN   22     POS                   REGISTERED ; OUTPUT
PIN   23     ZERO                  REGISTERED ; OUTPUT
PIN   24     VCC                              ;
```

```
;-------------------------------- Boolean Equation Segment --------------------------------
EQUATIONS

ZERO := /D7*/D6*/D5*/D4*/D3*/D2*/D1*/D0

POS := D7
CRY := C4

ZERO.TRST = VCC
POS.TRST = VCC
CRY.TRST = VCC
```

do? First, it fetches an instruction from memory; second, it interprets the instruction. To see this process in more detail, we need to know what the instruction code consists of. You might want to review the instructions we want our machine to perform at the beginning of this section before we see how to code them.

Since this is an 8-bit machine, we'll make the instructions consist of a sequence of 8-bit words. For any given instruction we'll make the first 8-bit word be the actual instruction code (op code). Following the op code, we'll have the data the instruction needs in the form of an 8-bit number, an address that tells us where to find an 8-bit number, or just an address if the instruction is a jump. There are only three types of arguments an operation can have—either a number (called *immediate*), an address to a memory location (called *address*), or an address for a jump. Figure 12.6-13 summarizes the coding of our instruction or op codes, and Table 12.6-3 lists all the valid codes. Notice that the two least significant bits of the op code tell us how many words follow and the next four bits tells us what to do.

Now, after this digression, back to the instruction processor. Let's try to make a rough list of what this processor must do. It must:

1. Fetch an instruction from memory using a given address
2. Store this instruction
3. Determine if the instruction is followed by nothing, an 8-bit data word, or an address (two 8-bit words)
4. Increment the address if another word is to follow
5. Get the next word if required and store it either as data, part of an address to data, or part of an address for a jump
6. Increment the address if another word is to follow
7. Get the next word if required and store it as part of an address to data or part of an address for a jump
8. Interpret the instruction and feed the proper control signals to the arithmetic processor to perform the operation requested
9. Increment the address unless a jump is indicated
10. Start over at 1 on this list

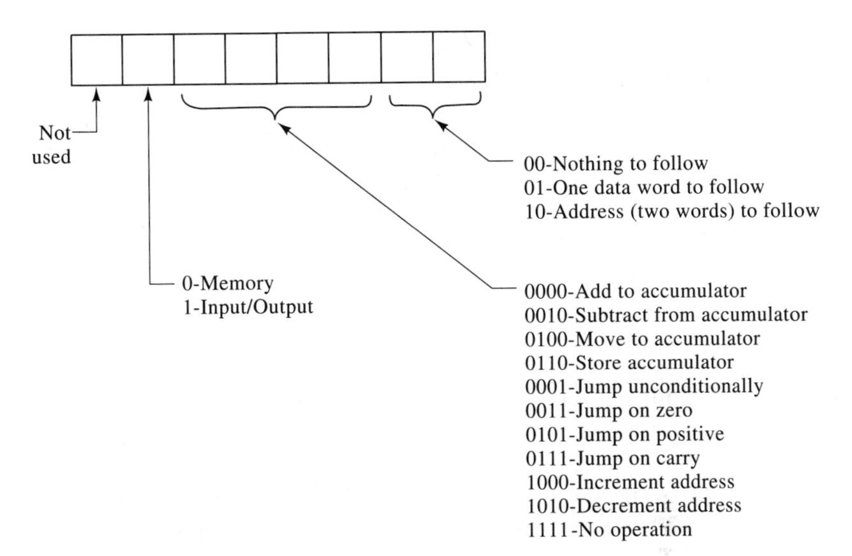

Figure 12.6-13 Instruction or op codes.

Table 12.6-3 Valid Instruction or OP Codes

OP Code	Meaning
0000001	Add immediate data to accumulator
0000010	Add addressed data to accumulator
0001001	Subtract immediate data from accumulator
0001010	Subtract addressed data from accumulator
0010001	Move immediate data to accumulator
0010010	Move addressed data to accumulator
0011010	Store data from accumulator to address
0000110	Jump unconditionally to address
0001110	Jump on zero to address
0010110	Jump on positive to address
0011110	Jump on carry to address
0100000	Increment address
0101000	Decrement address
0111100	No operation
1010010	Move addressed IO data to accumulator
1011010	Move data from accumulator to addressed IO

This is only a rough list. The order of operations may be imperfect, and some operations may be missing, but the list will get us started. Actually, it's not bad. First, we can see that we need to store an address to the current instruction and have provisions for incrementing this address. This requires a register of some kind. The register is normally called the **program counter** (**PC**) because it counts through the instruction addresses. Second, we need to store instruction codes. This requires a second register, which is normally called an **instruction register**. Third, we need to store addresses for data used in an instruction. This requires a third register which we'll call the **address register**. Finally, we need a system controller, which we've called the **instruction controller** (INS_CTL). Figure 12.6-14 shows the partitioning of the instruction processor we get from putting these subsystems together.

In this partitioning, we can complete the details of the interconnections without going any further down the hierarchy. We know what each of the subsystems must do. Both the program counter and the address register must be able to store an address that is presented to them as two 8-bit words—a high order word and a low order word. The program counter must be able to increment so it can count through the instructions in the program. The address register must be able to both increment and decrement as required by the specifications of the computer. Both must interface with the address bus using tristate outputs that can be enabled at the proper times. Finally, the program counter must have a provision for reset that resets the address to a known starting address. We'll make this address 0. The instruction register must be an 8-bit resetable register and little else. Finally, the system controller must furnish control for the three registers, the arithmetic processor, and the I/O and memory.

The address register and the program counter function almost identically. Since they must both increment (the address register must also decrement), they are essentially counters, but they must also have provisions for parallel load and tristate output. We've chosen to implement these registers (counters) in PALs. It's not difficult to make a counter using a PAL, and provisions for parallel load and tristate output are also easily added. The one problem is the number of pins available. Figure 12.6-15, which shows the two different PALs with functional labels on the pins, shows that there are just enough pins if we use two PALs in each register—one for each 8 bits in the 16-bit register. Actually, we could use one of the new multiple PALs (AMD MACHs) and use only one device, but we won't. Most of the pin labels are obvious—/INH is an inhibit function (asserted low) used for cascading as well as for holding the count when a load is made. /ENBL (asserted low) is the output enable used for bus control. TC is the terminal count, which is asserted when all the bits are 1s (all zeros for the up-down counter when it is counting down). The two PALs differ in that one has a reset and the other an up/down control (UP).

Figures 12.6-16 and 12.6-17 show circuit diagrams for the address register and program counter using these PALs. Notice the additional complexity of the program counter. This is because on careful consideration, we realize that when we execute a jump by loading an address into the program counter, we destroy the old address. Actually, it's worse than that; we destroy only half of the address. We'd better have provisions for getting the entire new address before we load any part of it. We do this with an extra 8-bit register, which stores the high order word of the address when LOADH is asserted and holds it until LOADL is asserted.

The designs for the PALs are not difficult if we don't try to design them as state machines using state transition diagrams. To design these PALs, which are essentially

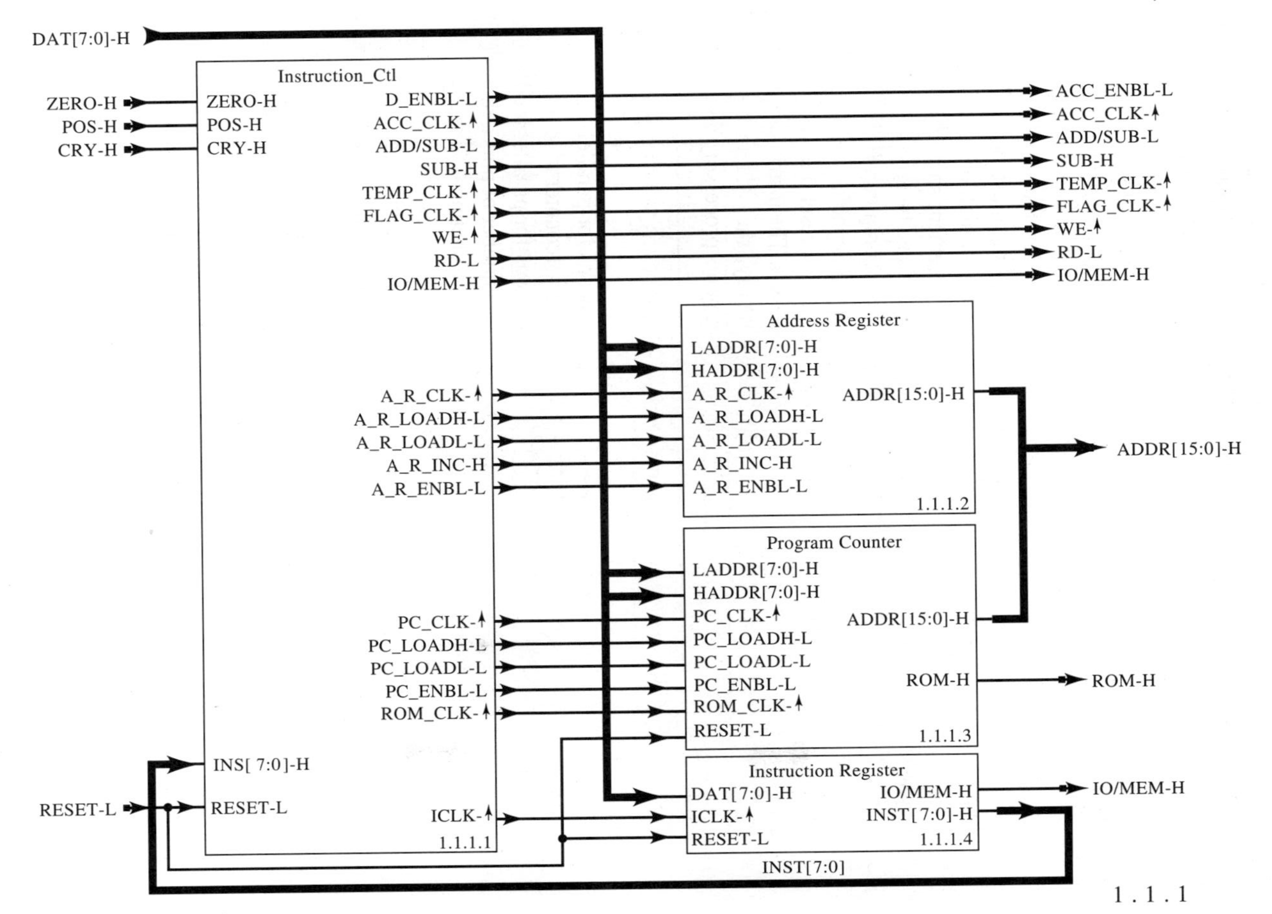

Figure 12.6-14 Instruction processor.

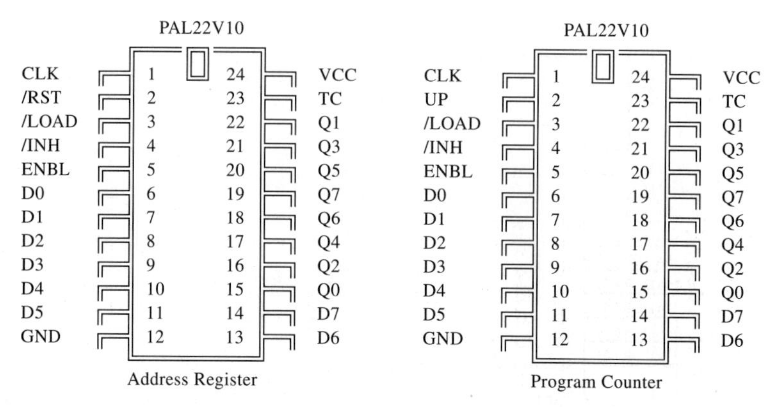

Figure 12.6-15 PALs used in the address register and program counter.

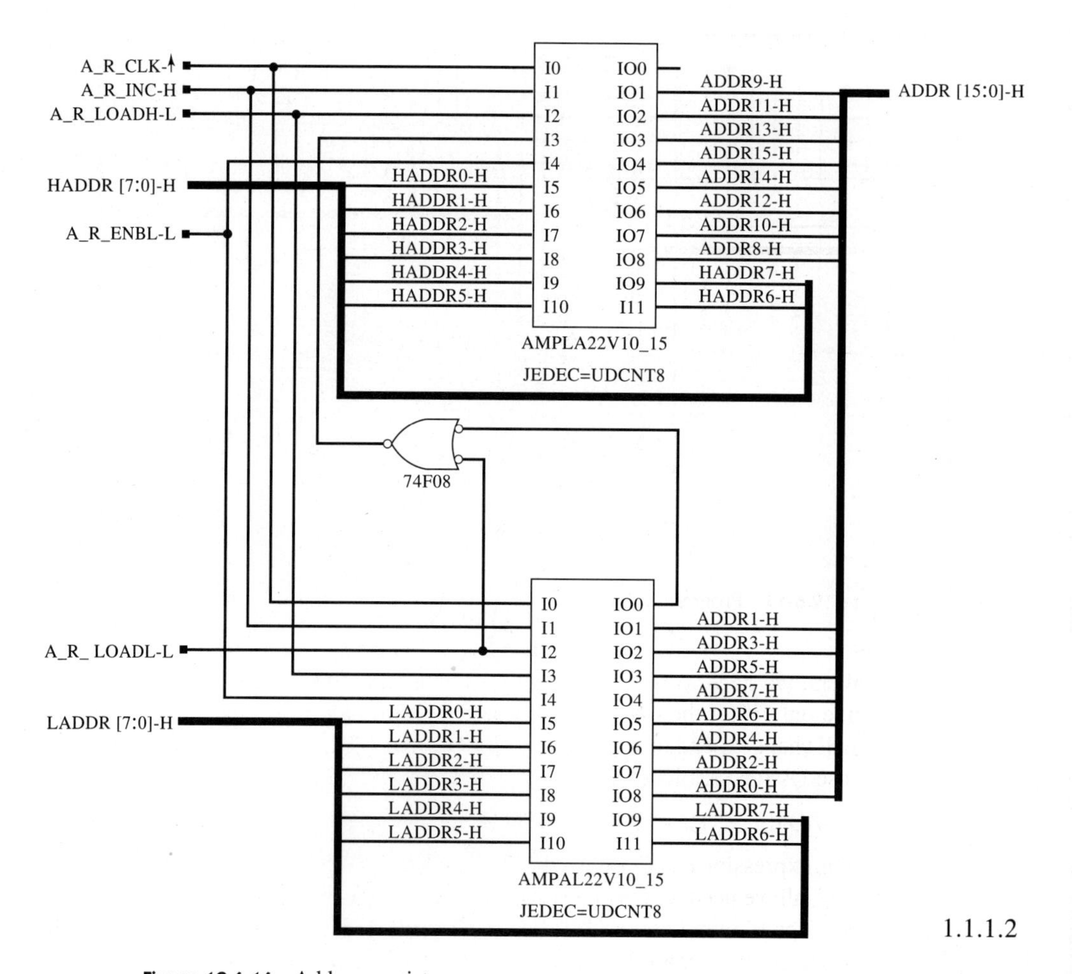

Figure 12.6-16 Address register.

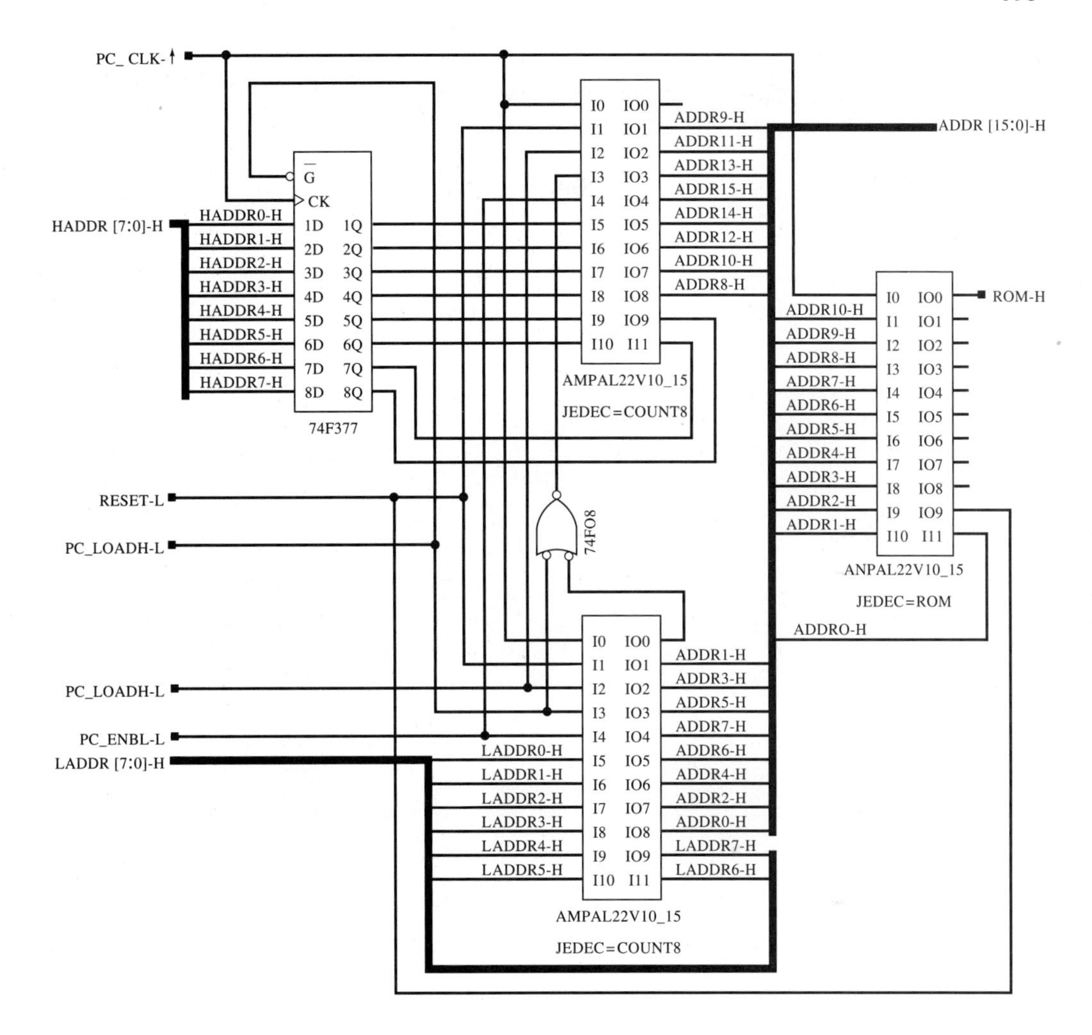

Figure 12.6-17 Program counter.

counters, we need to realize that when a binary counter counts, any given bit will change only if all the lower order bits are 1s. For instance, in an 8-bit counter, bit Q4 will change if and only if bits Q3, Q2, Q1, and Q0 are all 1s. (We bow to tradition and use Qs for the bits of a counter). This condition has a simple logic expression,

$$Q4(n + 1) = \overline{Q4} \cdot Q3 \cdot Q2 \cdot Q1 \cdot Q0 + Q4 \cdot \overline{Q3 \cdot Q2 \cdot Q1 \cdot Q0}$$

Similar expressions can be written for the changes in all the bits. For the program counter, all we need to do is add provisions for an inhibit, a parallel load, and tristate output. The terminal count (TC) follows the same idea. It's only asserted when all the bits in the counter are 1s. Table 12.6-4 shows the source program for programming the program counter PALs, the ones with the resets.

Table 12.6-4 Source Program for Program Counter PAL

```
;PALASM Design Description

;------------------------------------ Declaration Segment ------------------------------------
TITLE      8-BIT COUNTER
PATTERN
REVISION
AUTHOR     A W SHAW
COMPANY    UTAH STATE UNIVERSITY
DATE       09/21/92

CHIP    count8   PAL22V10

;--------------------------------------- PIN Declarations ---------------------------------------
PIN  1     CLK                          ; CLOCK
PIN  2     /RST                         ; INPUT
PIN  3     /LOAD                        ; INPUT
PIN  4     /INH                         ; INPUT
PIN  5     /ENBL                        ; INPUT
PIN  6     D0                           ; INPUT
PIN  7     D1                           ; INPUT
PIN  8     D2                           ; INPUT
PIN  9     D3                           ; INPUT
PIN  10    D4                           ; INPUT
PIN  11    D5                           ; INPUT
PIN  12    GND                          ;
PIN  13    D6                           ; INPUT
PIN  14    D7                           ; INPUT
PIN  15    Q0           REGISTERED      ; OUTPUT
PIN  16    Q2           REGISTERED      ; OUTPUT
PIN  17    Q4           REGISTERED      ; OUTPUT
PIN  18    Q6           REGISTERED      ; OUTPUT
PIN  19    Q7           REGISTERED      ; OUTPUT
PIN  20    Q5           REGISTERED      ; OUTPUT
PIN  21    Q3           REGISTERED      ; OUTPUT
PIN  22    Q1           REGISTERED      ; OUTPUT
PIN  23    TC           COMBINATORIAL   ; OUTPUT
PIN  24    VCC                          ;
NODE 1     GLOBAL

;-------------------------------- Boolean Equation Segment --------------------------------
EQUATIONS

Q0 := (/Q0*/INH + Q0*INH)*/LOAD + D0*LOAD
Q1 := ((/Q1*Q0 + Q1*/Q0)*/INH + Q1*INH)*/LOAD + D1*LOAD
Q2 := ((/Q2*Q1*Q0 + Q2*/(Q1*Q0))*/INH + Q2*INH)*/LOAD + D2*LOAD
Q3 := ((/Q3*Q2*Q1*Q0 + Q3*/(Q2*Q3*Q0))*/INH + Q1*INH)*/LOAD + D3*LOAD
Q4 := ((/Q4*Q3*Q2*Q1*Q0 + Q4*/(Q3*Q2*Q1*Q0))*/INH + Q4*INH)
      */LOAD + D4*LOAD
```

```
Q5 := ((/Q5*Q4*Q3*Q2*Q1*Q0 + Q5*/(Q4*Q3*Q2*Q1*Q0))*/INH + Q5*INH)
      */LOAD + D5*LOAD
Q6 := ((/Q6*Q5*Q4*Q3*Q2*Q1*Q0 + Q6*/(Q5*Q4*Q3*Q2*Q1*Q0))
      */INH + Q6*INH)*/LOAD + D6*LOAD
Q7 := ((/Q7*Q6*Q5*Q4*Q3*Q2*Q1*Q0 + Q7*/(Q6*Q5*Q4*Q3*Q2*Q1*Q0))*/INH
      + Q7*INH)*/LOAD + D7*LOAD
TC  = Q7*Q6*Q5*Q4*Q3*Q2*Q1*Q0

GLOBAL.RSTF = RST
Q0.TRST = ENBL
Q1.TRST = ENBL
Q2.TRST = ENBL
Q3.TRST = ENBL
Q4.TRST = ENBL
Q5.TRST = ENBL
Q6.TRST = ENBL
Q7.TRST = ENBL
```

The PALs for the address register are complicated because they have to count both up and down. Actually, the conditions for a bit changing in a down count are much the same as for a bit changing in an up count. In a down count, a bit changes if all the lower order bits are zero. This gives rise to the logic expression (using Q4 as an example) of

$$Q4(n + 1) = Q4 \cdot \overline{Q3} \cdot \overline{Q2} \cdot \overline{Q1} \cdot \overline{Q0} + Q4 \cdot \overline{\overline{Q3} \cdot \overline{Q2} \cdot \overline{Q1} \cdot \overline{Q0}}$$

In this counter, both the logic for up counting and that for down counting must be present with the choice conditional on the input UP. The terminal count (TC) must also be conditional on UP. Table 12.6-5 gives the source program for the address register PALs.

The program register has an additional function. It determines when to change from ROM to R/W memory. This is done in the ROM PAL. It is a simple 2-state state machine that enters state A on reset and stays in state A until address 003F is reached, then it enters state B and stays there until another reset. In state A, the output ROM is asserted. In state B, the output ROM is not asserted. Table 12.6-6 is the source program for the ROM PAL.

Figure 12.6-18 shows the circuit diagram for the final subsystem of the instruction processor, except the controller. It's simply an 8-bit register. The design of the controller is another matter. The approach as always is to determine what the controller has to do by considering the subsystems it has to control. This involves finding not only the proper control signal but also the timing of these signals.

In this design, if we tried to determine all of the timing constraints involved in the controller design we'd be overwhelmed, so we'll look at a representative set of timing constraints instead. We've already made some choices. We're going to choose a 25 MHz clock and allow two clock cycles for a memory read or write. With these

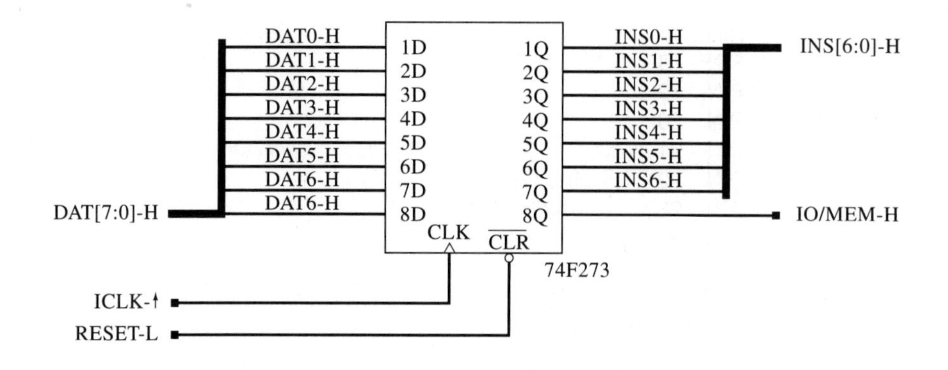

1.1.1.4

Figure 12.6-18 Instruction register.

Table 12.6-5 Source Program for Address Register PALs

```
;PALASM Design Description

;------------------------------------- Declaration Segment -------------------------------------
TITLE      8-BIT UP DOWN COUNTER
PATTERN
REVISION
AUTHOR     A W SHAW
COMPANY    UTAH STATE UNIVERSITY
DATE       09/21/92

CHIP    UDCNT8   PAL22V10

;----------------------------------------- PIN Declarations -----------------------------------------
PIN   1  CLK                ; CLOCK
PIN   2  UP                 ; INPUT
PIN   3  /LOAD              ; INPUT
PIN   4  /INH               ; INPUT
PIN   5  ENBL               ; INPUT
PIN   6  D0                 ; INPUT
PIN   7  D1                 ; INPUT
PIN   8  D2                 ; INPUT
PIN   9  D3                 ; INPUT
PIN   10 D4                 ; INPUT
PIN   11 D5                 ; INPUT
PIN   12 GND                ;
PIN   13 D6                 ; INPUT
PIN   14 D7                 ; INPUT
PIN   15 Q0  REGISTERED     ; OUTPUT
PIN   16 Q2  REGISTERED     ; OUTPUT
```

```
PIN  17     Q4                    REGISTERED ; OUTPUT
PIN  18     Q6                    REGISTERED ; OUTPUT
PIN  19     Q7                    REGISTERED ; OUTPUT
PIN  20     Q5                    REGISTERED ; OUTPUT
PIN  21     Q3                    REGISTERED ; OUTPUT
PIN  22     Q1                    REGISTERED ; OUTPUT
PIN  23     TC                 COMBINATORIAL ; OUTPUT
PIN  24     VCC                              ;

;---------------------------------- Boolean Equation Segment ----------------------------------
EQUATIONS

Q0 := (/Q0*/INH + Q0*INH)*/LOAD + D0*LOAD
Q1 := (((/Q1*Q0 + Q1*/Q0)*UP + (/Q1*/Q0 + Q1*Q0)*/UP)*/INH + Q1*INH)
      */LOAD + D1*LOAD
Q2 := (((/Q2*Q1*Q0 + Q2*/(Q1*Q0))*UP + (/Q2*/Q1*/Q0 + Q2*/(/Q1*/Q0))*/UP)
      */INH + Q2*INH)*/LOAD + D2*LOAD
Q3 := (((/Q3*Q2*Q1*Q0 + Q3*/(Q2*Q1*Q0))*UP + (/Q3*/Q2*/Q1*/Q0 + Q3
      */(/Q2*/Q3*/Q0))*/UP)*/INH + Q3*INH)*/LOAD + D3*LOAD
Q4 := (((/Q4*Q3*Q2*Q1*Q0 + Q4*/(Q3*Q2*Q1*Q0))*UP + (/Q4*/Q3*/Q2*/Q1
      */Q0 + Q4*/(/Q3*/Q2*/Q1*/Q0))*/UP)*/INH + Q4*INH)*/LOAD + D4*LOAD
Q5 := (((Q5*Q4*Q3*Q2*Q1*Q0 + Q5*/(Q4*Q3*Q2*Q1*Q0))*UP + (/Q5*/Q4*/Q3
      */Q2*/Q1*/Q0 + Q5*/(/Q4*/Q3*/Q2*/Q1*/Q0))*/UP)*/INH + Q5*INH)*/LOAD
      + D5*LOAD
Q6 := (((/Q6*Q5*Q4*Q3*Q2*Q1*Q0 + Q6*/(Q5*Q4*Q3*Q2*Q1*Q0))*UP + (/Q6
      */Q5*/Q4*/Q3*/Q2*/Q1*/Q0 + Q6*/(/Q5*/Q4*/Q3*/Q2*/Q1*/Q0))*/UP)
      */INH + Q6*INH)*/LOAD + D6*LOAD
Q7 := (((/Q7*Q6*Q5*Q4*Q3*Q2*Q1*Q0 + Q7*/(Q6*Q5*Q4*Q3*Q2*Q1*Q0))
      *UP + (/Q7*/Q6*/Q5*/Q4*/Q3*/Q2*/Q1*/Q0 + Q7*/(/Q6*/Q5*/Q4*/Q3*/Q2
      */Q1*/Q0))*/UP)*/INH + Q7*INH)*/LOAD + D7*LOAD
TC  = Q7*Q6*Q5*Q4*Q3*Q2*Q1*Q0*UP + /Q7*/Q6*/Q5*/Q4*/Q3*/Q2*/Q1*/Q0
      */UP

Q0.TRST = ENBL
Q1.TRST = ENBL
Q2.TRST = ENBL
Q3.TRST = ENBL
Q4.TRST = ENBL
Q5.TRST = ENBL
Q6.TRST = ENBL
Q7.TRST = ENBL
```

Table 12.6-6 Source Program for ROM PAL

```
;PALASM Design Description

;------------------------------------ Declaration Segment ------------------------------------
TITLE       ROM SELECTION
PATTERN
REVISION
AUTHOR      A W SHAW
COMPANY     UTAH STATE UNIVERSITY
DATE        10/31/92

CHIP     ROM   PAL22V10

;---------------------------------------- PIN Declarations ----------------------------------------
PIN   1        CLK                               ; CLOCK
PIN   2        ADDR10                            ; INPUT
PIN   3        ADDR9                             ; INPUT
PIN   4        ADDR8                             ; INPUT
PIN   5        ADDR7                             ; INPUT
PIN   6        ADDR6                             ; INPUT
PIN   7        ADDR5                             ; INPUT
PIN   8        ADDR4                             ; INPUT
PIN   9        ADDR3                             ; INPUT
PIN   10       ADDR2                             ; INPUT
PIN   11       ADDR1                             ; INPUT
PIN   12       GND                               ;
PIN   13       ADDR0                             ; INPUT
PIN   14       /RST                                ; INPUT
PIN   15       NC                                ;
PIN   16       NC                                ;
PIN   17       NC                                ;
PIN   18       NC                                  ;
PIN   19       NC                                  ;
PIN   20       NC                                  ;
PIN   21       NC                                  ;
PIN   22       B1                     REGISTERED ; OUTPUT
PIN   23       /ROM                   REGISTERED ; OUTPUT
PIN   24       VCC                                 ;
NODE 1         GLOBAL                              ;

;--------------------------------- State Equations Segment ---------------------------------
STATE

A = /B1*/ROM
B =  B1*ROM

A := FINISH → B
     + NOT_FINISH → A

B := VCC → B
```

```
CONDITIONS
NOT_FINISH = /(ADDR10*ADDR9*ADDR8*ADDR7*ADDR6*ADDR5*ADDR4
              *ADDR3*ADDR2*ADDR1*ADDR0)
FINISH = ADDR10*ADDR9*ADDR8*ADDR7*ADDR6*ADDR5*ADDR4*ADDR3
         *ADDR2*ADDR1*ADDR0

;-------------------------------- Boolean Equations Segment --------------------------------
EQUATIONS

GLOBAL.RSTF = RST
ROM.TRST = VCC
B1.TRST = VCC

;------------------------------------- Simulation Segment -------------------------------------
SIMULATION

SETF RST /ADDR10 /ADDR9 /ADDR8 /ADDR7 /ADDR6 /ADDR5 /ADDR4
/ADDR3
SETF /ADDR2 /ADDR1 /ADDR0
CLOCKF CLK
SETF /RST /ADDR10 /ADDR9 /ADDR8 /ADDR7 /ADDR6 /ADDR5 /ADDR4
/ADDR3
SETF /ADDR2 /ADDR1 /ADDR0
CLOCKF CLK
SETF /ADDR10 /ADDR9 /ADDR8 /ADDR7 /ADDR6 /ADDR5 /ADDR4 /ADDR3
/ADDR2
SETF /ADDR1 ADDR0
CLOCKF CLK
SETF /ADDR10 /ADDR9 /ADDR8 /ADDR7 /ADDR6 /ADDR5 /ADDR4 /ADDR3
/ADDR2
SETF ADDR1 /ADDR0
CLOCKF CLK
SETF /ADDR10 /ADDR9 /ADDR8 /ADDR7 /ADDR6 /ADDR5 /ADDR4 /ADDR3
/ADDR2
SETF ADDR1 ADDR0
CLOCKF CLK
SETF ADDR10 ADDR9 ADDR8 ADDR7 ADDR6 ADDR5 ADDR4 ADDR3
ADDR2 ADDR1
SETF ADDR0
CLOCKF CLK
SETF /ADDR10 /ADDR9 /ADDR8 /ADDR7 /ADDR6 /ADDR5 /ADDR4 /ADDR3
/ADDR2
SETF ADDR1 /ADDR0
CLOCKF CLK
SETF ADDR10 /ADDR9 /ADDR8 /ADDR7 /ADDR6 /ADDR5 /ADDR4 /ADDR3
/ADDR2
SETF ADDR1 /ADDR0

;---------------------------------------------------------------------------------------------
```

choices, let's consider what we need by way of control signals to do an add if the data in the add follow the op code in the instruction. We'll assume that the address for the instruction code is in the program counter, so we have to first get the op code, which requires us to:

1. Enable the program counter (PC_ENBL) so it will feed an address to the memory via the address bus
2. Assert a memory read (RD)
3. Wait two clock cycles
4. Assert the instruction register clock (ICLK) and disable the program counter

Now we have to increment the program counter and do another read into the temporary register. So we:

5. Assert the program counter clock (PC_CLK)
6. Enable the program counter
7. Assert a memory read
8. Wait two clock cycles
9. Assert the temporary register clock (TEMP_CLK) and disable the program counter

The data are now present for the add. We assume the accumulator already had data. So we:

10. Wait one clock cycle for the data to propagate through the adders
11. At the same time enable the data out of the adder (ADD/SUB_ENBL)
12. Assert the accumulator clock (ACC_CLK) to clock the results into the accumulator and the flag clock (FLAG_CLK) to clock the results into the flag register

Now we can go back and read a new instruction and execute it.

Figure 12.6-19 shows these timing steps. Note the cause-effect relation between the RD and the PC_ENBL as cause and the data on the data bus as effect. The PC_ENBL actually enables the address from the program counter to the address bus, and the address to memory causes the data to appear on the data bus. The timing on the diagram is to scale. For the memory chosen, the delay from enable to address (approximately 10 ns) to data out (45 ns) is 55 ns. The data are clocked by the ICLK into the instruction register and appear as a code and a size. The memory is actually faster than we need for this clock speed. Note the time between the data valid and the ICLK. Following this first read, the program counter is incremented by the PC_CLK and a new read is initiated. This read is the same as the first, but the data are clocked into the temporary register. Data are now present for an add after a delay. The data out of the adder are enabled by the ADD/SUB_ENBL and clocked into the accumulator by ACC_CLK and into the flag register by FLAG_CLK.

With these conditions, let's see what we can do about a controller. First, the controller does a number of read cycles, all with the same timing but different destinations. We can simplify our design by making a controller that controls the timing on the reads (also the adds and subtracts and the write) and a controller that controls the order of the reads. What we're proposing is to partition the controller into two as shown in Figure 12.6-20. The systems controller will control the cycling

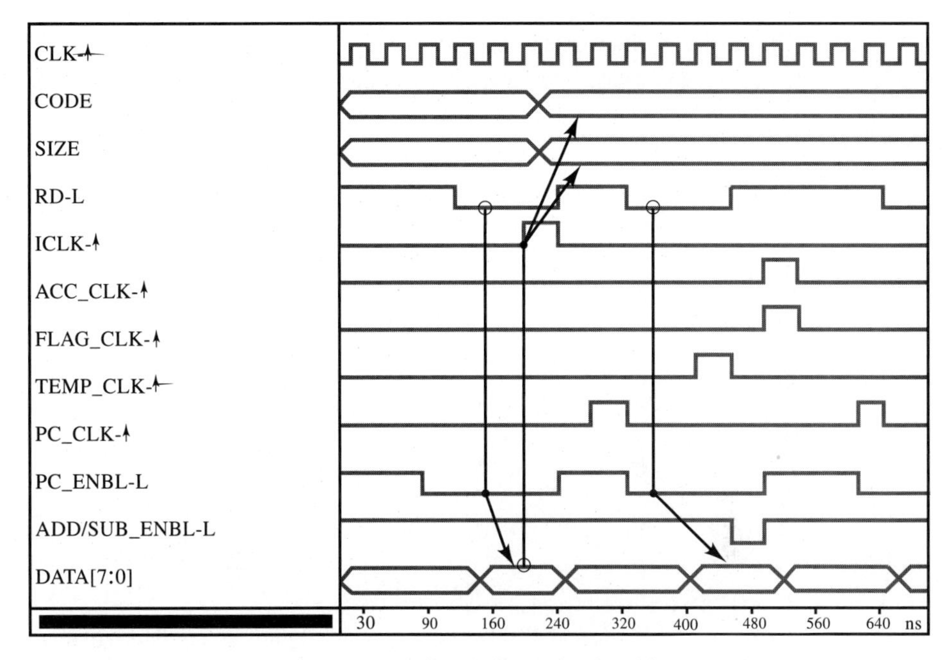

Figure 12.6-19 Timing for the controller during a simple add.

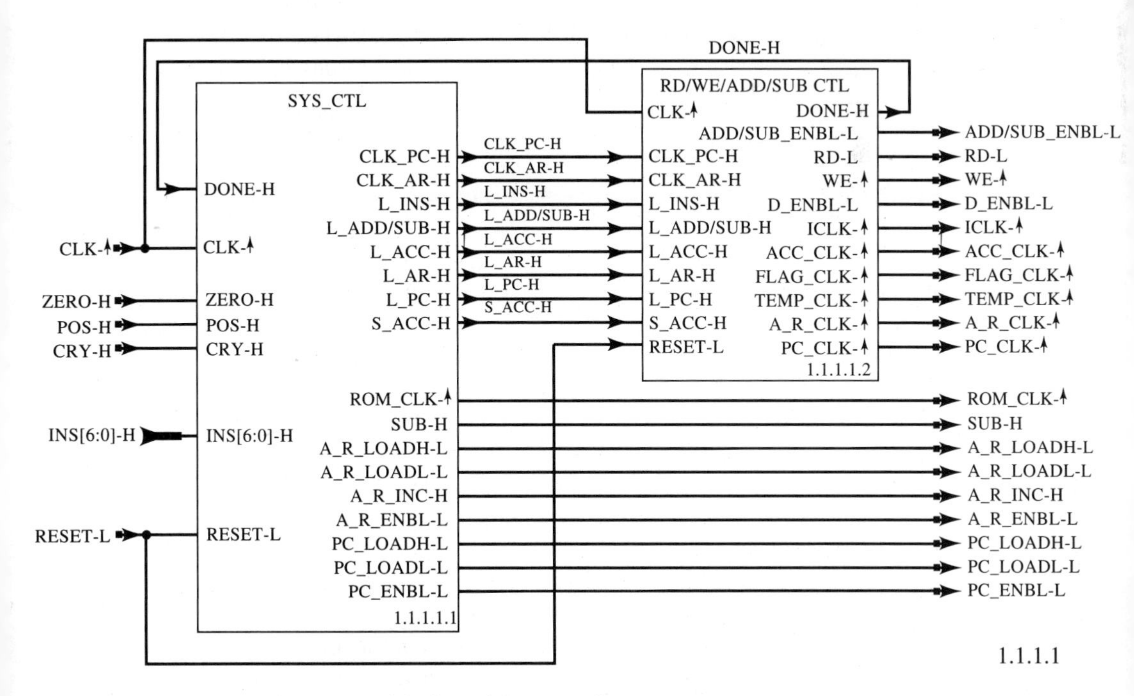

Figure 12.6-20 Partitioning of the controller.

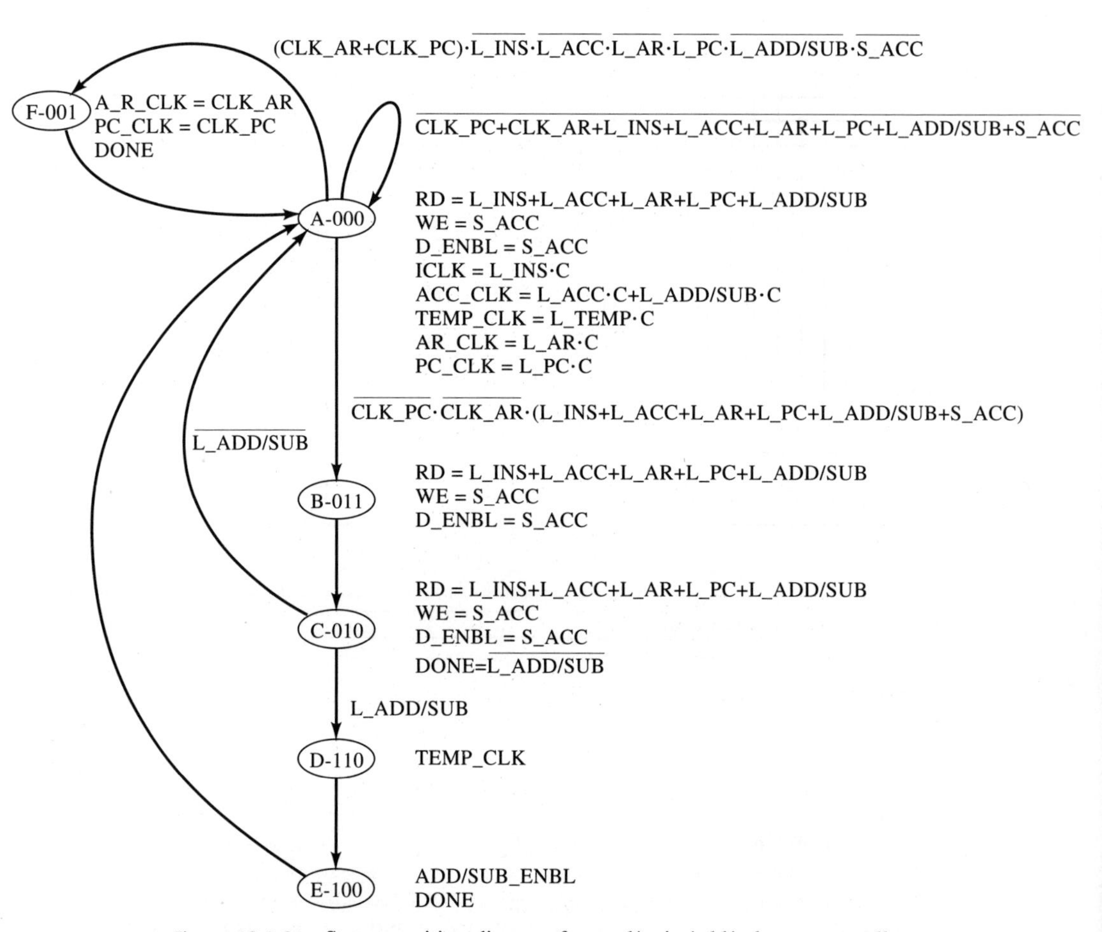

Figure 12.6-21 State transition diagram for read/write/add/subtract controller.

through the various read (and write) cycles to be performed in the execution of an instruction, including the reading of the instruction and any read/write/add/subtract directed by the instruction. The read/write/add/subtract controller controls the read cycle. The add and subtract cycles will be extended read cycles. We can see that we could have included this controller in the arithmetic processor. There's no reason to change at this time, because both ways will function correctly. The signals between the two controllers are essentially requests to the read/write/add/subtract controller and a reply (DONE) back. The one problem with partitioning the controller is slow timing. A single controller will be faster but more complex—the classic trade-off in many designs.

The controllers we'll design will be state machines in PALs. We can design these directly from the state transition diagrams. So let's construct the diagrams. First the read/write/add/subtract controller: it must wait in a state until it gets a request, then execute a read or a write cycle (two states). This is the smaller

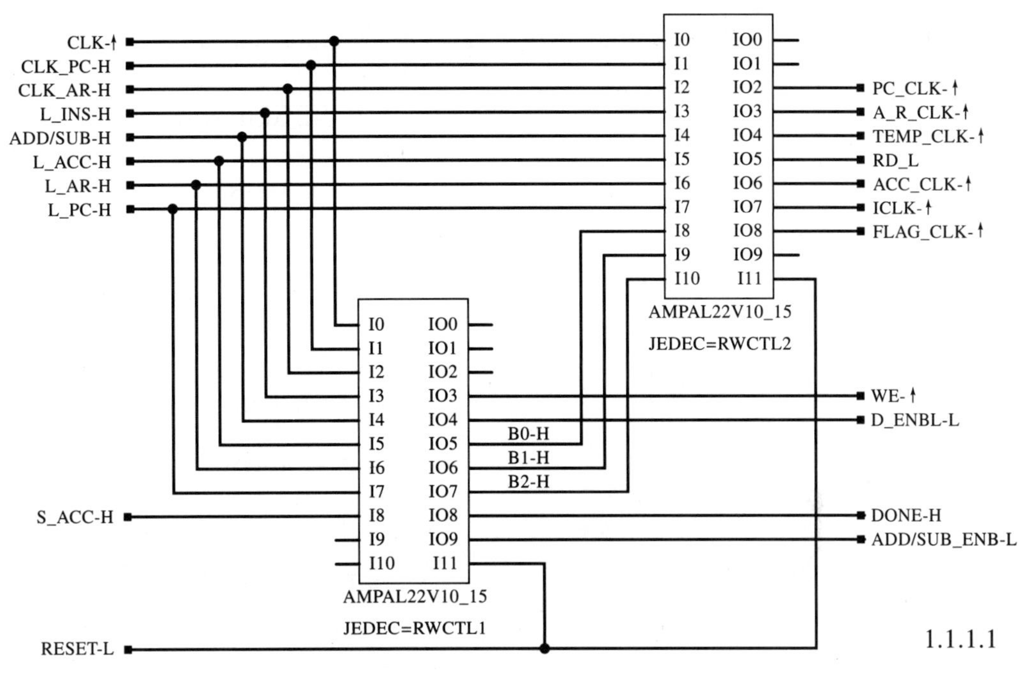

Figure 12.6-22 Circuit diagram for read/write/add/subtract controller.

cycle below state A in Figure 12.6-21. By extending the cycle through two more states, we can include an add or subtract. The single-state loop from A is the loop that produces a clock to increment the arithmetic register and the program counter. Notice how the same basic states are made to produce a variety of outputs depending on inputs. This state machine will not fit in a single PAL. It requires too many inputs and outputs and too many states. Figure 12.6-22 shows the circuit diagram for this controller using two PALs and Tables 12.6-7 and 12.6-8 give the source programs for the two PALs. The PALs are divided into a state machine proper and an output decoder. The state machine uses state transition language, but the output decoder must use basic logic equations that include the state variables.

The final part of the design is the system controller. It essentially steps the computer through an instruction as directed by the op code. Again, in this design, we find it convenient to divide it into two parts—a decoder to decode the instruction code into more meaningful control signals, and a state machine. The state machine actually had to be programmed into two PALs because of register and pin limitations. Figure 12.6-23 is the circuit diagram for the system controller. The decoded output from the decoder can be seen in this diagram. It has to do with what registers are involved in the instruction and how many words follow the op code in the instruction.

Table 12.6-7 Read/Write/Add/Subtract Controller (State)

```
;PALASM Design Description

;------------------------------------ Declaration Segment ------------------------------------
TITLE      RD/WE/ADD/SUB CTL (STATE)
PATTERN
REVISION
AUTHOR     A W SHAW
COMPANY    UTAH STATE UNIVERSITY
DATE       09/24/92

CHIP    RWCTL1   PAL22V10

;---------------------------------------- PIN Declarations ----------------------------------------
PIN  1     CLK                                    ; CLOCK
PIN  2     CLK_PC                                 ; INPUT
PIN  3     CLK_AR                                 ; INPUT
PIN  4     L_INS                                  ; INPUT
PIN  5     L_ADD_SUB                              ; INPUT
PIN  6     L_ACC                                  ; INPUT
PIN  7     L_AR                                   ; INPUT
PIN  8     L_PC                                   ; INPUT
PIN  9     S_ACC                                  ; INPUT
PIN  10    NC                                     ; INPUT
PIN  11    NC                                     ;
PIN  12    GND                                    ;
PIN  13    /RESET                                 ; INPUT
PIN  14    /ADD_SUB_ENBL           COMBINATORIAL  ; OUTPUT
PIN  15    DONE                    COMBINATORIAL  ; OUTPUT
PIN  16    B2                      REGISTERED     ; OUTPUT
PIN  17    B1                      REGISTERED     ; OUTPUT
PIN  18    B0                      REGISTERED     ; OUTPUT
PIN  19    /D_ENBL                 REGISTERED     ; OUTPUT
PIN  20    /WE                     REGISTERED     ; OUTPUT
PIN  21    NC                                     ;
PIN  22    NC                                     ;
PIN  23    NC                                     ;
PIN  24    VCC                                    ;
NODE 1     GLOBAL

;----------------------------------- State Equation Segment -----------------------------------
STATE
MEALY_MACHINE
```

```
A = /B2*/B1*/B0
B = /B2* B1* B0
C = /B2* B1*/B0
D =  B2* B1*/B0
E =  B2*/B1*/B0
F = /B2*/B1* B0

A := N_INPUT → A
     + L_INPUT → B
     + CLK_INPUT → F
B := VCC → C
C := A_S → D
     + N_A_S → A
D := VCC → E
E := VCC → A
F := VCC → A

CONDITIONS
N_INPUT = /CLK_PC*/CLK_AR*/L_INS*/L_ACC*/L_AR*/L_PC*/L_ADD_SUB
          */S_ACC
L_INPUT = /CLK_PC*/CLK_AR*(L_INS+L_ACC+L_AR+L_PC
          +L_ADD_SUB+S_ACC)
CLK_INPUT = (CLK_PC+CLK_AR)*/L_INS*/L_ACC*/L_AR*/L_PC*/
            L_ADD_SUB*/S_ACC
A_S = L_ADD_SUB
N_A_S = /L_ADD_SUB

;-------------------------------- Boolean Equation Segment --------------------------------
EQUATIONS
WE := (/B2*/B1*/B0+/B2*B1*B0)*S_ACC
D_ENBL := (/B2*/B1*/B0+/B2*B1*B0+/B2*B1*/B0)*S_ACC
ADD_SUB_ENBL = B2*/B1*/B0
DONE = B2*/B1*/B0+/B2*/B1*B0+/B2*B1*/B0*/L_ADD_SUB

GLOBAL.RSTF = RESET
B2.TRST = VCC
B1.TRST = VCC
B0.TRST = VCC
WE.TRST = VCC
D_ENBL.TRST = VCC
ADD_SUB_ENBL.TRST = VCC
DONE.TRST = VCC
```

Table 12.6-8 Read/Write/Add/Subtract Controller (Output)

```
;PALASM Design Description

;------------------------------------ Declaration Segment ------------------------------------
TITLE     RD/WE/ADD/SUB CTL (OUTPUT)
PATTERN
REVISION
AUTHOR    A W SHAW
COMPANY   UTAH STATE UNIVERSITY
DATE      09/24/92

CHIP      RWCTL2   PAL22V10

;---------------------------------------- PIN Declarations ----------------------------------------
PIN   1     CLK                        ; CLOCK
PIN   2     CLK_PC                     ; INPUT
PIN   3     CLK_AR                     ; INPUT
PIN   4     L_INS                      ; INPUT
PIN   5     L_ADD_SUB                  ; INPUT
PIN   6     L_ACC                      ; INPUT
PIN   7     L_AR                       ; INPUT
PIN   8     L_PC                       ; INPUT
PIN   9     B0                         ; INPUT
PIN   10    B1                         ; INPUT
PIN   11    B2                         ; INPUT
PIN   12    GND                        ; INPUT
PIN   13    /RESET                     ; INPUT
PIN   14    NC                         ;
PIN   15    FLAG_CLK        REGISTERED ; OUTPUT
PIN   16    ICLK            REGISTERED ; OUTPUT
PIN   17    ACC_CLK         REGISTERED ; OUTPUT
PIN   18    /RD             REGISTERED ; OUTPUT
PIN   19    TEMP_CLK        REGISTERED ; OUTPUT
PIN   20    A_R_CLK         REGISTERED ; OUTPUT
PIN   21    PC_CLK          REGISTERED ; OUTPUT
PIN   22    NC                         ;
PIN   23    NC                         ;
PIN   24    VCC                        ;
NODE 1      GLOBAL

;-------------------------------- Boolean Equation Segment --------------------------------
EQUATIONS

RD := (/B2*/B1*/B0 + /B2*B1*B0 + /B2*B1*/B0)*(L_INS + L_ACC + L_AR + L_PC
      + L_ADD_SUB)
ICLK := /B2*B1*/B0*/L_ADD_SUB*L_INS
```

```
ACC_CLK:= /B2*B1*/B0*/L_ADD_SUB*L_ACC+B2*/B1*/B0*L_ADD_SUB
TEMP_CLK := /B2*B1*/B0*L_ADD_SUB
A_R_CLK := /B2*B1*/B0*/L_ADD_SUB*L_AR+/B2*/B1*B0*CLK_AR
PC_CLK := /B2*B1*/B0*/L_ADD_SUB*L_PC+/B2*/B1*B0*CLK_PC
FLAG_CLK := B2*/B1*/B0*L_ADD_SUB

GLOBAL.RSTF = RESET

ICLK.TRST = VCC
ACC_CLK.TRST = VCC
TEMP_CLK.TRST = VCC
A_R_CLK.TRST = VCC
PC_CLK.TRST = VCC
FLAG_CLK.TRST = VCC
```

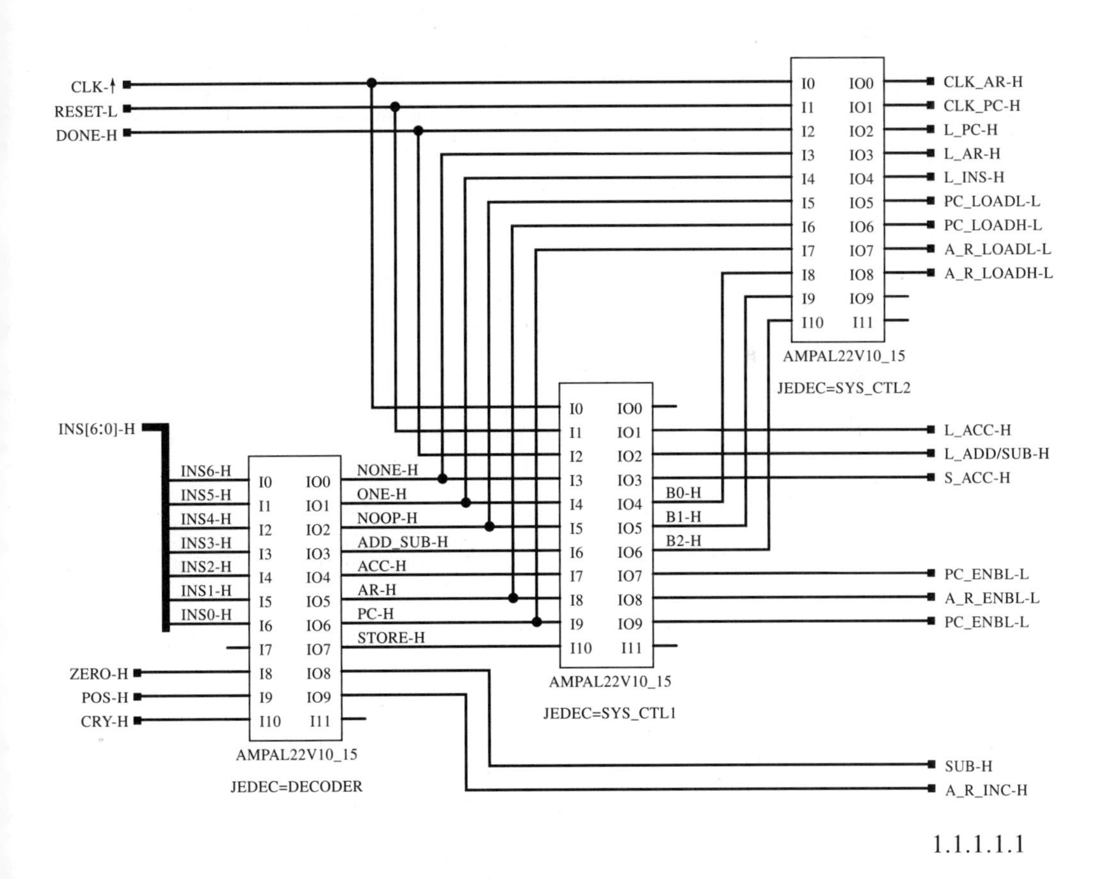

Figure 12.6-23 System controller circuit diagram.

The state machine to control the processing of the instructions must execute an initial read that gets the op code from memory (the fetch) and up to three additional read/writes. These reads occur in states C, E, and G in the state transition diagram of Figure 12.6-24. The fetch occurs in state A. Following most of these read/writes, the program counter needs to be incremented (states B, D, and F), the loop through state H is for an increment or decrement of the address counter. As with many controllers, this controller uses only a few states but has many input controlled outputs depending on the instruction to be performed in the computer. Notice how the machine remains in each of the states until the read/write/add/subtract controller asserts a DONE control. The source programs for the system controller PALs are given in Tables 12.6-9 through 12.6-11. The state machine was programmed using state transition language, but the output had to use logic equations because it was in a separate PAL. PALASM simulation of this and the previous two PAL decoders was not very helpful. In order to see that these controllers operated correctly, they were simulated using a circuit simulator. The results of this simulation, although not shown here, were similar to the timing of Figure 12.6-19. It's quite possible to do a simulation of the entire computer, but to do this takes a bit of preparation. First a small program must be written for the computer and stored in the simulated computer memory. This program must start in the memory location with address 0000, since that is where the computer starts when it's reset. We need to also remember that the computer starts in ROM and stays in ROM until the address location with address 003F is accessed.

We'll use the test program in Table 12.6-12 for our simulation. This table shows the address of each word in the program, the word, and an explanation of what the word does. All of the numbers are in hexadecimal. The first three words in the program, starting at address 0000 in ROM, are an instruction to jump to address 003D. The next three words, starting at address 003D, are another jump instruction. Notice that the last word of this instruction is stored in location 003F, so this jump, which is a jump to address 0000, causes a jump to address 0000 in R/W memory. Now starting in address 0000 in R./W memory, the first two words are an instruction to move 35 into the accumulator. The three words starting at address 0002 are an instruction to move the data word in the accumulator (35) to the memory location with address 1000. The two words starting at address 0005 are an instruction to move 13 into the accumulator. The three words starting at address 0007 are an instruction to add the data word in address location 1000 to the accumulator. Notice that this data word is the 35 we previously stored. The final three words starting at address 000A are an instruction to jump to address 011A.

Figures 12.6-25 through 12.6-28 show the timing results for a partial simulation of the computer as it executes the instructions of the test program. ADDR and DATA are the address on the memory address bus and the data on the data bus. RD-L and WR-↑ are the memory read and write. ROM-H is the signal indicating that ROM memory should be accessed. Notice how it's asserted when the computer is reset. STATE is the state of the system controller. Rather than show each bit of the state code, we've shown the state code as a hexadecimal number. For instance, state 011 is state 3. DONE is the signal indicating that the read/write/add/subtract

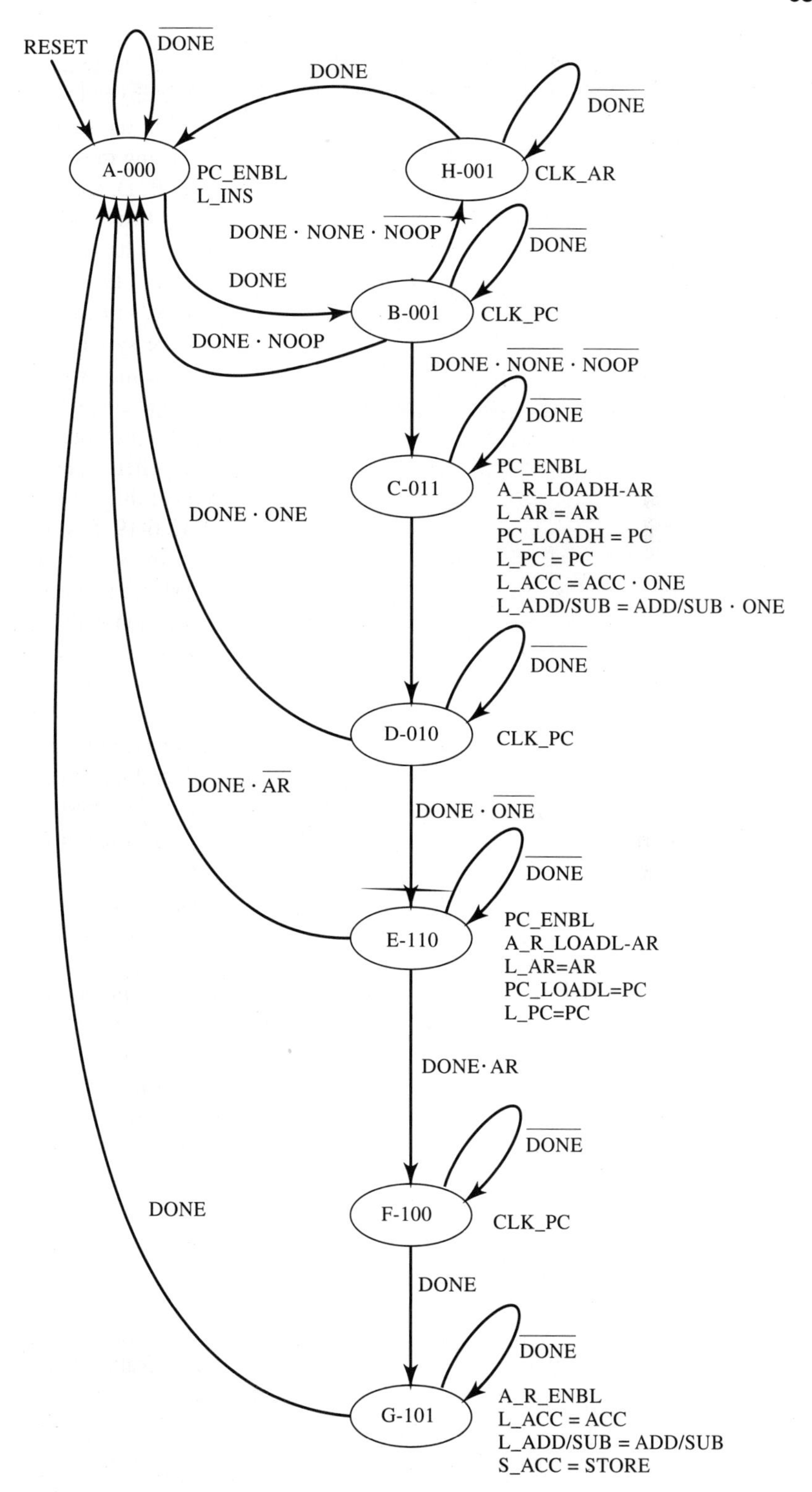

Figure 12.6-24 State transition diagram for system controller.

Table 12.6-9 Source Program for Decoder PAL

```
;PALASM Design Description

;------------------------------------ Declaration Segment ------------------------------------
TITLE      OP CODE DECODER 1
PATTERN
REVISION
AUTHOR     A W SHAW
COMPANY    UTAH STATE UNIVERSITY
DATE       09/25/92

CHIP     DECODER1   PAL22V10

;---------------------------------------- PIN Declarations ----------------------------------------
PIN  1     INS6                          ; INPUT
PIN  2     INS5                          ; INPUT
PIN  3     INS4                          ; INPUT
PIN  4     INS3                          ; INPUT
PIN  5     INS2                          ; INPUT
PIN  6     INS1                          ; INPUT
PIN  7     INS0                          ; INPUT
PIN  8     ZERO                          ; INPUT
PIN  9     POS                           ; INPUT
PIN  10    CRY                           ; INPUT
PIN  11    NC                            ;
PIN  12    GND                           ;
PIN  13    NC                            ;
PIN  14    A_R_INC       COMBINATORIAL   ; OUTPUT
PIN  15    SUB           COMBINATORIAL   ; OUTPUT
PIN  16    STORE         COMBINATORIAL   ; OUTPUT
PIN  17    PC            COMBINATORIAL   ; OUTPUT
PIN  18    AR            COMBINATORIAL   ; OUTPUT
PIN  19    ACC           COMBINATORIAL   ; OUTPUT
PIN  20    ADD_SUB       COMBINATORIAL   ; OUTPUT
PIN  21    NOOP          COMBINATORIAL   ; OUTPUT
PIN  22    ONE           COMBINATORIAL   ; OUTPUT
PIN  23    NONE          COMBINATORIAL   ; OUTPUT
PIN  24    VCC                           ;

;-------------------------------- Boolean Equation Segment --------------------------------
EQUATIONS

NONE = /INS1*/INS0

ONE = /INS1*INS0

ADD_SUB = /INS5*/INS4*/INS2
ACC = /INS5*INS4*/INS3*/INS2
```

```
AR = /INS5*/INS4*/INS3*/INS2*INS1*/INS0 + /INS5*/INS4*/INS3*INS2*INS1*/INS0
       + /INS5*INS4*/INS3*/INS2*INS1*/INS0 + /INS5*INS4*/INS3*/INS2*INS1*/INS0

PC = /INS5*/INS4*/INS3*INS2 + /INS5*/INS4*INS3*INS2*ZERO
       + /INS5*INS4*/INS3*INS2*POS + /INS5*INS4*INS3*INS2*CRY

NOOP = INS5*INS4*INS3*INS2 + /INS5*/INS4*INS3*INS2*/ZERO
       + /INS5*INS4*/INS3*INS2*/POS + /INS5*INS4*INS3*INS2*/CRY

SUB = /INS5*/INS4*INS3*/INS2

A_R_INC = /INS3

STORE = /INS5*INS4*INS3*/INS2

NONE.TRST = VCC
ONE.TRST = VCC
ADD_SUB.TRST = VCC
ACC.TRST = VCC
AR.TRST = VCC
PC.TRST = VCC
NOOP.TRST = VCC
SUB.TRST = VCC
```

Table 12.6-10 Source Program for System Controller (State)

```
;PALASM Design Description

;------------------------------------ Declaration Segment ------------------------------------
TITLE      SYSTEM CONTROLLER 1 (STATE)
PATTERN
REVISION
AUTHOR     A W SHAW
COMPANY    UTAH STATE UNIVERSITY
DATE       09/25/92

CHIP    SYS_CTL1    PAL22V10

;---------------------------------------- PIN Declarations ----------------------------------------

PIN  1         CLK                                  ; CLOCK
PIN  2         /RESET                               ; INPUT
PIN  3         DONE                                 ; INPUT
PIN  4         NONE                                 ; INPUT
PIN  5         ONE                                  ; INPUT
PIN  6         NOOP                                 ; INPUT
PIN  7         ADD_SUB                              ; INPUT
PIN  8         ACC                                  ; INPUT
PIN  9         AR                                   ; INPUT
```

```
PIN  10      PC                               ; INPUT
PIN  11      STORE                            ; INPUT
PIN  12      GND                              ;
PIN  13      NC                               ;
PIN  14      ROM_CLK              REGISTERED ; OUTPUT
PIN  15      /A_R_ENBL            REGISTERED ; OUTPUT
PIN  16      /PC_ENBL             REGISTERED ; OUTPUT
PIN  17      B0                   REGISTERED ; OUTPUT
PIN  18      B1                   REGISTERED ; OUTPUT
PIN  19      B2                   REGISTERED ; OUTPUT
PIN  20      S_ACC                REGISTERED ; OUTPUT
PIN  21      L_ADD_SUB            REGISTERED ; OUTPUT
PIN  22      L_ACC                REGISTERED ; OUTPUT
PIN  23      NC                               ;
PIN  24      VCC                              ;
NODE 1       GLOBAL                           ;

;-------------------------------- State Equations Segment --------------------------------

STATE

MEALY_MACHINE
A = /B2*/B1*/B0
B = /B2*/B1* B0
C = /B2* B1* B0
D = /B2* B1*/B0
E =  B2* B1*/B0
F =  B2*/B1*/B0
G =  B2*/B1* B0
H =  B2* B1* B0

A := C_DONE → B
     + C_N_DONE → A

B := C_DONE_NOOP → A
     + C_DONE_N_NN_NP → C
     + C_DONE_NONE → H
     + C_N_DONE → B

C := C_DONE → D
     + C_N_DONE → C

D := C_DONE_N_ONE → E
     + C_DONE_ONE → A
     + C_N_DONE → D

E := C_DONE_AR → F
     + C_DONE_N_AR → A
     + C_N_DONE → E
```

```
F := C_DONE → G
      + C_N_DONE → F

G := C_DONE → A
      + C_N_DONE → G

H := C_DONE → A
      + C_N_DONE → H

CONDITIONS

C_DONE = DONE
C_N_DONE = /DONE
C_DONE_NOOP = DONE*NOOP
C_DONE_NONE = DONE*NONE*/NOOP
C_DONE_N_NN_NP = DONE*/NONE*/NOOP
C_DONE_ONE = DONE*ONE
C_DONE_N_ONE = DONE*/ONE
C_DONE_AR = DONE*AR
C_DONE_N_AR = DONE*/AR

;--------------------------------- Boolean Equation Seqment ---------------------------------
EQUATIONS

PC_ENBL := /B2*/B1*/B0*/DONE + /B2*/B1*B0*DONE + /B2*B1*B0/DONE
            + /B2*B1*/B0*DONE + B2*B1*/B0*/DONE + B2*B1*/B0*DONE*
           /AR + B2*/B1*/B0*DONE + B2*/B1*B0*DONE

A_R_ENBL := B2*/B1*/B0*DONE + B2*/B1*B0*/DONE

ROM_CLK := B2*B1*B0*DONE

S_ACC := (B2*/B1*/B0*DONE + B2*/B1*B0*/DONE)*STORE

L_ACC := (/B2*/B1*B0*DONE*/NONE*/NOOP + /B2*B1*B0*/DONE)*ACC*ONE
         + (B2*/B1*/B0*DONE + B2*/B1*B0*/DONE)*ACC

L_ADD_SUB := (/B2*/B1*B0*DONE*/NONE*/NOOP + /B2*B1*B0*
             /DONE)*ADD_SUB*ONE + (B2*/B1*/B0*DONE
              + B2*/B1*B0*/DONE)*ADD_SUB

GLOBAL.RSTF = RESET

B2.TRST = VCC
B1.TRST = VCC
B0.TRST = VCC
PC_ENBL.TRST = VCC
```

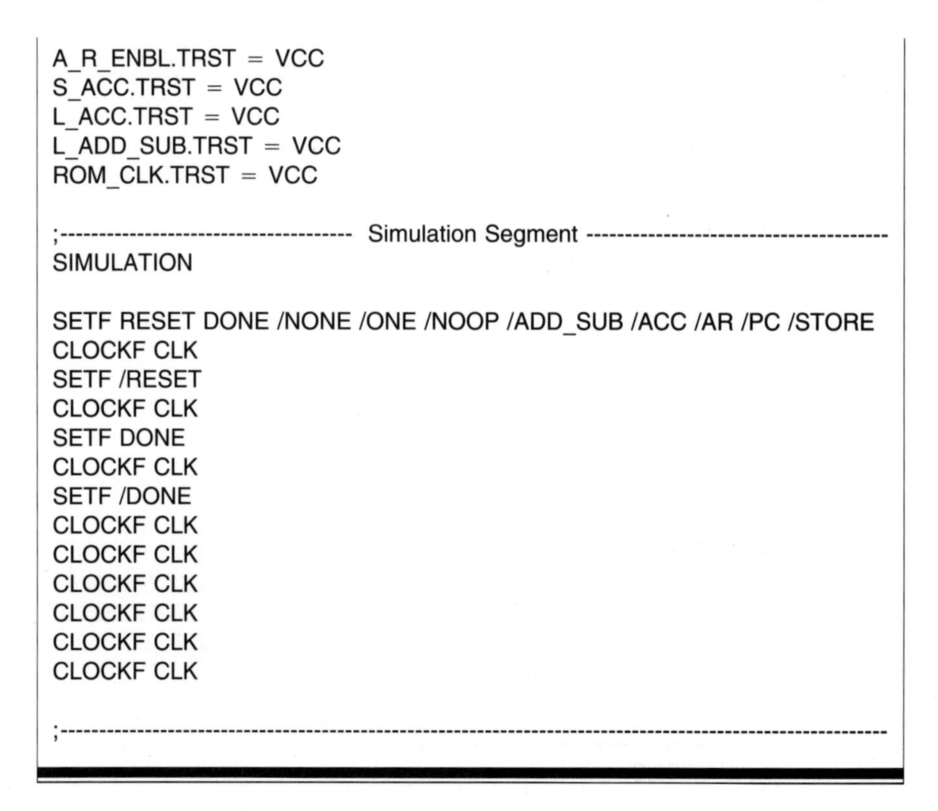

```
A_R_ENBL.TRST = VCC
S_ACC.TRST = VCC
L_ACC.TRST = VCC
L_ADD_SUB.TRST = VCC
ROM_CLK.TRST = VCC

;------------------------------------ Simulation Segment ------------------------------------
SIMULATION

SETF RESET DONE /NONE /ONE /NOOP /ADD_SUB /ACC /AR /PC /STORE
CLOCKF CLK
SETF /RESET
CLOCKF CLK
SETF DONE
CLOCKF CLK
SETF /DONE
CLOCKF CLK
CLOCKF CLK
CLOCKF CLK
CLOCKF CLK
CLOCKF CLK
CLOCKF CLK

;-------------------------------------------------------------------------------------------
```

Table 12.6-11 Source Program for System Controller (Output)

```
;PALASM Design Description

;------------------------------------ Declaration Segment ------------------------------------
TITLE      SYSTEM CONTROLLER 2 (OUTPUTS)
PATTERN
REVISION
AUTHOR     A W SHAW
COMPANY    UTAH STATE UNIVERSITY
DATE       09/25/92

CHIP    SYS_CTL2   PAL22V10

;--------------------------------------- PIN Declarations ---------------------------------------

PIN   1         CLK                                   ; CLOCK
PIN   2         /RESET                                ; INPUT
PIN   3         DONE                                  ; INPUT
PIN   4         NONE                                  ; INPUT
PIN   5         ONE                                   ; INPUT
```

```
PIN  6    NOOP                      ; INPUT
PIN  7    AR                        ; INPUT
PIN  8    PC                        ; INPUT
PIN  9    B2                        ; INPUT
PIN 10    B1                        ; INPUT
PIN 11    B0                        ; INPUT
PIN 12    GND                       ;
PIN 13    NC                        ;
PIN 14    NC                        ;
PIN 15    /A_R_LOADH     REGISTERED ; OUTPUT
PIN 16    /A_R_LOADL     REGISTERED ; OUTPUT
PIN 17    /PC_LOADH      REGISTERED ; OUTPUT
PIN 18    /PC_LOADL      REGISTERED ; OUTPUT
PIN 19    L_INS          REGISTERED ; OUTPUT
PIN 20    L_AR           REGISTERED ; OUTPUT
PIN 21    L_PC           REGISTERED ; OUTPUT
PIN 22    CLK_PC         REGISTERED ; OUTPUT
PIN 23    CLK_AR         REGISTERED ; OUTPUT
PIN 24    VCC                       ;
NODE 1    GLOBAL                    ;

;---------------------------------- Boolean Equation Segment ----------------------------------
EQUATIONS

A_R_LOADH := (/B2*/B1*B0*DONE*/NONE*/NOOP + /B2*B1*B0*/DONE)*AR
A_R_LOADL := (/B2*B1*/B0*DONE*/ONE + B2*B1*/B0*/DONE)*AR

PC_LOADH := (/B2*/B1*B0*DONE*/NONE*/NOOP + /B2*B1*B0*/DONE)*PC
PC_LOADL := (/B2*B1*/B0*DONE*/ONE + B2*B1*/B0*/DONE)*PC
L_INS := /B2*/B1*/B0*/DONE + B2*B1*B0*DONE + /B2*/B1*B0*DONE*NOOP
         + /B2*B1*/B0*DONE*ONE + B2*/B1*/B0*DONE*/AR
         + B2*/B1*B0*DONE

L_AR := (/B2*/B1*B0*DONE*/NONE*/NOOP + /B2*B1*B0*/DONE
        + /B2*B1*/B0*DONE*/ONE + B2*B1*/B0*/DONE)*AR

L_PC := (/B2*/B1*B0*DONE*/NONE*/NOOP + /B2*B1*B0*/DONE
        + /B2*B1*/B0*DONE*/ONE + B2*B1*/B0*/DONE)*PC

CLK_PC := /B2*/B1*/B0*DONE + /B2*/B1*B0*/DONE
          + /B2*B1*B0*DONE + /B2*B1*/B0*/DONE
          + B2*B1*/B0*DONE*AR + B2*/B1*/B0*/DONE

CLK_AR := /B2*/B1*B0*DONE*NONE*/NOOP + B2*B1*B0*/DONE

GLOBAL.RSTF = RESET
```

```
A_R_LOADH.TRST = VCC
A_R_LOADL.TRST = VCC
PC_LOADH.TRST = VCC
PC_LOADL.TRST = VCC
L_INS.TRST = VCC
L_AR.TRST = VCC
L_PC.TRST = VCC
CLK_PC.TRST = VCC
CLK_AR.TRST = VCC

;------------------------------------ Simulation Segment -------------------------------------
SIMULATION

SETF RESET  /B2  /B1  /B0  /DONE
CLOCKF CLK
CLOCKF CLK
SETF /RESET
CLOCKF CLK
CLOCKF CLK
CLOCKF CLK

;-----------------------------------------------------------------------------------------------
```

Table 12.6-12 Test Program

Location	Program	Meaning
0000	06	Jump to the address which follows
0001	00	First (high order) word of address
0002	3D	Second (low order) word of address
003D	06	Jump to address which follows
003E	00	First (high order) word of address
003F	00	Second (low order) word of address
0000	11	Move the data word which follows into the accumulator
0001	35	Data word to be moved
0002	1A	Move the date from the accumulator to the memory location with the address which follows
0003	10	First (high order) word of address
0004	00	Second (low order) word of address
0005	11	Move the data word which follows into the accumulator
0006	13	Data word to be moved
0007	02	Add the data word from the memory location with the address which follows to the accumulator
0008	10	First (high order) word off the address
0009	00	Second (low order) word off the address
000A	06	Jump to the address which follows

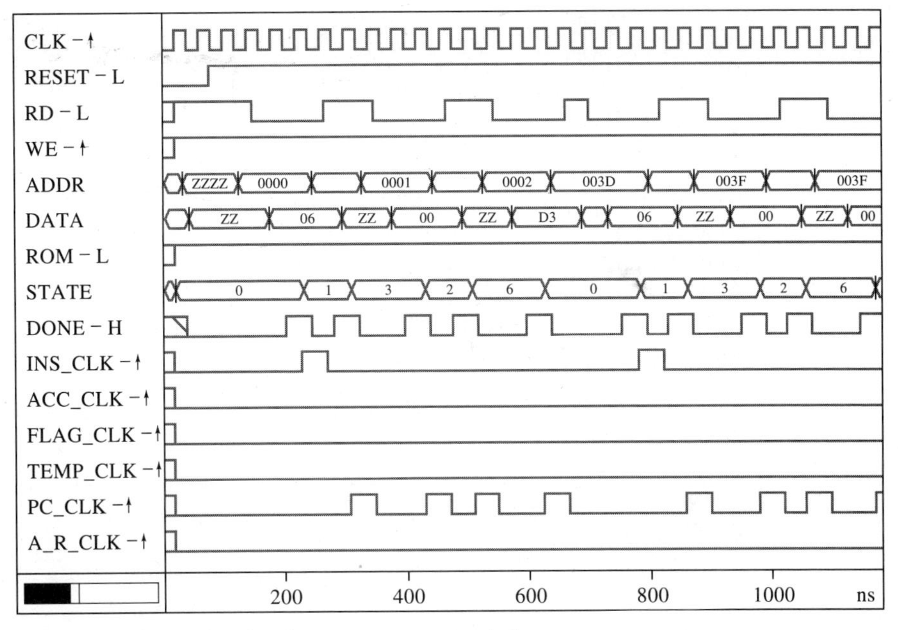

Figure 12.6-25 Timing for the test program simulation.

controller has completed an operation. The remaining signals are clocks to the instruction register, accumulator, flag register, temporary register, program counter, and address register.

The initial reset forces the initial state of the system controller and the initial memory address to zero. In state 0, the system controller requests the read/write/add/subtract controller to do a memory read. The completion of the memory read causes DONE to be asserted, as we see in the simulation. This first read places 06, the first instruction in the test program, on the data bus as shown on the simulation. ICLK clocks this instruction into the instruction register. A timing consideration that should be checked is the setup time of the data before this clock. If we measure the setup time in the simulation, we find it is much longer than the setup time required by the instruction register. When the system controller sees that DONE is asserted, it goes to state 1 just as it should. In state 1, the system controller requests the read/write/add/subtract controller to increment the program counter. The read/write/add/subtract controller does this by asserting the PC_INC input to the program counter (not shown in the simulation) and outputting one PC_CLK pulse as seen in the simulation. When the read/write/add/subtract controller has completed the increment it asserts DONE.

While the program counter is being incremented, the instruction in the instruction register is decoded in the instruction decoder and subsequent states of the system controller are based on the result of this decoding. Op code 06 causes the instruction decoder to assert PC (not shown in the simulation), indicating that the instruction uses the program counter (a jump). For op code 06, the system controller proceeds

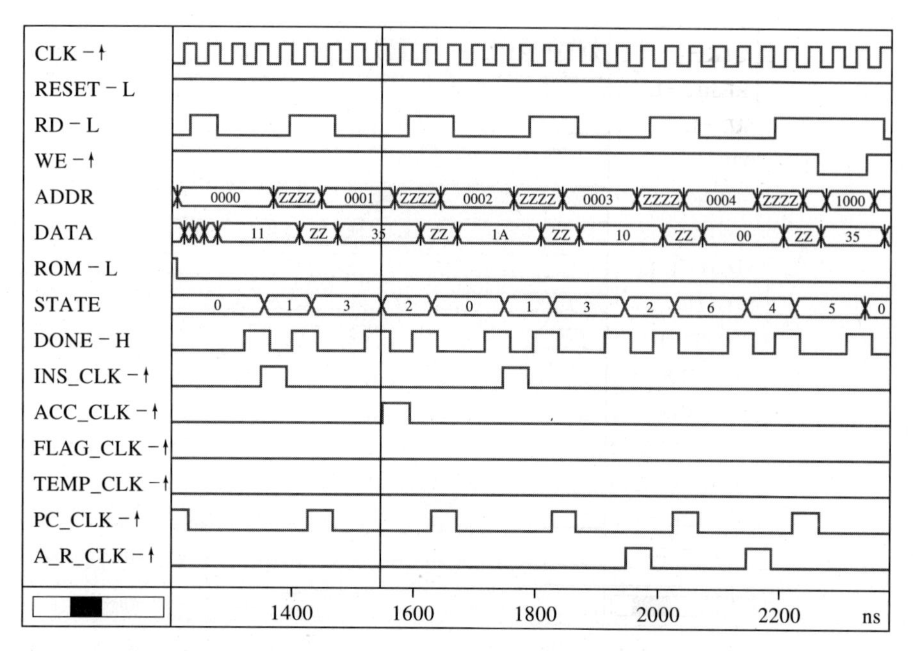

Figure 12.6-26 Timing for the test program simulation (continued).

through states 3, 2, and 6. In state 3, 00, the next word in the program, is read and stored in the holding register for the upper half of the program counter. If we store it directly in the upper half of the program counter, we'll destroy the upper part of the address we need to get the next word of the instruction. In state 2, the program counter is incremented. In state 6, 3D, the next word in the program is read and stored in the lower half of the program counter. At the same time the upper half of the program counter is loaded from the holding register. The program counter now contains the address we wish to jump to. We can see how the data bus first passes a value 00 and then a value 3D, and how each of these values is clocked into the program counter by PC_CLK. This completes the jump instruction and the system controller returns to state 0.

The next instruction is also a jump. Notice how it starts in address 003D, as we can see from the address bus in the simulation. In this jump instruction, the state machine again goes through the state 0, 1, 3, 2, and 6 with PC asserted. The sequence of address changes and data on the data bus as the instruction is executed can clearly be seen on the simulation. Since the memory read executed in state 6 of this sequence is from address 003F, ROM is deasserted and the subsequent memory reads and writes are to R/W memory.

The remaining part of the program is explained much more concisely, but you should nevertheless follow it through on the simulation. This part of the program starts at about 1200 ns in the timing diagram and adds two numbers in the accumulator. The first instruction (op code 11) loads the accumulator with the hex number 35

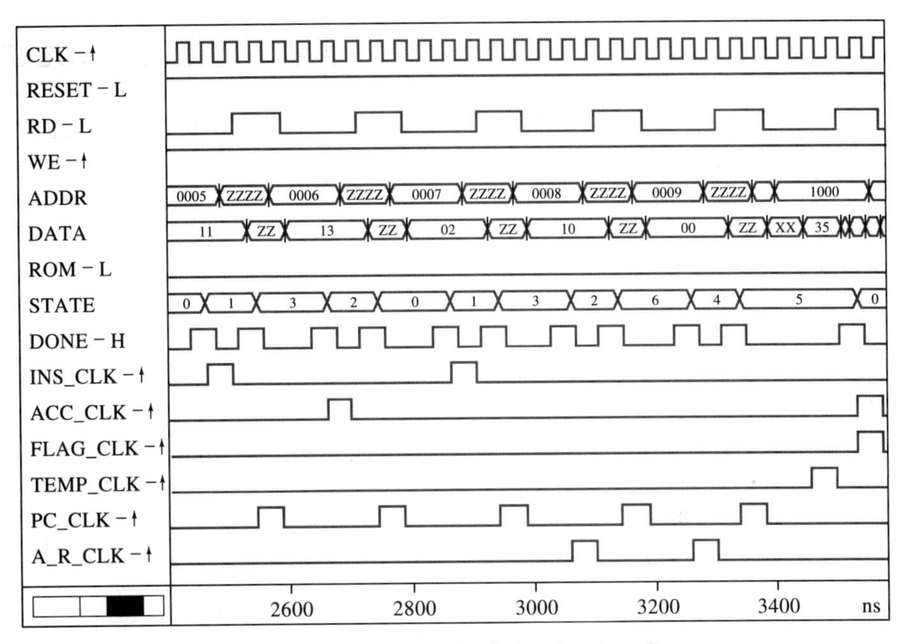

Figure 12.6-27 Timing for the test program simulation (continued).

(states 0, 1, 3, and 2 with ACC and ONE asserted). The next instruction (op code 1A) stores the 35 from the accumulator to memory location 1000 (states 0, 1, 3, 2, 6, 4, and 5 with AR and ACC asserted). Notice how A_R_CLK is used to clock the address into the address register at the end of states 3 and 6 (at approximately 2000 ns and 2200 ns). The next instruction (starting at approximately 2400 ns in the timing diagram) loads the accumulator again (op code 11). This time the number 13 is loaded (states 0, 1, 3, and 2 with ACC and ONE asserted). The next instruction (starting at approximately 2800 ns in the simulation) reads the data word stored in memory location 1000 and adds it to the accumulator (op code 02). This instruction again sends the system controller through states 0, 1, 3, 2, 6, 4, and 5 but with AR and ADD/SUB asserted. The purpose of the action in states 0, 1, 3, 6, and 4 is to get the 35 from memory. At about 3500 ns, TEMP_CLK clocks the data into the temporary register and after a delay to allow propagation through the adders, ACC_CLK, clocks the sum into the accumulator, and FLAG_CLK clocks the ZERO, POS, and CRY into the address flags. These clocks can be seen in the simulation.

This simulation is by no means complete. It's just a sample of the simulation needed to completely test the functions of the computer. We've only checked a jump, store to the accumulator, and add. Also missing from the description of this design is much of the iteration that was part of the design. A normal design sequence consists of a design followed by a simulation. The first simulation detects errors in the design and these errors are corrected. Then the design is resimulated. This process continues until a design which simulates correctly is produced. The correctly simulated design is then prototyped in hardware and may well exhibit further

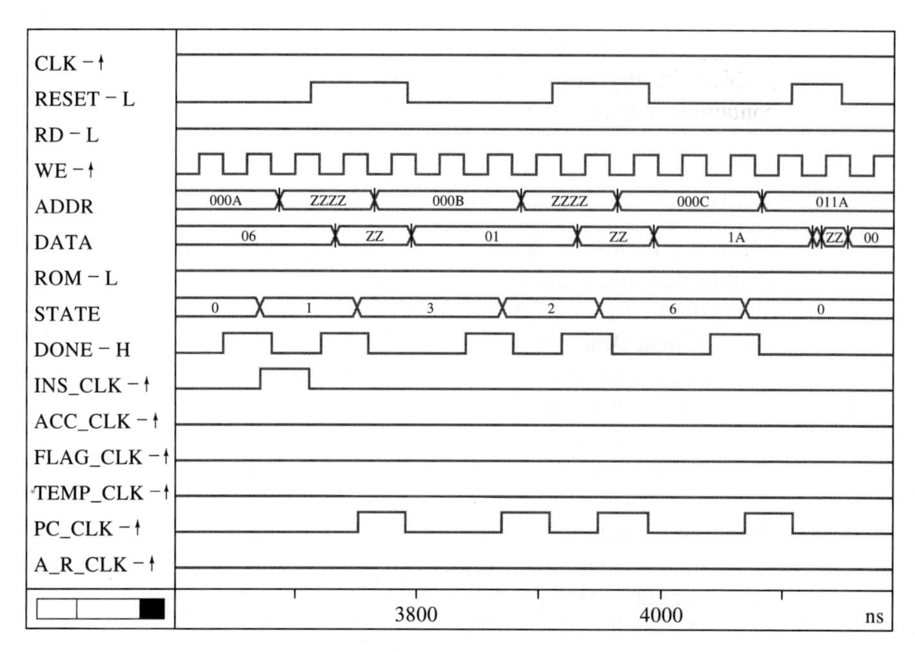

Figure 12.6-28 Timing for the test program simulation (continued).

problems which then must be corrected. The beauty of simulation is that the errors in the *prototype* are reduced or even eliminated.

In the course of designing this computer, only one of the subsystems was simulated. In a design of this complexity, it's well to simulate the subsystems as they are designed, and then simulate the entire system. Sometimes the simulation of the subsystems can be difficult because it's necessary to generate inputs produced by other interacting subsystems and make them have reasonable timing.

This design was rather complicated and could not have been undertaken successfully without a systematic plan of attack. Essentially, the plan in any design is to systematically break the system down into smaller and smaller subsystems until the subsystems are small enough to design from simple devices or as state machines. The design we've just completed could have been partitioned in other ways and still have been successfully designed. There are often many equally good designs for a given system.

Now, a final word about documentation. The design technique we've followed produces good documentation. The set of hierarchical drawings, including the top-level functionals, offers an easily followed map to the design approach. In a complete documentation, these should be supplemented with a verbal description of the system, timing diagrams of critical timing, and a signal dictionary listing all the signals with a description of what they do in the system. The description should contain any user selectable features of the system, such as switches and jumpers, and indicators. The timing diagrams can be conveniently generated as the system is simulated.

If you understand the concepts of the text, you should be able to get by that old problem in any design—What should I do first? You should also have a plan containing subsequent steps to complete the design.

KEY TERMS

Systems Perform a sequence of operations just like a state machine, but are more complicated to design.
Subsystem Result of partitioning of a system into more easily designed parts.
Hierarchical Design Structure A top down approach in which systems are repeatedly divided into smaller and smaller subsystems until the designer reaches subsystems which can be easily designed.
System Controller A subsystem that controls all other subsystems.
Handshake Interchange of signals used in communications between digital systems.
Partition To divide into subsystems.
Functional Partition or Functional A diagram showing the partitioning of a system or subsystem.
Flat A circuit in which the entire system is described without subdividing into subsystems.

EXERCISES

1. Design a digital system to change serial input to parallel output. Assume that you're given inputs DAT(H) and SYNC(L), and a clock CLK. DAT is synchronized so it changes on the negative clock edge. DAT consists of 4-bit data words with two bits of zero padding between each word. SYNC is asserted for one clock cycle at the beginning of the padding as shown in the figure below. Present the

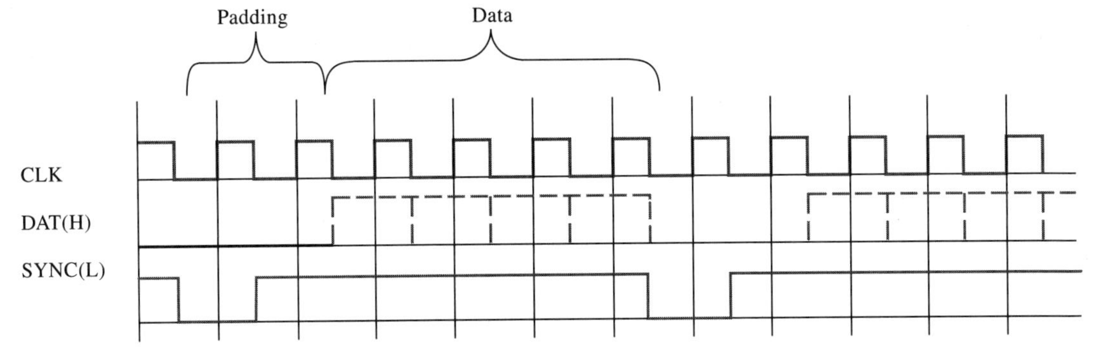

output on four parallel output lines as soon as it's formed, and hold it until the next parallel output is formed. Each time an output is formed, an output handshake line HDSHK(L) should be asserted for four clock cycles.

2. Design a digital circuit to average four 4-bit binary numbers and output the

average as a 4-bit binary number. The input is presented sequentially on four input lines DAT3(H) through DAT0(H). These data lines are under the control of two handshake lines. One handshake, which you must generate, signals the source of the input data that you need a new binary number. Call this handshake RDY(L). It should be asserted when a new number is needed. The other handshake signal, which is an input, is asserted to signal you that the new binary number is available on the inputs. Call it DAV(H) (Data Available). The average should be output on output lines OUT3(H) through OUT0(H). The output also has two handshake lines. The first, NRD(L) (Not Ready for Data), is an input that tells you not to send a new average when it's asserted. The second, which you must produce, is asserted when a new average is output. Call it ODAV(H) (Output Data Available). You must wait until NRD (Not Ready for Data) is deasserted and then output an average alone with ODAV. You must use an *adder* from Chapter 5.

3. Design a programmable digital timer that will assert a signal TIME1(H) for a programmable period of N sec, and then assert a signal TIME2 for a programmable period of M sec. These two output signals should alternate indefinitely. Assume that N and M are 8-bit binary numbers given in seconds. They are to be available on an 8-bit input IN7(H) through IN0(H). To get these numbers you must first assert an output N(H) and wait for an input ENTER(L). ENTER means that the number N is available at the input. To get the number M you repeat the process but assert an output M(H). Start the timing process when an input signal ST(L) is asserted and continue until it's deasserted.

4. Design a digital system that will either add or subtract two 8-bit binary numbers which are presented to it sequentially on eight parallel input lines DAT7(H) through DAT0(H) and will output the 8-bit sum or difference on eight output right lines RES7(H) through RES0(H). If the sum or difference results in an overflow or underflow, an output ERROR(L) should be asserted. Input and output handshakes are to be the same as those in problem 2.

5. Design an adaptive serial-to-parallel data conversion system. The serial data are input to the system on input line SER(H). A data clock is input on line DCLK, and a sync signal is input on line SYNC(H). The input data are 8-bit binary numbers with two bits of padding between them. Changes in SER are synchronous with the positive-to-negative clock edge. SYNC is asserted for one clock simultaneously with the first bit of padding. There are two parallel 8-line output busses, HI[7:0] and LO[7:0]. When the input binary number is greater than 01111111, it should be output on HI; when it's less than or equal to 01111111, it should be output on LO. When an output is present on HI, an output handshake HIAV(H) is to be asserted. When an output is present on LO, an output handshake LOAV(H) is to be asserted. These handshakes must be deasserted for at least one clock cycle when the output data changes. You will need an *adder* or ALU and a multiplexer from Chapter 5.

Flow Control Language

Fletcher, since the publication of his text, has evolved a method for describing a state machine using graphics similar to the flow diagrams used to describe computer programs. We will describe this *flow control language* by comparing it to a state transition diagram. Suppose we have a simple state machine described by the transition diagram we first introduced in Chapter 7, which we have repeated below in Figure A–1. The flow control diagram for describing this state machine is shown in Figure A–2. We can interpret this diagram using the key found in Figure A–3.

There are three different kinds of blocks in a flow control diagram. A state is represented by a rectangular block formatted as shown in the top of Figure A–3 and

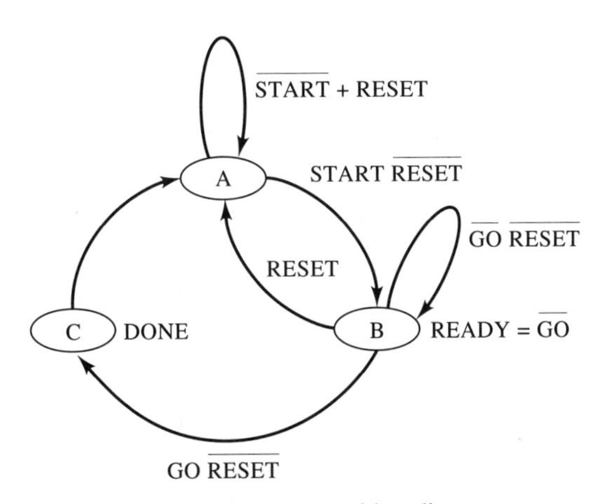

Figure A–1 Example state transition diagram.

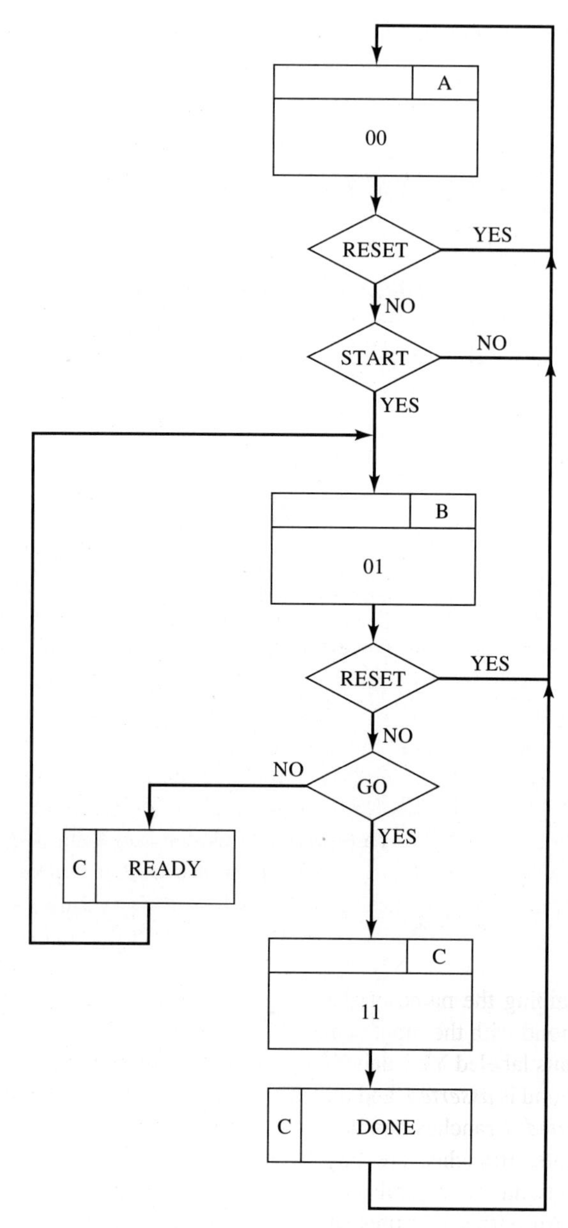

Figure A–2 Flow control language description of the state machine described by the state transition diagram of Figure A–1.

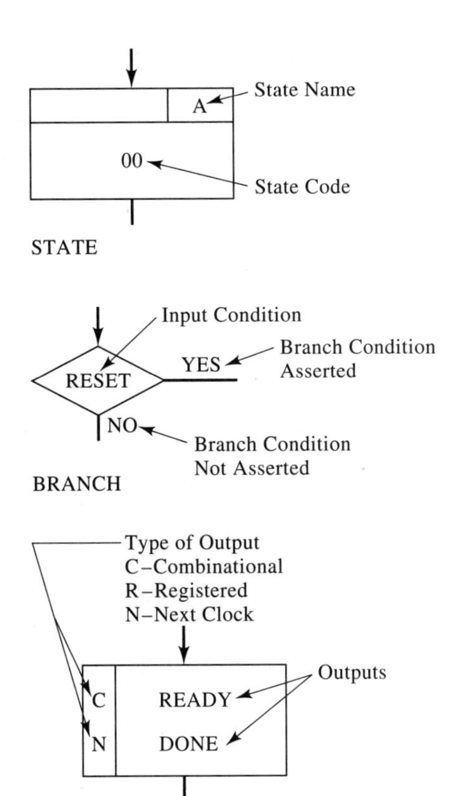

Figure A–3 Key for flow control language.

containing the name of the state and the state code. A branch is represented by a diamond with the input causing the branch shown in the diamond. A branch has two outputs labeled YES and NO. The YES output branch is taken when the input in the diamond is *asserted*, and the NO output is taken when the input in the diamond is *not asserted.* Branches handle both conditional transitions to new state and conditional outputs. Branches are binary (two way), so a transition or an output which is conditional on several inputs requires several branches.

An output is represented by a rectangle formatted as shown at the bottom of Figure A–3 and contains a code specifying the type and the name for each of the outputs. This code, which specifies that the output is either a conditional output (C), a registered output (R), or a registered output delayed to the next clock edge (N), contains information not normally found in the state transition diagram.

In Figure A–2, the rectangle at the top represents state A which has a state code 00. The branch just below this state shows that if RESET is asserted the machine goes back to state A. If reset is not asserted then there is a second branch and the

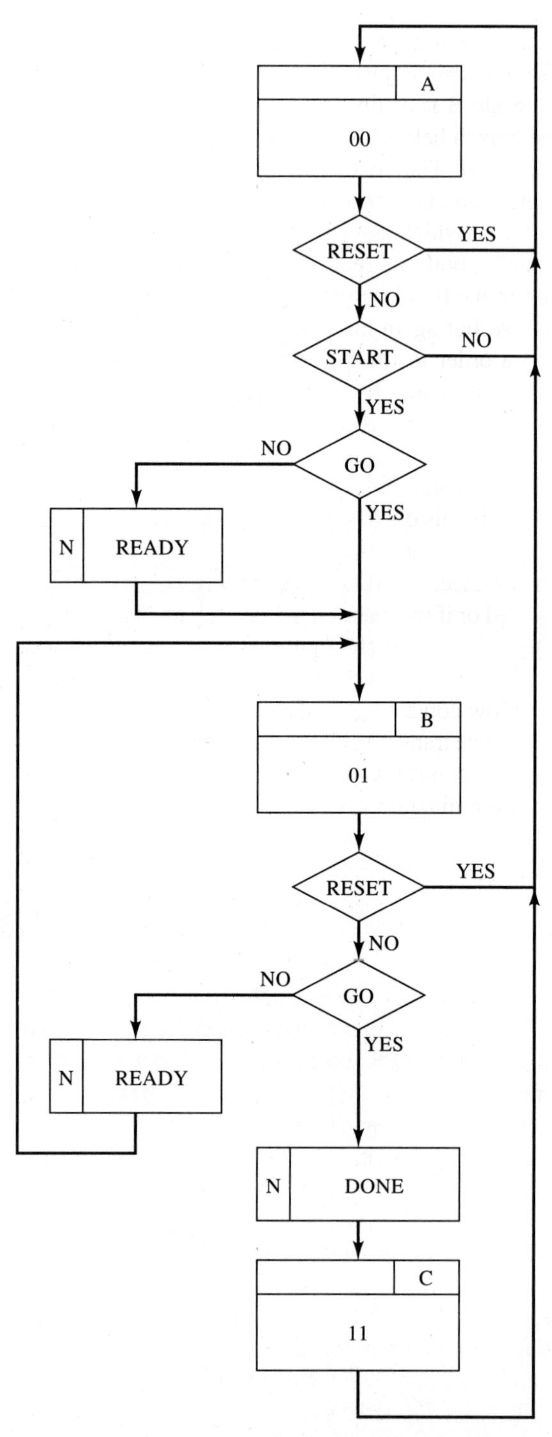

Figure A–4 Flow control diagram with registered outputs.

machine returns to state A if START is not asserted and goes to state B if start is asserted. This is just the same as the state transition diagram of Figure A–1.

State B is again represented by a rectangle. The state code for this state is 01. The branch below state B shows that the machine again returns to state A if RESET is asserted. The next branch shows that the machine stays in state B if GO is not asserted and goes to state C if GO is asserted, but notice how the flow goes through an output which asserts READY if RESET is not asserted and GO is not asserted. Actually, both the return to state B and the conditional output are described in one loop of the flow control diagram. The final state C makes an unconditional return to state A, but again the flow passes through an output which asserts DONE.

In order to finish the flow control diagram, it is necessary to assume that the outputs in states B and C are combinational outputs. Combinational outputs are asserted with only a small delay due to the propagation delay in the output decoders. However, the outputs can have serious glitches. If we want to register the outputs and not change the timing of the outputs, the flow control diagram of Figure A–4 results. In this diagram, all the outputs are register outputs that are delayed one clock cycle. Notice how they are placed in the flow so that they decode the previous state. For instance, READY is asserted if the state machine enters state B and GO is not asserted or if the machine stays in state B (GO not asserted). Register outputs without delay can also be shown in the flow control language, but these outputs do not fit the flow as well as delayed outputs—especially conditional outputs.

Flow control language is a very concise language, but it is not compact. Generally, a state transition diagram and flow control diagram specify the same information. The flow control language allows the types of outputs to be specified, while the state transition diagram does not.

Index